Belize Shelf—Carbonate Sediments, Clastic Sediments, and Ecology

The Association gratefully acknowledges
publication support in the form of a loan
from the American Petroleum Institute.

STUDIES IN GEOLOGY NO. 2

Belize Shelf—Carbonate Sediments, Clastic Sediments, and Ecology

and a paper on

Petrology and Diagenesis of Carbonate Eolianites of Northeastern Yucatán Peninsula, Mexico

edited by

Kenneth F. Wantland and Walter C. Pusey III

Published by
The American Association of Petroleum Geologists
Tulsa, Oklahoma, U.S.A., 1975

Published August 1975

Library of Congress Catalog Card No. 75-20995

ISBN 0-89181-001-3

Printed by Banknote Printing & Lithographing Co.
Tulsa, Oklahoma, U.S.A.

Contents

Preface

The continental shelf of Belize (formerly British Honduras) is an ideal natural laboratory in which to observe Holocene depositional processes. The area is small but it is rich in the diversity of tropical shelf settings. Best known for the shelf-edge barrier system and the abundant reefs, the Belize shelf includes an equally diverse array of marginal-marine and carbonate-island environments, as well as a complete transition from quartzose nearshore sediments to pure carbonate deposits across the narrow shelf lagoon. Interest in Belize is augmented by its proximity and accessibility to the United States and by its English-speaking culture.

The opportunity to study mixed terrigenous-carbonate sediments and the processes leading to the formation of impure limestones prompted the initial interest in Belize by E. G. Purdy in 1960. The project grew to include other phases of clastic sedimentation, carbonate sedimentation, and ecology, resulting in an integrated program of shelf studies. Nine doctoral dissertations at Rice University resulted over the period from 1961-1967. Seven of these studies, together with some original work by Purdy and William Ward's independent study of sedimentology at Isla Mujeres, are presented in this volume.

The 1960 reconnaissance study of the Belize shelf was financed by Jersey Production Research Company, now merged as part of Esso Production Research Company. Funds for the following 3 years were provided by Humble Oil & Refining Company, Jersey Production Research, and Imperial Oil Ltd. The work on Ambergris Cay was made possible through funds of the Petroleum Research Fund of the American Chemical Society and Research Grant 71-A of the American Petroleum Institute. Throughout all phases of the Belize shelf endeavor, the Geology Department of Rice University was a source of additional financial support, chiefly, but not exclusively, used for constructing equipment. In particular, the invaluable help of Elze Hemmen is gratefully acknowledged. E. William Behrens and J. Keith Evans were instrumental in endowing many of the equipment items with a flexibility and strength that required no improvement throughout the lifetime of the project.

W. Ford Young, of Belize City, facilitated shipments of equipment and samples to and from Belize. The British Comptroller of Customs, David Bradley, and his successor, Albert Johnson, expedited the import and export of all items. Those who have been at sea in reef environments will appreciate the importance of having a pilot with first-hand knowledge of navigational obstacles and an ability to "read the water." In this capacity, Jack Cabral saved the expedition more than once from an untimely end. Acknowledgment is also made to former graduate students who participated in the Belize field seasons but whose names do not appear as authors within this volume, either because their results have been published previously or because their participation was a training prerequisite to environmental investigations, recent or ancient, elsewhere. Specifically, we are indebted to the following, all of whom contributed significantly to an understanding of the subjects presented in this volume, and all of whom shared with the authors the adventure of discovery in Belize: E. William Behrens, D. W. Dalrymple, J. K. Evans, P. H. Heckel, W. C. Krueger, R. K. Matthews, and A. R. Troell.

The collection of these papers into a single volume has been an intermittent project since the work in Belize was completed. The manuscripts have been reviewed and revised and, in some cases, bear only partial resemblance to the originals because ideas have evolved over the past few years.

The editors are grateful to the American Petroleum Institute and The American Association of Petroleum Geologists for their support, patience, and perseverance in seeing this work through to completion. Special thanks go to AAPG Book Editor Peggy Rice and Assistant Editor Debby Zikmund for their efforts in finalizing the volume. Editors Wantland and Pusey also appreciate the assistance and indulgence of their respective corporations during the preparation of the book.

EDWARD G. PURDY
Exxon Exploration, Inc.
Walton-on-Thames, Surrey, England

WALTER C. PUSEY III
Continental Oil Company
Ponca City, Oklahoma

KENNETH F. WANTLAND
Cities Service Oil Company
Tulsa, Oklahoma

Continental Shelf of Belize—Regional Shelf Attributes[1]

EDWARD G. PURDY,[2] WALTER C. PUSEY III,[3] and KENNETH F. WANTLAND[4]

ABSTRACT The dynamic interaction of bathymetry and hydrography on the narrow, reef-bounded, continental shelf of Belize produces a complex pattern of Holocene sediment types and biotic assemblages. Basically, the Belize shelf consists of a seaward reef-cay complex and a shoreward lagoon system. The mainland of northern Belize is a low karst surface. The adjacent northern shelf is shallow and is mantled by a suite of almost wholly carbonate sediments. The southern Belize mainland is mountainous and drained by numerous streams transporting large volumes of clastic sediment seaward. The corresponding southern shelf is a well-developed karst surface characterized by considerable bathymetric relief evidenced by numerous steep-sided patch reefs and shelf atolls surrounded by deep channels. The diverse sediment types of the southern shelf range from nearshore quartzose sands and deltaic deposits to a suite of lagoonal marls distinguished from one another by the dominant biotic constituents, to wholly carbonate reef and platform bioclastic sediments. The overall distribution of Holocene sediment reflects a structural influence and is closely related to the historical development of the Belize shelf.

The diversity, abundance, and composition of the benthonic biotic assemblages of the shelf vary systematically in response to bathymetry, hydrography, and sediment types. This is most evident in the coral reef development. Reefs occur in a myriad of sizes, shapes, and environmental settings, but the most obvious and impressive constructional feature is the nearly continuous barrier reef developed on the seaward shelf edge. The diversity and distribution of mollusk assemblages demonstrate the strong influence of substrate, bathymetry, and hydrography on the benthonic fauna. The diverse epifaunal opportunities offered by marine grass (*Thalassia*) are especially evident.

LOCATION AND CLIMATE

Belize comprises the southeast margin of the Yucatán peninsula (Fig. 1). The country is bordered on the north by Mexico and on the west and south by Guatemala. The northern half of the country is physiographically a part of the low-lying limestone surface of the Yucatán platform. The southern half is dominated by the rugged Maya Mountains, which have elevations as great as 1,100 m (3,600 ft) and a terrain developed on a wide range of lithologies including metasedimentary rocks, granites, and intrusive rocks, as well as limestones, sandstones, and shales.

The Belize continental shelf is approximately 250 km (150 mi) long and 15-50 km (10-30 mi) wide. The shelf is terminated abruptly on the east by an escarpment on which an extensive fringing- and barrier-reef system is developed; Smith (1971, p. 29) described this system as the longest *continuous* barrier formation in the West Indies. The eastern escarpment descends precipitously into the Caribbean Sea to depths of more than 600 fm (1,200 m). These depths separate the shelf proper from three atolls—Turneffe, Lighthouse, and Glovers—located from 8 to 40 km to the east. During the Belize shelf study, over 500 bottom-sediment samples were collected from locations ranging from the Mexico-Belize border in the north to the southern part of the shelf near the Guatemalan coast. Figure 2 shows the sample locations for all major traverses, and Figure 3 shows sample locations from closely spaced traverses across the southern shelf lagoon.

The climate of Belize has been described by Wright et al (1959). The shelf lies wholly within the tropical climate belt. Annual rainfall increases in a southerly direction, from 124 cm (49 in.) at Chetumal,

[1]Manuscript received, October 3. 1974.

[2]Esso Exploration Inc., Walton on Thames, Surrey, England.
[3]Continental Oil Company, Ponca City, Oklahoma.
[4]Exploration and Production Research, Cities Service Oil Company, Tulsa, Oklahoma.

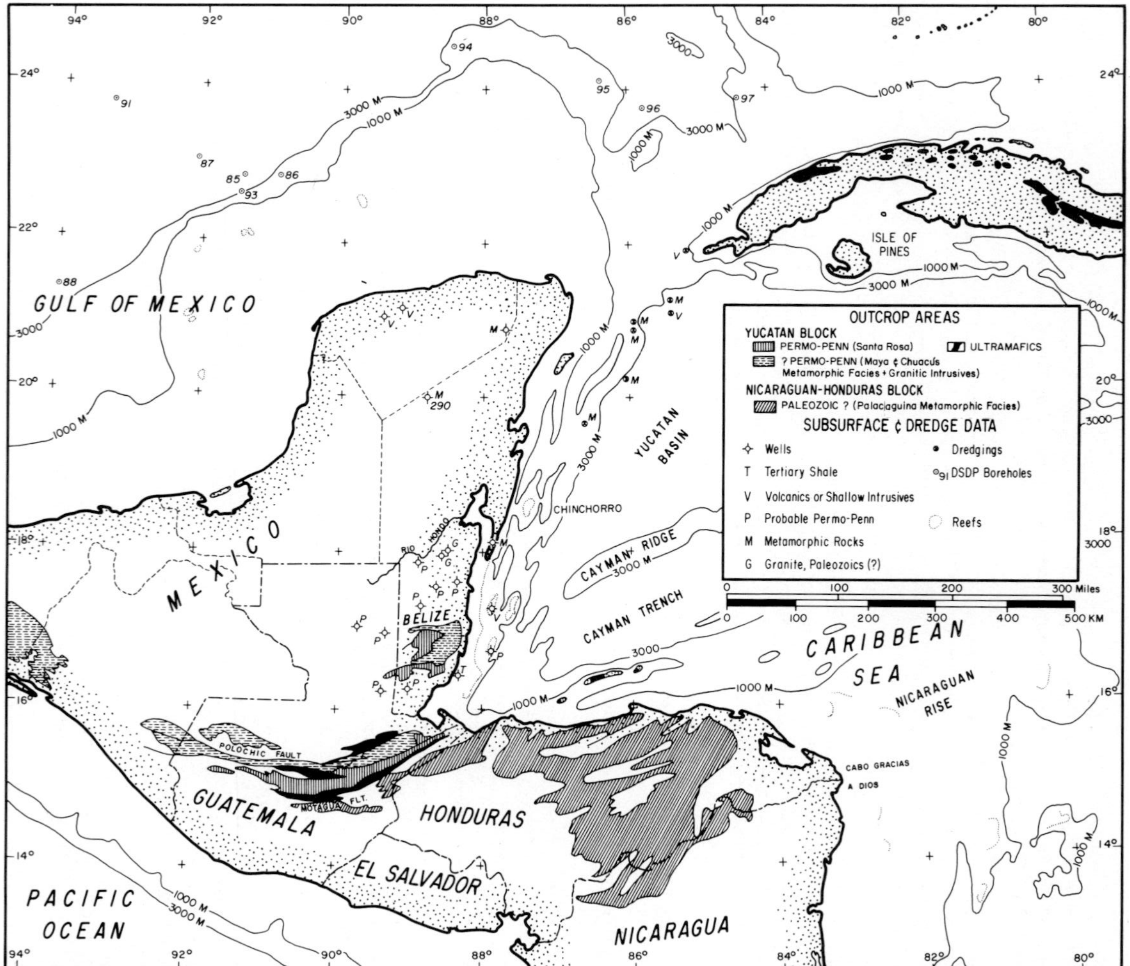

FIG. 1—Tectonic features of Northern Central America (modified from King, 1969, with data from Dengo, 1969, and Baie, 1970). Two basic Paleozoic blocks of continental crust are shown. Yucatan block is best known because sediments are only partially metamorphosed. Rocks of Macal are indisputably dated by fossils as Pennsylvanian and Early Permian. Palacaguina metamorphic rocks of Nicaraguan-Honduras block are not dated, but are presumed to be Paleozoic. Serpentinite bodies (in black) are associated with fault zones and extend ENE along south margin of Cayman Trench. Serpentinites are indicative of oceanic-type crust between blocks.

Mexico, to 178 cm (70 in.) at Belize City, and to more than 380 cm (150 in.) about 130 km south of Belize City near the Guatemalan border. The rainfall is unevenly distributed in an annual cycle consisting of a winter dry season lasting 3-5 months and a summer wet season lasting 7-9 months. Air temperatures range from an average of 27°C for the warm May-September period to 24°C for the cool November-January season.

The area is strongly influenced by easterly trade winds even though it lies near the tropical doldrums belt. The trade winds blow most consistently and strongest from the northeast and east octants at velocities of 12-15 knots (22-28 km/hr; U.S. Navy Hydrographic Office, 1955). Afternoon onshore breezes frequently reinforce these winds to 15-20 knots (28-37 km/hr).

Cold air masses from the United States frequently are moved into Belize during the October-January period by strong northerly winds referred to as "northers." The air temperatures accompanying these cold air masses can be as low as 10°C (Stoddart, 1962, p. 514), and the northers themselves are often associated with heavy rainfall. These phenomena, however, have only a slight effect on the reef water temperature. Hurricanes are fairly common in Belize, and Stoddart (1963) noted that, over the 30-year period from 1931 to 1961, the coast has been struck by a hurricane on the average of once every 6 years. He also provided an extremely useful account of the havoc wrought by one hurricane (Stoddart, 1963).

BATHYMETRY

North of the latitude of Belize City, shelf depths are generally less than 3 fm (5.5 m); south of Belize City, an areally extensive, north-south, elongated depression descends to depths as great as 15 fm (26 m; Fig. 4). The east side of this depression shoals abruptly to depths of 5 fm and less, and these shallow depths are maintained eastward to the crest of the barrier reef. In essence, these attributes define a shelf lagoon that is bordered on its seaward side by a barrier platform. North of Belize City a line of islands separates a northern shelf lagoon from an outer platform, a relation that recalls the barrier-rim separation of the outer platform and shelf lagoon in the Bahamas (Newell et al, 1959). Thus one can recognize a *barrier rim* that bounds a shallow *northern shelf lagoon* and a *barrier platform* that limits a much deeper *southern shelf lagoon* (Fig. 4). Toward the south the barrier platform becomes increasingly dissected and terminates, as does the barrier reef, in a peculiar hook. South of the hook, the more open circulation of the *Gulf of Honduras* prevails. Within these major provinces there are minor subdivisions such as the *Port of Honduras, Victoria Channel,* and *English Cay Channel* that do not concern us here except as convenient points of reference.

WATER CHARACTERISTICS

The water characteristics of greatest significance in the study of carbonate shelves are temperature, salinity, calcium carbonate saturation, circulation, and turbidity. On the Belize shelf, these parameters are imperfectly known, although gross aspects can be described.

Water Temperature

Surface-water temperatures measured at sediment sampling localities during the months of July and August, 1961 and 1962, range from 26.4° to

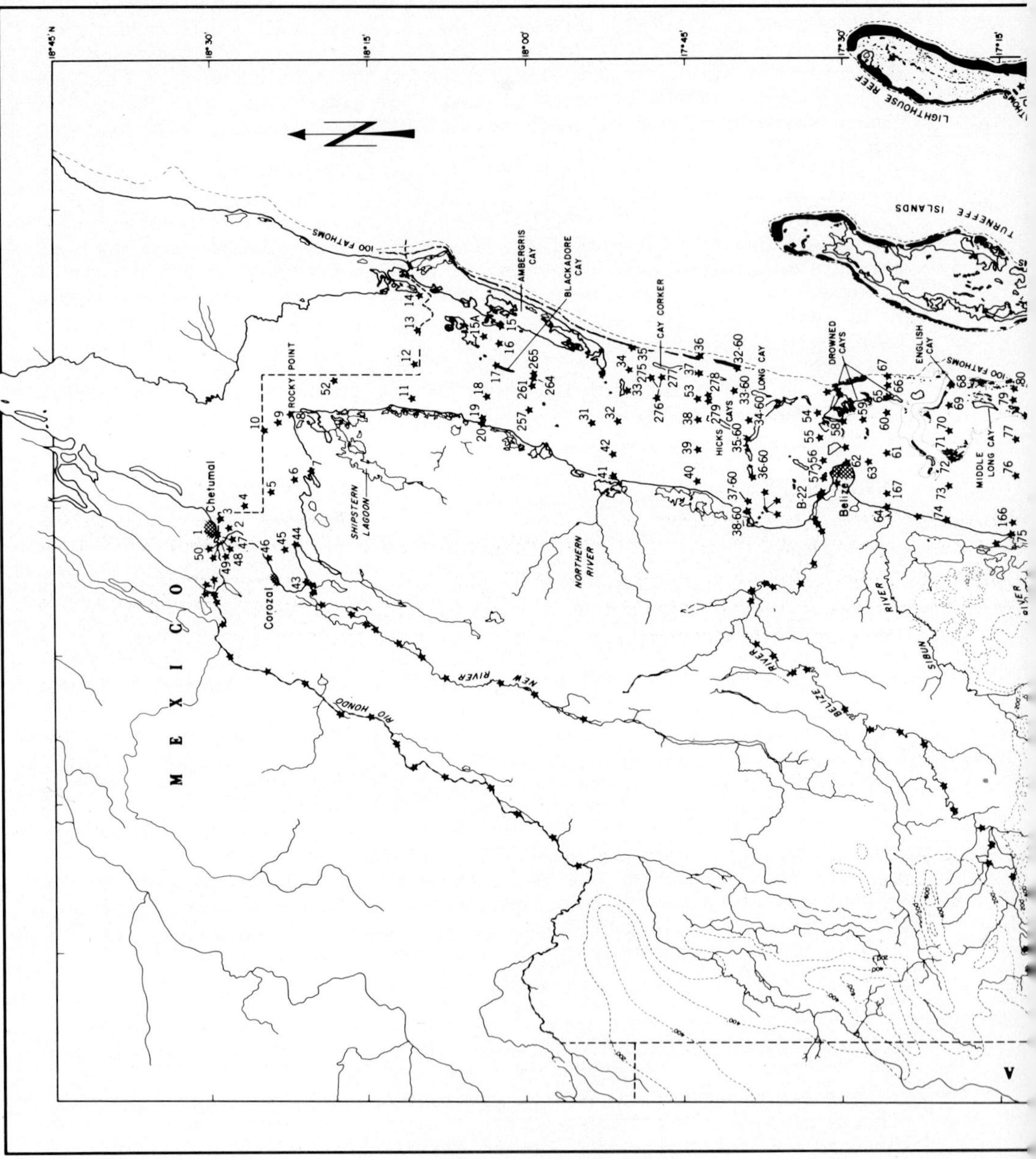
18°45'N
18°30'
18°15'
18°00'
17°45'
17°30'
17°15'
MEXICO
Chetumal
Corozal
RIO HONDO
NEW RIVER
NORTHERN RIVER
SHIPSTERN LAGOON
ROCKY POINT
AMBERGRIS CAY
BLACKADORE CAY
CAY CORKER
HICKS CAYS
LONG CAY
DROWNED CAYS
Belize
BELIZE RIVER
SIBUN RIVER
ENGLISH CAY
MIDDLE LONG CAY
TURNEFFE ISLANDS
LIGHTHOUSE REEF
100 FATHOMS
A

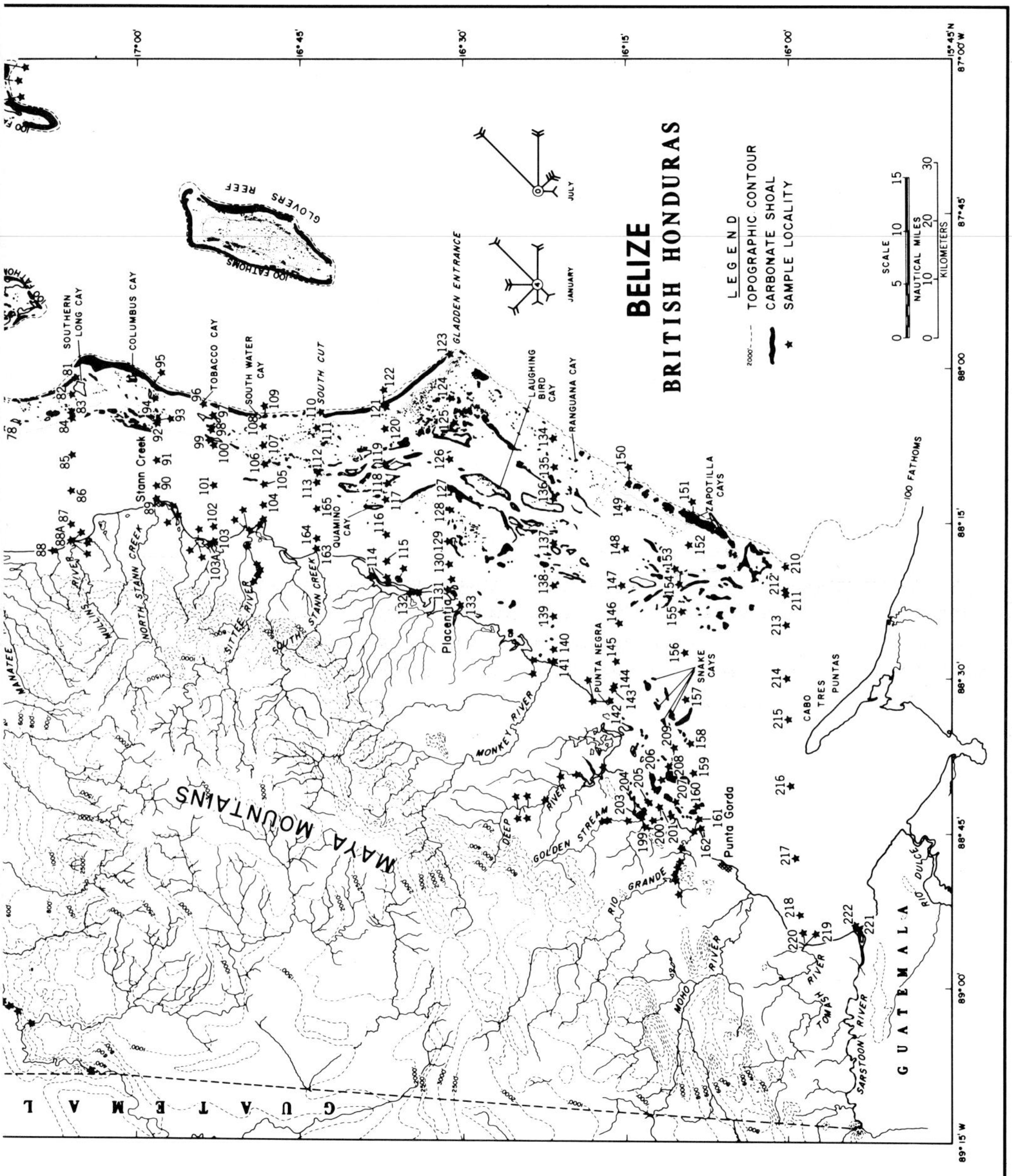

FIG. 2—Index map of Belize shelf. Sample numbers are shown for all sample locations except detailed traverses.

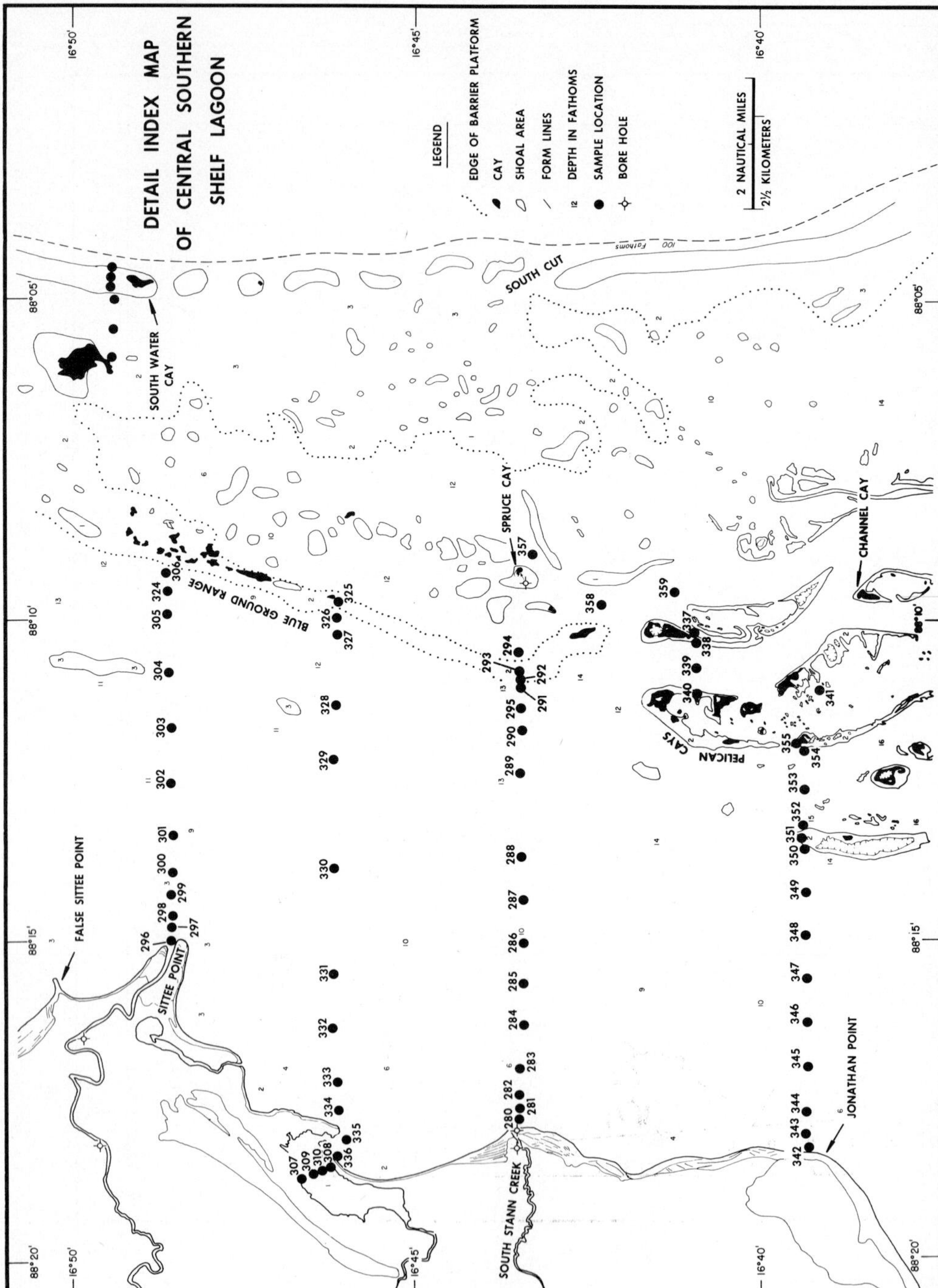

FIG. 3–Detailed index map of part of central southern Belize shelf showing sample locations on closely spaced traverses across southern shelf lagoon.

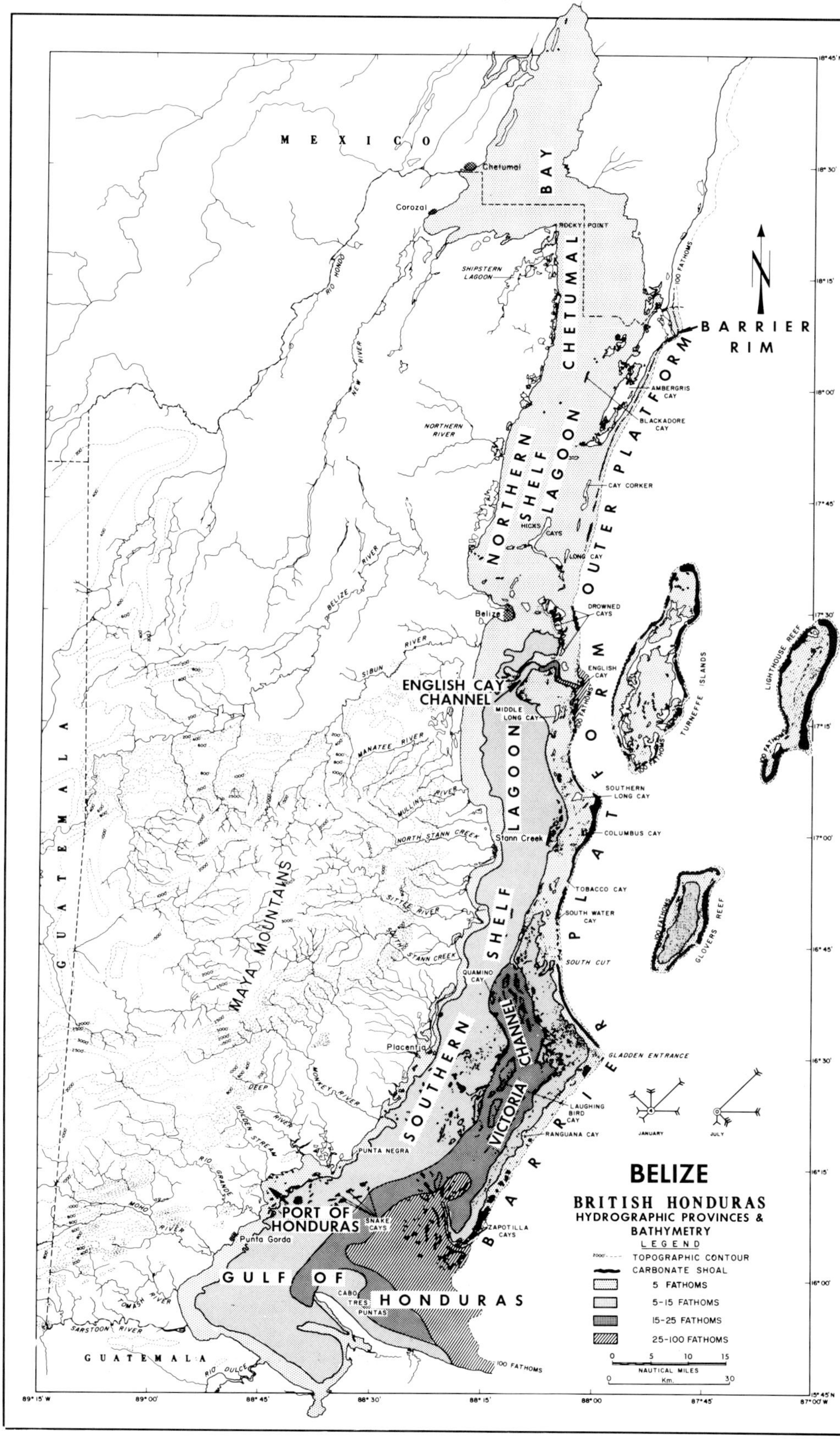

FIG. 4—Bathymetry and hydrographic provinces of Belize shelf.

30.9°C and average 28.9°C (N=215). Surface-water temperatures measured in January 1972 range between 24.6° and 27.0°C and have a mean value of 26.2°C (N=64). The January measurements were limited to the southern shelf, whereas the July and August determinations include both the northern and southern shelves. Both sets of values are available.[5] A few bottom-water temperature measurements showed as much as 1°C temperature difference relative to the warmer surface waters. This effect was minimized in shallow water, but even in the deeper water of the southern shelf lagoon the differences between surface- and bottom-water temperatures appear devoid of any overall pattern.

Salinity

Surface and bottom salinity measurements were made with a conductivity bridge during the summer months of June and August in 1961 and 1962. The results for the northern Belize shelf have been reported by Pusey (1964, Table 18); those for the southern shelf are available elsewhere.[6] These data are the basis for the bottom and surface salinity contours illustrated in Figures 5 and 6.

Two tendencies are apparent on the surface salinity map (Fig. 6). First is the apparent decrease in salinity northward into Chetumal Bay and southward into the Gulf of Honduras. The decrease northward is a consequence of the Ambergris Cay peninsular barrier to marine circulation in conjunction with mainland runoff into the bay. The decrease southward is a result of the substantial increase in fresh water entering the sea, chiefly from the Dulce, Sarstoon, Tomash, and Moho Rivers. The increase in fresh water, in turn, results from the general increase in rainfall to the south. The second tendency is for nearshore surface salinities to be higher in the northern shelf lagoon than in the southern shelf lagoon. This also is a result of the difference in rainfall.

A comparison of the bottom and surface salinity maps (Figs. 5, 6) shows essentially no difference for the northern shelf lagoon but a striking difference for the southern shelf lagoon. This difference reflects the prevalence of freshwater wedges in the south and the tendency for vertical salinity differences to be homogenized in the relatively shallow water of the north. The difference in density between fresh water and salt water causes wedges of fresh water that emanate from river mouths to float on the denser salt water. The fronts of the wedges are readily identified from the air by color contrasts and at sea by the large amount of flotsam that characteristically accumulates near their borders.

Some of these wedges have been observed as far east as the barrier platform. Local inhabitants claim that "sweet water" occasionally reaches the southern reef hook, and the surface salinity pattern in this area is certainly consistent with that observation. Wind-induced wave agitation destroys the density contrast by mixing the fresh water with the underlying salt water. The extent of this mixing is dependent on the depth to which agitation occurs; therefore, salinity homogenization increases as water depth decreases, if other factors are equal. Thus the deeper southern shelf lagoon shows significant salinity layering and the shallow northern lagoon does not. As might be anticipated, the salinity differences tend to increase landward. In one place off the mouth of the Monkey River in only 10 ft (3 m) of water (Fig. 2), a surface salinity of 3.3‰ and a bottom salinity of 32.5‰ were recorded, giving a difference of 29.2‰.

[5]Complete tabulation of values may be obtained from ASIS/NAPS, % Microfiche Publications, 305 East 46th Street, New York, New York 10017.

[6]See footnote 5.

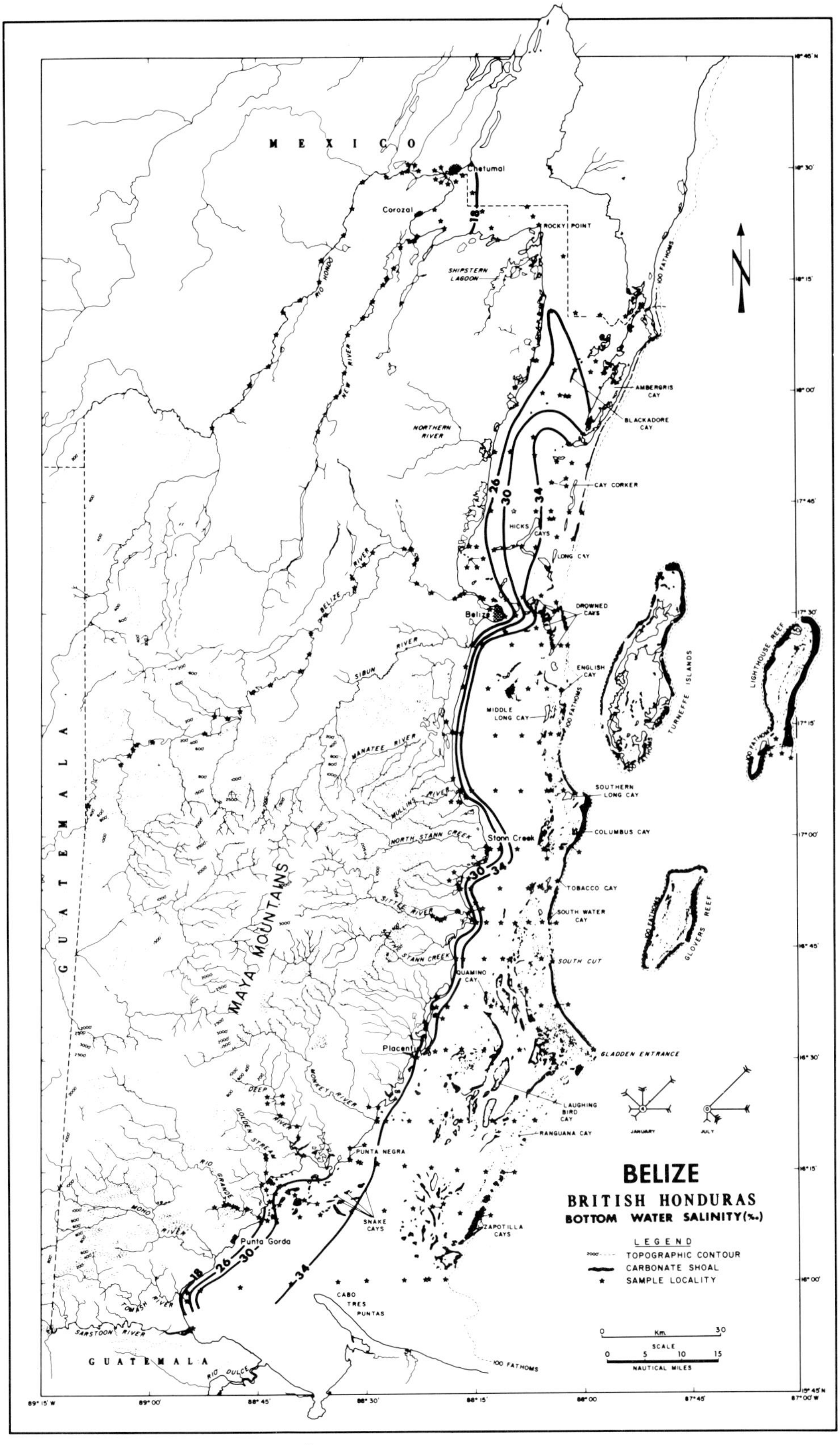

FIG. 5—Bottom-water salinity (°/oo) of Belize shelf waters. Data collected in summer months (June-August), 1961-62.

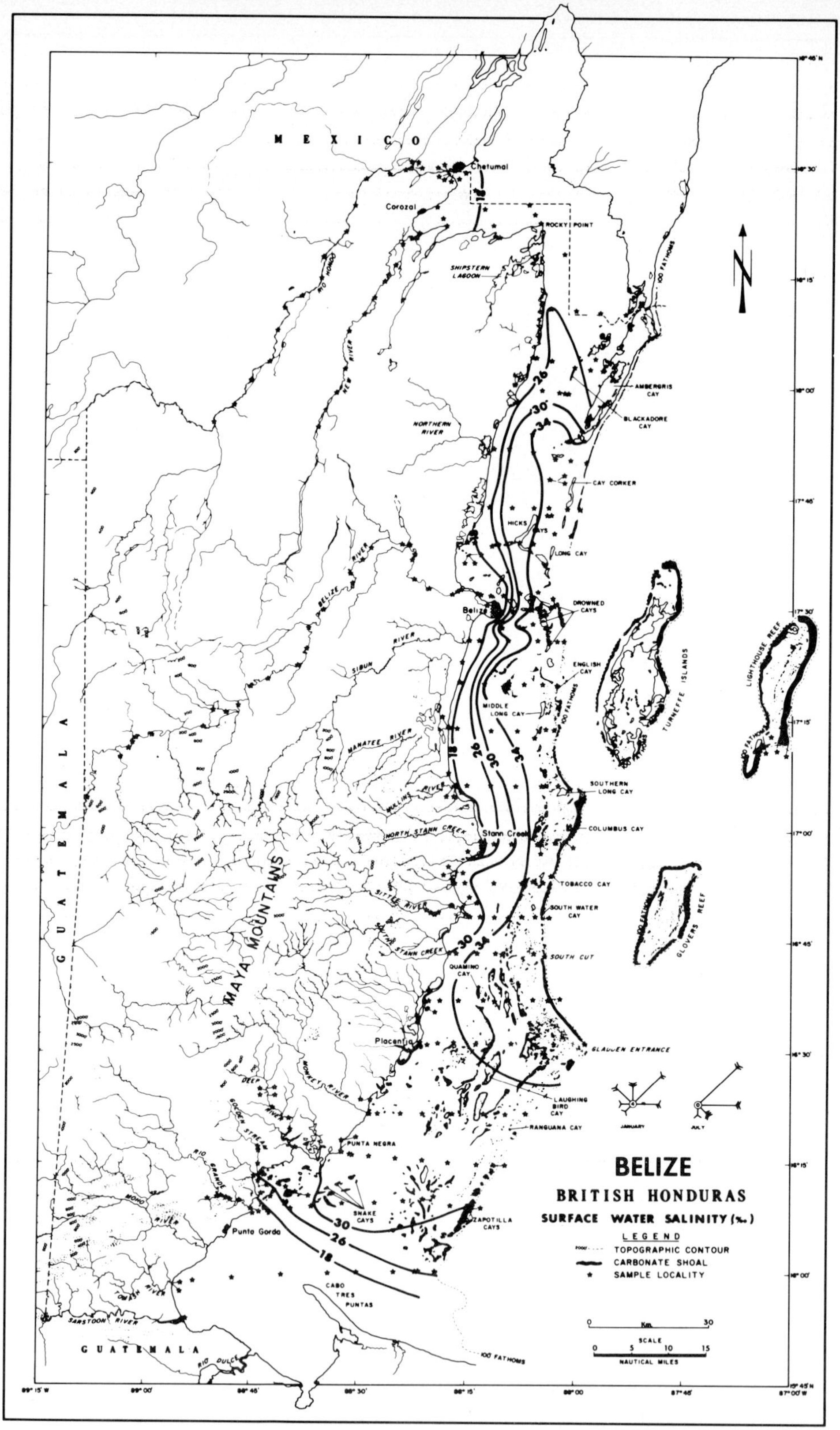

FIG. 6—Surface-water salinity (°/oo) of Belize shelf. Data collected in summer months (June-August), 1961-62. Comparison with Figure 5 demonstrates strong "wedge" effect of fresher surface waters overlying more marine bottom waters in relatively deep southern shelf lagoon; no such effect is apparent in shallow northern shelf lagoon and Chetumal Bay.

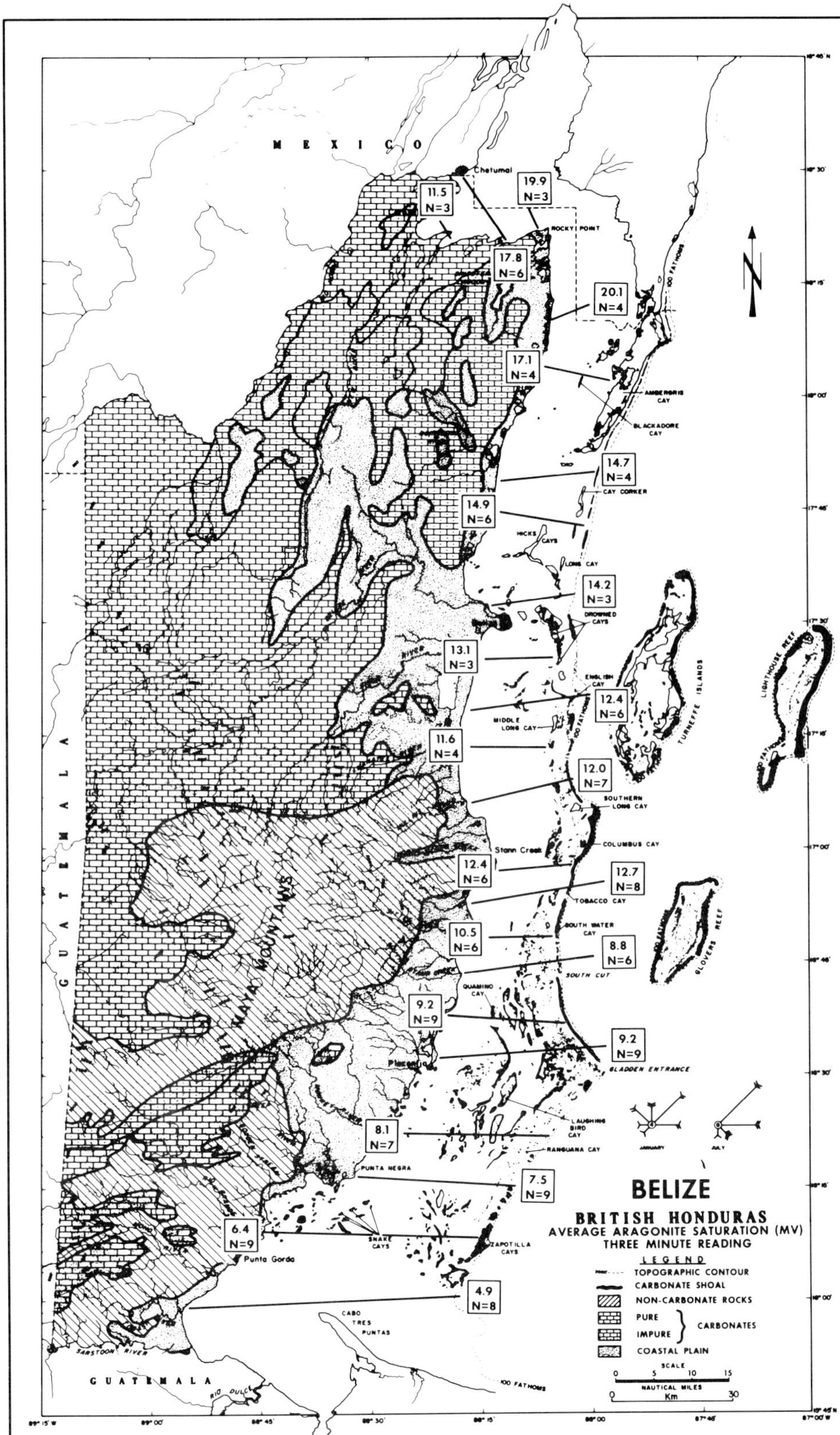

FIG. 7—Carbonate saturation (mv) of Belize shelf waters. Data shown are average values for each sampling traverse across shelf; extent of each traverse is illustrated by solid lines. Also shown are major lithologic suites of Belize mainland.

The ecological significance of this wedge effect is that "marine" bottom salinities occur extremely close to shore—and this is true even off the mouths of some of the larger rivers (e.g., Monkey River). Marine organisms follow suit, and therefore it is apparent that the occurrence of fossil marine organisms need not necessarily imply any great distance from shore, even when the mainland has considerable relief, as in the case of southern Belize.

Carbonate Saturation

Measurements of the degree to which the shelf waters are saturated with calcium carbonate were made with a Weyl saturometer (Weyl, 1961) using previously washed and dried Bahamian oolite (aragonite) as a standard. Readings were taken at 1-minute intervals up to and including 4 minutes. Empirically, the 3-minute readings proved to have the greatest reproducibility, and these are the values that were used to characterize carbonate saturation.[7] The saturometer results demonstrated that the shelf waters were consistently supersaturated with respect to the aragonite of Bahamian oolite; however, the absence of any consistent trend from locality to locality defied any simple generalizations (e.g., with respect to proximity to shore). Part of the problem may have been that we were measuring relative rather than absolute saturation, as we had no way of determining the partial pressure of CO_2 in the atmosphere. Day-to-day fluctuations in atmospheric PCO_2 would of course affect the saturometer readings, which were made on different days and, in some cases, in different years, and thereby obscure any trends that might be present. A second source of difficulty may have been the sensitivity of the instrument itself, as suggested by the fact that reproducibility was best, but far from perfect, 3 minutes after the Bahamian standard had been introduced. In any event, regional trends only became apparent after averaging the values for *each* saturometer *traverse* across the shelf. These averages, illustrated in Figure 7, demonstrate an overall tendency for the degree of supersaturation on the shelf to decrease in a southerly direction. The decrease in average values south of the latitude of Belize City probably reflects the increase in rainfall and mainland drainage in the same direction; thus carbonate saturation shows a general positive correlation with surface salinity (Table 1). In contrast, the increase in average supersaturation values north of Belize City is negatively correlated with surface salinity (Table 1). In large part this relation would seem to stem from the restricted circulation of Chetumal Bay in conjunction with the large amounts of dissolved carbonate introduced onto the shelf by the few rivers that drain the almost exclusively carbonate terrane of the mainland. Curiously, however, water samples from New River and Rio Hondo, which drain into Chetumal Bay, are markedly undersaturated with respect to the Bahamian oolite standard, and this is true even at the river mouths (-11 and -28 mv at the mouths of the Rio Hondo and New River, respectively). Perhaps the high degree of supersaturation in Chetumal Bay is largely a function of evaporation of fresh water after it enters the bay. The dry season would be the most likely time for saturation to increase; if this is the case, the present supersaturation values within the bay would reflect dilution from higher values attained during that season. This possibility is as yet untested.

Water Movements

Little information is available on this subject. U.S. Navy Hydrographic Office Publication 20 (1952, p. 187) indicates a spring tidal range of 0.5 m (1.5 ft) but also notes that northers commonly lower the

[7] See footnote 5.

TABLE 1. SELECTED CARBONATE-SATURATION AND SALINITY DATA

Sample Number	*Salinity (°/oo)*	*Saturation (mv) (3-minute reading)*
Traverse from mouth of Belize River to barrier platform.		
57	1.9	-16
62	1.6	-10
55	33.5	+21
54	35.2	+14
Traverse across Chetumal Bay from New River to barrier platform.		
43	0.8	-28
45	6.7	+10
5	19.7	+20
10	20.1	+20
52	18.5	+20
12	23.3	+22
18	27.9	+16
32	34.8	+12
35	36.2	+14

surface by 0.8 m (2.5 ft). No information on currents is available for the northern shelf lagoon. In Chetumal Bay, however, the fresh water draining into the bay, in conjunction with the insulating effect of the Ambergris Cay peninsula, insures a general surface-current movement southward. In the southern shelf lagoon there are indications of southerly currents with velocities of 0.9-2.3 km/hr (½-¼ knot), according to U.S. Navy Hydrographic Office Chart 1497 (1931). In the Gulf of Honduras, Chart 1498 (U.S. Navy Hydrographic Office, 1959) indicates easterly movement of water issuing from both the shelf lagoon on the north and the cul-de-sac on the south. Velocities of 1.6 km/hr (1 knot) are recorded. Apparently the circulation impediment of the barrier platform and the wind-induced currents combine to funnel shelf-lagoon waters southward into the Gulf of Honduras. The inevitable "pile-up" whereby a large quantity of fresh water drains directly into the Gulf of Honduras creates a hydraulic head that can only be dissipated by easterly flow against the prevailing wind direction. Such flow occurs south of the barrier platform. Data on bottom currents are nonexistent; nonetheless, there is no reason to believe that bottom-current directions are different from their surface counterparts, at least in the shelf-lagoon regions.

Turbidity

There are no quantitative measurements of turbidity on the Belize shelf, but qualitative observations suggest a relatively high turbidity in the southern shelf lagoon, where underwater visibility often approximates 3-5 m, and a relatively low turbidity for the outer and barrier platform areas, where underwater visibility has been estimated at 30-40 m.

SHELF BIOTA

Among the most impressive aspects of life on the Belize shelf are the coral reefs, but relatively little has been published about them.

FIG. 8—Oblique aerial view looking south along an uninterrupted portion of barrier reef. Line of breakers defines seaward margin of reef. Water depths across shallow barrier platform to right vary from a few centimeters on reef flat to a few meters on *Thalassia* meadows on far right. Just visible on reef tract in background is Columbus Cay.

Stoddart (1962) described some of the reef attributes of Turneffe, Glovers, and Lighthouse atolls. He also surveyed portions of the barrier reef and provided a lucid account of storm damage by a major hurricane. Thorpe and Bregazzi (1960) conducted some reef-growth experiments at Rendezvous Cay. York (*in* Wantland and Pusey, 1971) compiled an illustrated catalog of Belize shelf corals, and Purdy (1974a) was concerned largely with the genesis of the existing reef configurations.

Coral Communities

A profusion of coral-reef communities on the Belize shelf occurs in all shapes and sizes. The largest and most impressive development is the barrier reef of the southern shelf. From the air the crest of the reef appears as a relatively uninterrupted color band paralleling the shelf edge (Fig. 8). On the northern shelf the barrier reef appears less continuous and becomes a fringing reef bordering the windward lagoon of Ambergris Cay. From the air the fringing reef appears better developed than the northern-shelf barrier reef, as though the presence of the island itself were somehow responsible for the enhanced reef growth. Perhaps the island affords the reef protection from the hyposaline waters of the northern shelf lagoon in a manner analogous to that proposed for Bahamian reefs (Purdy and Imbrie, 1964). Whatever the cause, reef development on the northern shelf is virtually restricted to the area of the barrier rim

FIG. 9—Oblique aerial view toward southeast of Victoria Channel shelf atolls. Largest atoll in view is that of Pelican Cays, which is approximately 5.5 mi (8.8 km) long. To left, Elbow Cay atoll is partly truncated by frame of photograph. Other two visible atolls are unnamed as they are not associated with named islands. Tarpum Cay appears in background on a linear carbonate shoal. Note high degree of parallelism between sides of adjoining atolls. In upper left corner, barrier platform is identified by line of demarcation formed by its shoal-dominated border and deeper water of Victoria Channel. Barely visible in extreme background of same corner is a light band of color that identifies reef flat and apron of barrier reef.

and outer platform. On the southern shelf, however, coral reefs also appear within the shelf lagoon. There they occur not only as isolated shoals, but also as shelf atolls that commonly have a distinctive rhomboid outline (Fig. 9) and are largely restricted to the Victoria Channel area.

Surprisingly, coral communities also occur near the mainland coast of southern Belize. They are more prevalent south of the latitude of South Stann Creek, but there are occurrences as far north as False Sittee Point (see Figs. 2, 3 for location names). Many of the nearshore coral developments are demonstrably associated with a rock bottom; some, however, appear to have developed on a quartz-sand substrate (Fig. 10) that must have been stabilized sufficiently to permit coral colonization. The implication of a stabilized sand substrate is also suggestive of the relict nature of some of the adjacent coastal-plain sand bodies (e.g., Placentia peninsula; see High, this volume, for a different approach that leads to the same conclusion). *Acropora palmata*, the coral comprising the bulk of the barrier-reef crest, also is a member of some of the nearshore communities. The coastal occurrence of this stenohaline species is a forceful demonstration of the ecologic significance of normal-marine salinities under the freshwater wedge.

FIG. 10—Oblique aerial view of the mainland coast at Placentia, looking north. Town is built on the southern tip of a long, sinuous peninsula composed largely of quartz sand. Placentia Cay (lower right) is a carbonate island fringed by living coral. Open water on right is southern shelf lagoon; enclosed water on upper left isSeine Bight, which separates Placentia Peninsula from mainland proper.

Farther south in the Port of Honduras, focused mainland drainage lowers bottom salinities so that *A. palmata* apparently is restricted to shoals having normal-marine bottom salinities. Shoals with lower bottom salinities are dominated by large massive heads of *Siderastrea siderea* and much smaller colonies of *S. radians* associated with *Porites porites* vars.

The coral reefs obviously modify the environment markedly and influence the nature of water movement and sedimentation, both in the immediate vicinity of the reefs and to a considerable distance away from them. The faunal consequence of interest, however, is that the reefs provide a wide array of habitats for the diversification of groups other than the corals themselves. Some forms, such as the stalkless crinoids (Fig. 11) and some brachiopods, are found only within reef communities. More instructive, however, is the distribution of mollusks over the entire shelf.

Mollusk Communities

The basic distribution pattern of mollusks on the Belize shelf was determined by Robertson (1963) of the Philadelphia Academy of Natural Sciences. His taxonomic distribution of *live* mollusks has been supplemented by the identification of shells in subsequently collected sediment samples (His list of commonly occurring mollusks is given in the Appendix). The result is the map of mollusk communities in Figure 12. Robertson (1963) did not identify these groupings as communities, but implicit in his discussion is the fact that these are naturally recurring associations. He referred to the separate groupings by roman numeral rather than by name. The community names cited here simply reflect the most common member of each group among the live samples collected by Robertson. The only exceptions to this procedure are the Swamp and Reef-complex communities. The *Swamp* community is not well sampled and in essence refers to coastal freshwater and brackish-water habitats which are commonly, but not invariably, dominated by mangroves. This association also occurs on some of the larger islands on the shelf, but in large part it is restricted to the mainland coast. Only the mainland distribution has been indicated in Figure 12, and here its landward extent is in large part predicated on the mangrove-coastal-plain soil division of Wright et al (1959). The *Reef-complex* community is so-called because it involves large numbers of reef-dependent habitats on the outer and barrier platforms. Included within this category, however, are turtle grass (*Thalassia testudinum*) associations that do not appear to be particularly reef-dependent. Unfortunately, the complex alternation of environments on the outer and barrier platforms precludes finer subdivisions at the scale of the investigation; thus it is necessary to group this environmental potpourri under the Reef-complex heading.

FIG. 11—Stalkless crinoids (comatulids), such as *Tropiometra carinata* illustrated here, are restricted to reef community. They are most commonly found attached to branching corals (e.g., *Acropora cervicornis*) in protected, quiet-water environments. Width of specimen is approximately 10 in. (25.4 cm).

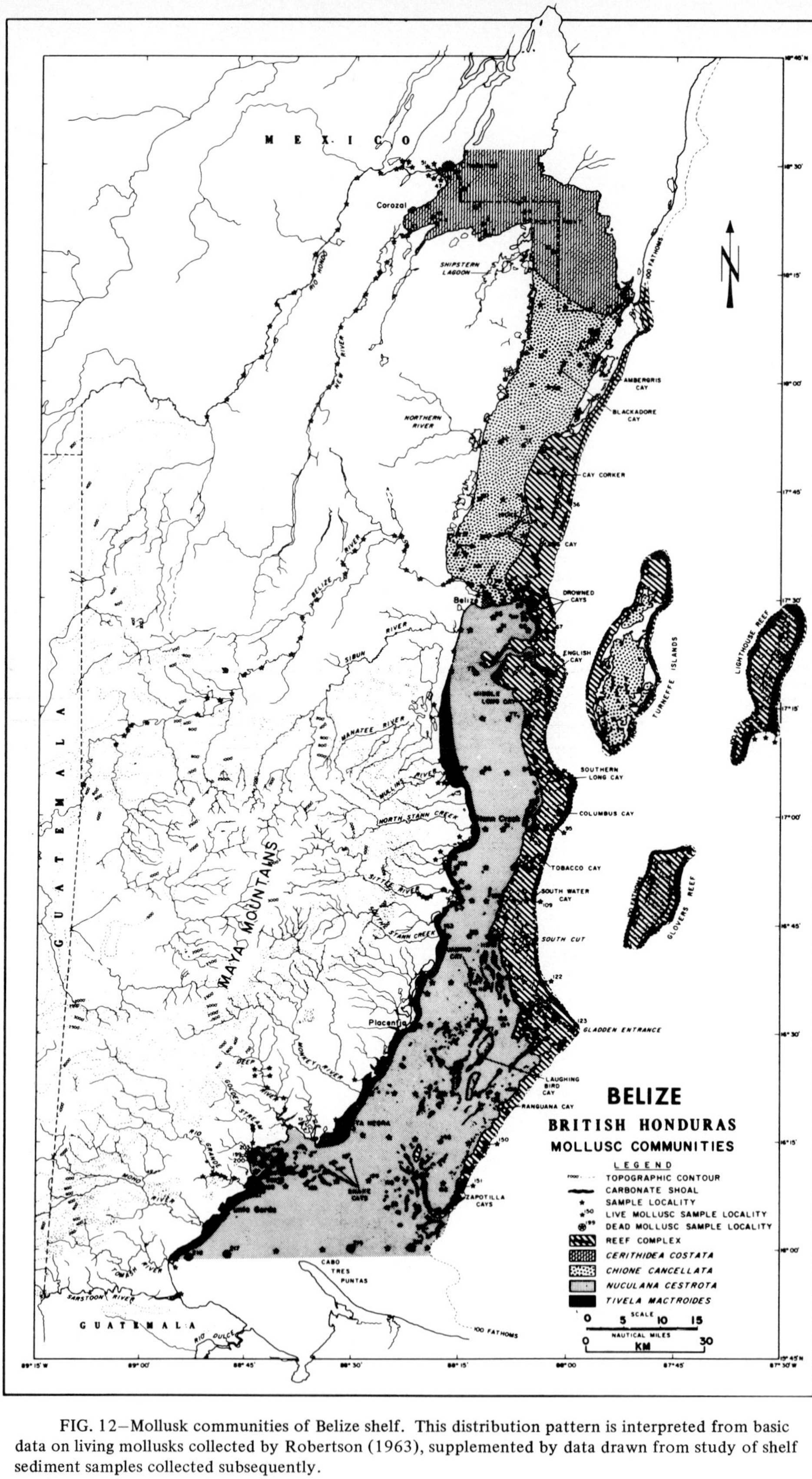

FIG. 12–Mollusk communities of Belize shelf. This distribution pattern is interpreted from basic data on living mollusks collected by Robertson (1963), supplemented by data drawn from study of shelf sediment samples collected subsequently.

It is clear that the Swamp and Reef-complex assemblages are habitat communities. In contrast, the remaining groups are in large part Petersen communities, although even they involve some habitat bias. Approximately 350 species of marine mollusks were collected by Robertson (1963). Of these, only 286 species occurred at two or more sampling stations with a frequency of five or more individuals per sample (Robertson, personal commun., 1965; see Appendix). Three of these species are either planktonic or nektonic, and therefore are not pertinent to an analysis of bottom communities. The other 283 species are distributed among the six communities as indicated in Table 2. As one would anticipate, the Reef-complex community, with its hodgepodge of environments, has the greatest number of species. The decrease in species from the Reef-complex community to the *Chione cancellata* community, and then to the *Cerithidea costata* group, is correlated with a decrease in bottom salinity in the same direction (Table 2). Considering the taxa involved (Appendix), it seems reasonable to presume a cause-and-effect relation with respect to salinity. The decrease in value of the same attributes into the southern shelf lagoon is not so readily explained, because the bottom salinity is essentially the same for both the Reef-complex and *Nuculana cestrota* communities. The major habitat difference there is the mud content of the *Nuculana cestrota* substrate (Table 2). It exceeds by a considerable margin that of any other habitat in Belize. A second characteristic of the *Nuculana cestrota* habitat is the notable lack of macroscopic flora. With the exception of scarce specimens of *Halophila baillonis* that were dredged occasionally from the shoaler parts of the habitat, the community is devoid of macroscopic plant life. This combination of soft substrate and no macroscopic flora dictates a preeminently infaunal mode of existence, as opposed to the dominant epifaunal nature of the Reef-complex mollusks (Table 2). What is surprising, however, is the high percentage of species that are specific to the *Nuculana cestrota* community, for its percentage of endemic mollusks is greater than that of any of the other mollusk associations (Table 2). The soft consistency of the substrate apparently requires special adaptive measures, resulting in the highly endemic mollusk population. Among other things, the thinness of the mollusk shells in this habitat affords a striking contrast with their relative thickness in the other molluscan communities. In this particular instance, a thin shell seems to be an obvious advantage in keeping the weight of an animal light in a substrate whose consistency is so soft that the organism constantly threatens to "drown" in it. If this reasoning is correct, then an abundance of thin-shelled fossil mollusks may be a clue to the consistency of a fossil substrate.

Calculations of affinities among the shelf mollusk communities (Table 3) demonstrate in yet another way the highly individual nature of the *Nuculana cestrota* community. The same table also documents an unexpectedly high affinity between the nearshore *Tivela mactroides* community and the Reef-complex assemblage. This relatively close relation results primarily from the fact that both areas are characterized by a wide range of common epifaunal habitats (e.g., marine grass, dead and live coral heads, etc). The major difference between these two habitats is the greater number of marine epifaunal opportunities offered by the Reef-complex environment. In fact, it is fair to say that a coral reef the size of the Belize barrier reef is the underwater equivalent of a tropical rain forest in that the variety of life forms undoubtedly is maximized for marine environments in a coral reef in the same way that it is for terrestrial environments in a tropical rain forest. The analogy can be extended further by comparing the gross habitat role of the corals in the reef to that of the trees in the rain forest.

Wantland (this volume) noted a similar phenomenon in the distribution of Foraminifera. Maximum diversity in this group is found in the *Thalassia*

TABLE 2. SOME ATTRIBUTES OF BELIZE MOLLUSK ASSEMBLAGES

Community	*Number of Species*	*Percent Diversity*[1]	*Percent Endemic*[2]	*Percent Epifauna*	*Percent Infauna*	*Bottom Salinity* ‰[3]			*Weight Percent Mud*[3]		
						N	$\overline{X}$	s	N	$\overline{X}$	s
Swamp	12	3	58	56	44	INSUFFICIENT DATA					
Cerithidea costata	8	2	25	13	87	14	17	4.7	6	47	9.5
Chione cancellata	15	4	40	21	79	3	29	29.5	13	36	4.3
Reef Complex	157	41	52	66	33	26	35[4]	1.9	28	13	20.0
Nuculana cestrota	70	19	70	17	83	15	35	.9	21	82	19.1
Tivela mactroides	119	31	31	54	56	8	32	2.5	11	14	26.8

[1]Diversity expressed as percentage of total number of species occurrences for all communities (N = 381). Thus for the swamp community 12/381 x 100 = 3 percent.

[2]Percentage of endemic species calculated as the percentage of the total number of species occurring in the community that are not found elsewhere. For the swamp community 7 species are found only within that assemblage; thus 7/12 x 100 = 58 percent.

[3]These data refer to samples taken at the same time and in the same place as Robertson's collection of live mollusks. The averages for bottom salinity and mud content of the substrate are in some instances based on a small number of samples, but they are broadly consistent with the bottom salinity pattern recorded previously (Fig. 5) and the textural trends illustrated subsequently (Fig. 14) for a much larger number of samples distributed over the entire shelf.

[4]Strictly speaking, this is a surface salinity average. Bottom salinities in general were not measured in the Reef-complex area because there seemed to be no appreciable difference between surface and bottom salinity where measured and because the general shoal nature of the habitat suggested rapid dissipation of any small differences that might occur from time to time.

TABLE 3. AFFINITIES AMONG BELIZE MOLLUSK COMMUNITIES[1]

	Swamp	*Cerithidea costata*	*Chione cancellata*	*Reef complex*	*Nuculana cestrota*	*Tivela mactroides*
Swamp	XX	.00	.09(.00)[2]	.03(.03)	.00	.01(.01)
Cerithidea costata	.00	XX	.17(.03)	.02(.01)	.00	.04(.01)
Chione cancellata	.09(.00)[2]	.17(.03)	XX	.05(.02)	.00	.06(.01)
Reef complex	.03(.03)	.02(.01)	.05(.02)	XX	.09(.05)	.36(.25)[3]
Nuculana cestrota	.00	.00	.00	.09(.05)	XX	.13(.06)
Tivela mactioides	.01(.01)	.04(.01)	.06(.01)	.37(.25)[3]	.13(.06)	XX

[1]Calculation example for affinity between Reef-complex and *Nuculana cestrota* communities. Total numbers of extra-community occurrences for species of the Reef-complex and *Nuculana cestrota* communities are 75 and 23, respectively. Number of species in common is 9. Affinity = $\frac{9}{75 + 23}$ = .09. Numbers in parentheses refer to epifaunal affinity; calculated in same way with respect to number of epifaunal species in common.

[2]One species that may be either epifaunal or infaunal has not been included in the epifaunal affinity value.

[3]Two species that may be either epifaunal or infaunal have not been included in the epifaunal affinity figure.

meadows of the barrier platform, and there is an affinity of this assemblage with that found in the *Thalassia* meadows along the mainland coast of southern Belize. The relatively high diversity of both assemblages is attributed to the diverse epifaunal opportunities available in the marine-grass community.

Other examples of the ecologic consequences of the reef environment in particular, and the ecology and distribution of shelf fauna in general, are provided by Teeter's study of ostracods (this volume) and Wantland's study of Foraminifera (Wantland, 1969; this volume).

SEDIMENTS

Composition

The carbonate content of the shelf sediments was determined by weight loss after treatment with 5-percent nitric acid solution. Replicate analyses indicate that data reproducibility is generally within 5 percent of the values obtained.[8] The few greater differences were due to the presence of one or more relatively large shells in the sample split, but even these differences are well within the 30-percent contour interval used to illustrate the areal variation of carbonate content in Figure 13. It is evident that the entire northern shelf and the southern barrier platform are carbonate provinces, whereas the southern shelf lagoon is transitional between nearshore deposition of terrigenous clastic material and barrier-platform deposition of calcium carbonate. The Gulf of Honduras is similar in this regard except that the areal extent of the terrigenous influence is considerably greater.

Texture

The weight-percent mud (silt and clay) in the samples provides a crude but generally reliable index of relative bottom-current intensity, particularly if the data are used to establish general trends rather than specific sample comparisons. The wet-sieve percentages[9] are contoured in Figure 14. Data reproducibility is again generally within 5 percent of the reported value. The illustrated contour patterns provide a further characterization of the shelf sediments. The barrier-rim, outer-platform, and barrier-platform sediments are sandy, as are the nearshore sediments. In contrast, the shelf-lagoon deposits are decidedly muddy, and the southern shelf lagoon is much muddier than its northern counterpart.

FACIES

The composition of the sediment provides a first-order facies distinction between terrigenous and carbonate sediments and also establishes a third category that is transitional between these two end members. Texture secondarily characterizes these primary subdivisions as sands, sandy muds, and muds; and the dominant or characteristic element of the coarse fraction (coarser than 1/16 mm) provides a third means of identification. These three parameters yield the facies map illustrated in Figure 15. In northern Belize the constituent composition of the coarse fraction has been determined by point-count analysis of thin sections (Pusey, 1964); in southern Belize the dominant elements composing the

[8] See footnote 5.
[9] See footnote 5.

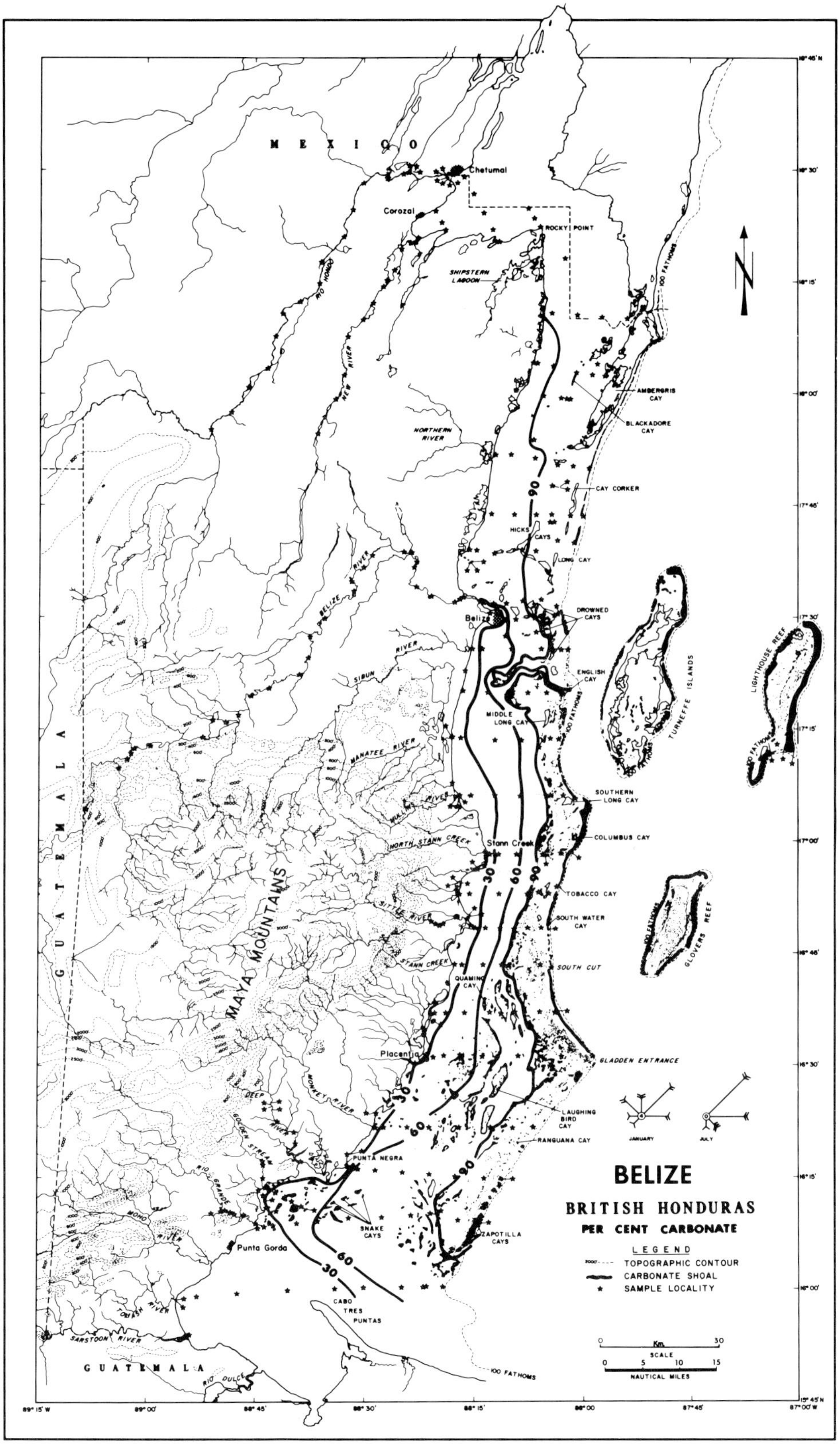

FIG. 13—Percent carbonate in Belize shelf sediments.

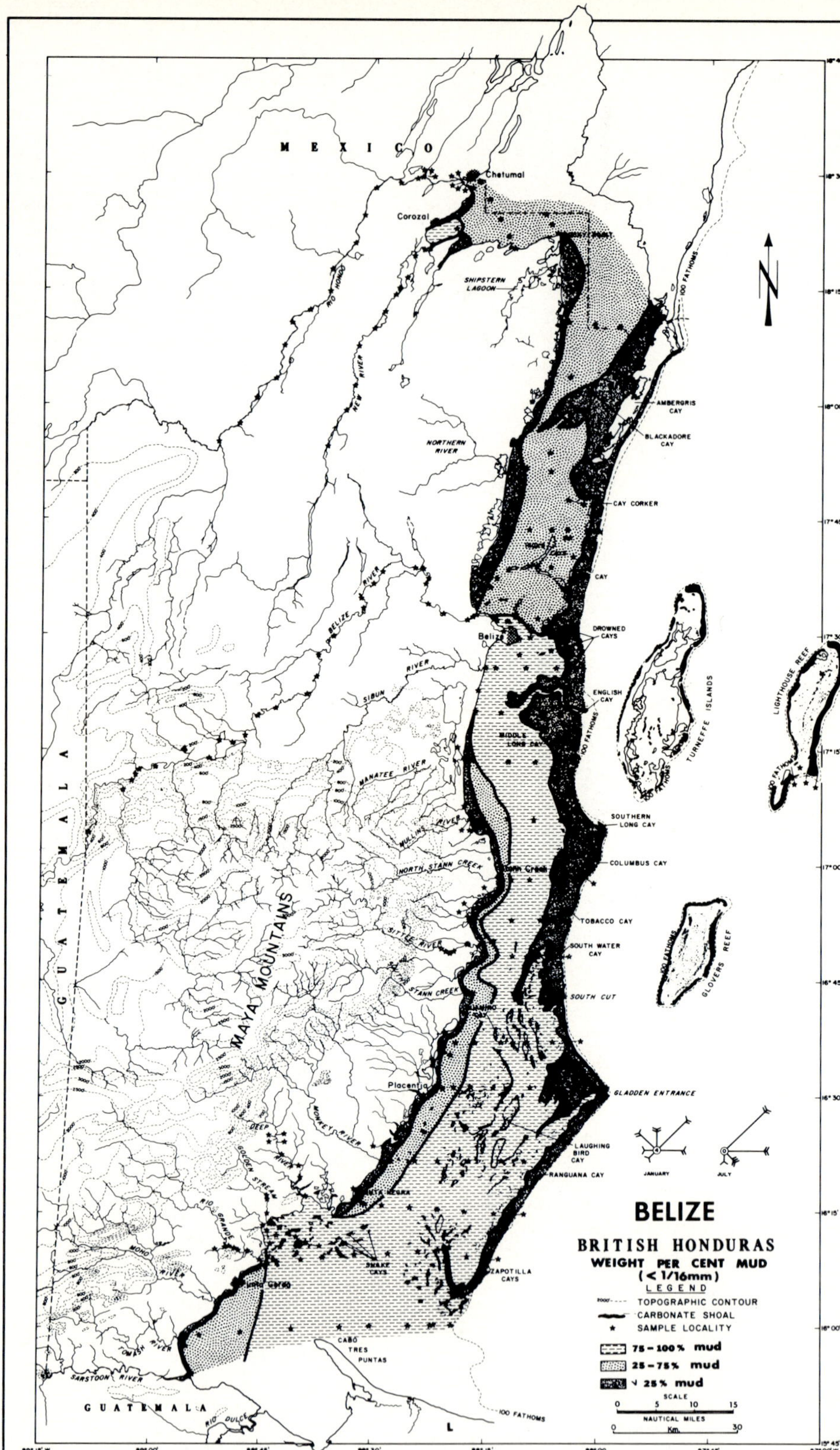

FIG. 14—Weight percent mud (<1/16 mm) in Belize sediments.

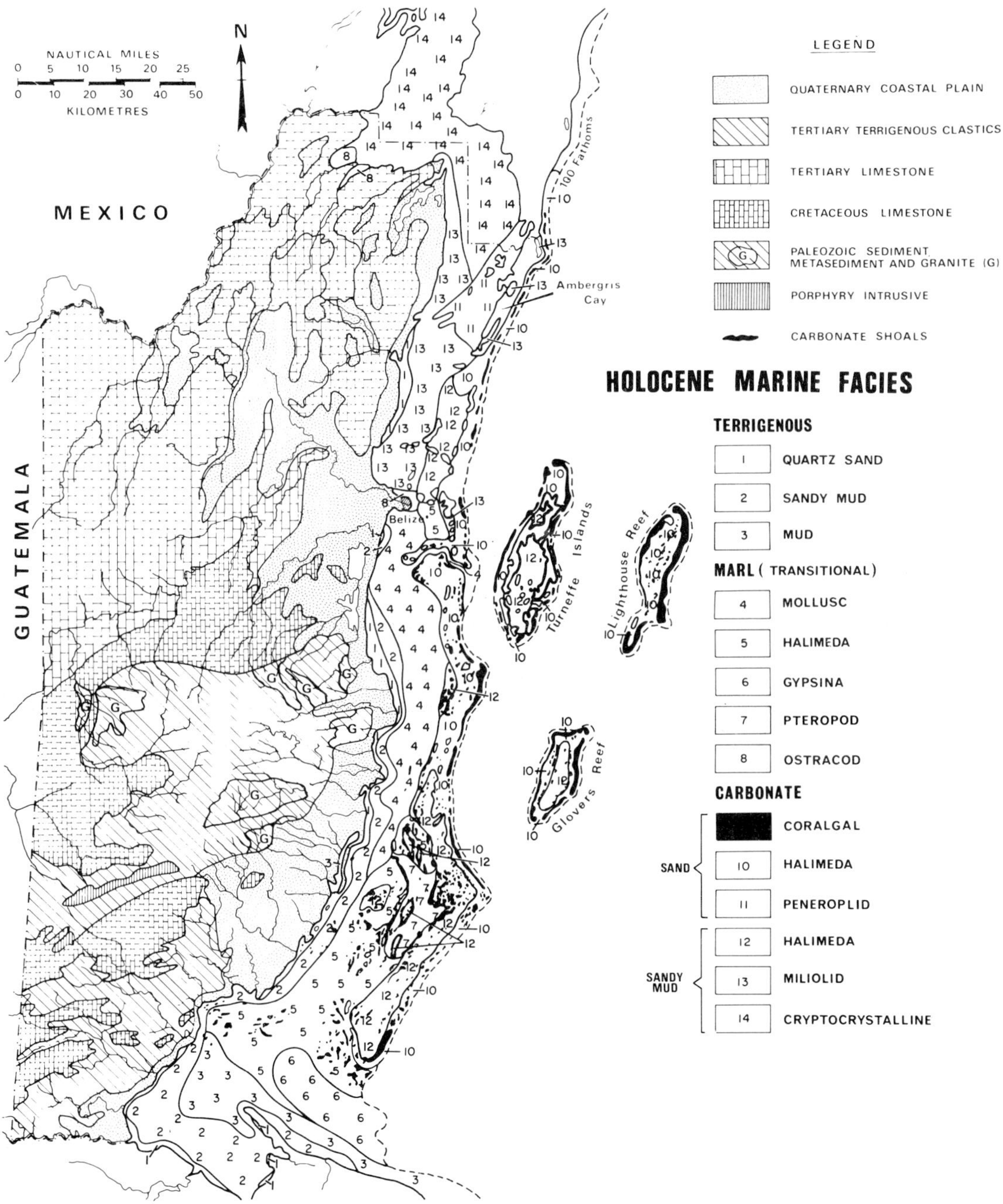

FIG. 15—Major sedimentary facies of Belize shelf. This map illustrates distribution of surficial sedimentary facies on marine shelf as well as age and lithologic types of major rock units outcropping on Belize mainland.

same fraction have been estimated by visual inspection of thin sections. This difference in mode of analysis is unfortunate, but it is doubtful that point-count analysis of the southern Belize sections would produce any meaningful differences with respect to the facies identified in Figure 15, because the differences among these facies are by no means subtle. Nevertheless, the pertinent quantitative and qualitative data are summarized in Table 4.

TABLE 4. CARBONATE CONTENT, GRAIN SIZE, AND CONSTITUENT COMPOSITION OF BELIZE FACIES (N=308)

FACIES	Mean Weight Percent $CaCO_3$[1]	Mean Grain Size[2]: Weight Percent 1/8-1/16 mm	Weight Percent less than 1/16 mm	Mean Volume Percent of Fraction Greater Than 1/16 mm[3]: Coralline Algae	*Halimeda*	Foraminifera	Corals	Mollusks	Pteropods	Ostracods	Cryptocrystalline Grains	Terrigenous Clastics	No. of Thin Sections	Total No. Samples[4]
Terrigenous														
Northern quartz sand[5]	79(4)	8(4)	10(4)	-	-	13	-	28	-	-	22	19	4	4
Southern quartz sand[6]	8(14)	6(16)	8(23)	-	-	-	-	-	-	-	-	99	14	23
Sandy mud[6]	30(28)	7(27)	61(24)	-	3[9]	-	-	14	-	-	-	76	17	28
Mud[6]	23(8)	3(8)	88(8)	-	-	23	-	12	10	-	-	50	3	8
Marl														
Ostracod[7]	73(5)	7(6)	89(6)	-	-	16	-	21	-	20	26	5	6	6
Mollusk[6]	54(36)	5(32)	83(37)	-	16[10]	11	-	37	-	-	-	32	22	37
Halimeda[6]	70(46)	7(41)	72(47)	-	43	5	21	14	-	-	-	4	28	47
Pteropod[6]	8(8)	3(8)	94(8)	-	22	14	4	8	39	-	-	-	8	8
Gypsina[6]	76(5)	11(5)	36(5)	12[11]	20[11]	52[11]	9	3	4	-	-	-	4	5
Carbonate														
Cryptocrystalline sandy mud[7]	90(15)	19(15)	34(15)	-	-	10	-	16	-	-	67	-	15	15
Miliolid sandy mud[7]	89(17)	20(17)	60(17)	-	-	34[12]	-	22	-	4	22	-	17	17
Halimeda sandy mud[8]	94(30)	14(30)	62(31)	-	40	6	22	17	-	-	-	2	33	33
Halimeda sand[8]	97(30)	8(34)	14(36)	2	50	10	18	16	-	-	-	-	35	36
Coralgal sand[8]	98(28)	3(24)	4(28)	14	26	6	37	9	-	-	-	-	28	28
Peneroplid sand[7]	95(13)	6(13)	10(13)	-	-	56[13]	-	18	-	-	20	-	13	13

[1]Percent carbonate determined from the amount of acid-insoluble residue left after treatment with 5 percent nitric acid.

[2]Numbers in parentheses refer to the number of samples analyzed.

[3]Only the major facies determining constituents are shown, and thus the indicated average constituent composition of each facies does not total to 100 percent. *Averages of less than 2 percent have been omitted.*

[4]In some cases data on carbonate content and grain size were available for samples from which no thin sections had been prepared. Usually these two parameters in conjunction with geographic proximity to thin-sectioned samples provided an unequivocal means of facies assignment. Thus the total number of samples within the facies sometimes exceeds the number of thin sections.

[5]The quartz-sand facies of Figure 15 has been subdivided here to show the considerable compositional difference between the area north and south of Belize City. This difference extends to the nature of the terrigenous material itself, for the quartz sand grains of the north are associated in the south with fragments of quartzite, which in places exceed the individual quartz grains in abundance, and with feldspar and scattered grains of microcline. Grain size and point-counted constituent composition of the northern quartz sand are from Pusey (1964).

[6]Constituent composition based on thin-section *estimates* of volumetric abundance.

[7]Grain size and point-counted constituent composition from Pusey (1964).

[8]Constituent composition is based largely on thin section *estimates* of volumetric abundance but some point-count data of Pusey (1964) are also involved. Similarly, some of Pusey's grain-size statistics were included in computing the indicated means.

[9]The *Halimeda* content reflects the sporadic occurrence of "carbonate communities" along the coast. Near Placentia these communities include the prominent barrier-reef coral, *Acropora palmata*.

[10]The *Halimeda* fragments have been derived in large part from the barrier platform.

[11]These grain types are associated with each other in a consortiumlike fashion. *Gypsina* is the only volumetrically significant foraminifer present.

[12]Miliolids constitute 22 percent of the coarse fraction (Pusey, 1964, p. 153).

[13]Peneroplids constitute 47 percent of the coarse fraction (Pusey, 1964, p. 134).

Terrigenous Facies

These sediments are confined largely to the southern shelf. The reason for this is that the northern Belize mainland is a low-lying limestone terrane, whereas southern Belize is a predominantly noncarbonate terrane of considerable relief (Fig. 15). Correspondingly, the northern coastal-plain sediments are represented by prograding carbonate units, and those of the south consist of prograding terrigenous clastic units.

The quartz-sand facies of northern Belize is only a few inches thick and reflects reworking of an underlying gray clay containing abundant sand- and silt-size quartz grains (Pusey, 1964, p. 189). High (1966) suggested that in part this basal clay may be a residual soil developed on top of the underlying, pre-Holocene limestone; but, whatever its origin, the quartz-sand facies is much more abundant in southern Belize. There it is represented not only by angular quartz grains but also by fragments of quartzite—which in some places exceed the individual quartz grains in abundance—and by feldspar including scarce microcline. This sand facies in southern Belize extends no more than a mile from shore, where it is succeeded by the finer grained sediments of the terrigenous sandy-mud facies. Aside from the distinction in texture, the only other significant difference between these two facies is the increasing amount of molluscan skeletal debris in the same direction. This increase probably reflects dissipation of the environmental stress condition imposed by the shifting, megarippled sand substrate as it becomes stabilized and thus muddier with increasing distance from shore. The increase in areal extent of terrigenous sandy mud in the Gulf of Honduras, as well as the presence of a terrigenous mud facies in the same area, is simply an expression of the increased land drainage that is focused seaward by the cul-de-sac that opposes the prevailing wind direction. Other than texture, there is no appreciable difference in the coarse fraction (>1/16 mm) of these two sediment types.

In several places, sand-size quartz grains are present in the shelf lagoon some distance from shore. The extremely soft mud in which the quartz grains are found, and which separates them from the mainland, precludes traction transport or saltation across the shelf lagoon. Some of the quartz grains may have been reworked by burrowing from the pre-Holocene terrigenous clastic sediments that are known to underlie the shelf lagoon in places. Piston cores from the southern shelf lagoon typically show a high degree of Holocene burrow mottling (Fig. 16), but the Holocene - pre-Holocene contact is usually distinct and therefore not suggestive of much reworking across the unconformity. Some other explanation seems to be required. Wolf (1961) described an instance of flotation of sand-size quartz grains into the Australian Gulf of Carpentaria by tidal rivers. Rising river water apparently is able to pick up sun-dried quartz and transport it as floating patches of sand that are supported by the surface tension of the water. Experimentally, surface tension has been shown to be capable of supporting quartz grains up to 5 mm in diameter (Wolf, 1961, p. 477).

Measurements of the maximum thin-section diameter of quartz grains from the shelf-lagoon muds can be as great as 32 mm (sample 286, Fig. 3), but the diameters are commonly much smaller than 5 mm within a given sample (e.g., samples 357 and 358). Because freshwater wedges occur throughout the length of the shelf lagoon, it seems plausible to assume that some of the smaller quartz grains were transported into the shelf lagoon by freshwater flotation. Ultimate wetting of the grain surfaces through wave agitation would then account for their deposition. The larger quartz grains most likely have been reworked from underlying rocks.

With regard to the terrigenous clay component of the sediment, Krueger (1963) noted that northern rivers transport chiefly montmorillonite,

FIG. 16—Slabbed piston core of sediments from southern shelf lagoon (Station 120). Core shows high degree of burrow mottling in fine-grained sediment.

whereas southern rivers carry relatively less montmorillonite and more kaolinite and illite. These differences are related to differences in provenance and climate and are reflected in the distribution of clays on the shelf (Scott, 1966; this volume). Within this frame of reference, however, the relative abundance of clay minerals appears to be strongly controlled by settling velocities. In the southern shelf lagoon, Scott (this volume) noted a consistent tendency for montmorillonite to increase in abundance, relative to either kaolinite or illite, in an offshore direction. She related this tendency to the apparent dependence of clay-mineral settling velocities on salinity as determined experimentally by Whitehouse et al (1960).

Carbonate Facies

It is apparent from Figure 15 that the coralgal and *Halimeda* sands and the *Halimeda* sandy muds are common to both northern and southern Belize. The coralgal facies includes the barrier reef and patch reefs of the outer and barrier platform, as well as the innumerable reef-dominated carbonate shoals in the southern shelf lagoon. The sediment is characterized by a relative abundance of corals and coralline algae. *Halimeda* fragments are commonly more abundant than either of these constituents but, collectively, corals and coralline algae are more abundant and, moreover, are relatively important only within this facies.

Where the barrier reef is continuous along the edge of the shelf, as in the vicinity of Columbus Cay, there is an exemplary lagoonward textural gradation from an algal-encrusted coral-rubble flat through a reef apron of coralgal sands (Fig. 17). The reef itself is the principal source of both the rubble and sand. The continuous buffeting by waves, including the action of hurricanes, provides reef debris that is transported lagoonward by wave-generated currents. The strength of these currents diminishes lagoonward from the point where they are created by the waves breaking on the reef, resulting in the textural gradation from rubble flat to sand apron. In many places the apron sand seems to be

FIG. 17—Oblique aerial view of barrier-reef sequence near Columbus Cay (upper right), looking south. Sequence from left to right shows well-developed spur and groove structure dipping steeply into Caribbean Sea perpendicular to reef trend, a narrow light band of coarse rubble directly shoreward of living barrier reef, a darker band of algal-coated rubble on reef flat, a broad reef apron of rippled carbonate sand, and dark-colored, shallow-water grass (*Thalassia*) meadows surrounding Columbus Cay.

FIG. 18—Oblique aerial view of reef apron near Columbus Cay. In this area reef apron is exceptionally broad as a result of wave refraction around seaward bend in reef trend. Width of reef flat and reef apron at this locality is approximately 0.5 mi (0.8 km).

definitely transgressing over the adjacent barrier platform; this effect is particularly pronounced at reef corners (Fig. 18), where wave refraction insures a wider-than-usual expanse of reef apron. Interestingly, the opposite tendency is apparent near reef passes, where the reef apron seems to narrow markedly (Fig. 19). The barrier reef shows no signs of growth inhibition on either side of these passes, and thus it is reasonable to believe that the production of reef debris is comparable to that observed elsewhere along the reef. The implication is that the return (seaward) flow of water from waves breaking on the reef transports most of this debris into the reef passes. Since the passes seem to be characterized in general by a relatively thin veneer of unconsolidated sediment, a further inference of transportation over the shelf edge seems warranted. Thus the relatively deep-water deposits at the base of the shelf should be associated with cones of shallow-water reef debris at the mouths of reef passes. Whether this is the case remains to be demonstrated. A variation on this theme concerns the lack of clearly defined apron sands in other barrier reefs such as the Bahamian Andros reef and the Florida reef tract. These reefs differ from the Belize example in that they are not nearly as continuous along their length. They are discontinuous or open reefs as compared to the relatively continuous or closed Belize barrier. The more open the reef barrier, the greater is the probability that a considerable quantity of potential apron sand will escape seaward. Consequently, open and closed barriers should differ not only with respect to the presence and/or integrity of the reef apron, but also with respect to the volume of potential apron debris that is transported seaward to be deposited in deeper water. These relations should also exist for fossil reefs.

FIG. 19—Oblique aerial view along trend of barrier reef near South Water Cay (center of photo). In this area linear reef trend is broken by numerous passes which provide pathways for seaward transport of reef debris. At this location reef apron is narrow, approximately 0.25 mi (0.4 km) in width.

The *Halimeda*-sand facies is, as the name suggests, dominated by fragments of this calcareous alga. The greater abundance of *Halimeda* skeletons in this facies than in the coralgal facies is largely related to the decrease in quantity of corals and coralline algae. The *Halimeda* sandy mud is simply a finer grained part of the same facies.

The other three carbonate facies are confined to the north and have been described in detail by Pusey (1964; this volume), although his emphasis and conclusions differ somewhat from those offered here. The abundant peneroplid foraminifers in the peneroplid sand are largely altered to cryptocrystalline carbonate, and there are few indications that they are living in the environment in which they are presently found (Wantland, 1967, p. 75). Taken together, these factors imply a "relict" deposit. The occurrence of similar current-winnowed accumulations of peneroplids along some of the western beaches of Ambergris Cay and along the borders of the intra-island lagoons of the same island (Ebanks, 1967) suggests further that these are drowned beach deposits, perhaps outlining the former extent of Ambergris Cay. The miliolid sandy-mud facies reflects the influence of both decreasing current intensity and decreasing salinity. The increase in mud content is correlated with the former, and the virtual absence of *Halimeda* and abundance of miliolid Foraminifera are correlated with the latter. In Chetumal Bay, even lower salinities further restrict the presence of skeleton-bearing organisms, and the result is the predominance of the nonskeletal cryptocrystalline grains that characterize the cryptocrystalline sandy mud. There appears to be very little *absolute* increase in cryptocrystalline grains as compared to the miliolid sandy mud; the relative change is due chiefly to the decrease in number of skeletal fragments, presumably because of the salinity effect. It is pertinent to note that the salinity change toward Chetumal Bay may not always be in the direction of lower salinities. The present salinity observations were made during the wet season, but during the dry season in February-April the Bay may become hypersaline, because halophilic red mangroves have been observed as much as 80 km (50 mi) upstream in the Rio Hondo and New River (Pusey, 1964, p. 36). Perhaps it is this extreme seasonal variation that is largely responsible for the relative paucity of organisms. It is also interesting to note that the mud fractions of the *Halimeda*, miliolid, and cryptocrystalline sandy muds show a facies change from aragonite to high-Mg calcite to low-Mg calcite, respectively (Pusey, 1964, p. 193, 209-218).

Marl Facies

The word "marl" is used to describe sediments transitional between the terrigenous and carbonate end members. Specifically, it refers to sediments in which, on the average, more than *35 percent* of the grains are finer than 1/16 mm and the total calcium content ranges between *50* and *85 percent*. In the north the only marls present are those dominated by ostracods off the mouths of the New and Belize Rivers. The virtual absence of most other organisms with hard parts can be ascribed chiefly to the highly variable salinity conditions in those places. The specific abundance of ostracods is perhaps directly or indirectly related to nutrients transported by the rivers. However, their absence off the mouths of other rivers is perplexing. Marl facies (except those dominated by ostracods) are virtually absent in the north but are predominant in the south.

Mollusk marl is a marl facies in which the coarse fraction is dominated by mollusk shells and also by Foraminifera. A few of the foraminifers are infilled and replaced by pyrite. Less commonly, a greenish material presumed to be glauconite infills and replaces skeletal fragments. Farther south the marl also becomes characterized by *Halimeda* plates (Fig. 20). The marl bottom is devoid of a macroscopic flora, so the source of the plates must be the large number of carbonate shoals that begin to appear in the shelf lagoon. These shoals are absent to the north where *Halimeda* fragments are not present in the mollusk marl. The plates within the small area of *Halimeda* marl at the northern extremity of the southern shelf lagoon (Fig. 15) were derived from the adjacent barrier platform and from part of the northern shelf lagoon. The same facies exists as a narrow zone lagoonward from the northern barrier platform, but it is too narrow to illustrate at the scale of the facies map in Figure 15. The *Halimeda* marls are thus primarily an index of bathymetric relief expressed in the form of carbonate shoals. This bathymetric diversity persists into the Gulf of Honduras, where the shoals are deeply submerged (150-225 ft or 46-69 m) and thus no longer visible from the air. Coupled with the increasing depth of shoal submergence is the appearance, on the shoals, of the foraminifer *Gypsina* encrusting extremely large *Halimeda* plates (Fig. 21). Coralline algae and the coral *Agaricia* have also been recovered by grab samples, and it would seem that these deposits are similar to, if not identical with, the Yucatán shelf *Agaricia-Montastrea* and *Gypsina-Lithothamnium* communities described by Logan (1969, p. 151-152). Certainly the depth relations are comparable, but the important point is that the shoals, like those of the *Halimeda* marl, are the local source for the sand-size encrusted *Halimeda* plates. Viewed in this context the *Gypsina* marl is a deeper water equivalent of the *Halimeda* marl.

The absence of *Halimeda* in particular and a macroscopic flora in general from the level bottom of the southern shelf lagoon is puzzling. The murky green color of the water in this region and the relatively

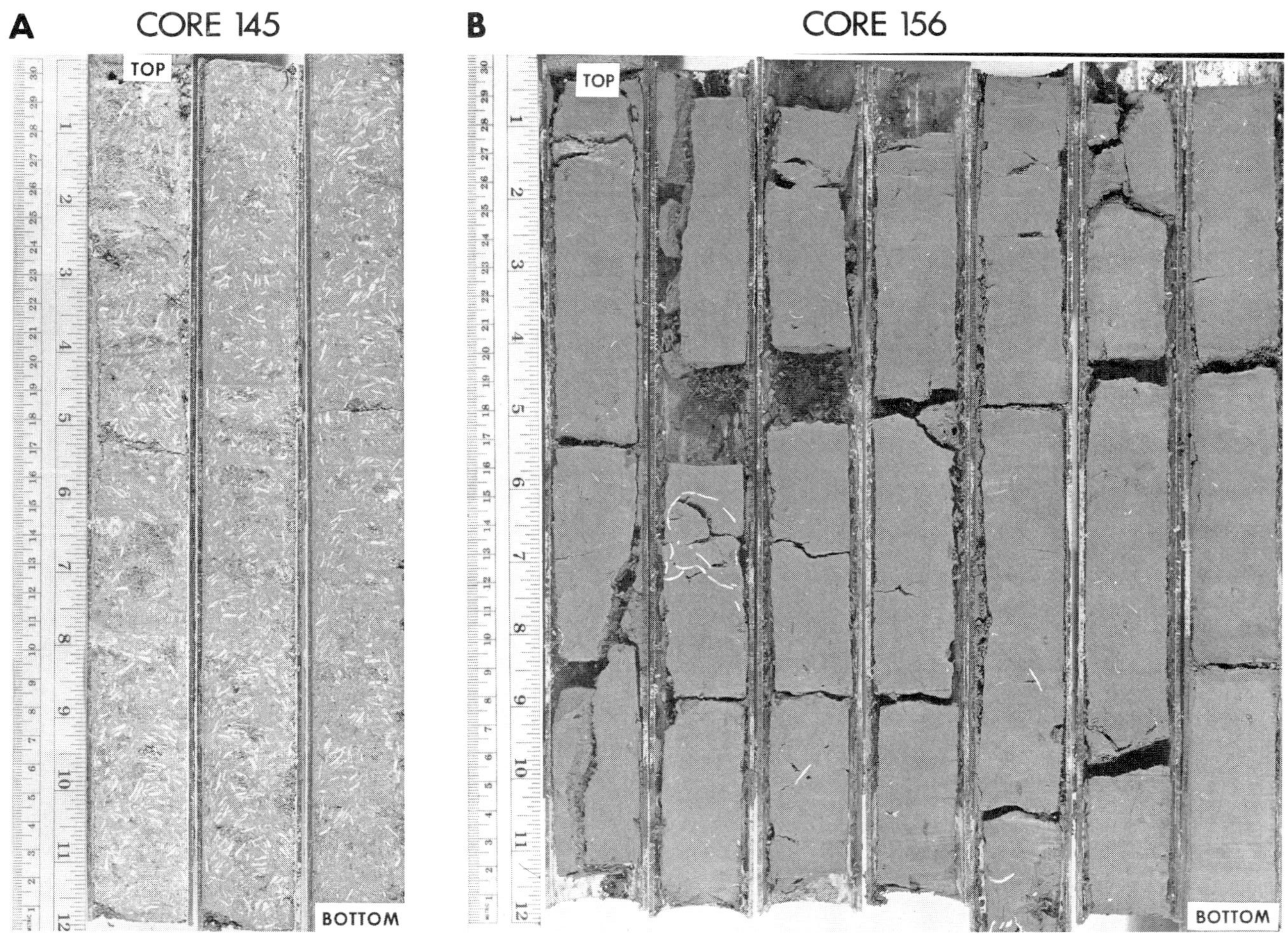

FIG. 20–Slabbed piston cores illustrating contrast in *Halimeda* content within *Halimeda* marl facies. Core in **A** (Station 145) is replete with *Halimeda* plates whereas core in **B** (Station 156) is virtually devoid of plates. Skeletal fragments visible in core 156 are those of an echinoid. To some extent this contrast may be a function of plate size, smaller plates being less apparent as well as less abundant than their larger counterparts. Alternatively, contrast may reflect significant differences in depositional history, because *Halimeda* marl designation is based on plate content of *surface* sediment samples. Plate-poor cores (**B**) are decidedly the norm rather than the exception.

A

CORE 213

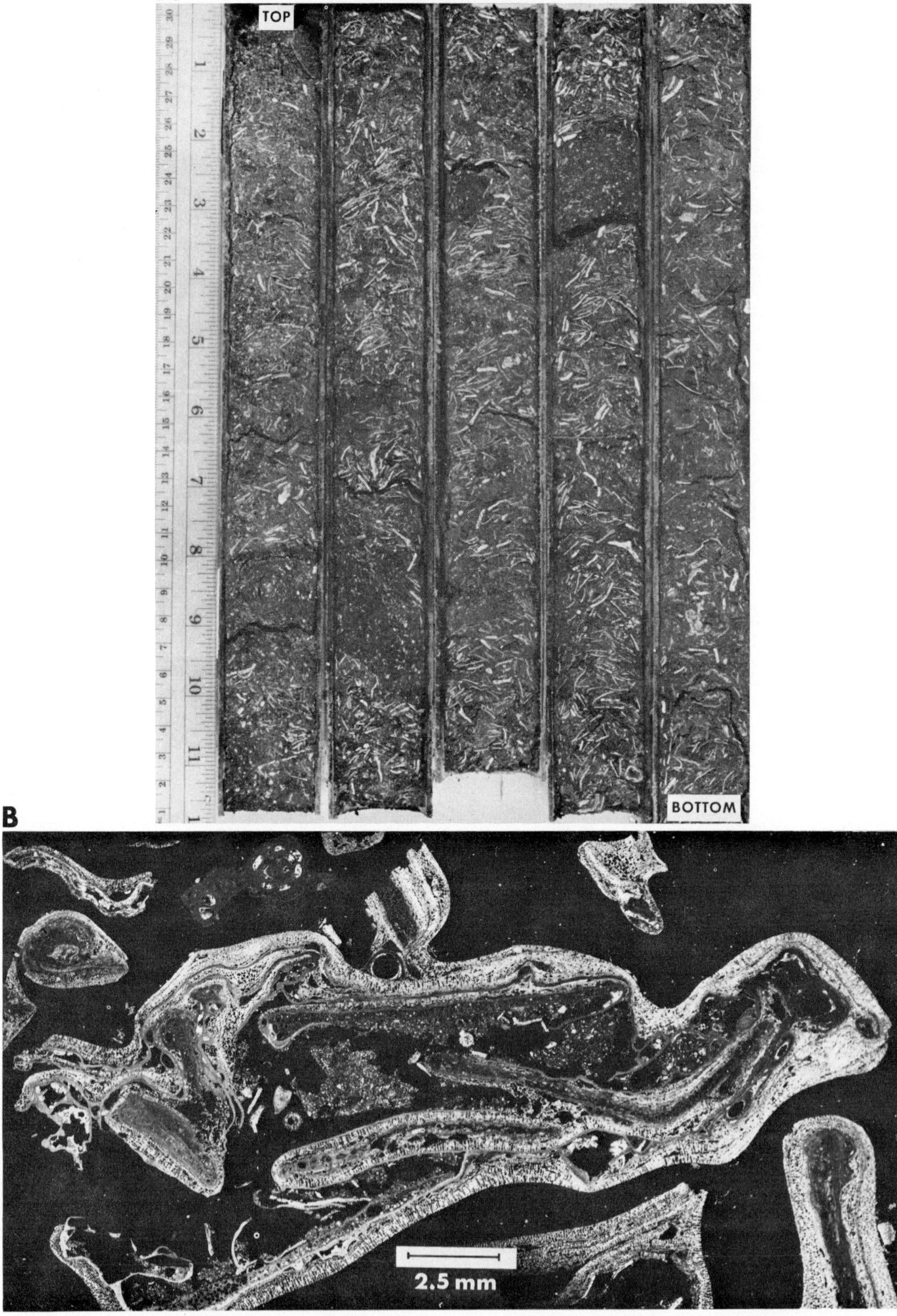

FIG. 21–Slabbed piston core and thin-section photomicrograph of sediments in *Gypsina* marl facies (Station 213) from Gulf of Honduras area. **A.** Large, shoal-derived *Halimeda* plates are abundant in core. Water depth of this station is 122 ft (37.5 m). **B.** Thin section of surface sediment demonstrates that *Halimeda* plates are coated by encrusting foraminifer, *Gypsina*. Polarized light.

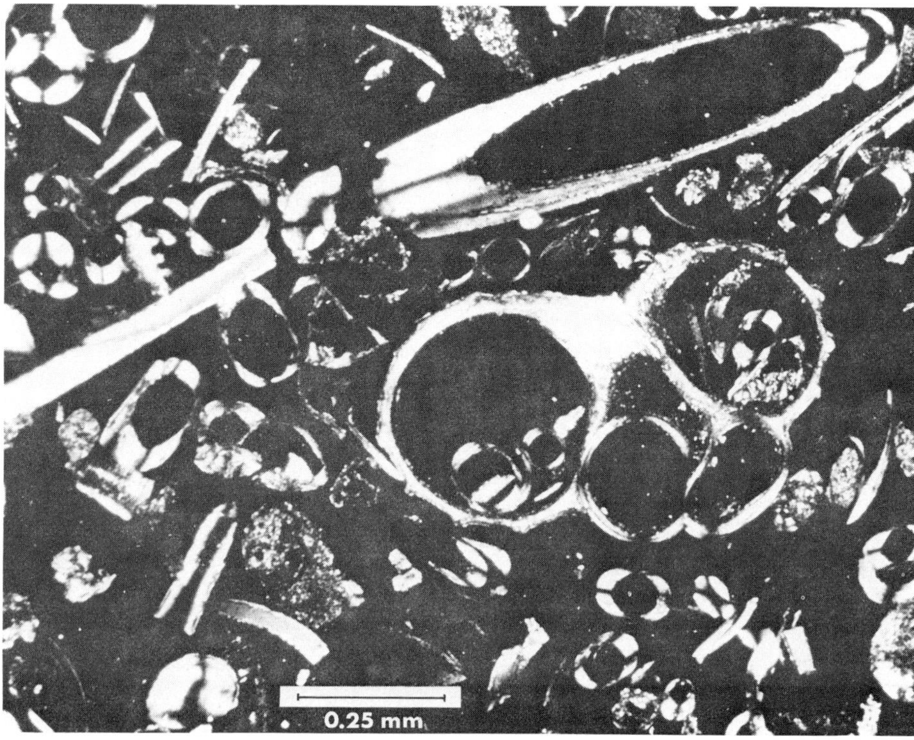

FIG. 22—Thin section of sediments from Victoria Channel near Laughing Bird Cay (Station 169) in pteropod marl facies. Polarized light.

great water column through which light has to pass suggest that the bottom may be below the zone of effective light penetration, resulting in the absence of flora. However, Bruce Purser (personal commun.) has pointed out that a similar situation prevails in the Persian Gulf, and the clarity of the water there demonstrates that turbidity is not the limiting factor. The problem is certainly deserving of further study because of the insight it may provide on paleoecologic reconstructions. Regardless of cause, the widespread distribution of *Halimeda* relative to other carbonate-shoal grain types is indicative of preferential transportation. The reason for this is probably the waferlike shape of the *Halimeda* plates in conjunction with their hollow internal tubes (utricles). Both attributes combine to produce a particle with a relatively large surface area and low bulk density; thus the plates are more readily transported than similar sized fragments that are less flattened and/or have a greater bulk density (e.g., corals and articulated coralline algae). With regard to transportability, *Halimeda* is very much the carbonate analogue of terrigenous mica.

The only remaining facies is the pteropod marl, which is largely restricted to Victoria Channel. Pteropods occur elsewhere in the southern shelf lagoon, particularly south of the Victoria Channel area; however, their abundance is diluted by the presence of *Halimeda*, and it is only in Victoria Channel that they constitute the major part of the sedimentary sand fraction (Fig. 22). In part, the abundance of pteropods in this particular region reflects the fact that the bottom is largely free from shoals and thus is free from sources of *Halimeda* dilution. This is not the only reason for their abundance, however; if it were, the level bottom of the mollusk marl should also be characterized by numerous pteropods, and this is not the case.

Surface salinity within the mollusk marl facies is generally less than 34°/₀₀, whereas in the Victoria Channel region it is generally greater than 34°/₀₀ (Fig. 6). Surface or near-surface salinity is apt to be an important limiting factor in the distribution of planktonic gastropods, and therefore it seems plausible that surface salinity differences are responsible for the observed contrast in pteropod abundance between the mollusk and pteropod marls. South of Victoria Channel the increase in shoal density—and thus in *Halimeda* abundance—and the decrease in surface salinity probably combine to limit further the relative abundance of pteropods in the sediment.

The carbonate mud in the southern shelf lagoon is considered by Matthews (1966) to be the product of autochthonous and allochthonous skeletal breakdown. Scholle and Kling (1972) agree with this general conclusion but question the magnitude of the effect. These authors estimate that between 5 and 20 percent of the bulk sediment sampled by them in the southern shelf lagoon consists of coccoliths, which they assume, with good reason, to be from an indigenous shelf-lagoon population. This occurrence was unknown to Matthews but primarily affects the amount of mud he presumed to be the result of *in situ* breakdown. It does not affect his estimate of the amount of mud derived from the shoals and barrier platform (Fig. 23). Scholle and Kling (1972, p. 203) also noted that the lagoonal carbonate muds could be loosely termed a "coccolith ooze" because no other single identifiable constituent is more abundant and because the general 10-15-percent coccolith content is not too different from the 10-21-percent values reported for some modern examples of *Globigerina* ooze and some ancient chalks. Their point that fossil coccolith-rich sediments should not be regarded as automatic indicators of deep water is especially well taken for potential ancient analogues of the mollusk and pteropod marl. In these facies the wide expanses of level bottom, lacking indigenous calcareous algae, would provide little evidence of how shallow the water really was. Indeed, a fossil equivalent of all the southern marl facies might well be interpreted as indicating a deepening of water from the *Gypsina* and *Halimeda* marl toward the pteropod and mollusk marl—exactly the opposite of the true relation.

The carbonate mud of the northern shelf lagoon contains virtually no coccoliths (Scholle and Kling, 1972, p. 202), and the aragonite mud of the *Halimeda* facies and the high-Mg calcite mud of the miliolid facies are said by Pusey (1964) to have been formed largely by abrasion. The origin of the low-Mg calcite muds in Chetumal Bay is enigmatic but may prove to be similar to that of the freshwater calcitic mud described by Scholl (1964) from the Whitewater Bay region of southwest Florida.

DISCUSSION

Fundamentally, the distribution of Belize facies reflects a structural bias (Purdy, 1974b). Therefore, it can be reasoned that the general distribution of terrigenous versus carbonate sediments dates from the origin of the presently observed structural fabric. Today Belize is part of one of two Paleozoic blocks composing northern Central America (Fig. 1). The basement of the northernmost or Yucatán block, to which Belize belongs, is known to be of Pennsylvanian and Permian age (Ross, 1962). The southern Nicaragua-Honduras block contains undated metamorphic rocks. A zone of faults, which contains serpentinite injections, separates the two blocks. This fracture zone is on a direct line with the offshore

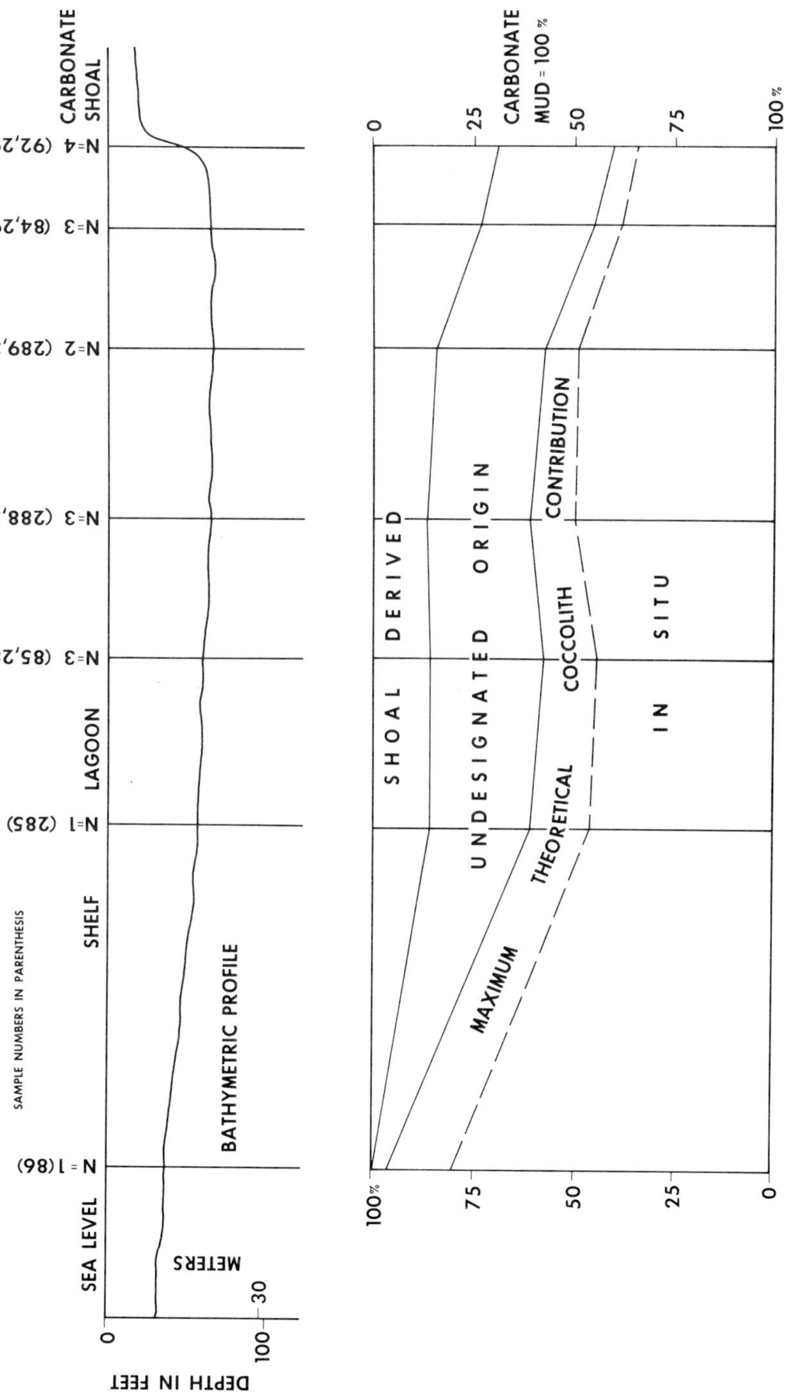

FIG. 23—Idealized derivation of muds (<1/16 mm) in sediments of southern shelf lagoon.

Cayman Trench. The history of the trench and adjacent continental blocks has been the subject of much speculation but, despite an increasing flow of data (e.g., McBirney, 1969; Vedder et al, 1971; Uchupi, 1973; Dillon and Vedder, 1973; Vedder et al, 1973; and Pyle et al, 1973), the history of the area is still poorly known.

Even so, the earliest record of the Yucatan block as a stable platform is Late Jurassic (Viniegra, 1971). Fragmentation of the east margin and westward tilting of the platform produced the NNE-SSW basement ridges which control the modern bathymetric provinces (Purdy, 1974b). The tectonic mechanism for this fragmentation is unknown. Strike-slip shearing has been proposed by advocates of plate tectonics (e.g., Freeland and Dietz, 1971), and normal faulting has been advocated by others (e.g., Viniegra, 1971). Subsurface data indicate that breakup began in Late Cretaceous time. Hole 95 of the Deep Sea Drilling Project (Fig. 1) encountered supratidal dolomites of Albian age (Beall et al, 1973) overlain by bathyal chalks of Santonian age (Worzel et al, 1973, p. 264-265).

Development of the *en échelon* ridge-and-trough system of east Yucatán may have occurred later, because samples of phyllite and volcanic rocks which were dredged from the flanks of ridges have yielded K-Ar dates of 33-92 m.y. (Vedder et al, 1973; Pyle et al, 1973). These Oligocene to Turonian dates should be regarded as minimal because of possible argon leakage (Dalrymple and Lanphere, 1969). Sedimentary facies in Belize also reflect an early Tertiary age. The Palmetto Cay well east of the Maya Mountains (Fig. 1) was still in Tertiary shale at the total depth of 16,005 ft (4,878 m), whereas wells 60 mi (97 km) to the east encountered Eocene shelf carbonate rocks (Dillon and Vedder, 1973). The Tertiary ridge-and-trough system is nearly stable at present. From 1931 to 1972 only three epicenters have been located in the western Yucatán Basin. In contrast, 52 epicenters have been located in the western Cayman Trough (Barazangi and Dorman, 1969; Gutenberg and Richter, 1954; Sykes and Ewing, 1965; Dillon and Vedder, 1973; Hypocentral Data File, Environmental Data Service, National Oceanic and Atmospheric Administration). Collectively, the inference that can be drawn from these data is that the gross sedimentary fabric of Belize was established in early Tertiary time. Pleistocene fluctuations of sea level have further modified the overall pattern of Holocene facies (Purdy, 1974b), but in essence this modification is a climatic variation on the tectonic theme.

REFERENCES CITED

Baie, L. F., 1970, Possible structural link between Yucatan and Cuba: AAPG Bull., v. 54, no. 11, p. 2204-2207.

Barazangi, M., and J. Dorman, 1969, World seismicity maps compiled from ESSA, Coast and Geodetic Survey, epicenter data, 1961-1967: Seismol. Soc. America Bull., v. 59, no. 1, p. 369-380.

Beall, A. O., et al, 1973, Sedimentology, *in* J. L. Worzel et al, Initial reports of the Deep Sea Drilling Project, v. 10, p. 699-716.

Dalrymple, G. B., and M. A. Lanphere, 1969, Potassium-argon dating: San Francisco, W. H. Freeman, 258 p.

Dengo, G., 1969, Problems of tectonic relations between Central America and the Caribbean, *in* Geology of the American Mediterranean: Gulf Coast Assoc. Geol. Socs. Trans., v. 19, p. 311-320.

Dillon, W. P., and J. G. Vedder, 1973, Structure and development of the continental margin of British Honduras: Geol. Soc. America Bull., v. 84, no. 8, p. 2713-2732.

Ebanks, W. J., 1967, Recent carbonate sedimentation and diagenesis, Ambergris Cay, British Honduras: Rice Univ., PhD thesis, 276 p.

Freeland, G. L., and R. S. Dietz, 1971, Plate tectonic evolution of the

Caribbean - Gulf of Mexico region: Nature, v. 232, July 2, p. 20-23.

Gutenberg, B., and C. F. Richter, 1954, Seismicity of the earth and associated phenomena: Princeton, N. J., Princeton Univ. Press, 2d ed., 310 p.

High, L. R., Jr., 1966, Recent coastal sediments of British Honduras: Rice Univ., PhD thesis, 199 p.

King, P. B., 1969, New tectonic map of North America: Geol. Soc. America Bull., v. 80, no. 10, p. 2039-2041.

Krueger, W. C., Jr., 1963, Mineralogical composition and textural properties of river sediments from British Honduras: Rice Univ., Phd thesis, 106 p.

Logan, B. W., 1969, Coral reefs and banks, Yucatán Shelf, Mexico, *in* B. W. Logan et al, sediments and reefs, Yucatán Shelf, Mexico: AAPG Mem. 11, p. 129-198.

Matthews, R. K., 1966, Genesis of recent lime mud in southern British Honduras: Jour. Sed. Petrology, v. 36, no. 4, p. 428-454.

McBirney, A. R., ed., 1969, Tectonic relations of northern Central America and the western Caribbean—the Bonacca Expedition: AAPG Mem. 11, p. 199-280.

Newell, N. D., et al, 1959, Organism communities and bottom facies, Great Bahama Bank: Am. Mus. Nat. History Bull., v. 117, p. 177-228.

Purdy, E. G., 1974a, Reef configurations: cause and effect: Soc. Econ. Paleontologists and Mineralogists Spec. Pub. 18, p. 9-76.

——1974b, Karst-determined facies patterns in British Honduras: Holocene carbonate sedimentation model: AAPG Bull., v. 58, p. 825-855.

——and J. Imbrie, 1964, Carbonate sediments, Great Bahama Bank: Geol. Soc. America, Guidebook, Field Trip No. 2, p. 31-37.

Pusey, W. C., 1964, Recent calcium carbonate sedimentation in northern British Honduras: Rice Univ., PhD thesis, 247 p.

——1975, Holocene carbonate sedimentation on northern Belize shelf: this volume.

Pyle, T. E., et al, 1973, Metamorphic rocks from northwestern Caribbean Sea: Earth and Planetary Sci. Letters, v. 18, no. 2, p. 339-344.

Robertson, R. R., 1963, The mollusks of British Honduras: Philadelphia Shell Club, v. 1, no. 7, p. 15-20.

Ross, C. A., 1962, Permian Foraminifera from British Honduras: Paleontology, v. 5, pt. 2, p. 297-306.

Scholl, D. W., 1964, Recent sedimentary record in mangrove swamps and rise in sea level over the southwestern coast of Florida: Marine Geology, v. 1, no. 4, p. 344-366.

Scholle, P. A., and S. A. Kling, 1972, Southern British Honduras: lagoonal coccolith ooze: Jour. Sed. Petrology, v. 42, no. 1, p. 195-204.

Scott, M. R., 1966, Distribution of clay minerals on the British Honduras shelf: Rice Univ., PhD thesis, 116 p.

——1975, Distribution of clay minerals on the Belize shelf: this volume.

Smith, F. G. W., 1971, Atlantic reef corals: Univ. Miami Press, rev. ed., 164 p.

Stoddart, D. R., 1962, Catastrophic storm effects on the British Honduras reefs and cays: Nature, v. 196, p. 512-515.

——1963, Effects of Hurricane Hattie on the British Honduras reefs and cays, October 30-31, 1961: Atoll Research Bull., no. 95, 142 p.

Sykes, L. R., and M. Ewing, 1965, The seismicity of the Caribbean region: Jour. Geophys. Research, v. 70, no. 20, p. 5065-5074.

Teeter, J. W., 1975, Distribution of Holocene marine Ostracoda from Belize: this volume.

Thorpe, J. E., and P. K. Bregazzi, 1960, Experiments and observations on the corals at Rendezvous Cay: Cambridge Exped. to British Honduras, 1959-1960, Gen. Rept., p. 22-28.

Uchupi, E., 1973, Eastern Yucatan continental margin and Western Caribbean

tectonics: AAPG Bull., v. 57, no. 6, p. 1075-1085.
U.S. Navy Hydrographic Office, 1931, Ranguana Cay to Columbus Cay including inner channel and Glover Reef: H.O. Chart 1497, 18th ed., scale 1:125,000.
——1952, Sailing direction for the east coast of Central America and Mexico: H.O. Pub. 20, 5th ed. (with changes through 1961), 274 p.
——1955, Pilot chart of Central American waters: Chart H.S. 3500, scale 1:6,120,000 (estimated).
——1959, Approaches to Belize: H.O. Chart 1498, 1st ed. (revised), scale 1:125,000.
Vedder, J. G., et al, 1971, USGS IDOE Leg 2: Geotimes, v. 16, no. 12, p. 10-12.
——et al, 1973, Age and tectonic implications of some low-grade metamorphic rocks from the Yucatan Channel: U.S. Geol. Survey Jour. Research, v. 1, no. 2, p. 157-164.
Viniegra, O. F., 1971, Age and evolution of salt basins of southeastern Mexico: AAPG Bull., v. 55, no. 3, p. 478-494.
Wantland, K. F., 1967, Recent benthonic Foraminifera of the British Honduras shelf: Rice Univ., PhD thesis, 300 p.
——1969, Foraminiferal assemblages of the coastal lagoons of British Honduras, *in* A. Ayala-Castañares and F. B Phleger, eds., Lagunas Costeras, Un Simposio, Mem. Simp. Intern. Lagunas Costeras, UNAM-UNESCO, Nov. 28-30, 1967: Mexico, D.F., p. 621-644.
——1975, Distribution of Holocene benthonic Foraminifera on the Belize shelf: this volume.
——and W. C. Pusey III, 1971, A Guidebook for the field trip to the southern shelf of British Honduras: Gulf Coast Assoc. Geol. Socs., 21st Ann. Mtg., 87 p.
Weyl, P. K., 1961, The carbonate saturometer: Jour. Geology, v. 69, p. 32-44.
Whitehouse, U. G., L. M. Jeffrey, and J. D. Debbrecht, 1960, Differential settling tendencies of clay minerals in saline waters, *in* A. Swineford, ed., Clay and clay minerals: London, Pergamon Press, 7th Natl. Conf., Proc., p. 1-79.
Wolf, K. H., 1961, The flotation phenomenon—a by-passing and sorting process: Jour. Sed. Petrology, v. 31, p. 476-478.
Worzel, J. L., et al, 1973, Site reports, *in* J. L. Worzel et al, Initial reports of the Deep Sea Drilling Project, v. 10, p. 7-336.
Wright, A. C. S., et al, 1959, Land in British Honduras: London, Colonial Research Pub. 24, 327 p.

APPENDIX

Systematic List of Commonly Occurring Marine Mollusks of Belize

Robert R. Robertson
(January 1965)

This list includes only those species found at two or more stations, with a total of five or more specimens collected or observed. Some species not yet identified (primarily minute) are omitted.

The following abbreviations are used to designate the communities in which each species is normally found (modified from Robertson, 1963).

I = Freshwater fauna (*Dreissena sallei* community)
II = Chetumal Bay fauna (*Cerithidea costata* community)
III = Bulkhead fauna (*Chione cancellata* community)

IV = Reef-complex fauna
V = Southern-lagoon fauna (*Nuculana cestrota* community)
VI = Southern-continental-shore fauna (*Tivela mactroides* community)

Epifaunal mollusks (i.e., epiphytic, epizoic, or rock inhabiting) are designated as *epi*. Infaunal mollusks (inhabiting mud or sand) are designated as *in*. The notes on habitats are based on observations made in Belize, supplemented with some from the Bahama Islands.

Class POLYPLACOPHORA

Family ISCHNOCHITONIDAE

Stenoplax limaciformis (Sowerby, 1832). IV (epi). Under rocks lying on sand.

Class GASTROPODA

Family FISSURELLIDAE

Hemitoma octoradiata (Gmelin, 1791). IV (epi). On or under rocks on reefs, etc.
Diodora minuta (Lamarck, 1822). IV (epi). Same as above, but more commonly in sheltered, deeper water.
Diodora cavenensis (Lamarck, 1822). IV, VI (epi). Under rocks.
Fissurella (Cremides) barbadensis (Gmelin, 1791). IV (epi). On rocks, in places intertidal.

Family ACMAEIDAE

Acmaea pustulata (Helbling, 1779). IV (epi). Under rocks, on *Thalassia* leaves, etc.
Acmaea albicosta (C. B. Adams, 1845). IV (epi). On *Lithothamnion*-covered rocks on reefs.

Family TROCHIDAE

Calliostoma (Elmerlinia) jujubinum (Gmelin, 1791). IV (epi). Under rocks.
Nerita (Nerita) versicolor Gmelin, 1791. IV, VI (epi). Same as above.
Nerita (Theliostyla) tessellata Gmelin, 1791. IV, VI (epi). Same as above.
Nerita (Theliostyla) fulgurans Gmelin, 1791. IV, VI (epi). Intertidal on rocks in brackish or slightly brackish water (sporadic in IV).
Neritina (Vitta) virginea (Linnaeus, 1758). I (in or epi). On mud in mangrove swamps, sometimes on prop roots.
Smaragdia viridis (Linnaeus, 1758). IV (epi). On *Thalassia, Cymodocea* leaves, etc.

Family LITTORINIDAE

Littorina (Melarapha) ziczac Gmelin, 1791. IV (epi). Intertidal on rocks, etc.
Littorina (Melarapha) lineolata Orbigny, 1840. IV, VI (epi). Same as above.
Littorina (Melarapha) nebulosa (Lamarck, 1822). IV (epi). Intertidal on wood or cement, sometimes in slightly brackish water.
Littorina (Littoraria) angulifera (Lamarck, 1822). I, IV, VI (epi). Intertidal on wood.
Tectarius (Cenchritis) muricatus (Linnaeus, 1758). IV, VI (epi). Supratidal or intertidal on rocks, etc.

Nodilittorina (Echinolittorina) tuberculata (Menke, 1828). IV, VI (epi). Same as above, but lower on shore.

Calliostoma (Elmerlinia) javanicum (Lamarck, 1822). IVB (epi). Under branched Acropora.

Calliostoma sarcodum Dall, 1927. IVB (epi). Same as above.

Tegula (Agathistoma) fasciata (Born, 1778). IV, VI (epi). Under rocks, on *Thalassia* leaves, etc.

Tegula (Agathistoma) lividomaculata (C. B. Adams, 1845). IV (epi). Under rocks on reefs.

Cittarium pica (Linnaeus, 1758). IV, VI (epi). On intertidal and shallow subtidal rocks (sporadic in VI).

Family TURBINIDAE

Astraea (Lithopoma) caelata (Gmelin, 1791). IV (epi). On and under rocks on reefs, primarily at reef crest.

Astraea (Distellifer) tecta (Lightfoot, 1786). IV (epi). On and under rocks on reefs, etc.

Astraea phoebia Röding, 1798. IV (epi). On *Thalassia* leaves, etc.

Family PHASIANELLIDAE

Tricolia affinis cruenta Robertson, 1958. VI (epi).

Tricolia thalassicola Robertson, 1958. IV (epi). On *Thalassia* leaves, etc.

Family NERITIDAE

Nerita (Nerita) peloronta Linnaeus, 1758. IV, VI (epi). Intertidal on rocks, etc (sporadic in VI).

Family TRUNCATELLIDAE

Truncatella (Truncatella) pulchella Pfeiffer, 1839 (*sensu* A. de la Torre, 1960) (epi). Semiterrestrial (above high-tide mark under rocks and in vegetation; larvae marine).

Family RISSOIDAE

Alvania sp. IV, VC (epi). Live in IV on *Laurencia*; dead from VC. A.N.S.P. 285033.

Barleeia sp. IV, VI (epi). In algae. A.N.S.P. 285024.

Family TURRITELLIDAE

Turritella sp. VB (in). A.N.S.P. 282326.

Vermicularia spirata (Philippi, 1836). IV (epi). In sponges.

Family VERMETIDAE

Dendropoma corrodens (Orbigny, 1842). IV (epi). Sessile on intertidal rocks and other shells, partly embedded.

Family CAECIDAE

Caecum (Fartulum) antillarum Carpenter, 1858. I, IV (epi). In algae.

Family PLANAXIDAE

Planaxis (Planaxis) lineatus (da Costa, 1778). IV (epi). Under intertidal rocks.

Planaxis (Supplanaxis) nucleus (Bruguière, 1789). IV (epi). Under intertidal rocks.

Family MODULIDAE

Modulus modulus (Linnaeus, 1758). IVD, VI (epi). On *Thalassia* leaves, etc.
Modulus carchedonius (Lamarck, 1822). VI (epi).

Family POTAMIDIDAE

Cerithidea costata (da Costa, 1778). II (in). In surface layer of sediment.
Cerithidea pliculosa (Menke, 1829). I (in). As preceding.
Batillaria minima (Gmelin, 1791). I, III (in). As preceding.

Family FINELLIDAE

Finella dubia (Orbigny, 1842). IV, V (epi). In algae; dead in V.

Family CERITHIIDAE

Litiopa melanostoma Rang, 1829. (Epi.) On floating *Sargassum*.
Cerithium eburneum Bruguière, 1792. IV, VI (in). In surface layer of sediment.
Cerithium litteratum (Born, 1778). IV, VI (epi). On *Thalassia* leaves, etc.
Cerithium sagrae Orbigny, 1842 = ?*C. lutosum* Menke, 1828 [*C. variabile* C. B. Adams, 1845 (not Deshayes, 1833)]. I, III (in and epi). In surface layer of sediment, in algae, etc.
Cerithium stantoni Dall, 1907. IV (in or epi?).
Bittium (Bittiolum) varium (Pfeiffer, 1840). IV, VI (epi). With sponges, etc.

Family EULIMIDAE

Balcis eburnea (Mühlfeld, 1824). IV, VI (epi). Ectoparasitic on holothurians.

Family HIPPONICIDAE

Cheilea equestris (Linnaeus, 1767). IV (epi). Under rocks on reefs, etc.
Hipponix antiquatus (Linnaeus, 1767). IV, VI (epi). Under rocks, etc.

Family CALYPTRAEIDAE

Calyptraea centralis Conrad, 1841. V, VI (epi). On shells, etc.
Crucibulum auricula Gmelin, 1791. IV, VI (epi). On shells, etc.
Crepidula convexa Say, 1822. IV, VI (epi). On shells, *Thalassia* leaves, etc.

Family STROMBIDAE

Strombus (Strombus) pugilis pugilis Linnaeus, 1758. V (in). On soft mud.
Strombus (Tricornis) raninus Gmelin, 1791. IV, VI (in). On sand in *Thalassia* beds, etc.
Strombus (Tricornis) gigas Linnaeus, 1758. IV (in). On sand, with or without *Thalassia*, etc, or on sand-veneered rock.
Strombus (Tricornis) costatus Gmelin, 1791. IV (in). On sand.

Family NATICIDAE

Sinum perspectivum (Say, 1831). V (in). In soft mud.
Polinices lacteus (Guilding, 1834). IV (in). In sand.

Family OVULIDAE

Cyphoma gibbosum (Linnaeus, 1758). IV (epi). On gorgonians.

Family ERATOIDAE

Trivia (Niveria) quadripunctata (Gray, 1827). IV (epi). Under corals, etc.

Family CYPRAEIDAE

Cypraea (Trona) zebra Linnaeus, 1758. IV (epi). Under corals, etc.

Family CASSIDIDAE

Cassis tuberosa (Linnaeus, 1758). IV (in). Partially buried in sand, with or without *Thalassia*, etc.
Cassis madagascariensis Lamarck, 1822. IV (in). As preceding.
Cypraecassis testiculus (Linnaeus, 1758). IV (epi). Under rocks, etc.

Family CYMATIIDAE

Charonia variegata (Lamarck, 1816). IV (epi). Under and on rocks, etc.
Cymatium (Gutturnium) muricinum (Röding, 1798). IV(?), VI (epi). Under rocks, etc.

Family TONNIDAE

Tonna maculosa (Dillwyn, 1817). IV (in?). On sand?

Family MURICIDAE

Murex (Phyllonotus) pomum Gmelin, 1791. IV (in). On sand, with or without *Thalassia*.
Murex (Murex) cabritii Bernardi, 1859. VB (in). On soft mud.
Murex (Murex) messorius Sowerby, 1841. III, VI (in). On sand among *Thalassia*, etc.
Murex (Chicoreus) florifer Reeve, 1846. IV, VI (in and epi). On sand, gravel, under rocks, etc.
Muricopsis oxytata (M. Smith, 1938). [*Murex hexagonus* Lamarck, 1822 (not Gmelin, 1791)]. IV, VI (epi). Under rocks, etc.
Morula nodulosa (C. B. Adams, 1845). IV (epi). Under rocks, etc.
Risomurex sp. IV (epi). Under corals. A.N.S.P. 281636.
Thais (Stramonita) haemastoma floridana (Conrad, 1837). VI (epi). On quartzite boulder near water level (1 station only—RR92).
Thais (Stramonita) rustica (Lamarck, 1822). IV (epi). On or under rocks, intertidal or shallow subtidal.
Thais (Mancinella) deltoidea (Lamarck, 1822). IV (epi). On rocks, etc, in shallow water.

Family CORALLIOPHILIDAE

Coralliophila abbreviata (Lamarck, 1816). IV (epi). With living corals.
Coralliophila aberrans (C. B. Adams, 1850). IV (epi). Under incrustations of zoanthids (*Palythoa*).
Coralliophila caribaea Abbott, 1958. IV (epi). Under gorgonian holdfasts.

Family COLUMBELLIDAE

Columbella (Columbella) mercatoria (Linnaeus, 1758). IV, VI (epi). On *Thalassia* leaves, under rocks, etc.
Anachis pulchella (Blainville, 1829). IV (epi). On *Thalassia* leaves, etc.
Anachis obesa (C. B. Adams, 1845). VI (epi).

Anachis avara (Say, 1822). IV, VI (epi). On *Thalassia* leaves, sponges, etc.
Anachis picta (Turton, 1825). IV (epi). Under corals.
Anachis (Costoanachis) hotessieriana (Orbigny, 1842). IV(?), VI (epi).
Nitidella nitida (Lamarck, 1822). IV, VI (epi). With sea anemone *Stoichactis helianthus*.
Mitrella lunata (Say, 1826). III, IV, VI (epi). On *Thalassia* leaves and in algae.
Mitrella (Columbellopsis) nitens (C. B. Adams, 1850). V (in). On or in soft mud.

Family BUCCINIDAE

Antillophos cf. *A. adelus* (Schwengel, 1942). V (in). On or in soft mud.
Engina turbinella (Kiener, 1835-36). IV (epi). Under rocks, etc.

Family MELONGENIDAE

Melongena melongena (Linnaeus, 1758). II (dead), III, IV (sporadic), VI (in). On sand, etc.

Family NASSARIIDAE

Nassarius (Hinia) albus (Say, 1826). IV, VI (in). On and in sand.
Nassarius vibex (Say, 1822). VI (in). On and in sand and mud.

Family FASCIOLARIIDAE

Leucozonia nassa nassa (Gmelin, 1791). IV (epi). Under rocks, etc.
Fasciolaria tulipa (Linnaeus, 1758). IV, VI (in and epi). On sand, under rocks, etc.

Family OLIVIDAE

Olivella (Dactylidia) mutica (Say, 1822). VI (in). In sand.
Olivella sp. V (in). In soft mud. A.N.S.P. 282336.
Oliva caribaeensis Dall and Simpson, 1901. VC (in). In soft mud.

Family MITRIDAE

Pusia (Pusia) albocincta (C. B. Adams, 1845). IV, VI (epi). On *Thalassia*, algae, etc.
Pusia (Pusiolina) hanleyi (Dohrn, 1861). IV (epi). In algae, etc.

Family VASIDAE

Vasum (Vasum) muricatum (Born, 1778). IV, VI (in). On sand, among *Thalassia*, etc.
Xancus angulatus (Solander, 1786). IV (in). Partly buried in sand with or without *Thalassia*.

Family MARGINELLIDAE

Bullata ovuliformis (Orbigny, 1842). IV(?), VC (dead). (Epi?).
Marginellopsis serrei Bavay, 1911. IV (epi). In algae.
Prunum (Leptegouana) apicinum (Menke, 1828). III (in). On mud, etc.
Prunum (Leptegouana) guttatum (Dillwyn, 1817). IV (epi). Under rocks, etc.

Family TURRIDAE

Polystira vibex (Dall, 1889). VC (in). On or in soft mud.
Crassispira (Compsodrillia) leucocyma Dall, 1883. IV, VI (in). On or in sand.

Crassispira (Crassispirella) fuscescens (Reeve, 1843). VI (in). On or in sand.

Crassispira (Monilispira) albinodata (Reeve, 1845). IV (epi). Under rocks and corals.

Crassispira (Monilispira) monilis Bartsch and Rehder, 1939. VI (in). On or in sand.

Family CONIDAE

Conus (Conus) regius Gmelin, 1791. IV (epi). Under rocks and corals on reefs.

Conus (Conus) mus Hwass *in* Bruguière, 1792. IV (epi). Under rocks and corals.

Conus (Lithoconus) spurius Gmelin, 1791. IV, VI (in). In sand.

Conus (Lithoconus) aff. *C. floridanus* Gabb, 1869. VI (in). On or in sand.

Conus (Ximeniconus) jaspideus Gmelin, 1791. IV, VI (in). In sand.

Conus (Ximeniconus) pygmaeus Reeve, 1844. VI (in). On or in sand.

Family ACTEONIDAE

Acteon punctostriatus (C. B. Adams, 1840). V (in). In soft mud.

Family BULLIDAE

Bulla striata Bruguiere, 1792. IV, VI (in). On and in sand, etc.

Family ATYIDAE

Haminoea elegans (Gray, 1825). V (in). On or in soft mud.

Haminoea succinea (Conrad, 1846). III, IV, VI (in). On or in sand.

Family SCAPHANDRIDAE

Acteocina recta (Orbigny, 1841). V (in). On or in soft mud.

Family CAVOLINIIDAE

Creseis acicula Rang, 1828. V (pelagic). In plankton. Dead shells in soft mud.

Family APLYSIIDAE

Aplysia dactylomela Rang, 1828. IV (epi). On *Thalassia*, algae, rocks, etc.

Phyllaplysia sp. IV (epi). On *Thalassia*. A.N.S.P. 282408.

Family ELLOBIIDAE

Pedipea mirabilis (Mühlfeld, 1818). I, IV (epi). In intertidal algae, etc.

Melampus coffeus (Linnaeus, 1758). I (in and epi). On mud under mangroves in brackish swamp; also on mangrove roots and under wood, etc.

Class SCAPHOPODA

Family DENTALIIDAE

Dentalium (Antalis) antillarum Orbigny, 1842. IV (in). In sand.

Dentalium (Antalis) cf. *D. (A.) ceratum* Dall, 1881. V (in). In soft mud.

Dentalium (Antalis) disparile Orbigny, 1842. V, VI (in). In sand and soft mud.

Dentalium (Graptaome) semistriolatum Guilding, 1834. IV, VI (in). In sand.

Class BIVALVIA (PELECYPODA)

Family NUCULIDAE

Nucula (Ennucula) sp. V, VI (in). In soft mud and sand. A.N.S.P. 283530 and 285417.

Family NUCULANIDAE

Nuculana (Saccella) vulgaris (Brown and Pilsbry, 1913). V (in). In soft mud.
Nuculana (Saccella) sp. V, VI (in). In soft mud and sand. A.N.S.P. 282304.
Nuculana (Costelloleda) cestrota (Dall, 1890). V (in). In soft mud.
Yoldia solenoides Dall, 1881. VB (in). In soft mud.
Adrana sp. VI (in). In black mud. A.N.S.P. 282387.

Family ARCIDAE

Arca imbricata Bruguière, 1789. IV, VI (epi). Attached under corals or to other shells, etc.
Arca zebra (Swainson, 1833). IV, VI (epi). With preceding.
Barbatia cancellaria (Lamarck, 1819). IV, VI (epi). Attached under corals, etc.
Arcropsis adamsi (Dall, 1886). IV, VI (epi). With preceding.
Anadara (Larkinia) notabilis (Röding, 1798). IV, VI (in). Half buried in sand in *Thalassia* beds.
Anadara (Cunearca) chemnitzii (Philippi, 1851). V, VI (in). In sand (and soft mud?).
Anadara (Lunarca) ovalis (Bruguière, 1789). VI (in). In sand.

Family GLYCYMERIDIDAE

Glycymeris (Glycymerella) decussata (Linnaeus, 1758). IV (in; dead only). In sand (Bahamas).
Glycymeris (Tucetona) pectinata (Gmelin, 1791). IV (in). In sand.

Family MYTILIDAE

Crenella divaricata (Orbigny, 1842). V, VI (in). In soft mud and sand.
Modiolus americanus (Leach, 1815). IV, VI (epi). Attached to dead shells, etc.
Brachidontes (Hormomya) exustus (Linnaeus 1758). II, III, IV, VI (epi). Attached to dead shells, etc [*B. citrinus* (Röding) perhaps also present].
Brachidontes (Ischadium) recurvus (Rafinesque, 1820). VI (epi). Attached to solid objects in slightly brackish water.
Musculus lateralis (Say, 1822). IV, VI (epi). In algae, etc.
Lithophaga (Lithophaga) antillarum (Orbigny, 1842). IV (epi). Boring in coral.
Lithophaga (Lithophaga) nigra (Orbigny, 1842). IV (epi). With preceding.
Lithophaga (Diberus) bisulcata (Orbigny, 1842). IV (epi). With preceding.

Family ISOGNOMONIDAE

Isognomon (Melina) alatus (Gmelin, 1791). IV, VI (epi). On mangrove roots, wood, old *Strombus* shells, etc.

Isognomon (Melina) radiatus (Anton, 1839). IV (epi). Attached under corals, on wood, etc.

Parimalleus candeanus (Orbigny, 1842). IV (epi). Attached under corals and in cracks.

Family PTERIIDAE

Pteria colymbus (Röding, 1798). IV, VI (epi). Attached to gorgonians (rarely, mangrove roots).

Pinctada imbricata Röding, 1798. IV, VI (epi). Attached to solid objects.

Family PINNIDAE

Pinna carnea Gmelin, 1791. IV, VI (in). Slightly buried in coarse sediment.

Atrina (Servatrina) seminuda (Lamarck, 1819). IV, VI (in). With preceding.

Family PECTINIDAE

Pecten (Amusium) laurentii (Gmelin, 1791). V (in). On soft mud.

Chlamys sentis (Reeve, 1853). IV (epi). Attached under corals.

Chlamys ornata (Lamarck, 1819). IV (epi). Attached under corals.

Chlamys benedicti Verrill and Bush (*in* Verrill, 1897). VC, VI (epi). Attached under corals, etc.

Chlamys imbricata (Gmelin, 1791). IV (epi). Attached under corals.

Chlamys muscosa (Wodd, 1828). IV, V, VI (epi). Attached to solid objects.

Chlamys (Argopecten) lineolaris (Lamarck, 1819). VC (in). On soft mud.

Chlamys (Argopecten) gibba (Linnaeus, 1758). IV, V, VI (in). On sand (and mud?).

"Lyropecten" antillarum (Récluz, 1853). IV (epi). Attached to solid objects.

Cyclopecten nanus Verrill and Bush (in Verrill, 1897). VC (in). On soft mud.

Family SPONDYLIDAE

Spondylus americanus Hermann, 1781. IV, VC (epi). Attached to solid objects.

Family LIMIDAE

Lima (Lima) lima (Linnaeus, 1758). IV (epi). Attached under corals, etc.

Lima (Limaria) hians (Gmelin, 1791). IV, VI (epi). Loose under corals, etc.

Lima (Ctenoides) scabra (Born, 1778). IV (epi). Attached under corals.

Family ANOMIIDAE

Anomia simplex Orbigny, 1842. IV, VI (epi). Attached to solid objects.

Family OSTREIDAE

Ostrea cristata Born, 1778. V, VI (epi). Attached to solid objects (dead shells, etc.).

Ostrea frons (Linnaeus, 1758). IV, VI (epi). On gorgonians (rarely, wooden dock pilings).

Crassostrea rhizophorae (Guilding, 1828). I (epi). Attached to mangrove roots or cement walls, etc.

Family CRASSATELLIDAE

Crassinella lunulata (Conrad, 1834). VI (in). In sand?

Family CORBICULIDAE

Pseudocyrena floridana (Conrad, 1846). I (in). Partially buried in mud near mangroves.
Corbicula (Neocorbicula) sp. I (in). With preceding? A.N.S.P. 282046.

Family SPORTELLIDAE

Sportella (Fabella) sp. IV (in). In sand? A.N.S.P. 283480.

Family DIPLODONTIDAE

Diplodonta (Diplodonta) punctata (Say, 1822). IV, VI (in). In sand.
Diplodonta (Felaniella) candeana (Orbigny, 1842). V, VI (in). In soft mud (and sand?).
Diplodonta (Felaniella) sp. VI (in). In sand. A.N.S.P. 282246.
Diplodonta (Phlyctiderma) semiaspera Philippi, 1836. IV, VI (in). In sand.

Family LUCINIDAE

Lucina (Lucina) pensylvanica (Linnaeus, 1758). IV (in). In sand.
Lucina ("Phacoides") pectinata Gmelin, 1791. II, III, VI (in). In sand.
Lucina (Lucinisca) muricata (Spengler, 1798). V, VI, (in). In soft mud (and sand?).
Lucina (Parvilucina) multilineata Tuomey and Holmes, 1857. IV (in). In sand.
Lucina sp. V (in). In soft mud. A.N.S.P. 284920.
Anodontia alba Link, 1807. VI (in). In sand.
Codakia (Codakia) orbicularis (Linnaeus, 1758). IV (in). In sand.
Codakia (Ctena) orbiculata (Montagu, 1808). IV, VI (in). In sand.
Codakia (Ctena) costata Orbigny, 1842. III, IV (in). In sand.
Codakia (Ctena) pectinella (C. B. Adams, 1852). V, VI (in). In soft mud.
Divaricella (Divalinga) quadrisulcata (Orbigny, 1842). IV, VI (in). In sand.

Family CHAMIDAE

Chama sinuosa Broderip, 1836. IV (epi). Attached to mangrove roots, etc.
Chama congregata Conrad, 1833. IV (epi). Attached to solid objects.
Chama florida Lamarck, 1819. IV (epi). Attached to solid objects.
Pseudochama radians (Lamarck, 1819). IV, VI (epi). Attached to dead shells, etc.
Echinochama arcinella (Linnaeus, 1767). V, VI (epi). Loosely attached to solid objects.

Family CARDIIDAE

Trachycardium (Dallocardia) muricatum (Linnaeus, 1758). IV, VI (in). In sand.
Papyridea soleniformis (Bruguière, 1789). IV, VI (in). In sand.
Microcardium tinctum (Dall, 1881). VC (in). In soft mud.
Laevicardium laevigatum (Linnaeus, 1758). IV, V, VI (in). In sand and soft mud (specimens from two substrates distinguishable).
Laevicardium mortoni (Conrad, 1830). II, VI (in). In sand (semibrackish water).
Laevicardium pictum (Ravenel, 1861). VC (in). In soft mud or shell gravel (?).
Trigoniocardia antillarum (Orbigny, 1842), *sensu* R. T. Abbott (1958). V (in). In soft mud.

Americardia guppyi (Thiele, 1910). IV (in). In sand.
Americardia media (Linnaeus, 1758). IV (in). In coarse sand or gravel.

Family VENERIDAE

Gouldia cerina (C. B. Adams, 1845). IV, V, VI (in). In sand and soft mud.
Pitar (Pitar) fulminata (Menke, 1828). IV, VI (in). In sand.
Pitar (Pitar) aresta (Dall and Simpson, 1901). V (in). In soft mud.
Pitar (Lamellichocha) circinata (Born, 1778). VI (in). In sand.
Macrocallista (Paradione) maculata (Linnaeus, 1758). IV, VI (in). In sand.
Transennella cubaniana (Orbigny, 1842). IV (in). In sand.
Tivela mactroides (Born, 1778). VI (in). In sand.
Dosinia (Dosinidea) discus (Reeve, 1850). V (in). In soft mud.
Cyclinella tenuis Récluz, 1852. V, VI (in). In soft mud and sand(?).
Ventricolaria rigida (Dillwyn, 1817). V (in). In soft mud.
Chione (Chione) cancellata (Linnaeus, 1767). III, IV, VI (in). In sand, most abundant where water slightly brackish.
Chione (Chione) intapurpurea (Conrad, 1849). VI (in). In sand.
Chione (Chione) granulata (Gmelin, 1791). VI (in). In sand.
Chione (Chione) pygmaea (Lamarck, 1818). IV (in). In sand.
Chione (Lirophora) paphia (Linnaeus, 1767). IV, V, VI (in). In sand and soft mud (?).
Anomalocardia cuneimeris (Conrad, 1846). II, IV, VI (in). In sand, most abundant where water brackish.

Family PETRICOLIDAE

Petricola (Petricola) lapicida (Gmelin, 1791). IV, VI (epi). Boring or nestling in corals and calcareous rocks.

Family MACTRIDAE

Mulinia lateralis guadelupensis (Récluz, 1852). VI (in). In sand.
Anatina anatina (Spengler, 1802). V (in). In soft mud.
Mactra (Mactrotoma) fragilis Gmelin, 1791. III, VI (in). In sand.

Family DONACIDAE

Donax striatus Linnaeus, 1767. VI (in). In intertidal and shallow subtidal sand.
Iphigenia brasiliensis (Lamarck, 1818). VI (in). With preceding (?).

Family SANGUINOLARIIDAE

Tagelus divisus (Spengler, 1794). III, IV, VI (in). In mud; most abundant where water slightly brackish.
Solecurtus cumingianus Dunker, 1861. V (in). In soft mud.

Family SEMELIDAE

Semele proficua (Pulteney, 1794). VI (in). In sand or mud (?).
Semele purpurascens (Gmelin, 1791). VI (in). With preceding.
Semele bellastriata (Conrad, 1837). IV (in). In sand.
Cumingia antillarum (Orbigny, 1842). III, IV, VI (epi). Nestled in crevices in sponges, rocks, etc.
Abra aequalis (Say, 1822). V (in). In soft mud.

Family TELLINIDAE

Arcopagia (Cyclotellina) fausta (Pulteney, 1799). IV, VI (in). In sand, mud, or gravel.

Tellina (Tellina) radiata Linnaeus, 1758. IV (in). In sand.
Tellina (Tellinella) listeri Röding, 1798. IV (in). In sand.
Tellina (Eurytellina) angulosa Gmelin, 1791. V (in). In soft mud.
Tellina (Eurytellina) punicea Born, 1778. VI (in). In sand.
Tellina (Eurytellina) alternata Say, 1822. III, V, VI (in). In soft mud and sand.
Tellina (Eurytellina) vespusiana Orbigny, 1842. V (in). In soft mud.
Tellina (Angulus) mera Say, 1834. IV (in). In sand.
Tellina (Angulus) euvitrea Boss, 1964. V (in). In soft mud.
Tellina (Angulus) exerythra Boss, 1964. VI (in). In sand.
Tellina (Angulus) consobrina Orbigny, 1842. V, VI (in). In soft mud and sand.
Tellina (Scissula) similis Sowerby, 1806. IV (in). In sand.
Tellina (Acorylus) gouldii Hanley, 1846. IV (in). In sand.
Tellina (Merisca) martiniscensis Orbigny, 1842. V, VI (in). In soft mud and sand (?).
Tellina (Phyllodina) persica Dall and Simpson, 1901. V, VI (in). In soft mud.
Quadrans lintea (Conrad, 1837). IV, VI (in). In sand.
Strigilla mirabilis (Philippi, 1841). IV, VI (in). In sand.
Strigilla pisiformis (Linnaeus, 1758). VI (in). In sand.
Macoma constricta (Bruguière, 1792). II, III, VI (in). In sand, most abundant in slightly brackish water.
Macoma tenta souleyetiana Récluz, 1852. V, VI (in). In soft mud and sand (?).
Macoma pseudomera Dall and Simpson, 1901. V (in). In soft mud.
Macoma aurora (Hanley, 1844). VI (in). In sand.
Macoma tageliformis Dall, 1900. V (in). In soft mud.
Macoma (Cymatoica) orientalis hendersoni (Rehder, 1939). V, VI (in). In soft mud and sand (?).

Family CORBULIDAE

Corbula (Corbula) limatula Conrad, 1846. V (in). In soft mud.
Corbula (Caryocorbula) caribaea Orbigny, 1842. V (in). In soft mud.
Corbula (Caryocorbula) sp. VI (in). In sand. A.N.S.P. 282751.
Corbula (Tenuicorbula) aequivalvis Philippi, 1836. V, VI (in). In soft mud and sand (?).
Corbula sp. VC (in). In soft mud. A.N.S.P. 284936.

Family GASTROCHAENIDAE

Gastrochaena hians (Gmelin, 1791). IV, V, VI (epi). Boring in shells, rocks, etc.

Family PHOLADIDAE

Martesia (Martesia) striata (Linnaeus, 1758). IV, V, VI (epi). Boring in wood.

Family TEREDINIDAE

Teredo sp. IV (epi). Boring in wood. A.N.S.P. 282065.

Family LYONSIIDAE

Lyonsia sp. VI (epi). Nestling in crevices. A.N.S.P. 284044.

Family PANDORIDAE

Pandora (Pandorella) bushiana Dall, 1886. V, VI (in). In sand and soft mud (?).

Family CUSPIDARIIDAE

Cuspidaria sp. VC (in). In soft mud. A.N.S.P. 283055.
Cardiomya ornatissima (Orbigny, 1842). V, VI (in). In soft mud and sand (?).

Class CEPHALOPODA

Family SPIRULIDAE

Spirula spirula (Linnaeus, 1758). Shells on beaches; animal nektonic, deep-water; shell floats.

Family LOLIGINIDAE

Sepioteuthis sepioidea (Blainville, 1823). IV (nektonic). In schools.

Family OCTOPODIDAE

Octopus sp. IV (epi). Under dead corals, etc. A.N.S.P. 281778.

NUMBER OF SPECIES LISTED

Polyplacophora	1
Gastropoda	132
Scaphopoda	4
Bivalvia	146
Cephalopoda	3
	286

Geomorphology and Sedimentology of Holocene Coastal Deposits, Belize [1]

LEE R. HIGH, JR.[2]

ABSTRACT The Belize coastal areas are largely covered by mangrove swamps, which mantle a variety of landforms. Assemblages of landforms define six physiographic provinces, controlled by the hydrographic regime and mainland geology. North of Belize City, the shelf is shallow and the coast is somewhat protected by numerous cays. Only a few streams drain the low, karst surface of the Yucatán platform of the mainland opposite Chetumal Bay. Consequently, little clastic material is brought to the coast. Barrier lagoons and marshes dominate the northern coast. South of Belize City, the shelf lagoon is deep and the coastline is exposed to heavy surf. The Maya Mountains dominate the southern mainland and are drained by abundant small streams which transport large volumes of sediment. Numerous small deltas characterize the coast east of the mountains. Farther south, marshes are again prevalent.

Holocene coastal sediments are present in a narrow, discontinuous band a few hundred feet wide. This band is bordered by Pleistocene limestone and clay inland, and by marine sediments of the shelf lagoon. Within this narrow zone, there are many environments in which sediments are accumulating. These environments are differentiated readily by means of morphology and vegetation.

Analysis of 105 sediment samples from numerous locations along the coast indicates no simple correlations between sediment type and environment. The sediments show continuous gradations in properties with no environmental restrictions. The differences in the physical settings of similar depositional environments, even within a limited area, and the large number of factors prevent a simple correlation between sediment type and environment.

The problem of relating sediment to environment is largely resolved when only the sediments from a local area are considered. Comparisons of sediment properties from adjacent sample stations indicate differences in environment. Furthermore, the changes in sediment properties between adjacent environments are similar along the entire coast.

The stratigraphy and history of the Holocene coastal sediments were determined from 95 piston cores. The coastal areas north of Belize City are underlain by a regular and persistent sequence: Pleistocene limestone, weathered Pleistocene limestone, sandy terrigenous clay, fibrous mangrove peat, and differentiated carbonate mud and sand. This sequence is interpreted to record the Flandrian transgression. South of Belize City, however, there is no consistent stratigraphic sequence. The greater relief on the Pleistocene surface prevented the flooding of wide areas. Consequently, widespread environments and lithologic units did not form.

INTRODUCTION

Foremost among the numerous advantages offered by the Belize coast for study of Holocene coastal sediments is the complex nature of the coastline. The variety of landforms and depositional environments permits investigation of possible stratigraphic and lithologic differences

[1]Manuscript received, October 15, 1969.

[2]Department of Geology, Oberlin College, Oberlin, Ohio 44074.

This paper is taken from a PhD thesis submitted to Rice University. The study was suggested by E. G. Purdy who, together with J. J. W. Rogers, acted as supervisor and gave valuable time and thought necessary for its completion. Financial support was provided by Rice University, the Society of Sigma Xi, and a grant to E. G. Purdy from the Humble Oil and Refining Company, the Jersey Production Research Company, and Imperial Oil Limited. Computer time was provided by Oberlin College. Spore-pollen analyses were performed by W. C. Elsik of Esso Production Research Company. Radiocarbon dates were provided by Shell Development Company. The field assistance of E. G. Purdy, W. J. Ebanks, J. W. Teeter, G. E. Tebbutt, and A. Matter is gratefully acknowledged.

Many persons in Belize provided helpful assistance. In particular, W. Ford Young expedited shipment of equipment to and from Belize. Mr. Johnson, the Comptroller of Customs, facilitated customs clearance of the equipment. A. A. Quinn of the Public Works Department provided information on dredging operations in the Belize River. To these individuals in particular and to the people of Belize in general, I gratefully extend my thanks for their courtesy and hospitality. I am also indebted to Alan Craig of the Florida Atlantic University for helpful information as well as aerial photographs of parts of the coast. S. S. Streeter provided assistance in computer operation.

The manuscript has been reviewed by E. G. Purdy, J. J. W. Rogers, R. R. Lankford, and M. D. Picard. It has benefited significantly from their collective criticism.

among related environments along a single coast. Also, because of variations in climate, bathymetry, and mainland geology, several important factors can be studied with some measure of control on their influence. Studies under the direction of E. G. Purdy have established the Holocene geologic framework from the source area to the edge of the continental shelf of Belize; these studies have provided much information pertinent to understanding coastal sedimentation.

The general objective of this study was a reconnaissance of the geomorphology, sedimentology, and stratigraphy of Holocene coastal deposits. The specific aim was to provide criteria useful for paleoenvironmental interpretation. Accordingly, field and laboratory techniques were selected to emphasize the features of Holocene sediments and the depositional environments which are likely to be preserved or reflected in sedimentary rocks.

GENERAL SETTING

Physiography

Belize may be divided into two parts on the basis of topography and geology. The boundary, which is located just south of Belize City, is reflected by drainage patterns (Fig. 1). The northern half is a low, featureless karst platform developed on Cretaceous and Tertiary limestones (Flores, 1952, p. 405). Surface drainage is poor and through-flowing streams are uncommon. The New River and the Rio Hondo form the major drainage systems. Their courses are remarkably straight and possibly are fault controlled (Dixon, no date, p. 10; Ower, 1928a, p. 497). Relief in this area is slight; maximum elevation is about 200 ft (61 m), and almost all the coastal area for several miles inland is less than a few feet above sea level.

This low-lying plain is terminated abruptly to the south by the Maya Mountains, which exceed 3,000 ft (914 m) in height and which are well drained by numerous small streams. Streams draining the western and northern slopes are largely tributary to the Belize and Siburn Rivers, which wander over the northern plain before debouching into the sea. The small streams draining the eastern and southern slopes flow directly into the sea. Drainage density is particularly high along the coast directly east of the Maya Mountains, where streams commonly terminate in small deltas. North of Deep River the streams drain metamorphic and igneous terranes and transport considerable sand and mud (Krueger, 1963). In contrast, south of Deep River all streams head in limestone terrane and transport little clastic material.

This pattern of stream distribution, controlled largely by the geology and related physiography of the interior, has profound effects on coastal sedimentation.

Bathymetry

Just as the mainland is divisible into northern and southern sections on the basis of topography, the shelf is divided into two similar areas on the basis of bathymetry. North of the latitude of Belize City, the shelf is shallow, and few depths are greater than 15 ft (4.6 m); south of Belize City the shelf lagoon deepens toward the Port of Honduras, where depths are greater than 200 ft (61 m).

Currents in the northern shelf lagoon flow southward, owing to the discharge of water into Chetumal Bay by the New River and Rio Hondo. Wave action is slight in the extensive shallow water and the coast is essentially surfless. Tides are slight, rarely exceeding 1.5 ft (0.5 m; U.S. Hydrographic Office, 1952, p. 187). In contrast, large waves are

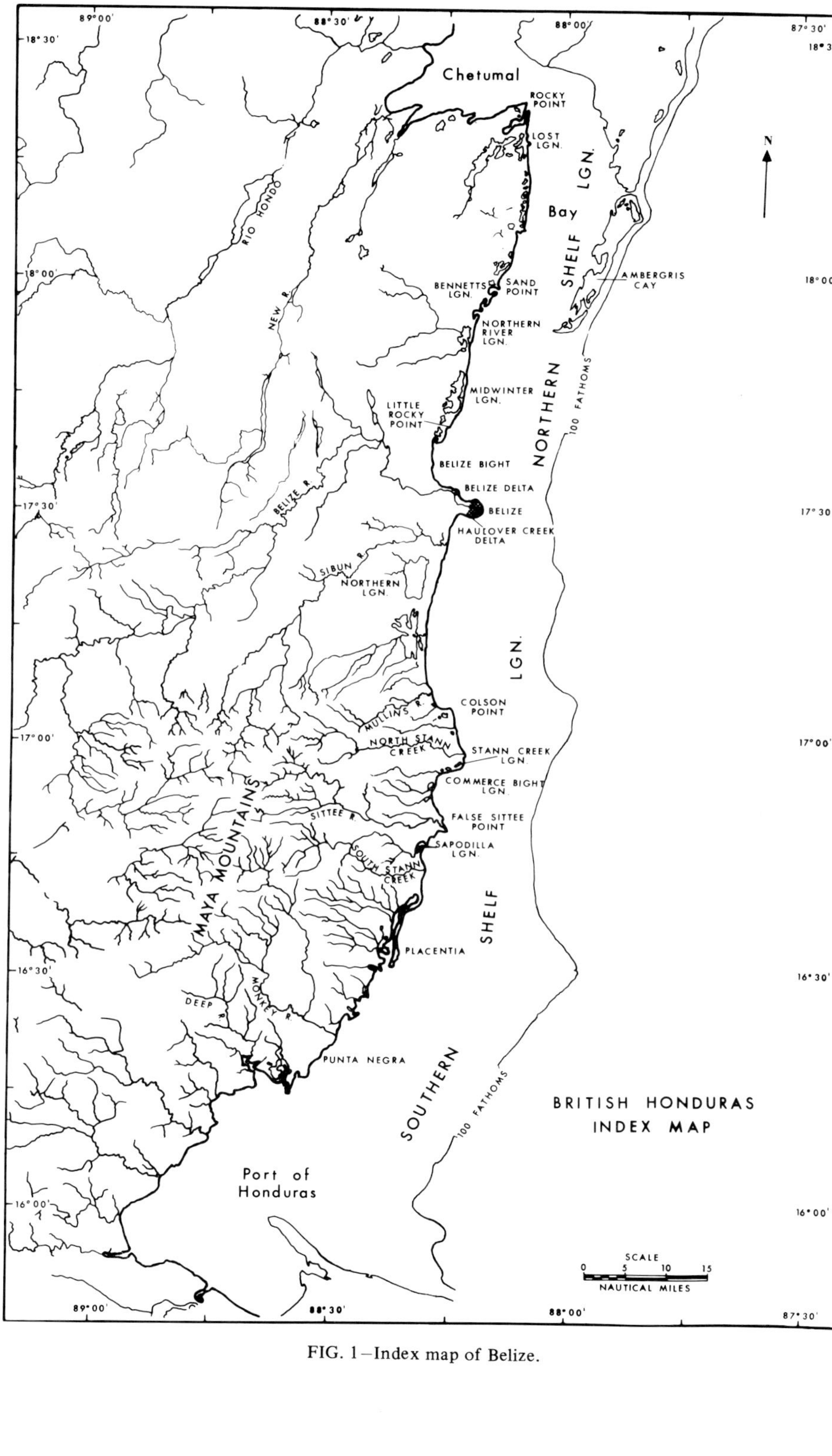

FIG. 1—Index map of Belize.

formed in the deeper water of the southern shelf lagoon, and moderate surf breaks on the beaches.

In contrast to the general quiet conditions which prevail along the northern coast, periodic storms bring heavy surf and abnormal water levels. The effects of these storms on coastal and offshore areas have been documented by Vermeer (1959, 1963), Stoddart (1962, 1963, 1969), and High (1969).

GEOMORPHOLOGY

Physiographic Provinces

Coastlines drowned by the Flandrian transgression and remaining essentially unmodified have been called "initial" (Guilcher, 1958, p. 153). All other coasts are considered to be "sequential." Subdivisions of sequential coasts may be based on degree of modification and dominant processes (erosion or deposition). In Belize all coastal areas visited by the writer are depositional; widespread marine erosion is not in evidence. The major coastal landforms, in order of increasing degree of depositional modification, are bights and marshes, lagoon-barrier systems, forelands, and deltas.

Assemblages of these major landforms permit division of the coastline of Belize into physiographic provinces. The assemblages are not rigid, exclusively defined groups; rather, they provide the means of description for basic subdivisions that are readily recognizable. There are six such segments of the coast: (1) northern initial coast, (2) northern lagoons, (3) Belize Bight, (4) central lagoons, (5) central deltas, and (6) southern coastal marshes.

Northern coast—From the mouth of Rio Hondo to Rocky Point, the coast is essentially initial and is controlled by the underlying rock structure, as reflected in the alignment of rivers, lagoons, and embayments. Recent coastal processes have done little to alter the coast. The only extensive Holocene coastal sediments are in the bight directly west of Rocky Point. A series of ridges, deposited as bayhead beaches, is tending to fill the bight and straighten the coastline. Otherwise, Holocene sediments are probably limited to a thin and discontinuous coastal zone.

Northern lagoons—The dominant features from Lost Lagoon to Belize Bight are coastal lagoons and barriers. Carbonate source areas, low relief, lack of surface drainage, shallow water offshore, and periodic storms are significant factors along this segment of the coast. Little clastic material from the interior reaches the coast, and offshore areas provide mainly carbonate mud. There are no river mouths to provide foci of deposition. Material winnowed from the bottom of Chetumal Bay is spread evenly along the coast by longshore currents. Mud flats behind the barriers and between the lagoons are extensive. The several forelands at Little Rocky Point mark the southern end of the northern lagoons province.

Belize Bight—Belize Bight is characterized by a featureless, marshy coastline. Deposition along the coast is slight, and carbonate and terrigenous muds are mixed offshore. The Belize River delta, at the southern end, is the most prominent feature along this segment.

Central lagoons—The dominant features of the fourth physiographic province, between Belize City and the mouth of Mullins River, are the Haulover Creek delta and the several large lagoons and beach-ridge barriers. The few rivers whose mouths are located along this segment have not constructed sizable deltas.

Features of this segment of the coast resemble the northern lagoons and barriers, but they are Pleistocene, rather than Holocene, as shown by an oxidized caliche zone at or near the surface.

Central deltas—The fifth province, extending from Mullins River to Placentia, is characterized by numerous small river deltas. Between the deltas are small lagoons and beach-ridge barriers. The many small rivers draining the eastern slope of the Maya Mountains transport considerable clastic material, which is deposited along the coast.

Southern marshes—Placentia sandbar, a unique feature of the coast, is included in the central delta province because of its beach ridges. Actually, the southern part of Placentia Lagoon belongs to the final segment of the coast—the marshes and estuaries in and around the Port of Honduras. Of the rivers draining this area, only the Monkey River transports significant amounts of sediment from the Maya Mountains, and only this river has constructed a delta. Consequently, sedimentation is most active in the low marshes that border the protected coast.

Initial Coasts

No strictly initial, or unmodified, coasts were observed, although some of the bight areas are little removed from the ria-type coasts described by Guilcher (1958, p. 153-157). The northern coast from Rocky Point to the mouth of the Rio Hondo was not visited and is poorly covered by the available aerial photographs; however, it may be concluded from the "Provisional Soil Map" of Wright et al (1959), the structure maps of Ower (1928a, Fig. 1, p. 497; 1928b, Fig. 1, p. 509) and Murray (1961, p. 199-201), and the lineaments displayed by the rivers, lagoons, and bays of this area (Fig. 1) that this is essentially an initial coast. For example, soils along the northern coast are directly influenced by the bedrock, and a thin mantle of Holocene coastal sediments is present only along the eastern coast. This general lack of Holocene sediments is suggested also by the irregularity of the northern coastline and the absence of typical sequential landforms on the available aerial photographs.

The straight courses of the Rio Hondo and New River closely parallel the dominant fault trends of the area, as do the several lagoons and embayments (Fig. 1). Thus, the evidence suggests that the northern coast is largely initial, having been modified little by either erosion or deposition as sea level rose to its present position.

Bights and Coastal Marshes

Bights and coastal marshes are the least modified of the sequential landforms. The bights, or large embayments, are bounded by a marshy estuarine coast. An example is the coast of Belize Bight, which is low-lying, regular, and featureless. Holocene sediments are limited to a narrow fringe along the coast. The ground surface inland resembles the Pleistocene surfaces elsewhere along the coast.

Belize Bight itself is an area of extensive shoaling. It is shallow and, as indicated by coring and probing, contains a thick section of Holocene sediments which were transported south along the coast by the longshore currents. Because of the prevailing southerly current, little sediment from the Belize River enters Belize Bight.

The fact that the Belize Bight coastline is roughly parallel with the western edge of the lagoons to the north (Fig. 1) suggests that Belize Bight represents a less modified coastal segment which is preserved because barriers have not formed. Sediment transported southward along the coast is trapped mainly at Little Rocky Point, where several extensive forelands have formed. Several small cays in the area of Little Rocky Point are now attached to the mainland by Holocene sediments. In the past, these cays probably served as the nuclei of deposition, initiating the forelands. Thus, the area directly to the south, Belize Bight, was deprived of sediment and barriers failed to form.

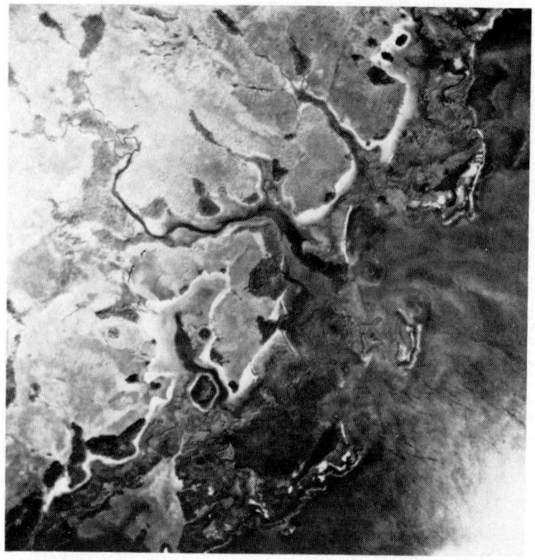

FIG. 2—Vertical aerial photograph of coast north of Monkey River. Estuaries have been filled and small creeks and marshes occupy valleys. White areas are stranded beaches formed from reworked Pleistocene sediment. Recent progradation has formed mud flats in front of beaches. Width of photograph, about 5 n. mi. North is toward top.

In contrast to Belize Bight, large bights south of Placentia have irregular and embayed coasts with numerous small cays offshore (Fig. 2). Streams that enter the sea along this part of the coast have not built deltas; instead, large, deep estuaries are present. The coast is low and marshy, having little relief. A large sand ridge, locally called "Pine Ridge," is present a short distance inland from these marshes. Although it cannot be traced continuously on the aerial photographs, Pine Ridge is probably the southern extension of Pleistocene features farther north on the Haulover Creek delta and along the central lagoon area.

The southern bights currently are being filled with sediment, and former cays can be seen in the marshes. Once the bights are filled and the coastline is straightened, the resulting regular wave pattern may construct beach ridges. Possibly, the final straightening will occur as barrier ridges form across the last embayments; new lagoon-barrier systems will be formed.

Lagoon-Barrier Systems

The most extensive landform, occupying approximately 50 percent of the coastline, is the lagoon-barrier complex. The general form is that of a shallow lagoon blocked by a narrow, elongate barrier. The barrier is constructed of a few parallel beach ridges backed by mud flats of varying widths; it is breached by a single tidal channel. This barrier shows only moderate depositional modification. Commonly, Pleistocene sediments are exposed on the floor or on the western edge of the lagoon.

Variations in this basic pattern are influenced by such factors as age, barrier width, lateral extent, and mud-flat development; these permit recognition of three different types of lagoon-barrier complexes, which are:

Barrier spit (e.g., Placentia Lagoon),

Interdelta barrier lagoon (e.g., Sapodilla Lagoon, Commerce Bight Lagoon, Stann Creek), and

Coastal lagoon (e.g., Northern Lagoon, Midwinter Lagoon, Northern River Lagoon, Bennetts Lagoon).

Barrier spit—Placentia sandbar constitutes the only large spit along the Belize coast. It is tied to the mainland on the north and separates Placentia Lagoon from the sea. The spit is constructed of a few beach ridges and is generally quite narrow, on the order of a few hundred feet. It widens toward the southern end, where the village of Placentia is located. When viewed from the air, the southern, free end does not show evidence of active growth such as recurved ridges and unvegetated spits. Rather, the eastern edge is eroded continually by heavy surf, and sediment is lost to deeper water. Also, sheetwash and wind activity are removing sand from the face of the bar and depositing it in Placentia Lagoon. Thus, the bar is a pre-modern feature that currently is being modified.

The lagoon edge of the bar is bordered by a low mangrove marsh. This marsh, which has its counterpart on the mainland, represents the gradual shrinking of Placentia Lagoon owing to the mangrove encroachment from both sides; indeed, the lagoon has almost been closed by mangroves in several places.

Interdelta lagoon barriers—Between the river-mouth deltas along the central part of the coast are small lagoons and barriers characterized by extreme mangrove encroachment and infilling. The construction of tidal deltas has greatly reduced the depth of the lagoons, with the exception of Sapodilla Lagoon.

The barriers are constructed of narrow beach ridges backed on the west by a generally wider zone of mangrove marsh and mud flat. Access to the lagoon is provided by a single tidal channel which remains relatively fixed in position.

Coastal lagoons—The most common lagoons are the coastal lagoons which occupy much of the mainland coast between Rocky Point and Colson

FIG. 3—Oblique aerial photograph looking south along coast of Chetumal Bay. Mouth of Northern River is in center of photograph. Midwinter Lagoon is in background. Beach ridges on barrier are heavily vegetated. Behind are barren high mud flats and heavily vegetated low mud flats. Offshore zone of shifting sediment is well developed. Note straight coast and relatively vegetation-free beaches in small bights.

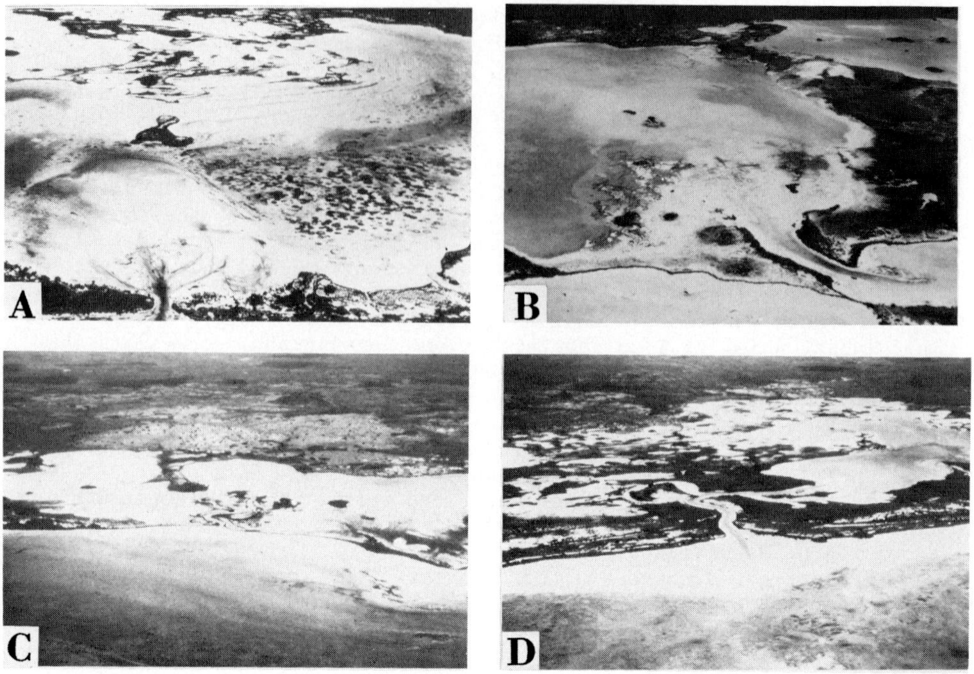

FIG. 4—Development of tidal deltas and coastal lagoons.

A. Submarine tidal delta, Midwinter Lagoon. Oblique aerial photograph looking west across barrier into lagoon. Tidal channel is simple breach through narrow barrier. Water depths over delta are generally less than 1 ft (0.3 m). Dark spots at bottom are *Thalassia* banks. At top are north-south ridges of Pleistocene limestone.

B. Tidal delta, Bennetts Lagoon. Oblique aerial photograph looking northwest. Sand Point is in lower right corner. Extensive tidal delta is dotted with mangrove islands.

C. Oblique aerial photograph of tidal delta at Will Edwards Lagoon, directly south of Bennetts Lagoon. View looking west. Much of tidal delta is exposed mud flat with dense mangrove cover.

D. Tidal delta, Northern River Lagoon. Oblique aerial photograph looking west. Original lagoon segmented by tidal delta with extensive mud flats.

Point. From Rocky Point to Little Rocky Point they are almost continuous.

The coastal lagoons bordering the northern shelf vary greatly in size and shape; most are larger than interdelta lagoons. They tend to be elongate and irregular, but many are rounded and roughly equidimensional. They are located near the coast and are separated from the sea by a narrow, well-defined and well-differentiated barrier (Fig. 3). The barrier is constructed of a few beach ridges with extensive mud flats developed on the lagoon side. Conspicuous bulges into the lagoon, representing both active and abandoned tidal deltas, add to the width of the barriers. In these areas the coastline is straight, constructed of a series of continuous beach ridges with few interruptions.

The barriers presently are growing at the expense of the lagoons. This sequence is illustrated clearly by several lagoons exhibiting different stages of development (Fig. 4).

The initial stage is the formation of a tidal channel through the narrow beach-ridge barrier. This channel probably is cut by storm washover (Stoddart, 1963, p. 101-111; Vermeer, 1963, p. 339-348; El-Ashry and Wanless, 1965, p. 404). Much of the material scoured from the new channel is deposited at either end as submarine deltas. Material transported by tidal currents is added to this. This stage of development

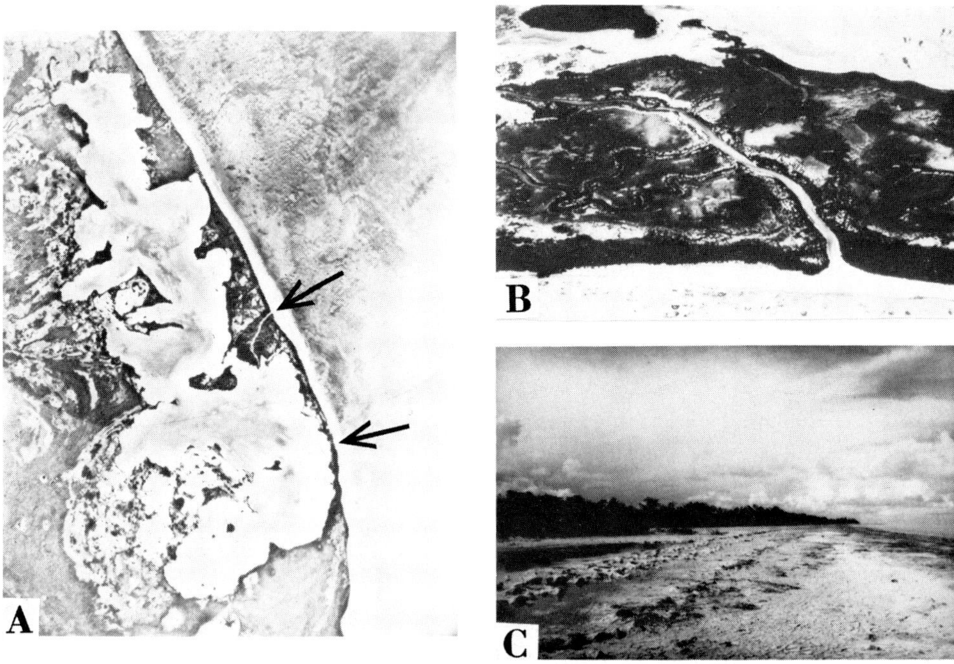

FIG. 5—Storm effects on tidal deltas, Midwinter Lagoon.

A. Vertical aerial mosaic, 1944. Arrows indicate two entrances. Abandoned tidal channels and deltas are north of open channel. Linear outcrops of Pleistocene limestone are evident on lagoon floor and are reflected in vegetation at left. These ridges form three arcuate systems. Two marine environments, coastal band of shifting sediment and stable bottom of Chetumal Bay, are clearly shown. Width of photograph, about 5 n. mi. North is toward top.

B. Northern entrance, 1965. Oblique aerial photograph looking west across barrier. Seaward end of tidal channel blocked by beach.

C. View north along new beach at northern entrance, 1965. Beach consists of several berms on crest of broad swell. Between swell and former shoreline is a flooded depression. Former channel mouth is marked by reentrant at center of photo.

is illustrated by the tidal channel of Midwinter Lagoon (Fig. 4A), where the submarine tidal delta is simply a bulge on either side of the barrier. The single tidal channel bifurcates, and several underwater channels drain, the submerged delta. Water depths are shallow—less than a foot on the banks and only a few feet in the distributaries. In the middle of the main channel the water is about 5-6 ft (1.5-1.8 m) deep. None of the banks are exposed; all are colonized by *Thalassia testudinum* (turtle grass), which effectively binds the sediment in place and prevents erosion. In addition, the dense grass acts as a baffle and promotes sedimentation (Ginsburg and Lowenstam, 1958, p. 312-314). This type of accretion is responsible for the continued expansion and shoaling of the banks until water over the banks is only about 6 in. (15 cm) deep. Then seeds of the red mangrove (*Rhyzophora mangle*) can become established in protected areas (Davis, 1940, p. 322-383; Vaughan, 1910; Chapman and Ronaldson, 1958). If the initial mangrove colonization is successful, the *Thalassia* bank subsequently will be covered by a thick stand of mangroves, the prop roots of which will trap additional sediment. This trapped sediment, together with the added vegetal matter, raises the surface until it is exposed, resulting in the formation of small, isolated mangrove islands (Fig. 4B), which are nuclei for further subaerial expansion. As the islands coalesce and new banks and islands form about the edges, the tidal delta continues to grow (Fig. 4C) until it perhaps stretches across the entire lagoon, resulting in lagoonal segmentation (Fig. 4D). Thus, the original large lagoon and narrow barrier will have been modified to a much wider barrier and smaller, more isolated lagoons.

If tidal channels were permanent features, the evolution of the barriers would end at this point. However, storm-constructed tidal channels are temporary features subject to subsequent storms which may destroy the open channel. The ephemeral nature of tidal channels is illustrated along the barrier at Midwinter Lagoon (Fig. 5).

Aerial photographs taken in 1944 show that there were two tidal channels connecting the sea with Midwinter Lagoon (Fig. 5A). The northern channel had a large, well-developed tidal delta. The southern channel (Fig. 4A) was small and had only a minor submarine tidal delta. By the summer of 1964, the northern tidal channel had been abandoned (Fig. 5B) and the southern channel provided the only access into the lagoon. During that 20-year interval, the drainage of Midwinter Lagoon had changed drastically.

The northern channel now is obstructed by a wide beach (Fig. 5C) deposited in a small bight to either side of the channel mouth. Within the tidal delta behind the beach ridges, no changes are seen and the channel remains open. The new beach, with a maximum width of about 100 ft (30 m), is constructed of a series of small berms which lie on the crest and west edge of a broad swell (Fig. 5C). Between the swell and the mangrove trees, which mark the old shoreline, a shallow depression is filled with water. Maximum relief is about 2 ft (0.6 m). The beach material is a soft, blue-black gel rich in water and hydrogen sulfide. Upon exposure to the air, the sediment rapidly oxidizes, suggesting that the pigmenting material is the hydrated iron monosulfide hydrotroilite ($FeS \cdot nH_2O$). Similar material is present in small patches along the mainland coast of Chetumal Bay and on Ambergris Cay.

The tidal channel at Midwinter Lagoon was blocked in 1959 by Hurricane Hattie, which deposited large amounts of mud and turtle grass on the coast (High, 1969).

The abandonment of old tidal channels and the formation of new ones increase the overall width of the barriers. The repeated construction of tidal deltas at new locations along the rear of the barrier is the major means of lagoonal restriction. Several abandoned tidal deltas at Midwinter Lagoon can be seen in Figure 5A. All are terminated abruptly seaward by beach ridges. The channel courses through the tidal deltas are little changed, except for infilling by heavy vegetation. Each records the sudden abandonment of well-established tidal drainage, and there is little or no degradation of the tidal deltas. Infrequent events, such as major storms, appear to cause the abandonment of tidal channels and the consequent evolution of the barriers.

The lagoons also are subjected to constant infilling by sediment from several sources, including stream and sheetwash runoff from the land side, sediment from offshore deposited in the lagoons by tidal currents, and windblown material from the beach ridges. This sediment is probably shifted about repeatedly on the lagoon floor until it becomes stabilized by *Thalassia*. Then the filling of the lagoon follows the general course described for evolution of submarine tidal deltas. The eventual fate of all the coastal lagoons is continued restriction and gradual infilling.

Two large, isolated lagoons between Belize City and Mullins River (Fig. 1) differ from other coastal lagoons in size and age. Cores taken from the barrier contain an oxidized zone indicative of a Pleistocene age. In general form, these Pleistocene barrier-lagoon systems resemble those just described, except that the Pleistocene features are several times larger and contain several sets of beach ridges. These differences are not sufficient to suggest major modification in the sequence outlined above for the development and evolution of barrier-lagoon complexes.

The Pleistocene beach ridges are remarkably parallel with the present coastal trend. They extend about 20 mi (32 km) from the southern edge of the Haulover Creek delta to just north of Mullins River, where they are truncated by the coast. The angle between the present coast and the Pleistocene coast, as determined by the beach ridges, is approximately 1°. The ridge reappears south of the Mullins River and, before disappearing again at Colson Point, changes in trend from generally north-south to northwest-southeast. The corresponding offset in the reef (Fig. 1) suggests structural control. Evidently, the offset was present during the Pleistocene and resulted in the observed deflection of the beach ridges. The Pleistocene ridges, which are truncated at Colson Point, are not seen again until the vicinity of Commerce Bight Lagoon. In the intervening area, they have been removed in places by post-Pleistocene erosion and, at Stann Creek, buried by Holocene sediments. South of Stann Creek the Pleistocene ridge system continues intermittently along the southern coast.

Forelands

Although they occupy only a small proportion of the Belize coastline, the cuspate and lobate forelands are important to an understanding of its evolution and genesis. Cuspate and lobate forelands differ in geometry as shown in Figure 6. Cuspate forelands are formed by the growth of two sets of beach ridges, or spits, which meet at a point. Other commonly used names are "triangular headlands" and "winged spits." Lobate forelands are constructed from only one set of beach ridges, which curve around the free end; they are essentially large, blunt, rounded stubby spits.

The orientations of the stranded beach ridges record the growth of the forelands. However, destruction of the older ridges by swamp encroachment (as described by Allen; 1965, p. 15-16) and erosion due to migration prevent complete preservation of the older form.

The construction of cuspate forelands is due to wave refraction around an offshore cay or irregular bottom. The refraction forms two wave sets that are nearly normal to each other (King, 1959, p. 93-100; Bascom, 1964, p. 70-77). Because coastal accretion is generally parallel with the dominant waves (Guilcher, 1958, p. 91-94, 182-186), the triangular shape is formed.

Lobate forelands form in similar environments where there are no offshore cays or irregular bathymetry to interrupt the onshore wave patterns. The lobate forelands start as normal spits, and the addition of new beach ridges leads to growth in length and width. Continued lateral growth, with no appreciable shift in the point of initiation where the beach ridges come together, results in a blunt lobate form, rather than an elongate narrow spit. Recurving of the beach ridges at the free end eventually isolates a large lagoon behind the foreland, as at Little Rocky Point.

A similar mechanism may have been responsible for forelands at Placentia and Punta Negra. They probably are the result of interference of more than one set of wind-driven waves, as it cannot be demonstrated that they contained former cays at their apexes. Their location at the widest points of the deep southern shelf (Fig. 1) supports this interpretation. Possibly, two or more permanent sets of wind-driven waves, generated in the Caribbean or on the shelf, interfere as they approach

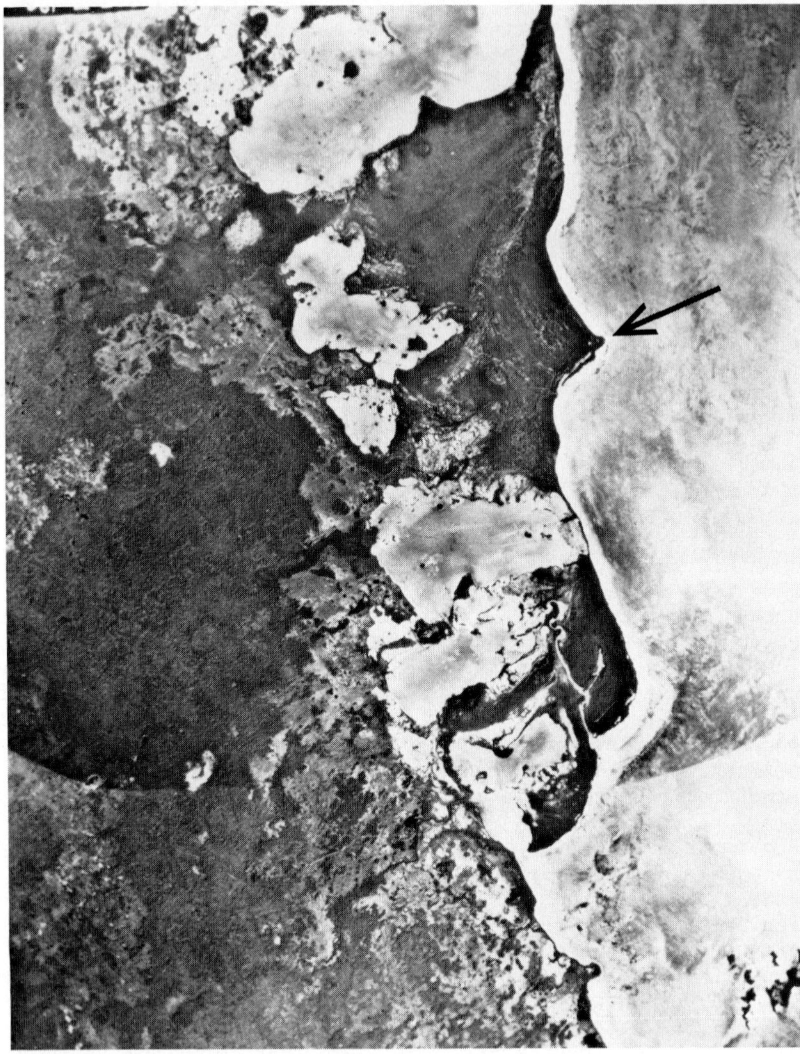

FIG. 6—Vertical aerial mosaic, Little Rocky Point area. Circular structures are Mayan lookouts or habitations. One at apex of cuspate foreland (arrow) is former cay. Wide beach-ridge system on northern side of cuspate foreland has been truncated at both ends. No such truncation is seen in two lobate forelands to south. Note alignment of coast of Belize Bight (bottom) and rear edge of lagoons. Southern end of Midwinter Lagoon is at top of picture. Width of photograph, about 5 n. mi. North is toward top.

the coast. Such a wave pattern is responsible for the large cuspate foreland at Dungeness along the English Channel (Guilcher, 1958, p. 91, 182-186).

The source of the large quantity of sand stored in the forelands is a problem. None of the forelands is adjacent to a river mouth from which the sediment required to continue growth could be derived. However, the general location of the forelands suggests that sediment is provided by longshore currents. With the exception of Placentia, the forelands are located at the southern end of straight reaches of coastline; to the south are indentations. Southward-flowing longshore currents are dispersed as they enter the large embayment. As a result, much of the material in motion along the straight stretches of coast comes to rest at the northern ends of the embayments. This material is then available for the construction of depositional features and constitutes a permanent contribution to a limited area of the coast. Thus, the coast is broken into compartments, each of which is approximately self sustaining. Exchange of sediment between coastal segments is probably negligible, as in southern California (Shepard, 1963a, p. 182-185).

Deltas

Deltas represent the most highly modified type of coast in Belize and record the adjustment of the coast to present-day conditions. Pre-Holocene topography is entirely subdued by the construction of a delta.

Deltas in Belize are of three types. The most common are the lobate and cuspate types of Bernard (1965, p. 334-335). Cuspate deltas, found at the mouths of most rivers south of the Belize River, are small. They are formed by a simple thickening of the beach-ridge system fronting much of the coast (Fig. 7). Each has a single distributary. In contrast, the larger delta of the Sittee River, which represents more active delta construction under the same general conditions, has built outward and changed from cuspate to lobate. The angle of the arc of the beach ridges has increased sharply, and the ridges have recurved at the tip, resulting in the final lobate form.

The Haulover Creek delta (Fig. 8), on which Belize City is built, is another lobate delta. However, this is not a Holocene feature, but is the partially destroyed remnant of the Pleistocene delta of the pre-Holocene Belize River. The most striking feature of the delta is its asymmetry. The wide southern side is marked by a large number of well-developed Pleistocene beach ridges, whereas the northern side is a low, featureless Holocene mud flat. The mud flat has almost been filled during the past few hundred years by material resulting from human occupation. Excavations on both sides of the river near the center of town indicate a Holocene section about 15-20 ft (4.5-6 m) thick (Ower, 1928a, p. 499).

The origin of the Haulover Creek delta has been questioned by Ower (1927, p. 376), who stated that the swamp behind Belize City indicates that this landform is a cay in the process of being attached to the mainland. That is, it is a tombolo rather than a delta. Present information helps to resolve the uncertainty concerning its origin. During the Pleistocene, when sea level stood at approximately its

FIG. 7—Cuspate delta, South Stann Creek. Oblique aerial photograph looking southeast. Well-developed beach ridges on northern side have been partly destroyed by stream meanders. Lunate or river-mouth bar is well developed.

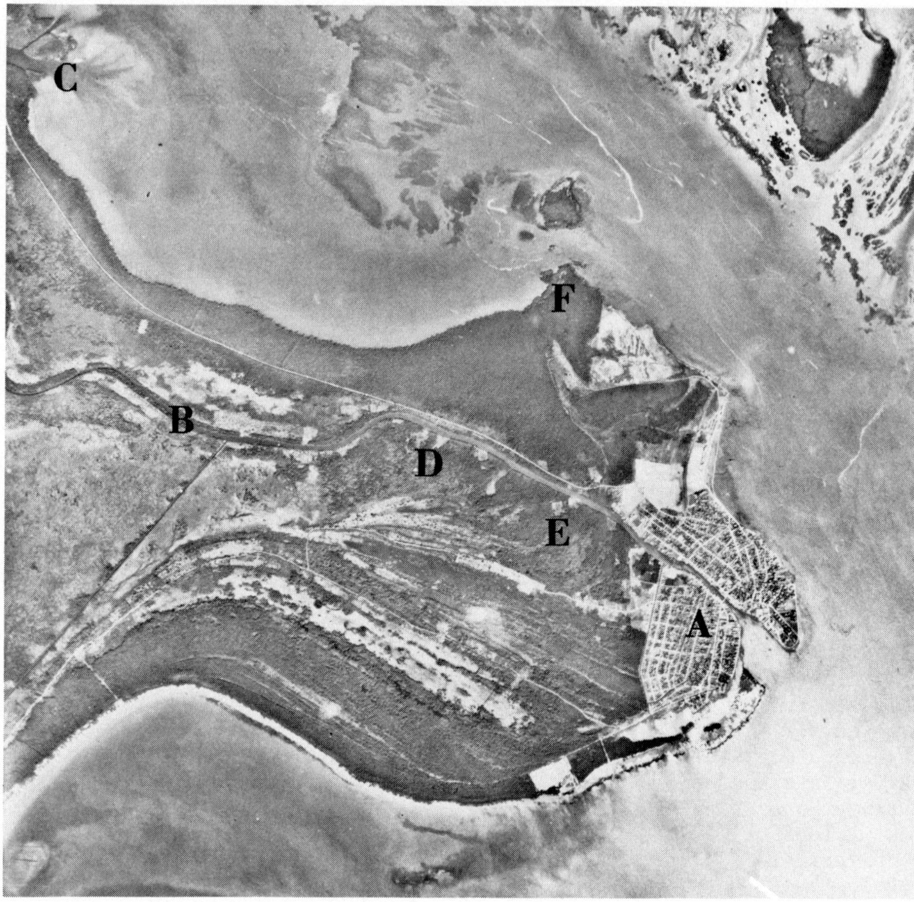

FIG. 8—Vertical aerial photograph, Pleistocene lobate delta of pre-Holocene Belize River. Width of photograph, about 5 n. mi. North is toward top. **A.** Belize City; **B.** Haulover Creek; **C.** mouth of Belize River; **D.** initial cuspate delta; **E.** lobate delta; **F.** former cay(?).

present position, the pre-Holocene Belize River constructed a large lobate delta. The several trends, or bundles, of beach ridges evident in the aerial photograph record different stages of delta growth (Fig. 8); each ridge records a former position of the shoreline. The initial form was a cuspate delta similar to those of South Stann Creek and Mullins River. Continued growth resulted in an elongate, lobate delta similar to the Sittee River delta. The recurving of the beach ridges indicates that the mouth of the pre-Holocene Belize River was located near the present course of Haulover Creek. Further growth resulted in the widening of the lobate delta as new bundles of beach ridges were added for each shift in orientation.

The absence of beach ridges on the north side is probably original. In the past, as today, the south side of the delta must have been exposed to the deeper waters of the southern shelf, whereas the north side was protected by sheltering cays and shallow water that prevented surf from striking the north side with as much force or frequency as the south side. Thus, the construction of beach ridges by storm waves (Price, 1945, p. 3; Russell and Howe, 1935, p. 458-461; Schofield, 1960, p. 470-476; High, 1969) was inhibited on the protected north side. A contributing factor was the southerly longshore current that deflected the load of the river southward.

The establishment of the modern outlet of the Belize River initiated the final phase in the construction of the Haulover Creek delta. The former course of the river was occupied by an overflow channel, Haulover Creek. The north side of the delta attained its final form as floodplains extended the low northern bank. The northern point of the delta probably formed as the shallow water between the mainland and small nearshore cays was filled with sediment from the Belize River (Fig. 8). Finally, material from human occupation rapidly raised the level to the present few feet above sea level.

The modern Belize River delta (Fig. 9) is an example of the birdfoot-type delta. Although small, it has three distributaries. The subaerial portion of the delta is essentially structureless, and there are no associated beach ridges. Low, flat mangrove swamps cover the surface.

The presence of these types of deltas in Belize can be explained by the hydrography of the shelf. Bernard (1965, p. 335) has ranked delta types in order of decreasing sediment influx and increasing energy of the waste mass as: Gilbert, birdfoot, lobate, cuspate, arcuate, and estuarine. Although there are no data on the sediment influx (the Belize River certainly carries a load far greater than that of other rivers in Belize), the birdfoot delta of the Belize River is located in a sheltered position, and the lobate and cuspate deltas of the rivers to the south are exposed to a much more rigorous hydrographic regime.

SEDIMENTOLOGY

The increasing number of studies of Holocene coasts permits the establishment of tentative models based on the type and mechanics of deposition. The most-studied coastal types include carbonate-evaporite, wave-dominated, and tide-dominated coasts. In contrast, little attention has been given to clastic deposition along quiet coasts, such as Belize, which are characterized by the lack of surf and tidal agitation and are dominated by clastic sediments.

Environments of Deposition

Depositional environments along the coast are listed in Table 1. Major environments listed correspond to physiographic features. Sub- and local environments are defined by local features, such as specific landform, elevation, and vegetation.

FIG. 9—Birdfoot delta of Belize River. Oblique aerial photograph looking northwest. Belize Bight is in right background. Entrance to Haulover Creek is just southwest of bridge. Vegetation on south side of delta is freshly cleared.

TABLE 1. CLASSIFICATION OF RECENT DEPOSITIONAL ENVIRONMENTS ALONG BELIZE COAST

Major Environment	*Subenvironment*	*Local Environment*	*Ref. No.*
Lagoon barrier	Lagoon complex	Lagoon	1
		Mangrove margin	2
		Mangrove island	3
		Windward beach ridge	4
	Mud-flat complex	Subaerially exposed tidal delta	5
		Interlagoonal	6
		Filled lagoon	7
		Back marsh	8
	Beach-ridge complex	Beach	9
		Beach ridge	10
		Swale	11
	Tidal channel	Channel	12
		Delta	13
		Abandoned & filled channel	14
	Nearshore	Shelf	15
		Shoreface	16
Delta	Prodelta		17
	Delta front	Lunate bar	18
		Other	19
	Delta plain	River channel	20
		Natural levee	21
		Interior swamp	22
		Beach	23
		Beach ridge	24
		Swale	25
	Delta margin	Lagoon	26
		Beach ridge	27
		Mud flat	28
		Nearshore	29
Bight	Coastal marsh		30
	Estuary		31
	Nearshore		32
	Cay	Shoreface	33
		Ridge	34
		Interior marsh	35
Cuspate and lobate foreland	Lagoon	Lagoon	36
		Marginal swamp	37
	Ridge and swale	Beach	38
		Beach ridge	39
	Nearshore		40

Lagoons—Lagoons are one of the characteristic coastal features, and are present along the entire coast in a variety of settings. Several of the largest, most extensive, and best-developed Holocene lagoons in the northern lagoons province were studied in detail. All are shallow; no depths greater than 5 ft (1.5 m) were found, and the average depth is estimated to be approximately 2 ft (0.6 m). Salinity varies widely within and between lagoons, ranging from almost hyposaline to hypersaline. Water enters the lagoons from the east by tidal exchange and from the west by runoff. In addition, the waters of the lagoons are freshened by infrequent heavy rain squalls. Dry periods, even of short duration, cause significant evaporation with a consequent increase in salinity. Water in the main lagoon is cool to tepid, whereas water in isolated pools and stagnant reaches may be warm or even hot.

The bottoms of the lagoons are sparsely vegetated. *Thalassia* is poorly developed, except in local patches near tidal channels. The marine alga *Batophora* is locally abundant where the bottom is hard.

The nature of the bottom varies widely. Pleistocene limestone is commonly exposed on the lagoon bottom. It may form small islands as, for example, at Lost and Midwinter Lagoons, where the irregular surface of the rock has as much as 3 ft (0.9 m) of relief locally. Holocene sediment accumulates in pockets between the rock projections. At Northern River and Bennetts Lagoons, Holocene sediment completely mantles the Pleistocene limestone. The sediment from different lagoons is generally a shelly calcareous mud or muddy shelly sand, but it varies greatly. However, within each lagoon the character of the sediment is fairly uniform.

Minor environments associated with the lagoons are mangrove fringe, mangrove island, and windward beach ridge. The mangrove fringe is a narrow zone of dense red mangrove growth that nearly encircles the lagoons. The sediment accumulating beneath the mangrove fringe is greatly influenced by the dense plant cover. It is a dark red-brown, soft mud rich in plant remains.

Mangrove islands are dense stands of large red and black (*Avicennia nitida*) mangroves that have colonized former *Thalassia* banks. The sediments formed are muddy peats.

The western, or windward, sides of the lagoons commonly have small, discontinuous beaches and beach ridges constructed of material winnowed from the bottom sediments. The beaches, located in small bights, are concave lagoonward. Colonization of the beach by grass or mangrove stabilizes the sediment and effectively retards erosion. This process was observed in several lagoons where the forest grows to the edge of the lagoon, making the windward beach ridge almost unrecognizable.

Although smaller and less well developed, lagoons of the cuspate forelands and interdelta areas are basically similar to the northern coastal lagoons. The same environments are present, although they are not so clearly differentiated. Several of the lagoons appear to be in the final stages of filling. The water is generally fresh or slightly brackish, undoubtedly reflecting the higher precipitation along the southern coast and the extensive surface drainage, as at Commerce Bight Lagoon. The bottoms may be covered with dense carpets of manatee and widgin grass.

Mud flats—Mud flats, the single most extensive environment along the coast, are located along embayed, estuarine coasts in association with lagoons.

Mud flats that form as tidal deltas surround coastal lagoons, and red mangroves colonize the fringes. Initially, the mud flat is low, flooded by lagoon water and covered by a thick growth of red mangroves. As sediment accumulates, the surface gradually is elevated until the mangroves no longer receive the needed brackish water and growth is inhibited. In this area of negligible tides, an elevation of about 1 ft (0.3 m) is sufficient. These higher surfaces are relatively bare except for a sparse growth of stunted red mangroves. Further sedimentation on the high mud flats is accomplished by storms. Erosion during storm flooding is inhibited by the cohesiveness of the sediment owing

FIG. 10—Environments on barriers.

A. Small red mangroves growing on low mud flat at Northern River Lagoon.

B. High, nearly barren mud flat on barrier at Northern River Lagoon, about 3 ft (0.9 m) above sea level. Three of four barrier vegetation zones can be seen. In background, palm trees and tall mangroves grow on beach ridges. Between these and barren mud flat is thick growth of seagrass. Fourth vegetation zone, not shown, is mangrove fringe.

C. Thick mature mangrove forest on barrier ridge at Midwinter Lagoon.

D. Beach ridge at Bennetts Lagoon, about 2 ft (0.6 m) high and 4 ft (1.2 m) wide, covered by thick growth of small red mangroves.

to subaerial exposure and by the protection of the mangrove fringe. The dense vegetation breaks the force of the wind and waves while the flood waters are temporarily ponded, allowing sediment to settle from suspension. Low parts of subaerial mud flats that catch local runoff are covered by tall, dense sawgrass.

Because of the differentiation of mud flats, the barriers commonly exhibit four vegetation zones (Fig. 10) which, from sea to lagoon, are: coconut palms and tall mangroves (on beach ridges); tall, dense sawgrass (in a band behind the oldest beach ridge); high, relatively bare mud flat; and mangrove fringe.

Mud-flat sediments are rather uniform, light-colored, sandy carbonate muds with scattered plant debris. Salt crystals are common on the surface but were not found in cores. Locally in the marshes on the mainland side of the lagoons, dark red-brown muddy peat is present.

A second major mud-flat environment is in the coastal marshes along estuarine coasts (Fig. 2). These marshes were investigated just north of the Monkey River. Superficially, the surface resembles parts of the subaerially exposed tidal deltas. It is low, flat, and covered with a sparse growth of small red mangroves. Flooding of the surface is probably common, and sedimentation is by mangrove entrapment and stabilization. However, the pattern of sedimentation is different from that of the tidal deltas. The coastal marshes are building outward from

Pleistocene sand ridges to fill embayments, whereas the tidal deltas are building inward to fill lagoons. The net result, a straight coastline, is the same, but the landforms and sediment distribution are markedly different.

Storm ridges—Much of the actual coastline is fronted by ridges and swales. The ridge systems are found on the barriers of coastal lagoons, on cuspate and lobate deltas and forelands, and in the bights between deltas.

The barrier-ridge system typically consists of a broad swell with two or three individual ridges on the crest. Discontinuous beaches on the seaward edge are deposited by Holocene storms (High, 1969). The ridges are all low. Maximum elevation above sea level is about 3-4 ft (0.9-1.2 m). The width is variable, ranging from a few feet to tens of yards. Vegetation on all but the youngest ridges is heavy. Large red mangroves and scattered coconut palms form a tangled, dense forest. In plan view, the ridges are straight to gently curved. Major ridges may be followed for several miles until they are terminated by tidal channels. Except for areas directly north of the forelands, truncation of the beach ridges by the present coastline was never observed.

Swales between ridges are low, marshy areas approximately at sea level. Where narrow, the swales show no appreciable difference in vegetation from the ridges. However, where well differentiated from the ridges, the swales support a much denser growth of larger mangroves and include both red and black varieties. No coconut palms were observed in the swales.

The formation of beach ridges by storms is indicated by the similarities of barrier ridges and modern storm deposits along the same segments of the coast (High, 1969, p. 239).

Although the initial sediment on the ridges is coarser than that of the swales, the addition of fine detrital material and plant remains causes significant modifications. The youngest ridges are constructed of nearly pure sand, whereas older ridges are less well sorted and contain varied amounts of plant debris.

Tidal channels—Each larger lagoon is connected to the sea by a deep tidal channel swept by moderate currents. Tidal channels of the northern coast have little sediment on the bottom, and Pleistocene limestone is exposed locally. After the channel has formed and the adjacent mud flat is colonized by mangroves, the position of the channel is stabilized. As a tidal delta grows, the initial break in the barrier becomes a long, sinuous creek with several distributaries at the lagoon end; the seaward end, which is exposed to surf and longshore currents, remains a single channel. As the surrounding mud flats grow, the channel banks become steep and support a dense growth of red mangroves. Eventually, further encroachment is checked as the increasing erosive power of the tidal currents balances the tendency for sedimentation to restrict the channel further. This stage in channel development can be recognized readily by the abruptness of the channel sides. An actual overhang may develop where the outermost mangrove plants are supported by the root mat and little or no sediment is packed around the prop roots.

Although the tidal range is small, strong currents in the restricted channels prevent the accumulation of much sediment. Bottom samples were difficult to obtain because of the current, water depth, and rocky bottom. One sample from the channel at Midwinter Lagoon was composed entirely of worn cobbles of Pleistocene limestone.

The associated submarine tidal-delta and filled-channel environments have been described. The sediments in the former range from poorly sorted muddy sand to well-sorted sand. Organic muds are present in the latter.

Deltas—Deltas are formed at the mouths of rivers that do not empty into lagoons or estuaries. Cuspate and lobate deltas are constructed of ridges and swales similar to those of the barriers. The ridges are about 3 ft (0.9 m) high, and the swales are approximately at sea level and are continually flooded. Widths of both are generally a few tens of feet. The pattern of vegetation is the opposite of that on the barriers; the ridges are heavily vegetated and the swales are relatively open. Both ridges and swales support large mangrove trees, but the ridges are covered also by a thick growth of lower shrubs and vines. Many of the larger trees near the coast are dead, having been defoliated by Hurricane Hattie in 1961.

As with the barrier ridges, the sediment of the ridges and swales on deltas is greatly modified by the heavy plant cover and is similar in both environments.

The delta ridges extend into the interdelta bights with little change. Each ridge and swale narrows and a few minor ridges may terminate. The distribution of sediment and vegetation in these areas is similar to that in the deltas.

The beaches fronting these ridges bear no resemblance to the storm-deposited beaches of the northern barriers. The former occupy a narrow but almost continuous band from Belize City to Placentia. Their maximum width is about 20 ft (6 m), and the steep profile reflects the heavy surf and coarse sediment. The beach sediment is clean, coarse sand, or even fine gravel, composed chiefly of quartz. The sources are the Maya Mountains and reworked Pleistocene sediment. In contrast, the beaches fronting the delta ridges are no more than a thin veneer thrown up on an eroding coast. They are transitory, migrating with the retreating shore. Older beaches may not be preserved but are redistributed as erosion proceeds.

The birdfoot delta of the Belize River (Fig. 9) contains a different suite of environments. Most of the subaerial part of the delta is a large mangrove swamp. A poorly developed natural levee is present along the river bank. The interior of the delta is low-lying and flooded. A mature black mangrove forest characterizes the heavy vegetation. As is common along the coast, many of the larger trees are dead. The sediments are soft, organic-rich muds.

The major submarine deltaic environment is the lunate river-mouth bar, which is present at all delta mouths (Fig. 7). The water over these bars is shallow, and the channel, if present, may shift. The sediments are composed of relatively clean sand with some sandy mud. Because of their position, the residence time of these sediments is relatively short, and they eventually enter the longshore-current system.

Nearshore marine environments—The final major environment is that of the nearshore marine area. The sediments of this environment are discussed elsewhere (Matthews, 1966; Pusey, this volume). Only the areas directly adjacent to the shore are of interest here. Extending seaward from the strandline for a few hundred feet is a zone of shifting sediment arranged in a series of giant ripples. This is the area of active longshore transportation. At Midwinter Lagoon the maximum relief across the first seven bars is about 2 ft (0.6 m); the average crest-to-crest distance is about 50 ft (15.2 m).

Seaward of this zone the sediments are relatively stationary and are partially covered by *Thalassia*. This zone of shifting sediment marks the seaward boundary of the study area.

Holocene Sediments

Methods—The sedimentary features selected for measurement in this study were chosen on the basis of their applicability to studies of ancient sediments. Commonly measured characteristics, such as water content, detailed textural measurements, plasticity, pH, Eh, etc., were rejected because it is not generally possible to make similar measurements on sedimentary rocks. Texture, carbonate content, organic content, and constituent-particle composition were judged to meet the requirements of applicability to sedimentary rocks.

Grab samples were collected at 128 stations during the summers of 1964 and 1965. Piston cores, up to 12 ft (3.7 m) long, were also collected at 95 of these stations. Sampling areas were selected from aerial photographs, taken by the U.S. Navy in 1944, on the basis of their potential for contributing to an understanding of the origin of particular types of coastal morphology; only minor consideration was given to uniform coverage of the coast. Where possible, sampling traverses extended across the entire belt of Holocene sediment. Stations along the traverse were selected by changes in the depositional environment, rather than by predetermined spacing.

One split of each dried sample was weighed and treated with commercial sodium hypochlorite (Clorox) to remove plant debris and organic matter. Any unlithified fecal pellets also were thus destroyed. The grain destruction reported by Pusey (1964, p. 49) following treatment with 4-percent hydrogen peroxide was not noted in the samples treated with sodium hypochlorite. After removal of organic matter, the remaining sediment was wet-seived and sand content was calculated. No other textural measurements were made.

The sand-size fraction of each sample was mixed with a polyester resin and thin sections were prepared. One hundred random points on each section were counted to estimate the constituent-particle composition of the sand fraction. Several independent counts of the same sample indicate that the results, reported as volumetric percent, are accurate to about ±5 percent of the common grain types. Other sample splits were used to determine carbonate and organic contents. Five-percent nitric acid and dilute sodium hypochlorate were used to remove carbonate and organic materials, respectively. The carbonate and organic contents were calculated by weight loss. Two splits were processed for each sample. Results, which agreed to within 5 percent, were averaged. Table 2 lists the results of these analyses.

Sample grouping—The initial problem in data evaluation is the recognition and definition of lithologies. It entails a critical appraisal of the properties measured in terms of their interdependence, sensitivity to environment, and reliability as environment indicators. Under favorable conditions (i.e., the properties are independent, sensitive, and reliable), discrete lithologies may be defined by one or more of the properties which are found to have a systematic variation among the samples. Such lithologic groups then may be related to the depositional environment, and the collective lithologic characteristics thus become the criteria for the recognition of that environment. This approach closely approximates that which would be applied to sedimentary rocks in an attempt to reconstruct paleoenvironments. A suite of samples would be studied, significant lithologies would be identified, and environmental interpretations would be based on the properties of these lithologies.

Possible causes of failure of this general procedure are that (1) the patterns may be present in the sample but pass unrecognized because of improper selection of sediment properties; (2) the patterns may have been present in the sediment at the time of deposition but be destroyed by subsequent reworking; and (3) the patterns never were present.

The first possibility is rejected because of the reliability of the basic sediment properties selected. Each is sensitive to certain aspects of the physical environment and may be expected to reflect them.

TABLE 2. PROPERTIES OF SEDIMENT SAMPLES

Environment (Table 1)	*Sample Number*	*Weight Percent*				*Constituent-Particle Composition (Percent volume of >1/16-mm fraction)*				
		Sand[1]	*Mud*[1]	*Organic*[1]	*Carbonate*	*Crypto-crystalline Carbonate*	*Shell Debris*	*Quartz*	*Volcanic Glass*	*Mica and Opaque*
Bennetts Lagoon										
5	Sd. Pt.	35	61	4	74	--	--	--	--	--
10	520	96	2	2	76	65	11	22	2	0
5	521	24	62	14	77	80	14	2	4	0
5	522	22	64	14	82	54	43	3	0	0
1	523	55	43	2	85	68	29	0	3	0
1	524	76	24	tr.	41	37	10	53	0	0
8	525	9	74	17	81	83	6	2	9	0
8	526	11	75	14	82	63	36	0	1	0
Lost Lagoon										
1	527	36	62	2	92	53	42	0	5	0
6	528	8	85	7	90	39	50	0	11	0
1	530	10	88	2	92	60	29	0	11	0
5	531	38	53	9	47	70	16	0	14	0
13	532	93	5	2	94	83	10	1	6	0
Northern River Lagoon										
8	533	3	64	33	14	12	54	33	1	0
4	534	14	45	41	13	0	0	96	4	0
1	535	47	50	3	58	34	49	12	5	0
3	536	5	21	74	20	--	--	--	--	--

14	537	4	52	44	33	--	--	--	--	--
14	538	3	30	67	24	15	46	39	0	0
14	539	5	72	23	41	51	40	8	0	0
1	540	36	60	4	40	29	21	43	7	0
5	541	24	68	8	52	35	22	36	7	0
5	542	32	53	15	77	47	31	10	12	0
11	543	26	64	10	79	58	42	0	0	0
11	544	63	30	7	88	69	29	2	0	0
10	545	34	46	20	86	74	24	2	0	0
9	546	20	53	27	72	48	35	12	5	0
15	41	95	3	2	82					
Midwinter Lagoon										
1	548	56	43	1	61	67	25	0	9	0
4	549	92	7	1	6	2	4	94	0	0
1	550	96	3	1	25	3	17	80	0	0
1	551	46	50	4	68	41	51	8	0	0
1	552	37	58	5	59	63	11	18	8	0
5	553	42	54	4	47	37	15	33	14	4
10	554	63	34	3	66	83	4	6	7	0
10	555	78	19	3	17	1	26	72	1	0
16	556	74	23	3	89	69	25	3	3	0
9	557	64	29	7	75	65	23	7	5	0
9	558	20	61	19	76	61	34	2	3	0
9	560B	39	44	17	81	73	19	0	8	0
16	561	66	28	6	83	80	14	1	5	0
16	562	98	1	1	25	36	4	58	3	0
13	563	73	23	4	45	47	17	35	2	0
13	565	91	6	3	38	54	10	31	5	0
15	40	96	0	4	65					
Belize Bight										
33	566	63	30	7	86	71	24	1	4	0
34	567	74	19	7	--	38	57	1	4	0
35	568	81	12	7	87	57	36	0	7	0

TABLE 2 (Continued)

Environment	Sample Number	Weight Percent				Constituent-Particle Composition (percent volume of >1/16-mm fraction)				
		Sand[1]	Mud[1]	Organic[1]	Carbonate	Crypto-crystalline Carbonate	Shell Debris	Quartz	Volcanic Glass	Mica and Opaque
35	568P	55	33	12	--	45	49	0	6	0
40	37-60	86	8	6	85					
40	38-60	75	17	8	88					
40	39-60	5	93	2	89					
Belize River Delta										
19	569	3	92	5	21	80	19	1	0	0
19	570	6	87	7	26	65	11	24	0	0
22	571	11	59	30	14	0	20	79	0	1
22	572	14	54	32	18	23	0	76	1	0
21	573	10	72	18	24	37	2	62	0	0
20	574	4	91	5	17	9	7	73	0	11
20	575	6	92	2	15	1	3	89	0	7
18	576	54	45	1	7	1	3	89	2	5
18	577	35	61	4	22	0	3	83	7	7
18	578	84	14	2	4	0	1	81	7	11
18	579	20	77	3	10	0	2	84	2	12
17	580	29	65	6	71	18	73	2	7	0
17	581	2	89	9	71	66	34	0	0	0
Haulover Creek Delta										
18	582	3	93	4	33	52	22	19	0	8
23	583	98	1	1	92	0	98	2	0	0

25	584	4	88	8	30	69	10	10	1	0
25	585	50	47	3	10					
24	586	3	89	8	27	0	0	96	4	0
24	587	35	35	30	33	0	0	98	1	1
25	588	3	62	35	26	20	28	50	0	2
24	592	62	31	7	4	0	0	93	7	0
22	595	3	89	8	25	34	14	50	2	0
Northern Lagoon										
7	621	37	62	1	6	13	10	74	3	0
10	622	83	15	2	1	0	0	94	6	0
11	623	57	38	5	12	0	4	93	0	3
10	624	88	10	2	11	0	3	76	9	12
16	625	60	39	1	5	0	2	85	1	12
Mullins River										
11	626	77	17	6	2	0	0	87	11	2
9	627	98	2	0	1	0	1	92	7	0
9	628	100	0	tr.	0	0	0	35	7	59
24	629	98	0	2	3	0	0	88	7	5
23	630	99	1	0	0	0	0	86	1	13
19	87	77	19	4	20					
19	88	95	4	1	1					
18	88A	100	0	0	0					
Colson Point										
26	631	46	50	4	3					
28	633	27	61	12	15	0	4	81	0	15
Stann Creek										
26	635	5	88	7	12	0	4	14	2	80
27	636	99	1	tr.	1	0	0	89	7	4

TABLE 2 (Continued)

Environment	*Sample Number*	*Weight Percent*				*Constituent-Particle Composition (percent volume of >1/16-mm fraction)*				
		Sand[1]	*Mud*[1]	*Organic*[1]	*Carbonate*	*Crypto-crystalline Carbonate*	*Shell Debris*	*Quartz*	*Volcanic Glass*	*Mica and Opaque*
Commerce Bight										
26	639	3	66	31	30	0	35	31	3	31
29	102	9	87	4	23					
29	103	95	4	1	6					
27	103A	99	1	tr.						
Sittee River										
29	643	99	0	1	2	0	0	86	9	5
Sapodilla Lagoon										
26	307	18	73	9	17					
26	308	1	91	8	17					
26	309	3	89	8	17					
29	335	35	59	6	32					
29	334	31	63	6	39					
Placentia Lagoon										
26	356	8	77	15	11					
Punta Negra										
40	142	100	0	tr.	4					
40	143	59	41	tr.	63					
40	144	94	4	2	92					

[1]Recalculated to 100 percent.

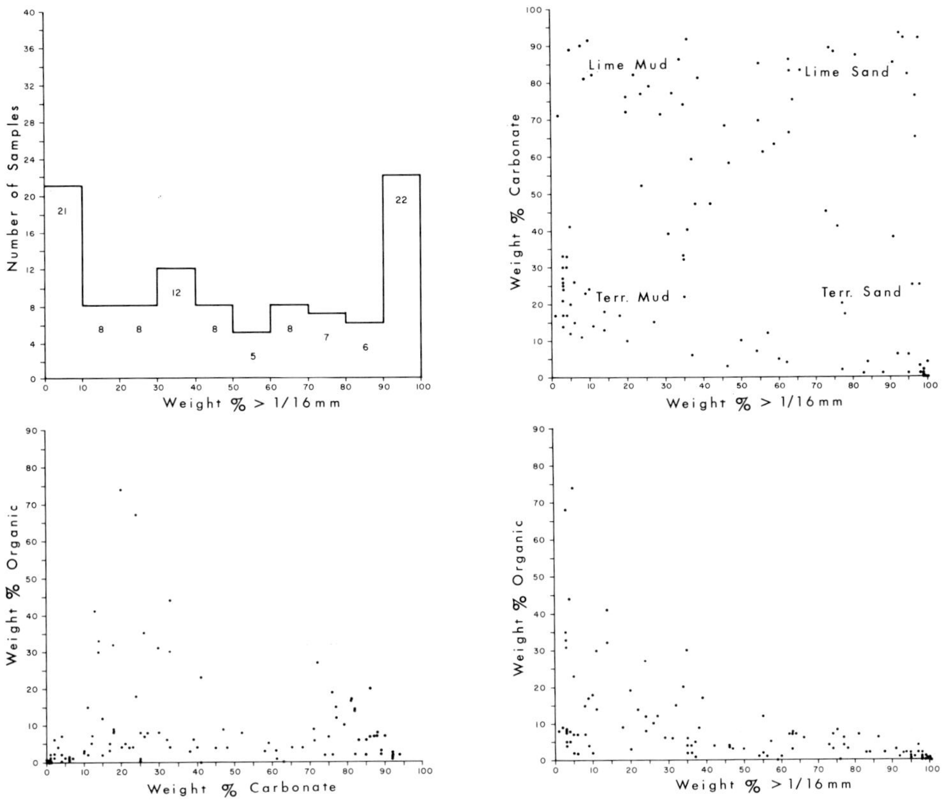

FIG. 11—Texture and composition of Holocene coastal sediments.

The second is likewise rejected because, although there is abundant evidence of reworking, it is of limited importance. Local reworking, such as by wave agitation and organic activity, generally affects only the sediments of a single environment. Except in the restricted case, where the sediments of a second environment are near the surface, no mixing will occur.

The third possibility is the most likely. The continuum of physical forces which exists in nature over any large area will be expressed in the sediment; distinct breaks between sediment types do not exist because there are no distinct breaks in the physical factors which control the environments. The complex interaction of all members of the system produces a vast number of differences in degree and direction. Thus, the end result for a single set of conditions cannot be predicted. Additionally, where controlling conditions are incompletely understood, results are even more complex.

The Holocene (and Pleistocene) coastal sediments of Belize show such a complex relation. Histograms of each of the nine sediment properties measured and cross-variable plots of all possible pairs of properties do not show any significant, consistent natural groupings (Fig. 11). When all samples are considered together, a continuum of properties is apparent (Fig. 12). Even the correlation of such related factors as sand content and quartz is uncertain. The correlation for quartz and sand content for all samples from northern Belize is 0.24. Thus, the scatter reflects the differences in hydrographic regime and source areas

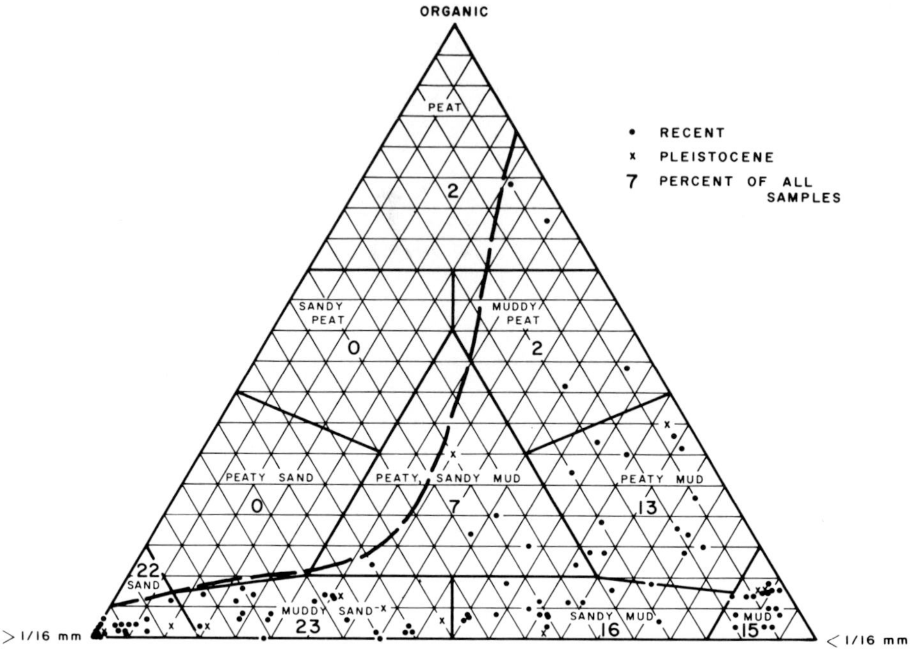

FIG. 12—Composition of Holocene coastal sediments.

between the northern and southern coasts; and these differences are sufficiently large to obliterate relatively minor effects of the depositional environment.

The data in Table 2 and studies along the coast suggest three major environments within the lagoon-barrier complexes: (1) lagoon, (2) mud flat - tidal delta, and (3) ridge-beach-nearshore. If samples from each group are identified on the plots, a moderate separation is apparent (Fig. 13). The considerable overlap is in part real and in part the result of sampling and local conditions. For example, the beaches are all recent storm deposits of organic muck which have not yet been winnowed. Similarly, some of the mud-flat samples may have been collected from unrecognized small ridges. A more detailed and systematic sampling of these environments, followed by statistical analysis of the data, might demonstrate more definite sediment-environment correlations.

Trends of sediment properties—The consideration of the sediment samples as isolated entities, without regard to other samples from the area, failed to define absolute criteria for each environment. However, if each traverse is considered separately, changes between adjacent environments may form a trend. If this trend can be generalized for several traverses, it can be used to identify the environments. To this end, the absolute values of each sediment parameter are not as important as the direction and general magnitude of change when compared to nearby samples from other environments. Purely local influences are recognized and eliminated.

As an example, the trend in sediment properties from Bennetts Lagoon is shown in Figure 14. Other traverses in northern Belize are similar. Comparison of these profiles indicates that there is a systematic variation in the sediment properties from one environment to another. Thus, barrier ridges and offshore deposits, when compared to other environments in the same area, are seen to be sandy, enriched in quartz, and poor in organic material. Similarly, mud-flat sediments are muddy with moderate amounts of plant remains; lagoon sediments are moderately sandy and deficient in organic matter. Carbonate content and grain composition, other than quartz, are variable and show no definite trends. Absolute values are not comparable from one area to another. For example, beach ridges at Northern River Lagoon contain about 50 percent sand. This figure is comparable to that for the lagoonal sediments at Bennetts Lagoon, where the beach ridges are nearly pure sand. Thus, the difficulty in obtaining definite characteristics for each environment is apparent. The local effects, even within a restricted area where all conditions are similar, are great enough to destroy the systematic pattern when the sediment samples are removed from their natural setting and grouped together. If, however, the samples are considered with respect to their relations within the local area, this problem is largely overcome.

Closer inspection of the traverse profiles reveals several irregularities in the actual trends. Many of these may be explained by considering sample position. Environments do not change abruptly along a definite border. Rather, there is a transition zone where each environment has a more or less marked influence on the other. Identification of the transition zone was difficult in places, and some samples were collected from the mixed zone by mistake. One example is sample 524 (Fig. 14), which has abnormally high sand and quartz contents in comparison to other lagoonal sediments along the same traverse. However, this sample was collected near the windward (west) shore of the lagoon where small beach ridges were present. Their proximity resulted in "contamination" of the lagoonal sediments.

Another probable source of deviations from the patterns arises from the attempt to fit a complex natural system into a simple conceptual scheme. However, in view of the large number of known and unknown variables, the unexplained irregularities seem minor.

In the past few years many attempts have been made to find universal, quantitative criteria by which the depositional environments of ancient sediments can be recognized. However, none of these attempts have produced the desired results, and recognition of depositional environments remains one of the most difficult tasks for the geologist. There are

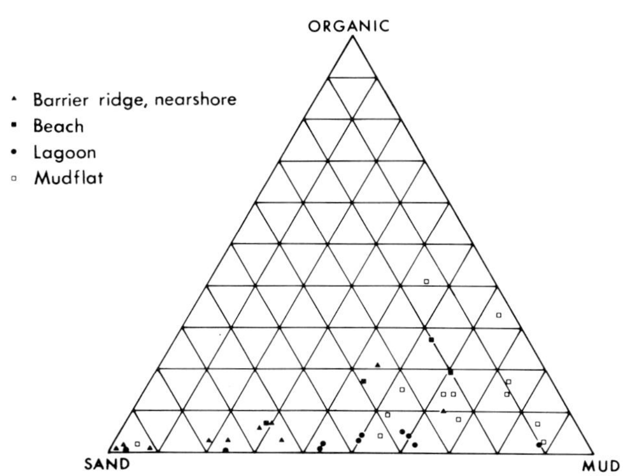

FIG. 13—Comparison of sediment composition of major coastal lagoon environments.

no absolute characteristics of the sediments deposited in a given depositional environment, regardless of local conditions. Natural conditions are too complex for such a simple approach. The continued failure of the attempts to find a "key" and the increasing divorcement of numerical descriptions from the actual sediments should indicate the need for a greater awareness of the roles of local conditions and the complexity of nature.

STRATIGRAPHY

Northern Belize

The Holocene sediments of northern Belize form a transgressive sequence overlying a poorly indurated Pleistocene limestone (Fig. 15). The basal unit consists of a stiff, light-colored, terrigenous clay similar to that presently accumulating on the low-lying, grass-covered meadows inland from the lagoons. Above the basal clay is a fibrous, massive peat, which is generally undisturbed, but in places shows signs of reworking and transportation. The uppermost unit is a complex of

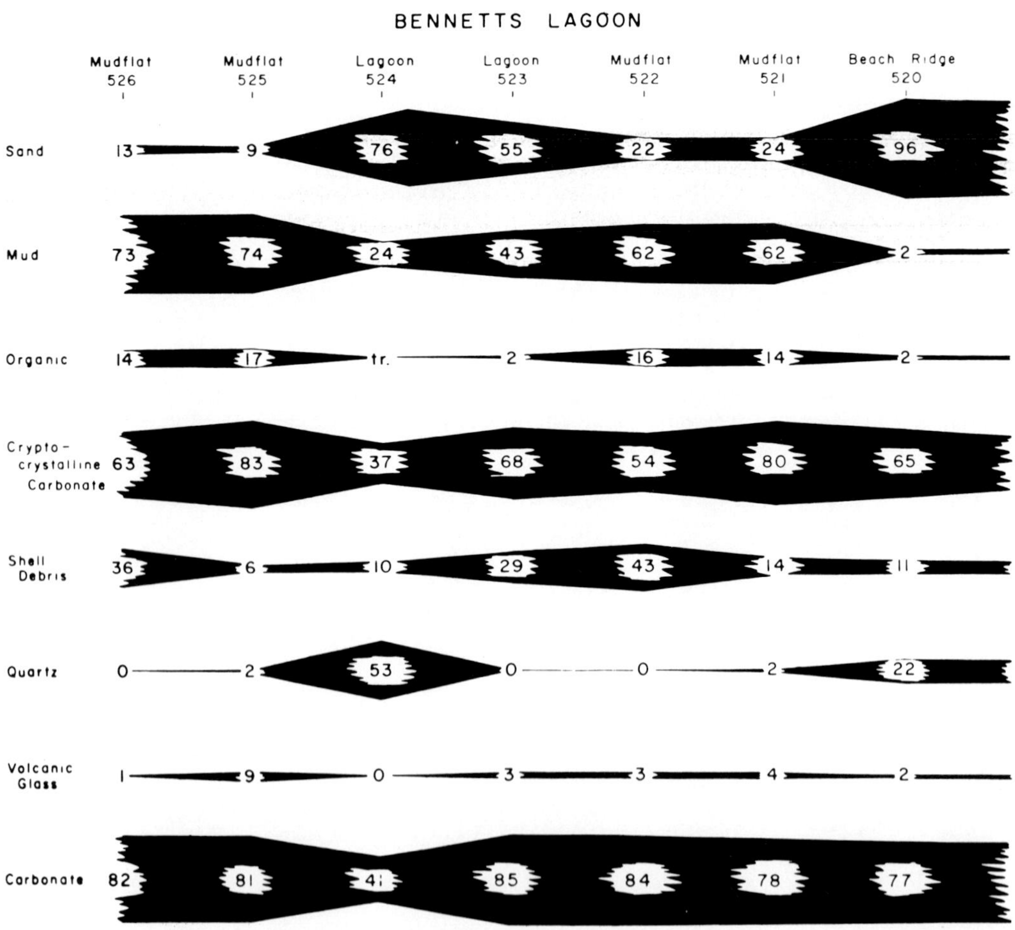

FIG. 14–Profile of changes in sediment properties at Bennetts Lagoon. Numbers represent weight percent (Table 2).

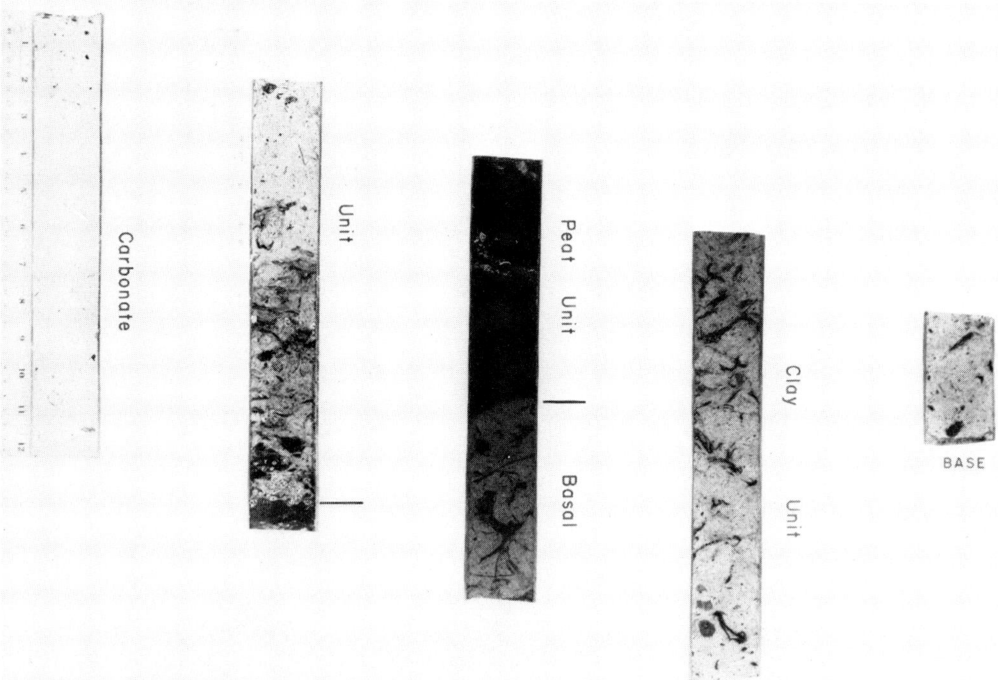

FIG. 15—General Holocene stratigraphic sequence. Core number 522, barrier mud flat, Bennetts Lagoon. Scale is in inches. Compaction is about 40 percent. Details of core are: carbonate unit (0-23 in. [58 cm])—light yellow-gray, silty carbonate mud; lower 6 in. (15 cm) contains abundant mangrove roots and peat fragments. Basal contact unconformable. Peat unit (23-31 in. or 58-79 cm)—dark red-brown, fibrous, massive mangrove peat. Basal contact unconformable. Basal clay (31-52 in. or 79-132 cm)—pale blue-green-gray, stiff, noncalcareous mud with abundant quartz sand and peat fragments; upper 8 in. (20 cm) is discolored by overlying peat. Pleistocene limestone—1 ft (30 cm) below base of core.

carbonate lithologies representing the surficial sediments. Although local irregularities exist in the form of partially eroded sections and additional units, this general sequence is persistent throughout northern Belize. Nowhere between Lost and Midwinter Lagoons were significant exceptions found.

Although most of the 44 cores from northern Belize do not contain a complete sequence, the general pattern can be determined when several cores are compared (Fig. 16). Color, general lithology, peat and carbonate content, fauna, and physical properties indicate the major units. The composition of the sand fraction proved to be diagnostic and was used where incomplete penetration or partial erosion was suspected.

The Holocene section along the northern coast is thin. The coring device and probe penetrated as deep as 12 ft (3.6 m), which was sufficient for this study because nearly one half of the cores or probes penetrated the entire Holocene section and bottomed in Pleistocene limestone. Based on these complete sections, it is estimated that the Holocene sediments average about 6 ft (1.8 m) in thickness. Maximum thickness is unknown.

Pleistocene surface—The contact between the Holocene sediments and the weakly consolidated Pleistocene limestone is unconformable. The limestone is considered to be Pleistocene because of its stratigraphic position, induration, and resemblance to Pleistocene limestone on Ambergris Cay. Although the exposures are limited and the limestone

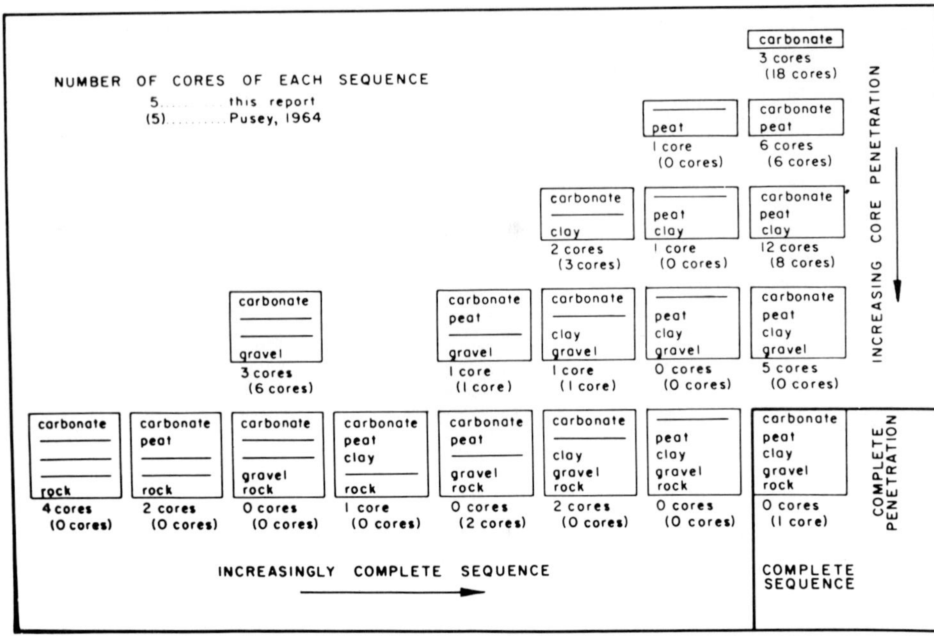

FIG. 16—Summary of core logs from northern Belize.

was not investigated in detail, two lithofacies were recognized. Dense, massive calcareous mudstone crops out in the Lost Lagoon area and is generally similar to the Pleistocene backreef facies on Ambergris Cay. The other lithofacies is present at Midwinter Lagoon, where Pleistocene exposures are extensive. There, the limestone occurs as thin, seaward-dipping beds (beachrock) arranged in large concentric arcs that are concave to the west. Three arcs of beachrock were identified.

Before deposition of the Holocene basal clay unit, solution by meteoric waters deeply pitted and etched the limestone and produced a highly irregular surface. Maximum local relief on the microkarst surface is at least 3 ft (0.9 m). Where exposed, this surface is similar to the more extensively exposed Pleistocene surface in the interior of Ambergris Cay. Available evidence indicates that relief on this surface does not control or influence Holocene physiography and topography. The presence and location of the coastal lagoons cannot be explained by flooding of a karst surface; there is no consistent correlation between topography and Pleistocene relief.

Basal clay unit—The basal clay unit overlying the eroded Pleistocene limestone consists mainly of a pale blue-gray, sticky, peaty, sandy (quartz), terrigenous clay with a distinctive greenish cast. In many places, the upper parts are stained reddish brown by water descending from the overlying peats (Fig. 15). Very coarse grains of quartz sand are scattered through the clay. Fine quartz sand is abundant and perhaps represents about 10 percent of the sediment. Small patches of bedded peat and peat fragments are common; mangrove roots and undisturbed peat are scarce. The stiff clay resists desiccation for several days after exposure. Because of the stiffness of the clay, it generally was not possible to penetrate more than 1-2 ft (0.3-0.6 m) with either the probe or core. Where the clay was thicker than this, binding and wall friction prevented complete penetration. These factors, rather than the length of the probe and coring tube (12 ft or 3.6 m), were responsible for most of the failures to penetrate the Holocene section completely.

The sand fraction of the basal clay unit is characterized by the abundance of quartz and the scarcity of carbonates. Small Foraminifera are notably sparse. In many cores, limestone rubble occurs in the lower parts of the basal clay unit (Fig. 16). These weathered fragments of limestone are a basal gravel marking the Pleistocene-Holocene unconformity.

The basal clay unit shows little variation between Lost and Midwinter Lagoons. The only trend noticed was a slight increase in quartz-sand content toward the south.

The basal clay unit is interpreted to have been deposited on a low, grass-covered inland meadow. In part, this unit may be a residual soil developed on the underlying Pleistocene limestone. The minimum elevation of this environment was probably a few feet above sea level. As the unit is now generally 4-6 ft (1.2-1.8 m) below sea level, it must have been deposited when the oceans stood at least 6 ft (1.8 m) below the present level. Thus, a minimum age of 3,000-4,000 years B.P. is indicated (Shepard, 1963b, p. 4; Scholl, 1964, p. 358).

The environmental interpretation on which this reconstruction is based is suggested by the general similarities in overall lithology between the basal clay and the sediment presently accumulating beneath a grassy meadow behind Bennetts Lagoon. In particular, the peculiar blue-greenish gray color characteristic of the basal clay was found only in this meadow environment. Also suggestive is the abundance of montmorillonite and other clay minerals which are not being deposited presently along the northern coast. A spore-pollen analysis of a basal clay sample in this area by W. C. Elsik, of the Esso Production Research Company in Houston, Texas, suggested "an open marsh, river channel, or lake environment that was not densely forested." The lack of shell material, a characteristic of present marine and barrier-lagoon sediments, is also suggestive of inland marshes.

The relation between the basal clay unit and the underlying Pleistocene limestone suggests little transportation of the material in the clay unit. The clay unit itself is massive and very poorly sorted; it contains abundant coarse sand and peat fragments. The peat fragments appear to be crudely bedded, but there are no thick-bedded peats which would indicate extensive reworking. The carbonate breccia layer, which directly overlies the Pleistocene limestone, appears to be rubble that accumulated in place as the surface weathered. Thus, it is probably the "C" horizon of a poorly developed pre-Holocene soil.

Peat unit—Dark red-brown, massive, fibrous peat directly overlies the basal clay unit. Generally it is similar to the peats presently accumulating beneath the dense stands of red and black mangroves and is presumed to be largely produced by debris from these plants. All recognizable plant remains are mangrove; no *Thalassia* peat was identified. The only observed difference between the surficial peats and those of this unit is that the surficial peats are less dense than the buried ones.

In many cores the peat is not well developed and the peat unit is represented by dark red-gray, peaty calcareous mud.

The peat or peaty mud is present throughout northern Belize. Where it is absent, as under some lagoons, it was probably removed by erosion. Average thickness of the peat unit is about 1 ft (0.3 m), and maximum thickness is about 4 ft (1.2 m). There is no apparent correlation between peat thickness and present water depth such as that reported for a similar sequence in Florida by Scholl (1964, p. 353).

The contact between the peat unit and the lower basal clay is generally gradational, but interbedding and erosional contacts are also

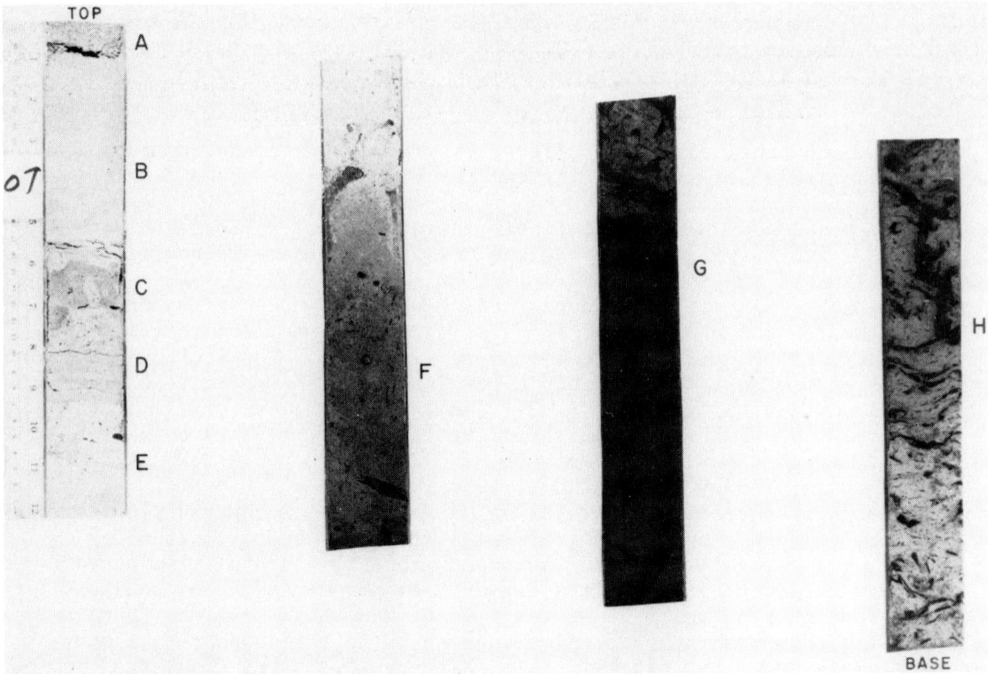

FIG. 17—Differentiation of carbonate unit. Core number 520, beach ridge, Bennetts Lagoon. Scale is in inches. Compaction is about 40 percent. Carbonate unit (0-25 in. or 0-63.5 cm)—**A,** 0-1 in. (0-2.5 cm) light green-gray, coarse, shelly sand; beach ridge; **B,** 1-6 in. (2.5-15.2 cm) light green-gray, laminated, shelly sand; nearshore; **C,** 6-7 in. (15.2-17.8 cm) dark blue-black mud; storm deposits; **D,** 7-9 in. (17.8-22.9 cm) light green-gray, laminated shell sand; nearshore; **E,** 9-13 in. (22.9-33.0 cm) light blue-gray, pelleted carbonate mud; high mud flat; **F,** 13-25 in. (33.0-63.5 cm) medium blue-gray, silty, peaty, carbonate mud; low mud flat. Peat unit—**G,** 25-34 in. (63.5-86.4 cm) dark red-brown, fibrous, massive, burrowed mangrove peat. Basal clay unit—**H,** 34-36 in. (86.4-91.4 cm) light blue-green-gray, sandy, peaty, noncalcareous mud.

present. Where the peat unit is well developed, the peat is generally massive; bedding, indicated by separation of layers during drying, is scarce.

Although red mangroves can colonize in waters a few feet deep, the plants only become dense, and thus contribute significant amounts of material to the sediment, at shallower depths or slightly above sea level. Similarly, although black mangroves can grow at elevations several feet above sea level, their greatest contribution to the sediment occurs at, or slightly below, sea level. Thus, the restricted habitat of mangroves and the very small tidal range in Belize combine to limit mangrove peat accumulation to a narrow zone at about sea level. Accordingly, mangrove peat is a reliable indicator of coastal environments.

The peat overlying the basal clay records the latest transgression in the area. As the sea rose, following the release of water from the retreating Wisconsin glaciers, the narrow zone suitable for abundant mangrove growth rapidly shifted westward over the land surface. Maximum transgression of the shoreline was to a position slightly west of the mainland margin of the present lagoons. Farther west the Pleistocene land surface is still exposed.

Carbonate unit—The upper transgressive unit is composed of various carbonate lithologies similar to those presently being deposited. Although diverse, these lithologies may be characterized as cream-colored calcareous muds or fine-grained skeletal sands. The sand fraction is largely composed of small Foraminifera and skeletal debris from mollusks and ostracods. Quartz is rare except locally; peat fragments are common, although they generally are finely divided and disseminated. These sediments are generally similar to the surficial sediments discussed under "Pleistocene surface."

The thickness of the carbonate unit is variable, but the maximum is 6 ft (1.8 m). Thickness variations correlate partly with surface relief; the carbonate unit tends to be thicker under the barriers than under the lagoons. The thickness also was influenced locally by erosion subsequent to the deposition of the peat layer and prior to the deposition of the carbonate sediments.

The carbonate unit is the most laterally extensive of the Holocene sequence. It was traced into Belize Bight, whereas the basal clay and peat units were not identified positively south of Midwinter Lagoon.

Cores of the carbonate unit commonly show two or more subdivisions (Fig. 17). If only the cores showing little evidence of internal truncation are considered, a possible sequence of carbonate lithologies can be identified. Based on these cores, the carbonate unit is divisible into two unequal parts. The lower, and generally thicker, part consists of relatively homogeneous calcareous mud. The upper part is represented by interbedded calcareous mud, calcareous sand, and shell debris.

The lower, muddy carbonate sediments are interpreted as the deposits of formerly more extensive mud flats. Within the lower calcareous mud there is a gradual increase in finely divided plant remains downward, and the contact between the carbonate and peat units is gradational. The sequence is interpreted as recording an increase in mud-flat elevation. From the initial mangrove swamp at sea level, continued sedimentation raised the surface until mangroves no longer flourished. As the

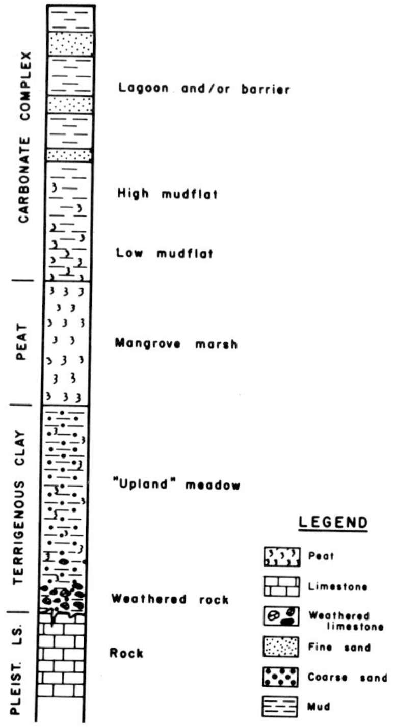

FIG. 18–Generalized Holocene stratigraphic sequence from northern Belize.

plant cover thinned, the sediment received less and less vegetal matter until virtually pure carbonate sediment was being deposited on barren mud flats.

The various carbonate lithologies in the upper part of the unit apparently record the establishment of the present conditions of small lagoons, barriers, and mud flats at the expense of the formerly more extensive mud flats. De Jong (1960, p. 643) has described a similar change in the pattern of sedimentation, from widespread units to highly differentiated local units, in the Netherlands. The change to modern, localized environments in Belize is attributed to a decrease in the rate of rise of sea level about 2,500 years ago. This date closely corresponds to a major decrease in the slope of the sea-level curve constructed by Scholl (1964, p. 358) for southern Florida.

Sequence and distribution—The generalized Holocene sequence, with environmental interpretations, is shown in Figure 18.

The Holocene section of Lost Lagoon (Fig. 19) is the most irregular of the four traverses in the northern coastal lagoons. Relief on the Pleistocene surface in this area is extreme, and abundant outcrops found behind the barrier mud flat tend to form small pockets in which the Holocene sediments collect. Generally the Holocene section is thin; Pleistocene limestone is widely exposed on the floors of the lagoons, and there is little more than a thin veneer of Holocene sediments. The barrier mud flat is underlain by a consistently thick Holocene section with no Pleistocene exposures. Pleistocene limestone does not reappear at the surface until rocky shoals are encountered a few hundred feet from the shoreline in Chetumal Bay.

Because the Holocene section is so thin and is interrupted by numerous Pleistocene outcrops at Lost Lagoon, the stratigraphic sequence is irregular. The three major lithologic units were recognized, but no core contains all three.

In contrast to Lost Lagoon, Bennetts Lagoon is characterized by its regular stratigraphic sequence (Fig. 19); six of the seven cores penetrated all three major units. The Pleistocene surface was reached by only one probe, at about 8 ft (2.4 m), and the configuration of this surface is not known. Directly in front of Sand Point, Pleistocene limestone crops out at sea level.

One striking feature of the Holocene section at Bennetts Lagoon is the nearly uniform thickness of the carbonate and peat units, regardless of surface topography. Because of this relation, there appears to be a correlation between surface topography and relief on the basal clay unit. This is especially true beneath the barrier where a well-defined high on the basal clay unit apparently controls the surface elevation. A less well-developed but similar correlation exists at Midwinter Lagoon (Fig. 20). However, no such correlation exists at Lost Lagoon (Fig. 19) and Northern River Lagoon (Fig. 21).

The surface topography at Northern River Lagoon (Fig. 21) apparently is the result of erosion after deposition of the peat unit and prior to deposition of the carbonate sediments. Beneath the lagoon, all older sediments have been removed and the thin carbonate beds lie directly on Pleistocene limestone. Beneath the mud flats, erosion was less effective and thicker carbonate beds cap the normal sequence. In the meadows and forests inland from the lagoon, peat accumulation has continued and surficial carbonate sediments are not present.

The Pleistocene high beneath the barrier at Northern River Lagoon is interpreted on the basis of the gravel present in one core. The only other evidence of relief is found west of the lagoon, where Pleistocene limestone that crops out extensively along the river banks is buried by at least 6 ft (1.8 m) of Holocene sediment only 100 yards (90 m) away.

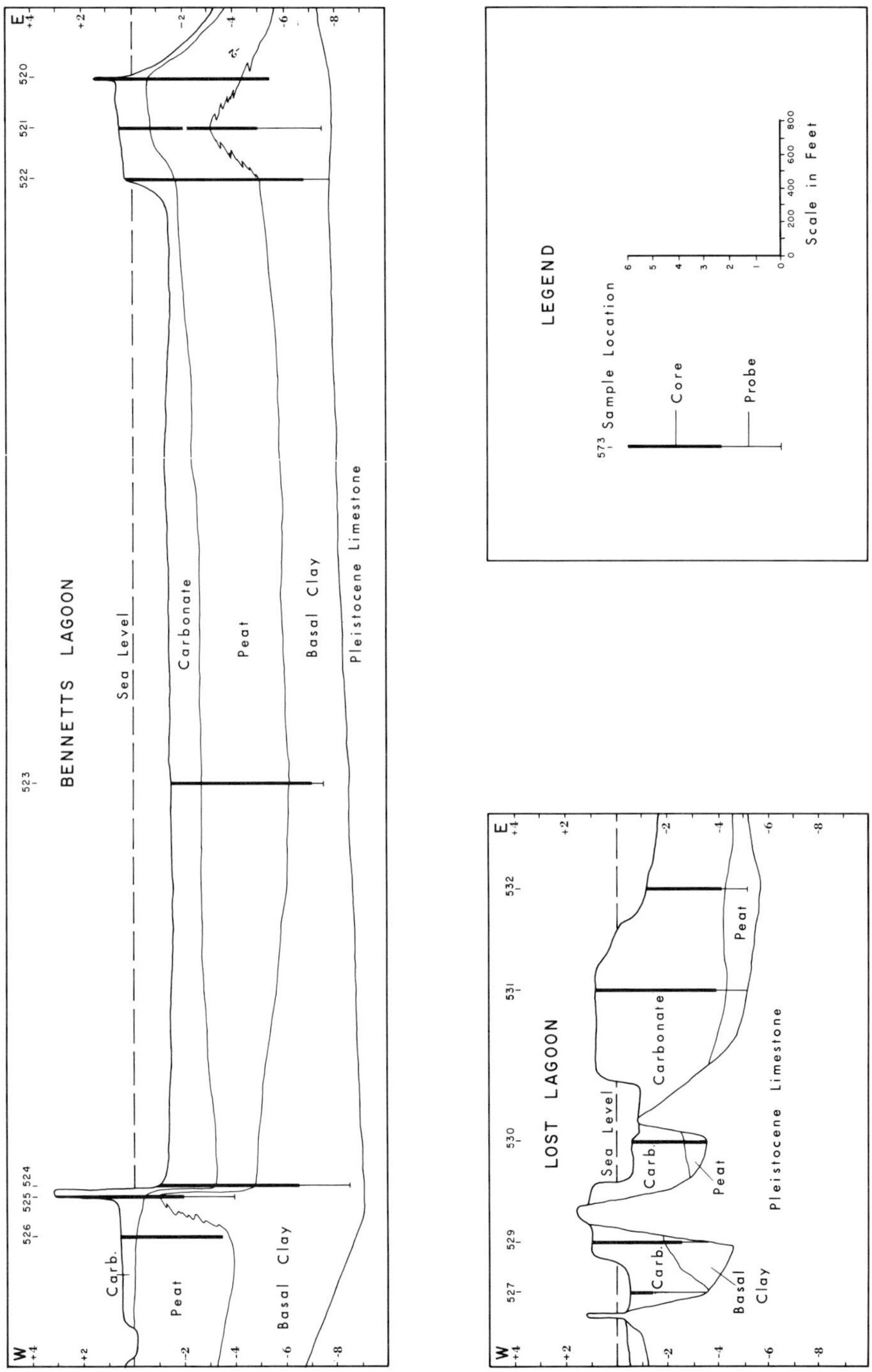

FIG. 19—Holocene stratigraphy at Lost and Bennetts Lagoons.

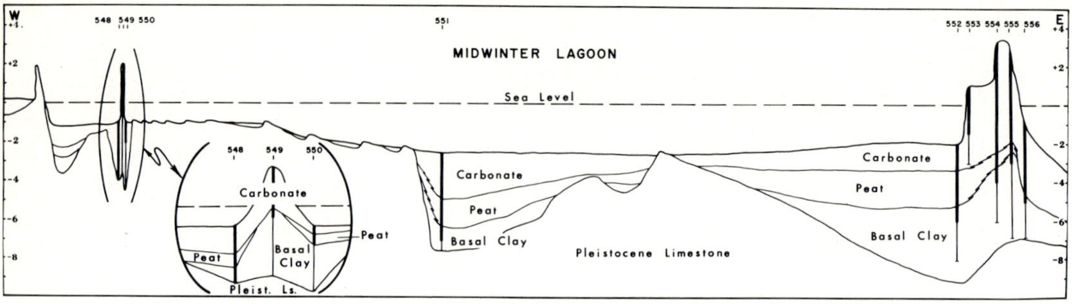

FIG. 20–Holocene stratigraphy at Midwinter Lagoon (legend same as Fig. 19).

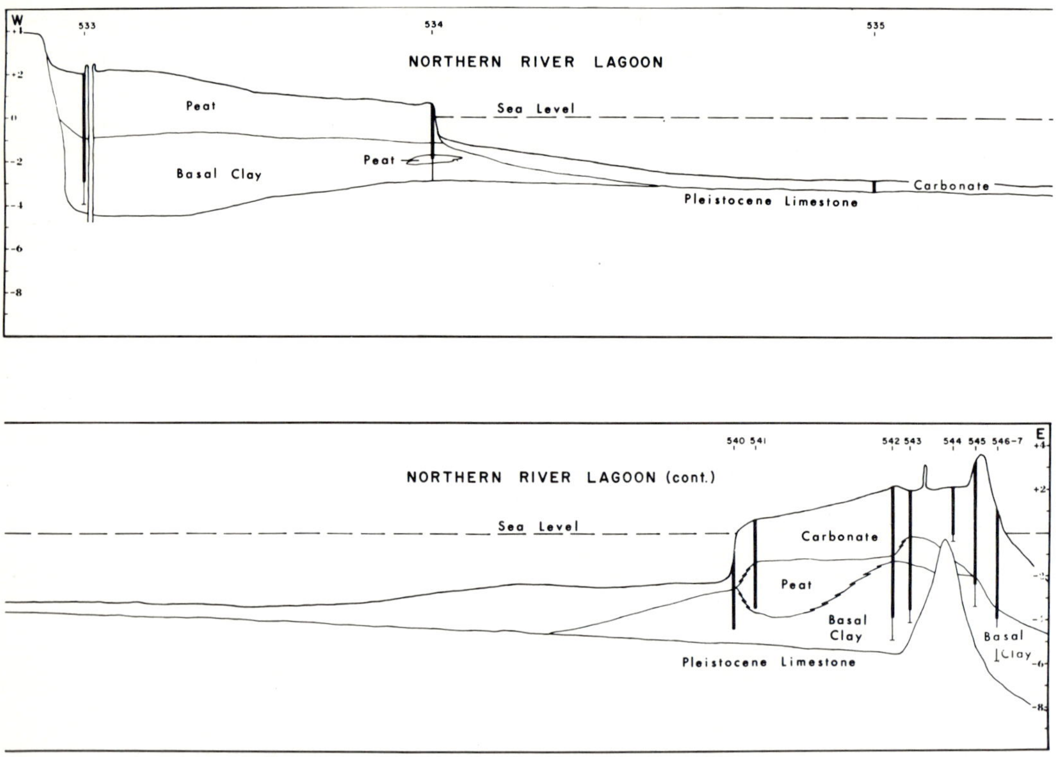

FIG. 21–Holocene stratigraphy at Northern River Lagoon (legend same as Fig. 19).

Probing between these two points indicated that the Pleistocene surface rises toward the outcrop, but in an irregular fashion.

The Holocene section at Midwinter Lagoon (Fig. 20) is similar to that at Lost Lagoon. It is, in part, irregular and discontinuous, and is strongly influenced by the underlying Pleistocene surface. Control on the configuration of the Pleistocene surface is best along this traverse. Much of the lagoon floor is composed of Pleistocene limestone with little or no sediment cover. The limestone forms platy, seaward-dipping beds arranged in concentric arcs.

The large depression in the Pleistocene surface midway across the lagoon suggests a pre-Holocene drainage channel. Although the north-south extent is not known, the east-west section is well defined by

Pleistocene exposures and by one core. Similar drainage channels in Pleistocene limestone beneath Chetumal Bay were reported by Pusey (1964, p. 219-220) and have been noted on the shelf by E. G. Purdy (1966, personal commun.) and Ebanks (1966, personal commun.).

The Pleistocene surface was not encountered beneath the barrier on the eastern side of Midwinter Lagoon. The Holocene section is thick and regular, having no unusual features.

Similarity to other areas—Similar sequences of Holocene sediments which underlie Chetumal Bay (Pusey, this volume) and Ambergris Cay (Ebanks, this volume) are believed to be correlative. A summary of Pusey's data (1964, p. 22) is included in Figure 16. Sequences representing modifications of this general pattern are believed to underlie the entire northern shelf.

The Holocene section from the Whitewater Bay and Ten Thousand Islands areas of southern Florida (Scholl, 1966, p. 350-356) is markedly similar to the Belize Holocene section. The base is the eroded surface of the Pleistocene Miami Oolite. Of the three Holocene units recognized, the oldest is a freshwater calcitic mud similar to that being deposited inland in the Everglades. Like the basal clay unit of northern Belize, it is believed to be nontransgressive. Above the basal calcitic mud are two transgressive units, a lower peat and an upper shelly sand. Because of the striking similarity in sequence and the similar physical settings, the sea-level curve for southern Florida (Scholl, 1964, p. 358) has been used to suggest tentative dates for similar events in Belize.

Southern Belize

In contrast to northern Belize, coastal areas south of Belize City are characterized by the lack of a persistent, regular sequence. Although a sequence may be established for a local area, it is not possible to extend it to adjacent regions.

The Holocene section at Commerce Bight Lagoon (Fig. 22) is composed of three units. The Pleistocene is a muddy quartz sand that is stained orange-red. The basal Holocene unit is a coarse, dark green-gray or blue-gray quartz sand that grades upward into dark red-brown peaty sediments. The peat unit varies from nearly pure unbedded peat to muddy and sandy bedded peat. On the far eastern side of the lagoon, where there is a thick growth of manatee grass, a thin layer of fine, micaceous, shelly sand overlies the peat. Throughout the rest of the lagoon, the peat unit is exposed on the bottom.

This sequence contrasts sharply with that of the next locality on the south. At False Sittee Point (Fig. 22), the Pleistocene was not encountered in any cores. The lowest Holocene lithology cored consists of green-gray, shelly, muddy, fine quartz sand with abundant coral debris (*Porites*). This unit grades upward into a similar sand with decreasing amounts of coral fragments and increasing peat content. A dark red-brown peat or peaty mud overlies the sand. The surficial sediments are sands on the beaches, shelly sand in shallow water, and peat with coral fragments in the mangrove swamps. The interpreted relations of the upper lithologies are shown in Figure 22.

At other localities in southern Belize, a general sequence was not identified. Rather, the entire cored section generally consisted of a single lithology.

Thus, southern Belize shows a distinctly poor Holocene stratigraphic section. At most places, lithologic units cannot be differentiated; where layering is present, it is a local phenomenon and cannot be extended beyond the area.

The stratigraphic differences between northern and southern Belize probably reflect basic differences in the physical setting, the Pleistocene surface, and the sediments available.

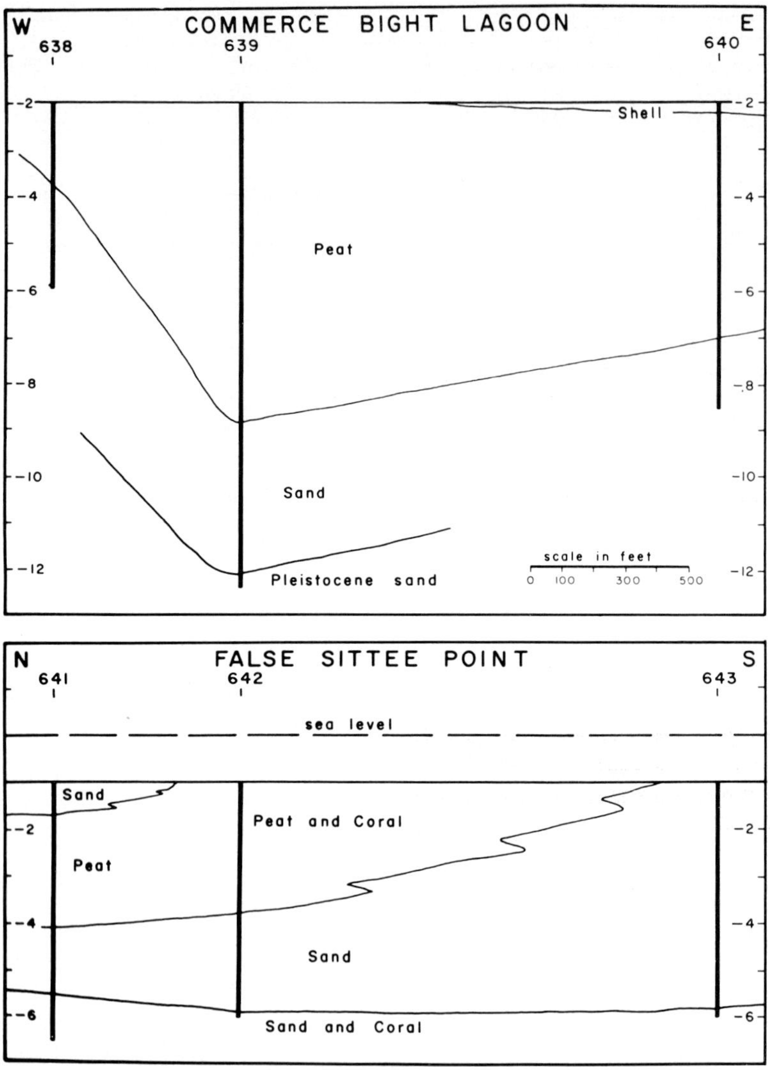

FIG. 22–Holocene stratigraphy at False Sittee Point and Commerce Bight Lagoon (legend after Fig. 19).

In northern Belize the Pleistocene surface is essentially flat, having no more than 20 ft (6 m) of relief in the 9 mi (14.5 km) between the coast and Ambergris Cay (Pusey, 1964, p. 180). In southern Belize the Pleistocene surface near the coast shows an appreciable dip to the east and south (E. G. Purdy, unpub. data). Therefore, the Holocene transgression in the south did not flood wide areas almost simultaneously as it did in the north; instead there was never more than a narrow zone subjected to coastal processes. The steeper surface prevented the formation of extensive mud flats and lagoons, which produced the regular stratigraphy of northern Belize, but favored the formation of a narrow beach subject to local conditions. Such beaches probably retreated with the transgressing sea. Even though the Flandrian transgression essentially ceased a few thousand years ago and has been replaced by a moderate regression along the Belize coast, these differences in sedimentation patterns between the northern and southern coastal areas are being maintained.

SUMMARY AND CONCLUSIONS

Genesis and Development of Coastline

The Belize coastline is the product of the post-Pleistocene Flandrian transgression. During the maximum advance of the Wisconsin glaciers, the entire continental shelf of Belize was exposed. As the glaciers retreated and water was returned to the oceans, the shelf was flooded progressively by the transgressing seas. No other event had as much influence on the present coastline as this transgression. There have been no significant structural movements, and no evidence for either rebound or subsidence was noted. The history of the coast is one of simple flooding of a stable platform.

The overall position and trend of the coastline are controlled largely by basement structure as indicated by the general parallelism of the coastline and the reef on the faulted edge of the shelf. However, the details of the coastal area, such as landform assemblages and sedimentation patterns, are the results of local conditions, geology of the source areas, and the hydrographic regime.

As sea level rose during late Pleistocene and early Holocene times, the first areas to be flooded were the deeper parts of the southern shelf. The present coastal area, along with the shallow northern shelf, did not begin to be affected by the transgression until about 5,000 years ago, when sea level stood at about -20 ft (-6 m; Shepard, 1963b, p. 3). Although local basins existed (Pusey, 1964, p. 178-180, 219-220), much of the northern shelf was flooded almost simultaneously. The restricted environment conducive to peat accumulation was widespread and migrated across the land surface rapidly. Continued sedimentation resulted in extensive mud flats. Several thousand years ago, when sea level was a few feet lower than at present (Scholl, 1964, p. 358), the rate of rise decreased markedly. Under the new conditions of virtual stillstand, the extensive mud-flat environments differentiated quickly as barriers and lagoons developed. If there are no further changes in sea level, which is highly improbable in view of the remaining glaciers, the lagoons will fill and the coastline will become straight.

The Holocene history of the southern shelf is not as clearly defined as that of the northern shelf. Because of the steep slope of the land surface, only narrow zones along the coast were affected at each stage of the transgression. Probably, only small, retreating beaches were constructed. Most of these beaches were destroyed as sea level continued to rise; some may have escaped erosion and either became buried beneath nearshore deposits or became stranded below effective wave base. With the attainment of the modern stillstand, minor progradation has occurred along the southern coast. Small deltas have been constructed at the mouths of most rivers. The bights between the deltas have been filling by the construction of barriers while peat and mud accumulation have extended the low-lying coasts. As with the northern coast, if sedimentation continues the coastline eventually will become straightened and will be fronted by long beach-ridge systems and small deltas.

The future development of the Belize coastline is dependent on changes in sea level. Melting of the existing glaciers would flood most of northern Belize, leaving the Maya Mountains as a mountainous island or peninsula. This process has occurred in the past and, barring tectonic activity, eventually will happen again. The immediate future is concerned with the present stillstand. If it persists for a few thousand years, most of the northern lagoons probably will be filled. A longer time is needed to fill in the coastal marshes and estuaries of the Port of Honduras. Minor progradation about the deltas will continue as long as the streams draining the Maya Mountains transport clastic material. However, renewed eustatic movements of sea level will impose a new set of conditions to which the coastline must adjust.

Coastal Sedimentation

This study has revealed features of coastal sedimentation which, although strongly influenced by the physical setting of Belize, are sufficiently general to have some applicability to other areas.

Role of storms—The importance of storms to coastal sedimentation has been documented from northern Belize (High, 1969). Far from being rare, destructive events, major storms are the means of progradation along much of the northern Belize coast. Other than slow filling of lagoons and marshes, little sedimentation appears to occur under "normal" conditions. In the absence of strong waves or tides, sediments deposited by storms are likely to be preserved.

Role of plants—In most environments plants have a significant role in initiating, controlling, and contributing to sedimentation. *Thalassia* and mangroves are the most important floral elements. Their influence is shown clearly in the construction of tidal deltas where they break the force of currents, entrap sediment and bind it in place, and contribute organic matter.

Coastal evolution—The major tendency on the Belize coast is the reduction of irregularities. This is a common process along many coasts (Guilcher, 1958, p. 179-191) where marshes and bights are filled, barriers are formed, and bay-head beaches are constructed. The only process increasing irregularity is the construction of river deltas. The final coastal form, which will develop if present conditions persist, appears to be a straight coast fronted by beach ridges and interrupted by small deltas.

Site of deposition—The most active sedimentation occurs on or directly adjacent to the coast. Material is contributed from both sides. Streams introduce sediment from the land, and sediment from the floor of the shelf is reworked. Much of the material is deposited within a few hundred yards of the coastline. Little sediment reaches deeper water (Matthews, 1966, p. 434, 443; Scott, 1966, p. 11; E. G. Purdy, unpub. data).

Similarity of Holocene and Pleistocene environments—The Holocene and Pleistocene deltaic and lagoonal environments differ mainly in size. The landform assemblages and sediments are basically similar. Thus, conclusions drawn from the present environments need not be restricted to Holocene sediments, but may find applicability to ancient analogues.

Correlation of landform assemblages and sedimentation patterns—Three major types of sedimentation are occurring along the coast of Belize. Each is characteristic of a given physiographic province—lagoon barrier, delta, and foreland. In the coastal lagoons of Chetumal Bay, the most active sedimentation is on the barrier mud flats. The bulk of the sediment is used in construction and extension of tidal deltas. Between Belize City and Placentia the coast is characterized by river deltas and beach ridges, although coastal mud flats are developed around the Monkey and Deep Rivers. The forelands contain abundant sediment that probably was carried in by southward-flowing longshore currents. The relation between landforms and sedimentation is indirect; neither strictly controls the other. Rather, both are controlled similarly by the physical factors of the environment.

Effects of longshore currents—In addition to transporting material along the coast, the longshore currents may explain several major coastal features. The area between Haulover Creek and Mullins River shows active erosion of Pleistocene deposits with no significant Holocene sediments. Possibly this erosion is due to the Belize Bight and Haulover Creek delta. The longshore currents that flow south along the coast of Chetumal Bay are interrupted and disorganized as they enter Belize Bight, causing deposition of much of the sediment load. South of the Haulover Creek delta, the longshore currents become reorganized but, having lost most of their load, must scour the Pleistocene sediments. The similarity of the Haulover Creek delta to a jetty, with deposition on the upcurrent site and erosion on the downcurrent side, is apparent.

In addition, longshore currents possibly provide a limit to progradation. As the coastline straightens, the currents become better organized and more powerful. Eventually, an equilibrium is reached where the erosive power of the currents checks progradation. Further coastal sedimentation then must overcome the longshore currents and thus probably is restricted to the unusual hydrographic conditions associated with major storms.

Indeterminacy—The final principle that may be drawn from this study is the application of indeterminacy to sedimentary petrology. The sediment-environment system is complex and poorly understood. It may never be possible to isolate a single inorganic feature of the sediment which will be diagnostic of the environment. Attempts to do so ignore the complexity of natural systems. Furthermore, individual samples cannot be isolated from their surroundings. Similar environments differ from place to place, and the physical forces can use only the material that is available. Accordingly, there are no universal criteria. Rather, the relation of each sample to other samples and environments must be considered.

REFERENCES CITED

Allen, J. R. L., 1965, Coastal geomorphology of eastern Nigeria: beach-ridge barrier islands and vegetated tidal flats: Geologie en Mijnbouw, v. 44, p. 1-21.

Bascom, W., 1964, Waves and beaches: Garden City, New York, Anchor Books, 267 p.

Bernard, H. A., 1965, A resume of river delta types: AAPG Bull., v. 49, p. 334-335.

Chapman, V. J., and J. W. Ronaldson, 1958, The mangrove and salt-marsh flats of the Auckland Isthmus: New Zealand Dept. Sci. and Indus. Research Bull. 125, 79 p.

Davis, J. H., Jr., 1940, The ecology and geologic role of mangroves in Florida: Tortugas Lab. Papers, v. 32, no. 16, p. 305-412.

De Jong, J. D., 1960, The morphological evolution of the Dutch coast: Geologie en Mijnbouw, v. 39, p. 638-643.

Dixon, C. G., no date, Geology of southern British Honduras: Belize, Government Printing Office, 85 p.

Ebanks, W. J., Jr., 1975, Holocene carbonate sedimentation and diagenesis, Ambergris Cay, Belize: this volume.

El-Ashry, M. T., and H. R. Wanless, 1965, Birth and early growth of a tidal delta: Jour. Geology, v. 73, p. 404-406.

Flores, G., 1952, Geology of northern British Honduras: AAPG Bull., v. 36, p. 404-413.

Ginsburg, R. N., and H. A. Lowenstam, 1958, The influence of marine bottom communities on the depositional environment of sediments: Jour. Geology, v. 66, p. 310-318.

Guilcher, A., 1958, Coastal and submarine morphology: London, Methuen and Co., 274 p.

High, L. R., Jr., 1969, Storms and sedimentary processes along the northern British Honduras coast: Jour. Sed. Petrology, v. 39, p. 235-245.

King, C. A. M., 1959, Beaches and coasts: London, Edward Arnold, 403 p.

Krueger, W. C., Jr., 1963, Mineralogical composition and textural properties of river sediments from British Honduras: Rice Univ., PhD thesis, 106 p.

Matthews, R. K., 1966, Genesis of Recent lime mud in southern British Honduras: Jour. Sed. Petrology, v. 36, p. 428-454.

Murray, G. E., 1961, Geology of the Atlantic and Gulf coastal province of North America: New York, Harper and Bros., 692 p.

Ower, L. H., 1927, Features of British Honduras: Geog. Jour., v. 70, p. 372-386.

——1928a, Geology of British Honduras: Jour. Geology, v. 36, p. 494-509.

——1928b, The silica lines of British Honduras: Geol. Mag., v. 65, p. 507-510.

Price, W. A., 1945, Environment and formation of the chenier plain: Texas A&M Univ., Research Found. Proj. 63, 15 p.

Pusey, W. C., 1964, Recent calcium carbonate sedimentation in northern British Honduras: Rice Univ., PhD thesis, 247 p.

——1975, Recent carbonate sedimentation on northern Belize shelf: this volume.

Russell, R. J., and H. V. Howe, 1935, Cheniers of southwestern Louisiana: Geog. Rev., v. 25, p. 449-461.

Schofield, J. C., 1960, Sea level fluctuations during the last 4000 years as recorded by a chenier plain, Firth of Thames, New Zealand: New Zealand Jour. Geology and Geophysics, v. 3, no. 3, p. 467-485.

Scholl, D. W., 1964, Recent sedimentary record in mangrove swamps and rise in sea level over the southwestern coast of Florida, Parts 1 and 2: Marine Geology, v. 1, p. 344-366; v. 2, p. 343-364.

Scott, M., 1966, Distribution of clay minerals on the British Honduras shelf: Rice Univ., PhD thesis, 61 p.

Shepard, F. P., 1963a, Submarine geology: New York, Harper and Row, 2d ed., 557 p.

——1963b, Thirty-five thousand years of sea level, *in* Essays in marine geology in honor of K. O. Emery: Los Angeles, Univ. Southern California Press, p. 1-10.

Stoddart, D. R., 1962, Catastrophic storm effects on the British Honduras reefs and cays: Nature, v. 196, p. 512-515.

——1963, Effects of Hurricane Hattie on the British Honduras reefs and cays, October 30-31, 1961: Atoll Research Bull., no. 95, 142 p.

——1969, Post-hurricane changes on the British Honduras reefs and cays: Re-survey of 1965: Atoll Research Bull., no. 131, 24 p.

U.S. Navy Hydrographic Office, 1952, Sailing direction for the east coast of Central America and Mexico: U.S. Navy Hydrog. Offic Pub. 20, 5th ed. (with changes through 1961), 274 p.

Vaughan, T. W., 1910, The geologic work of mangroves in southern Florida: Smithsonian Misc. Colln., v. 52, Quart. Issue, no. 5, p. 461-464.

Vermeer, D. E., 1959, The cays of British Honduras: Univ. Calif., Berkeley, M.A. thesis, 127 p.

——1963, Effects of Hurricane Hattie, 1961, on the cays of British Honduras: Zeitschr. Geomorphologie: v. 7, n.f., no. 4, p. 332-354.

Wright, A. C. S., et al, 1959, The pattern of land form and geology, *in* Land in British Honduras: London, Colonial Office, Colonial Research Pub. No. 24, p. 22-28.

Distribution of Clay Minerals on Belize Shelf[1]

MARTHA R. SCOTT[2]

ABSTRACT Recent sediments of the Belize continental shelf are characterized by a transition from detrital sediments near the shore to carbonate sediments near the barrier reef at the shelf edge. Studies of the less-than-2μ fraction of these sediments show that the major clay minerals are montmorillonite, kaolinite, and illite; chloritic mixed-layer clay and minor chlorite are also present.

The clay minerals of the shelf sediments are principally detrital in origin. Major differences in parent material and environment of soil formation are reflected in the clay minerals transported by the rivers and deposited on the adjacent shelf. Thus the montmorillonite-rich sediment of the northern shelf is derived from lime-enriched soils of northern Belize and the kaolinite-illite-montmorillonite sediments of the southern shelf reflect the more thoroughly leached soils of southern Belize.

Distribution of montmorillonite, illite, and kaolinite on the shelf is governed by the differential settling tendencies of these minerals, effects of longshore currents, and geographic distribution of shoals on the shelf. The relatively slow-settling montmorillonite is preferentially deposited in deep-water areas offshore. Relatively large concentrations of illite and kaolinite are found in the nearshore area and on the shallow barrier platform near the reef. On the barrier platform, agitation by currents and waves resuspends the sediment and permits montmorillonite to be transported away from the shoal into deeper water. Longshore currents transport montmorillonite-rich sediment southward from the northern shelf, creating a tongue of montmorillonite-rich sediment opposite rivers carrying sediment from an environment of kaolinite-rich, acid-leached soils.

Kaolinite is preferentially deposited in the nearshore area, although kaolinite and illite have approximately the same settling velocities. The formation of slower settling illite-montmorillonite settling entities may cause the preferential transport of illite farther from shore. Chlorite and chloritic mixed-layer clay are present in the sediments on and near the barrier platform, but are not found in sediments on other parts of the shelf or in the river sediments analyzed. The chlorite may be forming *in situ* by diagenesis of montmorillonite, or it may be brought to the area by ocean currents.

INTRODUCTION

The modern continental-shelf deposits of Belize grade from detrital sediments near the shore to carbonate sediments near the barrier reef at the shelf edge. This report is a study of the clay-mineral content of sediment grab samples from the Belize shelf lagoon. The clay-mineral content of the sediments varies both along and across the shelf. The objectives of this investigation were to delineate fully the distribution in terms of provenance, transport mechanisms, and possible diagenesis.

DESCRIPTION OF STUDY AREA

Mainland Features

The distribution of clays in the shelf sediments is strongly influenced by the climate, drainage, and bedrock lithologies of the adjacent

[1]Manuscript received, June 20, 1972.

[2]Department of Oceanography, Texas A&M University, College Station, Texas.

I am grateful to E. G. Purdy of Esso Production, Ltd., and John J. W. Rogers of Rice University, each of whom directed this study for part of its duration. Discussions of clay mineralogy with Richards A. Rowland were of great assistance. Karl K. Turekian of Yale University facilitated the completion of this work in New Haven, Connecticut. The samples used in this study were collected by other students at Rice University, several of whom are publishing their own work in this volume. The Belize research project was supported financially by the Humble Oil and Refining Company, the Jersey Production Research Company, and Imperial Oil, Ltd. A National Science Foundation graduate fellowship supported me for 2 years. This paper represents a major part of a doctoral dissertation submitted to Rice University in 1966.

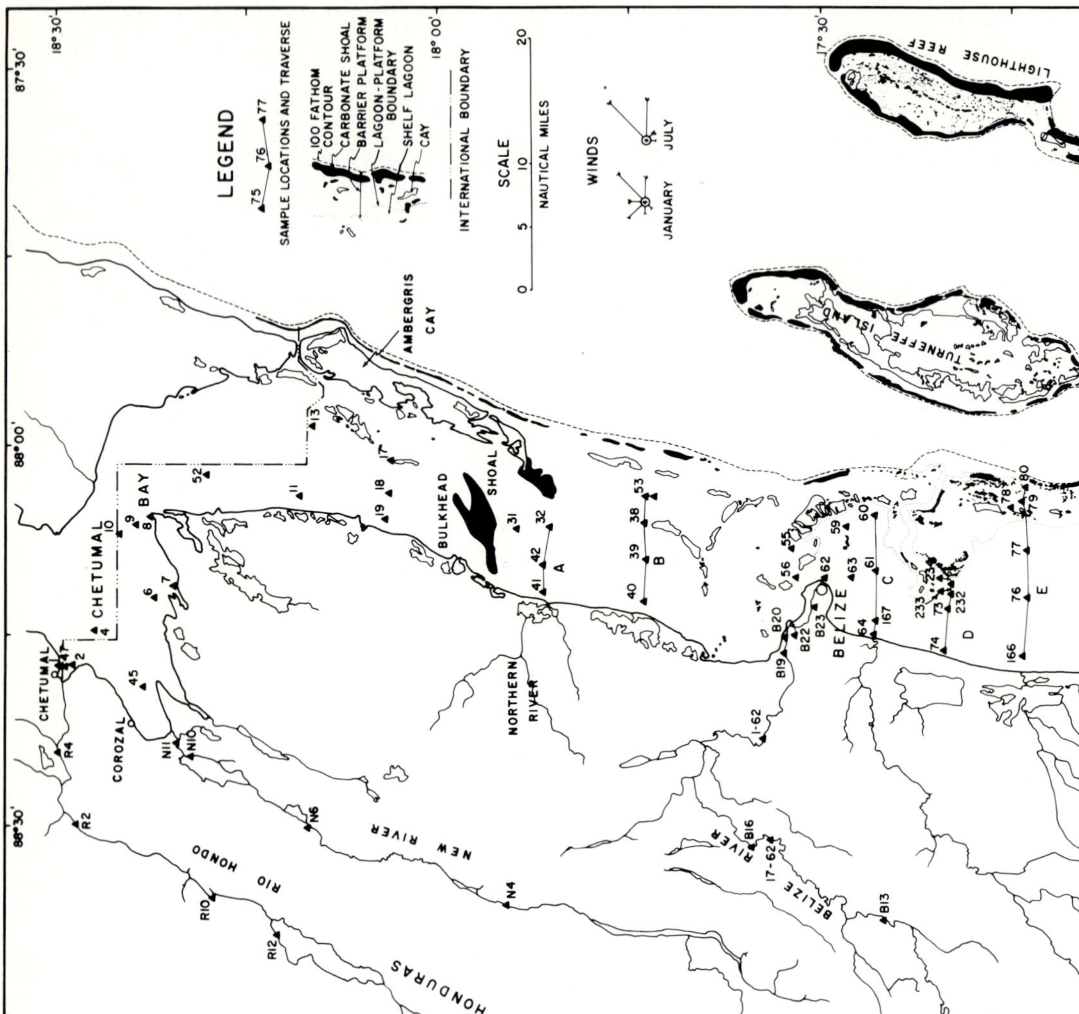
LEGEND
SAMPLE LOCATIONS AND TRAVERSE
100 FATHOM CONTOUR
CARBONATE SHOAL
BARRIER PLATFORM
LAGOON-PLATFORM BOUNDARY
SHELF LAGOON
CAY
INTERNATIONAL BOUNDARY
SCALE
NAUTICAL MILES
WINDS
JANUARY
JULY
LIGHTHOUSE REEF
TURNEFFE ISLAND
AMBERGRIS CAY
BULKHEAD SHOAL
CHETUMAL BAY
CHETUMAL
COROZAL
NORTHERN RIVER
NEW RIVER
RIO HONDO
BELIZE RIVER
BELIZE
MEXICO
BRITISH HONDURAS
87°30'
88°00'
88°30'
89°00'
18°30'
18°00'
17°30'

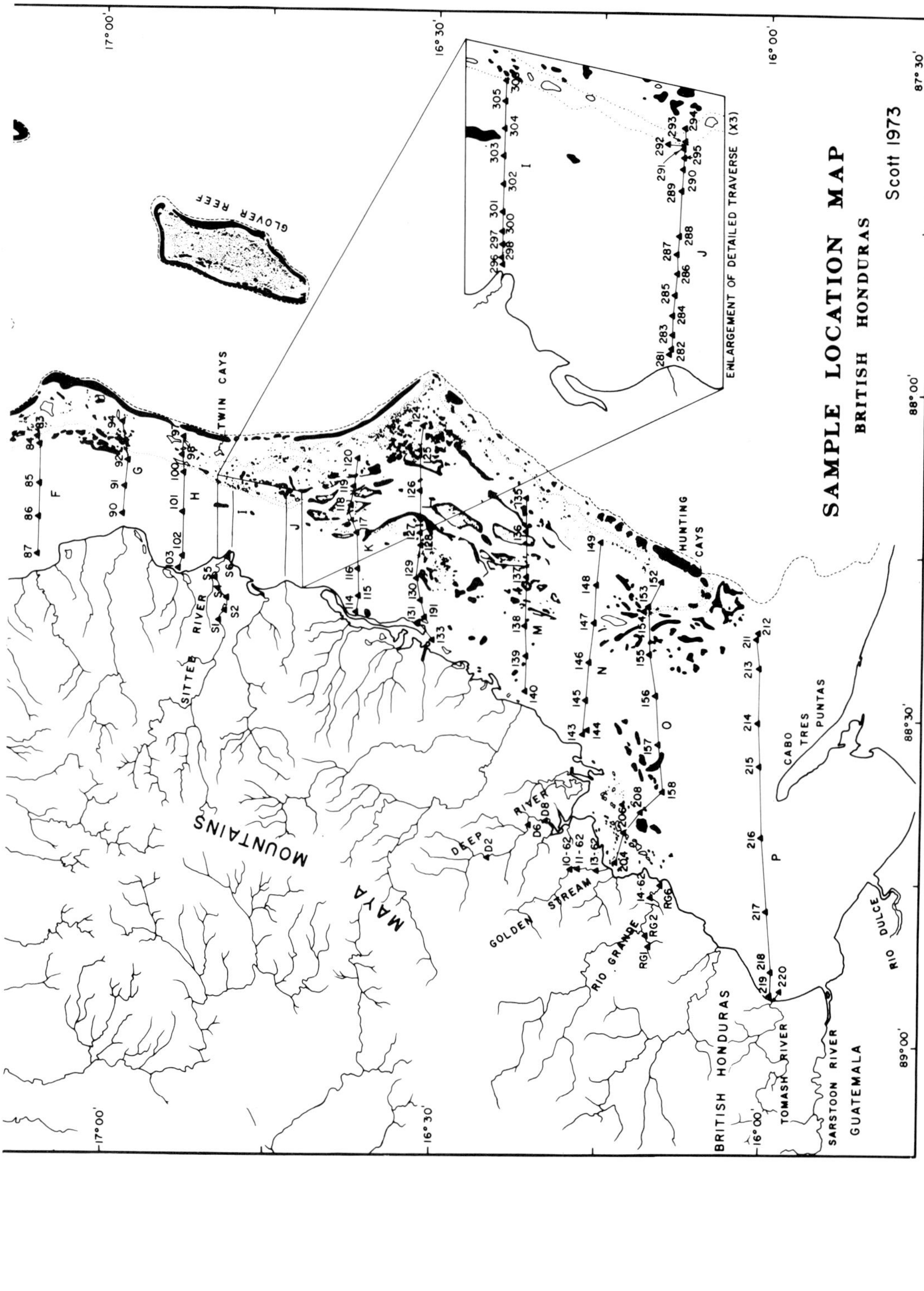

FIG. 1—Base map of Belize with locations of samples used in study.

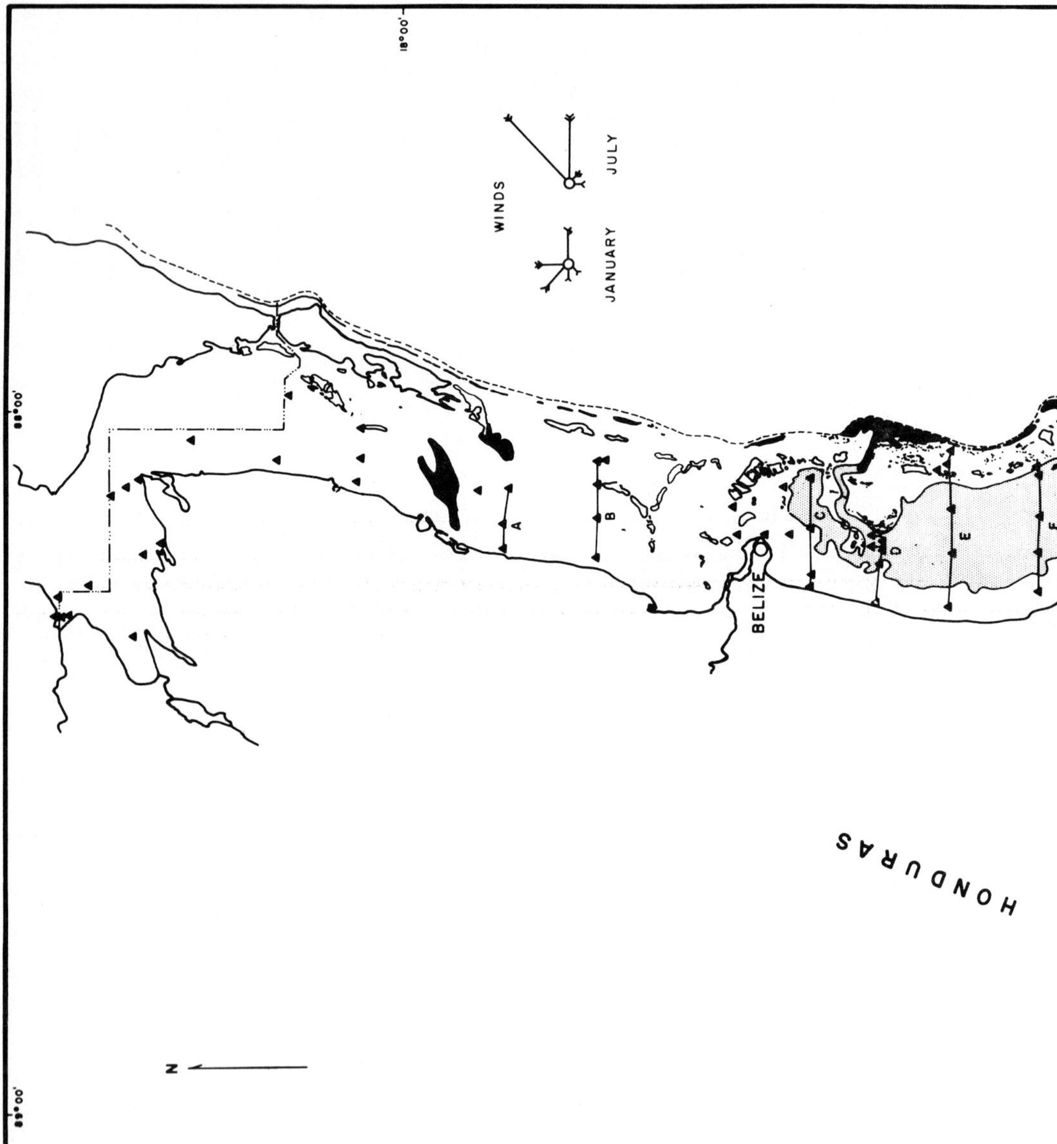
18°00'
WINDS
JANUARY
JULY
88°00'
A
B
C
D
E
F
BELIZE
HONDURAS
N
89°00'
18°00'

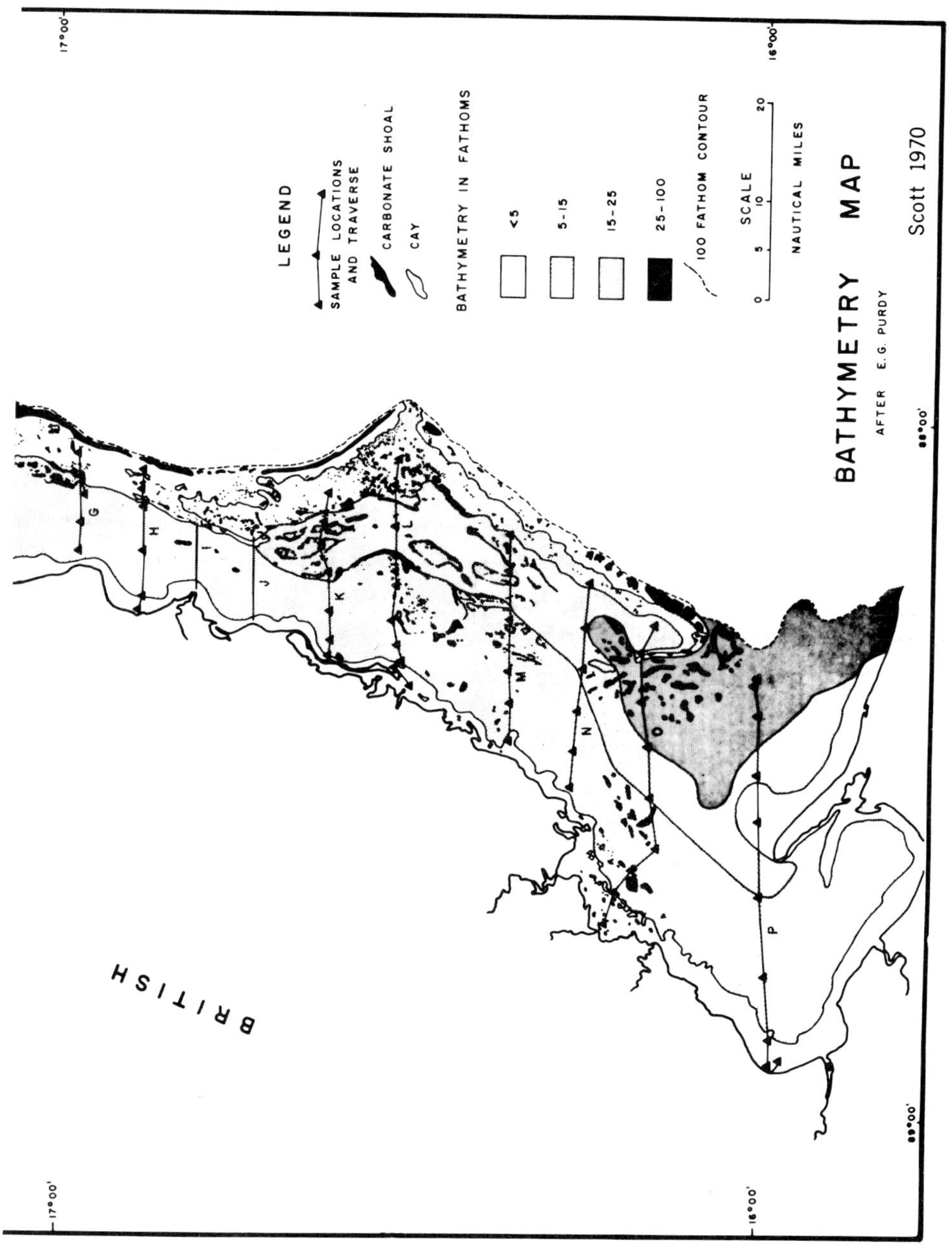

FIG. 2–Bathymetry of Belize shelf. Sample traverses referred to in text are indicated by letters.

mainland. South-central Belize is dominated by the Maya Mountains (Fig. 1), a deeply dissected, uplifted mass of Paleozoic sediments intruded by granites. The terrain in this area is quite rugged, the highest mountains rising to an elevation of more than 3,600 ft (1,100 m). Karst topography has developed on the remnants of Cretaceous limestone in the Maya Mountains. The mountains have the steepest declination on the north and east; almost all the rivers which drain these highlands empty onto the Belize continental shelf.

North of the Maya Mountains the terrain merges with the low, poorly drained karst landscape of the Yucatán Peninsula. The karst topography has developed on Cretaceous and Eocene limestones, which are the dominant rock type in northern Belize. The only significant noncarbonate outcrops in the area are the remnants of clastic sediments deposited on top of the Tertiary limestones during Pleistocene higher sea-level stands (Dixon *in* Romney, 1959). Another area where limestone is predominant lies south of the Maya Mountains and is characterized by low hills. The region is underlain by Cretaceous and Eocene limestones and Eocene shales.

Climatically, Belize is located in the transition zone between northern savanna and southern moist rainforest (Trewartha, 1961). The average annual rainfall ranges from 49 in. (125 cm) at Chetumal, Mexico, to 70 in. (178 cm) at Belize City, increasing to greater than 150 in. (381 cm) 80 mi (129 km) south of Belize City. The rainfall is unevenly distributed between a 3-5-month winter dry season and a 7-9-month summer wet season. The prevailing winds are the easterly trade winds, which blow most consistently from the east-northeast.

Shelf Bathymetry

The Belize shelf is about 150 mi (250 km) long and 15 mi (24 km) wide (Fig. 2). The shelf edge is characterized by a barrier reef which is present along almost the entire length of the shelf and is most continuous over an interval extending about 60 mi (97 km) south from the latitude of Belize City. On the eastern edge of the shelf, the bottom slopes very steeply into the Caribbean Sea. The barrier reef and the shoreline form the limits of the shelf lagoon; this feature is protected from wave action by the reef and forms a trap for both land-derived and reef-derived sediments.

North of the latitude of Belize City, the Belize shelf is extremely shallow; water depth generally is less than 15 ft (4.6 m). This part of the shelf, including Chetumal Bay, is referred to as the "northern shelf." The Bulkhead shoal, covered by 2-5 ft (0.6-1.5 m) of water, separates Chetumal Bay from the rest of the shelf.

The Belize shelf south of the latitude of Belize City is referred to in this report as the "southern shelf." The carbonate muds of this area have been studied by Matthews (1965, 1966). Along the eastern edge of the shelf is a shallow platform, the barrier platform. Water depths over the platform vary from a few feet to 5 fm. The eastern side of the barrier platform is bounded by the steep escarpment marking the edge of the Belize shelf. The western side of the barrier platform is marked by an abrupt descent into the southern shelf lagoon. The maximum depth of the southern shelf lagoon increases from 5 fm near Belize City to 35 fm at the southern end. The maximum depths are near the base of the barrier platform, and the water depth decreases gradually landward.

Water Characteristics

Very little is known about the current directions of the Belize shelf. Two large rivers, the Rio Hondo and the New River, empty into

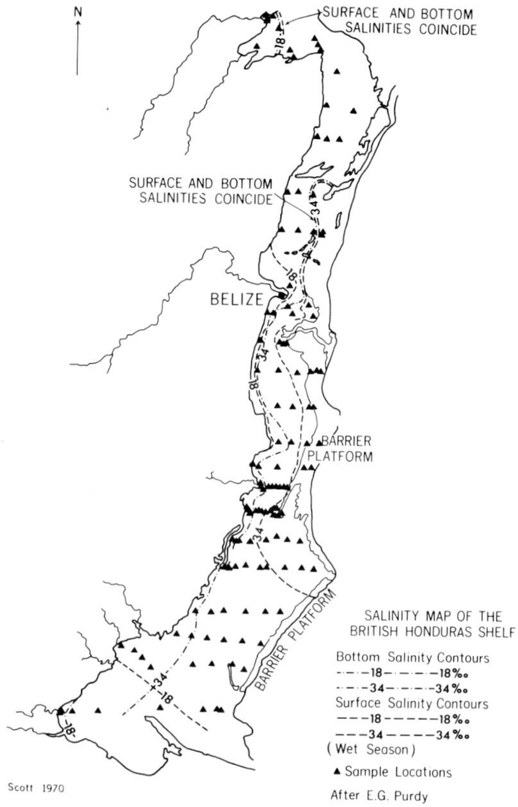

FIG. 3—Salinity map of Belize shelf.

upper Chetumal Bay, causing a net movement of water to the south. Trade winds push the water to the southwest, creating a persistent southerly longshore drift. Hydrographic Office Charts report a southward direction of flow of surface waters over most of the shelf but an easterly flow between Hunting Cays and Cabo Tres Puntas, Guatemala. The spring tidal range is 1.5 ft (0.5 m). During northers the tide commonly drops 2.5 ft (0.8 m); Pusey (1964) cited a report of a 7-ft (2.1 m) lowering of the tide induced by a hurricane in 1920. Pusey (1964) reported northeast-southwest-trending floral and depositional lineations which are visible in aerial photographs of the northern shelf; these lineations indicate the flow pattern of bottom currents and may be tidal in origin.

Surface and bottom salinity patterns are illustrated by Wantland and Pusey (this volume). These data were taken in the wet season and show that normal marine salinities prevail over most of the bottom of the southern shelf lagoon and the barrier and outer platforms (Fig. 3). A narrow band of brackish water is present along the southern mainland coast where fresh water dilutes seawater as a result of stream discharge. A broad salinity gradient is present across the shallow northern shelf lagoon, and waters in Chetumal Bay are strongly diluted by discharge of the Rio Hondo and the New River.

Surface salinities represent the interaction between freshwater wedges, northeastern winds, and tides. No information is available on the salinity distribution during the dry season. Saline bottom water penetrates a considerable distance up the rivers. Dixon (1957) reported halophilic mangroves 50 mi (80.5 km) upstream in the Rio Hondo and New River; Krueger (1963) reported mangroves 1-11 mi (1.6-17.6 km) upstream in the Belize River, Sittee River, Golden Stream, Deep River, and Rio Grande. The presence of halophilic plants in the rivers indicates that the saltwater penetration is not short-lived (Pusey, 1963).

General Features of Shelf Sediments

The sediments of the northern shelf are almost entirely carbonate (Fig. 4). Insoluble residues of the sediments from this region are characteristically less than 20 percent, and commonly less than 15 percent of the total sample. The high carbonate content of these sediments is primarily the result of a lack of terrigenous influx coupled with shallow-water conditions favoring the growth of organisms with carbonate skeletons. Many samples from the northern shelf have a high organic content; field notes indicate an organic film or scum present on the bottom in many areas. Pusey (1963) suggested the presence of the mucilaginous products of diatoms and other algae as a major factor in stabilization of the Bulkhead mud shoal.

The sediments of the southern shelf are influenced to a much greater extent by the influx of detrital materials. The rivers entering the southern shelf contribute clastic sediments from the sedimentary, igneous, and metamorphic rocks of southern Belize and northern Guatemala.

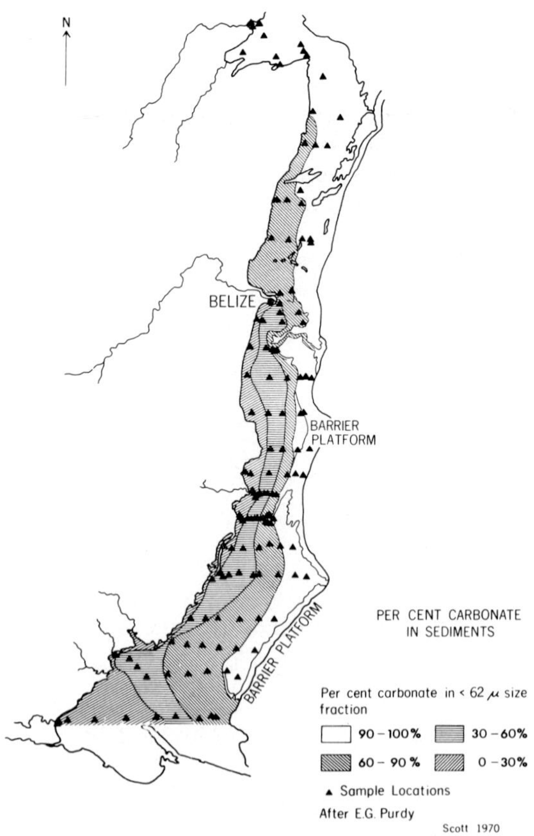

FIG. 4—Percent carbonate in less-than-62 μ size fraction of Belize shelf sediments.

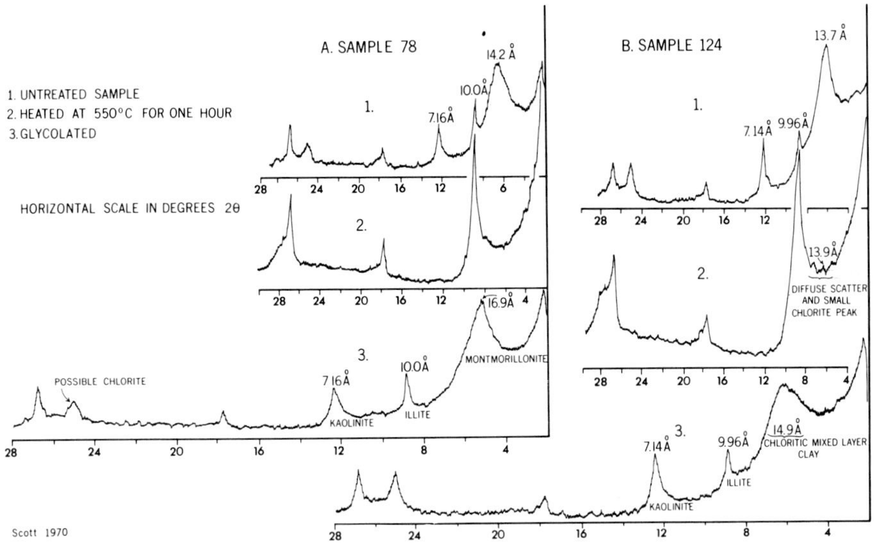

FIG. 5—Examples of X-ray diffraction patterns from samples studied. **A.** Sample 78: montmorillonite, illite, and kaolinite. **B.** Sample 124: chloritic mixed-layer clay, illite, and kaolinite.

The nearshore sediments consist of quartz sand that commonly contains muscovite or bleached biotite flakes. As a rule, the one or two samples closest to a river mouth in southern Belize contained so little fine-grained material that it was impossible to analyze their clay-mineral content. The terrigenous sand grades eastward into terrigenous mud, which grades into calcareous ("lime") mud toward the center of the southern shelf lagoon. According to Matthews (1965), the sediment of the southern shelf lagoon is typically 70-95 percent mud (<62μ), and about 50-90 percent of the mud fraction consists of calcium carbonate. *In situ* foraminifers and mollusks compose the bulk of the coarse fraction. Sediments of the reef complex consist almost entirely of sand-size skeletal debris. Skeletal sands predominate on the barrier platform and on the carbonate shoals in the southern shelf lagoon.

LABORATORY PROCEDURES

The fine fraction (<62μ) of the sediment was treated with 0.25 molar acetic acid to remove carbonate material, and then was washed repeatedly with deionized water until the clay was deflocculated. In a few cases a small amount of 0.20 molar sodium hexametaphosphate was used to complete deflocculation. The less-than-2μ size fraction of the samples was obtained by centrifugation; the slurry of this clay-size component was sedimented onto glass slides and dried. The prepared slides then were analyzed by X-ray diffraction.

A Norelco X-ray diffraction unit and Ni-filtered Cu Kα radiation were used in this study. An X-ray diffractogram was obtained for each untreated slide over the interval from 2° to 35° 2θ. Examples of the X-ray traces are shown in Figure 5. After X-ray patterns from the untreated slides had been recorded, the slides were subjected to glycolation by the method of Brunton (1955). The glycolated slides then were rerun at a goniometer speed of 1/2° 2θ/minute; these glycolated traces were used in the semiquantitative estimates of the relative abundance of clay minerals in the samples. In an attempt to verify the presence of small amounts of chlorite in some samples, the glycolated slides of these samples were rerun at 1/8° 2θ/minute over an interval from 24° to 26° 2θ to monitor the separation of a chlorite (004) peak from the kaolinite (002) peak (after the method described by Biscaye, 1964).

A slide prepared from each sample was heated at 550°C for 1 hour, and the dehydrated sample was then subjected to X-ray analysis. In many cases, the resulting X-ray traces were not entirely diagnostic for mineral identification; therefore, several samples were studied further with the aid of an oscillating-heating stage accessory to the Norelco diffractometer. The accessory equipment used in the oscillating-heating stage consisted of a diffractometer heating stage (Model XR-6 made by the Robert L. Stone Co. of Austin, Texas) and a Leeds-Northrup recorder. For each run, a portion of the less-than-2μ suspension was sedimented onto a platinum plate attached to a Pt-10-percent-Rh thermocouple. The sample plate was placed over the furnace, and the chamber was closed. Only the 15-Å peak, prominent in X-ray traces of almost every sample, was studied. The exact position of this peak was located, and the goniometer was set to oscillate over as small an angle as possible on either side of this position. The goniometer speed was set at 2° 2θ/minute. The minimum angle of oscillation was about 3° 2θ, rather than the optimum 1/2° 2θ suggested by Weiss and Rowland (1956a). The larger oscillation range prevented very close control on the reaction of the 15-Å peak to heat treatment, but it did not seriously hamper interpretation of the reactions. The sample was heated at a rate of 5°C/minute to at least 600°C. The temperature of the sample plate was recorded on a Leeds-Northrup recorder. Most of the samples were run in a dry nitrogen atmosphere in an attempt to prevent oxidation of any iron in the brucite layer of chlorite—a reaction which would inhibit the intensity increase of the chlorite (001) reflection (Weiss and Rowland, 1956a, p. 126) and thus hamper identification of trace amounts of chlorite. Examples of oscillating-heating X-ray traces are shown in Figure 6.

MINERAL IDENTIFICATION

Clay Minerals

The clay minerals in the Belize sediments were identified on the basis of the reaction of their basal reflections to glycolation and to heat treatment. The mineral identifications are based largely on the reference volume edited by Brown (1961). Because there were many clay minerals present, several of which had overlapping X-ray diffraction patterns, the minerals were identified only as belonging to one or another of the major clay-mineral groups. It is probable, for example, that the illite in these sediments is composed of two or more of the mica polymorphs; in such cases the X-ray diffraction pattern represents the average of the reflection from all the illite types present.

Montmorillonite—Material yielding a 15-Å reflection (001) from an untreated slide and a 17-Å reflection after glycolation was assigned to the montmorillonite group; this material collapsed to 10 Å after being heated for 1 hour at 550°C. Weak and diffuse (002) and (003) reflections were observed in X-ray graphs of a few samples which contained large amounts of montmorillonite. It is possible for vermiculite with a relatively low layer charge to be expanded by glycol (Walker, 1957). Therefore, any vermiculite in these sediments which expanded after glycolation with ethylene glycol would be included in the montmorillonite fraction.

Illite—The term "illite" in this report refers to material which gives an ordered sequence of reflections at 10, 5, and 3.3 Å and which

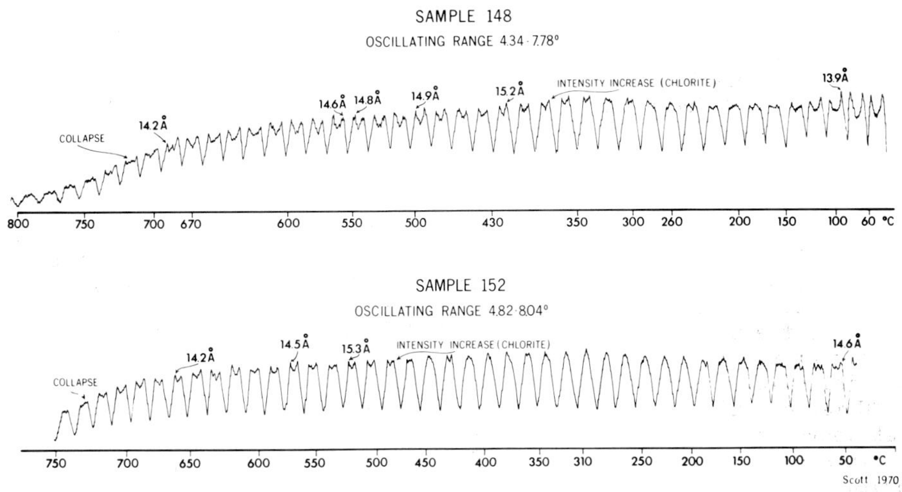

FIG. 6—Examples of oscillating-heating-stage X-ray traces.

does not react to glycolation or heat treatment. No attempt was made to distinguish any polymorphic varieties of illite. The mineral term is used as a "general term for the clay mineral constituent of argillaceous sediments belonging to the mica group," in accord with its definition by Grim et al (1937, p. 816).

Kaolinite—Material here referred to as "kaolinite" yielded a regular sequence of reflections at 7, 3.5, and 2.4 Å which showed no reaction to glycolation but collapsed to a nondiffracting state when heated at 550°C for 1 hour. The kaolinite diffraction maxima coincide with some of the chlorite reflections, and careful attention to detail was necessary in order to distinguish between the two minerals, especially since kaolinite was much more abundant in the samples than chlorite.

Chlorite—The identification of chlorite in Holocene sediments is commonly problematic, especially where other clay minerals are present. Well-ordered chlorite is characterized by a regular sequence of basal reflections at 14, 7, and 3.5 Å; the (001) peak does not shift on glycolation and becomes sharper and more intense after being heated at 550°C for 1 hour. In the Belize sediments, however, the chlorite present does not follow the above pattern of behavior, especially in the respect that in many samples its (001) peak collapses when the sample is heated at 550°C. Although chlorite is not a major constituent of the Belize sediments, it was studied extensively because many authors have reported the diagenetic formation of chlorite in Holocene sediments. The presence of poorly ordered chlorite was considered a possible key to determining whether diagenesis is an important factor in controlling the distribution of clay minerals in the Belize sediments.

Chlorite of a poorly ordered nature is quite common in Holocene sediments and has been described by several authors. For example, Griffin and Ingram (1955) described a chloritic mineral from the Neuse River estuary which collapsed to 10 Å after being heated to 400°C. Bradley (1954), Grim and Johns (1954), Grim et al (1949), and Powers (1957) discussed similar material, and attributed its anomalous reaction to heat treatment to poorly developed brucite layers, which may represent a lateral mixing of chlorite and vermiculite phases. Identification of the chloritic material was difficult, because only diffuse scatter remained in the 12-15-Å region after the sample had been heated at 550°C for 1 hour. Bradley (1954) and Biscaye (1964) pointed out that a distinction can be made between kaolinite and chlorite by the separation of the kaolinite (002) from the chlorite (004) on slow scans; that technique was used in this study.

Oscillating-heating-stage studies of Belize samples proved diagnostic for chlorite in several cases. Examples of the oscillation traces are shown in Figure 6. Oscillating-heating studies of ordered material by Weiss and Rowland (1956a, b) showed that the 14-Å reflection of chlorite shifts to 13.8 Å, increases markedly in intensity at about 575°C, and collapses at 850°C. In the Belize sediments, however, an increase in intensity of the chlorite peaks was observed in the range from 300 to 465°C, the temperature of collapse varied from 400 to 750°C, and the spacing of the (001) peak was higher, although it decreased with heat. Any material which maintained a spacing of 14 Å at temperatures greater than 300°C was considered to be chlorite, especially if the intensity of the peak increased during the oscillating-heat run (Fig. 6).

In several of the samples which retained a 14-Å reflection at temperatures greater than 300°C on the oscillating-heating stage, no increase in intensity of the 14-Å peak was observed. This phenomenon might be explained by the presence of iron in the brucite layer. Weiss and Rowland (1956a) found that an iron-bearing Vermont chlorite exhibited less increase in intensity of its 14-Å peak when heated than did an iron-poor clinochlore. In the presence of oxygen, iron in the brucite layer is oxidized, thus decreasing the amount of oxygen lost from the brucite layer during dehydration and diminishing the intensity change of the (001) reflection. In most of the oscillating-heating runs made on the Belize samples, a dry nitrogen atmosphere was used to prevent this effect of atmospheric oxygen. Sample 147 was run with and without nitrogen; both thermal reactions (intensity increase and collapse) occurred at lower temperatures in the presence of atmospheric oxygen. There was also less intensity increase of the (001) peak in the presence of atmospheric oxygen, implying the presence of iron in the brucite layers. The reaction of the chlorite lattice to heat is dependent on the rate of heating (Brindley, 1961, p. 252). Specifically, the rate of heating affects the degree of demarcation between the initial loss of water from the brucite layer—accompanied by an increase in intensity of the (001) peak—and the higher temperature loss of water from the talc layer. In the Belize sediments, several of the samples that showed only diffuse scatter in the 14-Å region after being heated rapidly to 550°C for 1 hour developed a distinct 14-Å peak which persisted at temperatures well above 550°C when heated slowly at the rate of 5°C/minute. In order to test this effect, a second heating-stage run of sample 152 was performed, and the temperature was held constant for 1 hour after it had reached 550°C. The 14-Å peak remained constant during this period, although only diffuse scatter was present in the 14-Å area of the trace from the rapidly heated slide. Poor demarcation between the two thermal reactions of chlorite may result in failure to identify chlorite, especially the poorly ordered sedimentary chlorites.

Mixed-layer clay—A survey of the distribution of mixed-layer clays by Weaver (1956) resulted in the conclusion that most sedimentary rocks contain some variety of mixed-layer clay. Diffraction patterns from the Belize sediment samples showed the following evidence of mixed 2:1 clays:

1. Glycolation of some samples resulted in an incomplete shift of the 15-Å reflection, either leaving a 14-16-Å shoulder on the expanded lattice reflection or expanding only a small percentage of the layers so that the major portion of the peak remained in the 14-16-Å region.

No expansion of the 10-, 5-, or 3.3-Å reflections was detectable after glycolation of the samples, indicating that the illite constituent is relatively free of contamination by expandable layers.

2. The same samples heated to 550°C for 1 hour commonly had X-ray traces showing varied amounts of diffuse scatter on the low-angle side of the 10-Å reflection. This scatter represents randomly mixed, uncollapsed layers which probably are present in varying amounts in individual grains. The mixed-layer clay characterized by incomplete expansion and incomplete collapse is referred to here as "chloritic mixed-layer clay." Figure 5B shows X-ray traces of a sample which contains large amounts of this type of mixed-layer clay. Because the mixed-layer clay in most of these samples exhibits a 14-Å reflection after glycolation, the writer concludes that much of the material consists of mixed chlorite and vermiculite layers. Each of these 2:1 minerals has a nonexpanding basal spacing of about 14 Å; the vermiculite layer collapses when heated, but the chlorite layer ordinarily does not. Illite and montmorillonite interlayering is probably also present. The partial expansion of the material may be due to the presence of montmorillonite or expandable vermiculite layers, or to laterally discontinuous brucite members of chloritic layers.

The Belize mixed-layer clay is similar to material described by Bradley (1954), Grim and Johns (1954), and Powers (1954, 1957) in that it contains mixed layers which remain uncollapsed by heat. The use of the term "chloritic" implies that the layers are prevented from collapsing by the presence of interlayered brucite. It is also possible for aluminum hydroxide to prevent the collapse of 2:1-layer silicates by heat. Aluminum-interlayered clay has been described by Rich and Obenshain (1955), Tamura (1957, 1958), Jackson, (1963), Rich (1968), and Sawheny (1968). Treatment with sodium citrate (Tamura, 1957, 1958) removes the interlayer material and permits the clay to collapse to 10 Å when heated. Citrate treatment was not performed on the Belize samples, so the presence of aluminum-interlayered clay is undetermined.

Accessory Minerals

The less-than-2μ size fraction of the Belize sediments is conspicuous in its lack of accessory minerals. Quartz is present in small amounts in all of the river samples and in many of the nearshore shelf samples; quartz was identified by reflections at 20.9 and 26.7° 2θ. Very small peaks at 9.2 and 9.5° 2θ were present in a few samples and were identified tentatively as talc or pyrophyllite. The accessory minerals were not included in the semiquantitative analysis listed in the Appendix.

QUANTITATIVE ESTIMATES OF CLAY MINERALS

To evaluate the relative distribution of a group of sedimentary constituents, it is necessary to quantify the amount of each constituent present. Clay-mineral data are particularly difficult to quantify. The ability of a clay mineral to diffract X-rays varies with chemical composition, crystal size, and the degree of ordering in the crystal lattice. Moreover, each of these parameters may vary from grain to grain of the same mineral in a sediment; internal standards are therefore of little use. Despite these serious sources of error, semiquantitative data are commonly computed for clay minerals by assigning appropriate weighting factors to basal X-ray peaks of the clay minerals being studied. In this report, semiquantitative data were computed from peak areas measured by planimeter on X-ray traces from the glycolated samples. The reflections used were the montmorillonite (001) at 17 Å, the illite (001) at 10 Å, and the kaolinite (001) and chlorite (002) at 7 Å. The weighting factors used were one times the montmorillonite peak and four times the illite peak (after Johns et al, 1954), and two times the chlorite-kaolinite reflection (after Biscaye, 1964, as modified from Weaver, 1961). In those samples which contained chlorite, the areas of the chlorite (004) peak at 3.54 Å and the kaolinite (002) peak at 3.58 Å were measured and directly compared in order to determine the contribution of each mineral to the 7-Å peak.

Several samples encountered in this study contained significant amounts of a chloritic mixed-layer clay (Fig. 8; Appendix). Because of the importance of this material, it was essential to include it in the semiquantitative analysis of the sediment, but there is no known appropriate weighting factor for material of such an indeterminate nature. For the sake of convenience, the large peak in the vicinity of 14-16 Å yielded by this material when glycolated was given the same weight as the 17-Å montmorillonite peak, and the resulting data are listed as the "percentage" of mixed-layer clay in the Appendix. The numerical values listed for mixed-layer clay are therefore not a reliable estimate of the absolute amount present, but they do have relative significance. The comparison of the montmorillonite peak area to four times the illite peak area is based on the relative scattering ability of three-layer clay minerals at angles corresponding to 17 Å and 10 Å, respectively (Johns et al, 1954). As the glycolated mixed-layer maximum lies between 14 and 16 Å in all cases, the use of the same weighting factor probably results in a lower estimate of the percentage than is realistic, owing to the effects of the polarization and Lorentz factors at the 2θ angles involved (Klug and Alexander, 1954).

The weighted peak areas were used to compute relative ratios and approximate percentages. The Appendix includes the montmorillonite/illite, montmorillonite/kaolinite, and kaolinite/illite ratios, as well as the clay percentages. The percentages were calculated by setting the sum of the weighted peak areas of the three major minerals equal to 100 percent, so that the fraction of the sum contributed by the weighted peak area of each mineral represents at least the relative amount of that mineral present. The numerical data in the Appendix are valuable

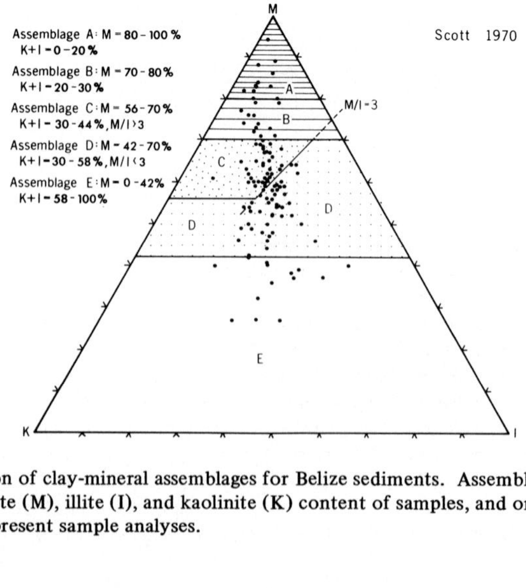

FIG. 7—Definition of clay-mineral assemblages for Belize sediments. Assemblages are defined on a basis of montmorillonite (M), illite (I), and kaolinite (K) content of samples, and on montmorillonite/illite ratios. Solid circles represent sample analyses.

in a relative sense only; Gibbs (1965) has shown that seven common techniques for quantitative analysis of clays have a precision of about ±10 percent. However, the accuracy of the Appendix data was much more variable. Diffraction patterns of samples prepared by particle settling yielded values for montmorillonite/kaolinite and montmorillonite/illite ratios which were too high; the slower settling montmorillonite was concentrated at the top of the clay film, and its peak was thus artificially enhanced because of nonpenetration of X-rays into the lower part of the film at the low 2θ angle corresponding to 17 Å. The pipette-on-glass method of Gibbs (1965), which corresponds to the one used in this study, resulted in montmorillonite/kaolinite and montmorillonite/illite ratios which were about 50 percent too high. It is therefore probable that the amount of montmorillonite in the Belize samples is not as high as indicated in the Appendix. The relative value of variation in ratios and percentages from sample to sample is not affected by the method of slide preparation, which was standard for all samples in this study, so that the validity of the trends shown in this report is not affected by the findings of Gibbs.

RESULTS OF SEMIQUANTITATIVE ANALYSIS

The major clay minerals in the Belize shelf sediments are montmorillonite, kaolinite, and illite. For the purposes of constructing a mineral-distribution map, five clay-mineral assemblages composed of different proportions of these minerals were defined by totaling montmorillonite, kaolinite, and illite to 100 percent. The assemblage boundaries are definied on a triangular diagram (Fig. 7) with the three major clay minerals as end members. In general these boundaries separate groups of samples on the basis of their montmorillonite content. Assemblage C, however, is bounded both by percentage lines and by a clay-mineral-ratio boundary (Fig. 7). This boundary was chosen because it delineates a pattern which has genetic significance. The geographic distribution of the assemblages is shown in Figure 8.

Montmorillonite is abundant and widespread on the shelf. The sediments of the northern shelf are characterized by an extremely high montmorillonite content (Figs. 7, 8), the average abundance being greater than 70 percent of the less-than-2μ size fraction of the insoluble residues. The sediments of the southern shelf, however, contain less than 70 percent montmorillonite in the clay fraction (Fig. 8). Contours showing the relative abundance of montmorillonite in the sediments of the southern shelf (Fig. 8) nearly parallel the coastline and bathymetric and insoluble-residue contours (Figs. 2, 4). Sediments from relatively deep parts of the southern shelf lagoon contain more montmorillonite than shallow-water sediments from this part of the shelf. Kaolinite and illite are present in about the same amounts in most samples, as shown by the distribution of sample analyses on the triangular diagram (Fig. 7). Both of these minerals are relatively scarce on the northern shelf (Fig. 8). In the southern-shelf sediments, illite and kaolinite are relatively abundant in the nearshore and barrier-platform areas and less abundant in the central part of the shelf lagoon. The distribution of these two minerals is in general complementary to the distribution of montmorillonite. A contoured map of kaolinite/illite ratios (Fig. 9) shows that kaolinite is slightly more abundant than illite in the sediments of the northern shelf and of the nearshore regions of the southern shelf, indicating a net decrease of kaolinite relative to illite in a direction away from these areas. Chlorite is present in only a few of the shelf samples, and only in very small

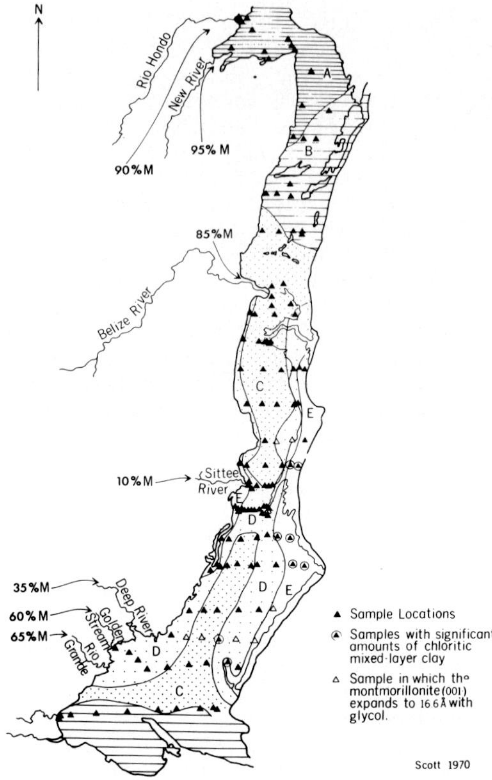

FIG. 8—Clay-mineral assemblage map for Belize sediments. Assemblages are defined in Figure 7. Percentages of montmorillonite (M) in clay size fraction of river sediments analyzed are indicated on map.

amounts (Appendix). Its presence in samples near or within the area characterized by abundant mixed-layer clay (samples 124, 125, 147-149, 152, 153) may be related to the accumulation of the mixed-layer clay. Significant amounts of mixed-layer clay were found in several samples. These samples are circled in Figure 8, but are not represented on the triangular diagram (Fig. 7). The distribution of the mixed-layer clay is confined to the sediments on or near the barrier platform on the southern shelf. In a few samples the montmorillonite peak expanded to 16.6 Å when glycolated, rather than to a more typical 16.9 Å. These samples are marked by open triangles in Figure 8; the incomplete expansion of the montmorillonite peak probably indicates the presence of intermixed nonexpanding layers.

DISCUSSION

The distribution of clay minerals in marine sediments may be controlled by three major factors: provenance, mechanisms of sediment transport in the depositional environment, and diagenetic transformations after deposition. Each of these factors is examined in the following discussion to determine its influence on the clays in this study. Extensive work on marine sediments (Weaver, 1958a, b; Griffin, 1962; Biscaye, 1965; Griffin et al, 1968; and others) has shown that provenance is the most influential factor in determining the mineralogy of marine sediments.

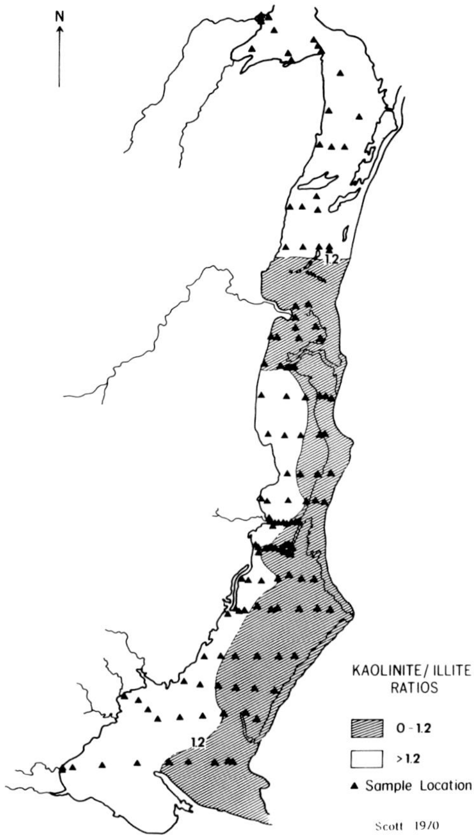

FIG. 9—Kaolinite/illite ratios for clay fraction of Belize sediments.

Provenance

Geologic reports on Belize have been published by Dickerson and Weisbord (1931), Flores (1952), Dixon (1957), Hoffstetter and Dixon (1960), Vinson (1962), and others. Northern Belize is underlain chiefly by limestone, much of it siliceous or tuffaceous. Even the unconsolidated coastal deposits of northern Belize are calcareous. In the Maya Mountains in central Belize, granitic intrusions and slightly metamorphosed terrigenous clastic rocks support the rugged topography. Limestone and calcareous shale crop out south of the mountains; south of the latitude of Belize City, the coastal deposits consist mainly of terrigenous detritus.

North of the Maya Mountains, Belize is a low-lying, poorly drained terrain which receives relatively little rainfall. Much of this area has an elevation of less than 200 ft (61 m) above sea level. According to Romney (1959), northern Belize is a dry tropical lowland with an average annual rainfall of less than 80 in. (203 cm). During 4 months of the year (the winter dry season), evaporation exceeds precipitation in this area. During the rainy season the poor drainage results in widespread swamps, especially in northernmost Belize (Murray, 1961). Romney (1959) has characterized the soils of most of Belize north of the Maya Mountains as being formed under conditions of "lime enrichment" (Fig. 10). The intermittent rainfall, poor drainage, abundance of

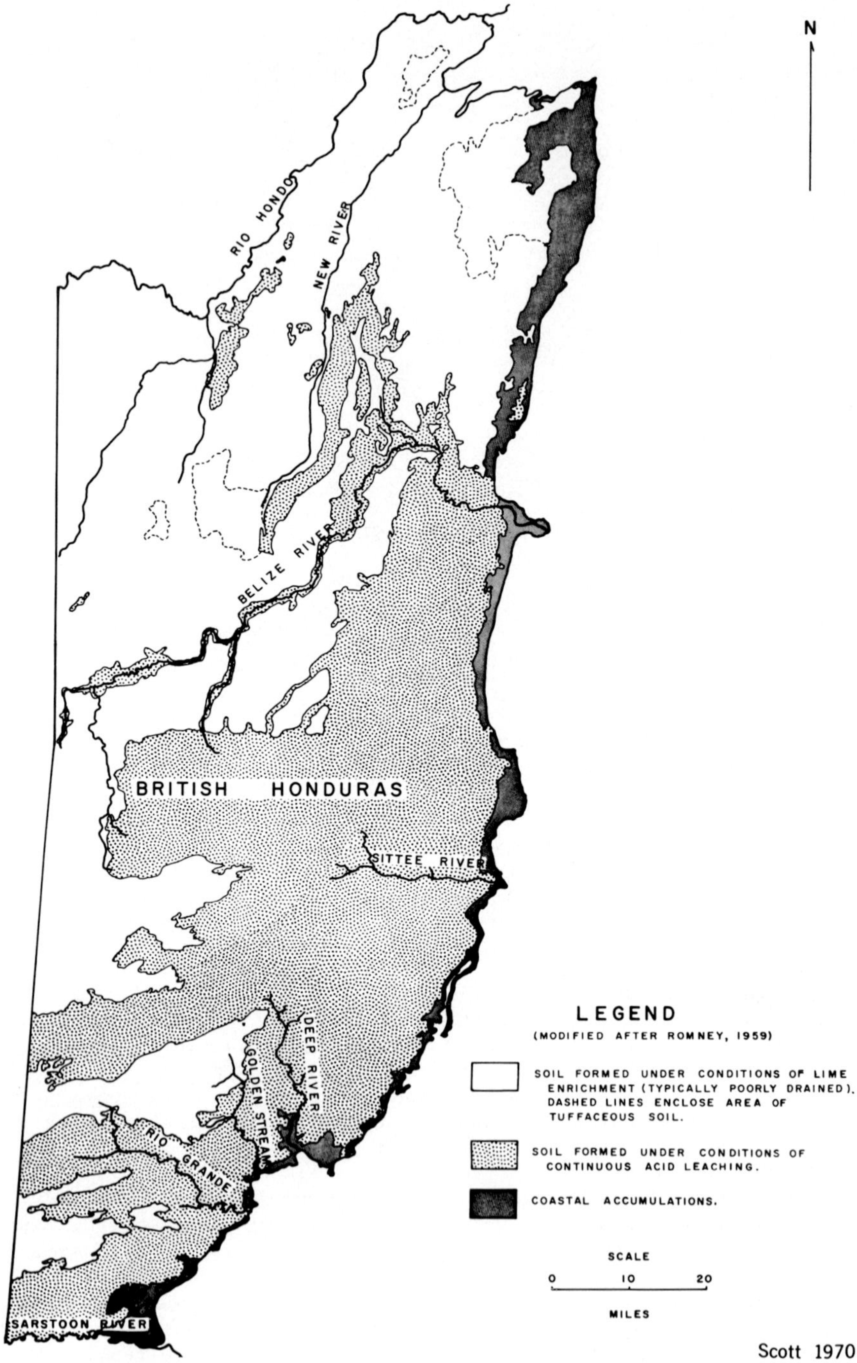

FIG. 10–Generalized soil map of Belize.

divalent cations (especially Ca), and availability of siliceous and tuffaceous limestone parent material correspond closely to the conditions under which montmorillonite is expected to form (Keller, 1962; Mohr and Van Baren, 1954). An X-ray analysis of a soil from the vicinity of Corozal by Fieldes (*in* Romney, 1959, p. 64) showed the presence of illite, traces of a kaolin mineral, and montmorillonoids.

The Maya Mountains and the area south of them contribute relatively large amounts of detritus to the shelf. The mountains have more than 3,000 ft (914 m) of relief, and annual rainfall is high—in some places more than 180 in. (457 cm) per year. The igneous intrusions and the slightly metamorphosed clastic rocks of this area are thus subjected to weathering conditions of acid leaching (Fig. 10). Such an environment is ideal for the formation of kaolinite. The rather poorly drained lowland areas at the base of the mountains are a less favorable environment for the formation of kaolinite. In fact, Romney (1959) stated that gleying, or weathering under conditions of waterlogging, is the prevalent mechanism of soil formation over a large part of the lowlands east of the mountains. X-ray analysis of one of these soils by Fieldes (*in* Romney, 1959, p. 96) indicated that the clay was predominantly kaolinitic but also contained a considerable amount of montmorillonoids.

Under the present environment of soil formation in Belize, at least two of the minerals found in the adjacent shelf sediments, montmorillonite and kaolinite, are the expected derivatives of the weathering process. Illite in soils is considered by many authors (e.g., Jackson, 1959) to be largely inherited from muscovite in the parent rock. Dixon (1957) stated that most of the granite in the Maya Mountains is a biotite granite, but that the adjacent metamorphic rocks have been "muscovitized." Therefore, the shales, slates, and phyllites are more probable sources of illite than are the igneous intrusives.

The clay-mineral content of the sediment carried by the rivers in Belize is the best index of the material being contributed to the adjacent shelf sediments (Appendix; Fig. 8). Krueger (1963) has studied bottom samples from seven of the major rivers of Belize. X-ray analysis of some of his samples by the writer showed that montmorillonite, kaolinite, and illite were the major clay minerals present; very small traces of chlorite were identified in some of the river samples (Appendix). The montmorillonite percentages of river clays reported here are consistently higher than those reported by Krueger; the discrepancy is probably the result of different methods of sample preparation and different peak-area weighting factors. The following conclusions may be drawn from the river-sediment data.

1. Bottom samples from the New River and Rio Hondo contain a very high percentage of montmorillonite, in agreement with the conditions of soil formation in their drainage basins.

2. The Belize River contains slightly less montmorillonite, in accordance with the fact that part of the acid-leached soil regime is included in its drainage basin, although the larger part of the area drained by this river is within the zone of lime-enriched soils (Fig. 10).

3. The southern rivers, which drain the Maya Mountains and the area south of them, contain less montmorillonite and more kaolinite than do the three northern rivers.

The Tomash and Sarstoon Rivers in the extreme southern part of Belize were not sampled, and the clay-mineral suites contributed to the shelf by these rivers are not known. Shelf samples adjacent to the mouths of the two rivers are montmorillonite-rich; it is probable that this material is derived largely from the Tomash and Sarstoon Rivers. The large freshwater wedge on the southern end of the study area (Fig. 3) is principally the result of fresh water from the Rio Dulce (E. G. Purdy, personal commun.). The Rio Dulce drains a region of metamorphic, ultramafic, and Tertiary volcanic and sedimentary rocks (Weyl, 1961), and undoubtedly contributes large amounts of sediment to the Belize shelf. X-ray graphs of samples 220-216 (traverse P, Fig. 2) show random scattering in the range of 2-10° 2θ, probably indicating a complete gradation of randomly mixed montmorillonite-illite present in addition to the discrete montmorillonite and illite found in the samples. The Tomash River, the Sarstoon River, or the Rio Dulce could be contributing this material to the shelf. The samples from the eastern part of this traverse (215-211) do not contain such mixed-layer material. Because of the complex and poorly known factors influencing the clay content of the sediment on this part of the shelf, traverse P is considered a special case and is not included in other discussions of clay-mineral distribution.

The clay minerals carried by the Belize rivers include substantial amounts of all the major clay minerals identified in the shelf sediments. The extremely high montmorillonite content of the sediments north of the latitude of Belize City (Fig. 8) is directly attributable to the high montmorillonite content of the New River and Rio Hondo, which contribute detritus to this part of the shelf. The Northern River probably supplies large amounts of montmorillonite to the shelf, but no data are available from this river. Montmorillonite-rich sediment is also contributed to the shelf by the Belize River, as evidenced by the Appendix and Figure 8. The relatively higher kaolinite and illite content of the southern rivers is reflected in a similarly higher proportion of these minerals in the sediments of the adjacent shelf area. Major differences in the parent materials and environment of soil formation between northern and southern Belize are thus reflected in differences in the clay suites transported by the rivers and in the clay minerals of the adjacent shelf sediments. The north-south variation in distribution of clays on the shelf is directly controlled by variations in the provenance. On a smaller scale, however, variations in the distribution of clays on the Belize shelf and similar areas cannot be explained solely on the basis of provenance. Such variations are obvious along traverses perpendicular to the coastline of Belize (Figs. 9, 10) and can be explained as functions of sediment-transport mechanisms operating in the depositional environment.

Distribution of Clay Minerals by Sedimentary Processes

Detrital clay minerals brought to a marine environment by rivers are subject to the same sedimentary processes as other mineral grains similarly introduced to the environment. The response of clay minerals to these sedimentary processes is different in one respect; the settling unit is not the discrete clay particle, but a floccule or coacervate consisting of many particles held together by van der Waals forces. Therefore, the ordinary concepts of the relation between particle size and settling velocity cannot be applied to clay minerals. Indeed, the reverse of the ordinary relation seems to be true: "In general, the larger solid particles are found in any slower-settling coacervates and the smaller particles in any faster-settling coacervates" (Whitehouse et al, 1960, p. 60). The differences in the settling velocities of kaolinite, illite, and montmorillonite (Fig. 11) are of primary importance in explaining the distribution of these minerals in sediments. The settling velocity of illite is slightly greater than that of kaolinite, and the settling velocities of both of these minerals are considerably greater than that of montmorillonite (Whitehouse et al, 1960).

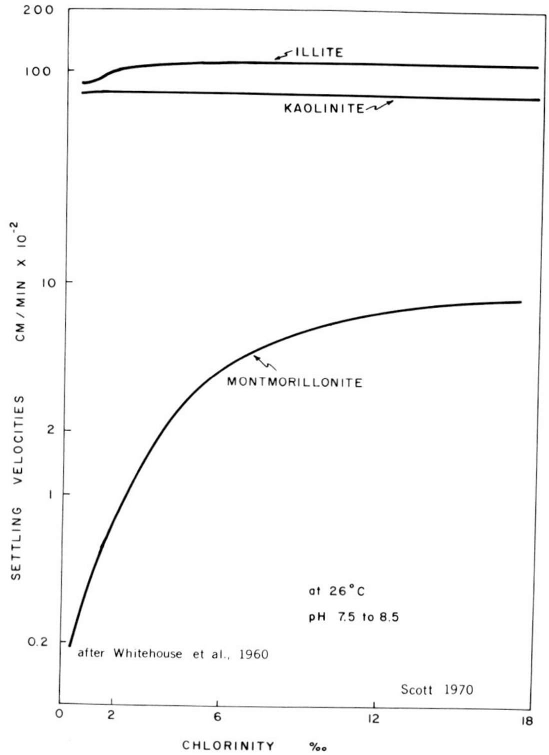

FIG. 11—Differential settling velocities of illite, kaolinite, and montmorillonite as a function of chlorinity.

The settling velocities of kaolinite and illite increase with chlorinity over a range from 0 to 2°/₀₀, but show little dependence on chlorinity at higher values. The increase in settling velocity of montmorillonite with chlorinity is more gradual, occurring over a range of 1-18°/₀₀. Figure 3 (salinity contours) shows that the surface and bottom salinities at almost all the sample locations are greater than 3.6°/₀₀, which is equivalent to 2°/₀₀ chlorinity. Thus the settling velocities of kaolinite and illite are constant over most of the shelf, but the montmorillonite settling velocity probably varies. The variation would not be of great enough magnitude to change the relative settling rate of the clay minerals.

A comparison of montmorillonite/illite and montmorillonite/kaolinite ratios to the bathymetric profiles of sample traverses (Fig. 12) shows a definite relation between montmorillonite content, distance from shore, and water depth. In all of the traverses the montmorillonite content increases with distance from shore and depth of water. This trend is in complete agreement with the expected result of the low settling velocity of montmorillonite determined by Whitehouse et al (1960) and with their experiments on differential transport of clay minerals. The faster settling illite and kaolinite tend to be concentrated in the nearshore, shallow-water area, whereas montmorillonite remains in suspension and is carried farther from shore and deposited in relatively deep water.

The montmorillonite content of relatively shallow-water samples at a relatively great distance from shore is characteristically low. Such samples are found near or on the barrier platform or on local shoals in

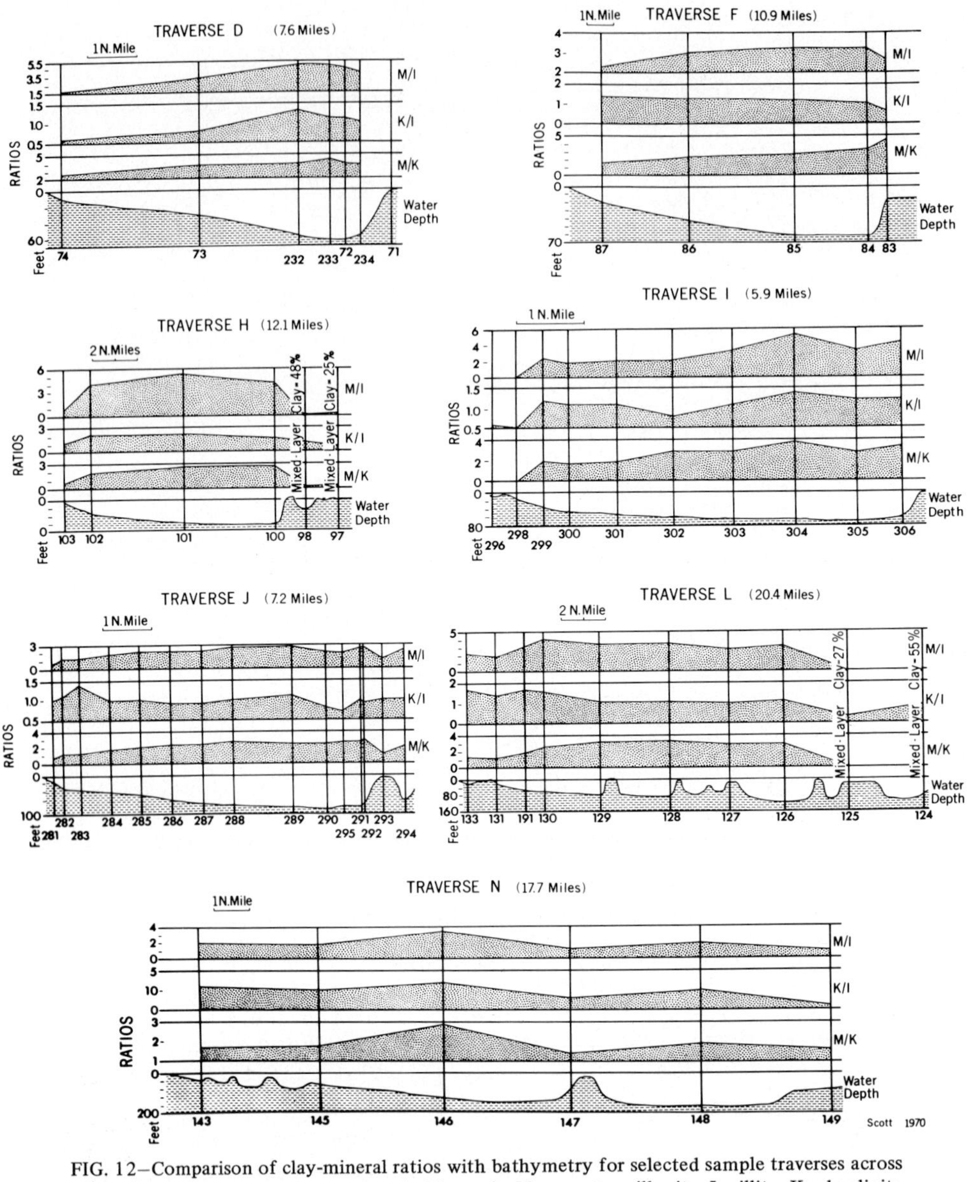

FIG. 12–Comparison of clay-mineral ratios with bathymetry for selected sample traverses across Belize shelf. Locations of traverses are given in Figure 2. M = montmorillonite, I = illite, K = kaolinite. Sample numbers are indicated below each bathymetric profile.

the southern shelf lagoon; these shallow-water sediments are more susceptible to the effects of currents and waves than are the sediments in nearby deeper water regions. The resuspension of the sediments by wave and current activity, combined with the net southward movement of the longshore current, results in a sorting out of montmorillonite from the shoal sediments. This relation is illustrated by a comparison of clay-mineral ratios with bathymetric profiles (Fig. 12). In several of these examples, the low-montmorillonite shoal samples contain a significant amount of chloritic mixed-layer clay; the origin of this material will be considered in the next section.

Additional evidence that the low-montmorillonite content of offshore shoal samples is the result of winnowing can be seen by comparing the clay-mineral ratios and bathymetric profiles of traverses J, L, and N (Fig. 12). In each of these traverses a deeper water sample was taken on the seaward side of one of the montmorillonite-poor shoal samples, and in each the deeper water sample contained more montmorillonite than the shoal sample. Such local variations are on too small a scale to appear on the clay-mineral distribution maps (Fig. 8), or they would appear as "islands" in the distribution pattern (e.g., sample 147, traverse N, Figs. 8, 12).

The relative distribution of kaolinite and illite (Fig. 9) seems anomalous when compared to the settling velocities determined for these minerals by Whitehouse et al (1960; Fig. 11). Kaolinite is more abundant nearshore, although illite was found to settle faster. However, Whitehouse et al found that mechanical mixtures of illite and montmorillonite tend to form settling entities which have settling velocities more nearly like those of montmorillonite. Thus the formation of illite-montmorillonite settling entities may be responsible for the offshore concentration of illite relative to kaolinite. Illite and kaolinite are present in about the same amounts in most samples from the southern shelf; the variation in the kaolinite/illite ratio is not of large magnitude, nor is it entirely consistent in direction from one traverse to the next (Fig. 12; Appendix).

Nearshore concentrations of kaolinite relative to illite are common in the geologic record (Weaver, 1959). In many cases, the transport of significant amounts of illite offshore by the formation of illite-montmorillonite settling entities may be cited as the probable cause of the distribution. In a significant number of cases, however, there is little or no montmorillonite present and such an inference cannot be made (e.g., Parham, 1964). Some authors have suggested that kaolinite is altered to illite or chlorite in the marine environment (Grim and Johns, 1954; Grim et al, 1949; Griffin and Ingram, 1955), and Swindale and Fan (1967) have observed alteration of gibbsite to chlorite, but such alteration is not found to occur in experimental work on diagenesis (Whitehouse and McCarter, 1958). It also seems unlikely from mineralogic considerations that illite is forming from kaolinite (Weaver, 1958b). The origin of this type of distribution is as yet uncertain.

Montmorillonite-rich sediment from the Belize River and the Northern River is carried southward by the longshore current and deposited as a long tongue of montmorillonite-rich sediment extending well south of the Sittee River (assemblage C, Fig. 8). Most of the streams south of the Belize River drain regions of acid-leached soil which would not be expected to yield large amounts of montmorillonite such as those which the northern rivers carry. A similar effect of a longshore current on the distribution of clay minerals in the northeastern Gulf of Mexico was described by Griffin (1962).

The montmorillonite-rich sediment added to the shelf by the northern rivers is progressively diluted southward by kaolinite from the rivers draining the Maya Mountains. The gap between the two areas of assemblage C (Fig. 8) reflects deposition of sediment from the Sittee River, which has an extremely low montmorillonite content and a low montmorillonite/illite ratio. The northward-pointing tongue of montmorillonite-rich sediment at the southern end of the shelf (assemblage C, Fig. 8) is at least in part the result of the preferential accumulation of montmorillonite in the relatively deep-water sediments of this part of the shelf (Fig. 2). The montmorillonite-rich sediment supplied to the area by the Golden Stream and Rio Grande (Appendix) undoubtedly serves to accent the concentration of montmorillonite in this area.

The most conspicuous effects of sedimentary processes on the distribution of clays on the Belize shelf are the preferential deposition of kaolinite and illite in the nearshore area and on local shoals, and the preferential deposition of montmorillonite in offshore quiet-water areas. Southward transport of montmorillonite-rich sediment by longshore currents is also a significant factor in determining clay distribution patterns. Similar cases of differential transport and deposition of clay minerals have been described by Parham (1964), Pryor and Glass (1961), Van Andel and Postma (1954), and Porrenga (1966). The high montmorillonite content of the abyssal plain in the Gulf of Mexico (Murray and Harrison, 1956; Pinsak and Murray, 1960) can also be considered as the result of differential transport of slow-settling montmorillonite away from nearshore, shallow parts of the Gulf.

Origin of Chloritic Mixed-Layer Clay

Chloritic mixed-layer clay is present in significant amounts (25-58 "percent") in samples of sediment on or near the barrier platform (Fig. 8). The confinement of such samples to one area of the shelf and the absence of similarly large amounts of the mixed-layer clay in other shelf samples and in river samples prompted an investigation of the origin of this material and its distribution. There are four possible explanations for the observed distribution.

1. The mixed-layer material could be present in all of the samples, but so greatly diluted by montmorillonite in most of them that it is only detectable in samples from areas of low montmorillonite accumulation. This interpretation seems reasonable, as the basal reflections of the two mineral types are close together on X-ray graphs. If this interpretation were valid, the nearshore sediments would be expected to contain detectable amounts of the mixed-layer material, because there is relatively little montmorillonite in these samples. Mixed-layer clay is not present in these samples and is present only in minute amounts in the river samples; therefore, this suggestion does not explain the observed distribution of mixed-layer clay.

2. The mixed-layer clay could be derived locally by submarine erosion of the Pleistocene surface on which the reefs are growing. This hypothesis seems unlikely, because the samples in question are located near reefs growing on Pleistocene carbonate surfaces rather than on terrigenous surfaces (E. G. Purdy, personal commun.). The carbonate Pleistocene sediments probably would not yield a significant amount of clay to the sediment now being formed. Also, the carbonate-based shoals are not limited to the barrier-platform area, whereas the mixed-layer clay is limited to that area.

3. The chloritic mixed-layer clay could be forming *in situ* by alteration of some other clay mineral. Many authors have reported the formation of chlorite, "chloritic" minerals, and illite from other 2:1 clay minerals in the marine environment (Grim et al, 1949; Grim and Johns, 1954; Griffin and Ingram, 1955; Powers, 1954, 1957; Pinsak and Murray, 1960; Nelson, 1960; Lynn and Whittig, 1966). These interpretations are based largely on the relative increase of the amount of chlorite (or illite) in the sediment in a seaward direction. In all of these cases the chlorite had diffuse X-ray peaks and was thermally unstable at temperatures of 300-500°C, as are the chlorite and chloritic mixed-layer clay from the Belize shelf.

The alteration of one clay mineral to another is commonly called "diagenesis," a term for which there is no universally accepted definition. Weaver (1958a, b) stated that cation adsorption by clay minerals introduced into the marine environment is not diagenesis; he reserves the term for changes in the basic clay-mineral lattice, and his definition is adhered to in this report. A strict separation between the process of ion exchange and "true" diagenesis may be misleading, however, because ion-exchange phenomena commonly may serve as the initiators of diagenesis (Whitehouse and McCarter, 1958). Experimental work by Whitehouse and McCarter has shown that montmorillonite aged in artificial seawater for 5 years is altered to as much as 18 percent chlorite. Illite and kaolinite showed almost no alteration when subjected to the same conditions. The modification of montmorillonite is relatively independent of total salt concentrations within the chlorinity range 7.2-19.4°/₀₀ (salinity 13-35°/₀₀). Surface and bottom-water salinities over nearly the entire Belize shelf are within this range (Fig. 3), so variation of salinity would not be a factor controlling diagenesis of the type studied by Whitehouse and McCarter (1958). Montmorillonite is the probable original material from which the Belize mixed-layer clay would form; montmorillonite is present in abundance on most parts of the shelf but not in samples containing large amounts of mixed-layer clay. The low montmorillonite content of these samples could be interpreted to indicate an alteration of montmorillonite to mixed-layer clay in addition to a winnowing out of the mineral by current and wave activity.

The alteration of montmorillonite to chlorite apparently is initiated by the preferential adsorption of magnesium by the clay when it enters the marine environment (Whitehouse and McCarter, 1958; Carrol and Starkey, 1960). Cation exchange with seawater approaches completion after 15 months. Aging of montmorillonite in seawater for 5 months to 5 years resulted in a decrease of a few percent in the SiO_2 content. Such a reaction indicates that "true" diagenesis took place. Whitehouse and McCarter (1958) found a slight decrease in the SiO_2/R_2O_3 ratio of montmorillonite aged in seawater; the reactions indicate possible replacement of tetrahedral Si by Al, which would produce a net negative charge, fixing Mg between the layers and producing a brucite sheet. The initially formed brucite layer would probably be water-rich and laterally discontinuous, resulting in partially expandable material much like the Belize mixed-layer clay. Material similar to this was observed by Whitehouse and McCarter (1958) during their experiments. The presence of a small amount of discrete chlorite in Belize samples containing chloritic mixed-layer clay may indicate that an alteration to chlorite is approaching completion in a certain percentage of the particles affected. If the Belize chlorite is formed by this type of diagenesis, it would be Mg-rich; but evidence that the chlorite in one sample may be Fe-rich has already been discussed in the section on mineral identification.

The presence of Fe-rich chlorite in deep-sea sediments of the Atlantic and Pacific Oceans (Biscaye, 1965; Gorbunova, 1963) indicates that formation of chlorite by magnesium adsorption is not a prevalent source of this material in deep-sea sediments. The chlorite in the Atlantic Ocean is probably detrital in origin, but this is not so clearly the case for the Pacific Ocean chlorite. However, Gorbunova (1963) stated that the montmorillonite in several Pacific Ocean sediment samples is Mg-rich (6-8 percent MgO in montmorillonite-rich samples).

Diagenetic origin of the chloritic mixed-layer clay and chlorite in the Belize sediments is a plausible hypothesis. The maximum amount of chlorite observed to form in 5 years by Whitehouse and McCarter (1958) was 18 percent, but some of the Belize samples contain more than 50 percent chloritic material. The reason for the concentration of the chloritic material on the barrier platform is not apparent. Whitehouse et al (1960) found that mixed-layer chlorite-montmorillonite settles

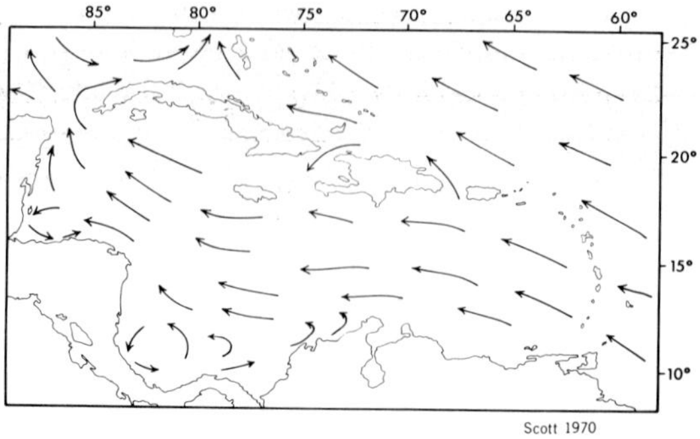

FIG. 13–Surface current directions in Caribbean Sea. After Wüst (1964).

slightly faster than pure montmorillonite, and chlorite settles at a rate near to that of kaolinite. It is therefore plausible that resuspension of the partially altered sediment and preferential net transport of the pure montmorillonite might tend to concentrate a faster settling portion of the clay which had been (or was being) altered to chlorite. Undetermined factors in the carbonate shoal environment of the barrier platform may be conducive to the type of diagenesis described.

4. The chloritic material may have been transported to the Belize shelf via the Caribbean Current. The Caribbean Current is driven westward by the northeast trade winds and turns northward as it approaches Central America (Fig. 13). Caribbean water contains about 8 mg/200 liters suspended material (Jacobs and Ewing, 1969a). Mineralogical analysis of suspended material in Caribbean water showed illite to be the major clay mineral and kaolinite and chlorite to be present in moderate amounts (Jacobs and Ewing, 1969b). Some of this chlorite may originate from rivers along the northern coast of Venezuela (Griffin and Goldberg, 1969) or even from the Amazon River. The Amazon River carries suspended sediment containing small amounts of chlorite (Gibbs, 1967), and its water mass is traceable into the Caribbean Sea as far as 80°W (Jacobs and Ewing, 1969b).

Sediments of the Caribbean Sea have been reported to contain 3-17 percent chlorite in the less-than-2μ fraction (Biscaye, 1965; Griffin and Goldberg, 1969; Jacobs and Ewing, 1969b). The chlorite content of sediments just east of the Belize shelf was found to be 13-16 percent (Griffin and Goldberg, 1969). Thus, ocean currents are a possible source of the chloritic mineral found on the outer edge of the Belize shelf. However, the chlorite reported in Caribbean water and sediments (Jacobs and Ewing, 1969b; Griffin and Goldberg, 1969) yields sharp, well-defined X-ray diffraction peaks and is a relatively minor component of the clay-size fraction. It is therefore difficult to identify this material positively with the Belize chlorite, which exhibits diffuse X-ray peaks and composes 25-60 percent of the less-than-2μ noncarbonate fraction of the sediment in some barrier-platform samples.

Of the four possible sources of chlorite in the barrier-platform sediments, the last two represent the most probable origins. The writer favors the hypothesis of diagenetic formation because of the abundance of the chloritic material and its apparent mineralogic differences from chlorite in Caribbean sediments.

CONCLUSIONS

The clay minerals of the Holocene sediments on the Belize shelf are principally detrital in origin. Major differences in parent material and environment of soil formation in the source area are reflected in the clay minerals transported by the rivers and deposited on the adjacent parts of the shelf. Variations in provenance are thus paralleled by the regional clay-mineral distribution pattern of the shelf sediments.

Distribution of montmorillonite, illite, and kaolinite on the shelf is governed by the differential settling tendencies of these minerals, the effects of the persistent longshore current, and the geographic distribution of shoal areas on the shelf. The relatively slow-settling montmorillonite is deposited preferentially in deep-water areas offshore. Concentrations of illite and kaolinite are found in the nearshore area and on the shallow barrier platform, where current and wave agitation resuspend the sediment and permit currents to transport montmorillonite away from shoals and into deeper water. The longshore current transports montmorillonite-rich detritus southward from the northern shelf, creating a tongue of montmorillonite-rich sediment opposite rivers carrying sediment from an environment of kaolinite-rich, acid-leached soils.

A concentration of chloritic minerals occurs in the sediments on and near the barrier platform. The chlorite may originate from westward-moving ocean surface currents or from diagenetic transformations of montmorillonite supplied from the adjacent landmass. Comparisons of the mineralogy and abundance of the material with the chlorite of Caribbean sediments lead the writer to prefer diagenesis as an origin.

The clay-mineral distribution patterns of the Belize sediments have analogues in other Holocene sediments and in ancient sediments. Preferential deposition of montmorillonite in offshore, relatively deep water and deposition of illite and kaolinite in nearshore, relatively shallow water are typical of such sediments. Low-montmorillonite sediments of offshore shoals are an exception to this generalized relation between clay-mineral content of sediments and basin geometry.

REFERENCES CITED

Biscaye, P. E., 1964, Distinction between kaolinite and chlorite in Recent sediments by X-ray diffraction: Am. Mineralogist, v. 49, p. 1281-1289.

——1965, Mineralogy and sedimentation of Recent deep-sea clay in the Atlantic Ocean and adjacent seas and oceans: Geol. Soc. America Bull., v. 76, p. 803-832.

Bradley, W. F., 1954, X-ray diffraction criteria for the characterization of chloritic material in sediments, p. 324-334 *in* A. Swineford and N. Plummer, eds., Clays and clay minerals: Natl. Research Council Pub. 327, 498 p.

Brindley, G. W., 1961, Chlorite minerals, Chap. 6 *in* G. Brown, ed., The X-ray identification and crystal structures of clay minerals, 2d ed.: London, Mineralog. Soc., p. 242-296.

Brown, G., ed., 1961, The X-ray identification and crystal structures of clay minerals, 2d ed.: London, Mineralog. Soc., 544 p.

Brunton, G., 1955, Vapor pressure glycolation of oriented clay minerals: Am. Mineralogist, v. 40, p. 124-126.

Carroll, D., and H. C. Starkey, 1960, Effect of sea-water on clay minerals, p. 80-101 *in* E. Ingerson, ed., Clays and clay minerals: London, Pergamon Press, 7th Natl. Conf. Proc., 369 p.

Dickerson, R. E., and N. E. Weisbord, 1931, Cretaceous limestone in British Honduras: Jour. Geology, v. 39, no. 5, p. 483-486.

Dixon, C. G., 1957, Geology of southern British Honduras: Belize City, Belize, H.M.S.O., 85 p.

Flores, G., 1952, Geology of northern British Honduras: AAPG Bull., v. 36, p. 404-409.

Gibbs, R. J., 1965, Error due to segregation in quantitative clay mineral X-ray diffraction mounting techniques: Am. Mineralogist, v. 50, p. 741-751.

——1967, The geochemistry of the Amazon River system: Part I; The factors that control the salinity and the concentration of the suspended solids: Geol. Soc. America Bull., v. 78, p. 1203-1232.

Gorbunova, Z. N., 1963, Clay minerals in the sediments of the Pacific Ocean: Litologiya i Poleznyye Iskopayemyye, v. 1, p. 28-42.

Griffin, G., 1962, Regional clay-mineral facies—products of weathering intensity and current distribution in the northeastern Gulf of Mexico: Geol. Soc. America Bull., v. 73, p. 737-768.

——and R. L. Ingram, 1955, Clay minerals of the Neuse River Estuary: Jour. Sed. Petrology, v. 25, p. 194-200.

Griffin, J. J., and E. D. Goldberg, 1969, Recent sediments of Caribbean Sea, *in* A. R. McBirney, ed., Tectonic relations of northern Central America and the western Caribbean—the Bonacca Expedition: AAPG Mem. 11, p. 258-268.

——H. Windom, and E. D. Goldberg, 1968, The distribution of clay minerals in the world ocean: Deep-Sea Research, v. 15, no. 4, p. 433-459.

Grim, R. E., and W. D. Johns, 1954, Clay mineral investigation of sediments in the northern Gulf of Mexico, p. 81-103 *in* A. Swineford and N. Plummer, eds., Clays and clay minerals: Natl. Research Council Pub. 327, 498 p.; reprinted (1957?) *in* Am. Petroleum Inst., Rept. Progress 1954-55, p. 145-167.

——R. H. Bray, and W. F. Bradley, 1937, The mica in argillaceous sediments: Am. Mineralogist, v. 22, no. 7, p. 813-829; Illinois Geol. Survey Rept. Inv. 44.

——R. S. Dietz, and W. F. Bradley, 1949, Clay mineral composition of some sediments from the Pacific Ocean off the California coast and the Gulf of California: Geol. Soc. Am. Bull., v. 60, no. 11, p. 1785-1808.

Hoffstetter, R., and C. G. Dixon, 1960, British Honduras, *in* Internat. Geol. Cong. Strat. Comm., Lexique stratigraphique international—v. 5, Amerique Latine—Fasc. 2a, Amerique Centrale: Paris, Centre Natl. Recherche Sci., p. 11-42.

Jackson, M. L., 1959, Frequency distribution of clay minerals in major great soils groups as related to the factors of soil formation, *in* A. Swineford, ed., Clays and clay minerals: Internat. Ser. Mons. Earth Sci., v. 2, p. 133-143.

——1963, Interlayering of expansible layer silicates in soils by chemical weathering, *in* Clays and clay minerals, 11th Natl. Conf., Ottawa, 1962, Proc.: New York, Macmillan, p. 29-46.

Jacobs, M. B., and M. Ewing, 1969a, Mineral source and transport in waters of the Gulf of Mexico and Caribbean Sea: Science, v. 163, no. 3869, p. 805-809.

——and——1969b, Suspended particulate matter; concentration in the major oceans: Science, v. 163, no. 3865, p. 380-383.

Johns, W. D., R. E. Grim, and W. F. Bradley, 1954, Quantitative estimations of clay minerals by diffraction methods: Jour. Sed. Petrology, v. 24, no. 4, p. 242-251.

Keller, W. D., 1962, Clay minerals in the Morrison Formation of the Colorado Plateau: U.S. Geol. Survey Bull. 1150, 90 p.

Klug, H. P., and L. E. Alexander, 1954, X-ray diffraction procedures—for polycrystalline and amorphous minerals: New York, John Wiley & Sons, 716 p.

Krueger, W. C., Jr., 1963, Mineralogical composition and textural properties of river sediments from British Honduras: Houston, Tex., Rice Univ., PhD thesis.

Lynn, W. C., and L. D. Whittig, 1966, Alteration and formation of clay minerals during cat clay development, *in* Clays and clay minerals: London and New York, Pergamon Press, 14th Natl. Conf., Berkeley, Calif., Proc., p. 241-248.

Matthews, R. K., 1965, Genesis of Recent lime mud in southern British Honduras: Houston, Tex., Rice Univ., PhD thesis.

——1966, Genesis of Recent lime mud in southern British Honduras: Jour. Sed. Petrology, v. 36, p. 428-454.

Mohr, E. C. J., and F. A. Van Baren, 1954, Tropical soils: New York, Interscience Publishers, 498 p.

Murray, G. E., 1961, Geology of the Atlantic and Gulf coastal province of North America: New York, Harper and Brothers, 692 p.

Murray, H. H., and J. L. Harrison, 1956, Clay mineral composition of Recent sediments from Sigsbee Deep: Jour. Sed. Petrology, v. 26, p. 363-368.

Nelson, B. W., 1960, Clay mineralogy of the bottom sediments, Rappahannock River, Virginia, p. 135-147 *in* E. Ingerson, ed., Clays and clay minerals: London, Pergamon Press, 7th Natl. Conf., Proc., 369 p.

Parham, W. E., 1964, Lateral clay-mineral variations in certain Pennsylvanian underclays, p. 581-602 *in* E. Ingerson, ed., Clays and clay minerals: London, Pergamon Press, 12th Natl. Conf., Proc., 692 p.

Pinsak, A. P., and H. H. Murray, 1960, Regional clay mineral patterns in the Gulf of Mexico, p. 162-177 *in* E. Ingerson, ed., Clays and clay minerals: London, Pergamon Press, 7th Natl. Conf., Proc., 369 p.

Porrenga, D. H., 1966, Clay minerals in Recent sediments of the Niger Delta, p. 221-223 *in* S. W. Bailey, ed., Clays and clay minerals: London, Pergamon Press, 14th Natl. Conf., Proc., 447 p.

Powers, M. C., 1954, Clay diagenesis in the Chesapeake Bay area, p. 68-80 *in* A. Swineford and N. Plummer, eds., Clays and clay minerals: Natl. Research Council Pub. 327, 498 p.

——1957, Adjustment of land-derived clays to the marine environment: Jour. Sed. Petrology, v. 27, p. 355-372.

Pryor, W. A., and H. D. Glass, 1961, Cretaceous-Tertiary clay mineralogy of the upper Mississippi embayment: Jour. Sed. Petrology, v. 31, p. 38-51.

Pusey, W. C., 1963, Recent carbonate shoal complexes in northern British Honduras (abs.): AAPG Bull., v. 47, p. 367.

——1964, Recent carbonate sedimentation in northern British Honduras: Houston, Tex., Rice Univ. PhD thesis.

Rich, C. I., 1968, Hydroxy interlayers in expansible layer silicates: Clays and clay minerals, v. 16, p. 15-30.

——and S. S. Obenshain, 1955, Clay mineral properties of a soil: Soil Sci. Soc. America Proc., v. 19, p. 334-339.

Romney, D. H., ed., 1959, Land in British Honduras, Report of the British Honduras land use survey team: Colonial Research Pub. No. 24, 327 p.

Sawheny, B. L., 1968, Aluminum interlayers in layer silicates—effect of OH/Al ratio of Al solution, time of reaction and type of structure: Clays and clay minerals, v. 16, p. 157-163.

Swindale, L. D., and P. F. Fan, 1967, Transformation of gibbsite to chlorite in ocean bottom sediments: Science, v. 157, p. 799-800.

Tamura, T., 1957, Identification of the 14 Å clay mineral component: Am. Mineralogist, v. 42, p. 107-110.

——1958, Clay minerals in acid soils: Jour. Soil Sci., v. 9, p. 141-147.

Trewartha, G. T., 1961, The earth's problem climates: Univ. Wisconsin Press, Madison, Wis., 334 p.

Van Andel, Tj., and H. Postma, 1954, Recent sediments in the Gulf of Paria, v. 1: Amsterdam, North Holland Publishing Co., 143 p.

Vinson, G., 1962, Upper Cretaceous and Tertiary stratigraphy of Guatemala: AAPG Bull., v. 46, p. 425-457.

Walker, G. F., 1957, Differentiation of vermiculites and smectites in clays: Clay Minerals Bull., v. 3, p. 154-163.

Weaver, C. E., 1956, The distribution and identification of mixed-layer clays in sedimentary rocks: Am. Mineralogist, v. 41, p. 202-221.

——1958a, A discussion on the origin of clay minerals in sedimentary rocks, p. 159-173 *in* A. Swineford, ed., Clays and clay minerals: Natl. Research Council Pub. 566, 360 p.

——1958b, Geologic interpretation of argillaceous sediments, Part I: Origin and significance of clay minerals in sedimentary rocks: AAPG Bull., v. 42, p. 254-271.

——1959, The clay petrology of sediments, *in* A. Swineford, ed., Clays and clay minerals: Internat. Ser. Mons. Earth Sci., v. 2, p. 154-187.

——1961, Clay minerals of the Ouachita structural belt and adjacent foreland, p. 147-162 *in* The Ouachita system: Austin, Texas, Texas Univ. Pub. 6120, 401 p.

Weiss, E. J., and R. A. Rowland, 1956a, Oscillating-heating X-ray diffractometer studies of clay mineral dehydroxylation: Am. Mineralogist, v. 41, p. 117-126.

——and——1956b, Effect of heat on vermiculite and mixed-layer vermiculite-chlorite: Am. Mineralogist, v. 41, p. 894-914.

Weyl, R., 1961, Die Geologie Mittelamerikas, Beiträge zur regionalen Geologie der Erde, Band 1: Berlin-Nikolassee, Gebrüder Borntraeger, 226 p.

Whitehouse, U. G., and R. S. McCarter, 1958, Diagenetic modification of clay mineral types in artificial sea water, p. 81-119 *in* A. Swineford, ed., Clays and clay minerals: Natl. Research Council Pub. 566, 306 p.

——L. M. Jeffrey, and J. D. Debbrecht, 1960, Differential settling of clay minerals in saline waters, p. 1-79 *in* A. Swineford, ed., Clays and clay minerals: London, Pergamon Press, 7th Natl. Conf., Proc., 369 p.

Wüst, Georg, 1964, Stratification and circulation in the Caribbean-Antillean Basins, Part I: New York, Columbia University Press, 201 p.

APPENDIX

Clay-Mineral Ratios and Weighted Peak-Area Percentages, Belize shelf[1]

	Weighted Peak-Area Percentages					*Ratios*		
Sample	M	K	I	C	ML	M/K	K/L	M/I
45	88	9	3	-	-	9.3	3.3	31.0
1	89	11	-	-	t	8.2	-	8.1
47	94	4	2	-	-	26.0	1.5	39.0
2	89	5	6	-	-	16.8	0.9	8.1
3	91	9	-	-	-	10.0	-	-
4	84	11	5	-	-	7.6	2.3	17.5
6	40	14	46	-	-	2.8	0.3	8.4
7	83	12	5	-	-	7.1	2.3	16.5
8	71	17	12	-	-	4.2	1.3	5.7
9	86	7	7	-	-	11.9	1.0	12.9
10	83	13	5	-	-	6.4	2.5	15.9
52	79	9	12	-	-	32.8	0.2	6.7
11	82	13	5	-	-	6.5	2.4	16.7
13	72	18	10	-	-	4.1	1.6	6.8
19	80	9	11	-	-	8.9	0.8	7.3
18	32	37	31	-	-	0.9	1.2	1.0
17	27	45	28	-	-	0.6	1.6	1.0
Traverse A								
41	73	14	13	-	-	5.2	1.1	5.8
42	79	14	7	-	-	5.4	2.0	10.6
32	82	14	4	-	-	5.8	3.2	18.4
31	76	17	7	-	-	4.5	2.4	10.7
Traverse B								
40	63	24	13	-	-	2.6	1.9	4.7
39	69	19	12	-	-	3.7	1.5	5.5
38	74	16	10	-	-	4.7	1.6	7.4
53	61	32	7	-	-	3.8	2.3	8.8
278	77	18	5	-	-	4.3	3.3	14.6
56	62	19	18	-	t	3.2	1.0	3.4
55	46	18	36	-	-	2.5	0.5	1.3
62	64	13	23	-	-	4.8	0.6	2.7
63	66	17	17	-	-	4.0	1.0	4.0
59	71	14	15	-	-	5.0	0.9	4.6
Traverse C								
64	37	27	36	-	t	1.3	0.8	1.0
167	64	16	20	-	t	3.9	0.8	3.2
61	67	18	15	-	-	3.7	1.2	4.4
60	64	14	22	t	t	4.7	0.6	2.9

[1]Sample numbers are listed according to their location on the shelf, from north to south. All samples except those in Chetumal Bay and those near the Belize Delta are listed in their respective order in the traverses from shore toward the shelf edge. Symbols: M = montmorillonite; K = kaolinite; I = illite; C = chlorite; ML = mixed-layer clay; t = trace.

APPENDIX (Continued)

	Weighted Peak-Area Percentages					*Ratios*		
Sample	M	K	I	C	ML	M/K	K/L	M/I
Traverse D								
74	50	19	30	-	-	2.6	0.6	1.6
73	63	17	20	-	-	3.7	0.8	3.1
232	68	18	14	-	-	3.8	1.3	4.9
233	69	16	15	-	t	4.4	1.1	4.7
72	67	18	15	-	-	3.8	1.1	4.4
234	65	18	17	-	-	3.7	1.0	3.8
Traverse E								
166	60	25	15	-	-	2.4	1.8	4.2
76	65	21	14	-	-	3.1	1.4	4.5
77	67	18	15	t	t	3.7	1.2	4.6
78	59	20	21	t	-	3.1	0.9	2.9
79	-	33	67	-	-	-	0.5	-
80	-	21	21	-	58	-	1.0	-
Traverse F								
87	49	30	21	-	-	1.6	1.4	2.3
86	57	24	19	-	-	2.4	1.3	3.0
85	59	22	19	-	t	2.7	1.2	3.2
84	61	20	19	-	-	3.1	1.0	3.2
83	62	14	24	-	-	4.5	0.6	2.6
Traverse G								
90	58	23	19	-	-	2.5	1.2	3.1
91	55	23	22	-	-	2.5	1.0	2.5
92	58	21	21	-	t	2.8	1.0	2.8
94	-	29	35	-	36	-	0.8	-
Traverse H								
103	27	40	33	-	t	0.7	1.2	0.8
102	56	30	14	-	-	1.9	2.1	4.1
101	65	24	11	-	t	2.6	2.1	5.4
100	61	24	15	-	-	2.6	1.6	4.2
98	-	29	23	-	48	-	1.2	-
97	4	29	42	-	25	0.1	0.7	0.1
Traverse I								
296	-	36	64	t	t	-	0.6	-
298	-	34	66	t	-	-	0.6	-
299	53	26	21	-	-	2.0	1.2	2.5
300	46	28	26	-	-	1.7	1.1	1.8
301	50	26	24	-	-	1.8	1.1	2.1
302	55	19	26	-	-	2.9	0.8	2.1
303	60	21	19	-	-	2.8	1.1	3.2
304	69	18	13	t	-	3.8	1.4	5.2
305	59	22	19	-	-	2.6	1.2	3.1
306	64	20	16	t	t	3.3	1.2	4.1
Traverse J								
281	27	35	38	-	-	0.8	1.0	0.7
282	40	31	29	t	-	1.3	1.1	1.4
283	40	31	29	-	t	1.3	1.4	1.4
284	48	26	26	-	-	1.8	1.0	1.8

APPENDIX (Continued)

	Weighted Peak-Area Percentages					*Ratios*		
Sample	M	K	I	C	ML	M/K	K/L	M/I
285	52	24	24	-	-	2.1	1.0	2.2
286	54	22	24	-	-	2.4	0.9	2.2
287	54	22	24	t	-	2.4	0.9	2.2
288	57	21	22	t	-	2.7	1.0	2.6
289	56	23	21	t	-	2.4	1.1	2.7
290	52	22	26	t	-	2.4	0.8	2.0
295	52	20	28	-	-	2.6	0.7	1.9
291	56	22	22	t	t	2.6	1.0	2.5
292	58	20	22	-	-	2.9	0.9	2.6
293	36	32	32	t	t	1.1	1.0	1.1
294	52	24	24	-	-	2.2	1.0	2.2
Traverse K								
114	42	34	24	-	-	1.2	1.4	1.7
115	54	28	18	-	-	1.9	1.6	3.0
116	62	24	14	-	-	2.6	1.7	4.4
117	64	20	16	-	-	3.2	1.3	4.1
118	54	22	24	-	t	2.4	0.9	2.3
119	-	26	22	-	52	-	1.2	-
120	-	29	28	-	43	-	1.1	-
Traverse L								
133	46	34	20	-	-	1.3	1.7	2.3
131	42	34	24	-	-	1.2	1.4	1.8
191	54	29	17	-	-	1.9	1.7	3.2
130	61	24	15	-	-	2.6	1.6	4.1
129	63	19	18	-	-	3.2	1.1	3.4
128	63	19	18	-	-	3.3	1.1	3.6
127	59	20	21	-	-	2.9	1.0	2.8
126	60	21	19	-	-	2.9	1.1	3.2
125	-	54	12	6	27	-	0.3	-
124	-	19	23	2	55	-	0.9	-
Traverse M								
140	44	29	27	-	-	1.6	1.1	1.6
139	60	23	17	-	-	2.6	1.4	3.5
138	62	19	19	-	-	3.2	1.0	3.2
137	57	19	24	-	-	3.0	0.8	2.4
136	59	18	23	-	-	3.2	0.8	2.5
135	38	27	35	-	-	1.4	0.8	1.1
Traverse N								
143	48	28	24	-	-	1.7	1.1	2.0
145	47	27	26	-	-	1.8	1.0	1.8
146	61	21	18	-	-	2.9	1.2	3.5
147	39	25	33	3	t	1.4	0.8	1.2
148	48	25	25	2	-	1.9	1.0	1.9
149	37	21	40	2	-	1.6	0.6	0.9
Traverse O								
204	40	42	17	-	-	1.0	2.4	2.3
206	50	30	20	-	-	1.7	1.5	2.5
208	49	32	19	-	-	1.5	1.7	2.6

APPENDIX (Continued)

Sample	*Weighted Peak-Area Percentages*					*Ratios*		
	M	K	I	C	ML	M/K	K/L	M/I
158	55	27	18	-	-	2.1	1.5	3.1
157	53	28	19	-	-	1.9	1.4	2.8
156	59	24	19	-	-	2.4	1.4	3.6
155	60	22	18	-	-	2.8	1.2	3.4
153	-	23	29	4	44	-	0.9	-
152	60	18	19	3	-	3.2	1.0	3.0
Traverse P								
220	76	16	8	-	-	4.7	2.1	9.5
219	80	14	6	-	-	5.7	2.2	12.1
218	83	9	8	-	t	9.5	1.0	10.1
217	74	19	7	t	-	3.9	2.8	10.6
216	73	17	10	t	-	4.3	1.6	7.0
215	67	17	16	-	-	4.1	1.0	4.3
214	68	14	17	1	-	4.4	0.9	3.4
213	70	17	12	1	-	4.0	1.5	6.2
211	54	21	24	1	t	2.4	0.9	2.2
212	56	19	24	1	-	2.9	0.8	2.4
RIVER SAMPLES								
Rio Hondo								
R-19	94	3	3	-	-	34.8	0.9	31.5
R-10	95	2	3	-	t	55.5	0.7	24.0
R-4	89	8	3	-	-	11.5	2.9	33.3
New River								
N-4	96	4	-	-	t	25.1	-	-
N-11	94	4	2	-	-	21.4	2.5	53.5
Belize River								
B-20	85	8	3	1	t	13.4	0.7	9.3
Sittee River								
S-2	-	38	58	-	4	-	0.7	-
S-6	9	43	47	-	1	0.2	0.9	0.2
Deep River								
D-8	36	52	12	-	t	0.7	4.4	3.1
Golden Stream								
13-62	64	29	7	-	t	2.2	4.4	9.5
Rio Grande								
RG-1	73	17	10	-	t	4.2	1.8	7.5
RG-6	64	30	6	-	-	2.2	5.2	11.5

Holocene Carbonate Sedimentation on Northern Belize Shelf [1]

WALTER C. PUSEY III[2]

ABSTRACT The continental shelf of northern Belize is mantled by calcium carbonate sediments deposited in environments characterized as fluvial, estuarine, and normal marine.

The composition, grain size, and carbonate mineralogy of these Holocene carbonate sediments were determined quantitatively by thin-section point-count analyses, sieve analyses, and X-ray diffraction techniques, respectively. Factor analysis of the accumulated data, supplemented with observations on the presence or absence of sedimentary structures, served to delineate eight facies: (1) reef facies—characterized by an abundance of coralline algae and corals welded into a wave-resistant frame; (2) *Halimeda* facies—characterized by abundant plates of the green alga *Halimeda*; (3) peneroplid-sand facies—typified by an abundance of large peneroplid Foraminifera; (4) miliolid-mud facies—distinguished by the association of many small miliolid Foraminifera with high-magnesium calcite mud; (5) laminated miliolid-mud facies—characterized by a relatively thick sequence containing laminations of foraminiferal sand intercalated with carbonate ("lime") mud; (6) terrigenous facies—characterized by the presence of relatively large quantities of quartz sand; (7) cryptocrystalline-grain facies—distinguished by the relative abundance of cryptocrystalline calcite grains; and (8) ostracod-mud facies—characterized by the association of ostracods and scattered diatoms and charophytes with "lime" and/or terrigenous mud.

The development of the coral-reef and *Halimeda* facies is limited to current-agitated marine waters of normal salinity, whereas the development of the other five carbonate facies is related to reduced current agitation and/or varying degrees of hyposalinity. The volume and mineralogy of the mud in each facies are controlled by current strength. They are determined by the transporting capacity of the currents in each of the depositional environments and, indirectly, by the capacity of the currents to induce the formation of mud-size particles through the abrasion of sand-size constituents.

INTRODUCTION

Systematic studies of modern carbonate environments have provided information that is directly applicable to the environmental reconstruction of ancient limestones. Investigations of carbonate sedimentation in Florida Bay (Ginsburg, 1956), the Bahama Banks (Illing, 1954; Purdy, 1963a, b), and Campeche Bank (Cann, 1962) have proved to be useful additions to our knowledge of carbonate depositional processes. Unfortunately, the information available is restricted to marine environments characterized by normal salinity or hypersalinity. With exception of the Florida Bay information (Ginsburg, 1956, 1957; Taft and Harbaugh, 1964), relatively little is known about carbonate sedimentation in hypo-

[1]Manuscript received, June 20, 1972.

[2]Senior Geologist, Continental Oil Company, Ponca City, Oklahoma 74601.

This paper is revised from the writer's PhD thesis, Rice University. This study was suggested and supervised by Edward G. Purdy, formerly of Rice University, and was financed by a grant from the Humble Oil & Refining Company (now Exxon), the Jersey Production Research Company (now Esso Research), and Imperial Oil Limited. Carbon-14 dates were provided through the courtesy of the Exploration Department and Geochemical Laboratory of the Humble Oil & Refining Company. The writer is indebted to the Geologic Research Center of the Humble Oil & Refining Company for suggesting the "freeze-lithification" method of preparing unconsolidated sediment cores for fabric studies. Flame-photometer analyses were made by Robert C. Harriss and David S. McKay, formerly of the Rice University Geology Department.

The field assistance of William A. Behrens, Philip H. Heckel, William C. Krueger, R. K. Matthews, Arthur R. Troell, and K. Frank Wantland is greatly appreciated. Robert Robertson, of the Philadelphia Academy of Natural Sciences, collected and identified the mollusks of the area. The enthusiastic cooperation of many people in Belize City during field operations is acknowledged. In particular, the writer is indebted to W. Ford Young of Belize City for his invaluable help in expediting field preparations and to David Bradley, Comptroller of Customs, for his courtesy in facilitating the import and export of scientific equipment. Finally, appreciation is expressed for the piloting services of Jack Cabral, who on more than one occasion prevented field operations from coming to an unfortunate and premature conclusion.

The constructive criticisms from John J. W. Rogers and Thomas E. Pulley, who served on the writer's thesis committee, are appreciated. The writer is indebted to the National Science Foundation for fellowships from 1960 to 1963.

saline environments or about the changes in carbonate sedimentation that accompany the transition from fluvial to normal-marine conditions. Such environmental transitions must have been commonplace among ancient carbonate environments; thus it seemed appropriate to study carbonate sedimentation where this transition is exemplified over a relatively great geographic area.

The area chosen for study was the northern part of Belize, where the environmental transition ranges from fluvial conditions through a rather large estuarine bay to a coral-reef environment. It is hoped that the information presented herein will prove useful in the recognition of ancient carbonate deposits that have accumulated under similar conditions.

Regional Setting

The regional setting of Belize has been discussed in detail by Purdy et al (this volume). Certain facets of the area are critical to the understanding of the modern sedimentary regime of the northern Belize shelf.

Belize forms the southeast side of the escarpment-bounded Yucatán platform. The Yucatán platform, like the Florida platform and the Bahama Banks, has been characterized by slow subsidence and attendant carbonate deposition throughout most of the Tertiary Period (Butterlin and Bonet, 1961).

The limestones of the Yucatán platform are eroded into a low-relief karstic surface with less than 1,000 ft (305 m) of elevation. The karst in the north and central parts of the platform is characterized by subsurface drainage. In contrast, the southeast side of the platform is surficially drained by rivers and elongate lagoons. Faulting, which accompanied the general uplift of the Yucatán platform, has produced a marked NNE-SSW lineation of the rivers and lagoons, as well as of the submarine escarpments which form the east rim of the Belize shelf.

The Belize shelf extends approximately 150 mi (240 km) north-south and averages 15 mi (24 km) in width. North of the latitude of the capital city of Belize, the shelf is less than 15 ft (4.6 m) deep, but south of Belize City the central portion of the shelf deepens to form an elongate depression with depths of almost 80 ft (24 m).

The Maya Mountains, which are approximately 40 mi (64 km) south of the study area, constitute a Paleozoic massif at the juncture of the stable Yucatán platform on the north and unstable Antillean mountain systems on the south. The mountains have elevations above 3,500 ft (1,067 m) and consist of Mesozoic and Paleozoic intrusive, metamorphic, and clastic rocks. The watershed of the Maya Mountains contributes large amounts of coarse siliceous clastic material to the southern part of the Belize coastal shelf but relatively small amounts to the shallow shelf on the north. This pattern, however, is relatively recent. During the Pleistocene higher sea-level stands, considerable amounts of clastic sediments were transported north and deposited over Tertiary limestones (Dixon *in* Romney, 1959, p. 22-28). These clastic sediments, which are not indurated, are the only significant source supplying terrigenous material to the northern Belize shelf.

RELATED INVESTIGATIONS

Prior to this study there were no reports dealing specifically with sedimentation in northern Belize. A few reports containing references to sediments have been concerned mainly with the geography of reefs and associated cays (islands) and with the geometry of the lagoon behind the barrier reef. Stoddart (1962a, p. 5) summarized these early geographic studies.

In 1842, Darwin (1889, p. 151-152), in his classic summary of coral reefs, commented (1) that the British Honduras (Belize) lagoon is a continuation of Chetumal Bay, (2) that large parts of the reef are sand or worn-down rock, and (3) that the sediment commonly is deposited in banks parallel with the shore. He further concluded that the barrier reef and the offshore atolls owed their peculiar barrier and atoll geometry to the shape of the underlying rock. In contrast with true atolls and barrier reefs, which are produced wholly by the growth of corals during subsidence, Belize reefs are more probably coral-fringed remnants of worn-down upheaval shoals. Darwin's conclusions were based on observations made from nautical charts and on information contained in the letters of Captain Bird Allen.

Ower (1927, 1928) discussed the general geology of the country and observed that "soft limestone and coral" can be seen on the lagoon bottom north of Belize City; he also recorded that Belize City is underlain by soft rock at a depth of 20 ft (6 m). He suggested that this may be a silt-covered reef. The presence of extensive low swamps west of the town caused him to conclude that Belize City lies not on a delta, but on a submerged cay which has almost become attached to the mainland by infilling with sediments from the Belize River. Ower (1928) also noted that there are lines of silica and chert nodules in the mainland Tertiary limestones and marls. He indicated that these were located along fault lines and believed that the nodules originated when water from silica springs mixed with seawater. Regarding the lagoon north of Belize City, Ower (*in* Mawson, 1929, p. 623) reported:

> ...limestone (is) now forming in a very shallow almost tideless area on the coast of Yucatan. The conditions of deposition appear very similar to those described by the Author (Mawson). The recrystallization here is also remarkably rapid, and the accumulations, which can be traced 50 miles inland, appear to have continued since Miocene times....The "biscuits" attain the size of slabs, which are obtained by diving, and are used as road material in Belize. Lines of shallow submarine siliceous springs have traversed the area, and it is suggested that the flints and limestones have been deposited by somewhat similar agencies.

Vermeer (1959) provided the first detailed geography of the northern Belize shelf. He classified the cays as: (1) coral-sand cays, (2) mangrove cays, and (3) mangrove-sand cays or transition cays. Exposed shallow-water cays are of the sand type; sheltered cays are of the mangrove type. He noted that all the cays north of Belize City are mangrove or mangrove-sand cays rooted in shoal banks of mud or sand. He described the sediments briefly as follows (p. 44):

> ...coral sand is common toward the seaward part of the lagoon while further toward the mainland the deposits are more apt to be mud. On the leeward side of the mangrove-sand cays the coral mud becomes mixed with decayed organic debris and silt from the rivers.

Vermeer (1963) also studied the cay damage produced by Hurricane Hattie, which struck 20 mi (32 km) south of Belize City in November 1963. Vermeer limited his reconnaissance to cays which lay north of the storm's center. He concluded that the storm was divided into a destructive waxing phase and a constructive waning phase. During the waxing phase, which was marked by a forerunner flood and a sudden hurricane surge, the floodwaters channeled low cays and deposited sand tongues on the leeward sides of cays. During the waning phase, which was marked by the recession of high water, the water built coral-rubble ridges on the emerging cays. The waning phase continued for several weeks after the storm, and

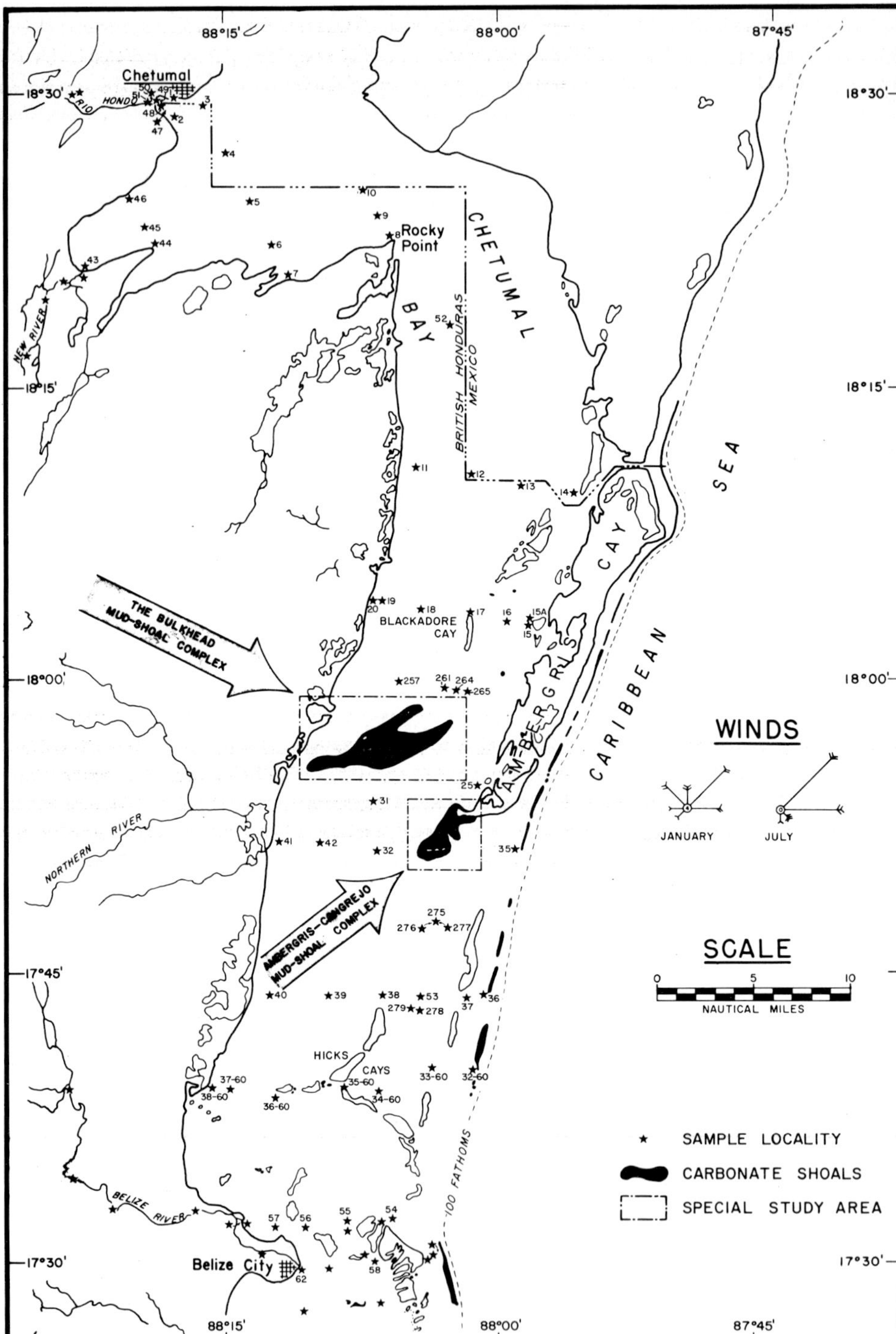

FIG. 1—Location of sample stations in northern Belize (only samples used in this study are numbered).

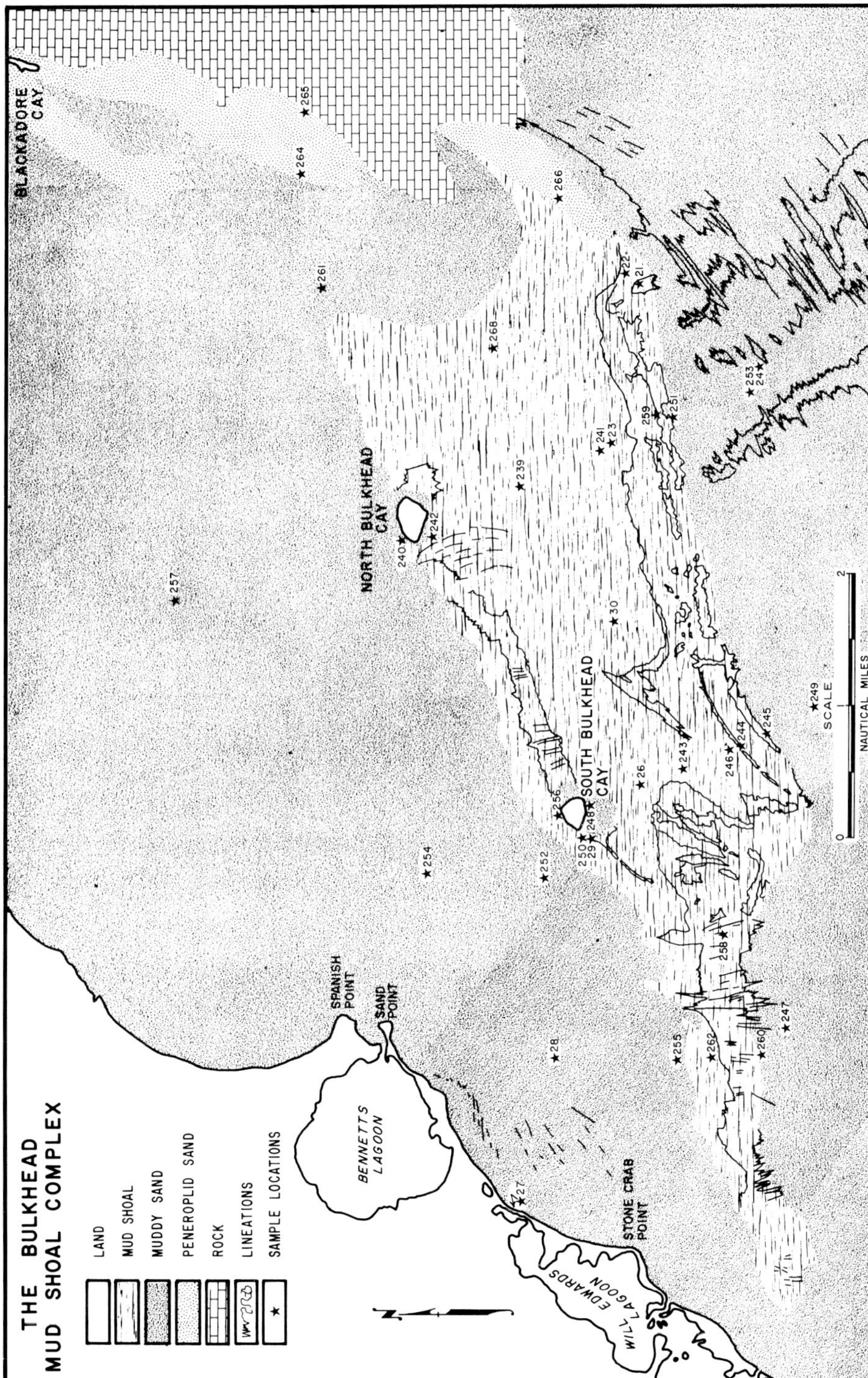

FIG. 2–Location of sample stations in Bulkhead area.

runoff water from the mainland produced outflowing currents strong enough to form deltas at the seaward ends of hurricane-cut channels.

Stoddart's (1960, 1962a, 1963) extensive work in Belize concerns the physiography of the offshore atolls and the barrier reef with its associated cays, especially south of Belize City. Stoddart (1962c, 1963) studied catastrophic storm effects on Belize reefs and cays by Hurricane Hattie and noted major vegetation damage but no significant cay destruction in northern Belize. South of Belize City, several cays disappeared, especially those which had been cleared of natural vegetation by man; and segments of the reef, including spur-and-groove systems, were destroyed. Much of the reef destruction is probably short-term.

Under the direction of Edward G. Purdy, Rice University graduate students began the systematic study of Belize in 1961. This writer's field work was conducted during July and August of 1961 and 1962. Subsequent to these field seasons, three workers (High, 1967; Ebanks, 1967; Tebbutt, 1967) sampled areas contiguous to the northern Belize shelf. Their results, which are discussed elsewhere in this volume, have modified and reinforced the writer's earlier conclusions (Pusey, 1964).

FIELD METHODS

All 105 sediment samples, with the exception of those from rocky bottoms and nearshore localities, were collected along predetermined traverses with a Petersen grab sampler. Sediment samples from rocky bottoms and nearshore localities were collected by diving. Station positions were determined by dead reckoning or by use of the Brunton compass for back-azimuth intersection from landmarks. Depending on the spacing of the samples and on the charts available, sample locations were plotted on U. S. Navy Hydrographic Chart 1498 (1959, 1:125,000), a base map constructed from U. S. Navy aerial photographs (1:40,000), or a British Honduras (Belize) Land Use Survey map (1:250,000; *in* Romney, 1959). Fathometer profiles were made along the lines of traverse. The positions of all samples used in this study are shown in Figures 1, 2, and 3.

At each sample station, salinity, water temperature, carbonate saturation, and thickness of sediment were measured. Salinity was determined with a conductivity bridge to an accuracy of ±0.2 parts per thousand. Saturation values were measured with a Weyl-type carbonate saturometer (Weyl, 1961). Sediment thicknesses were estimated with a 1-in.-diameter (2.54 cm) metal probe. Piston cores were obtained wherever sediment thicknesses exceeded 2 ft (0.6 m). A 150-lb piston corer was used to take 1 3/8-in.-diameter (2.85 cm) cores as long as 6 ft (1.8 m). In depths less than 5 ft (1.5 m), a tripod-piston corer (modeled after the bipod device of Ginsburg and Lloyd, 1956) was used to obtain 1 3/4-in.-diameter (2.85 cm) cores as long as 9 ft (2.7 m). Where the sediments were less than 2 ft (0.6 m) thick, punch cores were made by pushing plastic tubing into the sediment. General bottom observations were made regarding currents, ripple marks, flora, and fauna. In the bay and nearshore stations, turbid water made it necessary to dive in order to observe the bottom.

NORTHERN SHELF LAGOON AND CHETUMAL BAY

That part of the continental shelf between the latitude of Belize City and the Bulkhead mud-shoal complex is designated as the *northern*

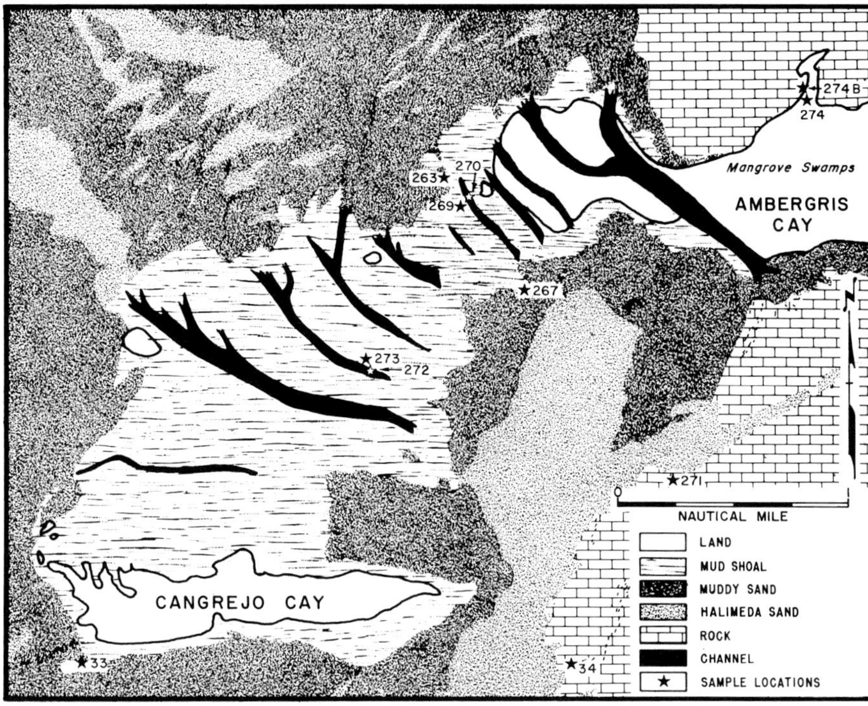

FIG. 3—Location of sample stations in Ambergris-Cangrejo mud-shoal area.

shelf (Fig. 1). The northern shelf consists of a relatively narrow *outer platform* which borders the Caribbean Sea. A discontinuous line of cays separates the outer platform from the *northern shelf lagoon*. The Bulkhead mud-shoal complex not only forms the northern limit of the northern shelf, but also serves as the southern limit of Chetumal Bay. The bay and the northern shelf include approximately 1,000 sq mi (2,590 km^2 of depositional area. This study does not include the bay area north of Chetumal, Mexico (18°30'N).

Bathymetry

Fathometer profiles and nautical charts show that the eastern edge of the northern shelf rises steeply from 500 fm to depths of 20 fm or less; the average slope approximates 40°. The seaward margin of the shelf is marked by a coral reef which gradually changes from a fringing reef along Ambergris Cay (Fig. 4B). The barrier reef is somewhat discontinuous and shelters approximately two thirds of the northern shelf. Figures 5 and 6 show the general bathymetry of the northern shelf and Chetumal Bay.

Local bottom relief on the northern Belize shelf is dependent on both Holocene sedimentation and Pleistocene erosion features superimposed on the structural fabric of the underlying Tertiary limestones. Relatively recent normal faulting has produced NNE-SSW topographic lineations (Sapper, 1896; Alvarez, 1954; Butterlin and Bonet, 1962). The Rio Hondo, New River, and numerous lagoons, as well as the shoals and deeps in Chetumal Bay, parallel this direction (Fig. 5). During the Pleistocene glacial stages, sea-level lowering accelerated erosion and produced a karstic surface. Numerous sinkholes are etched into the relatively

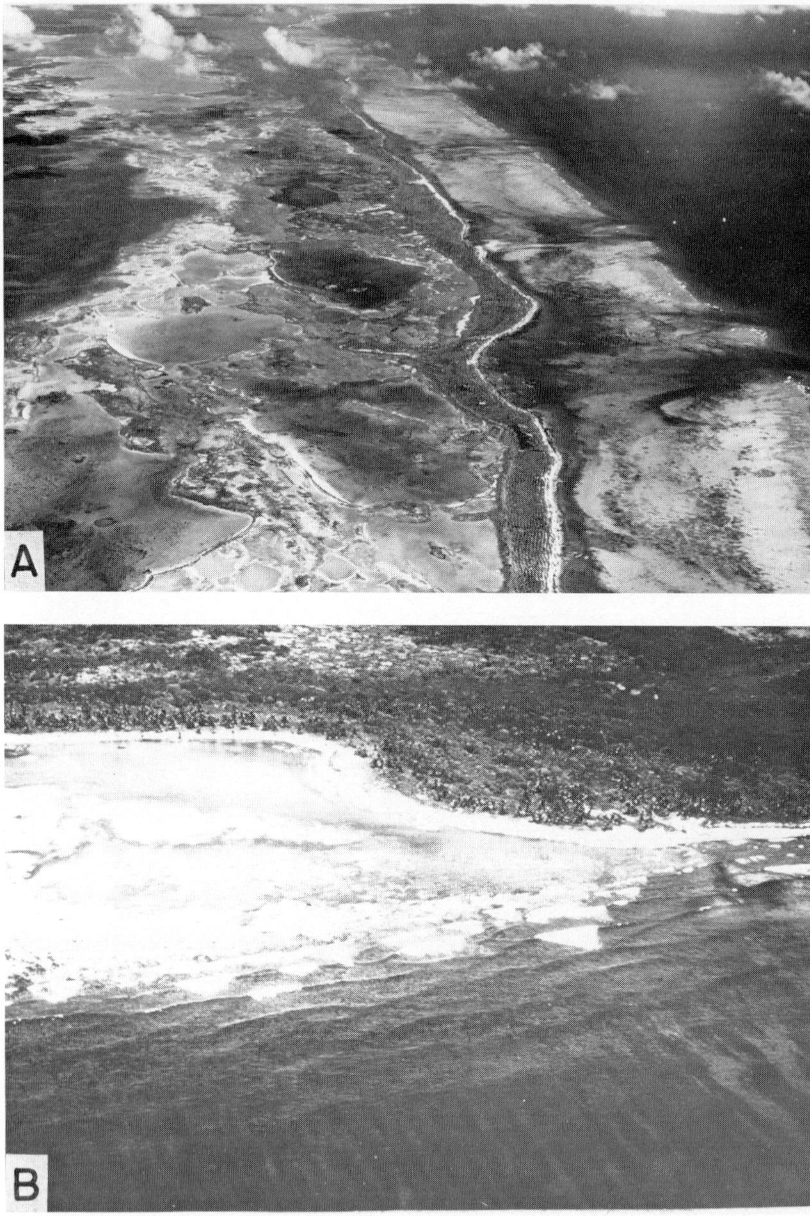

FIG. 4—Barrier reef of northern Belize.

A. Barrier reef along east side of Ambergris Cay. Oblique aerial photograph facing northeast; Caribbean Sea in upper right; backreef lagoon is approximately 1 mi (1.6 km) wide. Crest of reef is marked by white caps. Dark areas along reef trend are channels. Channel in right foreground has curved tributary channel in lagoon which suggests that channels may represent a drowned drainage system. Ambergris Cay consists of mangrove swamps, lakes (left side of photo), and seaward-facing beach ridges planted with rows of palms (center).

B. Barrier reef is becoming a fringing reef at north end of Ambergris Cay. Reef crest is in foreground, island in background. Oblique aerial photograph facing northwest. Spurs and grooves developed on seaward side of reef are barely evident.

flat limestones, giving parts of Ambergris Cay and the bottom of Chetumal Bay a cratered appearance. Drowned stream channels, which are sediment filled, are present on the rock high which separates the inner shelf lagoon from the outer shelf lagoon (Fig. 7B). The effect of the sink-holes and the incised stream channels on bathymetry is not great, because most have been filled with sediment.

Climate

The northern shelf lagoon lies within the tropical zone, where the coolest air temperatures average 75°F (24°C; Romney, 1959). The area is

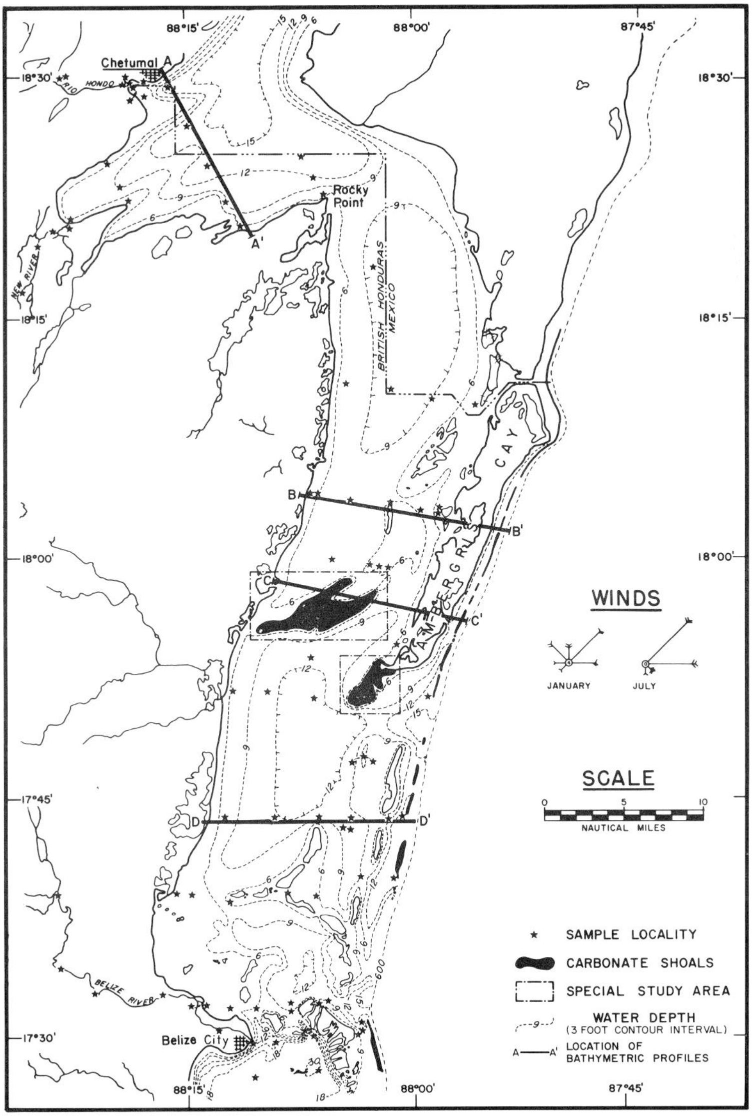

FIG. 5—Bathymetry of northern Belize (based on nautical charts and field work). Profiles are shown in Figure 6.

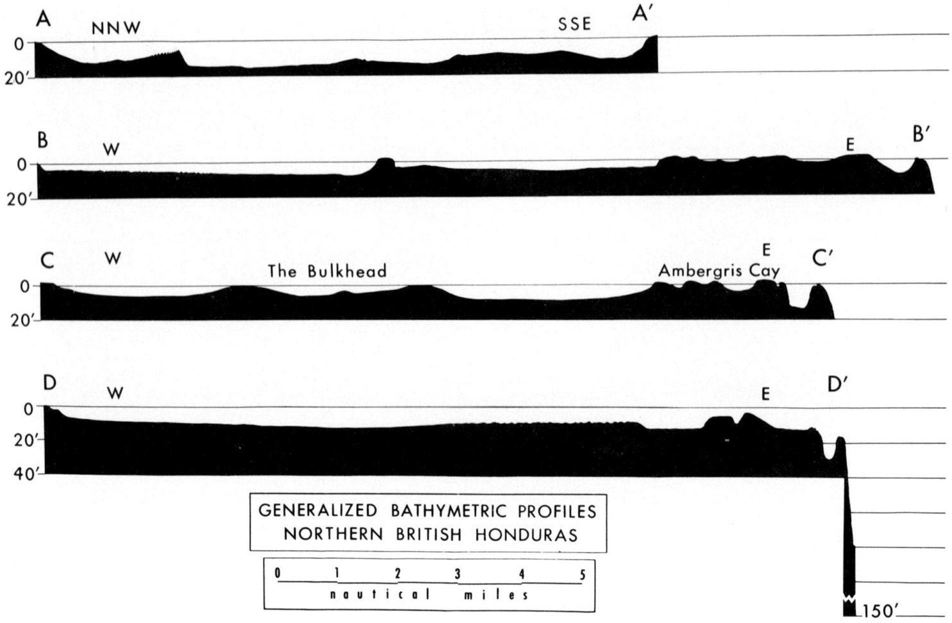

FIG. 6—Bathymetric profiles across northern Belize shelf (see Fig. 5 for location of profiles).

a transition zone between a northern savanna climate and a southern moist rain-forest climate (Trewartha, 1961). Annual rainfall increases from 49 in. (124 cm) at Chetumal, Mexico, to 70 in. (178 cm) at Belize City, and to more than 150 in. (380 cm) about 80 mi (128 km) south of Belize City. The annual cycle comprises a winter dry season lasting 3-5 months and a summer wet season lasting 7-9 months.

The area is strongly influenced by easterly trade winds, which blow most consistently and strongest from the northeast and east octants at velocities of 12-15 knots (U.S. Navy Hydrographic Office, 1955b). Afternoon onshore breezes frequently reinforce these winds to 15-20 knots. Strong northers—which bring rains, 30-knot winds, and air temperatures as low as 50°F (10°C)—have only a slight effect on reef water temperatures (Stoddart, 1962a, p. 14). The wind effect on the temperature of the more restricted and shallower bay water is unknown.

In the period 1931-1961, the Belize coast has been struck by hurricanes about once every 6 years (Stoddart, 1963). Although the population and manmade structures have been affected intensely by the hurricanes, alterations of the sedimentary environment have been minor and temporary, especially where natural vegetation remains uncleared (Stoddart, 1962c).

Water Characteristics

Within the milieu of carbonate sedimentation, water functions not only as an agent of transportation and deposition, but also has profound effects on primary carbonate-grain production and secondary modification of grain types. Physical and chemical properties of the hydrosphere may (1) indirectly affect the types of grains produced by controlling the distribution of carbonate-secreting organisms, and (2) directly affect, or even effect, carbonate grain types by causing carbonate precipitation or dissolution.

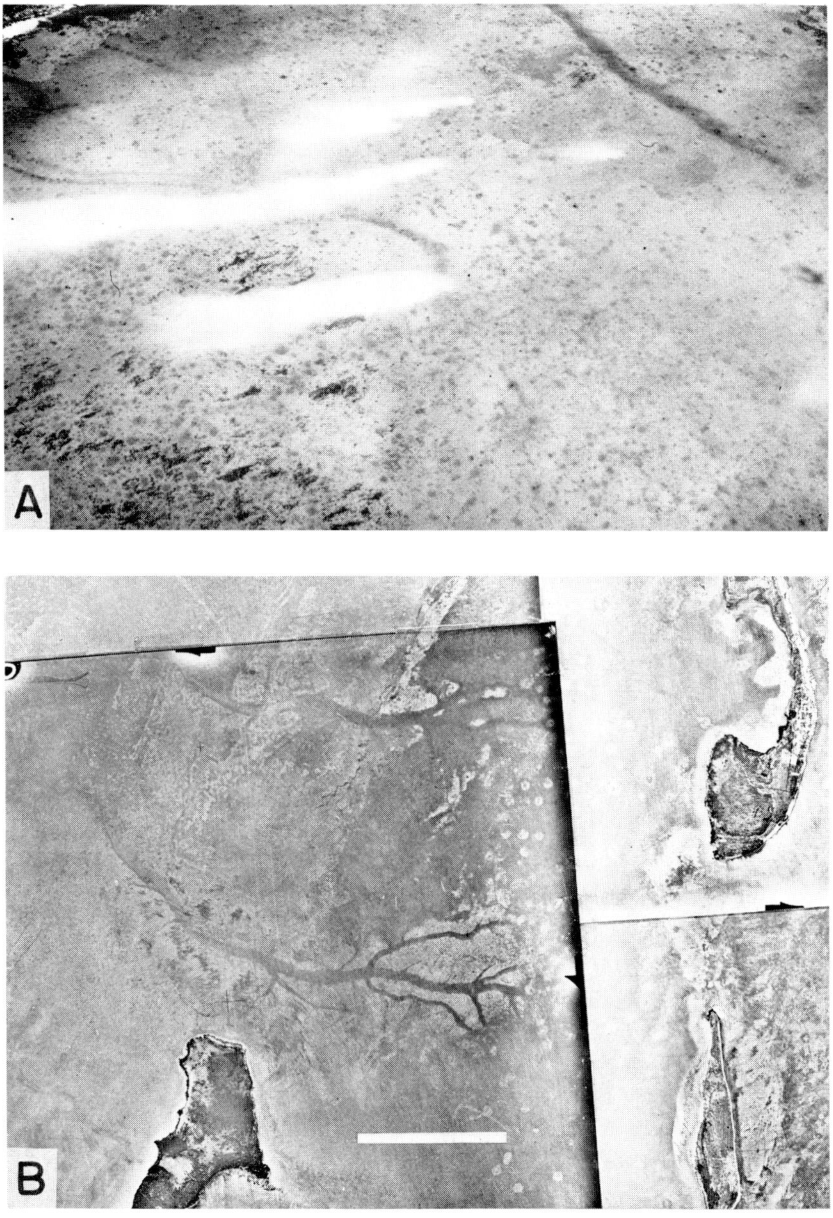

FIG. 7—Northern shelf lagoon.

A. "Whitings" near Hicks Cay. Oblique aerial photograph looking northwest. "Whitings" are elongate parallel with prevailing northeast wind, which blows from right side of photo to left. Boundary on windward side of each "whiting" is sharp and finely digitate, whereas leeward boundary is nondigitate and more gradational. "Whiting" in foreground is approximately 1,000 ft (305 m) long and 200 ft (60 m) wide.

B. Sediment-filled channels between Hicks Cay (lower left), Cay Caulker (upper right), and Cay Chapel (lower right). Vertical aerial mosaic; length of bar equals 1 n. mi. Light-colored areas are barren rock; darker areas are sediment-filled channels mantled by grass. Channels, cut into rock, formed during lower stands of sea level.

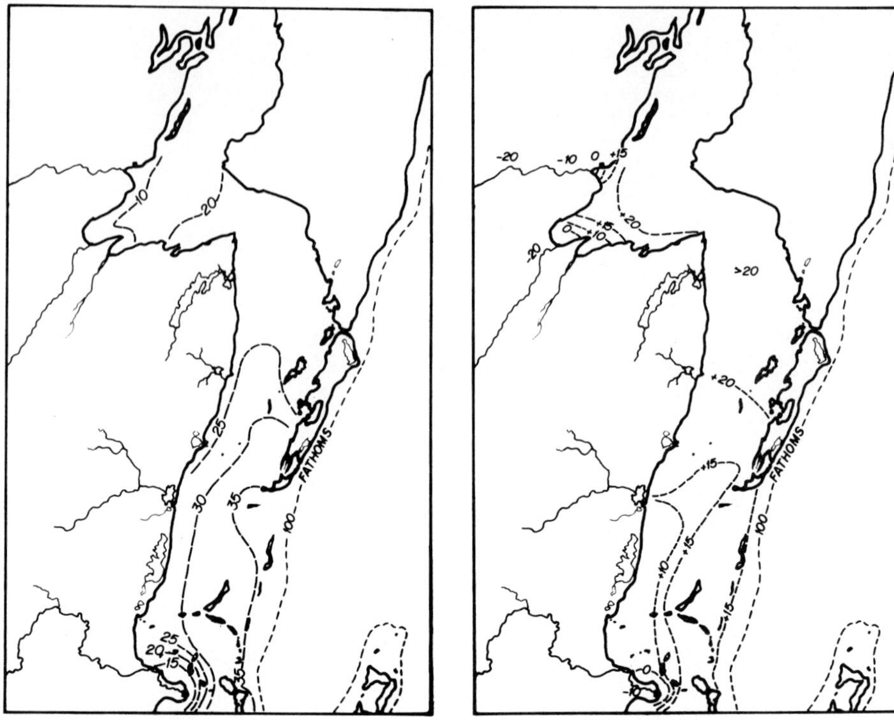

SURFACE-SALINITY (‰)
JULY-AUGUST 1961

ARAGONITE SATURATION (MV)
MID-AFTERNOON

FIG. 8—Water characteristics of northern Belize shelf (July and August, 1961).

A. Surface salinity (control points at each sample locality plus approximately 50 supplementary measurements).

B. Saturation of water relative to aragonite (control points at each sample locality plus 10 supplementary points).

Data and observations concerning salinity, temperature, carbonate saturation, and turbidity were gathered during the summers of 1961 and 1962. These data are summarized in Figure 8.

Salinity

The surface-salinity and bottom-salinity patterns illustrated in Figure 7 are based on more than 100 measurements made in July and August, 1961. Supplementary measurements made in 1962 indicate that these patterns are, in general, characteristic of the wet season. Surface waters are gradually diluted by freshwater runoff near the mainland and large cays (islands); however, rather high salinities persist well into Chetumal Bay. The distribution of surface salinities is largely the result of a complex interaction between freshwater wedges, northeast winds, and tides. The wedge effect is caused by less dense, low-salinity water overriding more dense, saline water—an effect that is most pronounced near rivers, where differences in salinity between surface and bottom waters may be as much as 14‰. In general, however, salinity differences between surface and bottom waters decrease to less than 1‰ a few miles from river mouths. Disruption of the freshwater wedges results from consistent winds, which almost daily disturb the water in the bay and lagoon sufficiently to cloud the water with suspended sediment.

Nothing is known of the salinities during the short but pronounced dry season from February through April. It is almost certain that Chetumal Bay becomes an evaporating basin as a result of decreased runoff coupled with continued evaporation by trade winds. Indirect evidence of this is afforded by the fact that salt water has been observed as far as 50 mi (80 km) upstream in the New River during the dry season (Dixon, 1958). The occurrence of halophilic red mangroves 50 mi (80 km) or more upstream in the Rio Hondo suggests that this saltwater penetration is not a short-lived occurrence.

Temperature

Average surface oceanic temperatures off northern Belize vary from 25.5°C in February to 28.5°C in August (Fuglister, 1947). During July 1961 and August 1962, surface temperatures near the reef approximated 29°C. Temperatures in the lagoon and bay were about 30°C, and daily warming of about 2°C was noticeable. Bottom temperatures were only 1/2° or, at most, 1° cooler than surface temperatures.

Carbonate Saturation

During the 1961 field season, a Weyl carbonate saturometer was used at all sediment stations. This instrument is basically a sensitive pH meter, which records changes in the hydrogen-ion concentration of an aqueous solution as carbonate is either dissolved or precipitated in the solution (Weyl, 1961). If a solution is supersaturated with respect to carbonate grains introduced into the solution, the grains will act as nuclei, and precipitation of carbonate will take place. Removal of carbonate ions from the solution increases the concentration of hydrogen ions, and consequently the potential of a pH electrode surrounded by the carbonate will become more positive.

The sensitivity of the instrument decreases with increasing pH (increasing amounts of the CO_2 in solution are present as the carbonate ion, rather than the bicarbonate ion); however, in the pH range of seawater (pH 7.5-8.3), the rate of change in sensitivity is negligible (Weyl, 1961, p. 43). Thus, if the pH of the waters measured is fairly uniform, direct comparisons of relative saturation of water samples may be made by comparing pH changes produced by the addition of carbonate to different water samples.

Saturometer response curves show that 2-60 minutes are required for equilibrium pH to be obtained (Weyl, 1961; Chave et al, 1962). Field conditions necessitated limiting saturation readings to 4 minutes. Within this time span the greatest reproducibility (±2 mv) was obtained at 3-minute intervals; the greatest rate of pH change occurred in the first 2 minutes. For this investigation a *saturometer value* for a water sample is defined as the pH change, expressed in millivolts,[3] 3 minutes after the introduction of calcium carbonate into the water sample. Saturation values for a given water sample vary according to mineralogy, impurities, grain size, and sample preparation (Weyl, 1961, p. 44). For this reason, an aragonite standard consisting of dry Bahamian ooids was chosen. The amount of oolite introduced in each water sample was sufficient to surround the glass electrode completely. In addition to relative evaluation of interstation water-saturation values, the intrastation saturation relation between a given sediment and the overlying water mass was measured by placing sediment instead of oolite into the saturometer.

[3]Since a positive millivolt value indicates supersaturation and a negative value indicates undersaturation, the writer prefers to use the raw millivolt readings rather than pH units. A calomel half cell was used, and delta millivolt readings therefore may be converted readily to delta pH by multiplying by -0.0169 (conversion tables may be found in Weast,

A significant diurnal variation of carbonate saturation has been well documented by Schmalz and Swanson (1969), in both Bermuda and at Eniwetok Atoll. This variation correlates with night and day and is apparently controlled by photosynthesis and respiration. During daylight, photosynthetic reactions remove carbon dioxide from seawater at a greater rate than respiration produces it; thus, the carbonate saturation increases (up to 10 mv above the average value); at night, respiration produces carbon dioxide and decreases carbonate saturation (up to 10 mv below the average value). Nearly all the saturometer measurements made in Belize were made between 10 a.m. and 6 p.m., when saturation was greatest. Two stations (no. 14 and no. 18) were measured at 9 a.m. and remeasured at midafternoon. In one case, the morning reading was higher; in the other, the midafternoon reading was higher. In both cases, the readings were within ±3 mv of the mean. Lateral differences in saturation were as much as 30 mv (Fig. 8B), greatly exceeding the expected diurnal variation for daylight hours.

In general, both bottom waters and surface waters were supersaturated with respect to subjacent sediment and Bahamian oolite. Within the fresh-water wedges less than a mile from the mouths of the Rio Hondo, New River, and Belize River, the surface water was undersaturated with respect to both the aragonite ooids and subjacent sediment, but bottom waters were supersaturated or barely undersaturated. This difference disappears within 2 mi (3.2 km) of the river mouths, even though pronounced salinity-wedge conditions persist.

With the exception of the above stations, there was no observed tendency for waters within a foot of the bottom to be closer to equilibrium with subjacent sediments. Within the reproducibility limits of ±2 mv, surface waters tended to give readings identical to near-bottom waters.

Saturation measurements of the water-sediment reaction vary with the mineralogy of the sediments. Bay waters tend to be more supersaturated with respect to indigenous low-magnesium calcite sediments than with the aragonitic oolite standard. Also, aragonite-rich and high-magnesium calcite sediments in the reef and lagoon produced saturation values comparable to those obtained with the Bahamian oolite standard. Muddy sediments, however, gave supersaturation readings 50-100 percent higher than those obtained with the oolite regardless of the mineralogy of the sediments. Without exception, these fine-grained gray sediments were reducing and evolved hydrogen sulfide when disturbed. The apparent saturation values of muddy sediment, then, are confounded in that they consist of hydrogen ions produced not only by carbonate reactions, but also by dissociation of hydrogen sulfide.

Saturometer data based on the standard of Bahamian oolite have a consistent areal pattern (Fig. 8B). Values obtained on and just seaward of the reef were similar to those obtained by the writer for waters of the Florida Straits and the northwest Gulf of Mexico. In contrast, saturation values are consistently higher inside Chetumal Bay. In the vicinity of the Bulkhead, they are only 20 percent higher than the reef values, but near Rocky Point they reach a maximum of 50 percent above the usual oceanic values. The cause of this pattern of increasing supersaturation is not fully understood. Measurements were consistent even though they were made at different times of day, at different temperatures, and during rain squalls. The interesting point is that the highest saturation values were recorded for brackish-water areas. Normally, one would anticipate a decrease in saturation to be associated with a decrease in salinity, other factors being equal. The reverse pattern that exists may be due in part to the high calcium content of runoff waters entering the bay. Few river data are available but, according to Dixon (1958), groundwaters in the limestones and marls of northern Belize tend to have

150-200 ppm of minerals that cause "temporary" hardness (calcium and magnesium). Water samples collected near the mouth of Chetumal Bay contain about 800 ppm calcium, which is almost twice as much as would be expected at the observed salinity. Other major anions are present in the expected amounts, indicating that the calcium is derived from runoff.

Little is known of the mechanism of interaction between carbonate grains and seawater. The equilibrium reaction is not adequately explained by prevalent theories of solution kinetics (Weyl, 1961; Schmalz and Chave, 1963). Regardless of what causes the supersaturated conditions, there can be little doubt that the carbonate grains in contact with the overlying waters tend to act as nuclei for precipitation. The saturometer reactions indicate that this applies not only to aragonitic Bahamian ooids but also to equivalent-sized calcite grains with magnesium carbonate contents of at least 10 mole percent.[4] During the dry season these supersaturated conditions are almost certainly intensified by evaporation.

Currents and Tides

Very little is known concerning water motions on the northern Belize shelf. According to the *Sailing Directions* (U.S. Navy Hydrographic Office, 1952, p. 187), the spring tidal range is only 1.5 ft (0.5 m), but during northers a lowering of 2.5 ft (0.8) is common. Metzgen (cited *in* Stoddart, 1963, p. 28) noted that in 1920 at Chetumal, Mexico, and at Corozal, Belize, about 10 mi (16 km) southwest, the sea receded several miles during a hurricane. This horizontal distance corresponds to a drop of at least 7 ft (2.2 m). Thus it is obvious that wind-induced currents are far more important than tide-induced currents. The addition of river waters to the bay produces a net flow from the bay into the lagoon. The trade winds modify this outflow by tending to push the waters to the southwest. Near shore the winds force the water southward, creating a persistent southerly longshore drift.

Strong currents associated with tidal rise and fall probably are sufficient in restricted channels behind the reef to sweep the bottom clean and maintain tidal channels in mud bars. Near the mouth of Chetumal Bay, depositional and floral lineations, plainly visible on aerial photographs (Fig. 9), indicate the flow pattern of bottom currents.

Turbidity

Underwater visibility ranges from over 50 ft (15.2 m) in the reef area to almost zero in Chetumal Bay. In the early morning, visibility is about 10 ft (3.0 m) in the bay, but by midafternoon the rising wind velocity has increased turbidity so that the waters are milky and visibility is reduced to less than 2 ft (0.6 m). Patches of white water, or whitings, were observed in flights over the lagoon in 1963 (Fig. 7A); however, none were noted during field sampling in 1961 and 1962.

SEDIMENT PROPERTIES

Sediments and sedimentary rocks have three fundamental properties: (1) texture, (2) composition, and (3) structure (Krumbein and Sloss, 1963, p. 93; Pettijohn, 1957, p. 13, 99, 157). Sedimentary texture traditionally consists of the size, shape, and arrangement of grains (Pettijohn, 1957, p. 13); however, with the notable exception of Dunham (1962, p. 108), texture in carbonate sediments has come to mean grain-size

[4]Calcites containing 10 mole percent magnesium carbonate are approximately equivalent in solubility to aragonite according to data presented by Chave et al (1962).

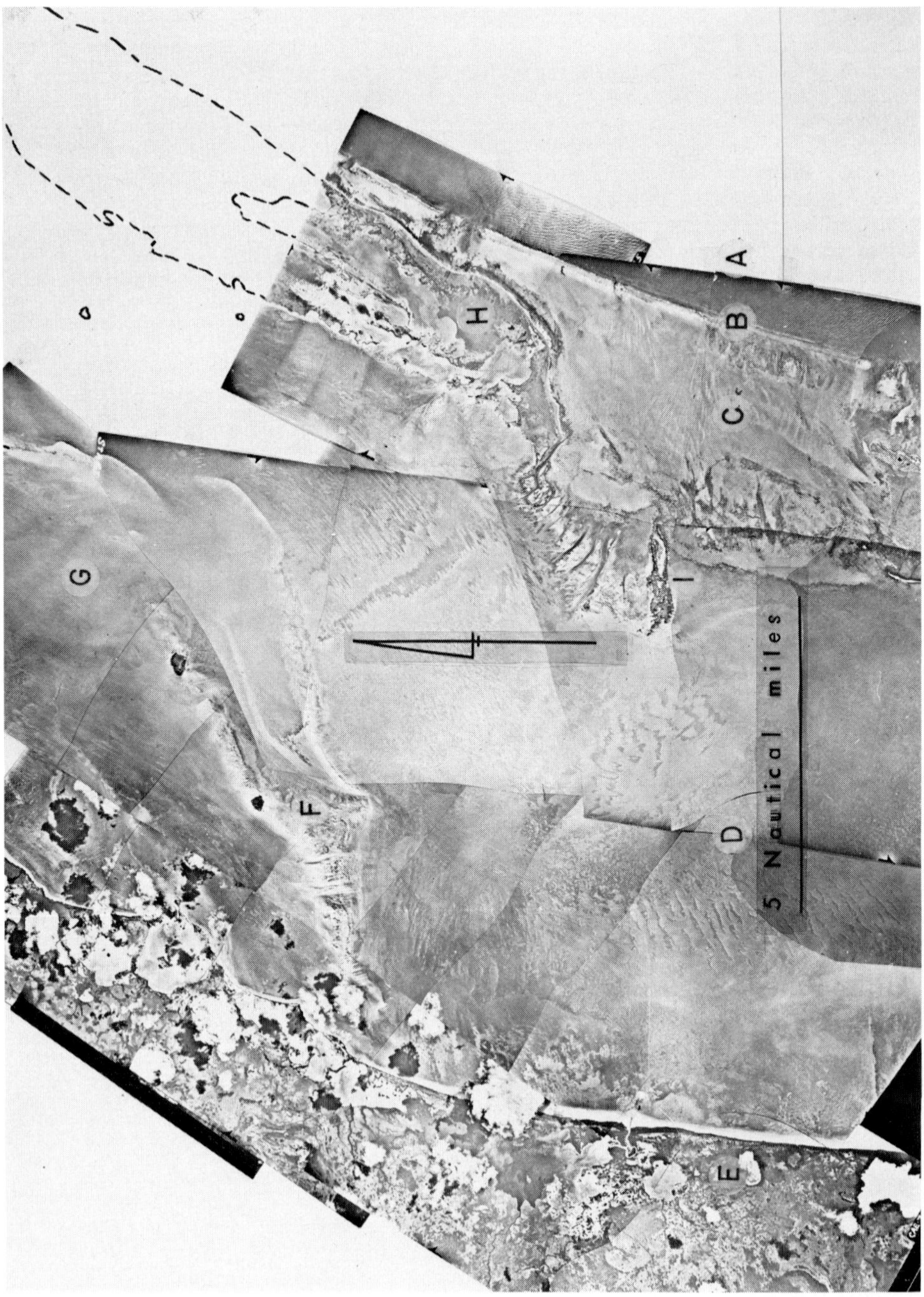

FIG. 9—Northern shelf lagoon near Bulkhead. Vertical aerial mosaic of southern Ambergris Cay and vicinity; arrow indicates north. Note lineations of floor of lagoon. These are believed to be current oriented. Dark areas are densely vegetated sediment of exposed rock. Letters designate following features: **A.** Caribbean Sea, **B.** barrier reef, **C.** backreef lagoon (outer shelf lagoon), **D.** inner shelf lagoon, **E.** mainland, **F.** Bulkhead mud-shoal complex, **G.** lower Chetumal Bay, **H.** Ambergris Cay, **I.** Cangrejo Cay.

parameters (Ham and Pray, 1962, p. 8). Unlike clastic sediments, the grain size of a carbonate sediment is only partially controlled by current intensity at the site of deposition. Coarse carbonate grains may be of local origin or may even have formed *in situ* as a product of chemical and/or organic processes. Thus, measures of finer grain sizes generally have been used as current-strength indicators in carbonate sediments (Ham and Pray, 1962, p. 8).

A further consequence of the local origin of many carbonate sediments is that their composition is related to the environmental conditions under which their constituent particles formed. For example, the environment is reflected not only in the distribution and abundance of the types of organisms that produce carbonate skeletons, but also is reflected in the distribution and abundance of inorganically produced grains such as oolite.

In its broadest sense, sedimentary structure includes large-scale "...features that in general are seen or studied best in the outcrop rather than hand specimen or thin sections" (Pettijohn, 1957, p. 157) and small-scale "...original lithologies or fabrics that are visible to the naked eye or with a hand lens...and are formed *on* or *just beneath* the depositional surface" (Moore and Scruton, 1957, p. 2724). Sedimentary structure reflects the shifting balance between sources of sediment, processes of deposition and diagenesis, and rates of deposition (Moore and Scruton, 1957, p. 2743).

Studies of carbonate sediments have stressed grain size and constituent-particle composition (e.g., Ginsburg, 1956; Cann, 1962; and Purdy, 1963a, b), but sedimentary structures are mentioned rarely. It is the purpose of the present study to consider all three features in order to determine how they reflect basic characteristics of depositional processes, grain provenance, and depositional rates.

Laboratory Procedure

Generalizations resulting from the study of Holocene sediments commonly are limited in their application to rocks by basic differences in techniques used in examining the sedimentary properties. In this investigation, an attempt was made to use techniques that are also suitable for the study of sedimentary rocks.

Sample Preparation

The sediment samples were dried to constant weight under heat lamps at less than 35°C. A subsample was separated with a Jones-type sample splitter and weighed. Sand-size grains were separated from the very fine sand and mud (silt + clay) fractions by wet-seiving through U.S. Standard 120-mesh (1/8 mm) and 230-mesh (1/16 mm) screens. Sand grains retained on the screens were dried and weighed, and the percentage of mud-size sediment was determined by weight loss. The initial drying process produced artificial aggregates in muddy sediments that might be confused with natural aggregates. As noted by Ginsburg (1956, p. 2423) and Purdy (1961, p. 30), treatment of the sample with 30 percent hydrogen peroxide destroys artificial aggregates and nonindurated fecal pellets. Unfortunately, however, this concentration of hydrogen peroxide almost totally destroys some common grain types, notably ostracods and certain pelecypod shells with high organic content. By trial and error, it was found that 4 percent hydrogen peroxide produced the required disaggregation of artificial aggregates in 1-2 days without appreciably destroying skeletal grains. Nonindurated fecal pellets were destroyed, however. To mitigate misidentification errors further, thin sections of artificial aggregates were studied petrographically.

The sand fraction of each sample was mixed thoroughly with a viscous polyester resin, which hardened to form the matrix of an artificial rock. Effects of differential grain settling in the plastic were countered (1) by stirring the mixture until hardening commenced, and (2) by cutting thin sections normal to the direction of any potential differential settling. Thin-section petrographic methods were used to determine types of constituent particles and to estimate the volumetric amount of each grain type.

Core Study

Sedimentary-fabric studies of the cores were successful only after a "freeze lithification" in dry ice. Cutting and careful scraping of soft unfrozen cores produced artificial "homogeneous" fabrics, whereas the freezing treatment disclosed the presence of several important fabric features. After freezing in dry ice, the cores were sliced longitudinally with a metal-cutting band saw. The thin slurry of cuttings, which formed and refroze during cutting, was removed by scraping to expose sedimentary structures. All the cores were photographed in black and white and in color; composition, color, texture, and minor structures were noted.

Carbonate Mineralogy

Carbonate mineralogy of the surface sediments and selected core intervals was determined to evaluate the possibility of diagenetic alteration. Subsamples of each sample were ground to sizes finer than 200 mesh (U.S. Standard) and analyzed with an X-ray diffractometer using Cu Kα radiation. Relative amounts of calcite and aragonite were determined using Chave's (1954) calibration curves, which are based on aragonite from *Strombus gigas* and low-magnesium calcite from *Ostrea lurida*. During the grinding process, pressures in excess of 3,000 bars cause partial conversion of calcite to its high-pressure polymorph aragonite (Burns and Bredig, 1956; Dachille and Roy, 1960). To determine if significant error was introduced by grinding, a sample of pure calcite (reagent-grade calcium carbonate) was ground continuously in an automatic mullite mortar and pestle. No measurable aragonite was produced until more than 10 hours had elapsed, and 80 hours were required for 20 percent inversion. Sample grinding preparatory to X-ray diffractometer analysis required no more than 5 minutes; therefore, it is concluded that the reported aragonite percentage for each sample is free of error ascribable to grinding.

In addition to aragonite inversion to calcite, lithification processes tend to change magnesium-rich calcite to relatively pure calcite or perhaps dolomite. Experimental work (Goldsmith et al, 1955) indicates that calcite containing more than 4 percent magnesium carbonate should be unstable at pressure and temperature conditions on the surface of the earth. Calcite containing more than 4 weight-percent magnesium carbonate is appropriately called "high-magnesium calcite" (Chave, 1954, 1962). The magnesium content of the calcite in the sediments was measured by the progressive shift of the major calcite diffraction peak toward smaller lattice spacings with increasing substitution of the smaller magnesium ions into calcium sites in the crystal (Chave, 1952). The calibration curve of Chave (1954) was used to compute the weight percentage of magnesium carbonate in the calcite lattice. Quartz, if naturally present, and silicon metal were used as internal standards.

The calcite in skeletal-rich samples is derived from organisms that may have different amounts of magnesium in their calcite (e.g., low-magnesium calcite from oysters and high-magnesium calcite from miliolid foraminifers). In practice, however, the calcites usually merged to form a symmetrical diffractogram curve. A few of the calcite peaks from the samples examined were asymmetric or even bimodal, but such curves invariably were restricted to samples containing fragments derived from underlying or nearby limestone.

Duplicate X-ray determinations of the weight percentage of aragonite relative to calcite were made for four samples. The maximum observed difference was 5 percent and the average observed difference was 3 percent. This reproducibility is considerably higher than the accuracy of ±10 percent assigned by Lowenstam (1954, p. 288) to the same calibration curve. The greatest potential source of error, of course, lies in assuming that all calcites have a similar ability to diffract X-rays. Taft (1962) showed that the diffraction intensity produced by peneroplids, which contain approximately 14 weight-percent magnesium carbonate, is only 75 percent of that of relatively pure calcite from mollusks. Thus a sample with a true composition of 25 percent aragonite and 75 percent peneroplid calcite would yield an apparent calcite value of 66 percent on a low-magnesium calcite-versus-aragonite calibration curve. Because calcites in recent sediments contain a spectrum of magnesium values, calcite values as much as 10 percent too low may result. For the sediments used in this study, the maximum error attributable to high-magnesium calcite is probably closer to 5 percent. The carbonate mineralogy for all sample stations is given in Appendix B.

Grain Size

The determination of grain-size distributions has long been considered a useful tool in interpreting transportational and depositional aspects of clastic sedimentation. Grain size in carbonate sediments, however, is affected not only by transportational and depositional conditions, but also by *in situ* changes such as physicochemical precipitation and solution or skeletal genesis and organic trituration of grains. Thus, the grain-size parameters of sorting, skewness, and kurtosis do not have the same environmental meaning for carbonate sediments as they do for terrigenous noncarbonate sediments.

Several workers studying Holocene carbonate environments (Ginsburg, 1956; Cann, 1962; Purdy, 1963a, b) have recognized these difficulties and have successfully used measures of the finer grained fraction of carbonate sediments as a crude current-strength index. This approach also has proved effective in the study of ancient carbonate rocks (e.g., Jaanusson, 1952; Klovan, 1963); therefore, a similar approach was used in the present study. The weight percentage of very fine sand (1/8-1/16 mm) and silt plus clay (<1/16 mm) was determined for each sample. The size finer than 1/16 mm was used for practical reasons. Particles smaller than 1/16 mm usually cannot be identified as to grain type. Several workers (Ginsburg, 1956, p. 2424; Purdy, 1963a, p. 339; Feray et al, 1962, p. 27) have noted difficulty in recognizing grains smaller than 1/8 mm; however, an abundance of miliolids and ostracods in the very fine sand fraction dictated the use of 1/16 mm as an upper size limit for mud. For foram-rich samples, use of 1/8 mm as the upper limit of mud meant classifying as much as a third of the recognizable grains as mud in 10 percent of the samples and as much as a fifth of the recognizable grains as mud in 25 percent of the samples. In order to compare grain-size data with those of earlier workers, who used 1/8 mm as the upper size limit of mud, the fraction finer than 1/8 mm was also measured. Insertion of a 1/8-mm sieve also facilitated sample processing by preventing sieve clogging.

Duplicate determinations of the weight percentage of grains smaller than 1/8 mm and smaller than 1/16 mm were made for five samples. The maximum observed difference in these duplicate determinations was 4 percent, and the average observed difference was 1 percent. Thus the grain-size determinations are probably accurate within ±4 percent of the recorded value.

Constituent-Grain Composition

Grain-size parameters are important clues to the clastic petrologist in deducing conditions of deposition, but the distribution and abundance of grain types are of paramount importance to the carbonate petrologist in identifying relevant attributes of the depositional environment. Until recently, few definitions of carbonate grain types based on thin-section petrography have been available. Descriptions directly applicable to this study have been published by Illing (1954), Ginsburg (1956), Hoskin (1963), and Purdy (1963a). The descriptions by these workers and a reference thin-section set of skeletons of common organisms facilitated the assignment of sedimentary constituents among 15 tabulation categories. The selection of tabulation categories was dictated more by the relative importance of a constituent than by the ability to distinguish it. Eleven major skeletal categories and four major nonskeletal categories were used. The following categories are described briefly to emphasize their diagnostic characteristics.

Skeletal Grains

Coralline algae—Close-packed rectangular cell walls distinguish this type of red alga (Fig. 10A) regardless of whether it is an encrusting or articulated form—although the articulated forms greatly outnumbered the encrusting forms. When fresh, the cell walls have a pale yellow-brown color in plane-polarized light. The walls are about 3μ thick and consist of 1 by 3μ crystals that have their long axes normal to the cell openings. Neveu (1961) and Lanlignel and Schmitt (1961) have published electron photomicrographs of *Lithothamnium* sp. showing these crystals in detail. The walls commonly are altered so that they appear dusty to the extent that their original crystal architecture is not discernible.

"Melobesia"—The red algal subfamily Melobesiae, which includes many multilayered coralline algae, also includes two genera (*Melobesia* and *Fosliella*) which have only a few cell layers (Taylor, 1960, p. 377ff). *Fosliella*, which formerly was included in *Melobesia*, is recognized by distinctive reproductive structures and hair-bearing cells. Because few such structures were encountered in the random orientations of grains within thin sections, both genera were tabulated as "*Melobesia*." It should be noted, however, that Taylor (1936, p. 123) reported *Fosliella farinosa* from Silk Cay, Belize; therefore, it is quite likely that many of the organisms tabulated as "*Melobesia*" are *Fosliella*.

In thin-section "*Melobesia*" appears yellowish brown in transmitted plane-polarized light, but is chalky white in reflected light. Between crossed nicols it appears dark with some orange highlights that are similar to those noted in peneroplids. The crustose thallus is usually less than a millimeter wide and is only a few tenths of a millimeter thick. Sections which are cut parallel with the crust near its base lack a regular cell arrangement, but sections slightly higher up show a reticulate pattern (Fig. 10B) that superficially resembles a cheilostome bryozoan. A few of the rectangular cells have small cover cells which resemble bryozoan opercula. Oltmanns' (1922-1923, p. 267) excellent illustrations of cover cells, however, permit one to distinguish readily between the two. A brown color in transmitted plane light, relatively thick walls (4-6μ), relatively small cells, and the absence of true opercula distinguish "*Melobesia*" from bryozoans. "*Melobesia*" also differs from encrusting foraminifers in cell arrangement and in the relatively small size of its cells.

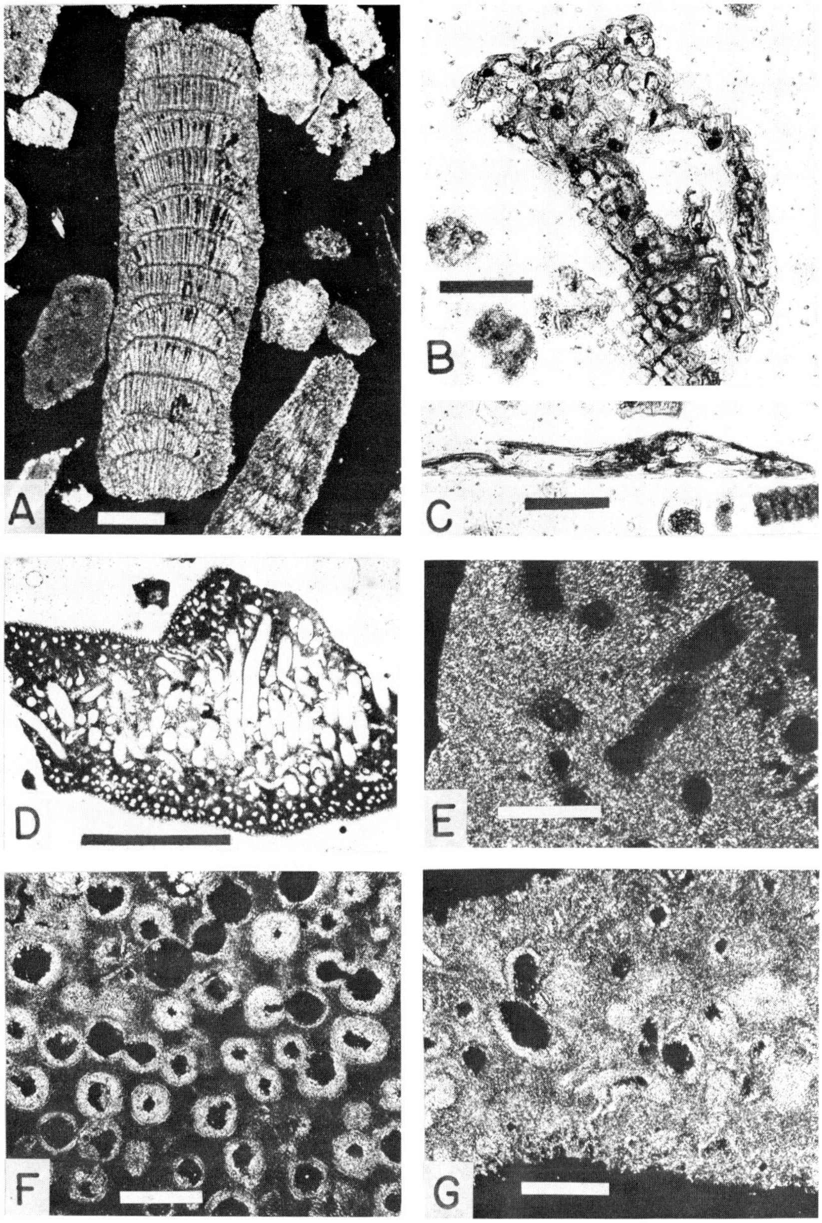

FIG. 10—Constituent grains—algae.

A. Red alga with long axis in plane of thin section. Most cell spaces are filled with calcite. Another red alga with long axis oblique to plane of thin section is located in lower right corner. Crossed nicols; bar length 100μ.

B. *"Melobesia"* with flat attachment surface in plane of thin section. Note rectangular arrangement of cells. Plane-polarized light; bar length 100μ.

C. *"Melobesia"* with flat attachment surface normal to plane of thin section. Plane-polarized light; bar length 100μ.

D. *Halimeda*. Typical structure of unaltered plate. Plane-polarized light; bar length 1 mm.

E. *Halimeda*. High magnification of very thin section. Note abundance of clay-size aragonite needles in wall. Crossed nicols; bar length 100μ.

F. *Halimeda* with void fillings. Acicular aragonite needles partly fill utricles. Note indistinct border between some infilled utricles and rest of grain; this indicates alteration of grain wall. Crossed nicols; bar length 100μ.

G. *Halimeda* partly altered to cryptocrystalline carbonate. Light-colored spots are infilled utricles; most of utricle walls are composed of cryptocrystalline material. Crossed nicols; bar length 100μ.

Thin sections cut perpendicular to the crust show a flat attachment surface (Fig. 10C). Under a binocular microscope, these flat surfaces commonly exhibit a fine rectangular meshwork having the same dimensions as the leaf cells of *Thalassia* grass. The epiphytic character of "*Melobesia*" from the tropical Atlantic Ocean has been noted by many writers (e.g., Thorne, 1954; Taylor, 1954, 1960). Although "*Melobesia*" rarely constitutes more than a few percent of the sand-size material, it invariably was found in relative abundance in sediments mantled by dense stands of *Thalassia*. Thus, even though marine grass is unlikely to be fossilized, "*Melobesia*" provides a valuable index of its existence.

Halimeda—In thin sections, disarticulated plates of this abundant codiacean green alga are a translucent brown in transmitted light. The plates are characterized by internal tubes (utricles) and small pits on the plate exterior. The latter appear in thin section as fine scallops on the edge of the plate (Fig. 10D). Under high magnification between crossed nicols, the plates appear granular (Fig. 10E). The plates commonly are altered to aggregates of randomly oriented, clay-size cryptocrystalline carbonate, which appears grayish brown in transmitted light (Fig. 10G). Alteration seems to begin near the periphery of the tube walls. In early stages of alteration the loss of brown color is accompanied by the appearance of dark brown clots (transmitted light) which effervesce in hydrogen peroxide, indicating that they are concentrations of organic matter. The alteration progresses inward until the entire particle consists of relatively large crystals which measure up to 5 by 20μ. Plate alteration to cryptocrystalline material is usually preceded or accompanied by the growth of aragonite needles into the tubes (Fig. 10F).

Miliolidae—This is the most abundantly represented and widespread foraminiferal family in northern Belize. Miliolid tests are readily identifiable by their chamber arrangement (Fig. 11A). In transmitted light the tests are transparent and are yellow to light brown in color; in reflected light, they appear chalk white. Between crossed nicols the test walls appear dark except for a line about 1μ thick on the wall surfaces. In electron photomicrographs, unaltered miliolid tests can be seen to consist of a mosaic of unoriented to poorly oriented calcite crystals 1/2 by 1μ in size or smaller (Hay et al, 1963). The surface of the test wall consists of slightly larger rhombs, which are well aligned in a shinglelike pattern. The dark appearance of the walls noted between crossed nicols in the optical microscope probably results from light scattering and compensation produced by transmission of polarized light through a relatively thick section (20-25μ) containing many layers of poorly oriented crystals.

Commonly the test walls are altered to semiopaque cryptocrystalline material which is grayish brown in transmitted light. Between crossed nicols the cryptocrystalline material seems to consist of crystals 2-5μ long (Fig. 11B). Individual crystal boundaries, however, cannot be resolved because of their small size, and what appears as single crystals possibly may be similarly oriented aggregates of submicroscopic crystals. The process of alteration may be recrystallization involving solution and reprecipitation of larger crystals.

Transformation of the test wall is commonly, but not invariably, accompanied by the growth of stubby calcite crystals that grow approximately normal to the wall surface (Fig. 11B, C). These fillings were noted in a few otherwise unaltered tests. The mineralogy of these crystal fillings was inferred from their blocky growth habit and their resemblance to X-ray-identified crystals growing inside peneroplids.

Peneroplidae—Tests of this family of Foraminifera are distinguished by their relatively large size, an orange-brown color in transmitted light, and their peculiar chamber arrangement (Fig. 11D). The presence

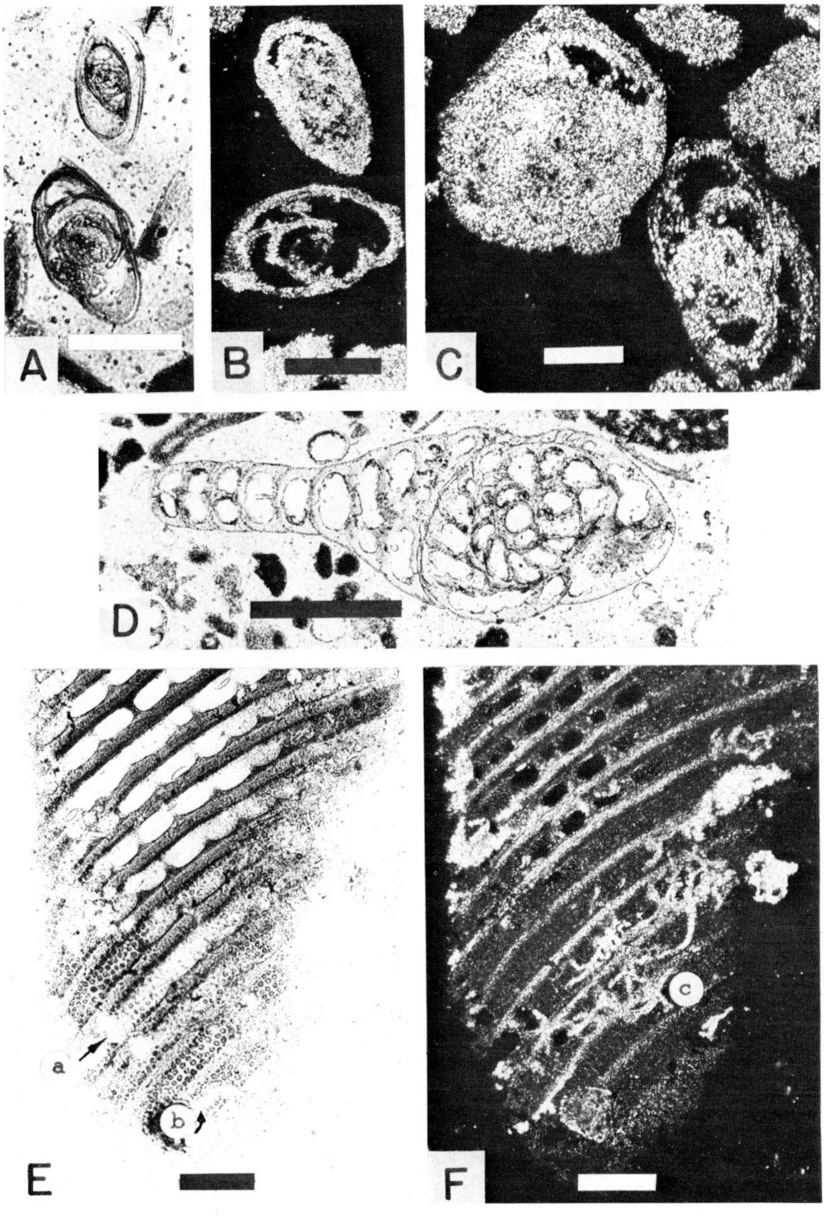

FIG. 11– Constituent grains– miliolid and peneroplid foraminifers.

A. Miliolids. Typical tests. Plane-polarized light; bar length 100μ.

B. Miliolids with altered wall structure. Lower test is partly altered to cryptocrystalline carbonate; upper test is completely altered and has some calcite crystals filling its chambers. Crossed nicols; bar length 100μ.

C. Miliolid almost completely altered to cryptocrystalline carbonate. Note that calcite crystals have partly occluded the chambers. Crossed nicols; bar length 100μ.

D. Peneroplid showing typical shape and transparency. Plane-polarized light; bar length 1 mm.

E. Peneroplid. Plane of thin section is nearly parallel with flat surface of test. Note widely spaced pits which are 6μ in diameter. A few large algal borings, **a**, and some smaller borings, **b**, are visible in test near bottom of photo. Plane-polarized light; bar length 100μ.

F. Same as **E.** Crossed nicols. Note that small borings, **c**, appear cryptocrystalline. Bar length 100μ.

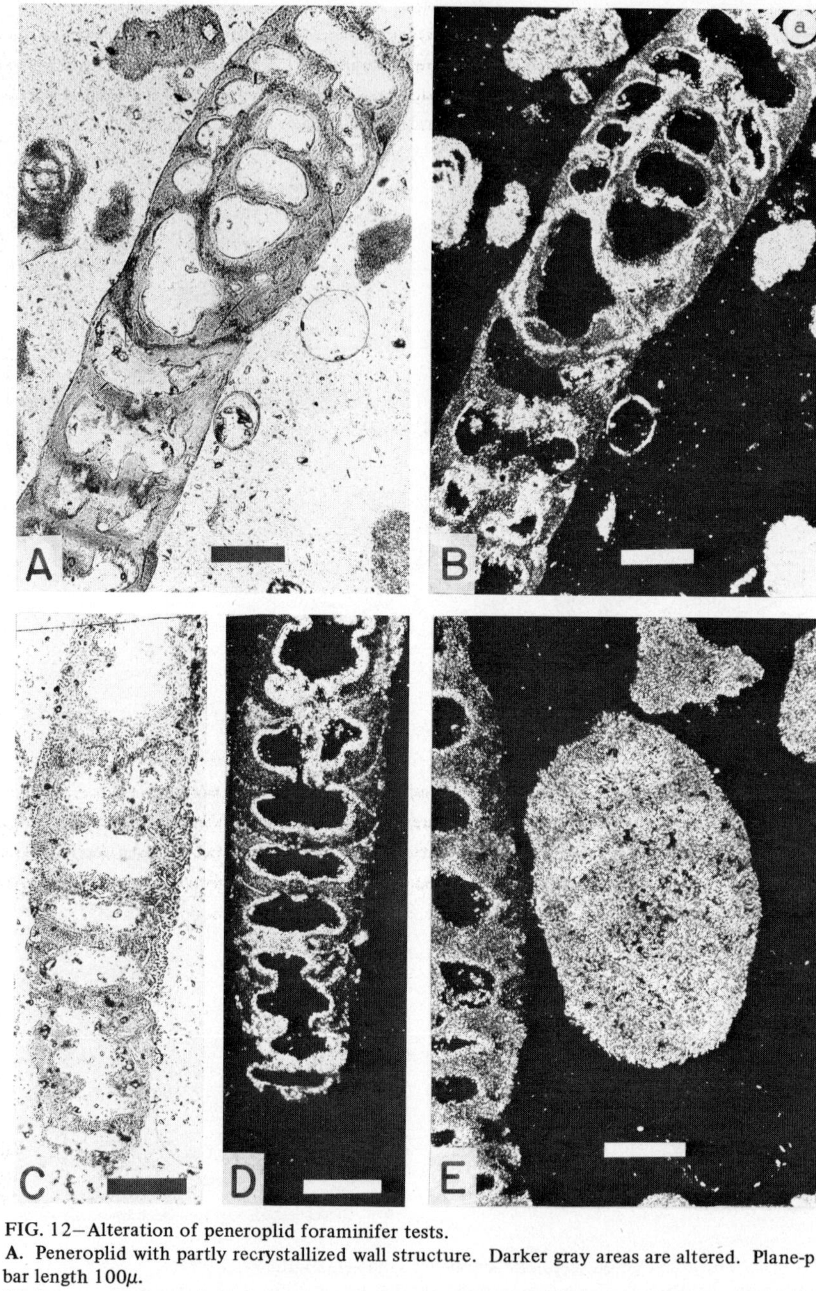

FIG. 12—Alteration of peneroplid foraminifer tests.

A. Peneroplid with partly recrystallized wall structure. Darker gray areas are altered. Plane-polarized light; bar length 100μ.

B. Same as **A.** Crossed nicols. Note that approximately one third of test wall is cryptocrystalline. Minor calcite void filling is visible in some chambers such as **a.** Bar length 100μ.

C. Peneroplid. Note abundant borings on right side of photo. Plane-polarized light; bar length 100μ.

D. Same as **C.** Crossed nicols. Calcite crystals line chambers, but walls are unaltered (compare with Fig. 12C). Note that borings do not appear cryptocrystalline. Bar length 100μ.

E. Recrystallized peneroplid in which filled chambers appear lighter than darker material of cryptocrystalline test wall because of relative coarseness of calcite fillings. Peneroplid on left side of photo is partly recrystallized to cryptocrystalline material. Crossed nicols; bar length 100μ.

of pits 1-7μ wide on unworn specimens (Fig. 11E) is characteristic of most genera (Hofker, 1950, 1951a, 1951b) and is useful in separating juvenile peneroplids from miliolids. *Archaias* and *Peneroplis* are important genera in northern Belize (Wantland, 1967). Between crossed nicols, the test wall generally appears dark ("isotropic"), but parts of the test commonly display undulate or fan extinction of oriented layers of submicroscopic calcite crystals. Tests in thin sections cut parallel with the axis of coiling commonly have an "anisotropic" rim and greatly resemble miliolids in wall structure. However, electron micrographs by Hay et al (1963) indicate that this resemblance is superficial, and there is apparently no outer shinglelike layer of crystals such as that found in miliolids. Like miliolids, peneroplids are chalky white in reflected light.

Peneroplids were noted in various stages of transformation to cryptocrystalline material (Fig. 12B). The transformation results in a loss of brownish color and imparts a more translucent gray appearance to the test in transmitted light. Alteration is gradual and, in early stages, selective (Fig. 12B). Between crossed nicols, cryptocrystalline tests appear to consist of randomly oriented crystals 2-5μ in diameter. Unaltered walls are composed of unoriented to poorly oriented crystals which average 1/4 by 1/2μ (Hay et al, 1963).

The cryptocrystalline appearance of the test wall is nearly everywhere associated with small borings into the test carbonate (Figs. 11E, 12C). These borings generally are 2-4μ in diameter and greatly resemble those of algae that penetrate mollusk shells in Bermuda (Humm and Taylor, 1961). Between crossed nicols the borings commonly appear cryptocrystalline (Fig. 13F).

Alteration of the test wall to cryptocrystalline material is usually associated with infilling of chambers by 5-15μ-long, high-magnesium calcite crystals, which were identified by X-ray (Fig. 12D). However, both wall alteration and crystal infilling do occur independently. The end products of wall alteration and chamber infilling are cryptocrystalline grains which, if abraded, can be identified only uncertainly as altered peneroplids (Fig. 12E).

Other Foraminifera—Although a great variety of other types of Foraminifera can be distinguished, they are not abundant and cannot consistently be differentiated from each other. The classification of Foraminifera by wall structure proposed by Reiss (1958) is an aid to identification but requires oriented sections for effective use. During tabulation the following general types of Foraminifera were noted as most abundant: colorless perforate (e.g., *Streblus* and *Elphidium*), colorless agglutinated (e.g., *Valvulina*), brown agglutinated (e.g., *Quinqueloculina*; Fig. 13A), and pink encrusting (e.g., *Homotrema*) foraminifers.

Aggregates bound by distinctly cellular brown carbonate were separated from grapestones and called "organic aggregates" by Purdy (1963a, p. 345) and were identified as an agglomerate foraminifer by Baars (1963, p. 108). Organic aggregates from northern Belize resemble the foraminiferal grains illustrated by Baars; however, other organic aggregates were encountered which may be *Quinqueloculina* sp. (Fig. 13A). In general, organic aggregates were scarce and were tabulated as "other Foraminifera."

Corals and hydrozoans—Coral skeletons can be recognized in thin section by their organization of aragonite crystals into fanlike bundles called "sclerodermites" (Wells, 1956, p. F337). Aligned sclerodermites form trabeculae and appear to share a center of calcification (Fig. 13B). The hydrozoan *Millepora* has a similar structure but has many pores

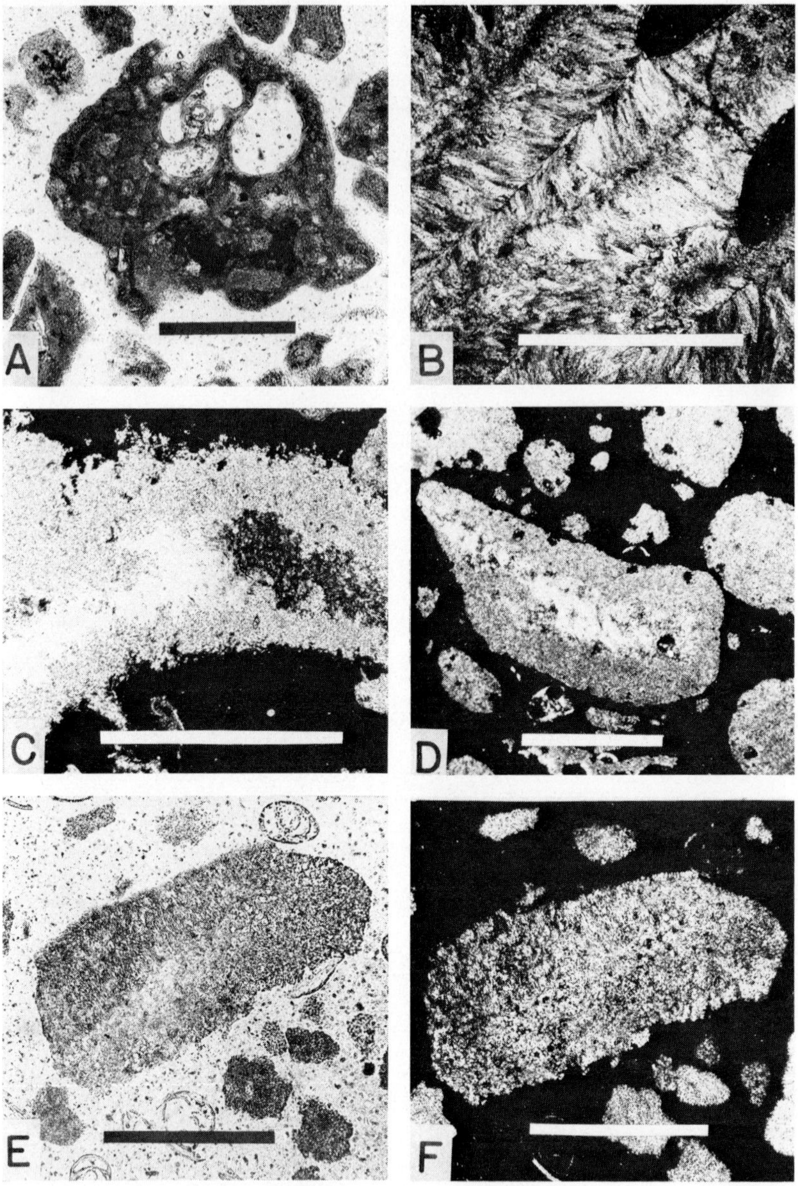

FIG. 13—Skeletal grains—other foraminifers, corals, and mollusks.

A. *Quinqueloculina* sp. showing initial quinqueloculine shape and later tendency to become irregular. Aggregate test consists of detrital carbonate grains cemented by an excess of brownish cement. This may explain origin of one type of organic aggregate. Plane-polarized light; bar length 200μ.

B. Coral (*Manicina areolata*) showing sclerodermites arranged in rows to form trabeculae. Crossed nicols; bar length 1 mm.

C. Mollusk shell with cryptocrystalline rim. Note irregular surface of grain and irregular but distinct boundary between rim and unaltered shell. Under higher magnification, numerous borings, about 5μ in diameter, can be seen. Grains in upper right corner are cryptocrystalline. Crossed nicols; bar length 500μ.

D. Mollusk(?) which is cryptocrystalline. Under higher magnification, this cryptocrystalline grain appears highly bored. Note great similarity between bored grain and cryptocrystalline grains in upper right corner. In addition to 5μ borings, there are a few (algal?) borings up to 60μ in diameter. Surface of bored grain is smooth, suggesting that much shell material may have been abraded to form mud. Crossed nicols; bar length 500μ.

E. Mollusk shell reduced to delicate mesh by 4μ-diameter borings. Mollusk structure has been destroyed almost completely. Plane-polarized light; bar length 250μ.

F. Same as **E.** Crossed nicols. Note cryptocrystalline appearance of grain. Bar length 250μ.

penetrating the skeleton which give it a spongy appearance. Fragments of *Millepora* were not commonly encountered.

Mollusks—Pelecypods and gastropods cannot be separated consistently on the basis of shell structure; only their shape is diagnostic. Because only fortuitous sections of large mollusk fragments could be categorized, no tabulation distinction was made between gastropods and pelecypods. Bøggild (1930) has published excellent illustrations showing the great variety of molluscan microstructure. Depending on the orientation of the thin section and the species of mollusk, grains may be recognized as mollusks by their shape and/or shell structure. Generally, mollusks appear either structureless or laminar in unpolarized light, but between crossed nicols the "structureless" shells are seen to consist of coarse crystals. Some common mollusks such as *Caecum* incorporate large amounts of organic material within their shell and superficially may resemble arthropod tubes in both color and structure. Between crossed nicols, however, the characteristically coarse crystals of mollusk structure can be seen.

Mollusks with partly altered wall structure that in transmitted light resembles grayish to brownish gray cryptocrystalline material are common. The boundary between unaltered shell and the "cryptocrystalline" shell is usually sharp. The "cryptocrystalline" material is represented by intensely bored shell material (Fig. 13C, D, E), whereas the unaltered shell contains few, if any, borings. These borings are 2-4μ in diameter and are similar in size and shape to those of the green algae *Ostreobium* and *Gomontia* (Humm and Taylor, 1961). Some "cryptocrystalline" grains manifest relicts of pelecypod microstructure between crossed nicols (Fig. 13D, E, F); consequently, it seems quite likely that continued boring and abrasional destruction of shell shape, coupled with recrystallization, can produce grains which are unrecognizable as altered mollusks.

Cryptocrystalline rims, which resemble carbonate-mud coatings, are almost identical with the "micrite envelopes" reported in ancient rocks by Bathurst (1966).

Ostracods—Paired valves of ostracods were recognized readily by the overlapping nature of the valves. Separate valves were identified by the presence of ornamentation, pores, straight hinge margins, or duplicature. The valves resemble juvenile pelecypods but, between crossed nicols, lack of molluscan microstructure is apparent. Fine lamellar microstructure, which has been cited by Cann (1962), Purdy (1963a), and Hoskin (1963) as diagnostic of ostracods, was rarely seen.

Miscellaneous skeletal grains—Identifiable skeletal grains, which were not abundant or which were found only locally, were tabulated in this category. These included diatoms, charophytes, sponge spicules, bryozoans, alcyonarian spicules, polychaete worm tubes, barnacles, and echinoderms.

Large benthonic diatoms were recognized by their characteristic box-like or saddlelike shapes and by their isotropic opaline skeletons. Among the genera illustrated by Patrick (*in* Ward and Whipple, 1959, p. 171-189), *Campylodiscus* and *Diploneis* were seen most commonly.

Charophyte oogonia were recognized by their shape and also by the thick, blocky calcite that comprises their walls (Fig. 14A). Charophytes were present locally near river mouths.

Siliceous sponge spicules were ubiquitous but not abundant. Monaxons of freshwater spongillids and tetraxons of brackish and marine sponges were identified by their shape and isotropism between crossed nicols. Calcareous sponge spicules were very scarce.

Bryozoans were recognized by their cell arrangement and their relatively large cell size as compared with red algae or Foraminifera.

Alcyonarian spicules have undulatory extinction and characteristic shapes, as illustrated by Bayer et al (1956, p. F187-227).

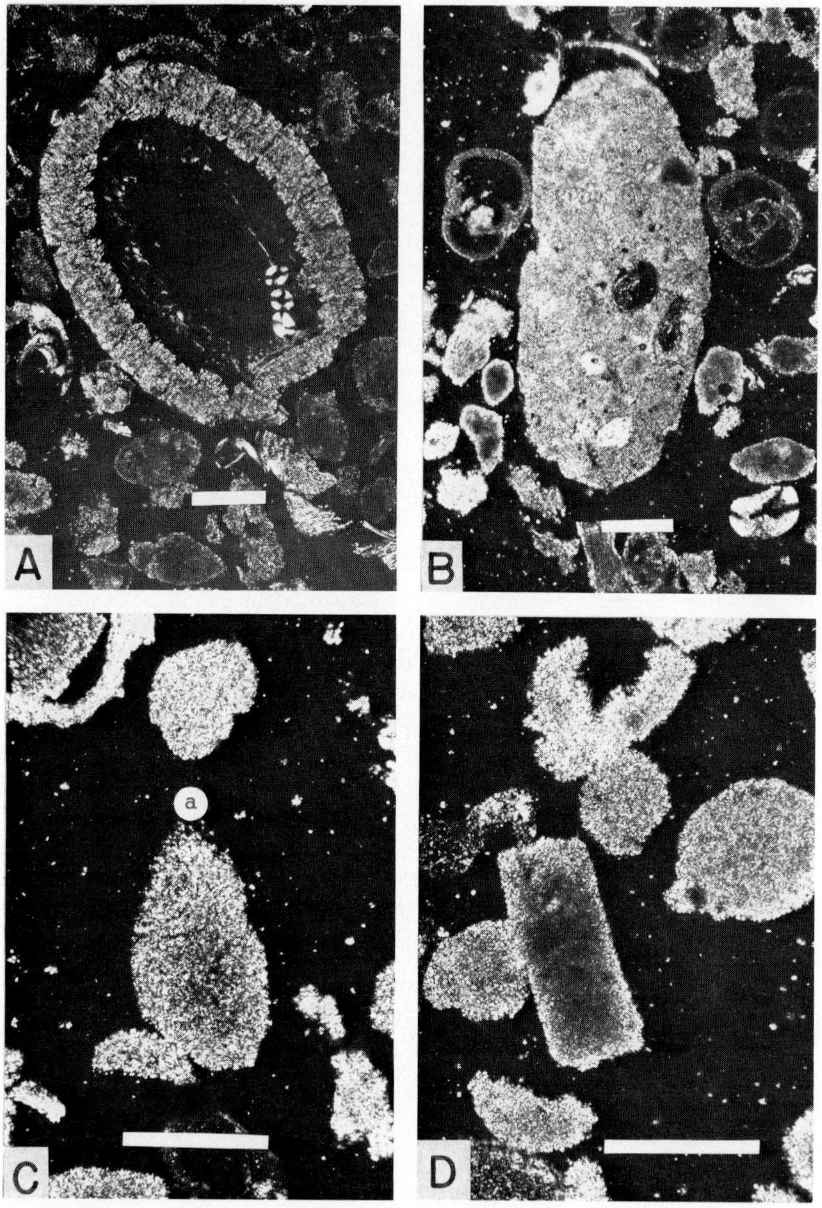

FIG. 14–Constituent grains–charophyte, fecal-pellet, and cryptocrystalline grains.

A. Longitudinal section of charophyte oogonium. Three spherical reproductive cells are visible inside oogonium. Crossed nicols; bar length 100μ.

B. Fecal pellet. Aggregate origin evidenced by sand- and silt-size skeletal fragments embedded in brownish matrix. Crossed nicols; bar length 100μ.

C. Cryptocrystalline grain which may be an indurated fecal pellet. Aggregate origin is suggested by tendency of irregular silt grains, **a**, to float in isotropic matrix near upper edge of grain. Note absence of recognizable skeletal grains. Center of grain is nearly opaque. Crossed nicols; bar length 100μ.

D. Cryptocrystalline grains. Rod-shaped grain resembles fecal pellets of some invertebrates, but texture of grain is cryptocrystalline. Opaque center of grain is typical of many cryptocrystalline grains and fecal pellets. Grains in upper right corner are cryptocrystalline. Crossed nicols; bar length 100μ.

Worm tubes were recognized by their circular shape in transverse section and their tube or spiral shape in longitudinal section. In transmitted light, the lamellar microstructure of their dusty yellow-brown walls is diagnostic. Coiled spirorbids were the most common type encountered.

Dusty gray and finely lamellate wall structure is diagnostic of balanid barnacle plates. Large fragments are further characterized by cross partitions in the porous plates. Some mollusks (oysters) have similar structure but are not finely lamellate.

Echinoderm plates, spines, and spicules have unit extinction between crossed nicols. Plates and spines also have a distinctive porous microstructure.

Unknown skeletal grains—All unidentifiable skeletal grains were tabulated as unknown. This category includes skeletal grains that have been so severely altered to cryptocrystalline material that the original skeletal type could not be determined with certainty.

Nonskeletal Grains

Fecal grains—Beginning with Moore's (1933) discussion of invertebrate feces, sand-size ellipsoidal sediment aggregates usually have been ascribed to organisms. This relation is assumed for ellipsoidal and rod-shaped aggregates in northern Belize. Two types of indurated fecal pellets, ellipsoids and rods, are present in northern Belize sediments. They differ in size, in shape, and to some extent in the types of sediment that form them. One type is similar to the pellets reported in the Bahamas (Illing, 1954; Purdy, 1963a) and Campeche Bank (Cann, 1962; Hoskin, 1963). These pellets are ellipsoidal grains which average 0.4 by 0.8 mm. Their aggregate nature is demonstrated by the presence of recognizable skeletal fragments in a matrix of finer grained particles (Fig. 14B). This type of pellet is present only locally in northern Belize and is rarely indurated sufficiently to survive wet sieving.

A second type of pellet is commonly present in many fine-grained samples from Chetumal Bay. These pellets are usually rod-shaped grains which average 0.1 by 0.3 mm. In thin section they consist of a brownish matrix (transmitted light) containing relatively few recognizable skeletal fragments. Rods that consist entirely of brownish matrix are common. At high magnification and between crossed nicols, fine silt-size and clay-size grains near the edge of the rod appear to be separated from grains deeper in the rod (Fig. 14C). This suggests that the rods are partially indurated fecal pellets.

With increasing induration, the fecal pellets tend to appear grayish brown and semiopaque in transmitted light. Rod-shaped fecal pellets commonly are more indurated than the ellipsoidal ones and are commonly difficult to distinguish from cryptocrystalline grains (Fig. 14D).

Rods and ellipsoids were assigned to the fecal-pellet category only if their aggregate nature was demonstrable.

Cryptocrystalline grains—These are the most abundant constituents in the sediments of northern Belize. In thin section the cryptocrystalline grains are translucent and vary from light brown to grayish brown in transmitted light. Between crossed nicols the grains appear to consist of randomly oriented crystals which are smaller than 5μ and average 2μ. X-ray analyses indicate that the component crystals are usually calcite, thus differentiating them from the aragonite cryptocrystalline grains (Purdy, 1963a, p. 347) from the Bahama Banks. The origin of cryptocrystalline grains will be considered in detail later, but it is important to emphasize now that this category is descriptive and includes grains which formed by more than one process. Cryptocrystalline grains from northern Belize include grains which form both by alteration of carbonate grains and by deposition of carbonate crusts on solid substrates.

Repeated reference has been made to the transitional alteration of skeletal and nonskeletal carbonate grains to cryptocrystalline material. Between crossed nicols, cryptocrystalline tests of miliolids and peneroplids appear to have recrystallized to more coarsely crystalline material (Figs. 10G, 11B, 12B). In contrast, cryptocrystalline skeletal grains such as mollusks, which are normally coarsely crystalline between crossed nicols, seem to have recrystallized to more fine-grained material (Fig. 13C, D, F).

Cryptocrystalline carbonate material is also present locally as irregular coatings on hard surfaces such as shells, wood, or rock (Fig. 15). Although these coatings in thin section have a tendency to be semi-opaque and appear dark gray in transmitted light, they resemble cryptocrystalline material formed by alteration of skeletal grains. Large fragments of cryptocrystalline coatings are laminated (Fig. 15B) and contain 5-8μ-diameter threads. The threads resemble algal threads found in algal concretions from Green Lake, New York (Bradley, 1929), and may be the remains of algae. A few small fragments of cryptocrystalline crusts can be identified by their laminations, but the majority of cryptocrystalline grains could not be assigned to the altered grain type or the crust type of cryptocrystalline grain. Therefore, all grains consisting of randomly oriented fine-grained carbonate that were not recognizable as altered grains were tabulated as "cryptocrystalline grains." Recognizable skeletal grains with cryptocrystalline texture were tabulated in their appropriate skeletal category. As this method tended to reduce the number of cryptocrystalline grains tabulated, skeletal grains showing more than 50 percent alteration to cryptocrystalline carbonate were tabulated both as their appropriate skeletal category and independently as "cryptocrystalline skeletal grains." This category then records the tendency for skeletal grains to alter to cryptocrystalline carbonate. "Total cryptocrystalline" refers to the sum of cryptocrystalline grains and cryptocrystalline skeletal grains.

Country rock—Particles of rock and caliche, collectively identified as country rock, are present near the shore and in places where the Holocene sedimentary column is thin. These constituents were identified by their spar cement. Included quartz grains and fossils are also characteristic. A few rock fragments were found to be intensely bored, and such fragments appeared to have cryptocrystalline rims.

Terrigenous minerals—This category includes all noncarbonate grains. Angular quartz and, to a lesser extent, chert and feldspar were common in some localities.

Tabulation Conventions

Each grain that came under the intersection of the microscope's cross hairs was identified as one of the 15 constituent-particle categories. Voids within a grain were counted as the grain only if the void was isolated from the exterior (e.g., foraminifers and closed ostracods). Voids which were open to the outside and thus did not fill sediment space were not counted as grains (e.g., open pelecypod shells). Grains encrusted by organisms (e.g., *Homotrema, Millepora,* or coralline algae) were counted, according to what came under the cross hairs, as either the substrate grain or the encrusting organism.

Point-Count Error

The constituent-grain composition of the fraction larger than 1/16 mm was determined for all sediment samples by point counts of thin sec-

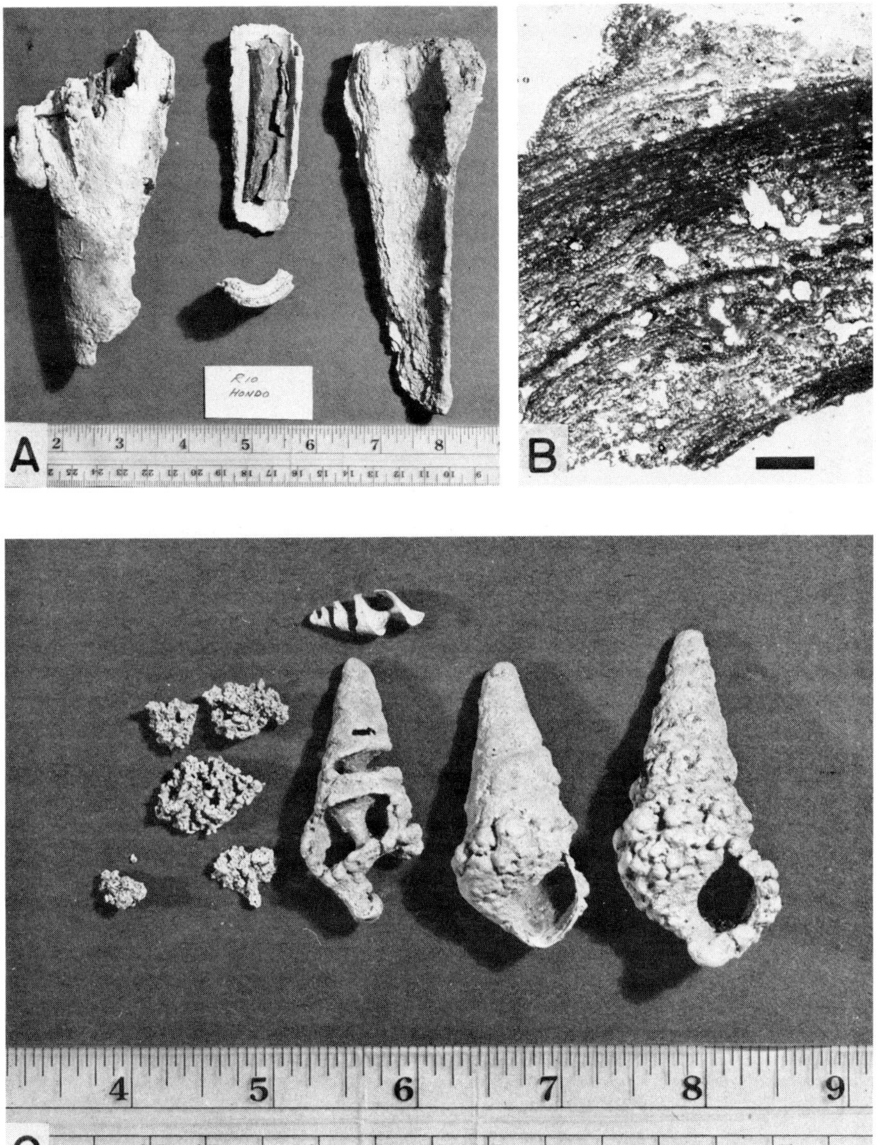

FIG. 15—Carbonate encrustations.

A. Carbonate encrustations on red mangrove roots at mouth of Rio Hondo. Scale in inches and centimeters.

B. Thin section of root encrustation transverse to length, showing laminations. Each layer contains numerous algal threads which are visible at higher magnification. Origin of large irregular holes is unknown. Plane-polarized light; bar length 1 mm.

C. Carbonate encrustations from lower reaches of Rio Hondo (Krueger locality 2-2). Irregular masses on left surround twigs and have gastropods trapped in carbonate. Large gastropods on right show successive stages of encrustation. Many shells are partly dissolved before lime encrustation begins. Shell at right shows typical botryoidal surface of well-developed crust. Scale in inches and centimeters.

tions. The reader is referred to Chayes (1956) for a detailed discussion of the various types of error associated with rock or, in this case, sediment composition from point-count analyses, and to Ginsburg (1956), Cann (1962), and Purdy (1963a) for a consideration of point-count analyses applied to carbonate sediments. A minimum number of 500 points was counted for each thin section. On the basis of data presented by Ginsburg (1956), Cann (1962), and Purdy (1963a), it seems conservative to state that the point-count estimates recorded in Appendix B are within 6 percent of the true value for the sample.

Sedimentary Structure

A sediment is characterized not only by grain size and constituent particle composition, but also by the spatial relations among its components. This last characteristic, of course, refers to sedimentary structure. For this study, sedimentary structure was divided into *internal structure*, which can be seen in a core, and *external structure*, which is mainly deduced by intercore correlation. This distinction is a matter of convenience rather than a reflection of a real dichotomy between end members of a spectrum of sedimentary structure.

Minor Internal Structure

Minor internal structures consist of small-scale variations in texture, composition, color, and/or grain orientation within a sediment (Moore and Scruton, 1957). In the present study the following types of minor internal structure were recognized (type names modified from Moore and Scruton, 1957): regular layering, irregular layering, mottling, and homogeneity.

Regular layering consists of contrasting, nearly parallel sediment layers of uniform thickness. The contrast between layers may result from both textural differences (Fig. 16C), which include graded bedding (Fig. 17B and C), and grain-orientation differences (Fig. 16D). Regular layering may result from initial deposition (Fig. 17A) or may be produced secondarily by variable current activity which winnows the surface of a previously deposited sediment (Fig. 16D). During coring operations, the core barrel sometimes entered the sediment at an angle or heeled over during penetration; this, of course, produced an artificial dip to the layering. True dip was ascertained by noting the orientation of grass blades within the cores. If the grass blades were oriented normal to a bedding surface which appeared to dip, it was assumed that the dip was an artifact of coring.

Irregular layers are represented by lenslike bodies surrounded by a contrasting matrix (Fig. 16B). As their name suggests, the lenses are irregular in outline; however, they do evidence elongation parallel with the depositional surface. Irregular layering may indicate the existence of irregular depositional surfaces such as that expressed by a ripple-marked bottom, but more commonly it probably records early stages of burrowing by organisms.

Mottles are irregular pockets, tubes, or lumps of sediment enclosed in a contrasting matrix (Figs. 16A; 17C, D, E). Mottles may indicate primary deposition of sedimentary aggregates but are more likely the work of burrowing animals.

Homogeneous sediments have no visible structure and may result from uniform depositional conditions or from the homogenization of a layered sediment by burrowing animals. No completely homogeneous sediments were found in the present study. Some uniformly textured and monotonously colored sediments appeared homogeneous at first glance, but close examination invariably demonstrated the presence of sedimentary structure.

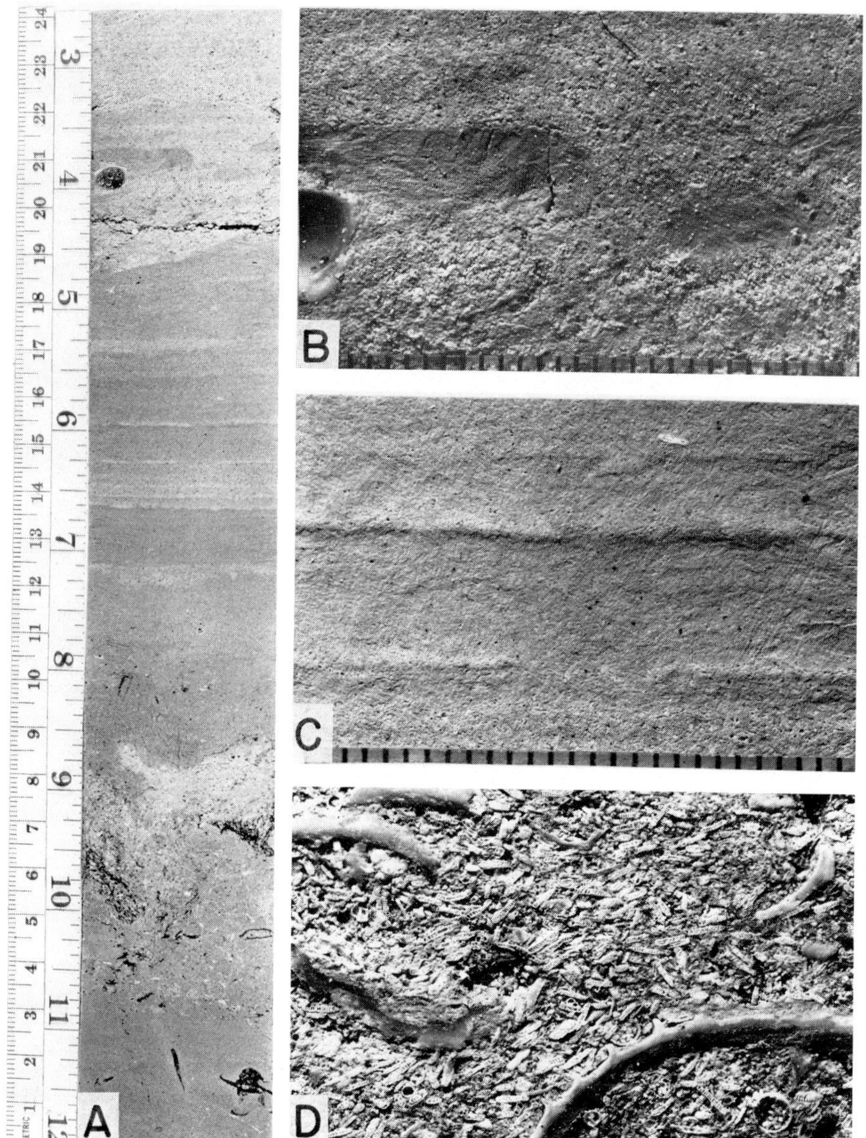

FIG. 16–Minor internal structures–sample 256 (near South Bulkhead Cay), water depth 2 ft (0.6 m).

A. Oriented segment of core containing minor internal structures 14.5-24.0 in. (36.8-61.0 cm) below sediment surface. Irregular layering is visible near top, regular layering near center, and mottles near bottom. Light-colored sand near top of core apparently truncates underlying mud layers. Truncation and formation of mudlumps may be by erosion or possibly by burrowing. Scale in centimeters and inches.

B. Detail of **A**, irregular layering. Dark lenses consist of carbonate mud in matrix of fine foraminiferal sand. Shell on left is immature *Chione cancellata*. Note elongation of lumps parallel with bedding. Scale in millimeters.

C. Detail of **A**, regular layering. Layers consist of fine miliolid sand in matrix of carbonate mud. Layers are parallel but vary slightly in thickness across core. Core was partly dried so sand layers would stand out. Elongate white grain near top is a peneroplid. Scale in millimeters.

D. Layering of elongate grains. Bedding direction is indicated by parallelism of grains set in fine sand matrix. Large grains are *Chione cancellata*; small grains are mostly peneroplids. Roundness of most shells plus grain-supported fabric indicates that layering was produced by currents strong enough to winnow away mud. Scale in millimeters.

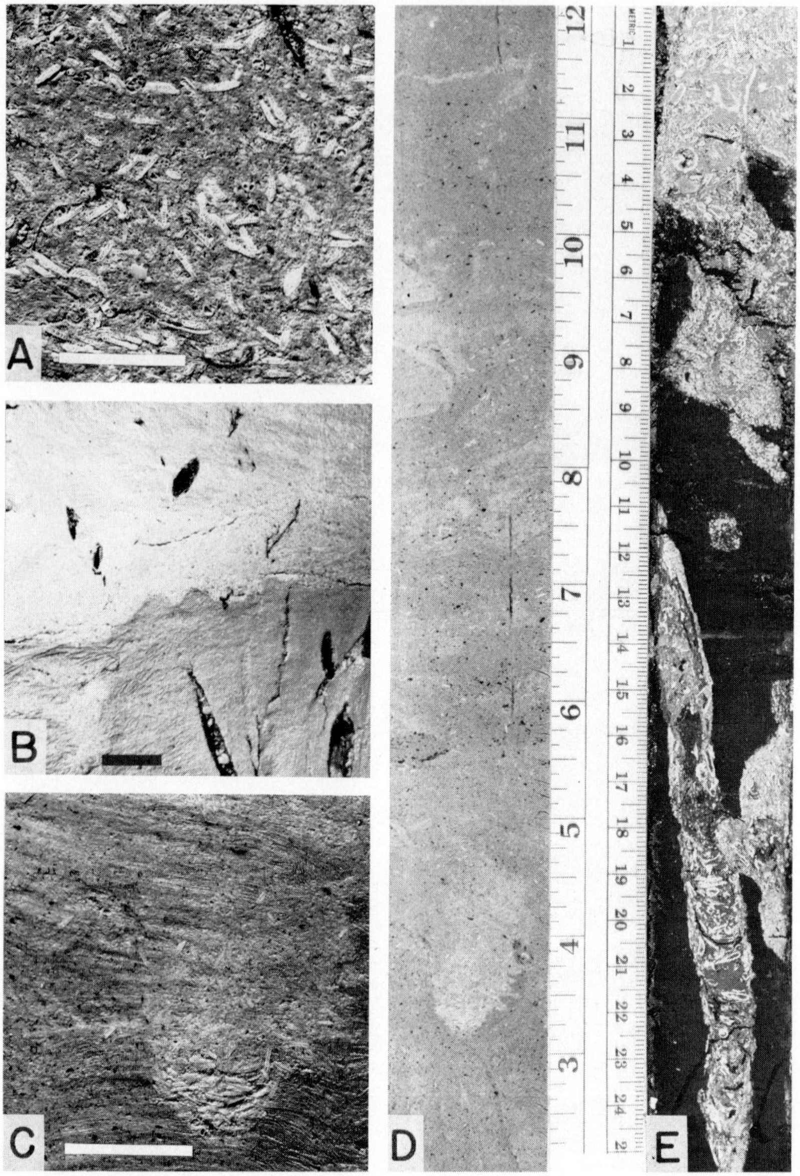

FIG. 17–Minor internal structures.

A. Regular layering. Oriented peneroplids and valvulinid foraminifers show bedding direction. Note equal distribution of sand grains in mud matrix. This indicates that mud and sand were deposited simultaneously, and sediment has not been winnowed. Bar length 5 mm (sample 256, 55-57 in. [140-145 cm] below sediment surface).

B. Sharp contact between dark-colored mud and lighter sand. Sand grades upward into mud (out of photo). Dark ellipsoids are leaves of *Syringodium* grass. Bar length 10 mm (sample 262, 43-45 in. [109-114 cm] below sediment surface).

C. Detail of lower part of **D.** Irregular sand pocket shows graded bedding and bedded grains. Dark specks are plant fragments; small lines are molds of ice crystals which formed during preparation of core. Bar length 10 mm.

D. Irregular layering and mottling. Irregular pockets and lenses of sandy mud (light colored) in carbonate mud (darker). These structures are gradational between irregular layering and mottling. Scale in inches and centimeters (sample 242, 38-46 in. [97-117 cm] below sediment surface).

E. Mottling. Burrows into mangrove peat are infilled with overlying sediment type (*Halimeda* sand). Note bedding of grains in vertical burrow. Circular horizontal burrow can be seen near center. Scale in inches and centimeters (sample 58, 50-61 in. [127-155 cm] below sediment surface).

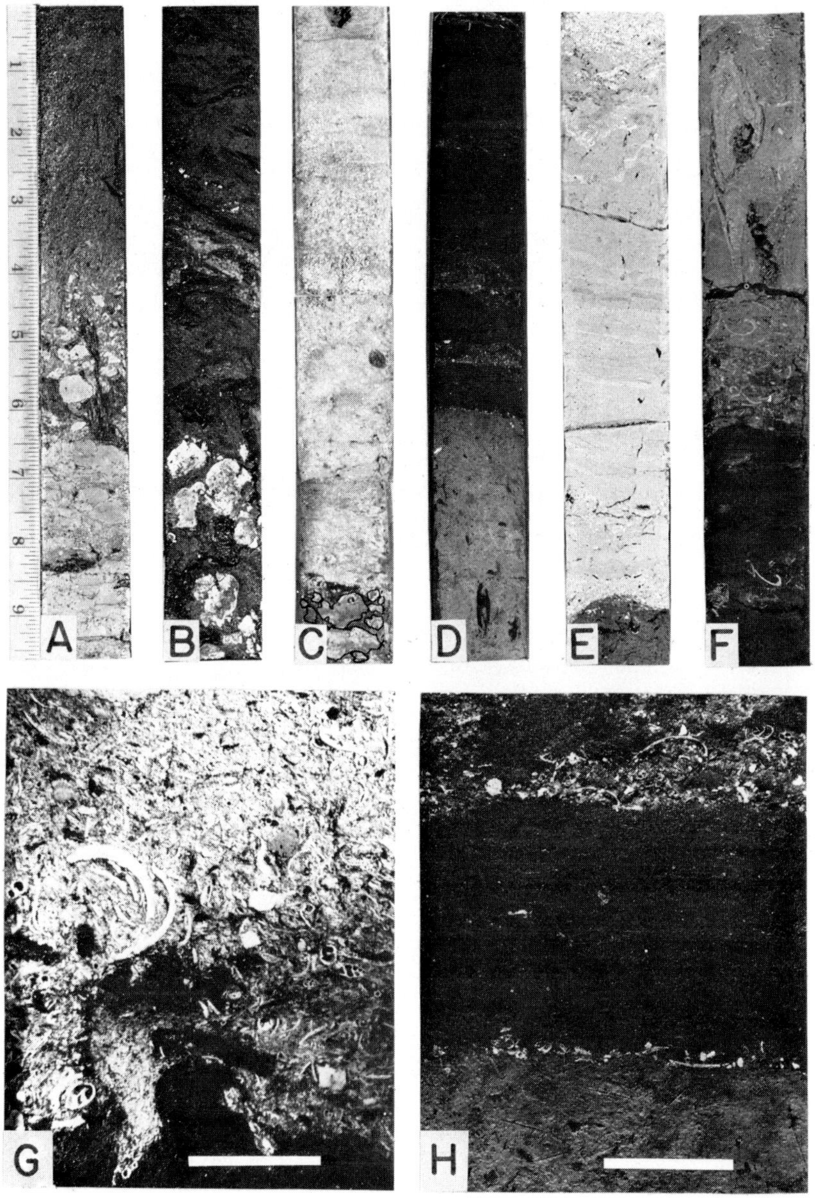

FIG. 18—Sedimentary contacts.

A. Pleistocene-Holocene contact; dark gray clay lying unconformably on weathered carbonate rock (light gray). Note basal gravel consisting of round rock fragments. Scale in inches (sample 262, 97-107 in. [246-271 cm] below sediment surface).

B. Bedded peat with carbonate-rock gravel near bottom. Gravel suggests that a rock surface lies just below core. Same scale as **A** (sample 239, 42-52 in. [107-132 cm] below sediment surface).

C. Carbonate sand with rock fragments near bottom (photograph retouched to show pebbles). Same scale as **A**.

D. Plane contact between gray clay and peat (see detail in **H**). Same scale as **A** (sample 256, 72-84 in. [183-213 cm] below sediment surface).

E. Contact between gray clay and laminated skeletal mud (light gray). Same scale as **A**.

F. Disturbed contact between peat (black) and miliolid mud (lighter gray). Note peat mixed into mottles in carbonate mud. Same scale as **A** (sample 31, 38-48 in. [97-122 cm] below sediment surface).

G. Irregular contact between peat (black) and carbonate sand. Rock fragments are common in sand. Bar length 10 mm (sample 248, 77-80 in. [196-203 cm] below sediment surface).

H. Detail of **D.** Note sharpness of peat-clay contact. Bar length 10 mm.

External Structure

External structure relates a sediment as a unit to surrounding sedimentary units. This includes lateral extent, thickness, and nature of the unit's contacts with surrounding units. Thickness and lateral extent of the surface sediments were determined, respectively, by core measurement and correlation of gross lithology between cores. Typical contacts between the sedimentary units encountered in this study include unconformable contacts (Fig. 18A, B, C) and conformable contacts which are commonly sharp (Fig. 18D, H) or disturbed (Fig. 18F, G).

SEDIMENTARY FACIES

The reconstruction of ancient depositional environments is a rapidly developing phase of sedimentary petrology. Obviously we cannot study ancient environments *per se*, but we can infer their characteristics from facies analyses. Unfortunately, the relation between sediments and their environments is not always a simple one. Within shallow-water bays and estuaries, for example, the sedimentary record may be biased in terms of recording mostly short-term environmental variations rather than the environmental norm of the area (e.g., the storm which displaced and concentrated mollusks along the margins of the Dutch Wadden Sea; Swennen, 1953, cited *in* Brongersma-Sanders, 1957, p. 987). The most important consequence of these complications is that the environmental attributes most commonly evaluated for an area may not explain some of the more important quantitative and qualitative characteristics of a facies. It seems appropriate to emphasize that the term "depositional environment" is used to mean the total effect of environment that produces general variations from place to place in sediments. These variations, of course, are facies. Thus a facies is "...any *areally restricted* part of a designated stratigraphic unit which exhibits characters *significantly different* from those of other parts of a unit" (Moore, 1949, p. 32; italics added).

Given a set of geologic observations on sedimentary samples, how does one recognize areally restricted samples that are significantly different from each other? There are three general types of techniques for processing such data: (1) visual grouping, in which the skill and intuition of the analyst are important, (2) arbitrary classification, in which there is an *a priori* assumption that the selected bases of classification are the most meaningful bases possible, and (3) statistical analyses, in which interaction and structure within the data are sought by mechanical procedures.

In this study the accumulated data were first interpreted by visual grouping and inspection. Inspection of the data matrix, in conjunction with plotting data maps and examining core photographs and aerial photographs, resulted in the facies pattern illustrated in Figure 19. This interpretation was based on the limited number of variables that could be manipulated mentally at any one time, and consequently the writer was not certain that a more meaningful classification could not be obtained by simultaneous consideration of all the variables by use of factor analysis.

The purpose of factor analysis is to seek structure in data. The reader is referred to Cattell (1952) for a nontechnical discussion of factor-analysis theory. The application of factor analysis to geologic situations has been discussed by Imbrie and Purdy (1962), and a detailed discussion of the techniques as applied to geologic data is presented by Imbrie (1963).

The philosophy behind factor analysis is well stated by Cattell (1952, p. 14): "In embarking upon the factor analysis one need have

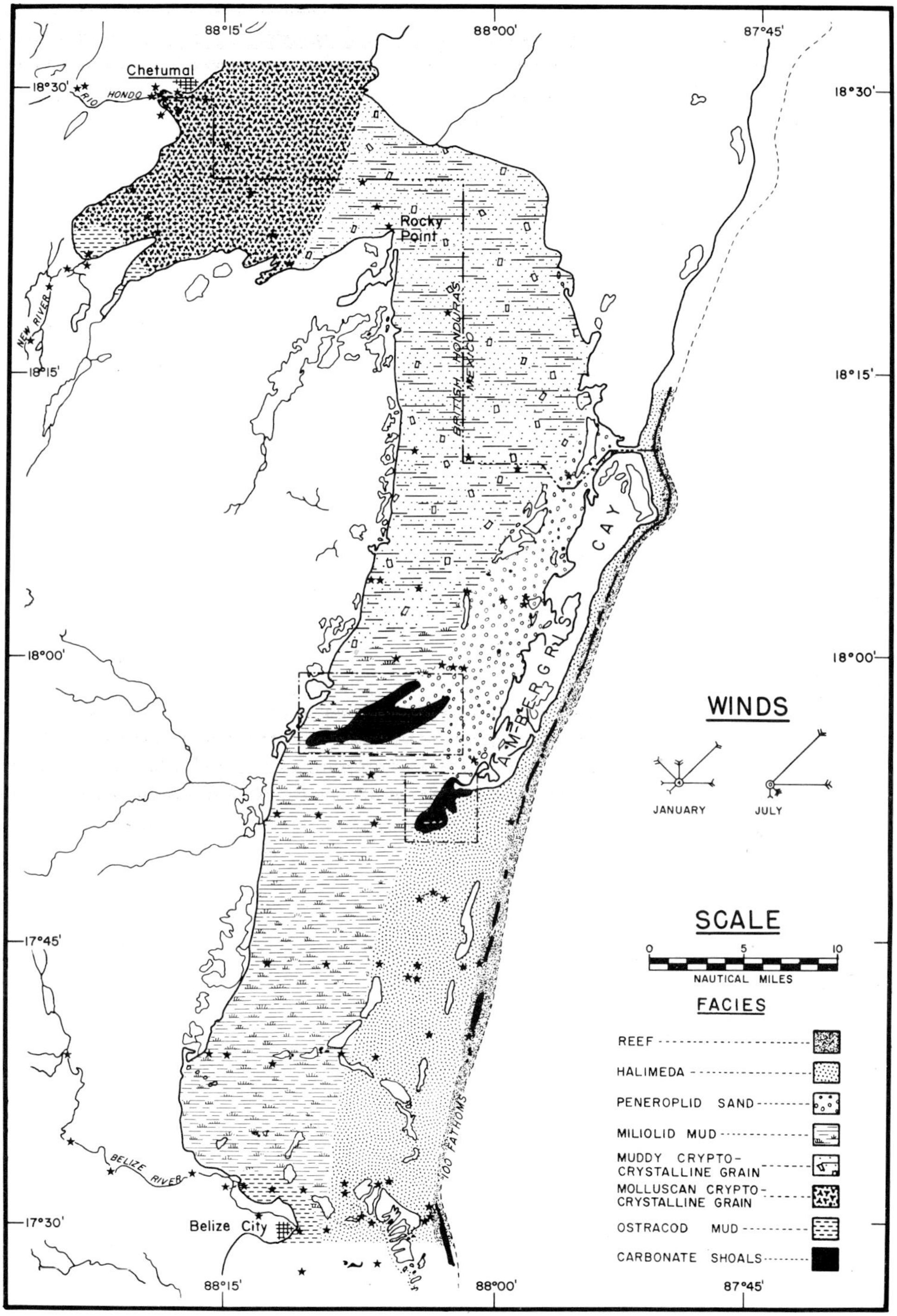

FIG. 19—Sediment facies distribution in northern Belize as determined by visual interpretation.

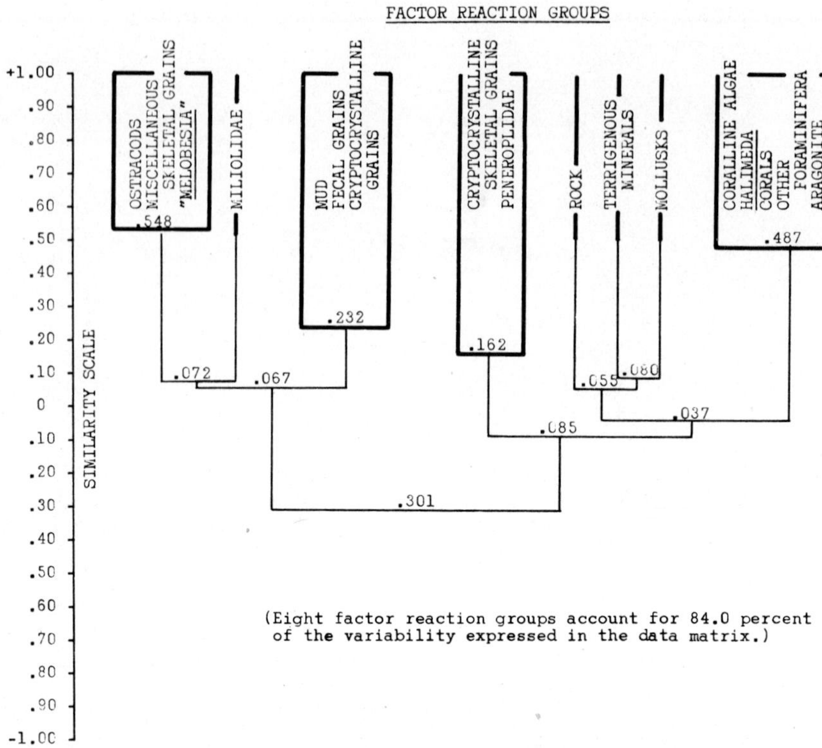

FIG. 20—Hierarchy diagram showing interaction of constituents in sediments of northern Belize.

no more definite idea than Columbus had of America in regard to what may be found. It is sufficient to hypothesize that *some* structure lies there." Factor analysis of a data matrix, such as in Appendix B, can find and delimit structure in the matrix. During visual interpretation of data, one may tend unconsciously to emphasize some attributes of the data or to disregard the importance of other data. Thus factor analysis can form a frame of reference by which the geologist can determine whether he has underexplained or overexplained the data.

Factor analyses in both the R-mode (between geologic variables) and the Q-mode (between geologic samples) were performed using the techniques of Imbrie (1963) and an IBM 7094 computer. A hierarchic representation of the R-mode factor analysis was constructed by computing mean similarities between the factor groups. The results of these computations are illustrated in Figure 20. The hierarchy represents interaction among the named variables. A similar hierarchy is presented in Figure 21 for geologic samples. In this case, however, the data were recomputed using the percent of observed range transformation (Imbrie and Purdy, 1962, p. 258) prior to factor extraction. The highest relative weighting of a sample on a factor in the rotated orthogonal matrix was used to collect the samples into six factor groups. Each of these factor groups has the areal distribution depicted in Figure 22, and thus the factor groups constitute sedimentary facies. Two nearshore samples consisting of rock fragments and one beach sand consisting of broken mollusk shells were extracted as factor groups but are not shown as facies because of their local extent. A comparison of the visually interpreted facies (Fig. 19) with the factor-delineated facies (Fig. 22) shows a general pattern similarity, but there are some significant differences. The equality of pattern of the reef facies in both illustrations stems from the fact that the reef was not sampled, and reef-apron sediments are not sufficiently different in composition from *Halimeda* sediments to be segregated by factor analysis. Thus the reef facies shown in Figure 22 is based on visual interpretation—which includes, of course, field observations. The terrigenous-sand facies delimited by factor analysis had been included previously in the miliolid-mud facies in spite of the fact that cores of this sediment contained large amounts of reworked material from the underlying sedimentary unit as well as miliolid mud. The amount of reworking in the area of these sediments is so intense, however, that the sediments themselves have a composition that is sufficiently diagnostic from a geologic point of view to warrant the erection of a separate facies. The factor analysis of cryptocrystalline-grain facies is the result of combining two visually interpreted facies: the muddy cryptocrystalline-grain facies and the molluscan cryptocrystalline-grain facies. The muddy cryptocrystalline-grain facies consists of 11 samples that contain mostly cryptocrystalline grains in a mud matrix. Factor analysis indicated that six, or over half, of these samples were most closely related to the miliolid-mud facies. In addition, visual comparison of the remaining five samples with samples in the cryptocrystalline facies convinced the writer that a single cryptocrystalline facies was more meaningful than the previous somewhat arbitrary division between the two visual cryptocrystalline-grain facies. Thus the facies delineated in northern Belize (Fig. 22) are based on a combination of both factor analysis and visual interpretation.

Reef facies

The barrier reef is limited to the seaward edge of the continental shelf and consists of a wave-resistant frame built largely by corals

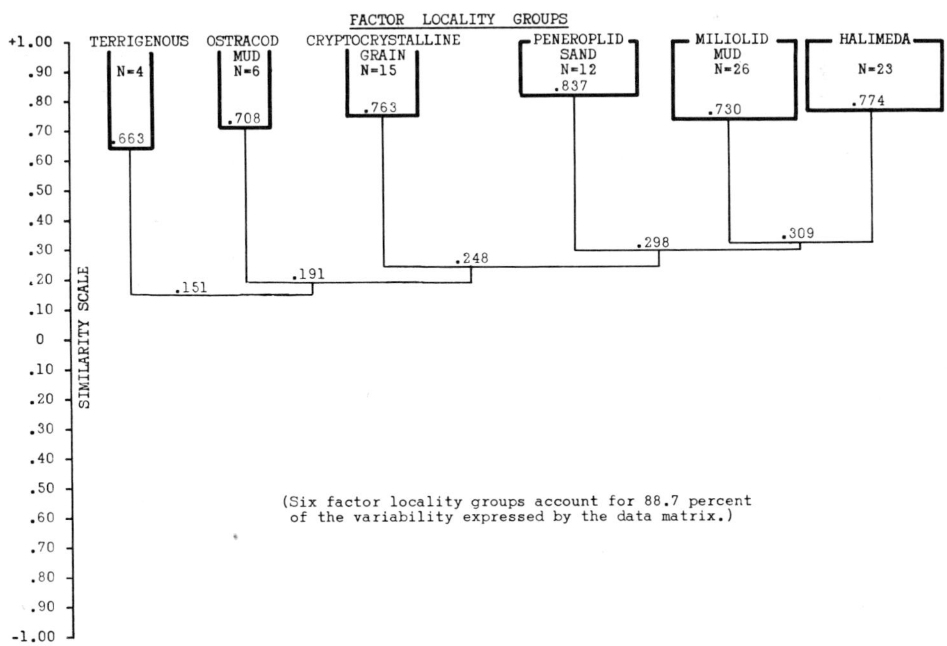

FIG. 21—Hierarchy diagram showing interaction of facies in sediments of northern Belize.

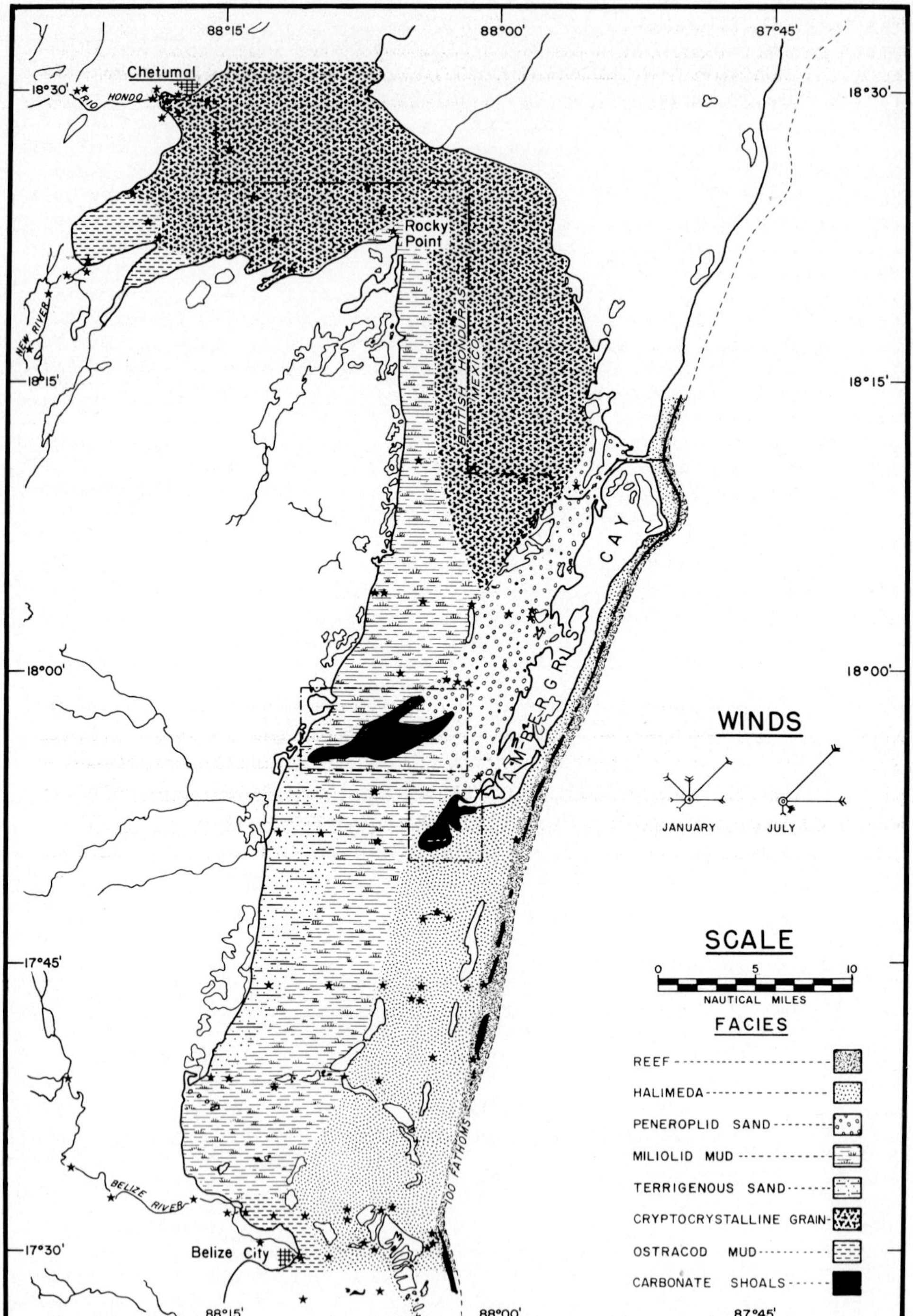

FIG. 22—Sediment facies distribution in northern Belize as determined by factor analysis.

and encrusting red algae. Because of its organically bound nature, the reef frame was not sampled. The distribution of the reef facies was determined by field observations and interpretation of aerial photographs. No detailed reef study was attempted. By correlation of aerial photographs, fathograms, and limited bottom observations, the reef can be divided into major zones which are similar to zones described by Thorpe and Bregazzi (1960) for the Belize barrier reef approximately 20 mi (32 km) southeast of Belize City. Their zones are also similar to those described by Goreau (1959) for Jamaican reefs.

The reef has colonized approximately three fourths of the shelf edge and is generally 300-400 yards (275-365 m) wide. A well-developed spur-and-groove system characterizes the windward side of the reef. The system is developed in depths of 10-60 ft (3-18 m) and is from 100 to over 200 yards (90-180 m) wide. Fathograms show that the buttresses dip seaward at about 2° or less and form a platform at the top of a steep slope of 20-30°. The spur-and-groove system is developed approximately normal to the reef trend, but in places intersects the reef crest at angles of 70°. Stoddart (1962a, p. 30) noted that elsewhere in Belize the axes of the systems tend to be developed at 90° to wave fronts. Wave refraction patterns on aerial photographs permit the extrapolation of Stoddart's observation to northern Belize. The origin of spur-and-groove systems is in debate, and the reader is referred to Cloud (1954), Newell and Rigby (1957), Goreau (1959), and Shinn (1963) for discussions of erosive versus constructive origins.

Along approximately half of the reef length, the spur-and-groove system is abruptly replaced at shallower depths by a narrow breaker zone which is characterized by wave-oriented elkhorn coral, *Acropora palmata*. Behind the breaker zone is the reef flat, which is usually 100-200 yards (90-180 m) wide and only a few feet deep. It is characterized by overturned and dead corals which are encrusted by *Millepora* and lithothamnid red algae. In places where the reef is particularly well developed, an apron of loose sand is present in deeper water behind the reef flat. This sand consists largely of reef detritus which has been swept across the reef flat.

Two sediment samples were collected from sand aprons adjacent to the reef flat. The constituent composition of only two samples is obviously insufficient to characterize the reef facies, but it does give some idea of the significant contributions of skeletal grains to the sediment (Table 1). Corals and coralline algae are distinctive of these sediments. Approximately half the red algae are encrusting melobesids and half are articulated forms. *Halimeda* plates are abundant but do not equal the combined volume of corals and red algae. Encrusting organisms, which are attached to hard skeletal substrates, are abundant and include some coralline algae, the red foraminifer *Homotrema*, bryozoans, and serpulids. Less than half of the *Halimeda* plates are partly altered to cryptocrystalline material. Unaltered plates are a uniform translucent brown in transmitted light and, with the exception of a few tiny needles, appear dark between crossed nicols. In transmitted light, partly cryptocrystalline plates have gray patches of cryptocrystalline material which in some cases surround islands of dark brown material. These brown islands may be organic material that has been excluded from the altered parts of the plate during the alteration process. The gray cryptocrystalline material appears to have an anisotropic fabric between crossed nicols and seems to be made up of crystals measuring 1-5μ. Uncommonly, acicular aragonite needles fill some of the pore spaces in the *Halimeda* plates. Although mollusks make their smallest contribution to the sediment in this facies, they have the greatest taxonomic diver-

TABLE 1. CONSTITUENT COMPOSITION, GRAIN SIZE, AND CARBONATE MINERALOGY OF TWO CORAL-REEF SAMPLES[1]

	Observed Range
Coralline algae	14.7-18.9
"Melobesia"	0.7- 1.2
Halimeda	24.1-38.9
Miliolidae	1.0- 1.9
Peneroplidae	2.2- 2.8
Other Foraminifera[2]	3.2- 8.8
Corals	13.3-20.1
Mollusks	11.1-15.8
Ostracods	0.4- 1.0
Miscellaneous skeletal[3]	2.8- 4.5
Unknown skeletal	5.0- 5.2
Total skeletal	98.7
Fecal grains	0.1- 0.6
Country rock	0
Terrigenous	0
Cryptocrystalline grains	0.5- 1.2
Cryptocrystalline skeletal	4.2-12.1
Total over 50 percent cryptocrystalline	5.4-12.7
Encrusting skeletal[4]	10.0-14.2
Weight percent <1/8 mm	9.6-15.6
Weight percent <1/16 mm	3.2- 3.4
Aragonite as weight percent of carbonate	63.0-72.0
Magnesium carbonate in calcite (weight percent)	13

[1]Samples from sand apron adjoining reef flat.
[2]Includes: *Homotrema*, many transparent calcareous species.
[3]Includes: echinoderms, serpulids, alcyonarian spicules, and bryozoans.
[4]Includes: *Homotrema*, serpulids, bryozoans, and *"Melobesia"*; tabulated independently.

sity here (Robertson, 1963). Typical species include borers (*Lithophaga* sp.), forms with shells cemented to a solid substrate (*Echinochama* sp. and *Hipponix antiquatus*), and many others found only on the reef.

In thin section, two distinct sizes of borings are seen in mollusks, coralline algae, and, rarely, coral fragments. The larger borings appear as 20-40μ holes at grain borders, whereas the smaller borings are displayed as tubes 3-6μ in diameter. The latter tend to occur well inside of grain borders. Boring is rarely intense, and the type of grain being bored is usually determined readily.

Miscellaneous skeletal grains include, in order of abundance: echinoderms, serpulids, alcyonarian spicules, and bryozoans.

Most of the sand grains are represented by aragonite skeletal fragments (corals, *Halimeda*, and many types of mollusks), thus explaining the high aragonite content of the samples. The calcite present in the bulk samples has a high magnesium carbonate content (approximately 13 weight percent) resulting mainly from the presence of numerous coralline algae and, to a lesser extent, from the presence of peneroplids, miliolids, and echinoderms. Less than 5 percent of the sediment is mud.

Depositional Environment

The typical environmental associations of reef corals have been summarized by Vaughan (1919), Yonge (1940), and Wells (1957). The location of the barrier reef on the windward edge of the shelf exposes it to open circulation of oceanic water. This circulation tends to decrease the severity of periodic temperature and salinity changes. Thus, although air temperatures on the mainland fall below 60°F (15°C), air temperatures on the reefs, which are only 20 mi (32 km) to the east, stay above 65°F (18°C); and water temperatures stay above 77°F (25°C; Stoddart, 1962a, p. 14). Water over the reef is very clear, and dark-colored coral heads can be seen at depths of 60 ft (18 m). Measured salinities on the reef were 35-36°/₀₀.

North of the latitude of Belize City, the barrier reef is not as well developed or as continuous as it is south of Belize City. This is true not only of northern Belize but also is true farther north with respect to Quintana Roo, Mexico, where the patchiness of reef development was mentioned by Vermeer (1959). The environmental reasons for the decrease in reef development are speculative at present. Vermeer (1959, p. 123) and Stoddart (1962a, p. 15) have suggested that upwellings of cold water are responsible. Temperature profiles in the Yucatan Channel (approximately 150 mi or 241 km north of Belize) support this hypothesis, as they show a tendency for relatively cool water to approach the surface near the Yucatán coast (Parr, 1937, p. 43). Conditions for upwelling on the east coast of Yucatán are not as regular as they are in the classic upwelling areas that are situated primarily off western coasts. The general associations of upwelling (Sverdrup et al, 1942, p. 140, 500) are: (1) steep coastal slope, (2) current direction such that the Coriolis force tends to move water away from land, and (3) steady winds which tend to blow surface water away from shore. The first two of these associations are present along the eastern Yucatán coast. The depth of water increases abruptly to more than 500 fm less than 10 mi (16 km) from the coast; the main Caribbean Current flows north parallel with the coast, but the prevailing trade winds are from the east. Periodically, however, northers with strong northwest and north winds blow almost directly offshore. Northers are associated with an increase in the velocity of a south-flowing countercurrent relative to the north-flowing Caribbean Current. Stoddart (1962a, p. 15) implied that such northers may produce upwellings; but the resultant south current should shift toward the coast by Coriolis effect and thus produce convergence and descent of surface waters near the coast (see Sverdrup et al, 1942, p. 500-501, for discussion of wind effects along coasts), which, of course, is the exact opposite of upwelling conditions. If upwelling does occur, it is more likely produced by south and southeast winds. Upwelling may also be associated with the axis of divergence of the Caribbean Current. The main Caribbean Current flows from southeast to northwest across the Caribbean Sea, turns north near the latitude of Ambergris Cay, and flows northward along the Yucatán coast. Part of the current, however, under the influence of the northeast trade winds, diverges and flows southward along the Belize coast. This axis of divergence is located approximately at the latitude of Ambergris Cay in the summer but, under the influence of northers, shifts northward approximately 50 mi (80 km) during the winter (U.S. Navy Hydrographic Office, 1955a).

The effect of upwelling on reef development is not understood. Studies of upwellings off California, Peru, and South Africa (Sverdrup et al, 1942, p. 402, 627, 724) indicate that upwelling water is derived from a depth of about 200 m. At this depth, water temperatures off the Yucatán coast are about 15-20°C (Parr, 1937, p. 36). Since 20°C is generally accepted as the minimum temperature for coral-reef development (Wells, 1957), it seems that upwelling, if it does indeed occur along the eastern Yucatán coast, may lower water temperatures sufficiently to prevent the formation of a well-developed reef.

The barrier reef is constructed largely by organisms which thrive only on hard substrates. With the exception of some reef-flat corals such as *Porites porites* and adult specimens of *Manicina areolata*, all the corals are attached. The hydrozoan *Millepora* sp. commonly encrusts dead skeletal material. Many of the red algae are represented as thin crusts which tend to cover older parts of coral colonies. A hard bottom and clear, warm water, however, are not the only factors limiting reef development, for hard-bottomed areas behind the reef support only scattered coral heads. Here, the obvious lack of turbulence as contrasted with reef conditions may be a third limiting factor.

Relatively deep-water reefs lacking a shallow surf zone are present along approximately one third of the reef trend in northern Belize. These areas of deep reef, which are up to 6 mi (9.7 km) long, commonly are adjacent to 100-1,000-yard-wide (91-910 m) channels through the reef trend. The channels continue into the lagoon, where their sinuosity suggests that they are abandoned stream channels (Fig. 4). There is evidence that the cays are localized on Pleistocene or older rock highs (Stoddart, 1963, p. 32, 36) and have grown through Holocene sedimentary accretion to these highs. Therefore, it is presumed by the writer that, during lower sea-level stands, the cays were low hills and that the reef channels were stream channels formed during glacial sea-level lowerings. Reef growth in the last 5,000 years has not succeeded in closing the channels. As mentioned, the longest stretches of deep reef are between the channels. The breaker zone may not have developed because of unfavorable environmental conditions for developing an *Acropora palmata* climax, or the breaker zone may have been destroyed by storms. Hurricanes certainly are capable of destroying the shallow-water parts of reefs. For example, Hurricane Hattie destroyed the shallow reef near St. George's Cay in 1961 and even destroyed deep-water spur-and-groove systems in southern Belize (Stoddart, 1962c, p. 513).

Much remains to be learned concerning reef development in northern Belize. The ideas developed above should not be considered as conclusions, but rather as working hypotheses for future research. The reasons for excellent reef development in some areas and poor development in others are not understood. The structure and origin of a major part of the reef, the spur-and-groove system, are unknown. The writer's inference that the barrier reef is only a veneer on an irregular rock surface should be tested by borings. Hydrographic data are too meager to prove or disprove the existence of upwellings off the Yucatán coast. Finally, comparative studies should be made of the relative abundance of major reef-forming corals in well-developed reefs versus poorly developed reefs.

Halimeda Facies

A high skeletal content characterized by abundant *Halimeda* plates is diagnostic of this sediment facies (Table 2). Corals and red algae are present but scarce. Partly cryptocrystalline *Halimeda* grains are common. "*Melobesia*" is common. Like the reef facies, the *Halimeda* facies consists predominantly of skeletal material which is aragonite or, less commonly, high-magnesium calcite. Miscellaneous skeletal grains are mostly serpulids but include echinoderms, bryozoans, alcyonarian spicules, and scarce radiolarians. Borings into grains are common but generally are not abundant; they consist of irregular small tubes 3-5μ in diameter and, less commonly, large holes 15-40μ in diameter. The amount of sediment smaller than 1/8 mm ranges between 2 and 81 percent, and the amount smaller than 1/16 mm ranges between 1 and 67 percent.

Nonskeletal grains are scarce and consist largely of country rock and cryptocrystalline grains. Indurated ellipsoidal fecal pellets are relatively abundant in the muddier sands. The common presence of partly

TABLE 2. MEAN CONSTITUENT COMPOSITION, GRAIN SIZE, AND CARBONATE MINERALOGY OF *HALIMEDA* FACIES[1]

	Mean	*Observed Range*	*Standard Deviation*
Coralline algae	0.5	0.0- 3.0	0.8
"*Melobesia*"	2.5	0.4-10.7	3.0
Halimeda	41.4	7.3-79.4	22.4
Miliolidae	3.2	0.6- 7.7	2.1
Peneroplidae	3.3	0.5-10.1	2.6
Other Foraminifera	4.5	1.4- 9.9	2.5
Corals	0.7	0.0- 2.3	2.1
Mollusks	26.5	11.7-49.0	11.4
Ostracods	1.6	0.0-11.8	2.9
Miscellaneous skeletal	2.7	0.6-10.1	2.8
Unknown skeletal	5.7	1.3-19.6	4.2
Total skeletal	92.7	80.9-99.8	6.2
Fecal grains	4.7	0.0-16.6	4.6
Cryptocrystalline grains	1.6	0.0- 6.1	2.5
Cryptocrystalline skeletal grains	6.5	0.0-47.7	11.6
Total over 50 percent cryptocrystalline	8.0	0.0-50.4	12.3
Country rock	0.8	0.0- 9.2	2.1
Terrigenous minerals	0.3	0.0- 3.5	0.8
Weight percent <1/8 mm	40.7	1.8-80.6	27.7
Weight percent <1/16 mm	26.8	1.2-68.9	23.6
Aragonite as weight percent of carbonate	71	54-89	10
Magnesium carbonate in calcite (weight percent)	12	7-15	2

[1]Nineteen samples used in calculation.

cryptocrystalline skeletal grains suggests that the cryptocrystalline grains in this facies are probably altered skeletal material. Samples containing small amounts of limestone country rock invariably have a bimodal calcite X-ray pattern that consists of a major high-magnesium and a minor low-magnesium peak. The low-magnesium peak is produced by the limestone, and the high-magnesium peak is produced by skeletal material such as peneroplids, miliolids, and red algae.

Between the reef and the line of cays, over half the bottom is represented by barren rock and the remaining half is represented by sediment of the *Halimeda* facies in the form of narrow bars. In contrast, the bottom west of the line of cays is mantled extensively by the *Halimeda* facies. Here, the average sediment thickness is approximately 4 ft (1.2 m), and channel fillings and bars are as thick as 8 ft (2.4 m). In both areas the *Halimeda* sediments are poorly bedded and/or burrowed (Fig. 17E).

Depositional Environment

Sediments of the *Halimeda* facies are located between the reef facies, characterized by considerable current agitation, and the miliolid-mud facies, characterized by negligible current agitation. The *Halimeda* facies is thus intermediate with respect to current conditions represented by the adjoining facies. It is also intermediate with respect to salinity; the observed bottom salinities within this facies range from normal marine (36°/₀₀) to slightly brackish (26°/₀₀).

The distribution of organisms within the *Halimeda* facies is related closely to edaphic conditions. In the outer lagoon, the large areas of rock bottom support a sparse population of sponges and alcyonarians. Corals (*Porites porites*) and green algae (*Halimeda, Penicillus,* and *Rhipocephalus*) live in small sand-filled depressions on the rock. The sandbars in the area are partly stabilized by sparse grass (*Syringodium* and less commonly *Thalassia*). The mollusk fauna of the sandbars is typified by burrowing types such as the pelecypods *Lucina pensylvanica* and *Codakia orbicularis* and the gastropods *Polinices lacteus* and *Terebra hastata* (Robertson, 1963).

In the inner lagoon where sediment constitutes most of the bottom, worm mounds and dense stands of *Thalassia* are common. The *Thalassia* provides cover and food for a vagile mollusk fauna including the large gastropods *Strombus gigas, S. costatus, S. raninus, Xancus angulatus,* and *Cassis madagascariensis,* and the several smaller gastropods *Modulus modulus, Astraea phoebia, Columbella mercatoria,* and *Cerithium litteratum* (Robertson, 1963). *Halimeda* plants are abundant. *Rhipocephalus* sp. and *Penicillus* sp., which are important producers of aragonite needles (Lowenstam, 1955, p. 271; Stockman et al, 1967), are common. Corals are not common, but where present are represented by *Manicina areolata, Porites porites,* and scarce *Siderastrea* sp. In addition to stabilizing the bottom and providing forage, *Thalassia* acts as a substrate for many encrusting and attached organisms. The green alga *Batophora,* which produces aragonite needles, is commonly attached to stout horizontal rhizomes of *Thalassia*; spirorbid worms, "*Melobesia,*" foraminifers, and bryozoans live attached to the leaves of the same plant. In dense *Thalassia* areas the remains of these organisms form as much as 5 or rarely 10 percent of the sediment. The cell pattern of *Thalassia* blades can be seen on the flat attachment surface of "*Melobesia*" and spirorbid tubes, so that their remains attest not only to their epiphytic habitat but also to the existence of a flora that has very little chance of being preserved in the geologic record. They thus become clues for recognizing fossil grass bottoms. Worm mounds are common in the *Thalassia* bottoms, and the large annelid worms encountered during coring probably account in part for the poor bedding of the sediment in these areas. Indurated fecal pellets, containing recognizable skeletal grains, are also abundant in the same grass-rich areas.

Grain Alteration

Halimeda plates in relatively pure sands invariably contain acicular aragonite crystals occluding the utricles (Fig. 10F, G). Less commonly, peneroplids in the same samples were observed to have their chambers partly filled with stubby calcite crystals. Such void fillings are most common in pure sands containing less than 10 percent mud. These sands are located in areas of relatively high current activity. In these sands, the constituent grains are exposed to the circulation of seawater that is supersaturated with respect to calcium carbonate (Fig. 8). Under such conditions it is likely that the crystals of a grain act as nuclei for the precipitation of mineralogically similar carbonate on free surfaces. Aragonitic *Halimeda* becomes infilled with aragonite; calcitic peneroplids become infilled with calcite. Competition for space favors the growth of crystals oriented normal to the growth surface. Crystals growing on the grain exterior are readily abraded, but void fillings are preserved.

Muddy *Halimeda* sands from quieter water areas, located leeward (west) of the line of cays, show little tendency for crystal infilling. The water of this area and of part of the adjacent miliolid-mud facies is slightly more supersaturated with respect to aragonite, however, than is water near the reef and nearer the mainland (Fig. 8). The higher supersaturation is probably related to the high rate of photosynthesis in these grass-rich areas. During the light reaction of photosynthesis, more carbon dioxide is removed from the surrounding water than is added from respiration; consequently, there should be an increase in supersaturation with respect to calcium carbonate, other factors being equal. Why, then, are crystal infillings scarce in these areas? Apparently it is a question of the relative length of exposure to circulating seawater. In muddy areas, sand grains are buried rather rapidly in a muddy matrix that impedes interstitial circulation of seawater, whereas, in sand areas, sand grains are buried more slowly and are reexposed frequently to seawater.

The environmental conditions favorable for void filling are also favorable for alteration of skeletal structure into cryptocrystalline material. In some samples, as high as 80 percent of the *Halimeda* grains are more than half altered to cryptocrystalline material. Unaltered *Halimeda* plates consist of interlaced aragonite needles which measure approximately 0.5 x 5μ (Lowenstam, 1955, p. 271). These needles are embedded in an organic matrix which is brown in transmitted light and is dark between crossed nicols (Fig. 10E). During grain alteration, the brown color becomes grayish in transmitted light and, under crossed nicols, areas which normally appear black are brownish-gray cryptocrystalline material. The cryptocrystalline material appears to consist of randomly oriented clay-size and fine silt-size aragonite. The process or processes which alter *Halimeda* skeletal structure to cryptocrystalline structure are unknown. The following two suggestions should be considered as working hypotheses:

1. The formation of cryptocrystalline grains may be inorganic; crystals within the grain may act as seed crystals, which tend to grow in the presence of supersaturated water. The preferential growth of patches of crystals may displace organic material and concentrate it as small dark islands in a cryptocrystalline matrix.

2. Microorganisms may be important in the formation of cryptocrystalline material. Although an environment with strong light, circulating water, and unstable coarse-sand substrate seems inhospitable for microorganisms, the microenvironment within *Halimeda* plates may be favorable for the growth of algae and bacteria. When *Halimeda* sands are stored in seawater, a strong hydrogen sulfide odor quickly develops, and mats of green algal filaments grow on lighted sides. Perhaps when the sands are cultured, the intergrain environment becomes, like the intragrain microenvironment, anaerobic and reducing. The hydrogen sulfide is probably produced by anaerobic sulfate-reducing bacteria which live by oxidizing organic material and reducing sulfates to hydrogen sulfide. Oppenheimer and Kornicker (1958), by careful *in situ* measurements of sediments, have shown that carbon dioxide is also produced and the sediment pH is lowered by sulfate-reducing bacteria. They propose the following reaction (p. 13):

$$2CH_3CHOHCOOH + 2H_2O + Ca^{2+} + SO_4{}^{2-} \longrightarrow$$

$$2CH_3COOH + 3H_2O + H_2S + CO_2 + CaCO_3.$$

Thus, sulfate-reducing bacteria not only destroy organic material, but also may cause precipitation of calcium carbonate.

Halimeda is one of the most abundant constituent grains associated with reef areas (see review by Hoskin, 1963, p. 54) and is commonly cited as the major source of aragonite needles in carbonate ("lime") muds. In laboratory experiments, chemical oxidation of the organic material binding the needles causes them to be released (Lowenstam, 1955, p. 270). In natural environments, however, heterotrophic bacteria and fungi are the main destroyers of organic material. It is possible that destruction of organic material by these organisms actually prevents needle release if it induces the formation of cryptocrystalline grains.

The reef has an important physical influence on the depositional environment of the *Halimeda* facies. The reef restricts water flow and creates the environment of the backreef lagoon. Goreau (1959, p. 89) considered the reef environment to be dynamic: early-stage growth by *Montastrea* spp. in relatively deep water forms a solid platform that grows upward into shallow water. The platform impedes circulation and produces turbulence in which the *Acropora palmata* climax stage develops. Thus the relatively quiet lagoon behind the reef in northern Belize owes its existence to reef growth.

The *Halimeda* facies occurs in the lagoon area of transition between the wave-agitated reef and the negligible currents of the inner lagoon. *Halimeda*-rich sediments have also been noted in similar areas of Florida Bay, the Gulf of Batabanó (Cuba), Funafuti Atoll, and Bikini Atoll (see review by Ginsburg et al, 1963). Ginsburg et al (p. 572) noted that the rate of production of *Halimeda* is high in reefs, but a higher rate of production of corals and coralline algae in reefs reduces the relative importance of *Halimeda*. Behind the reef, however, current conditions are unfavorable for a high production rate of corals and coralline algae, although they are still favorable for a high production rate of *Halimeda*. Thus, sands consisting almost entirely of *Halimeda* plates are produced.

Peneroplid-Sand Facies

Sediments of this facies are characterized by coarse sands containing abundant peneroplid foraminifers (Table 3). The only other important sand-size constituents are mollusks and cryptocrystalline grains. Most of the recognizable skeletal grains are more than half altered to cryptocrystalline material. Over four fifths of the sediment is calcite, evidencing a moderately high degree of magnesium carbonate substitution.

The peneroplids are nearly all of the same species—*Archaias angulatus* (Wantland, 1967, p. 75). Peneroplids account for most of the magnesium calcite in the peneroplid facies. Mollusks, which probably supply the relatively small amount of aragonite in the sands, are represented chiefly by *Chione cancellata* and *Brachidontes exustus*. The latter is commonly attached to rocks.

Microorganism borings are very common in all skeletal grains. For example, mollusks commonly are reduced to a spongy network of carbonate which in plane-polarized light superficially resembles cryptocrystalline grains (Fig. 13). Between crossed nicols, however, molluscan structure is clearly evident (Fig. 13F). Peneroplids and miliolids are also usually bored, and the more isolated borings commonly appear cryptocrystalline between crossed nicols (Fig. 11F). Grains with closely spaced borings appear cryptocrystalline between crossed nicols but commonly retain their original brown color in ordinary light. Few grains with patchy alteration to grayish cryptocrystalline material are evident (Fig. 12A).

The peneroplid facies is deposited unevenly on the windward side of lower Chetumal Bay. In many places the sands merely veneer an uneven rock surface. Elsewhere the sands form elongate sandbars as thick as 3 ft (0.9 m). Generally the sands directly overlie rock, but in a few

TABLE 3. MEAN CONSTITUENT COMPOSITION, GRAIN SIZE, AND CARBONATE MINERALOGY OF PENEROPLID-SAND FACIES[1]

	Mean	*Observed Range*	*Standard Deviation*
Coralline algae	0.0	0.0	-
"*Melobesia*"	0.3	0.0- 1.0	0.3
Halimeda	0.0	0.0	-
Miliolidae	7.1	1.0-16.3	5.1
Peneroplidae	47.3	19.7-75.7	16.0
Other Foraminifera	1.6	0.1- 3.8	1.2
Corals	0.0	0.0	-
Mollusks	18.5	6.1-48.9	12.0
Ostracods	0.7	0.0- 3.0	0.8
Miscellaneous skeletal	0.4	0.0- 2.1	0.6
Unknown skeletal	2.2	0.5- 3.5	1.3
Total skeletal	78.0	48.2-96.6	14.8
Fecal grains	0.2	0.0- 1.2	0.4
Cryptocrystalline grains	20.3	3.5-50.0	11.6
Cryptocrystalline skeletal grains	35.2	14.0-63.3	16.2
Total over 50 percent cryptocrystalline	55.4	21.7-76.3	18.2
Country rock	0.6	0.0- 1.7	0.6
Terrigenous minerals	1.0	0.0-12.6	3.5
Weight percent <1/8 mm	16.7	2.7-51.7	13.2
Weight percent <1/16 mm	10.2	1.8-39.8	10.6
Aragonite as weight percent of carbonate	16	3-48	14
Magnesium carbonate in calcite (weight percent)	11	7-13	2

[1]Thirteen samples used in calculations.

places they lie on peat deposited in sinkholes. On the north end of the Bulkhead (Fig. 9), the peneroplid sands grade downward into laminated carbonate muds. Laterally the sands grade into the finer grained sediments of the miliolid-mud and cryptocrystalline-grain facies. The peneroplid sediments are generally well bedded, and bedding is particularly well evidenced by the parallel orientation of the tabular peneroplids (Fig. 16D).

Depositional Environment

The peneroplid facies is deposited under rather strong current conditions, as evidenced not only by the virtual absence of mud and fine sand, but also by the large areas of rock bottom characterizing the depositional environment. The depth of water overlying the bottom generally ranges between 5 ft (1.5 m) on sandbars and 7 ft (2.1 m) on the rock areas. The filter-feeding clam *Chione cancellata* appears to thrive in the current-swept sandbars.

The observed salinities range from 35°/₀₀ to less than 20°/₀₀ and are influenced greatly by rainfall runoff from Ambergris Cay and numerous smaller cays. Narrow openings between Chetumal Bay and the sea, such as at the north end of Ambergris Cay, admit some marine water to the peneroplid depositional environment, but the amount is not sufficient to affect the general salinity pattern (Fig. 8). Weyl saturometer measurements indicate that the waters occupying the depositional environment are similar to seawater in their saturation relative to aragonite (+15 to +20 mv).

Origin of Cryptocrystalline Skeletal Grains

Cryptocrystalline skeletal grains are characteristic of the peneroplid-sand facies; therefore, it seems appropriate to consider their origin. The cryptocrystalline fabric of the grains in this facies is of comparatively recent origin. For example, sample 266, which contains approximately 70 percent cryptocrystalline skeletal grains (mostly peneroplids), 15 percent cryptocrystalline grains, and 15 percent skeletal grains, is dated as 450±105 years old by C-14 analysis. Moreover, the fact that all stages of transition were observed between partially cryptocrystalline skeletal grains and cryptocrystalline grains suggests that the latter have originated through alteration of skeletal carbonate. Therefore, the problem is reduced to finding the cause for skeletal alteration to cryptocrystalline carbonate. Fresh and altered peneroplids compose the bulk of this facies and were selected for special study in order to understand better the origin of cryptocrystalline grains.

Mineralogic changes accompanying alteration to cryptocrystalline material—Peneroplids with cryptocrystalline wall structure were recognized readily under the binocular microscope by their grayish appearance. The grayish appearance results from drusy void fillings that are usually present in cryptocrystalline peneroplids and compose an average of nearly 30 percent by volume of the altered tests (Table 4). As is shown in Table 4, cryptocrystalline peneroplids are represented by calcite with a magnesium content comparable to that of unaltered peneroplids. Chemical analyses for calcium ions and magnesium ions, converted to carbonate, confirm the high magnesium content of the cryptocrystalline peneroplid grains. Chemical determinations of magnesium, however, gave values consistently greater than X-ray determinations. The difference between chemical analyses and X-ray analyses of the magnesium content is slightly greater than the precision of the methods and is probably real. Chave (1954) and Goldsmith et al (1955, p. 220) noted in algal calcites that X-ray methods yielded magnesium values that were too low with respect to that determined by chemical analyses. The X-ray method of measuring magnesium is limited to the detection of magnesium present as magnesium carbonate replacing calcium carbonate in the calcite lattice. Thus the discrepancy between the two types of analyses suggests that magnesium is present in forms other than crystalline calcite—that is, as adsorbed ions (Goldsmith et al, 1955, p. 220) or in a finely divided mineral state. When algal calcites were heated under high temperature and pressure by Goldsmith et al, X-rays of interplanar spacing showed a shift toward spacings expected for the chemically determined magnesium content. It is evident, then, that one cannot positively describe the nature of the magnesium in the cryptocrystalline peneroplids from northern Belize, but one may safely conclude that the peneroplids do not undergo a significant mineralogic change during alteration to cryptocrystalline material.

Microorganic boring and alteration to cryptocrystalline material—Grain alteration to cryptocrystalline material is usually accompanied by microorganism borings. The most cited microborers are plants, especially green algae (Nadson, 1927; Humm and Taylor, 1961) and fungi (Pia, 1940). As shown in Figure 11E and F, microborings tend to be either

TABLE 4. COMPARISON OF PENEROPLID TESTS WITH AND WITHOUT CALCITE VOID FILLINGS

	Average Weight of Test (mg)[1]	*Carbonate Mineralogy (Weight Percent)*[2]			*Weight Percent Calcium and Magnesium Converted to Carbonate*[3]	
		Aragonite	*Calcite*	*$MgCO_3$*	*$CaCO_3$*	*$MgCO_3$*
White tests (without void fillings)	0.54	not detected	100	9.0	86	14
Gray tests (containing void fillings)[4]	0.76	?trace	>96	11.5	85	15

[1]Peneroplids were picked from sample 29. Many specimens intermediate between fresh white and dull gray were discarded. Weights of white tests are based on 314 individuals weighed in two lots. Weights of gray tests are based on 1,053 individuals weighed in six lots.

[2]X-ray techniques were used. Interplanar spacing was determined relative to a silicon-metal internal standard. Weight percent of magnesium carbonate determined from Chave's (1952, p. 191) curve of interplanar spacing versus magnesium carbonate. Reproducibility is about ±1 weight percent. Dolomite was not positively identified in any samples. The trace of aragonite found in altered tests is considerably less than the void-filling weight gain and probably is due to contamination.

[3]Flame-photometry method; analyst, David S. McKay of Rice University. Because small quantities (26-48 mg) were used, these values are probably only accurate to ±2 weight percent.

[4]The crystal fillings tend to absorb light and thus produce the darker appearance of the filled tests. Also, tests with void fillings tend to be worn so that their outer light-reflecting wall is gone.

relatively large (about 30μ) or very small (2-8μ). Small sinuous tubes which have extremely small irregularities are most common. These are similar in size and shape to the green algal genera *Ostreobium* and *Gomontia* as illustrated by Humm and Taylor (1961). Borings have been observed in opaline sponge spicules and coarse calcite spar but are generally found in carbonate skeletal grains.

In thin sections, skeletal material constructed of relatively large crystal units, such as are found in mollusks and echinoderms, commonly is seen to be surrounded by a rim of semiopaque cryptocrystalline material (Fig. 13C, D). Microorganism borings cannot be seen in the rim material because of its opacity, but a narrow transition zone of solitary borings is usually present between the cryptocrystalline rim and the unaltered skeletal material. Furthermore, when the carbonate is removed by acid, gray replicas of borings remain. This fact suggests that the cryptocrystalline rim is highly bored and that its semiopaque appearance is produced by small amounts of organic material in the tubes. A few intensely bored mollusks containing tubes with no opaque material are seen. These bored grains resemble finely cryptocrystalline grains in transmitted light (Fig. 13E), but between crossed nicols they display a coarse-grained molluscan architecture (Fig. 13F), especially if the stage is rotated. If opaque material is present in the tubes, however, it is difficult to determine whether or not recrystallization has produced the cryptocrystalline material. The correlation between microboring and cryptocrystalline alteration suggests that, if cryptocrystalline material in mollusks does form by recrystallization, the recrystallization is promoted by boring and is perhaps related to decomposition of organic material in the borings. This recrystallization apparently involves a decrease in crystal size.

In contrast, skeletal grains constructed of small crystals (less than approximately 2μ), such as miliolids or peneroplids (Hay et al, 1963), appear to alter to cryptocrystalline material that is more coarsely grained than the unaltered material. Alteration in this type of skeletal architecture, as exemplified by peneroplids, also is associated with microborings. Unlike mollusks, the peneroplid grain is rarely reduced to a meshwork of fragile carbonate (Fig. 12C) by intense boring. A few microborings that are not filled with opaque material are cryptocrystalline along their entire length (Fig. 11E, F). This suggests that recrystallization to coarser grain size may have occurred along the boring. Patchy alteration of miliolid and peneroplid wall structure is further evidence of recrystallization to produce cryptocrystalline material (Figs. 11B; 12A, B).

These cryptocrystalline patches within the test wall could be interpreted as sites at which there has been growth of calcite crystals by solid-state grain-boundary migration (grain growth in sense of Bathurst, 1958). Grain growth, of course, is a common metallurgical phenomenon at elevated temperatures, but at temperatures of 20-30°C it may be insignificant in calcium carbonate sediments. For example, in a review of experimental studies on the transformation of aragonite to calcite, Curl (1962, p. 69) concluded that solid-state transformation is inadequate to account for the phenomenon and suggested that recrystallization took place through the agency of a solvent. Although an increase in grain size of a single polymorph (in this case magnesium calcite) does not involve as great an ionic reorganization as changing to another polymorph such as aragonite, solid-state growth may still be too slow to be significant.

The common presence of cryptocrystalline patches in tests of peneroplids and miliolids without attendant borings indicates that microboring is not a prerequisite for recrystallization, but the high degree of correlation between cryptocrystalline material and microboring suggests that microboring may promote recrystallization to cryptocrystalline material. The mechanism of recrystallization is unknown but may be related to either the great increase of surface area in bored grains and/or the decomposition of organic material in the borings. The possible mechanism by which bacteriologic decomposition of organic material may induce precipitation was discussed previously in regard to the *Halimeda* facies. Regardless of what the mechanism of alteration to cryptocrystalline material is, recrystallization occurs at a size range that cannot be resolved with an ordinary petrographic microscope. Scanning-electron-microscopic studies of unaltered, partly altered, and fully altered grains should provide further insight into the formation of these particular cryptocrystalline grains.

Origin of Skeletal Void Fillings

Chambers of miliolid and peneroplid Foraminifera are commonly lined, or even filled, by clear calcite crystals (drusy growth of Bathurst, 1958). The blocky spar crystals vary from clay size to silt size (4 x 15μ maximum). They tend to grow normal to their attachment surface, but some deviate from this orientation by as much as 30°. Drusy growths are present on unaltered chamber walls (Fig. 12D) but much more commonly are associated with altered wall structure (Figs. 11B, C; 12E). Crystal growth appears to be limited to grain voids, but a few weakly cemented peneroplid-sand tubes were found (Fig. 23), suggesting that some intergrain cementing may occur. The sand tubes were buried beneath a foot of current-swept sand in water depths of 4-5 ft (1.2-1.5 m); consequently, their *in situ* orientation is unknown. The inside diameter of the hollow tubes is comparable to the size of *Thalassia* rhizomes, and the cementation may be related to decomposition of plant material.

Skeletal void fillings represent a carbonate weight increase of nearly 40 percent in peneroplids (Table 4). Thus the drusy crystals are a significant part of the grain mass, and any large mineralogic differences between them and the skeletal substrate should be detected on X-ray diffractograms. The data represented in Table 4 indicate that the void-filling crystals are high-magnesium calcite. It seems likely that the drusy crystals are the product of inorganic precipitation in a subsea environment. Skeletally produced high-magnesium calcite crystals may have acted as nuclei, with preferential growth into void spaces.

Void-filling and grain-cementing crystals of high-magnesium calcite of apparently inorganic origin are present in many shallow carbonate sediments of tropical areas such as the Persian Gulf (Shinn, 1969), the Mediterranean (Alexandersson, 1969), the Bermuda reefs (Ginsburg et al, 1967), the Jamaican reefs (Land and Goreau, 1970), and the Barbados reefs (Macintyre et al, 1968).

Prior to the discovery of inorganic high-magnesium calcite in sediments, several laboratory experiments and field observations suggested that magnesium-rich calcite crystals should precipitate from seawater. Rivière (1939a) found that powdered calcite, which was kept in recirculated seawater and ventilated with carbon dioxide-deficient air, contained up to 8 weight-percent magnesium carbonate in the calcite after 11 months. He further postulated (1939b) that this process of magnesium enrichment should be active in current-washed sediments. Taft (1962) found that high-magnesium calcite in Florida Bay sediment, which is up to 3,400 years old, shows no recrystallization to low-magnesium calcite. Furthermore, magnesium calcites with up to 15 percent magnesium carbonate give positive saturation values in seawater (Schmalz and Chave, 1963, p. 1207).

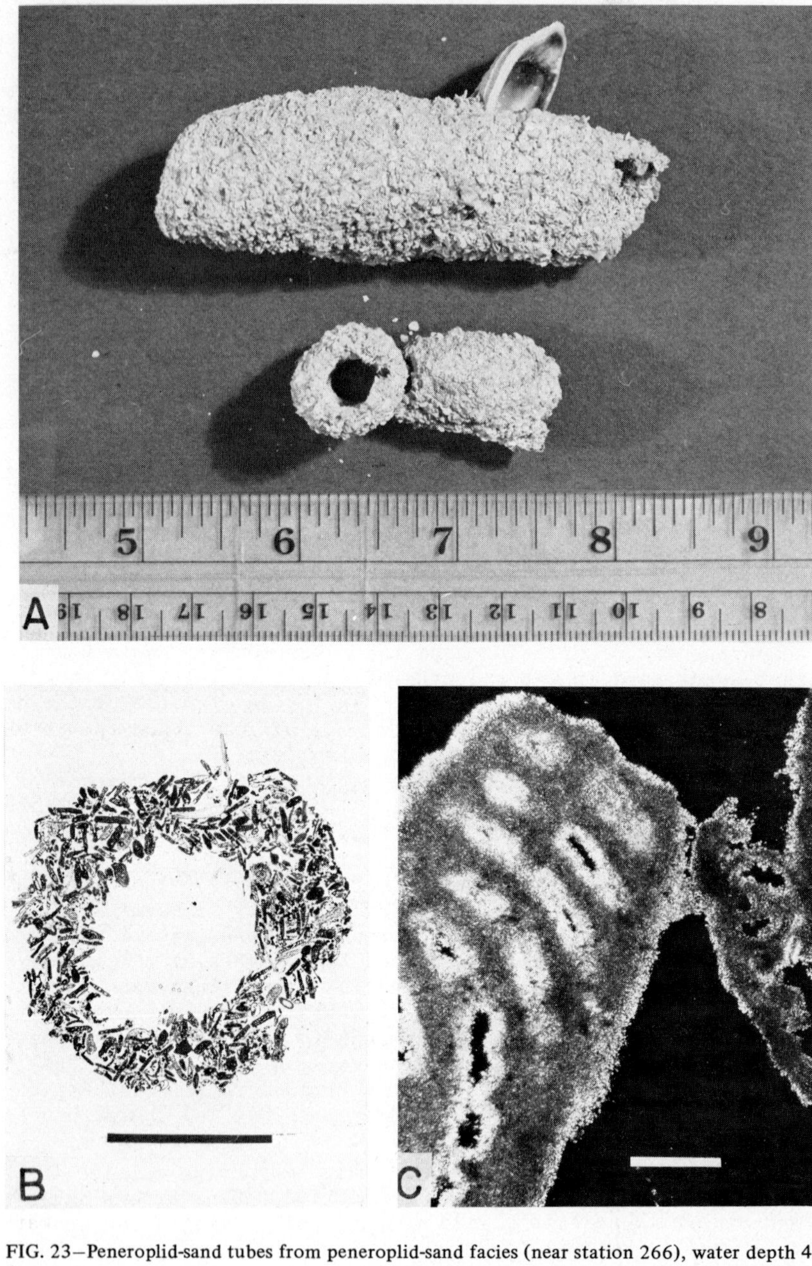

FIG. 23—Peneroplid-sand tubes from peneroplid-sand facies (near station 266), water depth 4.8 ft (1.5 m).

A. Peneroplid sand weakly cemented to form tubes. Large shell is *Chione cancellata*. *In situ* orientation of tubes is not known. Scale in inches and centimeters.

B. Thin section of peneroplid-sand tube. Most of grains are peneroplids varying from well preserved to cryptocrystalline. Carbonate cement is not visible at this magnification. Ordinary light; bar length 10 mm.

C. Thin section of peneroplid-sand tube showing cement. Surface of cryptocrystalline peneroplid is coated with clay-size crystals that presumably are calcite. Crystals cementing peneroplid to cryptocrystalline (mollusk?) grain in upper right are poorly oriented and thus resemble cryptocrystalline material. Bar length 100μ.

The most convincing explanation of the formation of high-magnesium calcites is provided by Winland (1969). He found that, in a mixture of seawater and Florida Bay sediment (which is predominantly aragonite and high-magnesium calcite), high-magnesium calcite can be alternately destroyed and formed by bubbling carbon dioxide and air, respectively, through the sediments. The amount of magnesium present is compatible with a distribution coeffieicnt for partitioning magnesium in calcite of 0.02 to as high as 0.06. In short, the partition theory (McIntire, 1963) predicts a magnesium carbonate content of 10-25 mole percent in calcites formed in equilibrium with seawater. The greatest mystery now is why low-magnesium calcite coexists with high-magnesium calcite in seawater.

There is a tendency for microorganic bores, crystal growths into voids, and cryptocrystalline skeletal material to occur in the same sample; yet these features may occur independently. Thus it is evident that these partly interdependent, but separate, processes tend to reach their maximum development under similar environmental conditions. Grain alteration to cryptocrystalline carbonate is most common in sediments deposited in current-swept areas. Sediments in these areas are thin and mobile, and grains are alternately buried and reexposed to circulating seawater and strong light. The water overlying the sands is slightly brackish but is equivalent to seawater in respect to aragonite supersaturation. These conditions apparently are favorable for the growth of boring microorganisms, the recrystallization of carbonate to cryptocrystalline material, and the precipitation of high-magnesium calcite. Microorganism boring possibly promotes recrystallization but is not clearly related to void-filling precipitation. Thus the sediments of the peneroplid-sand facies reflect their depositional environment not only by their coarse grain size and constituent-grain composition, but also by the attributes of the constituent grains themselves.

Miliolid-Mud Facies

As the name indicates, the miliolid-mud faciès is characterized by a relative abundance of mud in which miliolid grains are the dominant sand-size constituent (Table 5). Two thirds or more of the sand-size constituent grains are skeletal, and approximately half of these are represented by Foraminifera. Miliolids, especially unornamented species of *Quinqueloculina*, dominate the foraminifer population. Peneroplids are common but are dominated by one species, *Archaias angulatus* (Wantland, 1967, p. 75).

The nonskeletal grains are represented chiefly by fecal grains and cryptocrystalline grains. Both ellipsoid-shaped and rod-shaped fecal grains are present. Skeletal grains partly altered to cryptocrystalline carbonate are relatively scarce; however, they are abundant enough to be interpreted as the major source of the cryptocrystalline grains present.

The fact that the calcite content of the miliolid muds is more than twice that of the *Halimeda* facies is due to the abundance of calcite tests of Foraminifera and also to the absence of aragonite plates of *Halimeda*. This facies also differs from those previously considered, in that a small amount of quartz silt was evident in X-ray diffractograms of all the samples examined.

Depositional Environment

With the few exceptions noted subsequently, the miliolid muds generally are found in water depths of 6-12 ft (1.8-3.6 m). Observed bottom

TABLE 5. MEAN CONSTITUENT COMPOSITION, GRAIN SIZE, AND CARBONATE MINERALOGY OF MILIOLID-MUD FACIES[1]

	Mean	*Observed Range*	*Standard Deviation*
Coralline algae	0.1	0- 1.4	0.2
"Melobesia"	3.3	0.0-13.3	3.6
Halimeda	0.5	0.0- 3.6	1.1
Miliolidae	22.5	9.1-38.4	8.8
Peneroplidae	7.0	0.0-17.6	6.8
Other Foraminifera	4.4	1.7- 6.8	1.3
Corals	0.0	0.0- 0.2	0.0
Mollusks	21.7	7.0-36.4	9.0
Ostracods	3.9	0.4-18.6	4.3
Miscellaneous skeletal	2.9	0.0-11.0	3.5
Unknown skeletal	3.1	1.4- 8.4	1.9
Total skeletal	69.3	39.6-95.2	15.9
Fecal grains	5.8	0.2-16.7	5.8
Cryptocrystalline grains	22.3	0.6-52.3	17.3
Cryptocrystalline skeletal grains	5.4	1.1-12.1	3.6
Total over 50 percent cryptocrystalline	27.8	5.6-65.1	19.5
Country rock	1.9	0.0- 9.0	3.0
Terrigenous minerals	1.1	0.0- 8.3	2.2
Weight percent <1/8 mm	69.9	20.9-99.4	21.8
Weight percent <1/16 mm	50.4	14.1-92.7	25.0
Aragonite as weight percent of carbonate	27	6-55	15.5
Magnesium carbonate in calcite (weight percent)	9	5-12	2.9

[1]Seventeen samples used in calculations.

salinities varied between 28°/₀₀ and 33°/₀₀ in August 1961 but reached a high of 37°/₀₀ in August 1962 during a dry spell. The water is supersaturated with respect to aragonite, but the supersaturation pattern is not a simple one (Fig. 8). For example, saturation values tend to be positively correlated with salinity values between the *Halimeda* facies and the mainland, whereas an inverse relation is apparent between the same two variables in Chetumal Bay. The significance of this contrast will be considered subsequently.

Thalassia testudinum (turtle grass) is abundant and in places completely hides the bottom. Bands of dense *Thalassia* approximating 1/4 mi (0.4 km) in width and as much as 3 mi (4.8 km) in length are oriented normal to the main channel trend of the bay but are usually feathered parallel with the trend. As in the *Halimeda* facies, the *Thalassia* blades support epiphytic melobesids, spirorbids, and bryozoans that form a significant part of the skeletal constituents in grassy areas.

The water over the miliolid-mud area is typically clear in the morning, but by midafternoon is commonly so milky that underwater visibility is reduced to less than 3 ft (0.9 m). This increase in turbidity results from bottom agitation induced by northeast trade winds reinforced by a land breeze. At times patches of turbid water, resembling the "whitings" of the Bahamas, are present when the rest of the lagoon is clear. The white patches were photographed at several places in the lagoon between Belize City and the Bulkhead during an aerial reconnaissance by E. G. Purdy in August 1963 (Fig. 17A). The "whitings" are elongate in the direction of prevailing wind-induced currents. The part of the "whiting" facing into the current is in sharp contact with clear water along a sawlike border. The downcurrent end of each "whiting" is less opaque than its upcurrent counterpart and grades into clear water. No close examination of the "whitings" was made, but their shape indicates that they probably are similar to the "fish muds" reported in Florida Bay by Ginsburg (1956, p. 2398). The front edge of each patch may result from a school of fish that, through bottom-feeding activities, suspends fine-grained particles in the slow-moving current. The gradual settling and slow horizontal movement of this sediment then would combine to produce a cometlike tail. The possibility exists, of course, that northern Belize "whitings" are similar to Bahamian "whitings," which may represent either localized "clouds" of physicochemical carbonate precipitation (Cloud, 1962, p. 19ff) or "clouds" of stirred-up sediment that act as nuclei for additional carbonate precipitation (Purdy, 1963b, p. 488).

The miliolid-mud substrate supports a mollusk fauna characterized by a large number of individuals representing relatively few species, a characteristic that seems closely related to the variable salinity of the habitat. *Chione cancellata* is the most abundant mollusk present; the pelecypods "*Phacoides*" *pectinatus*, *Brachidontes exustus* (attached to shells and *Thalassia* rhizomes), and *Pinctada radiata* and the gastropod *Melongena melongena* are present in smaller numbers.

Ambergris-Cangrejo mud-shoal complex—One of the more striking bathymetric features of the miliolid-mud environment is a series of mud shoals located between Ambergris Cay and Cangrejo Cay (Fig. 3). These shoals generally are covered by less than 2 ft (0.6 m) of water and are separated from each other by relatively deep channels, some of which reach a depth of 10 ft (3.0 m). The surface of all the shoals is mantled by a dense carpet of *Thalassia testudinum* except where red mangroves have constructed mangrove cays.

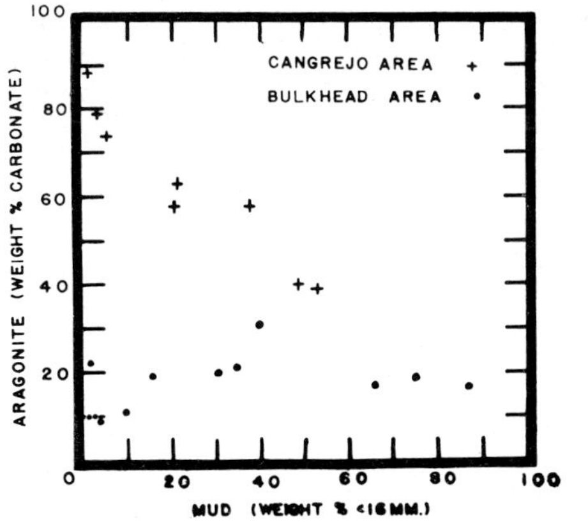

FIG. 24—Graph of aragonite content versus mud content in samples from Cangrejo Cay area and Bulkhead area (diagram based on data in Appendix II).

Sedimentologically, the shoals are composed of carbonate mud averaging approximately 10 ft (3.0 m) in thickness. Cores of the sediment supplemented by probes with a metal-tipped rod indicate that shoal sediment thicknesses vary from 5 ft (1.5 m) near Ambergris Cay to more than 15 ft (4.6 m) midway between Ambergris Cay and Cangrejo Cay. The sediment forming the shoals is somewhat transitional in constituent composition and mineralogic composition between the *Halimeda* facies to the east and the miliolid-mud facies to the west. The gradational change in mineralogy is illustrated in Figure 24, which shows that the percentage of aragonite relative to calcite in the mud fraction decreases from east to west. The average composition of the shoals, however, closely approximates that of the miliolid-mud facies; therefore, the Ambergris-Cangrejo mud-shoal complex is included within the miliolid-mud facies rather than identified as a separate facies.

The origin of the Ambergris-Cangrejo mud-shoal complex has been considered by Vermeer (1959). Water piled up between Ambergris Cay and the barrier reef by the wind and waves is funneled down a gradually deepening lagoon channel. The water flow sweeps the channel clean and carries coral-reef detritus. Vermeer (p. 48) wrote:

> As this flow moves southward it comes increasingly in contact with water flowing in through the break in the reef abreast of the southern tip of Ambergris Cay. As a result of the junction of the water flowing southward along the face of the cay with westwardly flowing water through the break in the reef, the southern portion of Ambergris Cay is curved toward the mainland.

Vermeer's hypothesis adequately accounts for the fact that the windward side of these mud shoals is somewhat similar in constituent composition and mineralogy to the *Halimeda* sands, but it does not explain the preponderance of calcite mud as opposed to aragonite sand on the leeward side of the shoals. In brief, if all the mud deposited on the shoals were derived from the *Halimeda* facies and/or the coral-reef facies, as required by Vermeer's hypothesis, there would be no obvious reason for the mineralogic differentiation of muds between windward and leeward sides of the shoal. Mud-size constituents derived from either the *Halimeda* or coral-reef facies should be largely aragonitic in composition like the mud-size constituents within each of these two facies. This apparent anomaly can be explained, however, by ebb-tidal currents that transport the calcite mud-size constituents of the miliolid-mud facies eastward until the ebb-tidal current velocities are checked by the longshore currents flowing southward from the east side of Ambergris Cay. Presumably, the result of this current interference would be the deposition of calcite mud in the present locale of the Ambergris-Cangrejo mud shoals.

Thus these particular mud shoals can be explained without recourse to the plant-baffle hypothesis advocated by Ginsburg and Lowenstam (1958) for the Florida Bay mud mounds. In the case of the Ambergris-Cangrejo mud shoals, the dense carpets of *Thalassia* seem to maintain the shoals but not to have caused their formation.

Laminated Miliolid Muds

Throughout the major part of the miliolid-mud facies, the sedimentary section, as evidenced by cores, is characterized by internal structures consisting either of poorly developed regular layering (Fig. 17A) or, less commonly, mottling (Figs. 17D, 18F) that is judged to be the product of burrowing organisms. In the area referred to as the "Bulkhead," however (see Fig. 2), the muds are distinctly laminated (Fig. 16C) and form mud shoals. The combination of these two attributes seemed

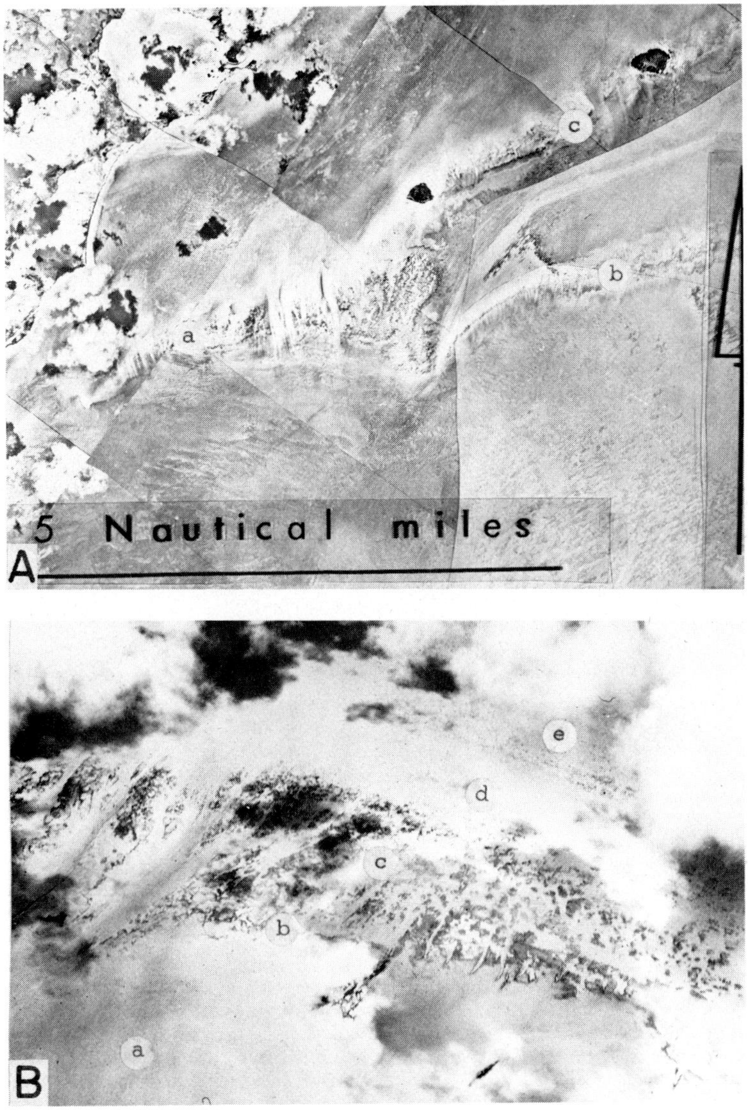

FIG. 25—Aerial photographs of Bulkhead mud-shoal complex.

A. Vertical aerial mosaic showing following features: **a.** Western lobe. Mud bar with many well-developed channels. Water depth at crest is 2-3 ft (0.6-0.9 m). **b.** Southeast lobe. Continuous mud bar with 2-4 ft (0.6-1.2 m) of water at crest. **c.** Northeastern lobe. Mud bar with few channels. Water depth at crest is 1 ft (0.3 m) or less.

B. Western lobe of Bulkhead mud-shoal complex. Oblique aerial photograph looking southeast from bay into lagoon. Letters designate following features: **a.** Slope from mud shoal into Chetumal Bay. Water depth 2-5 ft (0.6-1.5 m); no grass; covered with yellow-brown organic detritus. **b.** Crest of mud shoal. Water depth 2-3 ft (0.6-0.9 m), less than 1 ft (0.3 m) on *Chione*-peneroplid sand lenses. Several small channels barely interrupting crest can be seen in right part of photo; two large channels can be seen on left. **c.** Mud shoal. Water depth 3-4 ft (0.9-1.2 m); sparse grass. Large channel on left is 200 ft (60 m) wide and up to 7 ft (2.1 m) deep. **d.** Slope from mud shoal into lagoon. Water depth 4-6 ft (1.2-1.8 m); no grass, covered with yellow-brown organic detritus. **e.** Lagoon. Water depths 6-12 ft (1.8-3.6 m); moderate to dense grass cover on bottom.

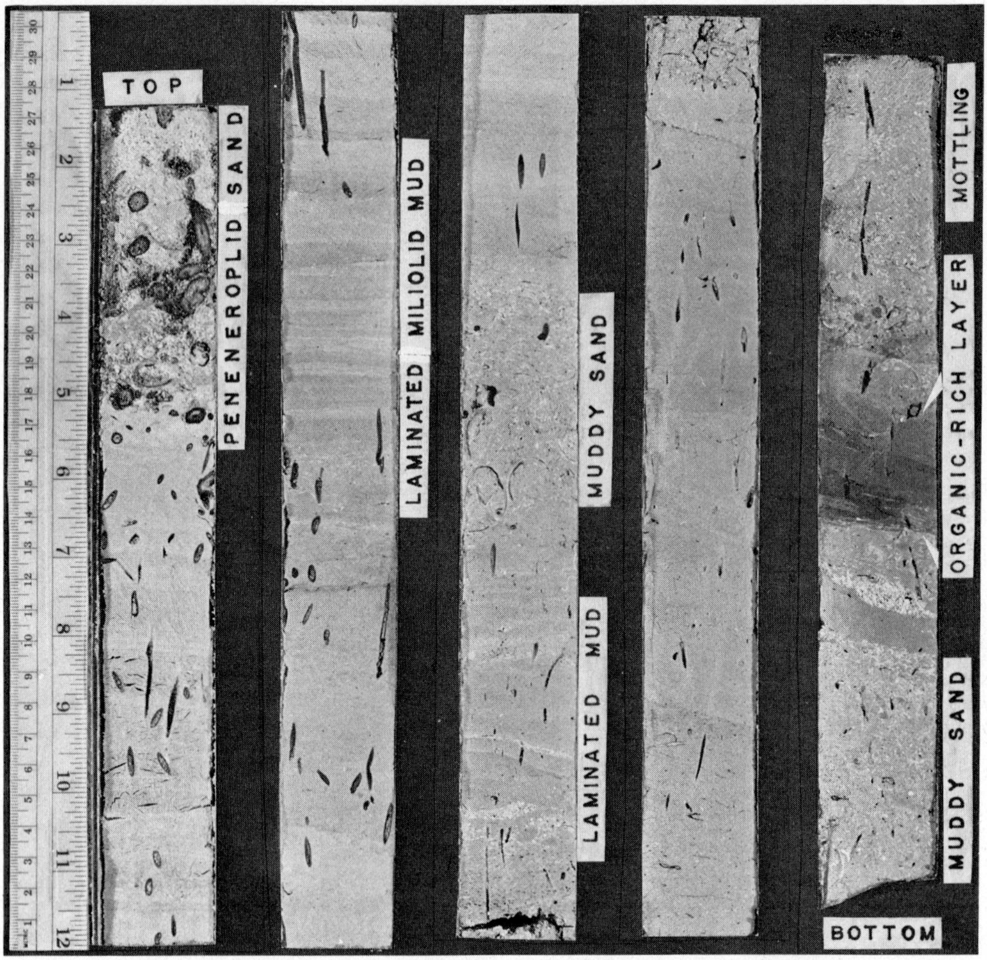

FIG. 26—Sedimentary structures in a core from crest of Bulkhead mud-shoal complex (core 244), water depth 3 ft (0.9 m). Incomplete core sequence; core compacted approximately 20 percent. Scale in inches and centimeters. Details of sequence, from top to bottom, are: (1) Peneroplid sand. Peneroplid sand with *Chione cancellata*. Horizontal rhizomes of *Thalassia* grass appear as 1/8-1/4-in. (3.2-6.3 mm) circles. (2) Laminated miliolid mud. Light-colored layers are rich in miliolids; darker layers are olive gray (5Y6/1) carbonate mud. Light layers grade upward into darker mud. Dark plant fragments are terete leaves of *Syringodium* grass. (3) Muddy sand. *Chione*-rich and peneroplid-rich layer. Note lack of laminations or other bedding features. (4) Laminated mud. Laminae of irregular peneroplid layers and lenses. Disseminated plant fragments are abundant. Change in bedding angle can be produced by bending of flexible coring tube during sampling, but parallellism of plant fragments with sides of core suggests that dip is depositional. (5) Mottled layer. Disrupted bedding and lumps of carbonate mud in peneroplid-mud matrix. (6) Organic-rich layer. Dark gray color is produced by fine plant material. (7) Muddy sand. Abundant peneroplids and *Chione*. Note lack of bedding in lowest part of core. Bottom of core is approximately 1 ft (0.3 m) above hard (rock?) surface.

sufficiently important to warrant the establishment of a separate facies, herein termed the "laminated miliolid muds." This distinction is based on criteria that were not used in the statistical grouping of samples, so the recognition of this facies does not reflect an inadequacy of the computer program in delineating facies. The establishment of this facies does convey correctly the impression that the statistically delineated

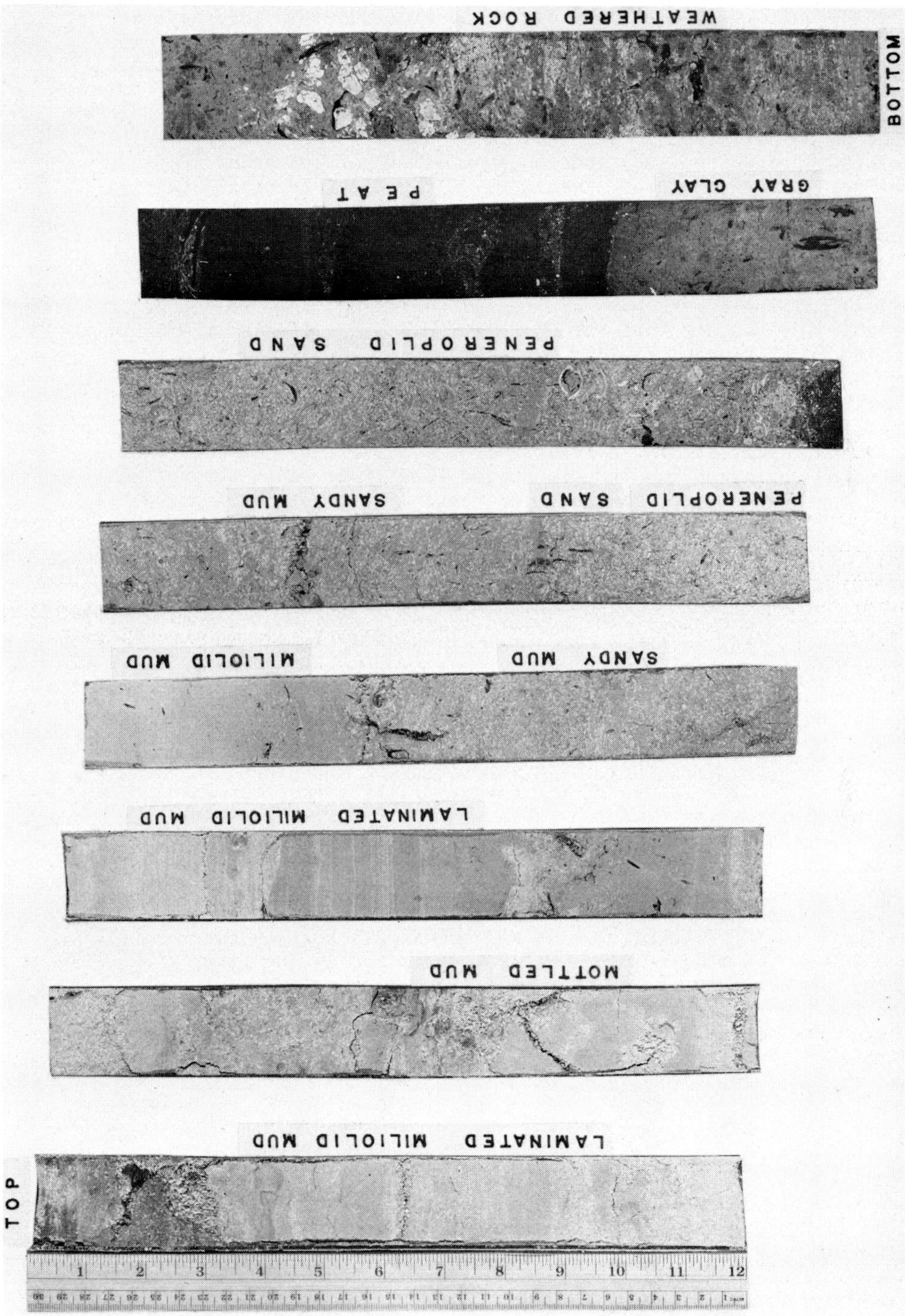

FIG. 27—Sedimentary structures in core from Bulkhead mud-shoal complex (core 256), water depth 1.8 ft (0.5 m). Complete sequence, core compacted less than 5 percent. Scale in inches and centimeters. Details of sequence are: (1) Laminated miliolid mud. Light-colored laminae of miliolids and peneroplids in carbonate-mud matrix. (2) Mottled mud. Mottles of peneroplid sands in carbonate-mud matrix. Mottles probably were produced by burrowing organisms. Shells are mostly *Chione cancellata.* (3) Laminated miliolid mud. Distinct laminae of miliolids alternating with carbonate mud. (4) Sandy mud. Poorly bedded peneroplids in carbonate mud. Note gradation into underlying peneroplid sand. (5) Peneroplid sand. Sand with abundant peneroplids and *Chione cancellata.* Rock fragments are abundant near peat contact. (6) Peat. Bedded mangrove peat. (7) Gray clay. Greenish-gray clay containing quartz silt and sand. (8) Weathered rock. Soft fossil-bearing rock. Note gravel at contact with gray clay.

facies should be viewed merely as a frame of reference for facies delineation.

The laminated carbonate muds average 6 ft (1.8 m) in thickness and locally are as thick as 9 ft (2.7 m). The mud is deposited in the form of an elongate shoal barring three fourths of the entrance to Chetumal Bay. This barrier to circulation is referred to as the "Bulkhead" on Hydrographic Office Charts and will be referred to by the same name in the succeeding paragraphs. The maximum measured slope from the mud shoal into surrounding sandy muds is about 11 ft/mi, or about 1/8°. The Bulkhead mud-shoal complex consists of two elongate shoals that converge on the west side of the complex to form a relatively wide shoal. The east and west parts of the mud-shoal complex are discussed separately.

Eastern Bulkhead shoal—The eastern two thirds of the Bulkhead, which consists of two narrow shoals, occupies the center of the bay entrance (Fig. 25A). The elongate bars are nearly continuous and are not deeply channeled. The southernmost bar is relatively deep (2-4 ft or 0.6-1.2 m) and is made up of several smaller bars which slope gradually to a grass-covered bottom. These bars are elongate, trending from west-southwest to east-northeast. *Thalassia* is present but is abundant only on deeper parts of the mud-shoal complex. The northern elongate shoal trends southwest-northeast and is in relatively shallow water. It crests at a 1-ft (0.3 m) depth and has been colonized locally by red mangroves to form two cays and several mangrove colonies. The crest is veneered by an inch or less of peneroplid sand overlying laminated mud. With the exception of the upper inch, the sediments of the northern and southern shoal are similar in composition, grain size, and structure (compare Figs. 26, 27).

Western Bulkhead shoal—The western third of the shoal complex differs from the rest of the Bulkhead by virtue of being interrupted by many well-developed channels (Fig. 25A, B). The channels apparently remain stable for more than a decade, because channels visible on aerial photographs taken in 1945 were also in evidence in 1963 despite the passage of several hurricanes across the area. The existence of a few filled channels shows, however, that their geometry does change over longer periods of time. The nature of the channel fillings was not studied, but it is known that the existing channels are mud floored.

The channels divide the mud shoal into a complex of shoals or what has been referred to herein as a "mud-shoal complex." On the south side of the complex, channels, which are small and numerous, coalesce to form a few large channels on the crest of the shoal (Fig. 28). The small channels are 30-70 ft (9-21 m) wide and mud shoals between the channels average 200 ft (60 m) in width. Natural submarine levees, less than a foot high, border many of the channels. Turtle grass is locally abundant on the levees and tends to delineate channel boundaries sharply.

The potential geologic record of the mud-shoal complex is shown by the detailed isopach map illustrated in Figure 28. The largest and deepest channels are on the shallow west side of the complex and are represented by enclosed depressions. The closed nature of these depressions results from the small channel deltas deposited at the north end of the channels. The channels themselves have characteristics typical of tidal inlets. The small channels on the south side of the mud-shoal complex resemble channels forming what Price (1963, p. 285) has designated as an "inlet drain." At the convergence of these channels, a large incised channel forms. Although winds and a hydrostatic head of river water tend to keep surface waters flowing southward, deeper currents of more saline flood tides can move northward over the shoal by means of the channels. This phenomenon was noted at station 254, where surface waters with a salinity of 18‰ were nearly motionless, but

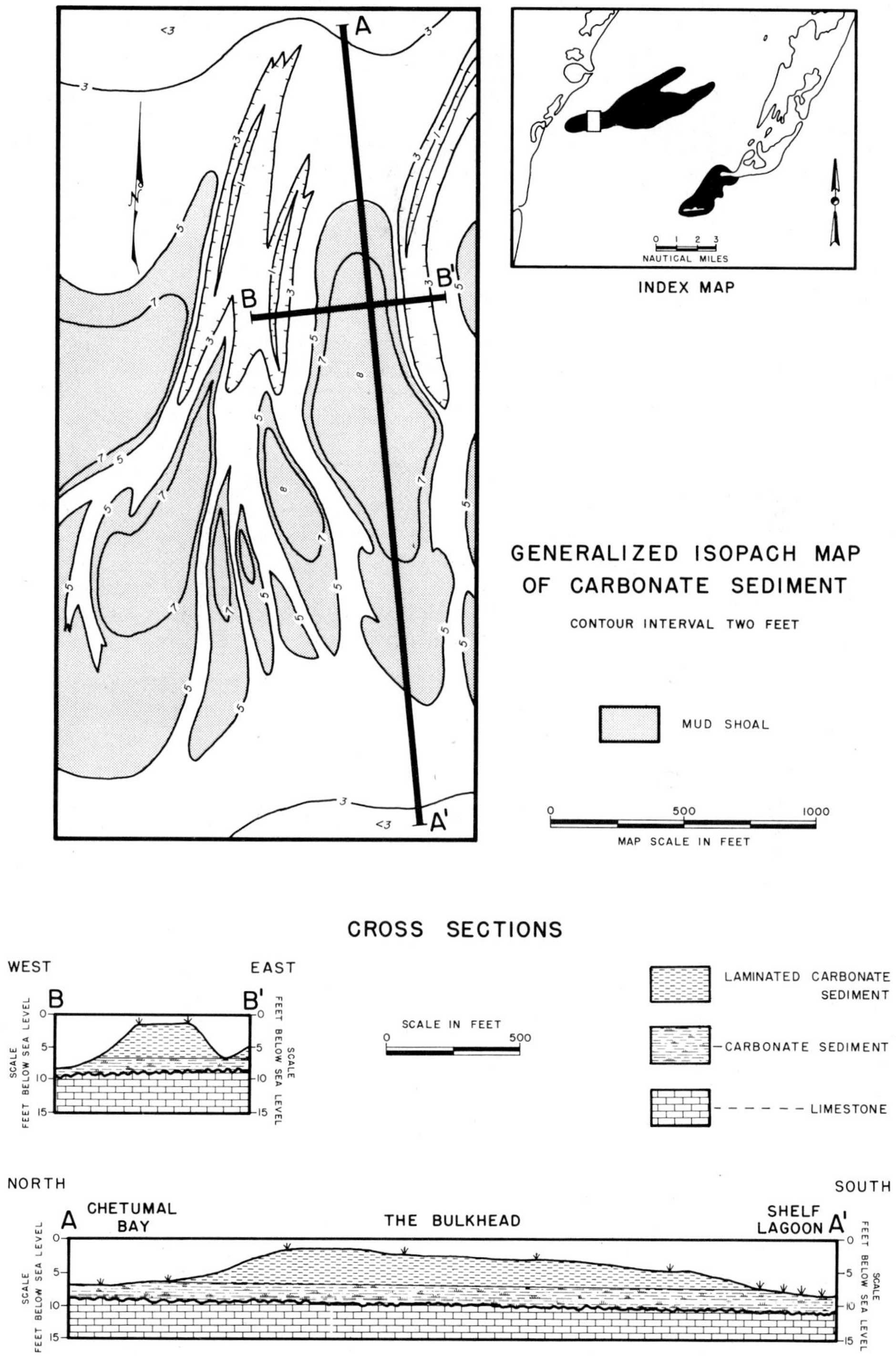

FIG. 28—Western part of Bulkhead mud-shoal complex.

bottom waters with a salinity of 26°/₀₀ were moving northward at a sufficient rate to hamper bottom sampling.

In places the channels are merely shallow depressions in the shoal crest (Fig. 25B). This fact, taken in conjunction with observed current flow, suggests the following sequence of channel development. As the western shoal accumulates, it restricts water circulation, and channels form that tend to maintain circulation. These initial channels are no more than low hollows developed in the top of the shoal. As the shoal grows and further restricts circulation, the channels grow larger and tend to drain a larger area of the shoal flat. Finally, numerous channels converge, forming a distinct flood tidal drain. Convergence of the channels results in a larger channel characterized by stronger currents (the tidal jet of Price, 1963) that slacken in the bay and form a small tidal delta at the north end of the channel.

It seems plausible that strong southerly-flowing currents are produced in the bay and in the channels during strong northers. These conditions do not persist for long periods, for the converging channels of tidal drains would tend to form on the north side of the shoal, whereas tidal deltas would tend to form on the south side. This would be the reverse of the existing pattern.

In contrast to the channels in the western Bulkhead, the channels in the Ambergris-Cangrejo mud shoal are relatively few in number but large in size (Fig. 3). Furthermore, tidal drains are developed on the northwest side of the shoal, indicating that the channels conduct ebbing tidal flow from Chetumal Bay. The Ambergris-Cangrejo mud shoal has reached maximum development in the sense that its flat-topped crest is approximately at mean low-water level. Consequently, water cannot flow over the shoal but must flow through the shoal by means of well-developed channels. Near Cangrejo Cay, deeper parts of the shoal contain channels that are poorly developed, suggesting that, like the Bulkhead channels, the channels in the Cangrejo-Ambergris shoal developed concomitantly with the growth of the shoal.

Sedimentary structures—The grain size and constituent composition of the laminated miliolid muds is similar to that of the other miliolid muds (compare Tables 5, 6); in fact, the only consistent difference appears to be the absence of laminations in the cores of the one facies and their presence in the other. The laminations are represented by alternate bands of dark and light color (Figs. 16A, 26). The darker, wider laminae consist of gray mud, whereas the light-colored laminae represent concentrations of Foraminifera. The laminae are generally 1/16-1/8 in. (1.6-3.2 mm) thick, but a few layers are as thick as 1/2 in. (13 mm). The density of laminae averages 1-2 laminae/inch, but as many as 20 laminae/inch have been noted (Fig. 16B). The lateral continuity of the laminae was not measured, but pinchouts within the 1 1/2-in. to 1 3/4-in. diameters of the cores indicate that foraminiferal laminae are lenses that, at best, are only a few feet wide. Most of the Foraminifera in the laminae are miliolids; however, laminae composed of peneroplids or mixtures of foraminifers are common.

In cores of the laminated muds the bottoms of relatively thick (1-2 in. or 2.5-5.1 cm) sand layers commonly lie in sharp contact with underlying mud and gradationally pass upward into mud again (Fig. 17B). In a few places the sharp contacts are irregular (Fig. 17B) or truncate older laminae (Fig. 16A). These sequences of graded beds lying on irregular or truncated surfaces are probably the record of the slow settling of suspended sediment after unusual erosive current conditions, such as might be produced by storms. Examination of sediment cores under a binocular microscope demonstrates that there is a gradation between relatively thick graded beds and laminae. In intermediate examples, thin light-colored layers appear to alternate with thicker dark-colored layers

TABLE 6. MEAN CONSTITUENT COMPOSITION, GRAIN SIZE, AND CARBONATE MINERALOGY OF LAMINATED MILIOLID-MUD FACIES[1]

	Mean	*Observed Range*	*Standard Deviation*
Coralline algae	0.0	0.0	-
"*Melobesia*"	1.9	0.5- 3.4	1.1
Halimeda	0.9	0.0	-
Miliolidae	34.0	17.8-52.1	10.4
Peneroplidae	12.0	2.6-30.3	10.2
Other Foraminifera	4.5	1.9- 7.1	1.7
Corals	0.0	0.0	-
Mollusks	16.9	7.5-27.9	7.0
Ostracods	4.3	2.2- 8.7	2.0
Miscellaneous skeletal	1.4	0.1- 3.5	1.0
Unknown skeletal	3.1	1.3- 5.4	1.2
Total skeletal	78.0	62.6-86.7	7.0
Fecal grains	3.5	1.0- 9.4	2.1
Cryptocrystalline grains	18.3	10.6-33.5	6.6
Cryptocrystalline skeletal grains	6.3	3.5-10.4	2.4
Total over 50 percent cryptocrystalline	24.6	15.6-43.9	7.9
Country rock	0.0	0.0- 0.2	0.1
Terrigenous minerals	0.0	0.0- 0.4	0.1
Weight percent <1/8 mm	78.6	58.8-93.1	13.2
Weight percent <1/16 mm	56.8	31.1-87.2	17.7
Aragonite as weight percent of carbonate	20	17-36	5
Magnesium carbonate in calcite (weight percent)	11	9-12	1

[1]Eleven samples used in calculations.

(Fig. 26, laminated miliolid mud). Examination of these layers under a binocular microscope, however, showed that the light-colored silt and sand laminae usually pass gradationally upward into mud over a distance of less than 2 mm. This fabric, in conjunction with the fact that laminae are best developed in the upper 2-3 ft (0.6-0.9 m) of sediment, suggests that the laminae are related to increased current activity associated with shoaling conditions. There is the possibility that some of the laminae represent productivity blooms of Foraminifera, in which case the laminated character of the sediment would result from a relatively rapid rate of foraminifer sedimentation relative to the rate of mud deposition. If this were the case, one would expect the foraminifers to be "floating" in a mud matrix. Many light-colored laminations, however, do not have a mud matrix. The general lack of laminations in deeper water sediments of similar miliolid-mud composition indicates that significant foraminiferal blooms are restricted to the shoals. Therefore, it is logical to conclude that the laminations do not owe their origin to productivity blooms.

The thick accumulations of gray mud in these cores do not appear bedded in core sections, but in thin section they were observed to be

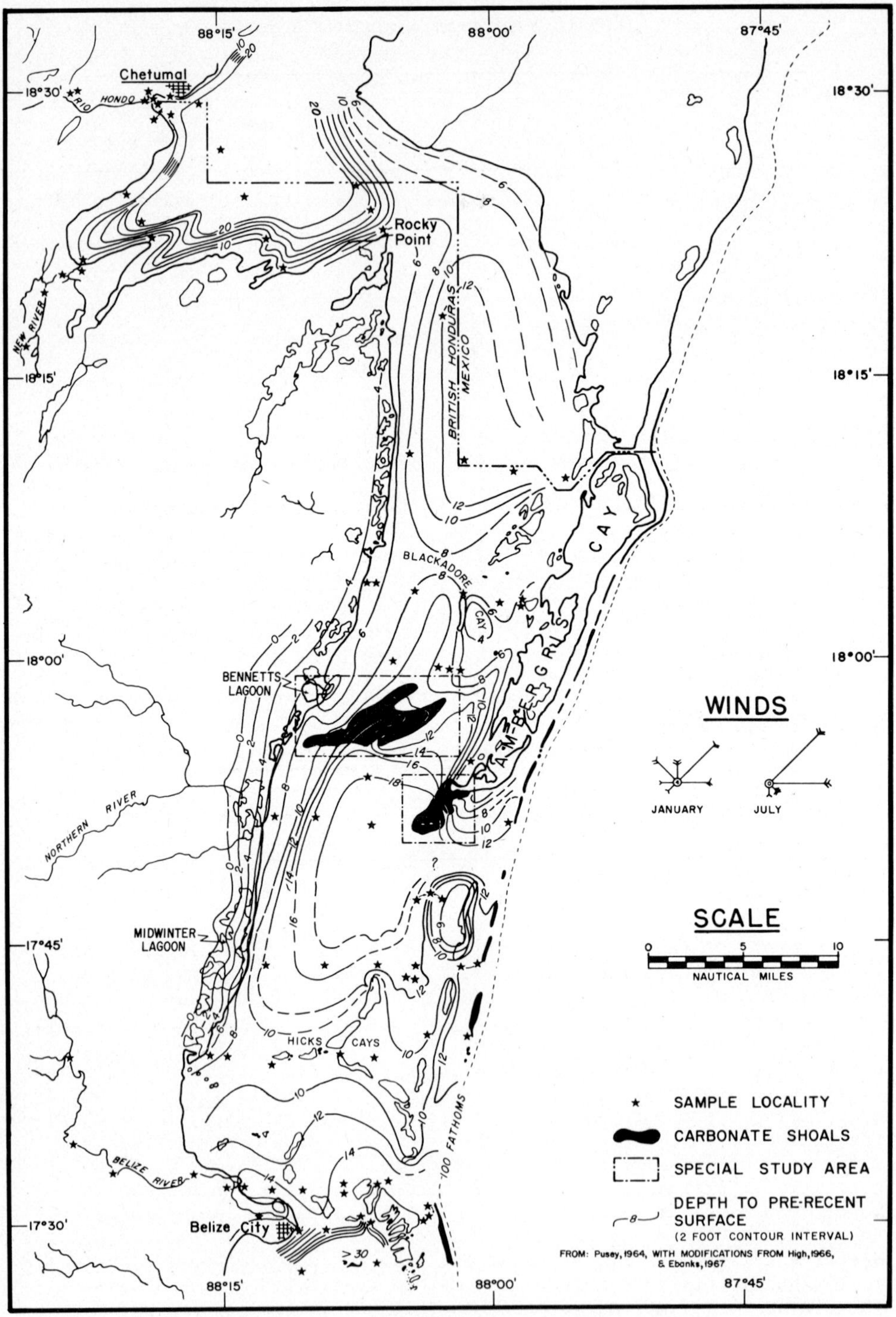

FIG. 29—Depth to pre-Holocene surface (sea-level datum).

well bedded. In the latter, elongate Foraminifera, diatoms, sponge spicules, and ostracods are aligned parallel with the depositional surface. Like the miliolid muds, the laminated miliolid-mud sediments contain leaves of *Syringodium* that are vertically oriented.

The thickness of the laminated miliolid muds averages 6 ft (1.8 m), and approximately 5 percent of this section is represented by sand layers 3-5 in. (0.2-1.5 m) thick (Figs. 26, 27). In most of these, the sands consist of mollusks (especially *Chione cancellata*) and peneroplids that do not "float" freely in a mud matrix. In these sections, plant fragments that may be rhizomes or roots are present as irregular dark streaks, and bedding is commonly disturbed or absent. Surface sediments with these same characteristics compose about 10 percent of the Bulkhead mud-shoal complex. The sand patches tend to occur in water depths of 1/2-3 ft (0.2-0.9 m) near the crest of the shoal, where they are 20-50 ft (6-15 m) wide. Dense *Thalassia* covers the sand patches and extends many pulpy rhizomes 4-5 in. (10-13 cm) into the sediment (Fig. 26, top). Red mangroves (*Rhizophora mangle*) are also present in the vicinity of the sand patches, and in two places they form mangrove cays. By analogy with the existing sand patches, shell-sand layers in the cores are probably a few tens of feet wide and tend to be elongate parallel with the length of the shoal complex.

Origin of the Bulkhead mud-shoal complex—Without detailed hydrographic data it is difficult to prove the origin of the Bulkhead mud-shoal complex. Plants that could baffle currents and allow suspended mud to drop out are inadequate to explain the shoal's origin, for, as noted, the only abundant marine grass in the area (*Thalassia*) is less abundant on the shoal than in surrounding areas. Furthermore, cores of the shoal show that grass-root remains are not common, and where they are present the roots are associated with *Chione*-peneroplid sands. *Thalassia* probably aids in stabilizing the muds but seems to have played a minor role as a current-reducing factor.

In order to determine the importance of preexisting topography in controlling the development of the mud-shoal complex, sediment-probe traverses were made. The data resulting from these traverses is illustrated in Figure 29. The Bulkhead tends to be associated with a break in bottom slope, but it is equally true that the mud-shoal complex transects the relief of the Pleistocene surface. Moreover, paleodrainage systems to the south (Fig. 7B) lack mud-shoal development. Thus, no dependency of shoal development on the preexisting topography is obvious.

The Bulkhead mud-shoal complex probably formed as the result of the combination of the geometry of Chetumal Bay and strong northeast winds. Lunar tidal rise and fall is less than 2 ft (0.6 m) and produces negligible currents, but wind-induced currents frequently disturb the shallow bay water and place fine sediment in suspension. Runoff water from two major rivers, the Rio Hondo and the New River, produces a hydraulic head in Chetumal Bay so that there is a net outward flow of water into the lagoon. This outward flow is accentuated by northeast trade winds and also, periodically, by lunar ebb currents. Currents bearing suspended sediment move toward the bay mouth, where the water deepens and the bay widens. At the bay mouth, the outflowing currents radiate outward and meet interference from wind-induced currents and tidal currents moving westward from the reef area. The outflowing currents disperse and lose velocity, and thus also lose their ability to carry suspended sediment. Sediment settles out locally in the bay-mouth area and gradually builds up a bay-mouth bar.

Reference has been made to the apparent stability of the Bulkhead geometry, a paradox in view of the fact that the mud-built structure has

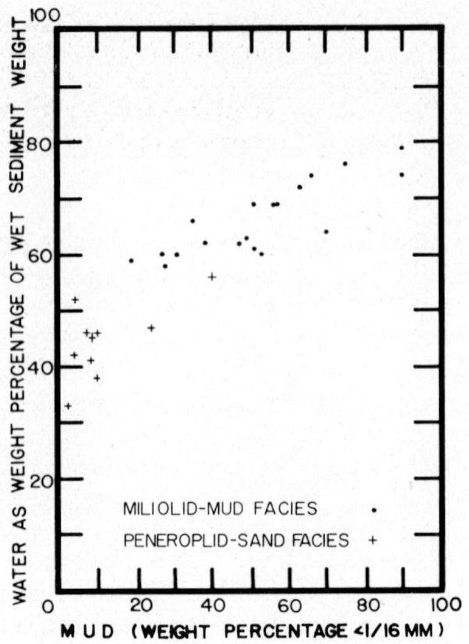

FIG. 30—Graph showing relation between mud content and water content of samples from miliolid-mud facies and peneroplid-sand facies (based on data in Table 7).

been subjected to severe storms, including hurricanes. It seems appropriate, therefore, to consider some of the physical characteristics not only of the muds composing the Bulkhead in particular, but also of the muds in northern Belize in general.

It is evident from Figure 30 and Table 7 that the water content of the sediment increases with the mud content. In muddy sediments the first 6 in. (15 cm) below the depositional interface contains 60-80 percent water by weight.[5] These figures correspond to 80 and 85 percent, respectively, of the sediment volume.[6] Below 6 in. (15 cm), the sediment appears to have more consistency, the water content having decreased to a value approximating 55 percent of sediment weight or 80 percent of sediment volume (Table 8). This tendency to retain water is also characteristic of carbonate muds in Florida Bay and the Bahamas (Table 8).

Much of the water in the sediment is contained within an organic slime. The organic content of the muds, as estimated by weight loss during complete oxidation in 30-percent hydrogen peroxide, is less than 5 percent, but much of this organic material appears to be expanded by included water to form a mucouslike slime. Carbonate grains, which form 20 percent or less of the sediment by volume, are obviously not in contact in three dimensions. They are supported by the watery slime.

Examination of undried samples with a binocular microscope reveals that nearly all the mud is aggregated into soft irregular ellipsoids and smooth rods that average 0.1-0.5 mm in diameter. The smooth grains

[5]Water content of the sediment is expressed herein as weight percentage of wet sediment weight. Sixty to seventy percent of the wet weight corresponds to 150-225 percent of dry sediment weight. The relation between percentage of dry weight and percentage of wet weight is:

$$\text{Water as percentage of wet weight} = \frac{\text{Water as percentage of dry weight}}{100 + \text{water as percentage of dry weight}}.$$

[6]If a density of 2.8 is assumed for the mixture of calcite and aragonite composing the sediment and a density of 1.0 is assumed for the water, the volume of water in the sediment, as a percent of the total volume, is:

$$100 \times \frac{2.8\ (\text{percent water as percent of wet weight})}{100 + 1.8\ (\text{percent water as percent of wet weight})}.$$

Because the water content of the sediment expressed as weight percent of dry weight varies between infinity and zero, it is difficult to visualize and results in a misleading overstatement of the water loss in the upper few inches of sediment.

are presumably fecal, but the irregular aggregates may be feces, pelecypod pseudofeces, or the aggregates of slime called "organic detritus" (Fox, 1950, 1957). The large naviculid diatom *Diploneis* and several other raphe-bearing genera, which are benthonic (Patrick, 1948) and produce mucilage (O'Colla, 1962), are common in the miliolid muds and may be responsible for much of the slime. Most of the detritus is a viscous organic film which contains carbonate silt, live diatoms, live miliolids, small nematode worms (40 x 500μ), unicellular green algae, sponge spicules, ostracods, partially decomposed plant fragments, and scarce thin threads of blue-green algae. In this regard, it is similar to the organic detritus illustrated by Kornicker (1958, p. 225) and Scoffin (1970) from Bimini Lagoon on the Great Bahama Bank. The organic film forms a light tan coating on most of the muds in northern Belize.

TABLE 7. MUD AND WATER CONTENT OF SELECTED SAMPLES FROM PENEROPLID-SAND AND MILIOLID-MUD FACIES

Sample	*Mud Content (weight percent <1/16 mm)*	*Water Content (percent of dry sediment weight)*[1]
Peneroplid Sand		
15	7	44
15A	4	52
16	24	47
21	2	28
25	8	41
241	40	56
254	9	45
261	10	46
264	10	38
265	4	42
266	2	33
Miliolid Mud		
23	66	74
24	27	60
26	56	69
28	57	69
30	75	76
31	47	62
32	90	74
39	49	63
239	51	69
243	51	61
245	71	64
246	63	72
250	87	79
251	38	62
253	28	57
257	19	59
259	31	60
268	35	66
272	53	60

[1]Determined for grab samples stored in sealed glass jars. Most values represent the mean of two determinations.

TABLE 8. WATER CONTENT OF CARBONATE-MUD SEDIMENTS[1]

Locality	*Interval (depth below depositional interface in feet)*				
	0-0.3	0-1	1-2	2-3	3-4
Belize stations[2]					
243	61	57	55	52	-
251	62	52	-	-	58
255	53	-	-	55	-
258	59	46	-	52	-
Average	59	52	55	53	58
Florida Bay[3]	70	58	-	-	-
Great Bahama Bank[4]	43	43	37	-	-

[1]Weight percent of wet sediment weight.

[2]Piston cores. Water content determined in 1-ft intervals for each half of longitudinally split core. Replicate analyses of core halves yielded values that were ±3 percent of reported mean. Analyses of the upper 0.3 ft were made from grab samples. Water volume loss over the first 1 ft is approximately 5 percent (decreases from 80 to 75 volume percent).

[3]Computed from data in Ginsburg (1957, p. 91). Piston core. The volume loss of water is approximately 10 percent (decreases from 85 to 75 percent).

[4]Data from Cloud (1962, p. 30). Piston core. The volume loss is about 6 percent (decreases from 67 to 61 percent).

The watery slime gives cohesiveness to muds such that the vertical walls of holes up to 1 1/2 ft (0.45 m) deep do not slump. To erode these mud sediments, then, currents not only must be strong enough to transport organic-slime aggregates, but also must be strong enough to overcome the cohesiveness imparted to the mud bottom by its close association with organic detritus. Flume experiments (Neumann et al, 1970) on subtidal mats from the Little Bahamas Bank show that the erosion of mats requires current velocities of two to five times that needed to erode unbound mud. Much of the silt-size carbonate material may be deposited initially as silt, but it can be eroded only as aggregates. Thus the cohesiveness of the muds helps to explain the resistance of the miliolid muds in general, and the Bulkhead in particular, to erosion by storms.

Terrigenous-Sand Facies

Skeletal sands with a relatively high quartz content characterize this facies (Table 9). The quartz grains are extremely angular and commonly show undulatory extinction between crossed nicols. Although quartz grains up to 1.5 mm are present, the average grain size is approximately 0.1 mm. The presence of angular chert fragments and feldspar grains is uncommon. Well over half the skeletal grains are represented by fragments of the pelecypods *Chione cancellata* and *Brachidontes exustus*. Many of these shells are smoothly rounded and have barnacles attached to them. Fragments of coarsely crystalline carbonate rock are common.

Although the terrigenous-sand facies is located close to the terrigenous alluvium of the mainland (Dixon *in* Romney, 1959, p. 22-28), the streams are not carrying quartz sand into the area at this time. This is not only a consequence of a low slope gradient in the drainage area, but also is the result of a settling-basin effect by small lagoons which separate the major streams from the shelf lagoon (Fig. 5).

The terrigenous-sand facies is only a few inches thick and is underlain by gray clay containing abundant sand-size and silt-size quartz grains. This clay is the source of most of the terrigenous minerals associated with the overlying mollusk grains. The gray clay and the molluscan shell debris are mixed by burrowing organisms to the extent that the contact between the predominantly molluscan sand and the terrigenous clay is irregular. This mixing also is evidenced by lumps of gray clay in the terrigenous sand and mottles of mollusk shells in the underlying clay. The worn appearance of the shells suggests that the sediments are frequently disturbed by currents. In brief, the terrigenous-sand facies is represented mainly by a residual shell hash containing reworked quartz sand from the underlying sediment.

Cryptocrystalline-Grain Facies

This facies, which is the most widespread facies in northern Belize, contains abundant cryptocrystalline grains (Table 10). In contrast to

TABLE 9. MEAN CONSTITUENT COMPOSITION, GRAIN SIZE, AND CARBONATE MINERALOGY OF TERRIGENOUS-SAND FACIES[1]

	Mean	*Observed Range*	*Standard Deviation*
Coralline algae	0.0	0.0	-
"*Melobesia*"	1.2	0.0- 2.4	1.1
Halimeda	0.0	0.0	-
Miliolidae	10.2	4.5-17.1	5.5
Peneroplidae	0.9	0.0- 2.0	1.1
Other Foraminifera	2.3	0.8- 4.9	1.8
Corals	0.0	0.0	-
Mollusks	27.8	22.1-33.9	5.6
Ostracods	1.0	0.0- 2.6	1.1
Miscellaneous skeletal	1.2	0.0- 2.3	1.2
Unknown skeletal	2.6	1.3- 3.6	1.1
Total skeletal	47.2	31.2-68.2	15.6
Fecal grains	0.8	0.0- 1.1	0.6
Cryptocrystalline grains	22.0	8.5-48.6	18.3
Cryptocrystalline skeletal grains	6.7	2.1-14.2	5.3
Total over 50 percent cryptocrystalline	28.7	13.0-62.8	23.4
Country rock	11.3	0.8-31.2	13.5
Terrigenous minerals	18.6	8.0-25.7	8.6
Weight percent <1/8 mm	28.4	10.6-49.7	19.7
Weight percent <1/16 mm	10.0	3.2-23.0	9.0
Aragonite as weight percent of carbonate	41	7-56	23
Magnesium carbonate in calcite (weight percent)	10	9-11	1

[1]Four samples used in calculations.

TABLE 10. MEAN CONSTITUENT COMPOSITION, GRAIN SIZE, AND CARBONATE MINERALOGY OF CRYPTOCRYSTALLINE-GRAIN FACIES[1]

	Mean	*Observed Range*	*Standard Deviation*
Coralline algae	0.0	0.0	-
"*Melobesia*"	0.2	0.0- 1.4	0.4
Halimeda	0.0	0.0- 0.1	0.0
Miliolidae	6.2	0.1-15.6	4.9
Peneroplidae	1.1	0.0- 8.5	2.9
Other Foraminifera	2.4	0.0- 5.4	1.9
Corals	0.0	0.0	-
Mollusks	15.5	4.6-41.6	9.5
Ostracods	1.5	0.0- 4.5	1.2
Miscellaneous skeletal[2]	0.1	0.0- 0.5	0.2
Unknown skeletal	1.4	0.0- 2.4	0.6
Total skeletal	28.3	6.7-58.5	13.1
Fecal grains	3.1	0.0-19.4	5.2
Cryptocrystalline grains	66.9	40.9-93.4	14.9
Cryptocrystalline skeletal grains	4.1	0.0-13.3	3.4
Total over 50 percent cryptocrystalline	71.0	43.1-96.3	15.0
Country rock	1.6	0.0-15.1	3.8
Terrigenous minerals	0.1	0.0- 0.5	0.2
Weight percent <1/8 mm	52.8	1.8-93.6	32.5
Weight percent <1/16 mm	34.0	1.3-82.4	26.1
Aragonite as weight percent of carbonate	7	2-19	4
Magnesium carbonate in calcite (weight percent)	3	0- 8	3

[1]Fifteen samples used in calculations.
[2]Includes: barnacles, bryozoans, echinoderms, sponge spicules, diatoms, and charophytes.

the miliolid-mud facies, skeletal grains are relatively unimportant in the cryptocrystalline-grain facies. Mollusks are the most abundant skeletal grains, but only a few species are represented (Robertson, 1963, p. 16). Mineralogically, the sediment is predominantly low-magnesium calcite. Skeletal grains commonly contain borings 3-5μ in diameter, and there commonly are large (40-60μ) holes near the periphery of the grains (Fig. 13D). In transmitted light some of these holes contain greenish material, probably representing green algae. On the periphery of many cryptocrystalline grains, rings of relatively coarse crystals are observed; these may be crystal-filled borings. Crystal fillings are present in foraminifer chambers, but they are not as well developed as in the peneroplid facies. Cryptocrystalline coatings are found on a few well-preserved miliolids. Country rock is locally abundant, and some cryptocrystalline grains are weakly cemented to form grain aggregates.

Sedimentary structure in the cryptocrystalline-grain facies is variable and is related to the local depositional environment. In lower Chetumal Bay the cryptocrystalline-grain sediments are thin and erratically distributed. Graded beds, many of which are less than an inch thick (Fig. 31A, B), characterize these sediments. Basal sands of the

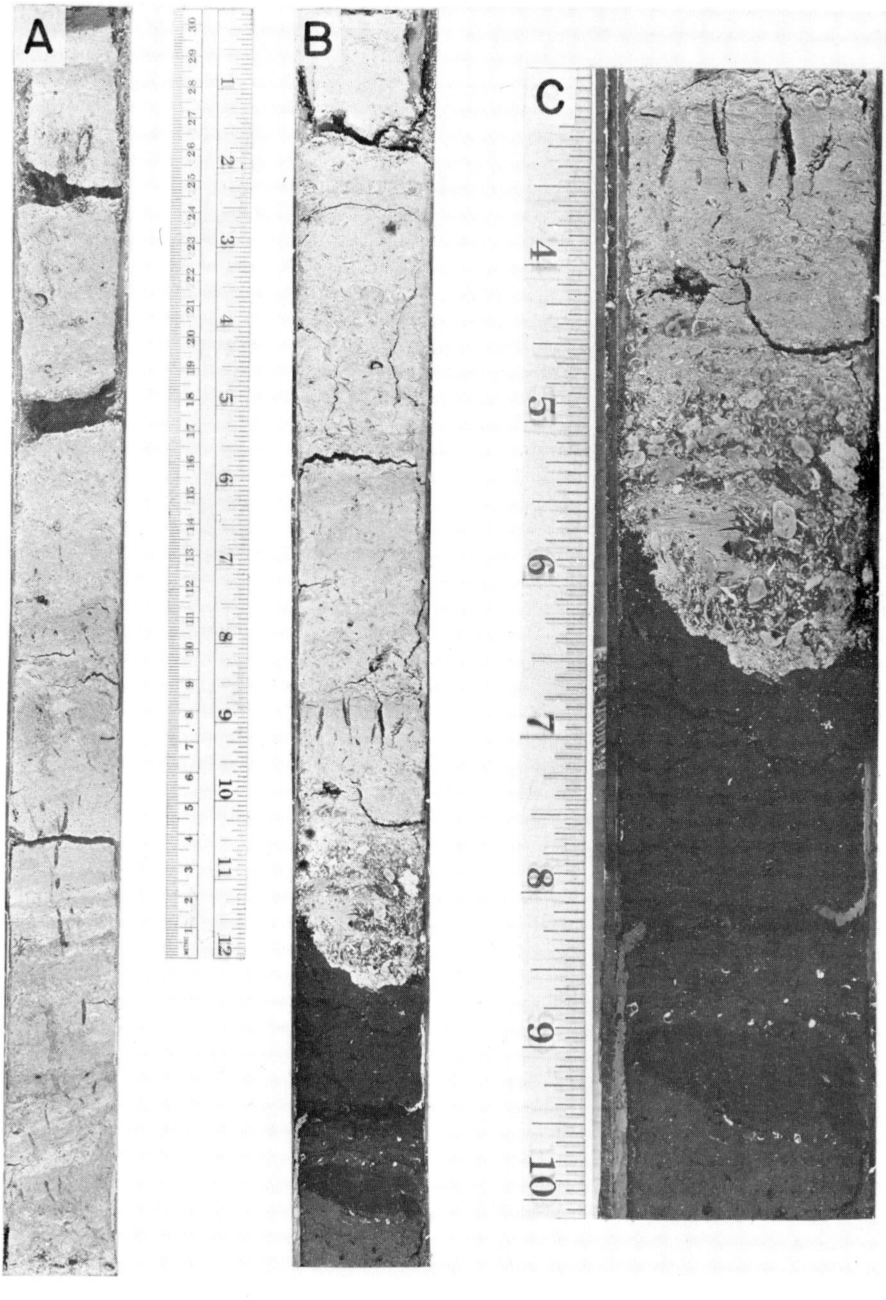

FIG. 31—Cores from cryptocrystalline facies in lower Chetumal Bay.

A. Core 13, water depth 9.5 ft (2.9 m), incomplete sequence. Light-colored layers are sand; darker layers are muddy sand. Note frequency of graded beds and sand lenses near bottom of core. *Thalassia* rhizomes can be seen near top of core. Scale in inches and centimeters.

B. Core 52, water depth 10.5 ft (3.2 m), complete sequence from top to bottom of graded beds, basal shell hash, peat, and gray clay. Scale same as **A.**

C. Detail of **B.** Near top of core, a graded bed is truncated by a sand layer. Note roots in truncated bed. Gravel at base of carbonate sequence contains mollusks and rock fragments. Peat is poorly bedded and contains layers of gastropods. Scale in inches.

graded beds commonly lie on an uneven surface which truncates older structures (Fig. 31C). High (1966, p. 161ff) has shown that these sediments are present under the coastal sediments and lagoons for at least a mile west of southern Chetumal Bay. Sediments in lower Chetumal Bay, including both miliolid muds and cryptocrystalline sediments, are deposited in the shallow depressions of an irregular rock surface. A few sinkholes are filled with more than 7 ft (2.1 m) of sediment; however, the average sediment thickness here is less than 2 ft (0.6 m), and approximately one third of the bottom is represented by bare rock. Sediment thicknesses in upper Chetumal Bay range from 1 ft (0.3 m) near shore to more than 7 ft (2.1 m) in deeper parts of the bay. The thick sediments consist of alternating muds and subordinate shell sands, which form graded beds in a few places. Mottling and burrows are scarce in both upper and lower Chetumal Bay sediments.

Depositional Environment

The depositional area of the cryptocrystalline facies encompasses a broad spectrum of environmental conditions that are variable from time to time. Chetumal Bay is affected greatly by runoff water from the Rio Hondo and the New River. Observed bottom salinities ranged from 1°/₀₀ at river mouths to 30°/₀₀ in lower Chetumal Bay. During the dry season (February to April), bottom salinities are higher. Dixon (1958) reported a dry-season saltwater wedge 50 mi (80 km) up the New River; and, on the basis of the distribution of halophilic red mangroves, a similar wedge which penetrates the Rio Hondo is indicated.

Bottom-water saturation with respect to aragonite varies from undersaturated conditions near river mouths to highly supersaturated conditions at Rocky Point (Fig. 8). Oceanic water penetrating from the lagoon into the bay enters a shallow area of restricted circulation where the water undergoes diurnal warming and evaporation and tends to become more saturated with respect to calcium carbonate. It was noted that water of the lower bay has calcium in excess of that expected from simple dilution of seawater. This calcium is brought in by the Rio Hondo and the New River, which drain limestone terranes (Dixon, 1958).

In addition to these conditions, biologic activities also may have an important role in increasing the aragonite supersaturation. Between the Bulkhead and Rocky Point and in parts of upper Chetumal Bay, there is a film of yellow-brown organic detritus, which is less than 1/4 in. (6.7 mm) thick but tends to erode in large patches which are several inches in diameter. Most of the film is yellow-brown, slimy, amorphous material. Blue-green algae, including gelatinous-walled *Gloeocapsa* and 3-4μ-diameter thread algae, are present but are not abundant. The photosynthetic activities of these algae remove carbon dioxide from the water and thereby tend to lower the solubility of calcium carbonate in the water.

The organic detritus forms a red stain in Rose Bengal; it contains a sparse living population of organisms including miliolids, rotalid foraminifers, and ostracods. Like the organic material in the miliolid-mud facies, the organic film contains many diatoms. The organic film, which covers nearly three fourths of the bottom, apparently is deleterious to mollusks, as no mollusks were found living on or under the film (Robertson, 1963, p. 16). The organic film, in conjunction with variable hydrographic conditions (e.g., salinity), probably accounts for the paucity of skeletal grains in the bay. The dominant skeletal group, mollusks, is restricted to five species (Robertson, 1963). The hard rocky bottom in lower Chetumal Bay provides an attachment substrate for *Brachidontes exustus* and *Pinctada radiata*, and the soft muddy bottom of upper

Chetumal Bay provides a suitable substrate for *Laevicardium mortoni*, *Cerithidea costata*, and *Anomalocardia cuneimeris*.

Wind-induced currents play a significant role in the deposition of the cryptocrystalline facies. The water in the bay is frequently milky and foam streaked parallel with the wind direction. As noted, sediments in relatively shallow, exposed areas such as lower Chetumal Bay occur as graded beds which are restricted to sediment pockets between rock highs. Fine-grained material tends to be deposited in deeper water such as is found in upper Chetumal Bay. Even the fine-grained, thick sediments of the upper bay contain graded beds which may record unusually severe storms such as hurricanes. It is known that several square miles of the bottom in Chetumal Bay were exposed during the 1920 hurricane (Stoddart, 1963, p. 28); this fact suggests that storm transport may be significant in explaining some of the primary structures of the cores from the cryptocrystalline facies. Certainly the importance of storm transport in Chetumal Bay cannot be evaluated properly unless sediment-distribution studies are made both before and after a hurricane.

Origin of Brownish Cryptocrystalline Grains

Despite the great abundance of cryptocrystalline grains, partly altered skeletal grains are scarce in the cryptocrystalline facies. This fact suggests that either the rate of recrystallization of skeletal grains is relatively rapid in this depositional environment or that the many cryptocrystalline grains are forming by processes other than the recrystallization of skeletal grains. The paucity of preservable organisms living in the area suggests that not many of the cryptocrystalline grains can be ascribed to the complete recrystallization of skeletal material. Excluding this possibility, the cryptocrystalline grains may represent (1) rock fragments, (2) fecal grains, and (3) algal crusts.

Rock fragments—A few rock fragments that are transitional to cryptocrystalline material were observed in thin sections. In some cases, coarse grains of limestone seem to be recrystallizing to cryptocrystalline grains; in other cases, fine grains of fossil-bearing rock seem to be altering to fine-grained material which lacks fossils. Since both rock fragments and partly cryptocrystalline rock fragments are scarce and are, in general, restricted to the nearshore area, rock alteration to form cryptocrystalline grains is thought to be unimportant in explaining the bulk of the cryptocrystalline grains in this facies.

Fecal grains—The mud-size fraction of the cryptocrystalline-grain facies consists predominantly of silt-size cryptocrystalline carbonate. Organic aggregation of this material forms fecal grains that can be identified in transmitted light as ellipsoidal or rod-shaped aggregates of silt-size and clay-size cryptocrystalline grains imbedded in a brownish matrix. Subsequent induration and recrystallization of the pellets produce cryptocrystalline grains that are uncertainly identified by their shape as altered fecal pellets. Rod- and ellipsoid-shaped cryptocrystalline grains are not abundant in the cryptocrystalline-grain facies; therefore, the alteration of fecal grains does not seem to contribute significantly to the cryptocrystalline-grain content of the facies.

Algal crusts—In the western part of Chetumal Bay, especially near the mouth of the Rio Hondo, laminated carbonate crusts are common on hard substrates such as shells or wood (Fig. 15). The carbonate composing these crusts is cryptocrystalline; however, both the encrusting habit and the composition (Table 11) of the crusts suggest that the crusts form in a different way from previously discussed cryptocrystalline carbonate. Thin sections of crusts at the mouth of the Rio Hondo contain a few gray tubes that greatly resemble the threads of blue-green

TABLE 11. CHARACTERISTICS OF CRYPTOCRYSTALLINE CARBONATE AT MOUTH OF RIO HONDO, BELIZE

Sample	*Carbonate Mineralogy*	*Carbon-14 Age (years)*[1]	*Acid-Insoluble (20% acetic)*	*Mineralogy of Insoluble Residue*[2]
Laminated crust surrounding dead root (Fig. 15)	Calcite with less than 4% magnesium carbonate substitution; no aragonite; no dolomite.	3,000±115	2.6%	Mostly montmorillonite, some kaolinite, trace of mixed-layer clay, no illite or chlorite; quartz present.
#49, bar sand with shells removed	Calcite with less than 2% magnesium carbonate substitution; no aragonite; no dolomite.	4,300±125	2.7%	Mostly montmorillonite, trace of kaolinite, trace of mixed-layer clay, no illite or chlorite; quartz present.

[1]Determined by courtesy of Exploration Department and Geochemical Laboratory of Humble Research Center, Houston, Texas.
[2]X-ray analysis of oriented slides of less-than-2μ fraction; analyses were made on both air-dried and glycolated samples.

algae described and illustrated from algal coatings in hard-water lakes and streams (Davis, 1900, 1901; Roddy, 1915; Bradley, 1929). When these crusts are acidified, the resulting grayish insoluble residue maintains the shape of the original carbonate crust. The residue effervesces strongly in 6-percent hydrogen peroxide, indicating that organic material is present. X-ray analyses of the acid-insoluble and hydrogen-peroxide-insoluble residues demonstrate the presence of detrital minerals (Table 11). Thus it is reasonable to conclude that the crusts are formed by (1) *accretion* of detrital grains such as quartz and clay minerals in sticky sheaths of blue-green algae, and (2) *precipitation* of calcium carbonate in connection with photosynthesis by the algae. During photosynthesis, the removal of carbon dioxide in excess of that liberated by respiration tends to raise the water pH and to precipitate calcium carbonate. After the death of the algae, bacterial decomposition of organic matter may induce precipitation of calcium carbonate (Oppenheimer and Kornicker, 1958), replacing the algal threads.

These crusts are poorly indurated and are readily abraded. A sandbar at the mouth of the Rio Hondo appears to consist almost entirely of fragmented and rounded crust material (Table 11). This cryptocrystalline sand yielded an apparent radiocarbon age of 4,300±125 years, which suggests that the material formed at the beginning of the stillstand of the sea and is not forming at present. Root incrustations a few yards away (Fig. 15), however, date 3,000±115 years. The accretionary nature of the crusts and their proximity to river water confounds a simple interpretation of the carbon-14 date. If carbonate is gained by both precipitation and accretion of detrital grains, the carbon-14 content may be diminished considerably. For example, Rubin and Taylor (1963) and Rubin et al (1963) have found that in hard-water rivers, living mollusks obtain up to 12 percent of their carbonate as "dead" carbonate from environmental water and, as a consequence, yield an apparent carbon-14 age of approximately 3,000 years. The carbon-14-deficient carbon is believed to be derived from dissolved limestone (Broecker, 1964) and to enter the organism as either food or carbon dioxide (Keith and Anderson, 1963). The presence of carbonate crusts on sticks on the river bottom indicates that the crusts are currently forming. If the crusts are actually 3,000 years old, the exposed wood should have disappeared long ago.

In thin section, algal crusts tend to be slightly more opaque and more brownish in transmitted light than partly cryptocrystalline skeletal grains. The writer was not successful, however, in establishing reliable petrographic criteria that would consistently distinguish algal-crust cryptocrystalline grains from other types of cryptocrystalline grains. Algal tubes, where preserved, are abundant in algal crusts, but generally preservation is too poor for them to be useful. A color criterion is not reliable because many cryptocrystalline grains are gradational between the brownish-gray color of algal crusts and the grayish-brown color of recrystallized skeletal grains. Unabraded algal crusts tend to be irregular, but abrasion of the soft cryptocrystalline carbonate renders shape criteria useless. Opaque threads, which are white in reflected light, are present near the center of many presumably algal grains at the mouth of the Rio Hondo. Similar threads are also present in partly cryptocrystalline skeletal grains. Despite the writer's inability to identify cryptocrystalline grains of different origins, indirect evidence suggests that cryptocrystalline grains of algal origin are more important than those of other origins in the cryptocrystalline-grain facies. Sediments near the Rio Hondo, on the basis of their brownish color in transmitted light and their insoluble residues, appear to be almost entirely abraded algal crusts. Sediments in the mid-bay are probably mixtures of transported

TABLE 12. MEAN CONSTITUENT COMPOSITION, GRAIN SIZE, AND CARBONATE MINERALOGY OF OSTRACOD-MUD FACIES[1]

	Mean	*Observed Range*	*Standard Deviation*
Coralline algae	0.0	0.0	-
"*Melobesia*"	0.0	0.0	-
Halimeda[2]	0.4	0.0- 2.7	1.1
Miliolidae	6.6	0.0-18.2	7.3
Peneroplidae	0.0	0.0- 0.2	0.0
Other Foraminifera	9.1	4.3-14.0	4.0
Corals	0.0	0.0	-
Mollusks	21.3	9.0-39.1	12.7
Ostracods	20.0	9.1-32.0	10.5
Miscellaneous skeletal[3]	6.8	1.2-25.5	9.2
Unknown skeletal	2.7	0.6- 5.9	2.0
Total skeletal	66.9	36.5-83.9	18.9
Fecal grains	0.5	0.0- 2.9	1.2
Cryptocrystalline grains	26.2	0.0-55.2	22.5
Cryptocrystalline skeletal grains	2.6	0.0-11.4	4.4
Total over 50 percent cryptocrystalline	28.8	0.0-60.1	25.4
Country rock	1.8	0.0- 8.2	3.2
Terrigenous minerals	4.6	0.0-18.0	7.5
Weight percent <1/8 mm	96.2	89.2-99.8	3.9
Weight percent <1/16 mm	89.0	64.0-99.2	13.5
Aragonite as weight percent of carbonate	10	0-41	15
Magnesium carbonate in calcite (weight percent)	4	0-11	3

[1]Six samples used in calculations.

[2]Found only in sample from Haulover Creek in Belize City. The sample is probably contaminated from boat cleaning.

[3]Consist predominantly of diatoms with lesser amounts of monaxon sponge spicules and charophytes.

algal crusts and altered skeletal material. Sediments in the lower bay are richer in skeletal grains, and most of the cryptocrystalline grains are probably altered skeletal material.

Regardless of the origin of the cryptocrystalline grains, the paucity of skeletal material in the cryptocrystalline-grain facies reflects the harshness of the brackish-bay depositional environment. Moreover, graded beds in even the deepest parts of the bay suggest that sporadic storms and possibly northers disturb and redistribute the sediments.

Ostracod-Mud Facies

This facies is restricted to the river-mouth areas of the Rio Hondo, New River, and Belize River and is characterized by ostracods and mud (Table 12). Large saddle-shaped diatoms (*Campylodiscus*) and opaline, monaxon sponge spicules are usually present, as is a rotalid foraminifer with finely perforate walls (probably *Streblus*). The sponge spicules resemble spicules of freshwater spongillids and are probably river borne. Charophytes are present in the mouths of the Rio Hondo and New River. Quartz and coarsely crystalline limestone fragments are common.

TABLE 13. MEAN COMPOSITION OF MUD FRACTION[1]

Facies	Number of samples counted	*Coarse Silt (15-62μ)*												*Fine Silt and Clay (<15μ)*
		Diatoms	Transparent forams	Miliolidae	Ostracods	Echinoderms	Tunicates	Miscellaneous[2]	Mollusks	Undifferentiated brown skeletal	Total skeletal	Terrigenous	Semi-opaque cryptocrystalline	Description
Halimeda	7	+	1	2	5	1	3	2	13	67[3]	99	1	+	mostly aragonite needles
Miliolid mud	4	0	2	6	6	+	2	1	7	21[4]	46	4	50	not examined
Laminated miliolid mud	11	1	3	5	15	1	1	1	5	12[4]	42	1	57	mostly calcite rhombs
Peneroplid sand	2	2	1	8	10	1	1	1	8	10[4]	41	1	57	not examined
Cryptocrystalline grain	5	+	1	3	8	+	+	+	2	3	17	1	82	not examined

[1] Based on counts of 100 grains per sample.
[2] Includes sponge spicules and rarely bryozoans or radiolarians.
[3] Includes fragments of *Halimeda*, "*Melobesia*," serpulids, peneroplids and coralline algae.
[4] Includes fragments of "*Melobesia*," serpulids, and peneroplids.

Fine-grained detrital sediments of the Belize River form the submerged portion of a delta within about 4 mi (6.4 km) of the river mouth. Sediments in this area and in the abandoned Belize River channel at Belize City are represented by black, foul-smelling mud. Quartz silt is common. X-ray analyses of the clays indicate that montmorillonite is the most abundant clay mineral present (Scott, 1966; this volume). Kreuger (1963) found a comparable clay-mineral suite in samples collected from the Belize River. Sand-size grains are characterized by the great abundance of ostracods. Among the mollusks, those containing considerable organic material within the shell wall structure (e.g., *Caecum* and *Neritina*) are particularly common. Iron sulfide spherules, commonly noted inside foraminifer tests and plant fragments, indicate that the sediments are reducing and iron is available. Thickness and structural features of the Belize delta were not determined.

The environment at the mouths of the rivers is inhospitable for most organisms. Mollusks and foraminifers apparently cannot tolerate the combination of low salinity, rapid deposition, and strongly reducing bottom conditions. Only vagile ostracods and, less commonly, planktonic diatoms are relatively abundant, and even these are greatly diluted by the great volume of river-borne silts and clays.

ORIGIN OF CARBONATE MUD

Silt-size and clay-size carbonate grains constitute the bulk of the miliolid-mud facies and a significant part of both the *Halimeda* and the cryptocrystalline-grain facies. Having considered the environmental significance of the diagnostic sand-size particles in each of these facies, it seems appropriate to discuss the origin of their respective carbonate muds. To investigate the origin of carbonate mud, several samples were studied in the following manner. The coarse-silt fraction (15-62μ), obtained by settling, was mounted on slides and point-counted. In addition, fine-silt-size and clay-size material from several samples was examined by Donald L. Gibbon of Rice University with an electron microscope. Table 13 summarizes the composition of the muds examined, and Figure 32 illustrates mud constituents from different facies.

In the *Halimeda* facies, the high frequency of mud formed from skeletal material suggests that the mud is largely skeletally derived. Matthews (1966), in a detailed petrographic and geochemical study of carbonate mud from southern Belize, reached essentially the same conclusion. Entire skeletons of microorganisms are minor, and those present are represented by diatoms, radiolarians, and foraminifers. Most of the silt grains are angular brownish grains which are translucent in transmitted light. Occasionally, these can be identified as *Halimeda* and, less commonly, as serpulids, "*Melobesia*," or peneroplids. Two possible modes of origin for the bulk of the skeletally derived carbonate mud are suggested: (1) skeletal disintegration and (2) skeletal abrasion. With regard to skeletal disintegration, it seems likely that boring organisms, such as algae or sponges, can effect particle-size reduction to the extent of producing silt- and clay-size grains from sand-size skeletons; but these organisms are not particularly abundant in the *Halimeda* facies. Another possibility is that decomposition of skeletal organic matter releases constituent silt-size and clay-size carbonate material to the surrounding sediment. In this way, major skeletal components such as mollusk prisms, tunicate spicules, sponge spicules, and aragonite needles contribute to the mud content of the sediment when the organic matrix in which they are embedded is destroyed. As noted, however, the decomposition of this organic material may be accompanied by

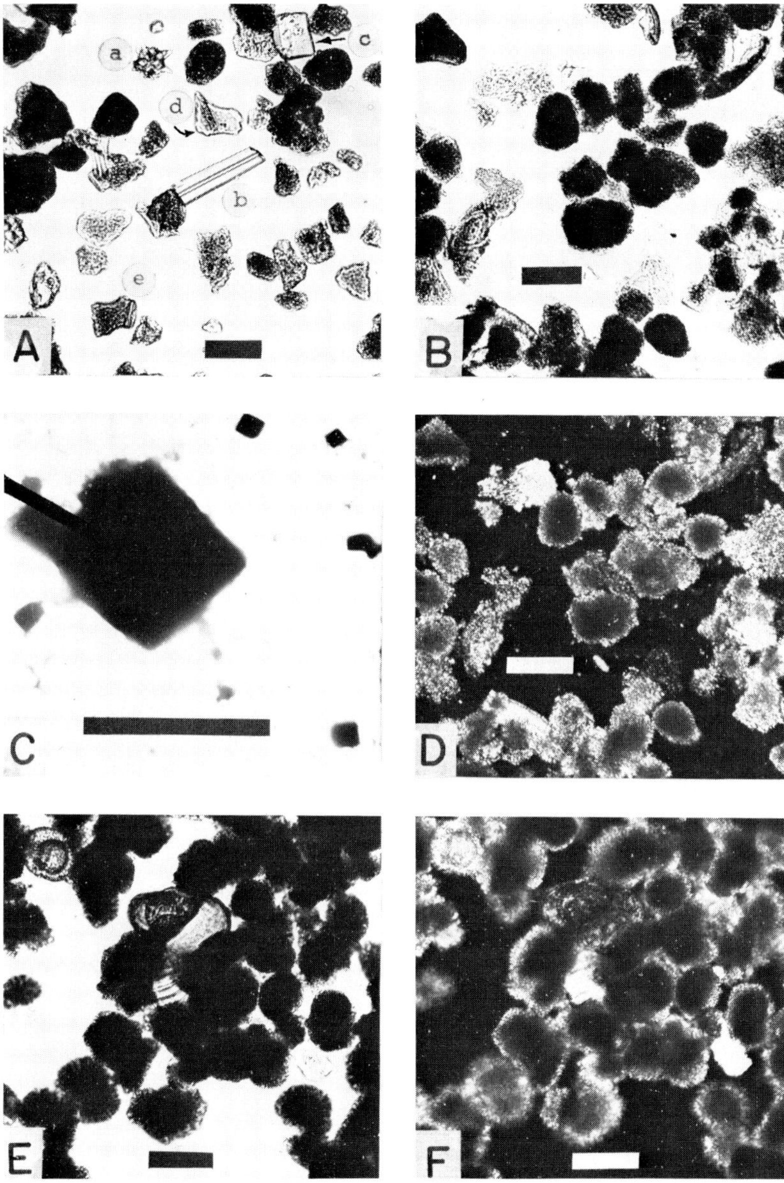

FIG. 32—Silt- and clay-size fraction from northern Belize sediments.

A. Coarse silt grains from *Halimeda* facies (sample 53). Predominantly skeletal grains including: **a**, tunicate spicule, **b**, siliceous sponge spicule, **c**, mollusk prism, **d**, ostracod, and **e**, brown skeletal grains. Many grains are bored. Plane-polarized light; bar length 50μ.

B. Coarse silt grains from miliolid-mud facies (sample 243). Approximately half skeletal (miliolids, ostracods, mollusks) and half semi-opaque cryptocrystalline grains. Note abundant borings and irregular grain borders associated with skeletal grains. Plane-polarized light; bar length 50μ.

C. Clay-size fraction from miliolid-mud facies (sample 250). X-ray diffractograms of fine-silt-size and clay-size material indicate that rhomb-shaped grains are high-magnesium calcite. Electron photomicrograph by Donald Gibbon; bar length 5μ.

D. Same as **B**. Crossed nicols. Note that cryptocrystalline grains appear as aggregates of silt- and clay-size carbonate. Bar length 50μ.

E. Coarse silt grains from cryptocrystalline-grain facies (sample 52). Less than one fourth of grains are recognizable as skeletal material (miliolids, mollusks). Cryptocrystalline grains compose bulk of sample. Plane-polarized light; bar length 50μ.

F. Same as **E**; crossed nicols. Note that miliolid in upper left appears partly cryptocrystalline. Bar length 50μ.

cementation of adjacent mud-size skeletal constituents rather than by release of these elements to the surrounding sediment. In muddy sediments, this type of cementation appears to be scarce, but in sandy sediments it appears to be common. It would seem, therefore, that abrasion is more important than skeletal disintegration in producing carbonate muds in sand areas. Of course, the converse would hold in muddy sediments where the chances for grain abrasion are obviously slight. In brief, the mud in the *Halimeda* facies is produced to some extent by skeletal disintegration (e.g., boring algae) and to a greater extent by abrasion.

The mud fraction of the miliolid-mud facies differs greatly from the mud in the aragonitic *Halimeda* facies with respect to both constituent composition and mineralogy (Tables 13, 14). The miliolid-mud fines are less than half skeletal, and are predominantly cryptocrystalline grains of high-magnesium calcite. The fine-silt and clay fraction of the miliolid muds is mineralogically similar to the coarst-silt and sand fraction of the same facies in that it consists mostly of calcite (Table 14). Thus it is unlikely that the mud is transported from the aragonite-rich *Halimeda* facies into the miliolid-mud facies. It is likely, however, that mud *from* the miliolid-mud facies is transported locally *into* the *Halimeda* facies, as exemplified by sediments of the Ambergris-Cangrejo mud-shoal complex located at the transition between the *Halimeda* facies and the miliolid-mud facies. As noted, in this area the percentage of aragonite composing the mud decreases as the mud content of the sediment increases (Fig. 24). This is due to a dilution effect by increasing contributions of those calcitic constituents such as foraminifers and cryptocrystalline grains that characterize the miliolid-mud facies.

Although skeletal silt grains with microboring and jagged borders are not scarce, less than half the mud fraction of the miliolid mud is directly ascribable to skeletal material. The bulk of the mud consists of translucent cryptocrystalline grains that are round but have a rough surface (Fig. 32B, D). These grains greatly resemble the sand-size and silt-size cryptocrystalline grains that typify the peneroplid and cryptocrystalline-grain facies. The silt-size and clay-size fraction of a miliolid-mud sample (no. 32) yielded a radiocarbon age of 1,000±105 years. The dated material contained a mixture of cryptocrystalline silt (35 percent) and skeletal (65 percent) material. It is unlikely that the cryptocrystalline material is fine-grained rock, for if the skeletal material is of recent origin and the cryptocrystalline material is "dead" carbonate from rock, an age of approximately 4,000 years should result.[7] Thus the silt-size cryptocrystalline material has been produced during the present stand of sea level and is probably still forming.

It was noted that the exact nature and mode of formation of sand-size cryptocrystalline material are poorly known. There is evidence, however, that silt-size cryptocrystalline grains in some areas are produced by breakdown of sand-size cryptocrystalline grains. For example, on the northeast end of the Bulkhead mud shoal, laminated miliolid mud grades into and is overlain by peneroplid sands, which are characterized by cryptocrystalline skeletal grains. There is negligible mineralogic difference between the sands and muds (Table 14); therefore, there is no clear relation between decreasing grain size and mineralogy such as was noted for the transition between the miliolid muds and *Halimeda* muds (Fig. 24). The size gradation in grains both *between* samples and *within* a sample indicates that the mud is produced by size reduction of sands.

[7]Radioactive decay is a first-order reaction, i.e., the rate of decomposition is proportional to the amount of decaying substance present. If only 65 percent of the carbonate contains measurable C-14, the amount of C-14 (assuming a decay rate of 50 percent in 5,500 years) in the total sample will be approximately 65 percent relative to recent material, giving an apparent age of just under 4,000 years.

TABLE 14. CARBONATE MINERALOGY OF DIFFERENT SIZE FRACTIONS OF SELECTED SAMPLES FROM MILIOLID-MUD AND PENEROPLID-SAND FACIES

	Sample	*Mud Content (% <1/16 mm)*	*Carbonate Mineralogy*					
			Aragonite Percentage			*Magnesium Carbonate in Calcite*[1]		
			Entire Sample	*<62μ*	*<5μ*	*Entire Sample*	*<62μ*	*<5μ*
Miliolid-mud facies	31	47	43	-	35	13	-	13
	32	90	26	26	-	11	11	-
	250	87	17	-	20	11	-	12
	257	19	15	16	-	13	13	-
	268	35	21	25	-	13	11	-
Peneroplid-sand facies	261	10	11	20	-	13	11	-
	264	3	10	15	-	13	13	-

[1]Weight percent; determined by X-ray diffraction. Reproducibility of ±1 percent; accuracy is probably ±2 percent.

Thus it seems reasonable to conclude that a major part of the mud in the miliolid-mud facies is probably transported into the depositional area from the peneroplid facies and the cryptocrystalline-grain facies.

Many areas of miliolid-mud deposition, however, are remote from cryptocrystalline-grain or peneroplid-sand areas, and yet they contain a high proportion of cryptocrystalline silt in their muds. Microboring and alteration of skeletal grains to cryptocrystalline material are not unusual in these samples but seems insufficient to account for the great volume of cryptocrystalline material. Direct physicochemical precipitation seems unlikely as these muds are largely high-magnesium calcite. It is possible that photosynthetic bacteria play an important role in the origin of the muds in these areas. LaLou (1957a, 1957b) demonstrated that anaerobic, but photosynthetic, sulfate-reducing bacteria can cause calcite, aragonite, and dolomite to precipitate in culture media. More recently, Oppenheimer and Master (1963) found that a mixture of algae and bacteria cultured in seawater and calcite sand contained up to 5 percent dolomite after 20 days. They implied that carbon dioxide-induced pH changes associated with photosynthesis and respiration caused the precipitation. Apparently, neither LaLou nor Oppenheimer and Master looked for high-magnesium calcite.

The dark-gray color of the miliolid muds less than an inch beneath the water-sediment interface indicates that the muds are reducing, thus satisfying the anaerobic environmental requirement of sulfate-reducing bacteria. The emission of hydrogen sulfide gas when the sediments are disturbed suggests that sulfate-reducing bacteria are actually present. Jars of miliolid mud that were stored wet developed green and/or purple films on their sides that may be bacterial in origin. Without additional points of comparison or experiments using Belize sediments, it would be mere speculation to ascribe a significant part of the cryptocrystalline mud grains to bacterial action.

DEVELOPMENT AND DISTRIBUTION OF FACIES

The unconsolidated sediments on the northern Belize shelf were deposited when sea level rose to near its present level after the Wisconsin glaciation. During the Wisconsin glaciation, sea level was lowered several hundred feet, exposing the entire shelf area to subaerial erosion. This erosion, which probably consisted largely of solution alteration of exposed Tertiary and Pleistocene limestones, produced an irregular rock surface, the general features of which are indicated in Figure 29. It should be noted that this map is based on sediment-thickness data obtained by coring and probing through the sediment section for the underlying rock surface. Although local relief, sinkholes, and possibly some valleys are not shown in this illustration, four drainage basins are delimited by the Rocky Point, Blackadore Cay, Hicks Cays, and Belize divides. It is evident that upper Chetumal Bay, lower Chetumal Bay, and the shelf lagoon did not become a continuous basin until sea level reached nearly its present position.

Detailed study of the basin between Blackadore Cay and Hicks Cays constitutes the basis for the discussion which follows. In the southern part of the basin, sediments are thin; however, filled drainage channels containing over 8 ft (2.4 m) of sediment show clearly on aerial photographs (Fig. 7B). In the north side of the basin, which includes the Bulkhead mud-shoal complex, the sediments are thick, but detailed probes indicate in a general way the location of channels beneath the sediment cover (Fig. 29). The configuration of the west side of the basin is poorly known because of progradation of the coast over marine sediments

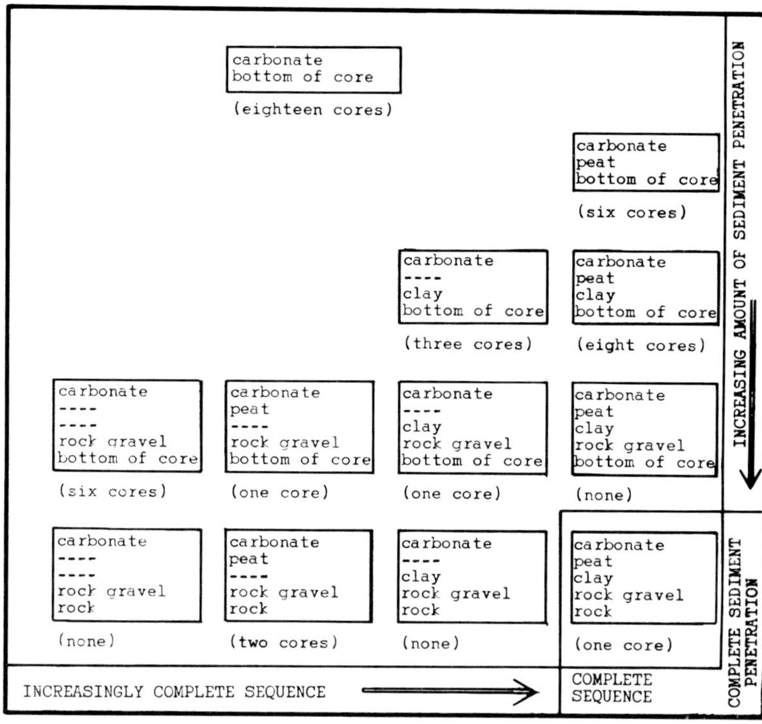

FIG. 33—Summary of sediment sequences in cores from northern Belize.

(High, 1966) and because of difficulty in penetrating the gray terrigenous clays that underlie the thin surface layer of carbonate sediments. Within this local basin, sediments were deposited in a lithologic sequence that varies consistently both vertically and laterally.

An idealized vertical section of the sediments, containing all the observed lithologic units, was determined by combining numerous partial, lithologic core sequences (Fig. 33) and by using one core that exemplified the entire lithologic sequence (Fig. 27). This sequence is analogous to the lower, or transgressive, part of classic Pennsylvanian cyclothems (Weller, 1960, p. 369-375). A basal unconformity is indicated by soil or gravel lying on an irregular limestone surface. The fossiliferous limestone is weathered and soft. The overlying gravel of limestone pebbles becomes increasingly diluted upward by a terrigenous gray clay matrix. The clay contains abundant plant fragments (about 10 percent of the sediment volume) that are vertically oriented except for a few inches of bedded fragments near the top. Some of the fragments are roots with pink low-magnesium calcite nodules inside. High (1966; this volume) found meadows onshore which are developed on clay soils that are nearly identical with the basal gray clays in the lagoon cores. Palynologic and stratigraphic evidence (High, 1966, p. 155), coupled with the present meadow-soil analogy, strongly suggests that the dense gray clays are soils which developed on the Pleistocene limestones just prior to flooding. Scarce miliolid foraminifers are found in the upper parts of these clays (Pusey, 1964, p. 223; High, 1966, p. 148). These foraminifers probably indicate some reworking of the soil as the lagoon transgressed the meadows.

The clay is overlain by bedded mangrove peat which commonly contains woody fragments. The clay-peat contact is gradational in places, but

TABLE 15. INSOLUBLE RESIDUES FROM CORES

Sample	*Depth in Feet Below Sediment Surface*	*Distance in Feet Above Contact with Noncarbonate*	*Percent Insoluble in Acetic Acid*[1]	*Description of Quartz in Insoluble Residue*[2]
31	3.0	1.5	29	trace of angular silt
243	6.0	unknown	16	none
	8.0	"	22	none
	8.7	"	23	trace of angular silt
244	4.0	4.5	10	none
	6.0	2.5	8	trace of angular silt
248	4.0	4.0	17	trace of angular silt
	6.0	2.0	23	none
	8.0	0.1	17	angular silt and sand (20 percent of residue)
250	0.0	8.2	18	none
	4.8	3.5	12	none
	7.2	1.0	22	angular silt and sand (1 percent of residue)
256	4.0	4.0	13	trace of angular silt
	6.0	2.0	11	silt and scarce sand (1 percent of residue)

[1] Three-to-seven-gram samples dissolved in 20 percent acetic acid.
[2] Most of the residues consisted of brownish plant(?) material, siliceous sponge spicules, diatoms, and scarce radiolarians.

more commonly is sharp (Fig. 18D) or burrowed (Fig. 18F) and is overlain by a coarse shell hash containing scattered limestone pebbles (Fig. 18A, B, C, G). The completeness and types of units present in the sequence at a given locality correlate well with the relief developed on the underlying rock surface. Sequences developed in rock lows tend to be more complete than sequences developed on rock highs. On the Bulkhead, sediments deposited in valleys contain a relatively complete sequence from rock to gravel, to clay, to peat, to carbonate mud; but sediments that were deposited on rock highs pass directly from rock into gravel and then into carbonate mud.

Carbonate sediments that are within a few feet of the contact with noncarbonate (clay or peat) material contain quartz sufficiently abundant to be detected by binocular examination of insoluble residues (Table 15). In contrast, quartz commonly is not present more than 3 ft (0.9 m) above this contact. The upward decrease in quartz content probably is due to retreat of the shoreline as the area flooded.

At the beginning of the post-Wisconsin rise in sea level, the shelf surface consisted of gentle hills and valleys. Local relief was 3 ft (0.9 m), with scattered sinkholes more than 10 ft (3.1 m) deep. Sea level rose from -300 ft (-91 m) to -20 ft (-6 m) at a rate of nearly 2 1/2 ft (0.8 m) per century, then it rose to its present level at a slower rate—averaging less than 1/2 ft (0.2 m) per century (Shepard, 1964). Approximately 6,000 years ago, the sea began flooding the basin, and marine deposition began with mangrove peats formed in swamps close to shore (Fig. 34A).

When sea level stood at -12 ft (-3.6 m), the area underlying the site of the Bulkhead was flooded first. A north-south valley, underlain by gray clay soil, was drowned, and a rock peninsula to the east was flanked and eventually covered by mangrove swamps. With further sea-level rise to -10 ft (-3.0 m), the peninsula was flooded, and only mangrove swamps remained (Fig. 34B). A second channel to the sea was flooded, south of Cay Corker, and marine circulation was established. In the deeper parts of the lagoon, miliolid muds containing small amounts of quartz sand that was reworked from exposed soil were deposited. The geometry of the depositional area attained its present configuration when sea level reached -8 ft (-2.4 m), and the shelf lagoon, lower Chetumal Bay, and upper Chetumal Bay became a continuous body of water. The commencement of water outflow from the bay into the shelf lagoon initiated deposition of the Bulkhead mud-shoal complex. Initially, the north and south parts of the Bulkhead were separated by a mangrove swamp, but the highest occurrence of peats in this area at -7 ft (-2.1 m) indicates that the swamp was soon flooded and covered by carbonate mud. At its maximum, approximately 2,000 years or less before present, the lagoon extended 1-2 mi (1.6-3.2 km) farther west (High, 1966) and covered parts of Ambergris Cay (Ebanks, 1967). Since that time, sea level has been nearly static and has allowed coastal and island sediments to prograde slowly over the lagoon. This process is less than 10 percent complete and will require thousands of years to destroy the lagoon.

In summary, the sequence of lithologies underlying the present facies is the result of a rising sea level which terminated subaerial erosion and soil formation and initiated marine deposition. As sea level rose, deposition changed from reworked clastic material (carbonate-rock fragments and terrigenous material) and locally produced mangrove peats to locally produced carbonate sediments.

SUMMARY AND CONCLUSIONS

The windward shelf edge of northern Belize is mantled by corals and coralline algae in the form of a barrier reef. In the protected lee of

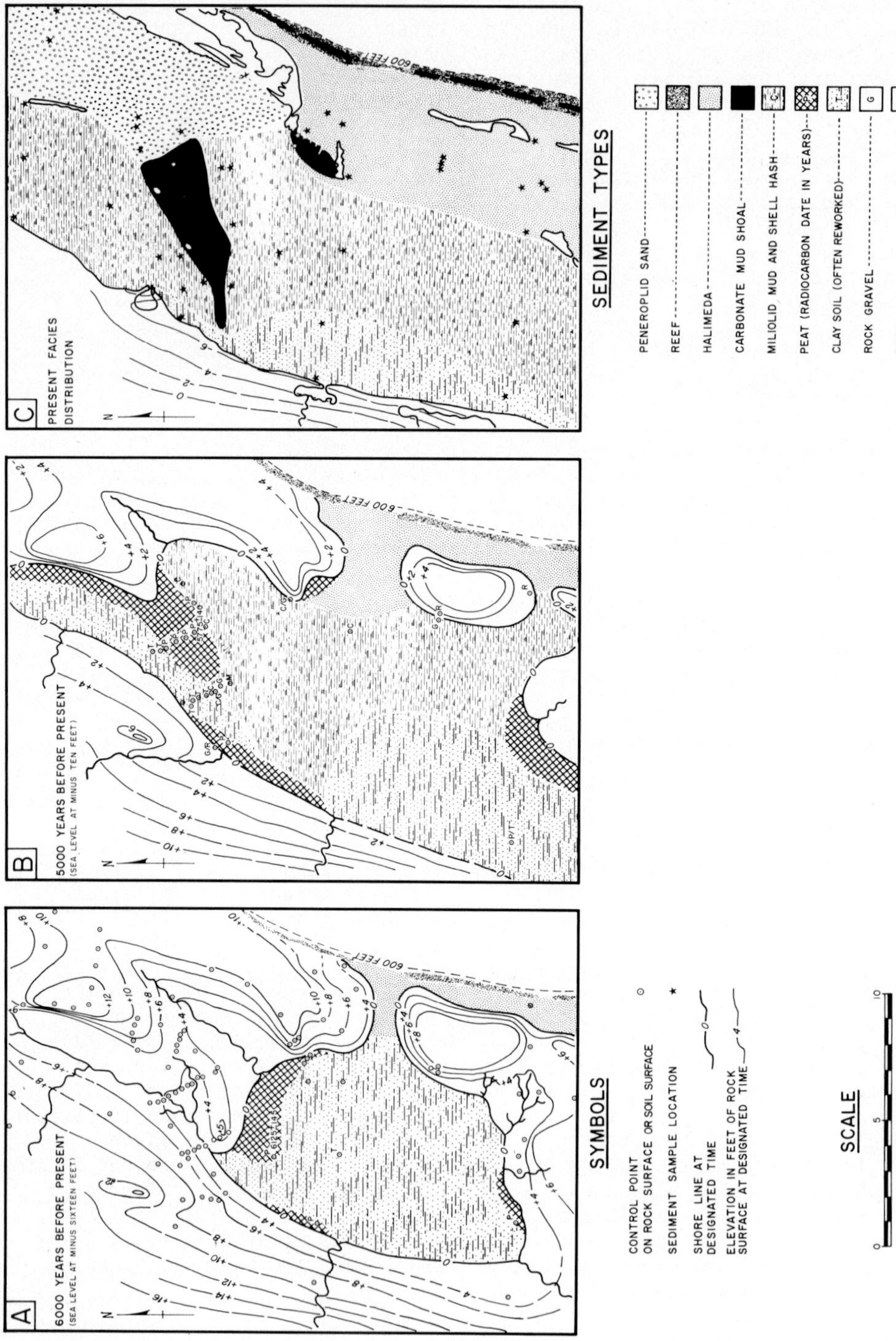

FIG. 34—Schematic representation of facies development during post-Wisconsin rise in sea level (from Pusey, 1964, with modifications from High, 1966).

the reef, sands composed of *Halimeda* predominate. Here, currents are unfavorable for reef growth but are strong enough to encourage the development of *Halimeda* meadows and to prevent mud deposition. In the deeper lagoon, farther west from the reef-rimmed shelf edge, currents are weaker and calcite muds, inhabited by organisms tolerant of brackish water, predominate. These are the miliolid muds and, although they are pelleted, the pellets are not indurated. Some of this calcite mud is formed within the depositional area as silt-size skeletal material, but most of the mud appears to have been transported into the miliolid-mud environment from areas of active abrasion, namely, the outer lagoon and the windward side of lower Chetumal Bay. In shallow water near the mainland, a thin sequence of molluscan carbonate sediments is mixed with the underlying terrigenous material by burrowing organisms, forming the terrigenous facies.

Two major rivers drain into upper Chetumal Bay, thereby insuring a net movement of water southward into lower Chetumal Bay. Currents leaving the lower bay slacken both by radiating out at the bay mouth and by meeting tidal and wind-induced currents. As a consequence of this localization of reduced current velocities, a mud shoal has formed at the bay mouth. Laminations in the mud-shoal sediments are produced by periodic but minor current disturbances. The mud shoal is transitional to the windward, current-swept areas of the peneroplid-sand facies. These high-magnesium calcite, peneroplid sands, as well as the aragonite *Halimeda* sands, are continuously bathed by circulating seawater. The water is supersaturated with respect to the carbonate mineralogy of the grains and, consequently, reacts with the grains so that the grains commonly are infilled with crystals of either high-magnesium calcite or aragonite corresponding to the mineralogy of the host grain. In these same areas of grain alteration, intensive microboring is common and promotes recrystallization, without major mineralogic change, to cryptocrystalline material.

The relatively restricted bay waters are subject to great seasonal variations in salinity, temperature, and related characters. These hydrographic conditions, in combination with a sediment-covering mat of organic detritus, limit the biota to sparse populations of few species. Sediments of this area are represented chiefly by the cryptocrystalline-grain facies. In the seaward part of the cryptocrystalline-grain facies, partly cryptocrystalline skeletal grains are common, and most of the cryptocrystalline grains are altered skeletal material. In the upper bay, however, cryptocrystalline grains are not associated with partly altered skeletal grains, and most of the cryptocrystalline grains are fragments of low-magnesium calcite crusts that are closely associated with blue-green algae. Sporadic storms produce currents that disturb the cryptocrystalline sediments and redeposit them as graded sand beds.

Near river mouths the environment is unfavorable for most marine organisms, and the sediments are characterized by ostracods in a mud matrix. Diatoms, woody organic material, and terrigenous minerals are locally abundant.

In summary, the development of the coral-reef and *Halimeda* facies is limited to current-agitated marine waters of normal salinity, whereas the development of the other facies is related to reduced current agitation and/or varying degrees of hyposalinity. The volume and mineralogy of the mud in each facies are determined both directly and indirectly by current strength. They are determined directly by the transporting capacity of the currents in each of the eight depositional environments, and indirectly by the capacity of currents to induce the formation of mud-size particles through the abrasion of sand-size constituents.

SELECTED REFERENCES

Alexandersson, T., 1969, Recent littoral and sublittoral high-Mg calcite lithification in the Mediterranean: Sedimentology, v. 12, p. 47-61.

Alvarez, M., Jr., 1954, Exploración geológica preliminar del Río Hondo, Quintana Roo: Asoc. Mexicana Geólogos Petroleros Bol., v. 6, no. 5-6, p. 207-213.

Baars, D. L., 1963, Petrology of carbonate rocks, *in* Shelf carbonates of the Paradox basin—a symposium: Four Corners Geol. Soc. 4th Field Conf., 1963; Durango, Colorado, Petroleum Inf., p. 101-129.

Barker, R. W., 1960, Taxonomic notes on the species figured by H. B. Brady in his report on the Foraminifera dredged by H. M. S. Challenger during the years 1873-1876: Soc. Econ. Paleontologists and Mineralogists Spec. Pub. 9, 238 p.

Bathurst, R. G. C., 1958, Diagenetic fabrics in some British Dinantian limestones: Liverpool and Manchester Geol. Jour., v. 2, p. 11-36.

——1966, Boring algae, micrite envelopes and lithification of molluscan biosparites: Geol. Jour., v. 5, no. 2, p. 89-109.

Bavendamn, W., 1932, Die mikrobiologische Kalkfällung in der tropischen, *in* International Expedition to the Bahamas, Contribution 7: Berlin, J. Springer, p. 205-276.

Bayer, F. M., and E. Montanaro-Gallitelli, 1956, Octocorallia with a section on Trachypsammiacea, *in* R. C. Moore, ed., Treatise on invertebrate paleontology, part F, Coelenterata: Lawrence, Kansas, Geol. Soc. American and Univ. Kansas Press, p. F166-F231.

Bøggild, O. B., 1930, The shell structure of the mollusks: Acad. Royale Sci. Danemark Mém., ser. 9, v. 2, p. 231-326.

Bradley, W. H., 1929, Algae reefs and oolites of the Green River Formation: U.S. Geol. Survey Prof. Paper 154-G, p. 203-223.

Broecker, W., 1964, Radiocarbon dating: a case against the proposed link between river mollusks and soil humus: Science, v. 143, p. 396-597.

Brongersma-Sanders, M., 1957, Mass mortality in the sea, *in* J. W. Hedgpeth, ed., Treatise on marine ecology and paleoecology: Geol. Soc. America Mem. 67, v. 1, p. 941-1010.

Burns, J. H., and M. A. Bredig, 1956, Transformation of calcite to aragonite by grinding: Jour. Chem. Physics, v. 25, p. 1281.

Butterlin, J., 1956, La constitucion géologique et la structure des Antilles: Paris, Centre Natl. Recherche Sci., 453 p.

——and F. Bonet, 1962, Les formations cénozoîques de la partie Mexicaine de la Presqu'île du Yucatán, *in* 3d Caribbean Geol. Conf., 1962: Kingston, Jamaica, 45 p. (mimeographed).

Cann, R. S., 1962, Recent calcium carbonate facies of the north-central Campeche Bank, Yucatán, Mexico: New York, Columbia Univ., PhD thesis, 121 p.

Cattell, R. B., 1952, Factor analysis; an introduction and manual for the psychologist and social scientist: New York, Harper & Bros., 462 p.

Cayeux, L., 1935, Les roches sédimentaires de France. Roches carbonatées (calcium et dolomies): Paris, Masson, 463 p.

Chave, K. E., 1952, A solid solution between calcite and dolomite: Jour. Geology, v. 60, p. 190-192.

——1954, Aspects of biogeochemistry of magnesium, 1; Calcareous marine organisms: Jour. Geology, v. 62, p. 266-283.

——1962, Factors influencing the mineralogy of carbonate sediments: Limnology and Oceanography, v. 7, p. 218-223.

——et al, 1962, Observations on the solubility of skeletal carbonates in aqueous solutions: Science, v. 137, p. 33-34.

Chayes, F., 1956, Petrographic modal analysis: New York, John Wiley & Sons, 113 p.

Cloud, P. E., Jr., 1954, Superficial aspects of modern organic reefs: Sci. Monthly, v. 79, p. 195-208.

——1962, Environment of calcium carbonate deposition west of Andros Island, Bahamas: U.S. Geol. Survey Prof. Paper 350, 138 p.

Curl, R. L., 1962, The aragonite-calcite problem: Natl. Speleol. Soc. Bull., v. 24, pt. 2, p. 57-73.

Dachille, F., and R. Roy, 1960, High-pressure phase transformations in laboratory mechanical mixers and mortars: Nature, v. 186, no. 4718, p. 34, 71.

Darwin, Charles, 1889, The structure and distribution of coral reefs, 3d ed.: New York, 344 p.

Davis, C. A., 1900, A contribution to the natural history of marl: Jour. Geology, v. 8, p. 485-497.

——1901, A second contribution to the natural history of marl: Jour. Geology, v. 9, p. 491-506.

Dixon, C. G., 1958, A hydrogeological survey in British Honduras: Gentbrugge, Belgium, Internat. Geod. Geophys. Union, Assoc. Sci. Hydrology Gen. Assembly, Toronto, 1957, v. 2, p. 230-235; Internat. Assoc. Sci. Hydrology Pub. 44.

Dunham, R. J., 1962, Classification of carbonate rocks according to depositional texture, *in* W. E. Ham, ed., Classification of carbonate rocks—a symposium: AAPG Mem. 1, p. 108-121.

Ebanks, W. J., Jr., 1967, Recent carbonate sedimentation and diagenesis, Ambergris Cay, British Honduras: Houston, Texas, Rice Univ., PhD thesis, 189 p.

Feray, D. E., E. Heuer, and W. G. Hewatt, 1962, Biological, genetic, and utilitarian aspects of limestone classification, *in* W. E. Ham, ed., Classification of carbonate rocks—a symposium: AAPG Mem. 1, p. 20-32.

Fox, D. L., 1950, Comparative metabolism of organic detritus by inshore animals: Ecology, v. 31, p. 100-108.

——1957, Particulate organic detritus, *in* J. W. Hedgpeth, ed., Treatise on marine ecology and paleoecology: Geol. Soc. America Mem. 67, v. 1, p. 383-390.

Fuglister, F. C., 1947, Average monthly sea surface temperatures of the western North Atlantic Ocean: Massachusetts Inst. Technology and Woods Hole Oceanog. Inst., Papers in Physical Oceanography and Meteorology, v. 10, no. 2, 25 p.

Ginsburg, R. N., 1956, Environmental relationships of grain size and constituent particles in some South Florida carbonate sediments: AAPG Bull., v. 40, p. 2384-2427.

——1957, Early diagenesis and lithification of shallow-water carbonate sediments in South Florida, *in* R. J. Le Blanc and J. G. Breeding, eds., Regional aspects of carbonate deposition—a symposium: Soc. Econ. Paleontologists Special Pub. 5, p. 80-99.

——and R. M. Lloyd, 1956, A manual piston coring device for use in shallow water: Jour. Sed. Petrology, v. 26, p. 64-66.

——and H. A. Lowenstam, 1958, the influence of marine bottom communities on the depositional environment of sediments: Jour. Geology, v. 66, p. 310-318.

——E. Shinn, and J. H. Schroeder, 1967, Submarine cementation and internal sedimentation within Bermuda reefs (abs.): Geol. Soc. America Spec. Paper 115, p. 78-79.

——et al, 1963, Shallow water carbonate sediments, *in* The sea, v. 3: New York, Wiley-Interscience, p. 554-582.

Goldsmith, J. R., D. L. Graf, and O. I. Joensuu, 1955, The occurrence of magnesian calcites in nature: Geochim. et Cosmochim. Acta, v. 7, p. 212-230.

Goreau, T. F., 1959, The ecology of Jamaican coral reefs, I—Species composition and zonation: Ecology, v. 40, no. 1, p. 67-90.

Ham, W. E., and L. C. Pray, 1962, Modern concepts and classifications of carbonate rocks, *in* W. E. Ham, ed., Classification of carbonate rocks—a symposium: AAPG Mem. 1, p. 2-19.

Hay, W. W., K. M. Towe, and R. Wright, 1963, Ultramicrostructure of some selected foraminiferal tests: Micropaleontology, v. 9, p. 171-195.

High, L. R., Jr., 1967, Recent coastal sediments of British Honduras: Houston, Texas, Rice Univ., PhD thesis, 199 p.

——1975, Geomorphology and sedimentology of Holocene coastal deposits, Belize: this volume.

Hofker, J., 1950, Recent Peneroplidae, Part I: Royal Micros. Soc. Jour., v. 70, p. 388-396.

——1951a, Recent Peneroplidae, Part II: Royal Micros. Soc. Jour., v. 71, p. 342-356.

——1951b, Recent Peneroplidae, Part III: Royal Micros. Soc. Jour., v. 71, p. 450-463.

Hoskin, C. M., 1963, Recent carbonate sedimentation on Alacran Reef, Yucatan, Mexico: Natl. Acad. Sci-Natl. Research Council Pub. 1089, 160 p.

Humm, H. J., and S. E. Taylor, 1961, Marine Chlorophyta of the upper west coast of Florida: Bull. Marine Sci. Gulf and Caribbean, v. 11, no. 3, p. 321-380.

Illing, L. V., 1954, Bahaman calcareous sands: AAPG Bull., v. 38, p. 1-94.

Imbrie, John, 1963, Factor and vector analysis programs for analyzing geologic data: Office Naval Research, Geography Branch, Northwestern Univ., Office Naval Research Task 389-135, Tech. Rept. 6, 83 p.

——and E. G. Purdy, 1962, Classification of modern Bahamian carbonate sediments, *in* W. E. Ham, ed., Classification of carbonate rocks—a symposium: AAPG Mem. 1, p. 253-272.

Jaanusson, V., 1952, Untersuchungen über Korngrösse der ordovizischen Kalksteine: Geol. Fören. Stockholm Förh., Bd. 74, h. 2., p. 124.

Keith, M. L., and G. M. Anderson, 1963, Radiocarbon dating; fictitious results with mollusk shells: Science, v. 141, p. 634-636.

Klovan, J. E., 1963, Facies analysis of Redwater reef complex, Alberta, Canada: New York, Columbia Univ., PhD thesis, 197 p.

Kornicker, L. S., 1958, Ecology and taxonomy of Recent marine ostracodes in the Bimini area, Great Bahama Bank: Inst. Marine Sci. Pub., v. 5, p. 194-300.

Krueger, W. C., Jr., 1963, Mineralogical composition and textural properties of river sediments from British Honduras: Houston, Texas, Rice Univ., PhD thesis, 106 p.

Krumbein, W. C., and L. L. Sloss, 1963, Stratigraphy and sedimentation, 2d ed.: San Francisco, W. H. Freeman & Co., 660 p.

LaLou, Claude, 1957a, Studies on bacterial precipitation of carbonates in sea water: Jour. Sed. Petrology, v. 27, p. 190-195.

——1957b, Étude expérimentale de la production de carbonates par les bactéries des vases de la baie de Villefranche-sur-Mer: Inst. Océanog. Annales, v. 33, no. 4, p. 201-266.

Land, L. S., and T. G. Goreau, 1970, Submarine lithification of Jamaican reefs: Jour. Sed. Petrology, v. 40, p. 457-462.

Lanlignel, A., and M. Schmitt, 1961, Examen de *Lithothamnium calcareum* au microscope électronique: Centre Natl. Recherche Sci., Colloques Internat., no. 103, p. 197-198.

Loeblich, A. R., Jr., and H. Tappan, 1964, Protista 2, *in* R. C. Moore, ed., Treatise on invertebrate paleontology, part C, Protista, v. 1: Lawrence, Kansas, Geol. Soc. America and Univ. Kansas Press, 511 p.

Lowenstam, H. A., 1954, Factors affecting the aragonite-calcite ratios in carbonate secreting marine organisms: Jour. Geology, v. 62, p. 284-322.

———1955, Aragonite needles secreted by algae and some sedimentary implications: Jour. Sed. Petrology, v. 25, p. 270-272.

Macintyre, I. G., E. W. Mountjoy, and B. F. D'Anglejan, 1968, An occurrence of submarine cementation of carbonate sediments off the west coast of Barbados, W.I.: Jour. Sed. Petrology, v. 38, p. 660-664.

Matthews, R. K., 1966, Genesis of Recent lime mud in southern British Honduras: Jour. Sed. Petrology, v. 35, p. 428-454.

Mawson, D., 1929, Some South Australian algal limestone in process of formation: Geol. Soc. London Quart. Jour., v. 85, p. 613-623.

McIntire, W. L., 1963, Trace element partition coefficients—a review of theory and applications to geology: Geochim. et Cosmochim. Acta, v. 27, p. 1209-1264.

Moore, D. G., and P. C. Scruton, 1957, Minor internal structures of some recent unconsolidated sediments: AAPG Bull., v. 41, p. 2723-2751.

Moore, H. B., 1933, Faecal pellets from marine deposits: Discovery Repts., v. 7, p. 17-26.

Moore, R. C., 1949, Meaning of facies, *in* Sedimentary facies in geologic history: Geol. Soc. America Mem. 39, p. 1-34.

Murray, G. E., 1961, Geology of the Atlantic and Gulf coastal province of North America: New York, Harper & Bros., 657 p.

———and A. Weidie, 1962, Regional geologic summary of Yucatan Peninsula, *in* Field trip to Peninsula of Yucatan: New Orleans Geol. Soc. Guidebook, p. 5-51.

Nadson, G., 1927, Les algues perforantes, leur distribution et leur rôle dans la nature: Paris, Acad. Sci. Comptes Rendus, v. 184, p. 896-898, p. 1015-1017.

Neumann, A. C., C. D. Gebelein, and T. P. Scoffin, 1970, The composition, structure and erodability of subtidal mats, Abaco, Bahamas: Jour. Sed. Petrology, v. 40, no. 1, p. 274-297.

Neveu, M., 1961, Contributions a l'étude du traitement des eaux d'alimentation par le *Lithothamnium calcareum*: Centre Natl. Recherche Sci., Colloque 103, p. 63.

Newell, N. D., and J. K. Rigby, 1957, Geological studies on the Great Bahama Bank, *in* Regional aspects of carbonate deposition: Soc. Econ. Paleontologists and Mineralogists Spec. Pub. 5, p. 15-72.

O'Colla, P. S., 1962, Mucilages, *in* Physiology and biochemistry of algae: New York, Academic Press, p. 337-356.

Oltmanns, F., 1922-1923, Morphologie und Biologie der Algen: Jena, Verlag Gustave Fischer, v. 2, 439 p.

Oppenheimer, C. H., and L. S. Kornicker, 1958, Effect of the microbial production of hydrogen sulfide and carbon dioxide on the pH of Recent sediments: Inst. Marine Sci. Pub., v. 5, p. 5-15.

———and I. M. Master, 1964, Transition of silicate and carbonate crystal structure by photosynthesis and metabolism (abs.): Geol. Soc. America Spec. Paper 76, p. 125.

Ower, L., 1927, Features of British Honduras: Geog. Jour., v. 70, p. 372-386.

———1928, Geology of British Honduras: Jour. Geology, v. 36, p. 494-509.

Parr, A. E., 1937, A contribution to the hydrography of the Caribbean and Cayman Seas: Yale Univ. Peabody Mus. Bull., v. 5, art 4, p. 1-110.

Patrick, R., 1948, Factors affecting the distribution of diatoms: Bot. Rev., v. 14, no. 8, p. 473-524.

Pettijohn, F. J., 1957, Sedimentary rocks: New York, Harper & Bros., 526 p.

Pia, J., 1940, Die kalklosenden Thallophyten: Archiv Hydrobiologie Beihefte—Ergebnisse Limnologie, v. 31, pts. 2-3, p. 364-392.
Price, W. A., 1963, Patterns of flow and channeling in tidal inlets: Jour. Sed. Petrology, v. 33, p. 279-290.
Purdy, E. G., 1961, Recent calcium carbonate facies of the Great Bahama Bank: New York, Columbia Univ., PhD thesis, 170 p.
——1963a, Recent calcium carbonate facies of the Great Bahama Bank—1, Petrography and reaction groups: Jour. Geology, v. 71, p. 334-355.
——1963b, Recent calcium carbonate facies of the Great Bahama Bank—2, Sedimentary facies: Jour. Geology, v. 71, p. 472-497.
——W. C. Pusey, and K. F. Wantland, 1975, Continental shelf of Belize —regional shelf attributes: this volume.
Pusey, W. C., 1964, Recent calcium carbonate deposition in northern British Honduras: Houston, Texas, Rice Univ., PhD thesis, 247 p.
Reiss, Z., 1958, Classification of lamellar Foraminifera: Micropaleontology, v. 4, p. 51-70.
Rivière, A., 1939a, Sur la dolomitisation des sediments calcaires: Paris, Acad. Sci. Comptes Rendus, v. 209, p. 597-599.
——1939b, Observations nouvelles sur le mecanisme de dolomitisation des sédiments calcaires: Paris, Acad. Sci. Comptes Rendus, v. 209, p. 691-692.
Robertson, R., 1963, The mollusks of British Honduras: Philadelphia Shell Club Proc., v. 1, no. 7, p. 15-20.
Roddy, H. J., 1915, Concretions in streams formed by the agency of blue-green algae and related plants: Am. Philos. Soc. Proc., v. 54, no. 218, p. 246-258.
Romney, D. H., ed., 1959, Land in British Honduras, report of the British Honduras land use survey team: London, Colonial Research Pub. 24, 327 p.
Rubin, M., R. C. Likens, and J. Berry, 1963, On the validity of radiocarbon dates from snail shells: Jour. Geology, v. 71, p. 84-89.
——and D. W. Taylor, 1963, Radiocarbon activity of shells from living clams and snails: Science, v. 141, p. 637.
Sapper, K. T., 1896, Geology of Chiapas, Tabasco and Yucatan: Jour. Geology, v. 4, p. 938-947.
Schmalz, R. F., and K. E. Chave, 1963, Calcium carbonate: factors affecting saturation in ocean waters off Bermuda: Science, v. 139, p. 1206-1207.
——and F. J. Swanson, 1969, Diurnal variations in the carbonate saturation of seawater: Jour. Sed. Petrology, v. 39, p. 255-267.
Scoffin, T. P., 1970, The trapping and binding of subtidal carbonate sediments by marine vegetation in Bimini Lagoon, Bahamas: Jour. Sed. Petrology, v. 40, p. 1-28.
Scott, M. R., 1966, Distribution of clay minerals on the British Honduras shelf: Houston, Texas, Rice Univ., PhD thesis, 77 p.
——1975, Distribution of clay minerals on Belize shelf: this volume.
Shepard, F. P., 1964, Sea level changes in the past 6000 years; possible archeological significance: Science, v. 143, p. 574-576.
Shinn, E. A., 1963, Spur and groove formation on the Florida reef tract: Jour. Sed. Petrology, v. 33, p. 291-303.
——1969, Lithification of Holocene carbonate sediments in the Persian Gulf: Sedimentology, v. 12, p. 109-144.
Stockman, K. W., R. N. Ginsburg, and E. A. Shinn, 1967, The production of lime mud by algae in south Florida: Jour. Sed. Petrology, v. 37, no. 2, p. 633-648.
Stoddart, D. R., 1960, The reefs and sand cays of British Honduras: Cambridge Exped. British Honduras 1959-1960, Gen. Rep., p. 16-22.
——1962a, Three Caribbean atolls: Turneffe Islands, Lighthouse Reef,

and Glover's Reef, British Honduras: Atoll Research Bull., no. 87, 151 p.
——1962b, Physiographic studies on the British Honduras reefs and cays: Geog. Jour., v. 128, p. 161-171.
——1962c, Catastrophic storm effects on the British Honduras reefs and cays: Nature, v. 196, p. 512-515.
——1963, Effects of Hurricane Hattie on the British Honduras reefs and cays, October 30-31, 1961: Atoll Research Bull., no. 95, 142 p.
Sverdrup, H. U., M. W. Johnson, and R. H. Fleming, 1942, The oceans; their physics, chemistry, and general biology: New York, Prentice-Hall, 1087 p.
Taft, W. H., 1962, Unconsolidated carbonate sediments of Florida Bay, Florida: Stanford, California, Stanford Univ., PhD thesis, 60 p.
——and J. W. Harbaugh, 1964, Modern carbonate sediments of southern Florida, Bahamas, and Espíritu Santo Island, Baja California—a comparison of their mineralogy and chemistry: Stanford Univ. Pubs. Geol. Sci., v. 8, no. 2, 133 p.
Taylor, W. R., 1936, Marine algae from the Yucatan Peninsula, *in* Botany of the Maya area, paper 7: Carnegie Inst. Washington Pub. 461, p. 115-124.
——1954, Sketch of the character of marine algal vegetation of the shores of the Gulf of Mexico, *in* Gulf of Mexico, its origin, waters and marine life: U.S. Dept. Interior Fishery Bull. 89, p. 177-192.
——1960, Marine algae of the eastern tropical and subtropical coasts of the Americas: Ann Arbor, Michigan, Univ. Michigan Press, 870 p.
Tebbutt, G. E., 1967, Diagenesis of Pleistocene limestone on Ambergris Cay, British Honduras (abs.): Dissert. Abs., sec. B, Sci. and Eng., v. 28, no. 5, p. 2002B.
Thorne, R. F., 1954, Flowering plants of the waters and shores of the Gulf of Mexico, *in* Gulf of Mexico, its origin, waters and marine life: U.S. Dept. Interior Fishery Bull. 89, p. 193-202.
Thorpe, J. E., and P. K. Bregazzi, 1960, Experiments and observations on the corals at Rendezvous Cay: Cambridge Exped. British Honduras, 1959-1960, Gen. Rept., p. 22-28.
Trewartha, G. T., 1961, The earth's problem climates: Madison, Wisconsin, Univ. Wisconsin Press, 334 p.
U.S. Navy Hydrographic Office, 1952, Sailing directions for the east coasts of Central America and Mexico: H.O. Pub. 20, 5th ed. (with changes through 1961), 274 p.
——1955a, Atlas of pilot charts, Central American waters and South Atlantic Ocean: H.O. Pub. 106, 16 charts.
——1955b, Pilot chart of Central American waters: Chart H.S. 3500, scale 1:6,120,000 (estimated).
——1959a, Approaches to Belize: Chart H.O. 1498, 1st ed. (revised), scale 1:125,000.
——1959b, Punta Herro to Cabo Gracias á Dios: Chart H.O. 394, 67th ed. (revised 1959), scale 1:936,430.
——1963, Yucatan channel and approaches: Chart H.O. 966, 67th ed. (revised), scale 1:906,530.
Vaughan, T. W., 1919, Corals and the formation of coral reefs: Smithsonian Inst. Ann. Rept. 197, p. 189-276.
Vermeer, D. E., 1959, The cays of British Honduras: Berkeley, Univ. California, M.A. thesis, 127 p.
——1963, Effects of hurricane Hattie, 1961, on the cays of British Honduras: Zeitschr. Geomorphologie, v. 7, no. 4, p. 332-354.
Walper, J. L., 1960, Geology of Cobán-Purulhá area, Alta Verapaz, Guatemala: AAPG Bull., v. 44, p. 1273-1315.
Wantland, K. F., 1967, Recent benthonic Foraminifera of the British

Honduras shelf: Houston, Texas, Rice Univ., PhD thesis, 308 p.
Ward, H. B., and G. C. Whipple, 1959, Fresh-water biology, 2d ed.: New York, John Wiley & Sons, 1248 p.
Weast, R. C., ed., 1968, Handbook of chemistry and physics: Cleveland, Ohio, Chemical Rubber Co., 49th ed.
Weller, J. M., 1960 Stratigraphic principles and practice: New York, Harper & Bros., 725 p.
Wells, J. W., 1956, Scleractinia, *in* R. C. Moore, ed., Treatise on invertebrate paleontology, part F, Coelenterata: Lawrence, Kansas, Geol. Soc. America and Univ. Kansas Press, p. F328-F444.
——1957, Coral reefs, *in* J. W. Hedgpeth, ed., Treatise on marine ecology and paleoecology: Geol. Soc. America Mem. 67, v. 1, p. 609-632.
West, R. C., 1962, Physical geology of the Yucatan platform, *in* Field trip to Peninsula of Yucatan: New Orleans Geol. Soc. Guidebook, p. 58-63.
Weyl, P. K., 1961, The carbonate saturometer: Jour. Geology, v. 69, p. 32-44.
Winland, H. D., 1969, Stability of calcium carbonate polymorphs in warm, shallow seawater: Jour. Sed. Petrology, v. 39, p. 1579-1587.
Yonge, C. M., 1940, The biology of reef-building corals: British Mus. (Nat. History), Great Barrier Reef Expedition, Sci. Repts., v. 1, no. 13, p. 353-390.

APPENDIX A

Water Data From Northern Belize, July-August 1961, August 1962

Station	Water Temperature (°C.)	Salinity (°/oo)	Saturometer (mv.) 1 min.	2 min.	3 min.	4 min.	Carbonate Standard Used	Date
1 surface	28.5	1.4	+ 12	+ 11	+ 10	+ 10	oolite	27 Jul'61
repeat	-	-	+ 12	+ 12	+ 12	+ 12	"	"
2 surface	29.5	1.5	+ 16	+ 16	+ 16	+ 15	oolite	"
bottom	-	15.1	+ 14	+ 14	+ 14	+ 14	"	"
repeat	-	-	+ 14	+ 14	+ 14	+ 14	"	"
3 surface	29.6	15.1	+ 18	+ 19	+ 20	+ 20	oolite	"
repeat	-	-	+ 18	+ 19	+ 19	+ 19	"	"
repeat	-	-	+ 20	+ 20	+ 20	+ 21	sediment	"
4 surface	28.7	17.9	+ 18	+ 20	+ 20	+ 20	oolite	"
surface	-	-	+ 33	+ 34	+ 35	+ 36	sediment	"
5 surface	29.0	19.7	+ 19	+ 20	+ 20	+ 20	oolite	"
surface	-	-	+ 28	+ 31	+ 31	+ 32	sediment	"
6 surface	28.5	20.5	+ 20	+ 22	+ 22	+ 22	oolite	"
surface	-	-	+ 30	+ 32	+ 34	+ 34	sediment	"
bottom	27.9	20.6	-	-	-	-	-	"
7 surface	28.5	20.7	+ 17	+ 18	+ 18	+ 19	oolite	28 Jul'61
bottom	28.5	20.7	+ 18	+ 18	+ 20	+ 20	oolite	"
surface	-	-	+ 26	+ 28	+ 29	+ 30	sediment	"
bottom	-	-	+ 22	+ 24	+ 26	+ 28	sediment	"
8 surface	28.9	20.6	+ 18	+ 18	+ 18	+ 18	oolite	"
surface	-	-	+ 23	+ 26	+ 28	+ 30	sediment	"
9 surface	28.9	19.9	+ 20	+ 21	+ 22	+ 22	oolite	"
bottom	29.0	19.8	+ 19	+ 20	+ 20	+ 21	oolite	"
surface	-	-	+ 24	+ 28	+ 30	+ 31	sediment	"
bottom	-	-	+ 28	+ 30	+ 32	+ 32	sediment	"
10 surface	29.0	20.1	+ 20	+ 20	+ 20	+ 20	oolite	"
surface	-	-	+ 29	+ 30	+ 32	+ 32	sediment	"
11 surface	29.7	26.0	+ 18	+ 20	+ 20	+ 21	oolite	"
surface	-	-	+ 36	+ 38	+ 39	+ 40	sediment	"
12 surface	29.4	23.3	+ 20	+ 20	+ 22	+ 22	oolite	"
surface	-	-	+ 26	+ 26	+ 27	+ 28	sediment	"
13 surface	29.2	23.8	+ 18	+ 19	+ 20	+ 20	oolite	"
surface	-	-	+ 29	+ 33	+ 34	+ 36	sediment	"
14 surface	28.0	19.4	+ 17	+ 18	+ 19	+ 19	oolite	29 Jul'61
surface	-	-	+ 18	+ 20	+ 20	+ 20	sediment	"
15 surface	-	21.6	+ 18	+ 20	+ 20	+ 21	oolite	5 Aug'61
surface	-	-	+ 23	+ 24	+ 26	+ 26	sediment	"
16 surface	28.5	20.4	+ 16	+ 17	+ 18	+ 19	oolite	"
surface	-	-	+ 16	+ 19	+ 22	+ 23	sediment	"
17 surface	27.9	27.0	+ 16	+ 18	+ 18	+ 18	oolite	29 Jul'61
surface	-	-	+ 17	+ 22	+ 24	+ 26	sediment	"
18 surface	27.9	27.4	+ 14	-	+ 16	+ 16	oolite	"
surface	-	-	+ 20	+ 29	+ 30	+ 31	sediment	"
19 surface	27.8	21.8	+ 14	+ 14	+ 15	+ 15	oolite	"
surface	-	-	+ 20	+ 21	+ 26	+ 27	sediment	"
bottom	27.4	21.9	-	-	-	-	-	"

APPENDIX A (Continued)

Station	Water Temperature (°C.)	Salinity (°/oo)	Saturometer (mv.) 1 min.	2 min.	3 min.	4 min.	Carbonate Standard Used	Date
23 surface	28.2	30.5	+ 16	+ 18	+ 18	+ 19	oolite	29 Jul'61
surface	-	-	+ 22	+ 26	+ 28	+ 28	sediment	"
24 surface	27.2	32.0	+ 10	+ 10	+ 10	+ 10	oolite	30 Jul'61
repeat	-	-	+ 14	+ 14	+ 13	+ 12	oolite	"
surface	-	-	+ 18	+ 20	+ 20	+ 20	sediment	"
25 surface	26.4	31.0	+ 14	+ 14	+ 14	+ 14	oolite	"
surface	-	-	+ 15	+ 16	+ 16	+ 17	sediment	"
26 surface	27.2	27.9	+ 17	+ 18	+ 18	+ 18	oolite	"
surface	-	-	+ 8	+ 12	+ 14	+ 16	sediment	"
repeat	-	-	+ 9	+ 12	+ 13	+ 14	sediment	"
27 surface	-	19.1	+ 16	+ 16	+ 19	+ 20	oolite	"
surface	-	-	+ 18	+ 19	+ 20	+ 20	sediment	"
28 surface	-	28.4	+ 18	+ 21	+ 22	+ 22	oolite	"
surface	-	-	+ 12	+ 15	+ 16	+ 18	sediment	"
31 surface	28.2	33.9	+ 12	+ 11	+ 10	-	oolite	"
repeat	-	-	+ 11	+ 10	+ 10	+ 10	oolite	"
surface	-	-	+ 22	+ 24	+ 26	+ 27	sediment	"
32 surface	28.2	34.8	+ 12	+ 12	+ 12	+ 13	oolite	"
surface	-	-	+ 26	+ 27	+ 28	+ 29	sediment	"
33 surface	28.2	35.8	+ 15	+ 16	+ 17	+ 18	oolite	"
repeat	-	-	+ 17	+ 18	+ 20	+ 20	oolite	"
repeat	-	-	+ 15	+ 16	+ 17	+ 18	oolite	"
surface	-	-	+ 19	+ 20	+ 20	+ 20	sediment	"
34 surface	27.8	36.0	+ 14	+ 16	+ 16	+ 16	oolite	31 Jul'61
surface	-	-	+ 14	+ 14	+ 16	+ 16	sediment	"
35 surface	27.7	36.1	+ 13	+ 14	+ 14	+ 14	oolite	"
surface	-	-	+ 10	+ 10	+ 11	+ 10	sediment	"
36 surface	28.2	35.8	+ 12	+ 13	+ 14	+ 14	oolite	"
surface	-	-	+ 9	+ 10	+ 10	+ 10	sediment	"
37 surface	28.7	19.3	+ 14	+ 14	+ 15	+ 15	oolite	"
repeat	-	-	+ 15	+ 15	+ 15	+ 14	oolite	"
surface	-	-	+ 12	+ 12	+ 14	+ 14	sediment	"
38 surface	28.7	20.1	+ 12	+ 14	+ 16	+ 16	oolite	"
surface	-	-	+ 10	+ 10	+ 11	+ 11	sediment	"
bottom	28.7	20.1	-	-	-	-	-	"
39 surface	28.9	33.4	+ 16	+ 18	+ 19	+ 20	oolite	"
surface	-	-	+ 16	+ 19	+ 30	+ 31	sediment	"
repeat	-	-	+ 26	+ 30	+ 31	+ 32	sediment	"
bottom	28.6	33.2	-	-	-	-	-	"
40 surface	29.2	26.1	+ 10	+ 10	+ 8	+ 6	oolite	"
repeat	-	-	+ 12	+ 12	+ 12	+ 13	oolite	"
surface	-	-	+ 14	+ 16	+ 18	+ 19	sediment	"
bottom	29.0	26.9	-	-	-	-	-	"
41 surface	30.4	20.9	+ 14	+ 16	+ 16	+ 16	oolite	"
surface	-	-	+ 10	+ 10	+ 12	+ 12	sediment	"
bottom	29.0	23.9	-	-	-	-	-	"
42 surface	28.4	29.5	-	+ 16	+ 16	+ 17	oolite	"
surface	-	-	+ 27	+ 30	+ 32	+ 34	sediment	"

APPENDIX A (Continued)

Station		Water Temperature (°C.)	Salinity (°/oo)	Saturometer (mv.) 1 min.	2 min.	3 min.	4 min.	Carbonate Standard Used	Date
43	surface	28.5	0.8	- 22	- 26	- 28	- 30	oolite	1 Aug'61
	repeat	-	-	- 21	- 26	- 28	- 28	oolite	"
	bottom	28.5	1.0	- 21	- 24	- 26	- 28	oolite	"
	repeat	-	-	- 21	- 24	- 25	- 26	oolite	"
	surface	-	-	+ 20	+ 26	+ 28	+ 28	sediment	"
	bottom	-	-	+ 20	+ 21	+ 22	+ 22	sediment	"
44	surface	30.4	5.9	+ 7	+ 7	+ 6	+ 6	oolite	"
	surface	-	-	+ 10	+ 12	+ 14	+ 14	sediment	"
	bottom	29.6	9.2	-	-	-	-	-	"
45	surface	30.3	6.7	+ 10	+ 10	+ 10	+ 9	oolite	"
	bottom	29.8	14.0	+ 18	+ 18	+ 18	+ 18	oolite	"
	surface	-	-	+ 16	+ 17	+ 16	+ 14	sediment	"
	bottom	-	-	+ 27	+ 29	+ 30	+ 31	sediment	"
46	surface	31.4	8.4	+ 13	+ 16	+ 18	+ 19	oolite	"
	bottom	-	8.6	+ 14	+ 15	+ 14	+ 15	oolite	"
	surface	-	-	+ 8	+ 10	+ 11	+ 12	sediment	"
	bottom	-	-	+ 9	+ 12	+ 13	+ 14	sediment	"
50	surface	28.6	0.5	- 5	- 8	- 10	- 11	oolite	4 Aug'61
	surface	-	-	+ 1	+ ½	- ½	- ½	sediment	"
52	surface	30.4	19.5	+ 18	+ 20	+ 20	+ 21	oolite	5 Aug'61
	bottom	30.5	19.5	+ 16	+ 18	+ 18	+ 18	oolite	"
	repeat	-	-	+ 16	+ 18	+ 19	+ 20	oolite	"
	surface	-	-	+ 16	+ 20	+ 22	+ 24	sediment	"
	bottom	-	-	+ 14	+ 16	+ 17	+ 18	sediment	"
53	surface	29.0	35.0	+ 14	+ 15	+ 16	+ 16	oolite	6 Aug'61
	repeat	-	-	+ 12	+ 12	+ 14	+ 13	oolite	"
	surface	-	-	+ 14	+ 16	+ 16	+ 17	sediment	"
54	surface	28.7	35.2	+ 12	+ 14	+ 14	+ 15	oolite	7 Aug'61
	surface	-	-	+ 9	+ 10	+ 12	+ 14	sediment	"
55	surface	29.2	33.5	+ 16	+ 18	+ 21	+ 22	oolite	"
	surface	-	-	+ 16	+ 18	+ 20	+ 21	sediment	"
56	surface	-	12.2	+ 6	+ 6	+ 8	+ 8	oolite	"
	repeat	-	-	+ 6	+ 6	+ 7	+ 7	oolite	"
	bottom	-	26.0	+ 16	+ 18	+ 18	+ 18	oolite	"
	surface	-	-	+ 10	+ 11	+ 12	+ 12	sediment	"
	bottom	-	-	+ 24	+ 26	+ 28	+ 30	sediment	"
57	surface	-	1.9	- 14	- 15	- 16	- 16	oolite	"
	bottom	-	5.2	- 8	- 7	- 6	- 6	oolite	"
	surface	-	-	- 10	- 10	- 9	- 9	sediment	"
	bottom	-	-	- 4	- 2	- ½	- 1	sediment	"
58	surface	30.2	34.2	+ 14	+ 16	+ 16	+ 16	oolite	"
	surface	-	-	+ 16	+ 18	+ 20	+ 21	sediment	"
62	surface	29.2	1.6	- 9	- 12	- 12	- 14	oolite	9 Aug'61
	repeat	-	-	- 4	- 6	- 10	- 10	oolite	"
	surface	-	-	+ 28	+ 42	+ 42	+ 44	sediment	"
	repeat	-	-	+ 6	+ 9	+ 12	+ 14	sediment	"
	repeat	-	-	+ 8	+ 10	+ 12	+ 14	sediment	"
239	surface	30.0	37.1						17 Aug'62
	bottom	30.0	36.9						"
240	bottom	-	36.4						"
241	surface	30.3	37.1						"
	bottom	-	37.3						"

APPENDIX A (Continued)

Station	Water Temperature (°C.)	Salinity (°/oo)	Saturometer (mv.) 1 min.	2 min.	3 min.	4 min.	Carbonate Standard Used	Date
243 surface	30.7	36.3						17 Aug'62
bottom	30.5	37.0						"
245 surface	29.7	37.0						"
bottom	-	36.7						"
248 bottom	-	27.3						18 Aug'62
254 surface	-	17.7						"
bottom	-	26.4						"
263 surface	29.9	35.2						19 Aug'62
bottom	29.8	36.0						"
267 surface	29.4	35.9						"
bottom	-	36.0						"
269 bottom	-	35.9						20 Aug'62
271 bottom	-	35.7						"
272 bottom	-	35.7						"
274 pond	-	36.0						"
278 surface	30.2	29.9						21 Aug'62

APPENDIX B

Composition, Grain Size, and Carbonate Mineralogy of Northern Belize Samples

Sample Number	Volume Percentage of Fraction Greater Than 1/16 mm.																	Grain Size		Carbonate Mineralogy		
	Coralline Algae	"*Melobesia*"	*Halimeda*	Miliolidae	Peneroplidae	Other Foraminifera	Corals	Mollusks	Ostracods	Miscellaneous Skeletal	Unknown Skeletal	Total Skeletal	Fecal Grains	Cryptocrystalline Grains	Cryptocrystalline Skeletal-Grains	Total Cryptocrystalline Grains	Country Rock	Terrigenous Minerals	Weight Percentage Less Than 1/8 mm	Weight Percentage Less Than 1/16 mm	Aragonite as Weight Percentage of Carbonate	Magnesium Carbonate in Calcite (Weight Percentage)	Weight Percentage Magnesium Calcite in Total Calcite
1	0.0	0.0	0.0	4.4	0.0	9.2	0.0	19.4	9.1	3.4	5.9	51.3	0.0	48.7	11.4	60.1	0.0	0.0	89.2	64.0	3	0	0
2	0.0	0.0	0.0	0.6	0.0	1.0	0.0	19.7	0.0	0.0	1.6	22.9	0.0	76.6	7.1	83.7	0.0	0.2	16.8	11.1	5	0	0
3	0.0	0.0	0.0	7.5	0.0	5.4	0.0	13.2	2.3	0.2	2.1	30.6	0.0	69.5	3.8	73.3	0.0	0.0	76.2	31.5	5	0	0
4	0.0	0.0	0.0	3.2	0.0	5.2	0.0	11.8	4.5	0.5	1.7	26.9	19.4	52.0	1.9	53.9	1.7	0.0	93.2	82.4	7	1	0
5	0.0	0.0	0.0	10.5	0.0	1.7	0.0	14.7	2.9	0.0	2.4	32.2	2.0	65.7	2.4	68.1	0.0	0.0	88.3	60.7	3	5	0
6	0.0	0.0	0.0	2.4	0.0	0.4	0.0	14.2	0.6	0.0	0.9	18.5	1.2	79.6	2.6	82.2	0.6	0.2	32.7	18.6	7	7	100
7	0.0	0.0	0.0	9.4	0.0	4.5	0.0	41.6	1.6	0.0	1.4	58.5	0.4	40.9	2.2	43.1	0.2	0.0	75.0	50.0	19	5	0
8	0.0	0.0	0.0	12.4	1.0	4.1	0.0	30.1	2.4	0.0	1.4	51.4	0.3	39.4	3.2	42.6	9.0	0.0	91.2	85.0	10	5	0
9	0.0	0.0	0.0	8.0	0.0	0.7	0.0	5.2	1.9	0.0	1.3	17.1	8.7	74.3	2.9	77.2	0.0	0.0	93.6	75.3	3	6	100
10	0.0	0.0	0.0	11.0	0.3	1.5	0.0	11.4	1.3	0.0	1.0	26.5	6.7	66.7	5.4	72.1	0.0	0.0	79.2	44.9	7	6	100
11	0.0	1.3	0.0	18.6	3.3	1.7	0.0	12.9	3.6	1.0	1.8	44.2	2.0	53.1	10.0	63.1	7.4	0.0	75.8	46.8	7	10	100
12	0.0	1.4	0.0	9.9	8.5	1.9	0.0	10.5	0.8	0.0	2.4	35.3	3.1	60.2	13.3	73.5	1.4	0.0	33.4	15.2	6	7	100
13	0.0	0.1	0.1	15.6	7.9	0.4	0.0	12.9	2.2	0.0	1.9	41.1	2.9	55.9	6.7	62.6	0.0	0.0	68.4	49.1	9	7	100
14	0.0	0.0	0.0	1.3	19.7	0.2	0.0	31.3	0.0	1.1	0.9	54.5	0.6	44.6	14.2	58.8	0.3	0.0	12.8	9.9	20	7	100
15	0.0	0.0	0.0	1.7	36.8	0.3	0.0	7.8	0.8	0.0	0.8	48.2	0.0	50.0	26.3	76.3	1.7	0.0	11.4	7.0	3	9	100
15A	0.0	0.0	0.0	1.0	62.1	1.9	0.0	6.1	0.3	0.0	0.6	72.0	0.0	27.0	34.2	61.2	1.0	0.0	5.8	3.7	11	10	100
16	0.0	0.4	0.0	6.4	53.0	0.8	0.0	10.3	0.5	0.1	0.5	71.6	0.0	28.2	39.3	67.5	0.4	0.0	30.3	24.0	12	9	100
17	0.0	1.1	0.2	9.1	14.2	2.5	0.0	19.8	0.9	1.5	2.0	51.2	4.6	36.7	9.0	45.7	7.7	0.0	53.7	32.5	29	7	100
18	0.0	0.6	0.0	38.4	1.7	3.8	0.0	10.9	2.9	0.6	1.4	60.2	3.0	36.5	6.5	43.0	0.3	0.0	75.7	44.2	10	7	100
19	0.0	1.0	0.0	20.2	0.0	4.4	0.0	9.7	2.9	0.0	1.5	39.6	1.6	57.3	7.8	65.1	1.5	0.0	91.7	73.2	6	1	0
20	0.0	0.2	0.0	1.9	2.9	3.2	0.0	66.5	0.3	2.4	1.8	79.3	0.0	10.7	2.3	13.0	10.0	0.0	6.2	2.5	10	4	0
21	0.0	0.1	0.0	3.9	75.7	0.1	0.0	13.6	0.1	0.5	2.5	96.6	0.0	3.5	63.3	66.8	0.0	0.0	4.8	2.4	22	9	100
23	0.0	2.9	0.0	38.2	3.4	1.9	0.0	17.9	4.1	1.1	2.5	72.0	3.0	24.7	4.8	29.5	0.0	0.4	92.6	66.0	17	9	100
24	0.0	5.4	0.0	21.9	17.6	4.6	0.0	26.7	4.5	1.0	2.7	84.4	2.7	12.8	5.0	17.8	0.0	0.0	47.3	26.6	32	11	100
25	0.0	0.5	0.0	12.2	24.7	1.0	0.0	48.9	0.8	2.1	1.7	91.9	0.2	6.5	15.2	21.7	1.5	0.0	16.4	7.5	48	0,12	98
26	0.0	3.4	0.0	27.3	15.6	4.2	0.0	23.8	2.6	1.2	2.3	80.4	4.5	14.9	8.7	23.6	0.0	0.0	69.3	55.8	20	9	100
27	0.0	0.0	0.0	11.6	2.0	1.5	0.0	22.1	0.7	0.0	3.6	41.5	1.1	48.6	14.2	62.8	0.8	8.0	49.7	9.5	7	11	100
28	0.0	0.4	0.0	28.3	10.0	4.2	0.0	19.8	3.5	0.9	2.5	69.6	0.7	25.3	11.1	36.4	2.4	2.0	74.3	56.9	11	9	100
29	0.0	0.5	0.0	8.2	48.5	1.4	0.0	27.2	0.9	0.6	2.8	90.1	0.5	9.4	35.4	44.8	0.0	0.0	18.9	10.9	21	12	100
30	0.0	1.0	0.0	43.5	1.9	4.8	0.0	24.9	5.2	1.2	1.9	84.4	1.0	14.6	3.5	18.1	0.0	0.0	88.6	74.7	19	10	100

APPENDIX B (Continued)

Sample Number	Volume Percentage of Fraction Greater Than 1/16 mm.																	Grain Size		Carbonate Mineralogy		
	Coralline Algae	"*Melobesia*"	*Halimeda*	Miliolidae	Peneroplidae	Other Foraminifera	Corals	Mollusks	Ostracods	Miscellaneous Skeletal	Unknown Skeletal	Total Skeletal	Fecal Grains	Cryptocrystalline Grains	Cryptocrystalline Skeletal Grains	Total Cryptocrystalline Grains	Country Rock	Terrigenous Minerals	Weight Percentage Less Than 1/8 mm	Weight Percentage Less Than 1/16 mm	Aragonite as Weight Percentage of Carbonate	Magnesium Carbonate in Calcite (Weight Percentage)	Weight Percentage Magnesium Calcite in Total Calcite
31	0.0	8.3	0.5	26.6	2.7	6.3	0.0	25.0	2.7	1.4	4.3	77.7	6.6	14.0	2.8	16.8	1.1	0.4	75.5	46.6	43	11	100
32	0.0	5.2	0.0	37.2	2.0	5.4	0.0	14.3	18.6	7.3	5.2	95.2	0.2	4.5	1.1	5.6	0.0	0.0	99.4	89.6	26	9	100
33	0.1	2.6	58.1	5.8	4.2	6.3	0.0	11.7	2.9	0.8	4.6	97.1	2.0	0.7	3.6	4.3	0.0	0.0	42.0	24.5	63	7	100
34	3.0	1.4	47.7	2.5	7.9	5.7	1.1	16.4	0.8	1.9	8.2	96.6	0.0	2.7	47.7	50.4	0.0	0.0	6.0	3.7	79	0,13	97
35	14.7	0.8	38.9	1.0	2.8	3.2	13.3	15.8	0.4	2.8	5.0	98.7	0.6	0.5	12.1	12.6	0.0	0.0	9.6	3.2	72	12	100
36	18.9	1.2	24.1	1.9	2.2	8.8	20.1	11.1	1.0	4.5	5.2	98.7	0.1	1.2	4.2	5.4	0.0	0.0	15.6	3.4	63	13	100
37	0.7	0.5	66.4	0.8	3.3	5.8	1.5	13.1	0.3	1.2	3.1	96.7	3.0	0.3	11.3	11.6	0.0	0.0	9.3	6.9	77	13	100
38	0.0	0.6	19.9	3.1	10.1	6.8	1.1	30.5	0.0	2.4	6.4	80.9	4.6	5.0	8.9	13.9	9.2	0.4	16.6	8.8	62	0,12	92
39	0.0	7.2	0.0	30.8	0.4	3.9	0.0	24.3	2.2	1.6	2.7	73.2	18.0	7.2	1.3	8.5	0.2	1.6	73.5	49.3	55	0, 9	95
40	0.0	1.8	0.0	7.5	0.0	2.1	0.0	31.1	0.8	2.3	2.2	47.9	0.8	18.9	6.1	25.0	6.8	25.4	12.8	3.2	56	0,11	87
41	0.0	0.4	0.0	4.5	0.0	0.8	0.0	23.9	0.0	0.3	1.3	31.2	0.0	11.9	2.1	14.0	31.2	25.7	10.6	4.5	50	0,11	70
42	0.0	2.4	0.0	17.1	1.7	4.9	0.0	33.9	2.6	2.1	3.5	68.2	1.2	8.5	4.5	13.0	6.5	15.4	40.4	23,0	50	0, 9	91
43	0.0	0.0	0.0	0.2	0.0	5.6	0.0	12.5	27.5	25.5	0.6	71.9	0.0	27.1	0.8	27.9	0.8	0.2	99.8	99.2	9	0	0
44	0.0	0.0	0.0	0.5	0.0	0.2	0.0	30.2	0.2	0.0	1.0	32.1	0.0	10.5	1.2	11.7	57.3	0.2	9.2	7.0	17	0	0
45	0.0	0.0	0.0	12.7	0.0	7.9	0.0	39.1	10.1	4.3	1.7	75.8	0.3	22.2	1.7	23.9	1.5	0.2	94.8	92.6	5	5	0
46	0.0	0.0	0.0	0.1	0.0	0.0	0.0	9.2	0.4	0.0	0.0	10.2	0.0	3.6	0.0	3.6	86.0	0.1	6.2	4.2	7	0	0
47	0.0	0.0	0.0	4.2	0.0	4.2	0.0	30.9	1.7	0.0	1.1	42.1	0.0	42.4	1.1	43.5	15.1	0.5	50.6	38.9	12	0	0
48	0.0	0.0	0.0	0.1	0.0	0.0	0.0	13.7	0.0	0.3	0.0	14.1	0.0	82.8	0.0	82.8	3.0	0.1	1.8	1.3	4	0	0
49	0.0	0.0	0.0	0.1	0.0	0.8	0.0	4.6	0.2	0.0	1.0	6.7	0.0	93.4	2.9	96.3	0.0	0.0	3.6	1.8	2	0	0
50	0.0	0.0	0.0	0.6	0.0	3.8	0.0	10.8	1.5	0.2	1.2	18.1	0.0	79.6	2.3	81.9	1.8	0.5	19.7	5.1	4	0	0
51	0.0	0.0	0.0	4.0	0.0	4.3	0.0	9.0	15.1	1.2	2.9	36.5	0.0	55.2	1.7	56.9	8.2	0.0	96.1	83.6	3	0	0
52	0.0	0.8	0.0	9.7	0.3	3.8	0.0	17.5	0.8	0.0	1.2	34.1	1.5	63.9	6.3	70.2	0.5	0.0	60.2	24.6	12	8	100
53	0.2	2.0	45.8	3.8	1.6	1.7	0.2	35.6	0.2	1.1	4.0	96.1	3.3	0.7	1.5	2.2	0.0	0.0	65.7	49.3	66	0,12	98
54	0.0	0.5	79.4	0.2	0.6	1.4	1.9	13.1	0.0	1.1	1.6	99.8	0.3	0.0	6.4	6.4	0.0	0.0	11.4	9.3	82	13	100
55	0.0	9.5	12.0	3.5	0.7	3.9	0.0	30.1	6.3	10.1	9.3	85.4	10.9	3.6	6.1	9.7	0.0	0.2	70.2	56.7	62	12	100
56	0.0	0.4	7.3	5.1	1.0	8.1	0.2	45.3	11.8	2.0	4.9	86.1	7.7	0.6	1.4	2.0	2.2	3.5	73.5	68.9	54	4,13	93
57	0.0	0.0	0.0	0.0	0.0	14.0	0.0	35.0	26.0	3.0	4.0	82.0	0.0	0.0	0.0	0.0	0.0	18.0	99.5	99.1	0	9	100
58	0.0	0.5	73.6	2.0	0.7	2.3	0.0	13.0	1.5	4.6	1.3	99.6	0.2	0.0	0.0	0.0	0.0	0.2	71.2	67.6	72	1,12	63
62	0.0	0.0	2.7	18.2	0.2	13.5	0.0	12.9	32.0	3.1	1.3	83.9	2.9	4.0	0.0	4.0	0.0	9.4	97.5	95.5	41	2,11	76

239	0.0	1.1	0.0	20.4	30.3	2.3	0.0	27.9	2.8	0.6	1.3	86.7	2.3	10.9	7.2	18.1	0.1	0.0	67.3	51.5	19	10	100
240																			36.6	20.4	45	3,11	92
241	0.0	1.0	0.0	16.3	38.9	3.8	0.0	23.9	3.0	0.0	3.2	90.0	1.2	8.8	14.0	22.8	0.0	0.0	51.7	39.8	31	10	100
242																			15.5	6.3	29	12	100
243	0.0	2.1	0.0	43.0	3.6	4.8	0.0	7.5	8.7	3.5	4.1	77.2	2.9	19.8	4.1	23.9	0.0	0.0	90.2	51.1	18	11	100
244																			20.4	8.1	36	12	100
245	0.0	3.3	0.0	38.4	13.0	6.3	0.0	13.7	5.5	2.9	3.0	86.1	3.3	10.6	5.0	15.6	0.0	0.0	84.8	71.4	36	12	100
246	0.0	3.3	0.0	52.1	2.6	7.1	0.0	4.5	5.9	1.2	4.3	81.0	2.2	16.7	3.8	20.5	0.2	0.0	89.0	63.2	19	11	100
247																			91.6	83.0	15	11	100
248																			13.5	6.3	39	12	100
249																			46.1	28.1	42	11	100
250	0.0	0.5	0.0	26.8	8.0	4.5	0.0	15.1	2.2	0.1	5.4	62.6	3.9	33.5	10.4	43.9	0.0	0.0	93.1	87.2	17	10	100
251	0.0	1.6	0.0	31.0	10.8	6.7	0.0	16.3	3.7	1.6	3.3	74.9	9.4	15.7	7.1	22.8	0.0	0.0	66.7	38.4	19	11	100
252																			47.8	25.1	14	11	100
253	0.0	4.0	0.0	19.2	15.5	4.6	0.0	33.6	0.9	0.8	1.4	80.0	8.4	11.4	4.3	15.7	0.2	0.0	54.4	28.3	32	11	100
254	0.0	0.5	0.0	11.0	36.6	2.9	0.0	14.8	1.6	0.2	3.5	71.2	0.0	15.2	24.9	40.1	0.9	12.6	16.8	9.0	23	12	100
255																			20.1	11.5	16	11	100
256																			97.0	83.8	20	11	100
257	0.0	1.2	0.0	21.9	17.1	5.3	0.0	14.6	1.8	1.7	3.3	66.8	15.8	17.5	12.1	29.6	0.0	0.0	44.0	19.0	15	11	100
258																			33.0	19.0	30	11	100
259	0.0	1.2	0.0	35.2	12.7	4.7	0.0	15.3	4.1	1.9	2.2	77.2	3.7	19.0	5.3	24.3	0.0	0.0	64.6	31.1	20	11	100
260																			85.8	65.6	18	10	100
261	0.0	0.2	0.0	12.6	42.2	2.4	0.0	20.7	0.5	0.0	1.4	80.0	0.2	19.6	35.6	55.2	0.3	0.0	25.7	10.4	11	11	100
262																			7.4	3.6	38	2,11	95
263	0.2	10.7	8.9	7.4	1.8	6.5	0.2	40.9	0.7	6.1	10.9	94.3	4.8	0.6	1.2	1.8	0.4	0.0	49.0	21.2	58	3,12	97
264	0.0	0.2	0.0	10.8	60.6	0.9	0.0	13.9	0.7	1.0	3.0	91.0	0.0	7.8	50.7	58.5	1.2	0.0	10.5	2.8	10	12	100
265	0.0	0.0	0.0	3.4	55.6	2.0	0.0	9.3	0.3	0.0	2.5	73.1	0.0	26.9	45.9	72.8	0.0	0.0	9.5	3.8	9	12	100
266	0.0	0.3	0.0	3.4	60.1	2.6	0.0	12.4	0.0	0.0	5.0	83.8	0.0	16.2	58.0	74.2	0.0	0.0	2.7	1.8	10	13	100
267	1.1	2.9	30.0	4.1	5.7	9.9	1.1	34.0	0.2	1.2	7.2	97.3	1.9	0.5	0.7	1.2	0.2	0.0	24.0	6.3	74	3,14	98
268	0.0	0.6	0.0	17.8	29.8	2.7	0.0	19.0	2.2	0.3	3.6	76.0	2.7	21.2	9.3	30.5	0.1	0.0	58.8	34.8	21	12	100
269																			85.3	48.9	40	5,12	98
270																			55.1	38.1	58	4,13	95
271	1.9	0.8	31.5	2.2	3.9	5.1	2.3	24.5	0.0	0.6	19.6	92.4	1.2	6.1	22.1	28.2	0.0	0.0	1.8	1.2	89	6,15	96
272	1.4	13.3	3.4	22.0	0.9	6.8	0.2	7.0	3.2	10.2	8.4	76.8	16.7	6.5	2.3	8.8	0.0	0.0	89.5	53.2	39	4,11	97
273																					36	3,12	98
274																			32.5	18.4	56	2,12	98
275	0.6	5.1	38.5	7.7	3.4	7.2	1.0	16.8	1.1	3.4	5.4	90.2	9.8	0.0	1.6	1.6	0.0	0.0	80.6	38.5	62	13	100
276	0.6	0.8	63.4	0.6	4.9	1.4	0.6	23.3	0.6	0.7	2.7	99.6	0.3	0.1	1.1	1.2	0.0	0.0	12.7	3.6	83	3,13	97
277	0.4	3.3	56.9	2.4	3.1	3.1	1.0	21.2	0.0	1.6	4.3	97.2	1.8	0.0	2.4	2.4	1.2	0.0	27.2	8.5	79	4,11	98
278	0.5	1.9	49.8	2.1	1.9	2.2	0.3	33.2	0.5	1.1	2.4	95.8	3.1	0.3	0.7	1.0	0.8	0.0	22.4	7.1	81	3,13	97
279	0.6	2.8	37.7	1.6	0.5	3.1	0.5	31.4	0.3	1.7	2.5	82.7	16.6	0.6	0.2	0.8	0.2	0.0	49.5	27.6	73	2,13	97
33-60	0.0	0.5	49.4	1.0	2.7	3.2	0.5	20.3	0.8	1.3	3.8	83.4	11.1	4.8	3.4	8.2	0.6	0.2	63.8	45.8	67	12	100
34-60	0.0	0.9	11.2	4.4	4.4	2.2	0.2	49.0	1.5	9.0	6.8	89.4	6.1	3.0	2.8	5.8	0.7	0.6	76.9	53.5	60	12	100
35-60	0.0	2.7	3.6	11.6	7.2	5.4	0.0	30.7	9.0	11.0	6.3	87.5	7.8	0.6	1.8	2.4	1.5	2.7	78.6	74.2	49	12	100
36-60	0.0	2.5	0.2	31.9	5.0	4.0	0.0	29.6	2.2	6.7	2.7	84.7	1.3	13.8	1.7	15.5	0.0	0.2	95.7	92.7	43	11	100
37-60	0.0	0.6	0.4	10.8	17.4	3.7	0.0	36.4	0.4	2.4	2.3	74.4	5.3	11.1	7.4	18.5	0.9	8.3	20.9	14.1	33	12	100
38-60	0.0	1.0	0.0	21.8	2.4	3.5	0.0	22.9	4.7	1.7	2.7	60.6	3.2	32.2	4.7	36.9	0.3	3.6	47.6	25.0	25	12	100

Holocene Carbonate Sedimentation and Diagenesis, Ambergris Cay, Belize [1]

WILLIAM J. EBANKS, JR.[2]

ABSTRACT Carbonate islands are obvious features of modern shelf seas, and the same was probably true of the extensive shallow seas of the past. The purpose of this investigation was to recognize and describe those attributes of island sediments that distinguish them from adjacent submarine deposits and to indicate the processes responsible for these features.

The most important characteristics of shelf islands are that (1) they strongly influence the depositional environments of adjacent submarine areas, (2) they comprise distinctive and potentially large bodies of carbonate sediment, and (3) sediments of such islands may be the key to environmental interpretation of much larger areas because of their definite connotation of water depth and position near a shelf edge.

Ambergris Cay, Belize, is a modern shelf island that has formed by Holocene sediment accumulation on and around a partly exposed, karstic limestone surface under conditions of rising sea level. The configuration of the limestone surface has had an important effect on the distribution of the Holocene sediment.

This island consists of a narrow, windward, sand beach ridge, leeward of which is a broad expanse of supratidal flats interspersed with intra-island lagoons and areas of exposed rock. The windward ridge is a storm deposit of coralgal sand derived from the adjacent reef lagoon. Its geometry and the importance of coral and calcified algal constituents distinguish this sand deposit from other island sediments. The supratidal-flat sediments are predominantly magnesium calcite, and the most important constituents are foraminifers and mollusks. These deposits are further characterized by their lateral and vertical variability, relatively high content of mud-size carbonate material, and assemblages of the following features: laminations, stromatolites, desiccation cracks, intraformational conglomerates, and root casts. Intra-island-lagoon sediments are typically nonlaminated, contain a high percentage of mud-size carbonate material, and have a restricted faunal assemblage.

In addition to these primary characteristics, island sediments are distinguished by early diagenetic modifications that reflect their subaerial exposure. Solution destruction and early cementation are results of exposure of the sediments to meteoric water. In contrast, early dolomitization affects supratidal-flat sediments associated with ponds of hypersaline water.

Thus, island sediments form a distinctive, locally "regressive" sequence within a transgressive marine situation. That is, the island's land area is growing at the expense of its submarine area, increasing its sheltering effect on surrounding submarine environments.

INTRODUCTION

Carbonate islands are among the most apparent features of modern carbonate-shelf seas, and yet they are not commonly identified in the stratigraphic record. This recognition problem results from a lack of adequate criteria, for studies of modern carbonate sediments have concentrated on submarine rather than subaerial accumulations. Most studies of carbonate islands heretofore have been limited to physiography or concerned with one particular aspect of supratidal sedimentation, such as beach sands, diagenesis, or algal mats. Notable exceptions are the brief

[1] Manuscript received, June 20, 1972.

[2] Department of Geology, Rice University, Houston, Texas. Present address: Kansas Geological Survey, Lawrence, Kansas.

This paper is based on a thesis submitted in May 1967 to the graduate school of Rice University in partial fulfillment of the requirements for the degree of Doctor of Philosophy. Financial support for this study was provided by the American Petroleum Institute's Research Grant 71-A and by the Petroleum Research Fund of the American Chemical Society. The Department of Geology, Rice University, provided excellent facilities and equipment and supplementary funds when needed. Esso Production Research Company provided for preparation of the manuscript.

E. G. Purdy suggested the study and supervised it through completion. His guidance in the field was invaluable. L. R. High, Jr., J. W. Teeter, and G. E. Tebbutt also were of great assistance in the field. W. Ford Young, of Belize City, furnished critical logistical help.

description by Purdy and Imbrie (1964) of the supratidal flats of Williams Island, Bahamas, and the recent papers by Roehl (1967) and Shinn et al (1969) concerning sedimentation on Andros Island, Bahamas.

The purpose of this investigation was to focus attention on carbonate islands as environments of deposition and, by study of a modern carbonate island, to understand better their origin and subsequent development. Particular emphasis was placed on identifying criteria for recognizing carbonate islands in the stratigraphic record.

REGIONAL GEOGRAPHY AND GEOLOGY

Belize is located on the southeast side of the Central American Yucatán Peninsula. It is bordered on the west and south by Guatemala, on the north by Mexico, and on the east by the open Caribbean Sea.

Belize comprises elements of both the intensely deformed Chiapas system of Guatemala and Mexico and the relatively stable Yucatán platform. South of the capital, Belize City, the landscape is dominated by the Maya Mountains, an uplifted mass of granite-intruded Paleozoic terrigenous rocks, now deeply dissected but having a maximum elevation of 3,000 ft (900 m). North of the latitude of Belize City, the Yucatán stable shelf, on which Ambergris Cay is situated, is developed on younger limestone with only 200 ft (60 m) of local relief (Ower, 1928; Dixon *in* Romney, 1959).

This platform was an area of slow, periodic subsidence with attendant carbonate deposition throughout much of Cretaceous and Tertiary time (Butterlin and Bonet, 1962; Flores, 1952); in this regard, it resembles the Floridan and Bahamian carbonate platforms. In Pliocene time the Yucatán platform was tilted northwestward. A system of north-northeast-trending, down-to-the-southeast, normal faults developed on the upwarped eastern margin. Numerous present topographic and bathymetric features in Belize are related to these faults (Ower, 1928; Stoddart, 1962a).

The Belize submarine shelf (Fig. 1) is approximately 150 mi (240 km) long from north to south and ranges from 8 to 30 mi (13-48 km) in width. Paralleling the physiographic features of the mainland, the shelf is divisible into bathymetrically different north and south portions. Southward from Belize City, the shelf's troughlike center deepens from 5 fm to more than 30 fm in the Gulf of Honduras (U.S. Navy Hydrographic Office, 1959). Seaward of this central depression is a narrow bathymetric prominence, the barrier platform; east of this, the steep escarpment marks the edge of the shelf. Along the barrier platform and on numerous carbonate shoals in the southern shelf are small sand cays. The crest of the shelf-edge escarpment is formed by a well-developed barrier reef.

In contrast, north of Belize City, the Belize shelf is nowhere more than 15 ft (4.6 m) deep. The barrier reef is less continuous in the north than in the south, but here, too, it marks the crest of the shelf escarpment. One to two miles (1.6-3.2 km) leeward of the barrier reef in the north is a line of mangrove-sand cays, including Ambergris Cay, situated on a discontinuous, pre-Holocene limestone ridge that is apparently the northern continuation of the buried ridge underlying the barrier platform farther south (Purdy, personal commun., 1966).

NORTHERN BELIZE SHELF

The continental shelf north of the latitude of Belize City (17°30'N) is termed the "northern Belize shelf" (Fig. 2). The narrow water body

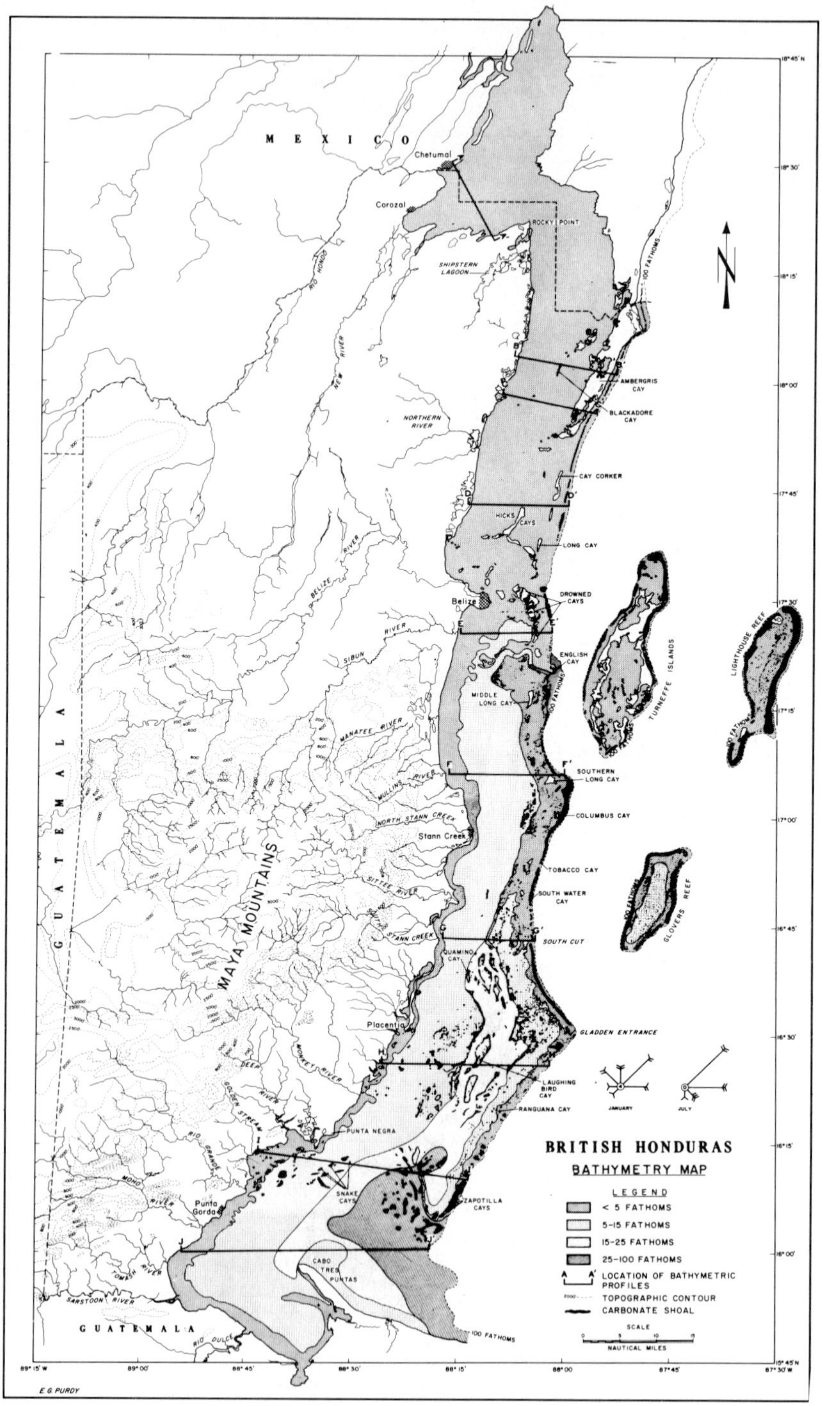

FIG. 1—Belize physiographic and bathymetric map; Ambergris Cay is northernmost island on shelf edge. After Purdy (personal commun., 1967).

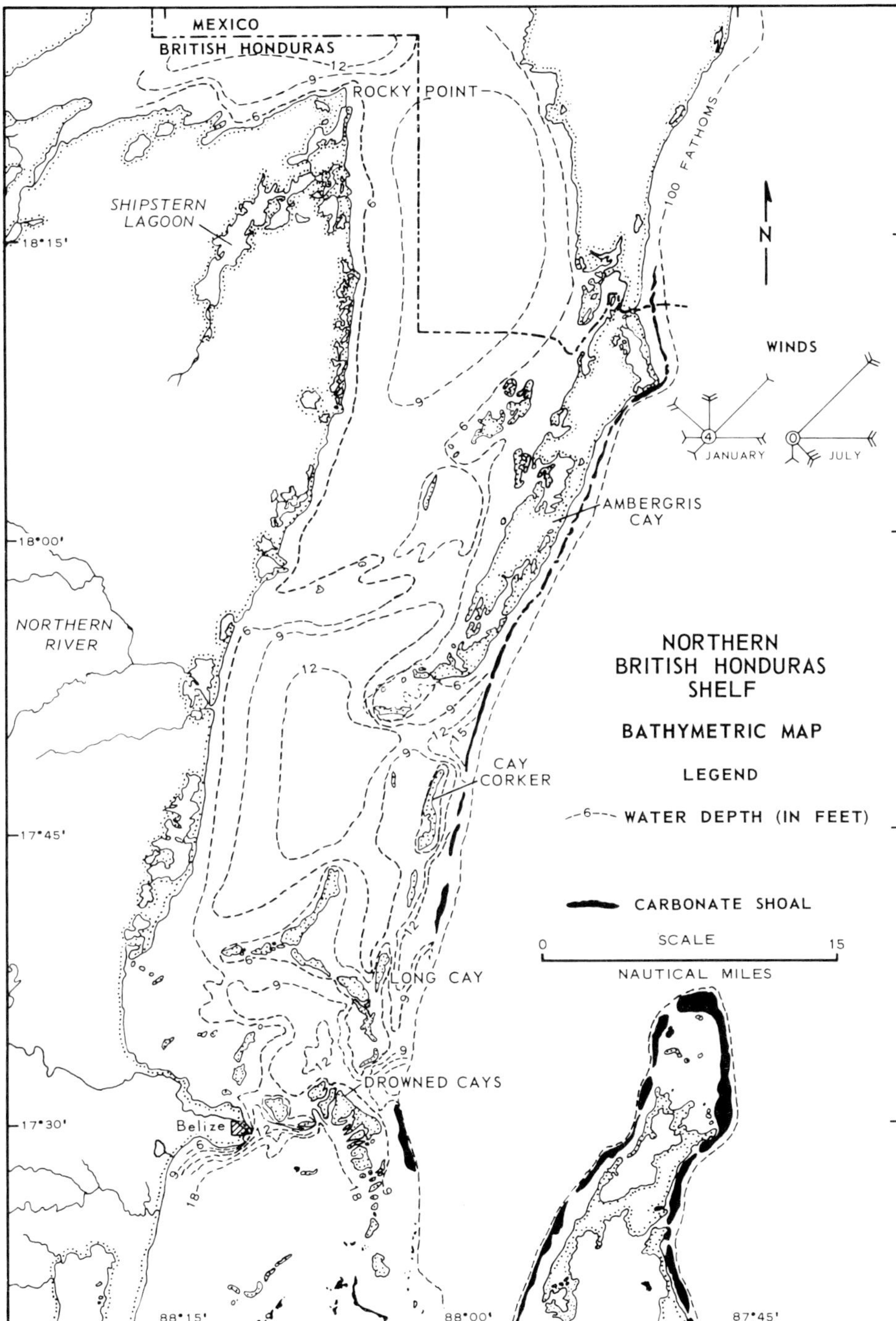

FIG. 2—Bathymetric map of northern Belize shelf. Modified from Pusey (1964).

between the barrier reef on the shelf margin and the line of mangrove-sand cays is referred to here as the *reef lagoon*, and the part of the submarine shelf leeward of these islands is termed the *shelf lagoon*. These contiguous water bodies include approximately 1,000 sq mi (2,590 km^2) of depositional area. The *shelf lagoon* is subdivided further into the *northern shelf lagoon* (between the latitude of Belize City and the Bulkhead mud shoal) and *Chetumal Bay* (from the Bulkhead northward). This terminology differs somewhat from that of Pusey (1964), from whom many of the data in this section are taken.

Bathymetry

The continental slope off the northern Belize shelf rises steeply from depths of more than 500 fm (900 m) to less than 20 fm (36 m; Vermeer, 1959; Pusey, this volume); the average slope approximates 40°. An elongate reef lagoon separates the reef from a line of mangrove-sand cays to the west by a distance of 1-2 mi (1.6-3.2 km). Bathymetrically, the reef lagoon is a troughlike depression whose central axis is nowhere more than 15 ft (4.5 m) deep.

Passages between the cays lead to the northern shelf lagoon and Chetumal Bay. The northern shelf lagoon is generally less than 12 ft (3.6 m) deep, and its average width is 10 mi (16 km). Chetumal Bay is somewhat shallower, with a depth generally less than 9 ft (2.7 m) west of Ambergris Cay.

Climate

The northern shelf lagoon lies wholly within the tropical zone, and the coolest air temperatures average 75°F (24°C; Romney, 1959). Annual rainfall increases from 50 in. (127 cm) at Chetumal, Mexico, at the north of Chetumal Bay, to 70 in. (178 cm) at Belize City, and to 166 in. (421 cm) at Punta Gorda, 80 mi (129 km) south of Belize City. On Ambergris Cay there is probably an annual rainfall of 50-60 in. The rainfall is unevenly distributed in an annual cycle of 3-5 months of winter dry season and 7-9 months of summer wet season.

Belize is strongly affected by easterly trade winds even though it is near the tropical doldrum belt. The trade winds blow from the east and northeast octants at velocities of 12-15 knots (22-28 km/hr; U.S. Navy Hydrographic Office, 1955). Afternoon onshore breezes reinforce these winds to 15-20 knots (28-37 km/hr). Morning offshore breezes are felt on the mainland, but only produce temporary calms at the cays. During the winter months, northers bring rains, 30-knot (55 km/hr) winds, and air temperatures as low as 50°F (10°C), but have only a slight effect on the seawater temperature over the reef (Stoddart, 1962a, p. 14). The wind effect on the temperature of the more restricted and shallower bay water is unknown. The very shallow water in the intra-island lagoons of the cays is probably noticeably affected, but only for short periods at a time.

During the late summer months and in the fall, hurricanes occur frequently along the Belize coast. Within the 30-year period 1931-1961, the coast has been struck, on an average, every 6 years (Stoddart, 1963). The submarine depositional environments are not drastically altered, but subaerially exposed cays and the mainland may be strongly affected, especially where natural vegetation has been cleared by humans (Stoddart, 1962c; Vermeer, 1963).

Hydrography

Currents and tides—There is very little recorded information on the tidal fluctuations on the northern Belize shelf. In *Sailing Directions*

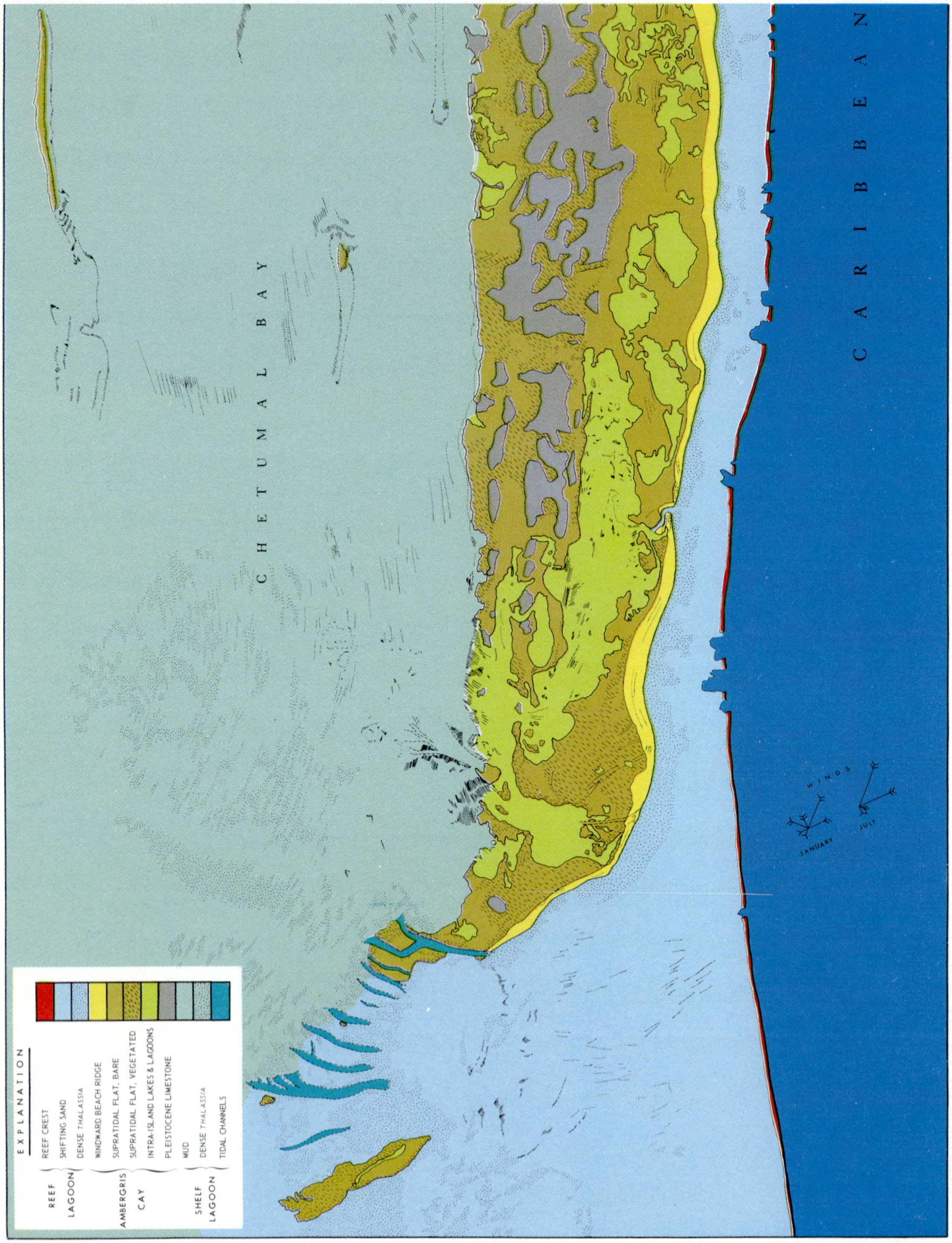
CHETUMAL BAY
CARIBBEAN
EXPLANATION
REEF
REEF CREST
LAGOON
SHIFTING SAND
DENSE THALASSIA
AMBERGRIS CAY
WINDWARD BEACH RIDGE
SUPRATIDAL FLAT, BARE
SUPRATIDAL FLAT, VEGETATED
INTRA-ISLAND LAKES & LAGOONS
PLEISTOCENE LIMESTONE
SHELF LAGOON
MUD
DENSE THALASSIA
TIDAL CHANNELS
WINDS
JANUARY
JULY

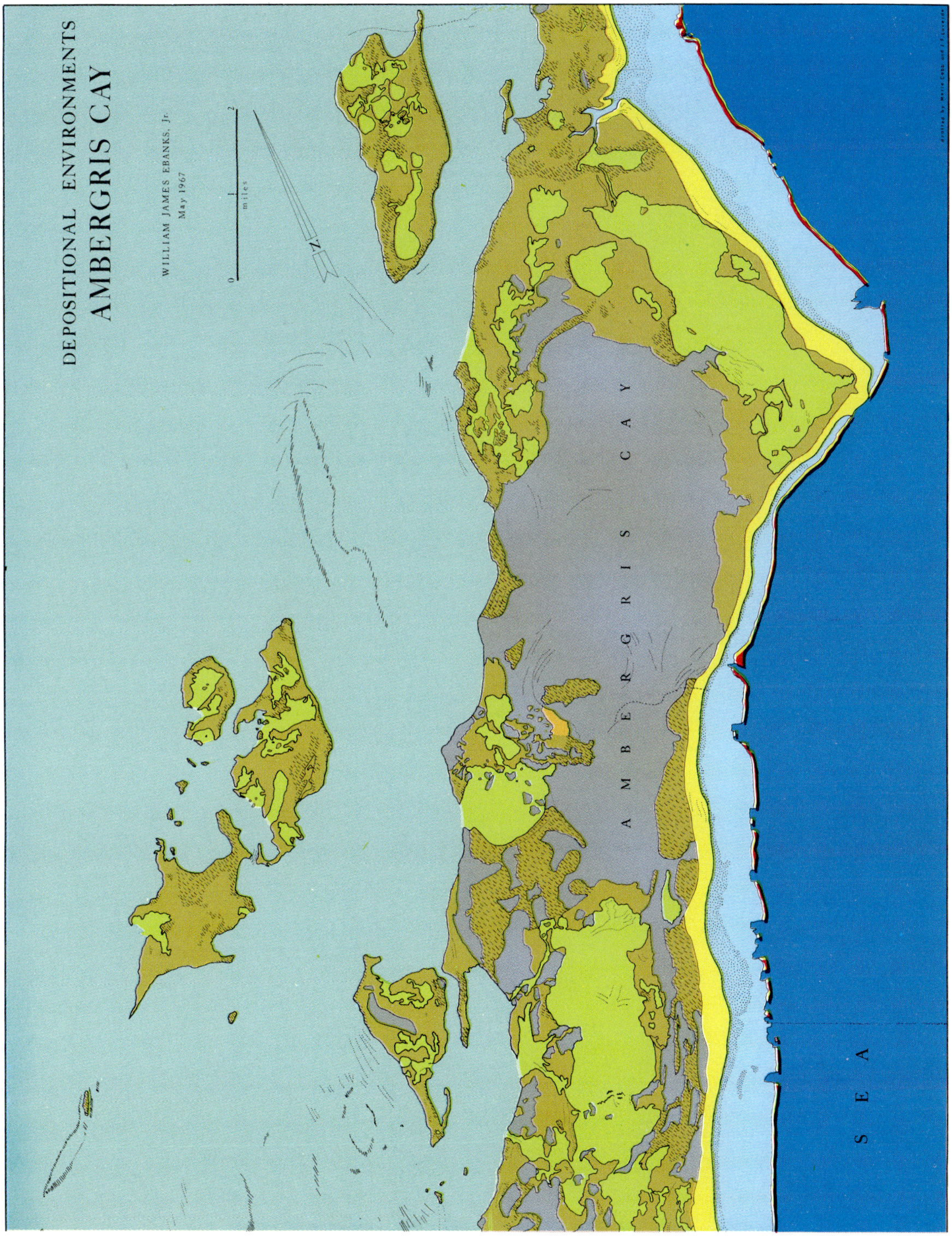

FIG. 3—Depositional environments of Holocene sediments.

(U.S. Navy Hydrographic Office, 1952, p. 187), the spring tidal range is listed as 1.5 ft (45 cm). The writer observed a maximum tidal range of 1.25-1.5 ft (37-45 cm) at Ambergris Cay on exposed, seaward shores. On exposed, leeward shores the range was somewhat less; and, on the shores of intra-island lagoons whose exchange with open water is restricted to narrow channels, the maximum tidal range was 0.5 ft (15 cm), and more often, almost nil.

During northers, water in Chetumal Bay commonly falls 2.5 ft (75 cm). Reports of hurricanes (Stoddart, 1963) indicate that the changing wind directions result in alternate draining and flooding of the shallow shelf. Thus, wind-induced currents are much more important for transporting sediment than are the weak tidal currents. Even so, the persistent presence of the tidal currents is reflected in channels through mud shoals on the shelf and in lineations of marine-grass beds near the mouth of Chetumal Bay (Fig. 3).

Currents in the reef lagoon windward of Ambergris Cay are rather complex, being caused by combined forces of wind and water. Water forced over the reef by incoming (westward-moving) sea swells "piles up" in the shallow lagoon. Surface longshore currents set up by the constant northeast wind displace some of this water southward and into the shelf lagoon. Most of the water returns seaward as bottom currents through numerous breaks in the reef.

Salinity—Purdy, Pusey, and Wantland (this volume) demonstrated that the northern shelf water is mainly hyposaline. Salinity ranges from 25°/₀₀ near the mainland coast to 35°/₀₀ at the line of islands on the shelf margin. Surface and bottom salinities are the same because of mixing.

Water temperature—Average surface temperatures of the seawater off the northern Belize shelf vary from 25.5°C in February to 28.5°C in August (Fuglister, 1947). Pusey (this volume) noted temperatures of approximately 29°C near the reef on the northern shelf and 30°C, with daily warming of 2°C, in the shelf lagoon and bay.

Carbonate saturation—Using a Weyl-type saturometer, Pusey in 1961 found values of water saturation for the shelf to be areally consistent with respect to Bahamian oolite. The interesting point he noted was that the highest saturation values were recorded for brackish-water areas. Pusey attributed this relation to the high calcium content of runoff waters from rivers draining the limestone terrain. In general, both bottom and surface waters were supersaturated with respect to subjacent sediment and the oolite standard.

PREVIOUS STUDIES

There are no previous, comprehensive studies of carbonate island sediments in Belize. The few works that have been published deal with the geography and physiography of the islands and reefs or with storm damage to the island's coast.

Vermeer (1959), studying the geography of the Belize cays, grouped them with respect to vegetation and degrees of exposure to the open sea. He identified them as (1) sand cays, (2) mangrove cays, and (3) mangrove-sand cays. The sand cays are most exposed; they are perched on the barrier rim near breaks in the reef, and are usually small and ephemeral. The mangrove cays are the least exposed; they occur in a sheltered position near the mainland. They consist of red mangrove (*Rhizophora mangle*) rooted in sand or mud banks. The mangrove-sand cays, including Ambergris Cay, are the largest of the three types of cays. As the name implies, they consist of a windward sand deposit and leeward mangrove mud flats. Stoddart (1963), in a report on storm damage to the reefs and cays of Belize, gave a brief and accurate physiographic description of Ambergris Cay, but did not dwell on the geology of the island.

The most recent work on the northern shelf of Belize is that of Pusey (1964; this volume), whose study of the sediments there is a natural background for study of Ambergris Cay. Pusey's data on the water and submarine sediments windward and leeward of Ambergris Cay provide the contrast necessary to accentuate the island's distinctive deposits.

AMBERGRIS CAY

Physiography and Hydrography

Ambergris Cay is the northernmost and largest cay in a long chain of islands on the outer continental shelf of Belize (Fig. 1). It lies just south of the Yucatán limestone platform and faces the open Caribbean Sea eastward across a barrier reef (Fig. 2). West of the island is the broad, shallow Chetumal Bay, which isolates the island physically and geologically from the mainland by a distance of 8-10 mi (13-16 km). The next cay south, Cay Corker, is 5 mi (8 km) away across a shallow tide-swept strait. In simplest terms, Ambergris Cay consists of a narrow, windward, beach ridge of sand and a leeward, broad complex of muddy supratidal flats and intra-island lagoons interspersed with areas of exposed rock (Fig. 4).

FIG. 4—Air photo from above San Pedro, Ambergris Cay, looking north along island's eastern shore. Barrier-reef crest (a) separates shallow reef lagoon (b) from deep Caribbean water east of island. Island's eastern shore is a windward beach ridge (c), cultivated in coconut palms. West of ridge is an expanse of supratidal flats (d) and intra-island lagoons (e). Note "cratered" appearance of island's surface, as Holocene sediments conform to underlying karst limestone. Dark area (f) in island's interior is heavily vegetated limestone terrain.

TABLE 1. SALINITY OF BOTTOM-WATER SAMPLES FROM AMBERGRIS CAY

Environment	*No. of Samples*	*Salinity (°/oo)* Range	*Salinity (°/oo)* Average
Beach-ridge water table	2	3.8- 4.5	4.1
Shelf lagoon	4	24.0- 38.8	33.5
Reef lagoon	1	—	37.2
Intra-island lagoons	14	25.5- 55.1	40.9
Supratidal ponds	7	41.5-123.8	76.3

The windward beach ridge, whose crest is 4-20 ft (1.2-6.0 m) above sea level, is constantly exposed to fresh, onshore winds. The fishing village of San Pedro overlooks the beautiful, clear water of the reef lagoon from the highest area of the ridge. A shallow, freshwater lens in this sand supports the cultivation of coconut palms.

In contrast, the low-lying supratidal flats in the island's sheltered interior are rarely refreshed by sea breeze. These flats adjoin stagnant or restricted, shallow lagoons and commonly are covered by ponds of hypersaline water. The saline sediment supports no useful crop, but is densely colonized by mangroves.

Some areas of the interior are underlain by karst limestone. These less accessible areas are covered by a thin, humic soil and dense, junglelike vegetation. Recent supratidal-flat sediments overlap the margins of these rocky areas.

The hydrographic conditions are extremely variable among the island's many separated lagoons and ponds. This factor strongly affects the distribution of the various lime-secreting organisms and the types of carbonate grains produced. It likewise affects the preservation, alteration, and even precipitation of sediment grains in various parts of the island. Table 1 illustrates the range of salinities and their grouping by environment on Ambergris Cay. It is significant that, whereas most of the water on the shelf is either of normal marine salinity or slightly hyposaline during the wet season, water of the intra-island lagoons and supratidal flats is distinctly hypersaline during the same period.

Scattered temperature observations in the lagoons and ponds ranged from 30°C in the deeper, larger, intra-island lagoons to a high at midday of 42°C in shallow supratidal ponds. Island waters in general are slightly supersaturated with respect to a Bahamian oolite standard, but the few measurements made do not permit interpretation of the slight differences between samples. The observed values of pH range from 7.1 to 8.4. The lowest value was for a hard, freshwater sample; the highest was for seawater in the reef lagoon; and intermediate values were obtained from samples of restricted-lagoon surface waters.

The water-chemistry data cited here, though spotty and incomplete, indicate that highly variable hydrographic conditions exist in the restricted waters of an island environment because of dilution and evaporation effects. The ecologic conditions in island waters must also be highly variable and must represent stress conditions for most marine organisms. Indeed, this fact is reflected in the relatively low taxonomic diversity of the fauna observed on Ambergris Cay.

The physiography of Ambergris Cay reflects the depositional environments that are important in the following description of island sedimentation. In order of discussion, these environments are (1) the windward beach ridge, (2) the supratidal flats, and (3) the intra-island lagoons. Transitions between these environments typically are abrupt, and each environment includes numerous sediment types. Thus, island sediments are highly variable horizontally and in vertical sequence. This variability emphasizes the fact that not one, but an *assemblage* of criteria may be necessary for the recognition of island sediments.

Besides these primary characteristics, island sediments show several diagenetic effects that are consequent upon subaerial exposure. The diagenetic processes to be discussed are (1) dissolution, (2) early cementation and pore filling, and (3) replacement (dolomitization). None of these processes is new but, in this context, they may be suggestive of a subaerially exposed sediment unit.

The submarine environments of deposition adjacent to Ambergris Cay are distinctly different from those of the island itself. They are the sources for much of the detrital carbonate sediment on the island. Pusey (this volume) discusses these environments in the context of a facies analysis of all the northern-shelf marine sediments. The discussion of these submarine environments is necessarily brief here, but it emphasizes the distinctiveness of the island sediments and indicates some of the processes involved in their accumulation.

Reef and Reef Lagoon

Environment—The coral barrier reef, 1-2 mi (1.6-3.2 km) windward of Ambergris Cay, exhibits a zonation similar to that described by Thorpe and Bregazzi (1960) for the reef at Rendezvous Cay, farther south. The breaker zone at the reef crest is formed largely by "trees" of *Acropora palmata* and round heads of *Montastrea annularis*. Leeward of the crest, a zone of patch reefs occurs on a floor of coral rubble in water 3-7 ft (1.0-2.1 m) deep; much of the rubble is encrusted by lithothamnid red algae. Broken sticks of *Acropora cervicornis* are particularly conspicuous in this rubble. An apron of continually shifting sand succeeds the patch-reef zone leeward and merges with a subsea meadow of sparse codiacean green algae and *Thalassia*, which colonize a sand bottom pocked with innumerable, low, crustacean- and worm-burrow mounds.

Along the western margin of the reef lagoon, and adjacent to the windward beach of the island, is a narrow band of densely vegetated sediment (Figs. 3, 4). *Thalassia testudinum* blankets this deposit and contributes large amounts of organic debris to it. The root system of these plants stabilizes the sand. The surface of the grass beds near shore is usually awash at low tide; and the seaward margin is only 3 or 4 ft (1-1.3 m) underwater. The *Thalassia* zone ranges from 100 to 500 yards (90-460 m) in width. The widest parts of the zone are usually opposite gaps in the reef.

There is great diversity in the fauna and flora of this shallow, "open-marine" environment, as indicated by samples of the bottom sediment (Table 2). No attempt will be made here to list all of the reef organisms observed, but Thorpe and Bregazzi (1960) reported a coral fauna that was second in diversity only to that of the Jamaican reefs in the Caribbean (Goreau, 1959).

Sediment—The importance of skeletal particles in this environment of deposition is apparent from Table 2, which shows an average particle

TABLE 2. AVERAGE CONSTITUENT COMPOSITION AND GRAIN SIZE OF AMBERGRIS CAY AND NEARBY SAMPLES

Constituent	*Average composition of Samples*				
	Windward Beach Ridge	*Supratidal Flats*	*Intra-island Lagoons*	*Shelf Lagoon*[1]	*Reef Lagoon*
Corals	8.6	0.0	0.5	0.0	13.1
Halimeda	35.5	0.1	6.2	0.2	34.3
Coralline algae	14.8	0.1	0.9	0.0	13.5
Soritidae	3.2	20.6	19.8	18.5	4.8
Miliolidae	0.8	18.8	18.9	11.9	0.6
Other Foraminifera	3.3	3.8	4.6	2.8	5.5
Mollusks	18.0	20.9	25.4	18.6	13.7
Ostracods	0.1	0.9	1.1	2.0	0.1
Miscellaneous skeletal	2.6	2.2	4.3	2.4	3.5
Total skeletal	90.8	67.0	84.0	58.5	87.4
Composite grains	1.2	1.3	1.2	--	5.6
Mud aggregates	1.7	18.6	12.8	--	0.0
Limestone fragments	0.0	2.3	1.3	1.4	0.0
Dolomitic crust fragments	2.2	9.4	0.3	--	0.0
Fecal grains	1.5	0.6	0.4	3.0	1.6
Unidentified	2.1	0.6	0.5	2.2	2.7
Terrigenous minerals	0.3	0.1	0.1	0.7	0.0
Cryptocrystalline grains[2]	5.1	29.3	14.3	36.5	5.6
Cryptocrystalline skeletal	25.9	15.2	18.0	14.9	26.4
Weight percent finer than 1/16 mm	12.3	29.0	23.1	31.5	1.4
Number of samples	19	45	40	45	43

[1]Data from Pusey (1964).

[2]Includes "composite grains," "mud aggregates," and "dolomitic crust fragments" of this study.

composition for the samples from the reef lagoon compared to those from island environments. None of the nonskeletal particles are of any importance. No ooids were observed in these sediments, whereas such grains have a very important status in similar shallow, current-swept, sandy areas of the Bahamas.

Reef-lagoon samples are a mineralogic mixture of aragonite and magnesium calcite, reflecting the varied contributions of the fauna and flora. Aragonite is important here because of the presence of so much *Halimeda* and coral debris. Local changes in the percentage of coralline algae are usually reflected in changes in the percentage of magnesium calcite. Figure 5 illustrates the changes in abundance of three of the most common sediment constituents from the reef to the island shore: (1) coral fragments are produced in great numbers at the reef and transported leeward; (2) *Halimeda* fragments become increasingly abundant with distance from the reef, as the substrate becomes more stable, and reach a maximum in the grass beds near shore, where *Halimeda* thrives among the *Thalassia* plants; (3) mollusk fragments are surprisingly constant in abundance. This uniformity may reflect constant production or compensating differences in rate of production and rate of breakdown across the lagoon. The reef-lagoon samples contain very little fine material, but what is present is concentrated in the grass-covered zone near shore.

Swinchatt (1965) studied the sediments between the line of keys and the reef arc in south Florida, an environment analogous to the reef lagoon at Ambergris Cay. The Florida situation and the processes affecting sediments near Ambergris Cay are very similar despite differences in depth and width of the reef lagoons, and the consequently more effective transport of material leeward at Ambergris Cay.

Windward Beach Ridge

Environment—The windward beach ridge on Ambergris Cay is a long, narrow, fairly straight sand body extending almost unbroken from Boca Chica in the south to the Mexico border in the north, a distance of approximately 26 mi (42 km; Fig. 3). The ridge is 100-500 yards (90-460 m) wide and ranges between 4 and 8 ft (1.2-2.4 m) in elevation along most of its length. The village of San Pedro is 22 ft (6.6 m) above sea level. The beach ridge is actually a compound feature with two or three ridges separated by very shallow swales. Vegetation on the beach ridge is a tangle of vines and thicket, typical of Caribbean sand cays (Stoddart, 1963). Mangroves have colonized swales leeward of the beach-ridge crest, where there is standing water, making the western extent of the beach-ridge sand indistinct. Washover fans extend the contact between the windward ridge and leeward mud flats westward. The Ambergris Cay beach ridge differs somewhat from the beach ridges on many of the Bahama Islands (Newell and Rigby, 1957; Doran, 1955) in that bedrock crops out in only one place. The rock surface rises under the ridge, but only at Reef Point does it appear at the surface.

Sediment—The sediment of the windward beach ridge is a deposit of sand and cobbles heaped by storm waves on a shoaling bottom. Vermeer (1959) and Stoddart (1963) described the effectiveness of Caribbean storms in modifying the shelf islands in northern Belize. Most of the effects they describe are destructive, such as leeward movement of sand and channeling of beach ridges, but there is evidence of construction as well, in the form of thin sheets of sand added to beach ridges and multiple beach berms formed above high-tide level. The normal tidal range of 12-18 in. (30-46 cm) is inadequate to explain these features, nor is wind an active agent of sediment transport here except as a part of storm effects. Only in areas stripped of natural vegetation by humans is the surface rippled or otherwise affected by the steady, but moderate,

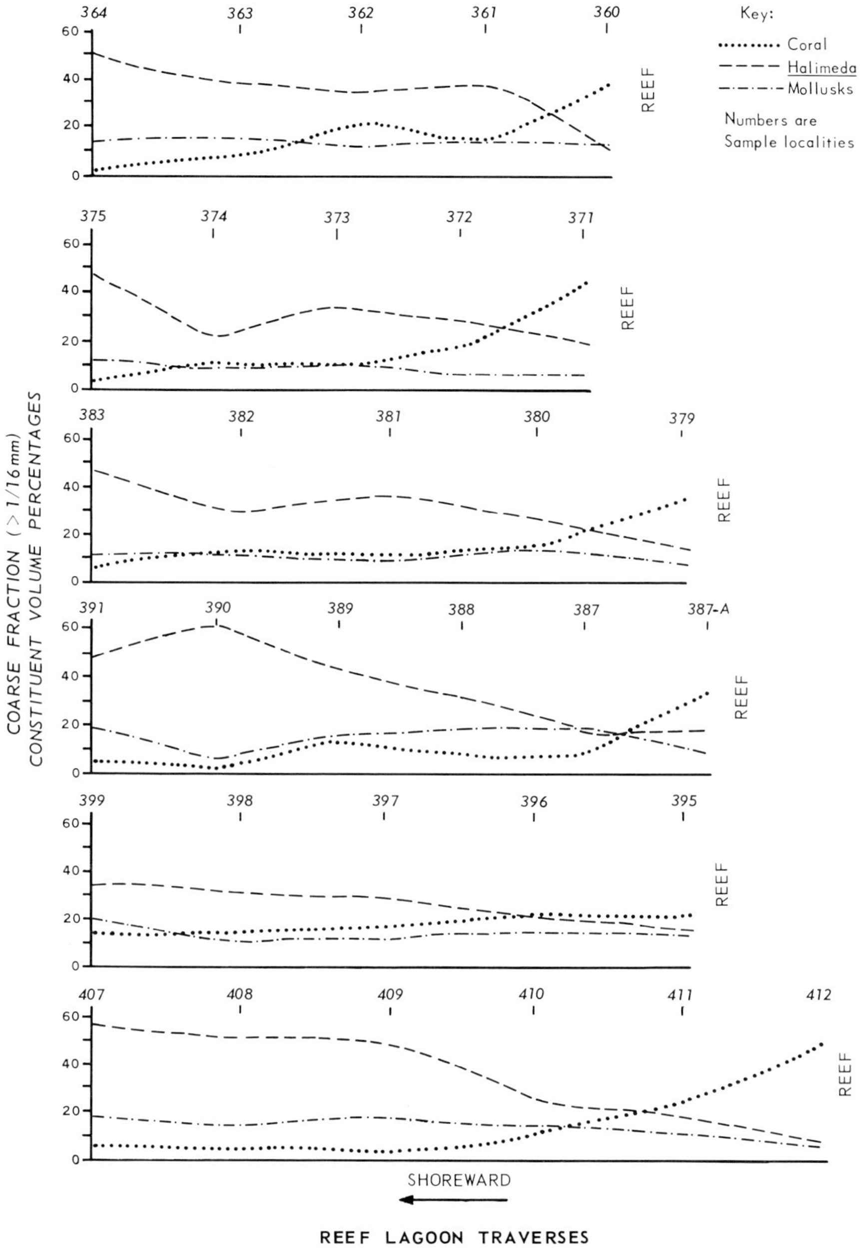

FIG. 5—Variations in abundance of three major constituents of coralgal sand along traverses across reef lagoon. Numbers are sample localities.

onshore wind. Conversely, tropical storms are a regular, periodic element in the weather pattern of the Belize coast and need not be considered a catastrophic mechanism (Stoddart, 1963). The observation that the coarseness of the beach-ridge material varies directly with its proximity to the reef is indirect evidence of the importance of storm waves in building and modifying the beach ridge (Fig. 6).

FIG. 6—Windward beach ridge, Ambergris Cay.

A. South of Reef Point; view is south toward Robles Point. Tall (10 ft or 3 m), dark gray ridge on extreme right is built of storm-thrown cobbles of modern coral. White sand and gravel seaward of ridge is younger sediment overlapping foot of ridge. Dark material on white sand surface is debris of wave-thrown seaweed marking highest "normal" high tide.

B. North of San Pedro, Ambergris Cay. Crest of ridge here is only 5 ft (1.5 m) above low tide level. Beach ridge is low and wide, with no cobble ridge.

No well-developed bedding was noted in the beach-ridge sand, either in cores or in pits, except for the narrow swash zone, where layers dip slightly seaward. In the upper third of the swash zone, where the beach sand is fairly even textured and not covered by seaweed, small pockmarks dot the surface. These are caused by air bubbles escaping from the sand surface after the passage of each wavelet (Hoyt and Henry, 1964; Dunham, 1970). Plant roots and burrowing crustaceans homogenize the sediment higher on the ridge.

The sediment of the windward beach ridge is very similar to that of the adjacent reef lagoon, from which it obviously has been derived. Its geometry and position adjacent to the leeward swamps are the main criteria distinguishing this deposit from the submarine deposits nearby. Table 2 shows the average composition of windward beach-ridge samples. The sands are largely abraded remains of coral, *Halimeda*, mollusks, coralline algae, and foraminifers. The first three contribute the large amount of aragonite that distinguishes the windward beach-ridge sands from other supratidal island deposits. The last two contribute most of the magnesium calcite to these sands. There is the same scarcity of nonskeletal particles in the beach sand as in the reef-lagoon deposits. The bulk of the beach-ridge sediment is poorly sorted, but very porous and uncemented, except for scattered patches of surficial, calcite-cemented sand. This porous beach sand contains lenses of fresh water tapped by several wells. The depth of the water table and the quality (hardness) of the water differ slightly from well to well, indicating the lenticular nature and incomplete communication between depositional units within the beach ridge.

Diagenesis—Purdy (1965) noted that, in subaerially exposed, unconsolidated carbonate sediments, carbonate dissolution, replacement, and precipitation occur, in that order. Evidence of the first and last of these processes, dissolution and precipitation (cementation), is present in the windward beach-ridge sediments. Wherever coral cobbles or heads are found on the beach above high tide, as at Reef Point, some of them can be broken apart by hand along lines of growth, as if "rotten." This characteristic may reflect decay of the organic matrix and etching of framework aragonite with consequent weakening of the skeletal structure. This process could be important in the production of fine carbonate debris without either abrasion or precipitation (Purdy, 1968).

Elsewhere on the windward beach, smaller skeletal particles are being destroyed by the percolation of fresh rainwater and its dissolution of carbonate material (Fig. 7). The obvious evidence of dissolution of skeletal carbonate is apparent as large vugs transecting architectural lines. In addition to the dissolution, fine, fragmental carbonate is produced within the skeleton being attacked.

The opposite effect occurs on the surface of the beach ridge, where patches of lightly cemented coralgal sand stand in low relief. Distinct from beachrock, which apparently is limited to the intertidal zone (Ginsburg, 1957), this sand is cemented by calcite and is well above the intertidal zone.

Between rainfalls, meteoric water, saturated with respect to calcite, evaporates from the beach-ridge surface, leaving rims of small, 20μ, bladed calcite crystals coating grains and filling interstices in a few places. Figure 9B shows an exceptionally well-developed example of this. More commonly, the calcite is discernable only at grain contacts ("meniscus cement" of Dunham, 1969). The cement is inferred to be calcite from comparison of X-ray diffractograms of cemented sand and adjacent uncemented sand which contains no other obvious source for the calcite.

There is a weak correlation between the occurrence of cemented beach sand and the presence of vegetation. Cementation apparently does not occur below a depth of 6 in. (15 cm), and typically is confined to the upper 2 in. (5 cm) of an exposed surface. This fact suggests that a certain degree of stability of the sand is necessary, and explains why the cemented patches are most common near the ridge crest, where plants minimize sand movement and cause a higher rate of moisture transpiration.

In brief, dissolution of carbonate, nonabrasional production of fine carbonate, and precipitation of cementing carbonate all occur within the windward beach ridge. There is, however, no evidence of mineralogic replacement of aragonite or magnesium calcite by calcite as suggested by Purdy (1965).

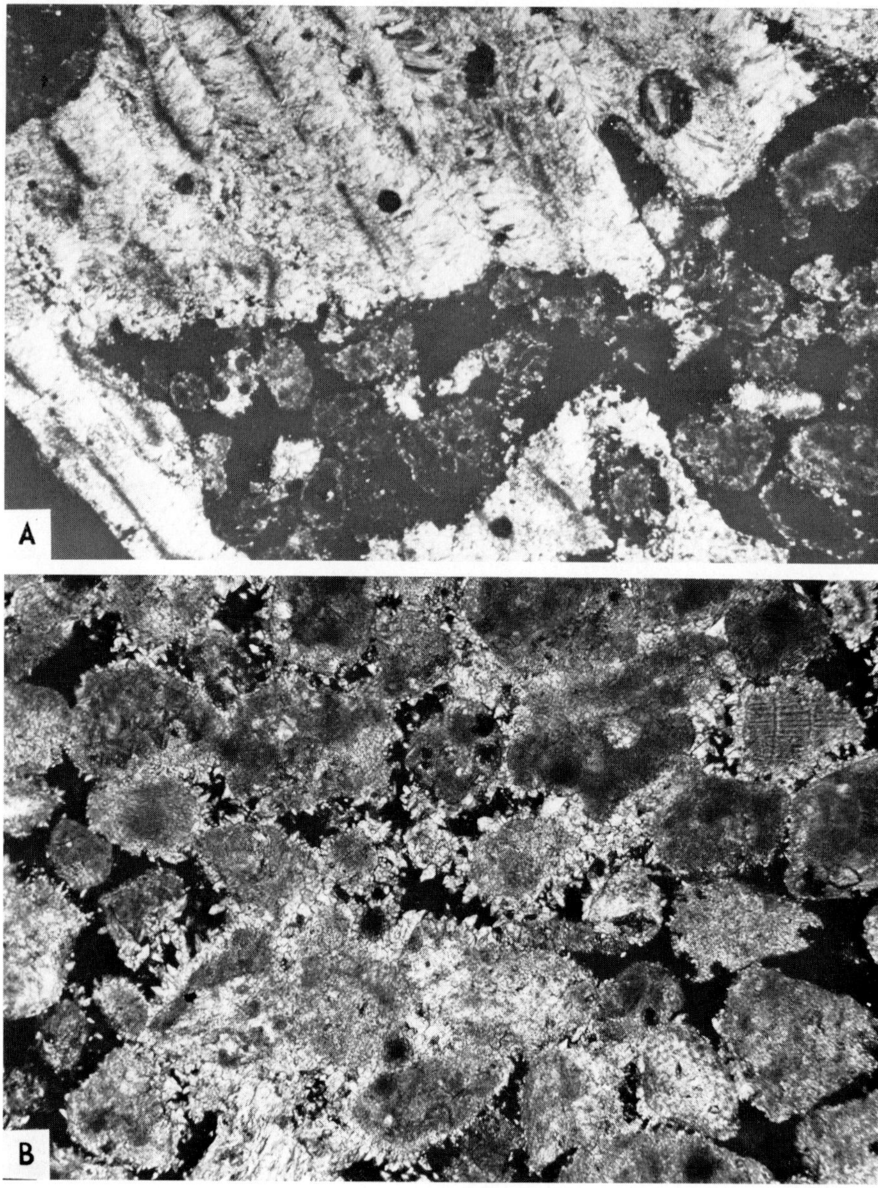

FIG. 7—Early diagenesis in windward beach-ridge sediment.

A. Photomicrograph of a fragment of coral, *Manicina areolata*, from within beach ridge, above fresh-water table. Dissolution of some of original, skeletal aragonite leaves large vugs transecting growth lines. Some particles inside vug are silt-size sediment produced during dissolution process. Such solution removal of carbonate minerals is common in uncemented sands subjected to percolation of fresh meteoric water. Dissolved carbonate thus derived may be important in cementing lower sediment. Crossed nicols (x32).

B. Photomicrograph of lightly cemented coralgal sand from surface of beach ridge at location 405. Cement is bladed calcite crystals developed in interstices, especially at grain contacts. This cemented sand is in discontinuous patches on surface of beach ridge, well above high-tide line. Crossed nicols (x10).

FIG. 8—Supratidal flats (light colored) surrounding intra-island lagoons on Ambergris Cay. View is eastward, near island's northern boundary. In background is windward beach ridge and, beyond, reef lagoon and open sea. Supratidal deposits here are mostly bare; in other localities they are densely vegetated. Note numerous shallow, oval ponds on flat surfaces.

Supratidal Flats

West of the windward beach ridge is a broad expanse of supratidal flats interspersed with intra-island lagoons and areas of exposed rock (Fig. 3). Predominantly fine-grained sediment which has accumulated in these environments forms some of the most distinctive deposits on the island (Fig. 8). The most significant feature of these deposits is the fact that they have effectively "welded" together the somewhat separated rock highs on the drowned topography into a continuous area of land, thereby greatly increasing the island's sheltering effect on surrounding submarine environments.

Environment—The areas of supratidal deposition on Ambergris Cay are not a single environment, but a complex of related environments. In some cases, it is difficult to assign such areas to a marine or to a subaerial category, for they are transitional between the two. A generalized map (Fig. 3) cannot begin to disclose the true complexity of these parts of the island (Shinn et al, 1969).

Much of the surface of the supratidal flats is covered by vegetation, mainly red mangrove (*Rhizophora mangle*) or black mangrove (*Avicennia nitida*). The shelly sand ridges on some shores support a palmetto - salt grass - mangrove assemblage. The red mangroves colonize marginal areas, near brackish or normal-marine water, where the island is frequently flooded. Black mangrove is the more common shrub on supratidal flats where hypersaline water stands most of the year (Fig. 9A). In areas of thick vegetation, the sediment may include a large amount of waterlogged, leafy and woody organic matter. In some parts of the island where storms have defoliated and destroyed mangrove colonies (Fig. 9B), a Holocene peat has formed.

Typical of sediment above an unconformity, the supratidal deposits of Ambergris Cay vary greatly in thickness from place to place because of the irregularity of the underlying limestone surface. Depressions in the limestone surface, presumably of solution origin, localize the sites of the many shallow, round or oval-shaped ponds on supratidal flats (Fig. 9C, D). These ephemeral water bodies are not distinguished from the flats on Figure 3, because they are actually part of the supratidal environment. The supratidal ponds are encircled by raised, levee-like rims of sediment. These rims are a few inches higher and slightly coarser than the sediment on the adjoining mud flat. Thus, the ponds are like small reservoirs of impounded water on supratidal-flat surfaces. The unusual and important feature of these stranded, ephemeral ponds is that their water chemistry varies drastically with the seasons. This chemistry has important diagenetic effects on the sediment.

Sediment—The variability of supratidal deposits is one of their most striking attributes. Some supratidal flats result from infilling of lakes and lagoons; some form after occlusion and filling of tidal channels; some form by accretion of sediment around an exposed bedrock knoll or a sediment bar or spit; and some form by simple addition of storm deposits to a submerged sediment mound. Whatever their origin, these flats represent growth of the island's land surface at the expense of its wet surface. Each of the situations mentioned leads to a common result: a subaerially exposed sediment surface with similar sedimentary structures and diagenetic effects. The lateral and vertical variability of the supratidal-flat deposits contrasts with the relative homogeneity of nearby submarine deposits.

Supratidal deposits on Ambergris Cay are commonly less than 2 ft (0.6 m) thick. They are typically a tan or gray, muddy carbonate sand varying greatly in texture. They are not characterized by a unique assemblage of constituent particles because they derive their sediment from all of the surrounding subsea areas. Exceptions to this may be cases where land organisms (e.g., terrestrial gastropods) are included in the sediment; but even these may not be definitive if transported seaward. There are no currents smoothing the surface and "painting" facies with different combinations of current velocity and sediment supply. The supratidal environment receives its deposits in spasmodic bursts, first from one direction, then from another, in highly variable quantities. Even the organic constituents vary with direction of supply, for the hydrography of the intra-island lagoons is so variable that different suites of organisms may thrive in unconnected lagoons bordering a single mud flat. Figure 10 graphically portrays the variability of composition between several supratidal-flat samples on the basis of the most important constituents. For comparison, Table 2 indicates that much of the island sediment is poorly differentiated from nearby submarine deposits on the basis of composition alone. Mineralogy is predictably variable also, but magnesium calcite is usually predominant because of the large contribution to these sediments by Foraminifera. From comparison of the constituents and mineralogy, it is apparent that most of the supratidal surface sediment of the island is derived from the intra-island lagoons or from the shallow shelf to the west rather than from the reef lagoon. This fact emphasizes the effectiveness of the windward beach ridge as a blockade to cross-island circulation.

Curved, shell-sand beach ridges are present along much of the leeward (western) shore of Ambergris Cay and on the margins of intra-island lagoons. They range from 1 to 4 ft (0.3-1.3 m) in elevation. These ridges are drastically different from the windward ridge in composition, consisting of skeletal remains of peneroplid foraminifers and a few genera of mollusks, such as cerithid gastropods or the pelecypods *Anomalocardia*, *Chione*, or *Pseudocyrena*.

FIG. 9—Ambergris Cay supratidal flats.

A. View east from location 487 across supratidal flats and ponds partly filled with hypersaline water. Black mangrove pneumatophore roots in foreground perforate sediment, destroying lamination but leaving distinctive root casts.

B. Area once heavily colonized by mangroves (location 615), but since denuded by a hurricane. Sediment here is 1 in. (2.5 cm) of shelly, organic-rich mud, overlying 2 ft (0.6 m) of modern mangrove peat.

C. Bright white supratidal flat near location 600, where only very sparse vegetation and little floodwater occupy surface. Beyond exposed flat is an intra-island lagoon and, beyond that, a dark strip that is windward beach ridge.

D. Supratidal flat where samples 487-497 were taken. Note flooded surface with numerous ponds and areas of algal mat (dark areas on pond margins). Algal mats are not present *in* supratidal ponds on Ambergris Cay. Dolomitized sediment is found on margins of many of oval ponds in this area and that in preceding photo

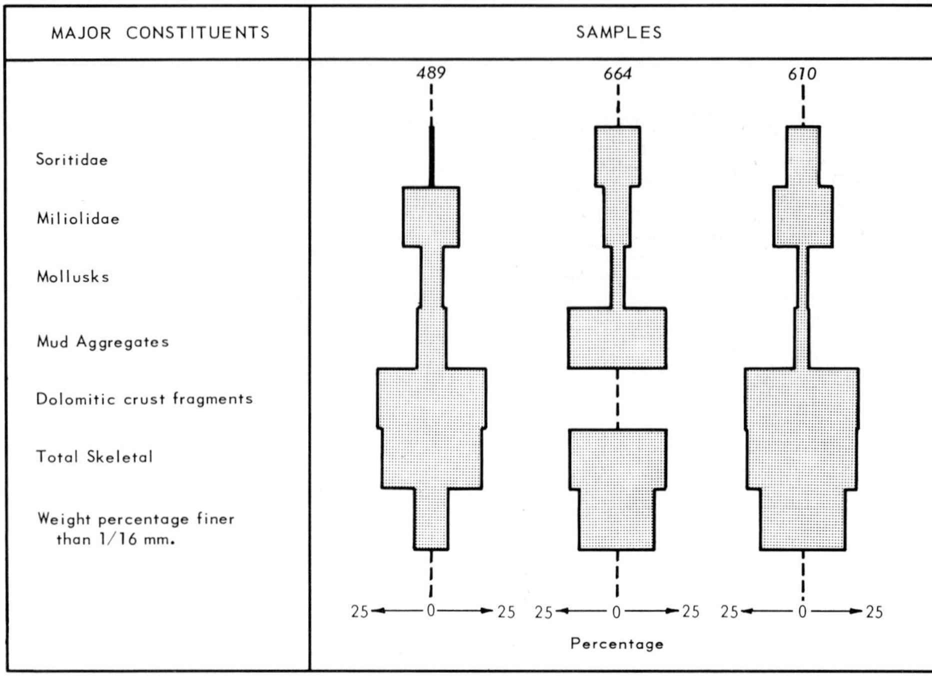

FIG. 10—Percentage diagram of abundances of major grain types in three supratidal-flat samples, demonstrating variability of samples from that environment.

Narrow intertidal areas are transitional between the higher supratidal flats or marginal beach ridges and lower intra-island lagoons (Fig. 8). The shelly carbonate mud accumulating there is usually very dark gray or black, and reeks of hydrogen sulfide gas when disturbed. These intertidal deposits retain much of their original interstitial water long after burial (intertidal beds beneath 2-3 ft [0.6-1.0 m] of sediment in cores shrink as much as 30 percent on drying). This water may be held in the mucouslike organic matter which is so abundant in these sediments (Pusey, this volume). The transitional nature of this intertidal sediment is emphasized by the fact that lagoonal deposits are softer, whereas supratidal sediments are much firmer, probably because of the desiccation and compaction which result from subaerial exposure.

Supratidal-flat deposits usually are distinctly laminated. In some places the lamination is textural—layers of mud alternate with thin layers of skeletal sand. In other places, difference in content of water and organic detritus of alternating layers produces color lamination in almost pure mud sequences. Color and texture lamination is also common in stromatolitic sediments (Fig. 11A).

Lithologic variability and lamination contrast strongly with the homogeneity of sediments from normal-marine environments. Pusey (this volume) describes textural lamination of sediments on a large mud-shoal complex in the shelf lagoon; elsewhere the submarine sediments of the northern shelf are essentially structureless except for vague mottling. Supratidal deposits rarely are mottled. The contrasting sedimentary structures in these two environments probably are caused by differences in abundance of burrowing organisms. In the marine environment, potential stratification of sediments is obliterated by burrowing organisms.

The virtual absence of these organisms in the supratidal environment precludes destruction of the original lamination.

Shelly sand layers in cores commonly truncate underlying mud laminae abruptly, but grade upward into more muddy material. Such small-scale graded bedding (Fig. 11) is produced when sediment settles out of seawater inundating the flats during a storm or other unusually high water. Inverse grading, as observed in some cores, may result from inertial flow in newly deposited sediment on supratidal flats that remain flooded for a while (Sanders, 1965). The lateral extent of storm-flood laminae is only a few meters in any direction, and muddy laminae are more laterally extensive than sandy laminae. This explains the difference in appearance of cores taken only a few meters apart. Desiccation polygons

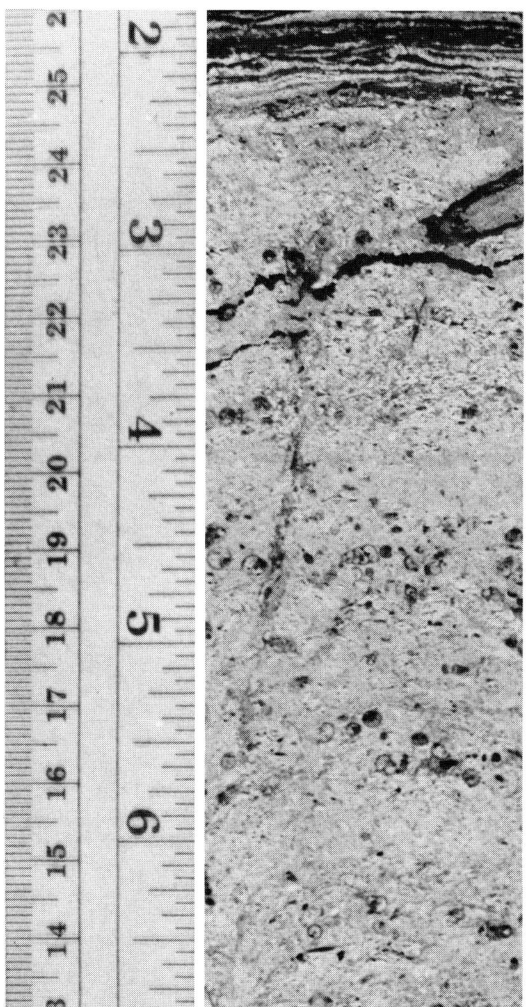

FIG. 11—Core of supratidal-flat sediment (incomplete), showing two types of lamination. Strong color lamination in upper 1/2 in. (1.27 cm) is stromatolite structure; dark algal-mat layers alternate with light sediment layers. Lower in core are less obvious textural laminae of alternating coarser and finer sediment. Scale is in inches and centimeters.

form everywhere that muddy sediment is exposed to drying on the supratidal flats, especially on the margins of intra-island lagoons. These cracks penetrate completely through the last muddy lamina deposited. Their complexity is proportional to the degree of desiccation, and the more complete polygons form farthest from the intertidal zone (Fig. 12A, B).

Subsequent washing of sediment over these polygons results in filling of the cracks, commonly by slightly more permeable, or sandy, sediment. Thus, these desiccation features are preserved after burial and are apparent in cross sections as intraformational, flat-pebble conglomerates. No fenestral laminae (Tebbutt et al, 1965) or desiccation structures (Fischer, 1964; Shinn, 1968) were observed in Ambergris Cay deposits.

Approximately 40 percent of the surface of Ambergris Cay supratidal deposits is covered by blue-green algal mats 1/8-1/2 in. (3-12 mm) thick (Fig. 12C, D). No mats were noted in the intertidal or subtidal zone (Gebelein, 1968). The characteristic, laminated, stromatolitic structure so common to algal-mat deposits (Black, 1933; Ginsburg et al, 1954; Illing et al, 1965) is apparent in Ambergris Cay supratidal sediments only to a depth of less than 1 in. (2.5 cm) and, more commonly, is not present below the surface mat. Stromatolitic structure is generally a color-banding of light-colored sediment layers with dark-colored organic layers. In addition to color-banding, there is a textural lamination of finer and coarser sediment layers which is less apparent than the color-banded interval but which persists to depths of 6-8 in. (15-20 cm; Fig. 11).

Apparently, the infrequently flooded, often desiccated supratidal flats on Ambergris Cay are such an oxidizing environment, at least on the surface, that the delicate algal threads deteriorate after a period of growth and are lost from the sediment. When an area is reflooded, it is rapidly recolonized by algae. Whether these algae have survived the dry period or are newly established algal colonies is uncertain. Likewise, whether the textural laminae mentioned above are actual testimony to the sediment-trapping capacity of algal mats or simply represent storm-tide deposits is questionable. The writer believes that, in areas not often inundated by tides or storms, desiccated algal mats (Fig. 12C, D) simply mark a supratidal environment and are not important to sediment accretion.

Dalrymple (1964, p. 41) described "algal carbonate grains" in intertidal algal mats in Baffin Bay, Texas. These grains are structureless masses of cryptocrystalline aragonite 1/16 mm to more than 10 mm across. The smaller grains are globular, whereas the larger ones are flakelike aggregates of the smaller grains. Apparently, the aragonite of these algal carbonate grains is precipitated directly from seawater, and precipitation possibly is induced by the decay of the organic material in the mat. In thin sections of algal mat from Ambergris Cay, the cryptocrystalline carbonate present as platy or irregular grains within algal-mat-carbonate sequences is optically indistinguishable from the "algal carbonate" grains of Dalrymple (1964), which are present in algal-mat-terrigenous-grain sequences. Further suggestion that there may be algae-associated precipitation of carbonate on Ambergris Cay is that blocky crystals of gypsum and halite are present in the uppermost layer of some mats. These scattered, ephemeral deposits are the only evaporites noted in the island sediment. They are not preserved as part of the permanent sediment record but, rather, are redissolved during each succeeding flood.

Diagenesis—Annual rainfall on Ambergris Cay averages 50-60 in. (127-152 cm), concentrated in a 3-4-month summer wet season. Periodic storms with accompanying flood tides inundate the low-lying flats with seawater.

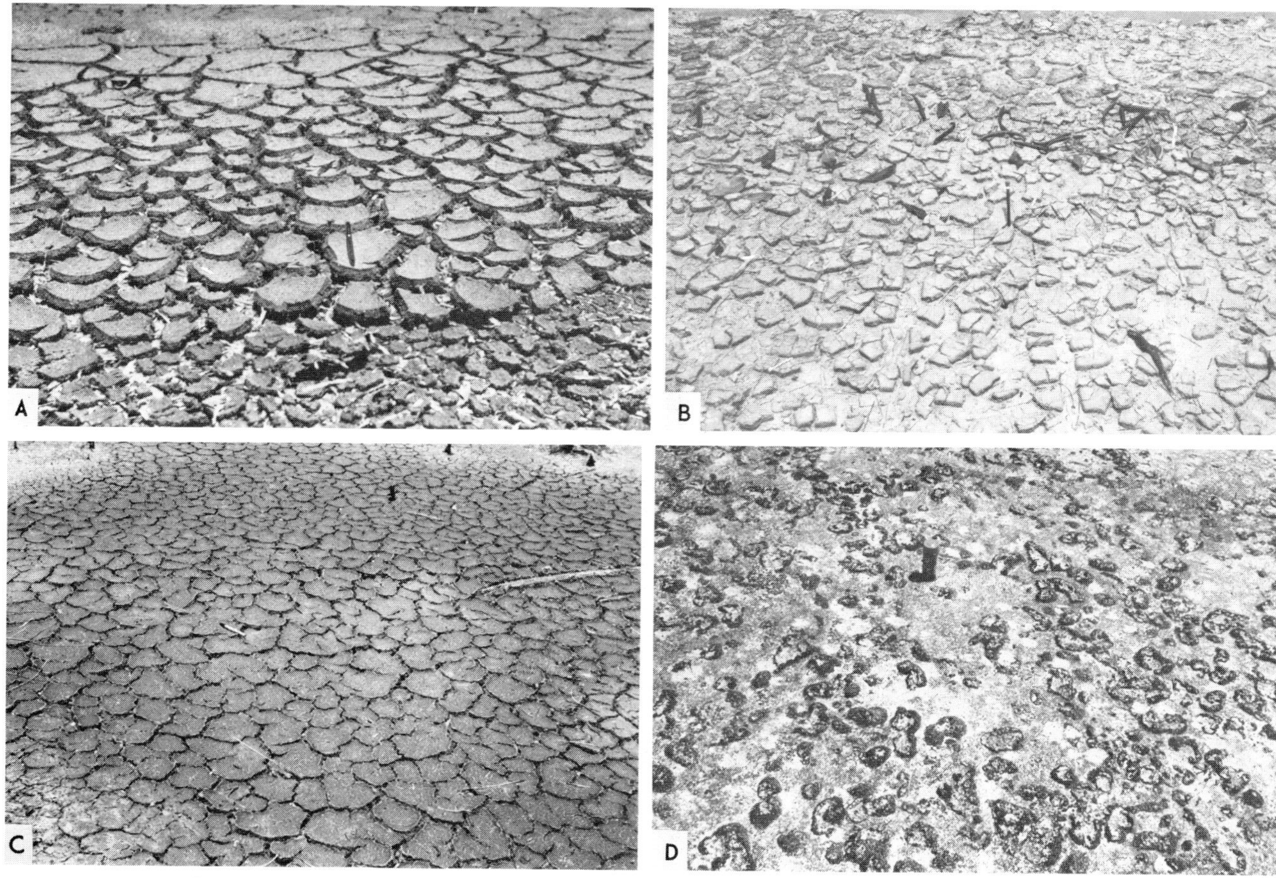

FIG. 12–Surface sedimentary structures.

A. Desiccation cracks in intertidal sediment on margin of intra-island lagoon. Shoreline is at top. Note that polygons separate from underlying sandy layer. White ribbonlike strips in foreground are bleached blades of marine grass *Thalassia*.

B. Mud polygons in intertidal sediment, disarranged by wave swash and partly buried by later sediment. Long, dark objects are mangrove seed pods. Pencil in center is for scale.

C. Dry, cracked mat of organic detritus and algae on a supratidal flat near location 615. Where organic matter is thoroughly mixed with fine sediment, polygons are strongly concave upward. Pencil is for scale.

D. Algal-mat surface structure on a supratidal flat, where bulbous, hollow, truncated blisters form. Can of insect repellant is for scale.

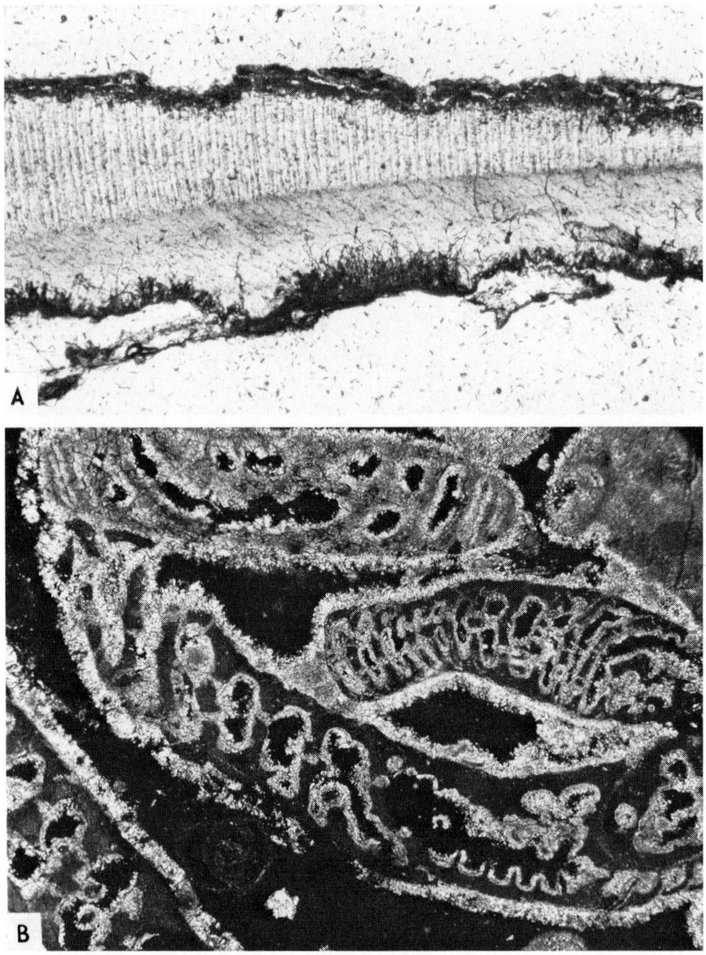

FIG. 13—Early diagenesis in supratidal-flat sediment.

A. Thin-section photomicrograph of a gastropod fragment. Surfaces of aragonite shell have been leached by fresh water and attacked by a "felt" of blue-green boring algae. Note hairlike borings perforating shell material and creating a cryptocrystalline, semi-opaque rind around grain. Plane-polarized light (x33).

B. Photomicrograph of intertidal beachrock from location 618. Beach-sand particles, predominantly peneroplid foraminifers, are coated by radially oriented magnesium calcite crystals. This is a particularly well-developed zone of cementation. Elsewhere in sample, cement rind is well developed only at grain contacts. Crossed nicols (x50).

Thus, the waters standing on the supratidal flats vary seasonally from cool, almost fresh, and distinctly undersaturated with respect to carbonate minerals, to hot, hypersaline, and capable of dolomitizing sediment. This variable hydrology, with accompanying periods of dryness, imposes chemically harsh conditions on exposed sediment and results in distinctive early diagenetic effects.

The dissolution of modern carbonate particles is limited to the subaerial environment. It results mainly from periodic exposure of the sediment to the initially acidic effect of rainwater and the boring activity of blue-green algae. Previously dry supratidal flats become flooded during periods of heavy rainfall. The raised margins of the flats prevent most of this water from running off, increasing its residence time over the sediment on the flats. The water quickly redissolves halite and any other soluble salts on the surface, but remains undersaturated with respect to carbonates for a much longer time.

Shells exposed on the flats are fresh on the underside, where partly buried, but blackened and leached on the upper side. The blackening results from blue-green algae that chemically bore the shell material. These algae facilitate solution diminution of the particles by increasing the surface area exposed to weathering (Fig. 13A). If these large, obvious particles are dissolved, smaller sand-size material must be affected similarly. Thus fine carbonate sediment may be produced *in situ*, without abrasion, by etching and breakdown of larger particles.

Early cementation occurs in island sediment as a ledge of beachrock in the intertidal zone of beaches on the western (leeward) shore of Ambergris Cay. These beaches are predominantly magnesium calcite sand, in contrast to the aragonite-rich windward beaches. The ledge has all the

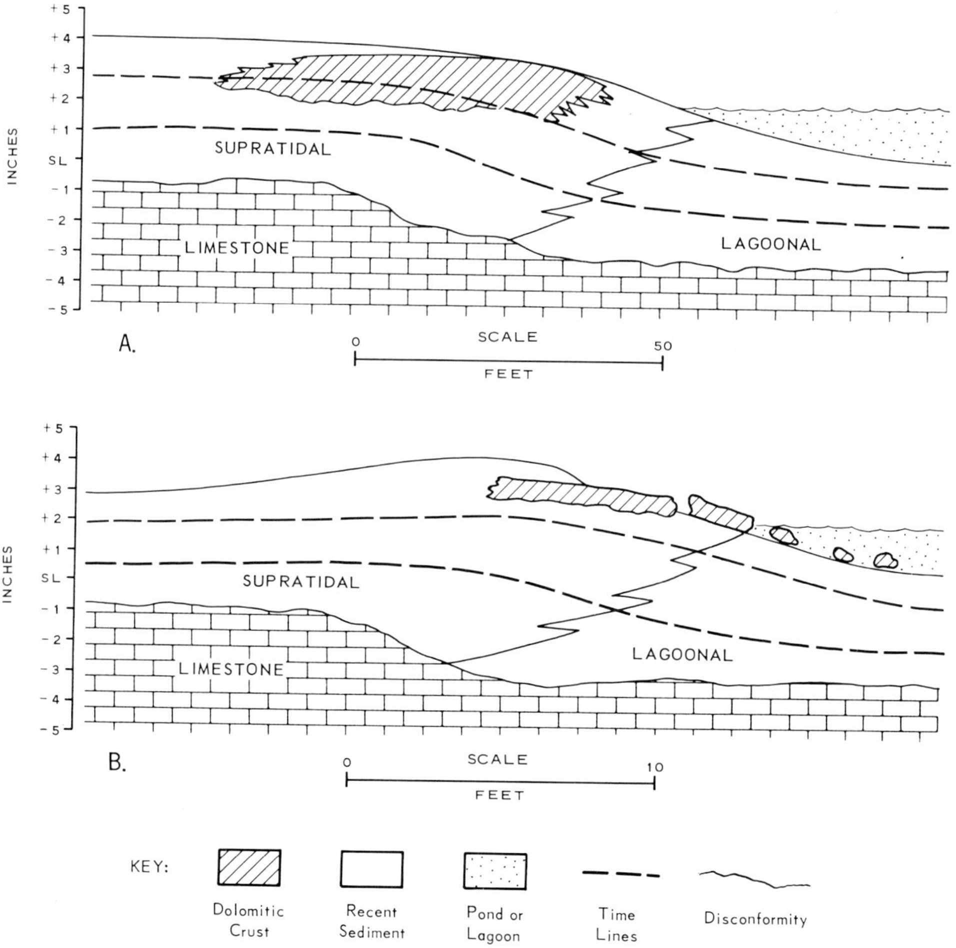

FIG. 14—Stratigraphy of dolomitic crust in Holocene sediment bordering supratidal ponds or intra-island lagoons. Note differences in horizontal scales. **A**. Pavementlike crust. **B**. Ledgelike crust.

properties of the platy, friable, skeletal sand around it, but is cemented by small (30μ) bladed crystals of magnesium calcite (Fig. 13B). In this respect, it differs from beachrock described by Ginsburg (1953) and Russell (1962).

The beachrock is present as 2-3-in.-thick (5-8 cm) ledges in the intertidal zone, and as scattered and patchy exposures or loose blocks on the surface. This may be explained if the beachrock is forming within the sand body and is exposed only where the shoreline is cut back by wave erosion (Russell, 1962). Indeed, the ledge does extend a few feet from the water line under a sand ridge (as inferred from probing). It rises slightly from the shore and is independent of the underlying rock surface. The beachrock is lightly cemented where it is buried; but, when exposed, it develops a rindlike casing and becomes better indurated. An exposed ledge is commonly encrusted by spirorbid worms and becomes discolored to light grayish green because of blue-green algae on the surface. No direct correlation could be established between the beachrock and a freshwater table—if a water table was present in the low, narrow sand ridge.

These observations serve only to confuse rather than clarify the question of beachrock formation. One point is clear, however; whatever the mechanism of its formation, beachrock is commonly noted on carbonate sand beaches and represents subaerial lithification and pore filling of an otherwise friable sand. Thus, beachrock is evidence of an intertidal or supratidal environment and may supplement other evidence of island sedimentation.

Dolomitic sediment is even more important than the solution and cementation phenomena in recognizing subaerially exposed horizons and the supratidal-flat environment. No radiocarbon age is available for dolomitic Holocene sediment on Ambergris Cay but, in manner of occurrence, physical appearance, and chemistry, this material appears very similar to the crustose dolomitic sediment reported by others from Bonaire (Deffeyes et al, 1964, 1965), Florida (Shinn, 1964), and the Bahamas (Shinn et al, 1965).

The Ambergris Cay crustose dolomitic sediment is present as an indurated ledge or as a partly buried pavement in the deposits marginal to ephemeral, hypersaline supratidal ponds (Figs. 9C, D, 14, 15A). Completely buried crusts also are present beneath mud flats in the intertidal zone adjacent to restricted intra-island lagoons (Fig. 15B). In each case, the dolomite forms in a zone of alternate wetting and drying where interstitial water is concentrated. Cores demonstrate that the sediments above and below the dolomitic layer are identical—that is, laminated, shelly, pelleted mud or muddy sand typical of the supratidal deposits of Ambergris Cay.

The dolomite crusts are almost everywhere white or cream colored, and may vary in thickness from 1/2 to 4 in. (1-10 cm), depending on the circumstances of their formation.

Sedimentary structures are usually well preserved in the lithified crusts. Textural laminations, small burrows, root casts, and algal-mat crenulations are the most common structures seen. None of the desiccation polygons noted by Shinn (1964) were observed in Ambergris Cay crusts. Moreover, completely massive (unlaminated) crusts are not uncommon. The most striking feature of these crusts is the many small voids that result from leaching of skeletal particles, chiefly foraminifer and mollusk fragments. Nonskeletal grains (i.e., mud aggregates and pellets) lose their identity in this dolomitized sediment and are rarely distinguishable from the matrix. The overall texture of most crusts is grumous, or clotted, with vuggy porosity (Fig. 15C, D).

The dolomite itself is optically indistinguishable from original cryptocrystalline matrix because of its fine grain size (1-5μ) and the

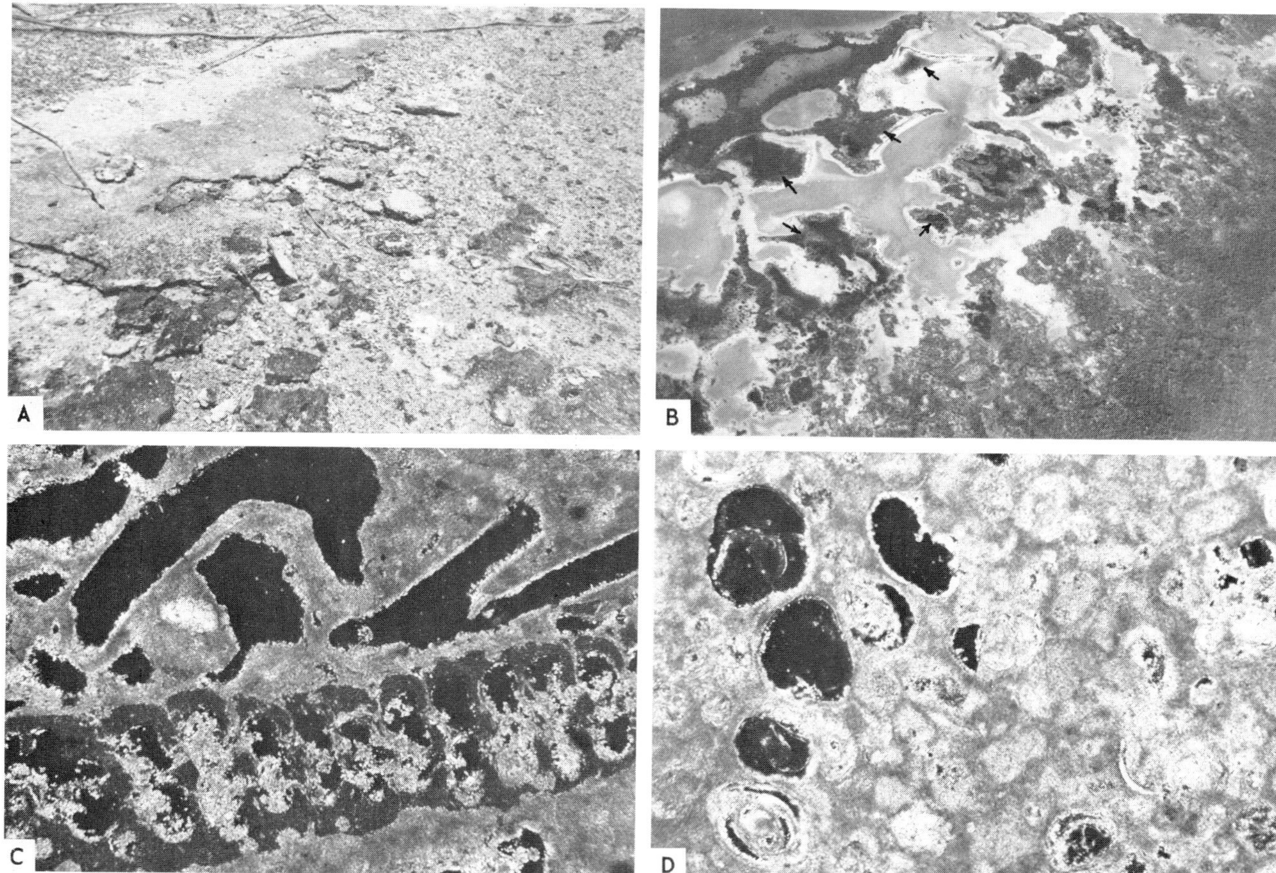

FIG. 15–Dolomitized supratidal sediment.

A. Indurated ledge of supratidal dolomite partly eroded from a raised rim of sandy sediment on margin of a small pond at location 612. Dolomitic sediment is being broken into clasts that will be included in next deposit covering area. Pencil is for scale.

B. An area where sediment accretion is filling in intra-island lagoons, resulting in formation of low-lying supratidal flats that are colonized by dark blue-green algal mats (arrows). Dolomitic sediment is present as crusts within Holocene deposits under these mats. See also Figure 9C, D.

C. Thin-section photomicrograph of dolomitic sediment crust from near location 600. Note preservation of magnesium calcite foraminifer and preferential removal (solution) of aragonite mollusk fragments (apparent as large voids in upper half of photo). Crossed nicols (x75).

D. Thin-section photomicrograph of dolomitic crust showing a muddy lamina in which matrix has been dolomitized and almost only porosity remaining is secondary, resulting from leaching of some miliolid foraminifer tests. Crossed nicols (x50).

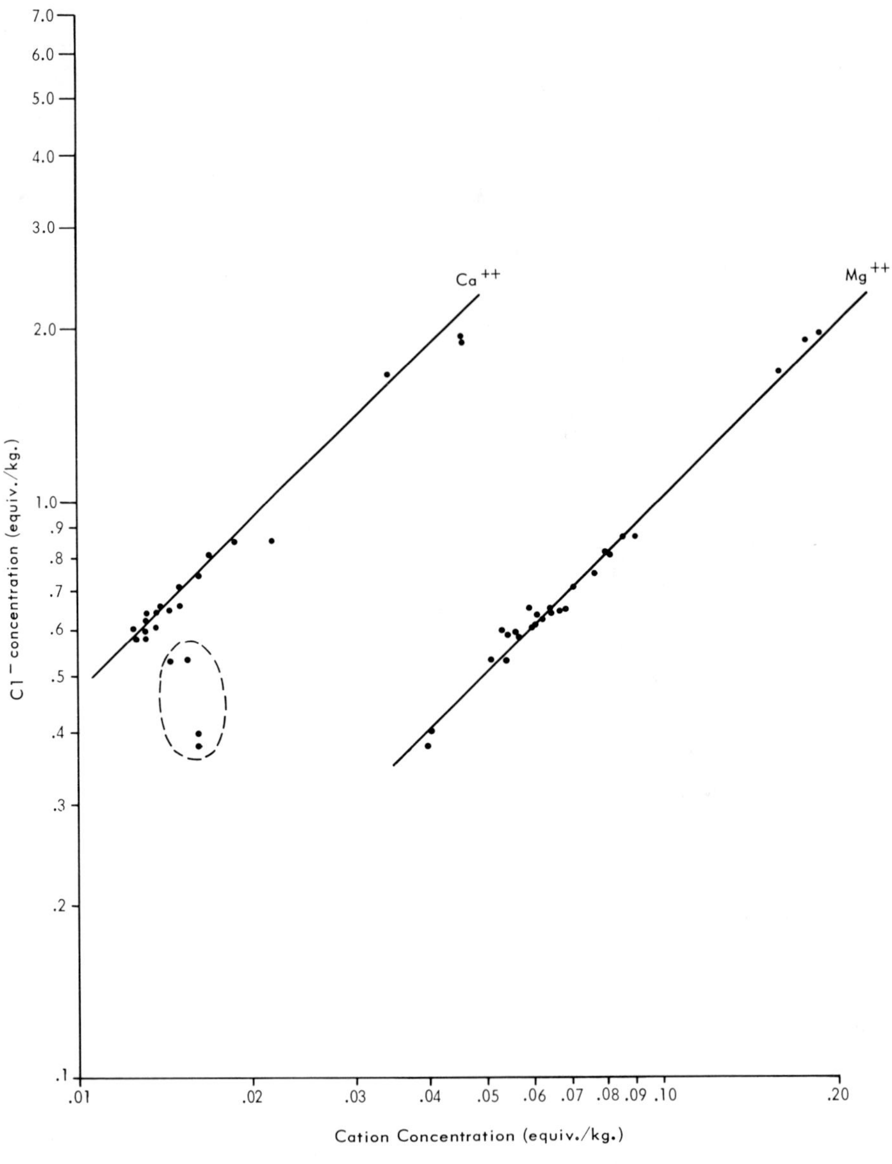

FIG. 16—Chemistry of island waters. Curves placed by inspection.

lack of good rhombic crystal shape. In fact, the presence of dolomite can only be inferred from the clotted texture and the bulk mineralogy. Selective sampling and X-ray analysis of different parts of the same crust indicate that the dolomite occurs mainly in the extremely fine-grained matrix, only replacing foraminifer tests in a few places. Furthermore, there is a direct correlation between percent matrix and degree of dolomitization in the bulk samples. No void-filling and cementing calcite as reported by Shinn et al (1965) was noted in these samples, but sampling may have been inadequate. Alternatively, the dolomite itself may act as a cement.

Ambergris Cay dolomite is somewhat distinct from that at other reported localities, in that the bulk of the material being replaced is magnesium calcite rather than aragonite. Chemically and structurally, the dolomite is non-ideal, with Mg/Ca ratio varying from $Mg_{40}Ca_{60}$ to $Mg_{46}Ca_{54}$, as determined by X-ray diffraction analysis (Goldsmith and Graf, 1958). Also apparent from X-ray patterns is the fact that "ordering" peaks are very weak or entirely missing. These characteristics indicate that the supratidal dolomite is very poorly crystallized, having excess Ca in the lattice and imperfect arrangement of the alternating Ca and Mg layers. It is similar to the early-forming, imperfect "protodolomite" of Graf and Goldsmith (1956) and to early diagenetic dolomite from Bonaire, Florida, and the Bahamas.

The consistent association of dolomitic sediment on Ambergris Cay with hypersaline supratidal ponds and intra-island lagoons forcefully suggests that the dolomitization is effected by concentrated water moving laterally outward from the ponds through the porous sediment or soaking downward through the sediment after each period of flooding. Dark algal mats commonly cover the sediment on the supratidal flat and probably facilitate evaporation of capillary moisture from the surface, further concentrating it interstitially. In the sabkhas bordering the Persian Gulf, an analogous situation, on a much larger scale, has been described by Kinsman (1964) and Butler (1969). Shinn et al (1965) suggested that periodic up-and-down movement of the saline groundwater table occurs beneath every mud flat in response to the tides, and "pumps" air in and out of the overlying sediment, thereby facilitating evaporation of capillary water from the surface. Either process offers a mechanism for moving saline water through the sediment and concentrating evaporative salts there. The high annual rainfall on Ambergris Cay would preclude preservation of the more soluble salts, halite and gypsum, in the sediment.

No samples of interstitial water were obtained from tidal-flat sediment, but numerous samples of water were taken from ponds and lagoons adjacent to dolomitic sediment. Discouragingly, these water samples have Mg/Ca ratios almost as low as that of nearby seawater (Fig. 16), although their salinity may be as high as 123°/₀₀. Water in completely restricted ponds apparently is diluted by rainfall or concentrated by evaporation, whereas water in interconnecting lagoons with slight tidal exchange is buffered against these changes. In either case, the cation ratios are unchanged. Where cation ratios are divergent (circled in Fig. 16), the waters are probably mixed with calcium-rich shelf-lagoon water because of their locations. Atwood and Bubb (1968) suggested that increased surface activity of carbonate sediment particles that are alternately wetted and dried, such as those in a supratidal-flat environment, may enhance the tendency of the grains to absorb magnesium from seawater (Berner, 1966). If this mechanism is operative on Ambergris Cay, there may be no need for contact of sediment with water having a high Mg/Ca ratio in order to effect dolomitization in the short time available.

Intra-island Lagoons

Environment—The intra-island lagoons of Ambergris Cay cover approximately one third of the island's surface. The larger lagoons range in depth from 4 to 6 ft (1.3-2.0 m) in their deep centers. Others may be nearly filled with sediment, or may have a shelflike, shallow area with a water depth of less than 2 ft (0.6 m). These partly enclosed water bodies have only limited exchange with the reef lagoon or Chetumal Bay through narrow mouths or sinuous tidal channels (Fig. 17).

Variable hydrography is characteristic from one lagoon to the next, or even within the same lagoon at different seasons. It results from

FIG. 17—Intra-island lagoons.

A. Large intra-island lagoon north of Reef Point, Ambergris Cay. View is eastward, with island's windward shore and reef lagoon in background. Water depth in center of lagoon is 4 ft (1.3 m). Baylike, shallow reach in foreground is only 1 ft (0.3 m) deep. Note arcuate shape of lagoon's leeward shore and less regular windward shore.

B. Shallow, highly restricted intra-island lagoons with only narrow tidal channels connecting them to larger water bodies. These lagoons are being filled with sediment as mud flats on their margins expand. Areas such as **a** and **b** are transitional between land and water; i.e., they are areas of lagoon that are filled almost completely with sediment.

(1) varying exchange between different lagoons and extra-island waters having different salinities and (2) seasonal fluctuation of high-rainfall and evaporative conditions in different-sized lagoons. The three largest intra-island lagoons (Fig. 3) illustrate this point well. North of Reef Point the large lagoon is open to exchange with the normal-marine environment only through a very narrow pass leading to Boca Bacalar Chico. West of Palmero Point, a second lagoon has very limited communication with Chetumal Bay through two tortuous channels that are barely maintained. San Pedro Lagoon, in contrast, has almost constant, though weak, circulation from the reef lagoon through a winding channel and with Chetumal Bay through a wide mouth. Local exchange with smaller, often hypersaline, very restricted water bodies further complicates the hydrographic pattern.

Chetumal Bay water west of the island varies from brackish to slightly hypersaline (24.0-37.4°/₀₀) near the island, in contrast to the normal-marine surface salinity of 37.2°/₀₀ measured east of Ambergris Cay. Intra-island lagoon waters on Ambergris Cay vary from 25.5°/₀₀ where recent rainfall runoff obviously affected the value, to 55.1°/₀₀ in a highly restricted arm of a larger lagoon.

It is not surprising, then, that the fauna and flora of these lagoons reflect their variable ecologic conditions. Invertebrate populations are almost exclusively molluscan and foraminiferal (Ebanks, 1967, p. 126). No attempt was made to study the distributions of the various living forms, but they are noted by Robertson (1963) to characterize habitats ranging from brackish mangrove swamps, to semibrackish bay and shoal areas, to normal-marine *Thalassia* beds. Qualitative observation in the field suggests that the lagoonal populations vary in specific composition with variations in hydrography. Likewise, the sand beaches bordering these lagoons vary in constituent composition with variations in lagoonal assemblages.

Wantland (1967) noted low-diversity, miliolid-dominant, foraminiferal populations in the Ambergris Cay intra-island lagoons, with major contributions of peneroplids, notably *Archais angulatus*. Lagoon faunas are characterized by relatively low taxonomic diversity, enormous numbers of individuals, and considerable intraspecific variability. *Thalassia* occurs as stunted, widely separated plants in most lagoons. Where there is fairly good exchange with marine water, as west of San Pedro or at Boca Bacalar Chico, this marine grass forms dense beds. *Cymodocea* is inconspicuously present with *Thalassia*. The dasycladacean algae *Batophora* and *Acetabularia*, the latter of which is partly calcified, colonize patches of bedrock or other hard substrate in the lagoons and locally form dense beds. Fragmental remains of these algae were not recognized in the sediment.

Sediment—Degree of exchange of water between intra-island lagoons and the bay or reef lagoon affects sedimentation both directly and indirectly. The greater the exchange with outside sources of sediment, the more material is transported into the island lagoons. Likewise, the greater the exchange with nutrient-bearing water, the more hospitable is the environment for organisms and the greater is the potential for adding skeletal debris to the sediment. Thus, in a lagoon such as that west of Palmero Point (Fig. 3), with very limited exchange, sediment is less than 1 ft (0.3 m) thick and is distributed in pockets on an otherwise bare rock bottom. In contrast, the well-ventilated lagoon west of San Pedro has at least 2 ft (0.6 m) of sediment, and commonly more, depending on depth to the pre-Holocene surface.

In intra-island lagoons, which rarely undergo tidal fluctuation, wind-induced currents are the most important mechanism of sediment transport. The fetch of prevailing winds from the east and northeast is only a few hundred yards (less than 1 km) in the largest lagoons, but these are never deeper than 6 ft (2 m). Evidently, wind waves are capable of disturbing and transporting sediment leeward, for a lens of shelly, winnowed sediment and a nearshore bar lend an arcuate shape to many leeward lagoonal shorelines (Fig. 17).

Deposits of modern sediment in intra-island lagoons are typically black or gray, fetid, extremely soft, shelly muds or muddy sands. These deposits are very rich in organic matter and reek of hydrogen sulfide when disturbed. The abundance of fecal pellets in lagoonal muds is evidence of the organic activity that flourishes there periodically. These elliptical or rod-shaped grains consist of silt- and clay-size carbonate embedded in an organic matrix. They rarely survive wet sieving and thus are not accurately represented in sediment thin sections. Lagoonal sediments have the consistency of mud despite the high weight-percentage of skeletal grains in some samples. Apparently, the very large amount of gelatinous organic matter helps to retain original interstitial water; thus there is little compaction of the sediment.

Though, as a whole, intra-island lagoonal deposits appear "massive" and heterogeneous, in detail numerous sedimentary structures are apparent. Irregular layers and mottles of texturally contrasting sediment (Fig. 18A) probably represent current-disturbed and burrowed material (Moore and Scruton, 1957). Graded beds within an otherwise massive deposit (Fig. 18B) suggest periodic storm mixing and redeposition (Sanders, 1965). In some places, however, apparent grading may simply reflect changing ecologic conditions and corresponding changes in suites of skeletal particles produced locally (Fig. 18C). The depositional fabric of such "pseudo-graded" deposits, with nonabraded skeletal particles "floating" in a finer matrix, suggests that there was simultaneous deposition of grains and matrix and argues against a storm-deposit origin.

Diagenesis—A process of pelletoid grain formation affecting intra-island lagoon deposits of Ambergris Cay is worthy of mention because of the common occurrence of pelleted limestones in the geologic record. Fecal pellets are only one type of pelletoid grain and cannot account for all of the material in this category. The grains referred to here are mud aggregates, which are subrounded to subangular, spherical or slightly elongate, 0.15 mm in diameter on the average, and composed of silt- and clay-size carbonate particles bound by a waxy organic matrix.

It was noted that intertidal deposits which form a narrow band at the shorelines of intra-island lagoons consist of laminated coarse and fine layers. Wavelets disturb and erode particles of this sediment. Skeletal sand is concentrated into very well-sorted lenses in the shallow water near shore, and the muddy layers, more cohesive and more difficult to erode, are "chipped away," forming small, soft mud-aggregate grains. These aggregates are microscopic analogs of the larger mud clasts reported by Ginsburg (1957) to have formed by waves cutting cohesive mud on the shores of Florida Bay keys.

On Ambergris Cay, these incipient pellets are suspended in a narrow zone of wavelet swash near shore. During occasional periods of high wind, these suspended particles are washed up on the supratidal flats, where they become semilithified by desiccation of the water contained in their amorphous matrix. The pellets may become a permanent part of the supratidal deposit or may be reintroduced into the lagoonal sediment by a later flood. Sediment consisting of particles of this type would compact into a packstone fabric on a supratidal flat, whereas in a lagoonal deposit they would more likely be part of a wackestone fabric (Dunham, 1962). These grains may be distinguished from fecal pellets by their lack of a microscopically recognizable aggregate nature. However, both mud aggregates and fecal pellets may be recrystallized to cryptocrystalline carbonate and thus may appear identical in rocks.

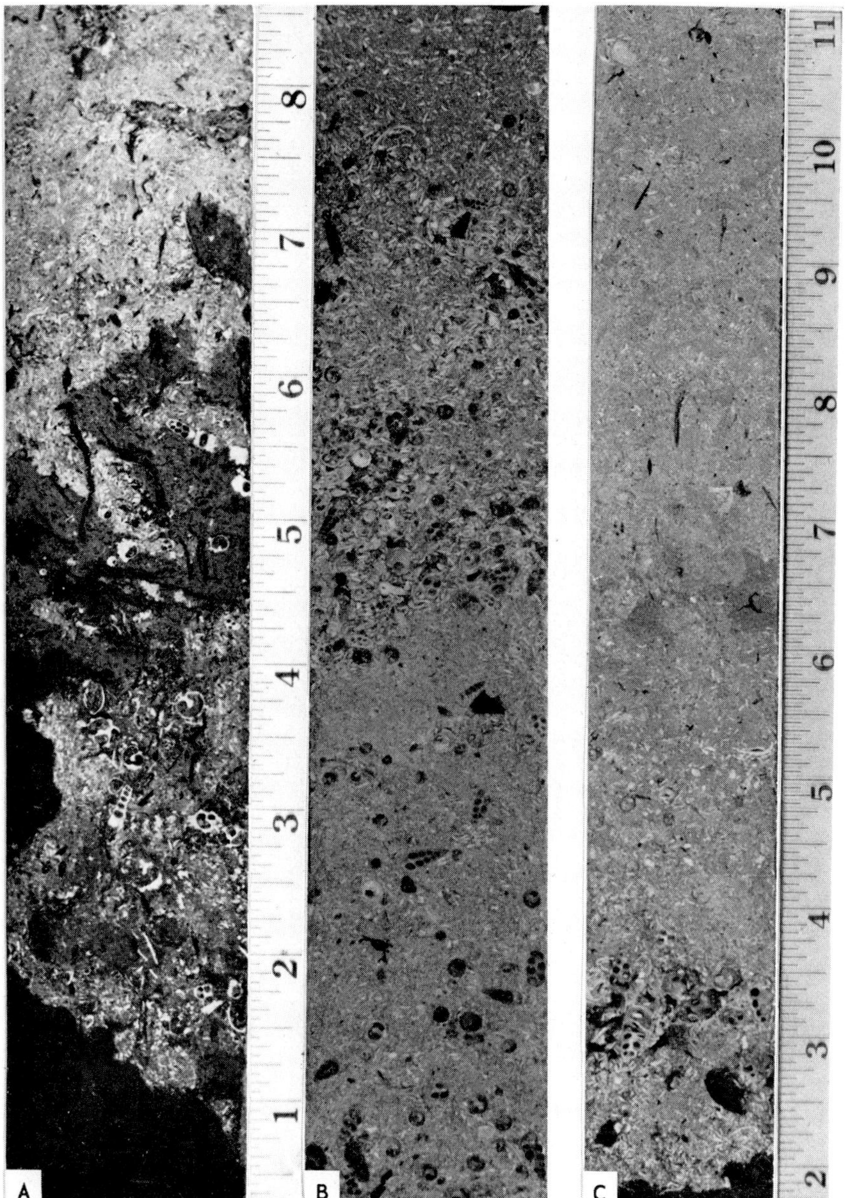

FIG. 18—Cores of intra-island lagoon sediment.

A. Core (incomplete) from location 485, showing textural and color mottling of shelly carbonate mud with organic-rich mud from peat layer at base. Mottled texture in this case was probably caused by burrowing organisms. This sequence is overlain by supratidal deposits. Scale is in inches.

B. Core (complete) from location 666, showing winnowed, graded, muddy sand (above 5-in. mark) in an otherwise structureless shelly mud. Graded sediment, composed of same constituents as sequence below, minus some of finer particles, is probably a product of current disturbance and redeposition. In this area, where good circulation has been maintained since modern sedimentation began, sediment shows no change in types or proportions of constituent organisms. Water is 1 ft (0.3 m). Scale is same as **A.**

C. Core from location 466 (incomplete). Increasing restriction of this lagoon is reflected in vertical change in sediment from molluscan-foraminiferal muddy sand at base to a miliolid-rich shelly mud above. Sequence is overlain by 10 in. (25 cm) more of restricted-lagoon sediment. Scale is in inches.

STRATIGRAPHY OF HOLOCENE SEDIMENT

Ambergris Cay sediments have evolved as a locally "regressive" sequence within an overall transgressive pattern. That is, this island has grown vertically and laterally through the accumulation of supratidal deposits within the larger framework of a progressively inundated shelf area. Recent sediment overlying the weathered limestone surface on Ambergris Cay comprises three widely recognizable units: (1) a stiff, gray, terrigenous basal clay; (2) a dark brown, fibrous peat; and (3) a complex of carbonate deposits representing several environments and including the present surface sediments. The similarity and proximity of these sedimentary units on Ambergris Cay to those described by Pusey (1964) and High (1966) from the shelf lagoon and mainland coastal complex, respectively, are taken as evidence of correlation among the three areas. With increasing thickness and maturity of the Holocene section, as on Andros Island, Bahamas, the "carbonate complex" becomes far more important than underlying units, and the areal and vertical distribution of various units within this complex becomes more predictable (Shinn et al, 1969).

Limestone Surface

During the Wisconsin glaciation when the entire Belize coastal shelf was exposed to subaerial weathering, a karstic surface was developed on the exposed limestone. On Ambergris Cay, local relief has had a pronounced effect on relative positioning of modern sedimentary environments. Depressions in the limestone have, in general, become intra-island lagoons, whereas intervening high areas received accretions of supratidal deposits. Completeness of the Holocene section correlates with relief on the underlying surface; the lowest units most commonly are present above depressions and absent on nearby highs, but the longest sediment cores are not everywhere the most complete. Weathered pebbles of this limestone form a "basal conglomerate" in Holocene sediment above the erosional unconformity.

Basal Clay

The gray to greenish-gray clay at the base of the Holocene sediment column is a distinctive unit consisting largely of terrigenous clay minerals with minor amounts of quartz sand and plant remains. This mud layer is noticeably more dehydrated than surface carbonate muds. It contains small, 2-3 mm, ferruginous concretions that commonly stain the clay orange or yellow.

The basal clay on Ambergris Cay was probably a coastal soil deposit. Its presence at higher elevations than the clay under the shelf lagoon, the absence of microfauna, the lack of evidence of resedimentation, and the presence of scattered black, bedded peat lenses and iron concretions all tend to support this contention. This deposit either never completely covered the underlying limestone or was stripped from higher surfaces by erosion, because weathered fragments of the limestone are found well up in the clay layer.

Mangrove Peat

A dark brown peat consisting mainly of woody fragments and fibrous remains of mangroves overlies the basal clay and is the oldest unit of the transgressive sequence. The contact between the two units is typically gradational but may be erosional. This peat differs little from the peat presently forming at the surface in certain areas, except that the older peat appears more compressed and contains less water.

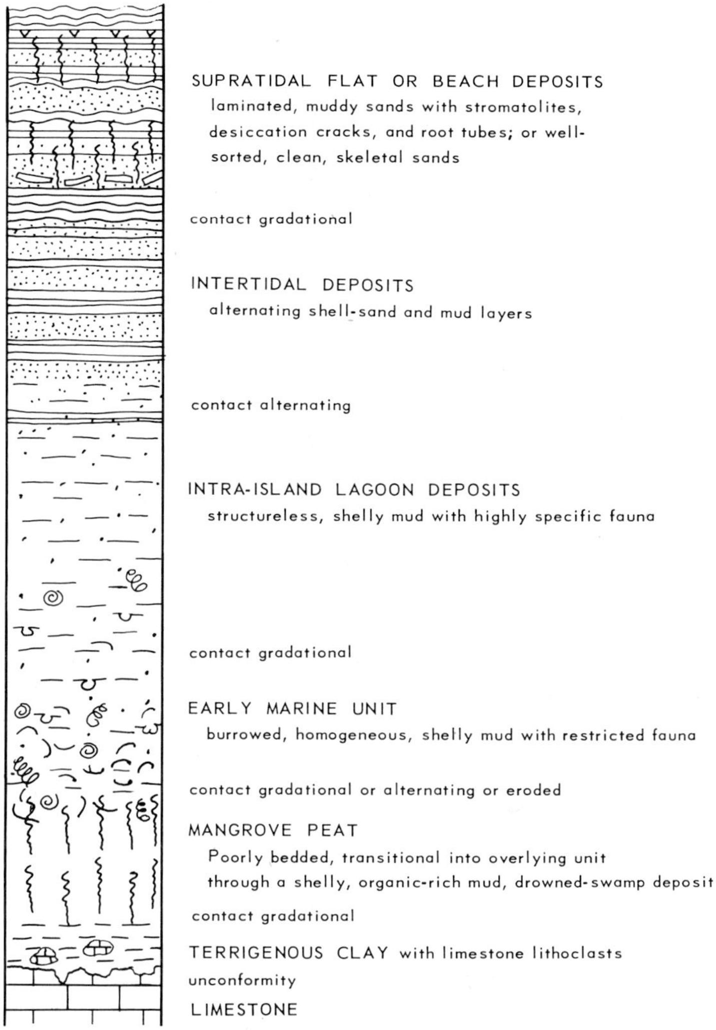

FIG. 19–Schematic log of an idealized "complete" core of Holocene section on Ambergris Cay, showing all sediment types in their proper vertical sequence. Only one core actually had this entire sequence of sediment types (Fig. 22). No vertical scale.

Holocene Carbonate Complex

With the widespread onset of marine conditions on the northern shelf of Belize about 5,000 years ago (Pusey, 1964), Ambergris Cay became more and more isolated from the mainland. The rising sea imposed a new set of ecologic conditions on the area, drowned the widespread mangrove swamps that had flourished early in the transgression, and was accompanied by increasing carbonate sedimentation. Essentially pure carbonate deposition is occurring at present. A trace of terrigenous material is derived from weathering of the underlying limestone or erosion of the basal clay in a few localities.

The transgressive history of the modern sea has left its record in this carbonate sequence, but Ambergris Cay sediments clearly demonstrate local shoaling and emergence of an island within this record. Though

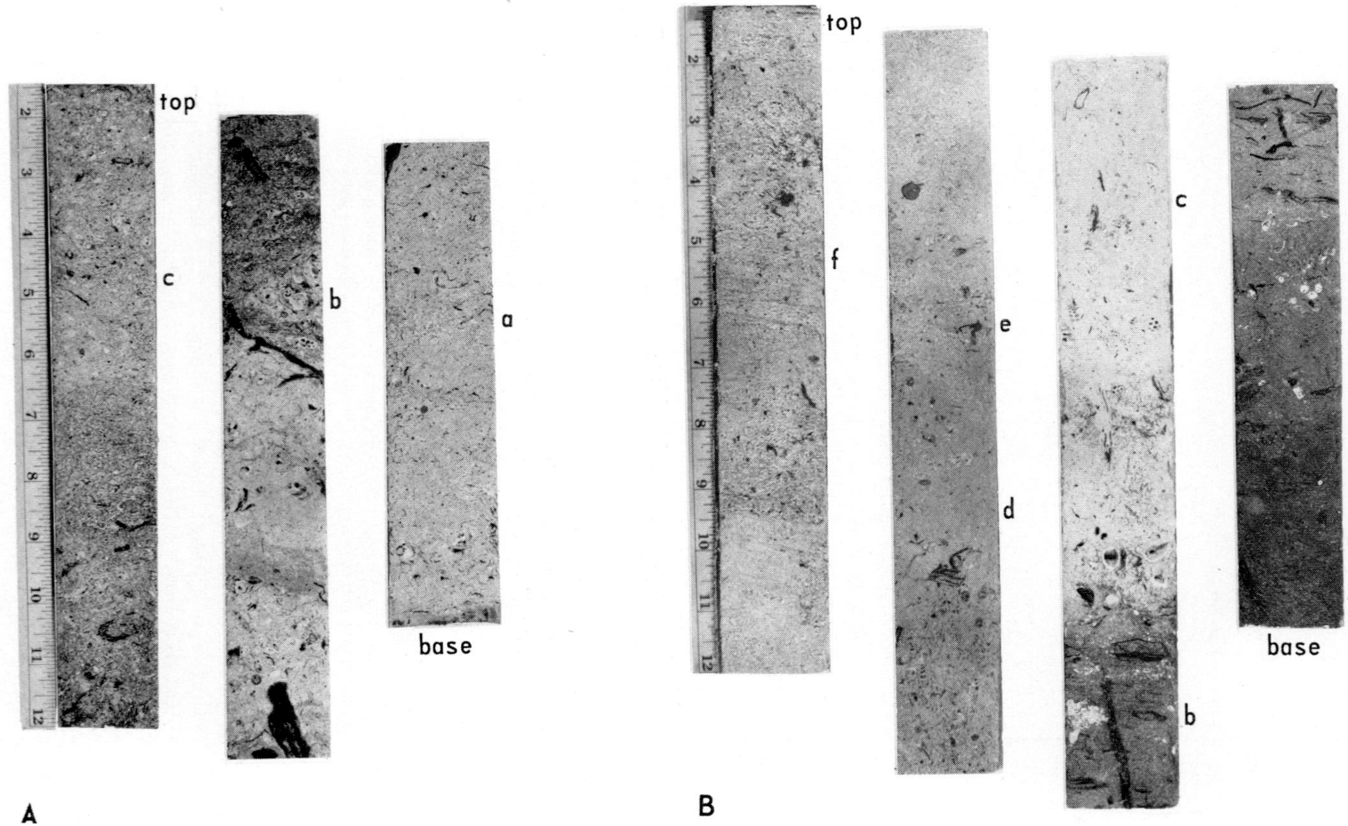

FIG. 20—**A.** Core from location 448, a supratidal flat, illustrating an *incomplete* sequence of island sediments. Rock was encountered at base of core. Sequence, from base upward, is: (**a**) intra-island-lagoon or shelf-lagoon shelly mud; (**b**) intertidal, interbedded shelly and muddy layers with some transported plant fragments; (**c**) supratidal-flat deposits of poorly sorted shell sand with much plant debris (swale deposit), overlain by better sorted shell sand with little plant debris or fine sediment (beach-ridge deposit). This typical sequence, from submarine to subaerial deposits, is found repeatedly around island. Scale is in inches.

B. Core from location 663, a supratidal flat, illustrating a *complete* sequence of Holocene island sediment (except basal clay), demonstrates sequence of events leading to island development during transgression. Sediments, from base upward, are: (**a**) mangrove peat, reflecting earliest marine incursion in area; (**b**) swamp deposit of muddy sediment with sparse fauna and much plant debris; (**c**) early marine, shelf-lagoon sediment of shelly mud with marine-grass root hairs in growth position; (**d**) intra-island lagoon, unlaminated, shelly mud; (**e**) intertidal, interbedded shell-sand and mud layers; (**f**) supratidal, laminated, muddy sands with good alignment of platy grains and inclined bedding (lee slope of low beach). Sequence reflects increasingly marine conditions (drowning of exposed rock terrain) followed by apparent regression or shallowing that results from growth of island. Scale is in inches.

very few cores contain the complete section, an ideal sequence of sediment types may be inferred from comparison of numerous incomplete cores (Figs. 19, 20).

Early marine unit—The oldest carbonate unit is a marine deposit. Its depositional fabric and sedimentary structures are identical to those of presently forming shelf-lagoon deposits (Pusey, 1964). This unit is not well bedded but, rather, has lenticular, irregular layers and textural mottling. These features denote current disturbance and resedimentation and organic burrowing (Moore and Scruton, 1957) that are scarce in supratidal deposits. This marine deposit ideally overlies mangrove peat, but may contact bedrock or clay where the peat is missing. In the latter case the contact is usually abrupt, but in the former there may be a gradation from peat to carbonate sediment as a result of extensive burrowing and intermixing of the two or of continuous sedimentation during the transition.

The lowest part of the early marine unit is typically a shell and rock gravel with very little mud matrix. The limestone fragments in this gravel are well rounded and probably were derived from nearby exposed limestone outcrops at the time of inundation. Above this gravel is a shelly carbonate mud with skeletal particles "floating" in the fine matrix. Platy particles in this sediment are not foliated, but are randomly oriented. This matrix-supported depositional fabric indicates simultaneous deposition of grains and matrix (Dunham, 1962). Any initial stratification has been destroyed by burrowing organisms. Another factor contributing to the homogeneity of this sediment is hydrographic restriction of the overlying water and consequent adverse ecologic conditions. In some areas, poorly sorted sediment results from accumulation of a variety of invertebrate remains; whereas in others, ecologic conditions limit the types of organisms that can survive, and a more well-sorted, less mixed sediment results from their local accumulation. This early marine deposit is not as variable from place to place as the later intra-island lagoon sediments in this aspect, presumably because of increasing restriction of the island's water bodies with time.

Intra-island-lagoon unit—Next in sequence above the early marine unit is either intra-island - lagoon sediment or supratidal-flat sediment. As mentioned, the former simply results from continuation of submarine deposition and usually is distinguishable, if at all, only on the basis of a decrease in taxonomic diversity of the organisms contributing shells to the sediment.

Sedimentation on Ambergris Cay is most active along the shores of intra-island lagoons, where skeletal particles and mud-size debris are continually transported, reworked, and deposited. These processes of shoreline accretion and stabilization are reflected in core traverses from a shallow lagoon, across a shoreline, and onto a supratidal flat (Fig. 21).

Supratidal-flat unit—Supratidal-flat sediments may be found overlying any of the older units—limestone, clay, peat, or lagoon carbonates. Ideally, transition from underlying beds is through an intertidal deposit, though this sequence is not present everywhere. Ambergris Cay supratidal deposits are extremely variable from place to place. Cores of sediment with stromatolitic layering, graded bedding, or massive structure commonly are taken within a few meters of each other, indicating the discontinuity of each of these types of features (Fig. 22).

Texturally laminated intertidal deposits interfinger with supratidal sediment in one direction and lagoonal deposits in the other and are between the two in vertical sequence. These alternating layers of shell sand and desiccation-cracked carbonate mud accreting to lagoon shores form the base on which supratidal deposits prograde lagoonward.

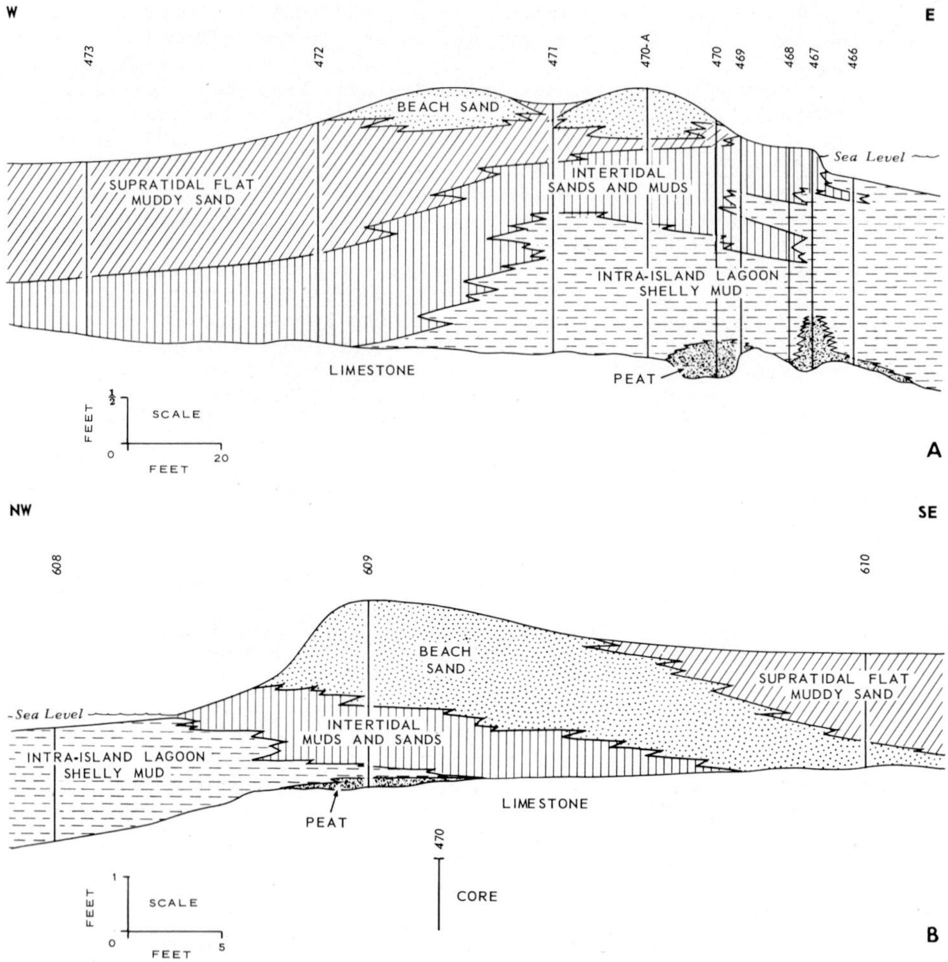

FIG. 21—Core traverses demonstrating progradation of intertidal and supratidal sediment over subtidal deposits on margin of intra-island lagoon. Note differences in scale in **A** and **B**.

The low beach ridges on the island's leeward shore and around intra-island lagoons usually form thin veneers or lenses of shell sand over intertidal or supratidal deposits. The shingled appearance of these winnowed peneroplid-foram and pelecypod-shell sands is quite distinctive in cores.

Reef-lagoon and windward-beach-ridge units—The descriptions given above are for the deposits west of the windward beach ridge. Only two cores taken in the marine environment east of the ridge penetrated to underlying rock. In both cores the sediment is unlaminated coral-algal sand and gravel very similar to deposits forming in the reef lagoon at present. There is no peat or terrigenous clay at the base of the Holocene section. The carbonate sediment is very poorly sorted throughout and contains a few indistinct burrow mottles. *Thalassia* rhizomes and debris are most apparent in the uppermost 6 in. (15 cm; Fig. 23). Very likely, the encroaching sea stripped any early clay or peat deposits off the island's windward limestone surface before much carbonate sand accumulated there. Subsequent deposits have been homogeneous in composition because the environment in which they accumulated has been relatively unchanging.

The small-diameter (2 in. or 5.1 cm) cores are an inadequate sample of the windward, current-swept deposits. Thus the possibility remains that current ripples and crossbedding characterize much of this sediment seaward of the *Thalassia* beds. Moreover, it is impossible to subdivide these windward Holocene sediments on the basis of the few incomplete cores taken there, and therefore no simple correlation between these and the very different leeward deposits can be made. The single core through the windward beach ridge contains no obviously marine beds. Transition

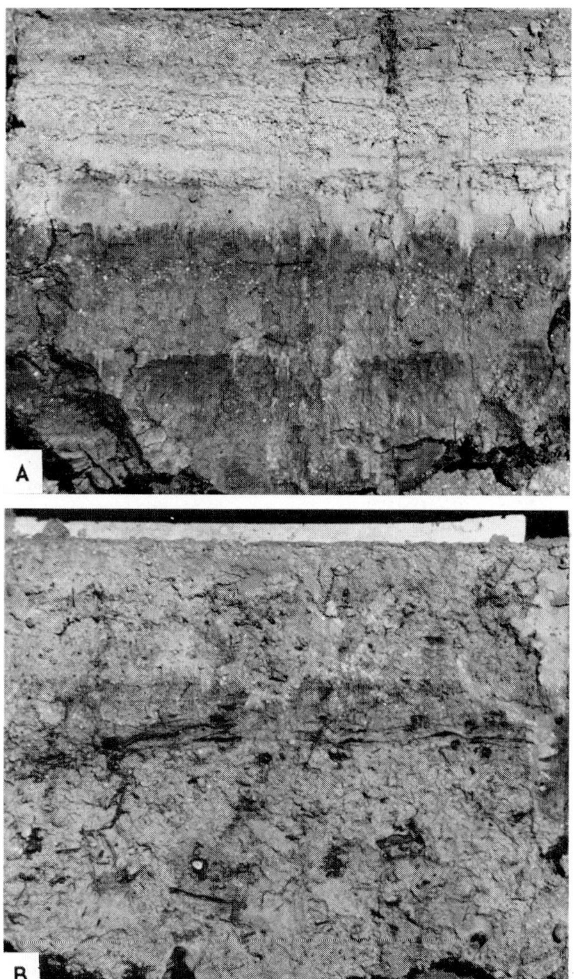

FIG. 22—Box cores of supratidal deposits.

A. Supratidal (light) sediments above intertidal (dark) deposits from location 471. Supratidal laminae are part of a low, raised rim on margin of a lagoon. They are more desiccated and cohesive than underlying intertidal sediment. Box core is 18 in. (46 cm) wide.

B. At location 472, a few meters from location 471, core emphasizes destructive effect of mangrove roots on lamination in somewhat finer supratidal sediment. Lower 6 in. (15 cm) of this box core are stratigraphically equivalent to intertidal deposits of **A**. Box core is 18 in. (46 cm) wide.

between the windward and leeward areas of the island is signified by interfingering of their sediments on the west side of the windward beach ridge (Fig. 24).

ORIGIN AND DEVELOPMENT OF AMBERGRIS CAY

Modern Islands

Ambergris Cay resembles many modern carbonate islands in its setting and physiography. Its situation on a shelf or bank edge is typical of most of the large islands in the Bahamas (Newell and Rigby, 1957), as is its isolation from terrigenous sedimentation. The barrier reef a mile or two windward of the island is common on modern carbonate banks; but some islands form on the leeward side of banks where no reefs grow. Indeed, even the foundation of older rock on which Ambergris Cay has been built has many counterparts in modern seas. This last point suggests a distinction among types of islands which has some bearing on their origin, surface area, and influence on surrounding sedimentation.

Many cays, or islands, on the southern Belize shelf form as storm deposits perched on reef flats and surrounded on all sides by deep water

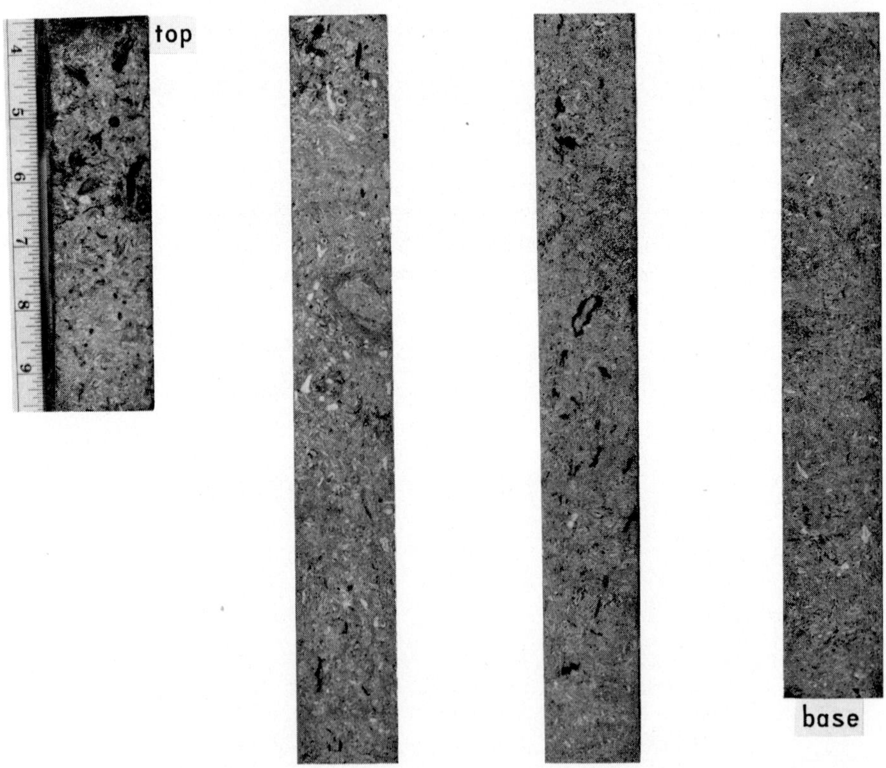

FIG. 23—Sediment core from location 383, in reef lagoon a few meters east (windward) of Ambergris Cay. Water depth is 1.5 ft (0.5 m). This core is typical of all those taken in reef-lagoon sediment. It consists, from base to top, of coralgal sand, with varied amounts of fine sediment and little plant debris except in top 8 in. (20 cm; reflecting dense *Thalassia* colonizing surface of sediment). Sediment is burrow mottled, but otherwise structureless. White, gravelly particles are fragments of coralline algae. *Halimeda* and coral fragments are also conspicuous. Scale is in inches.

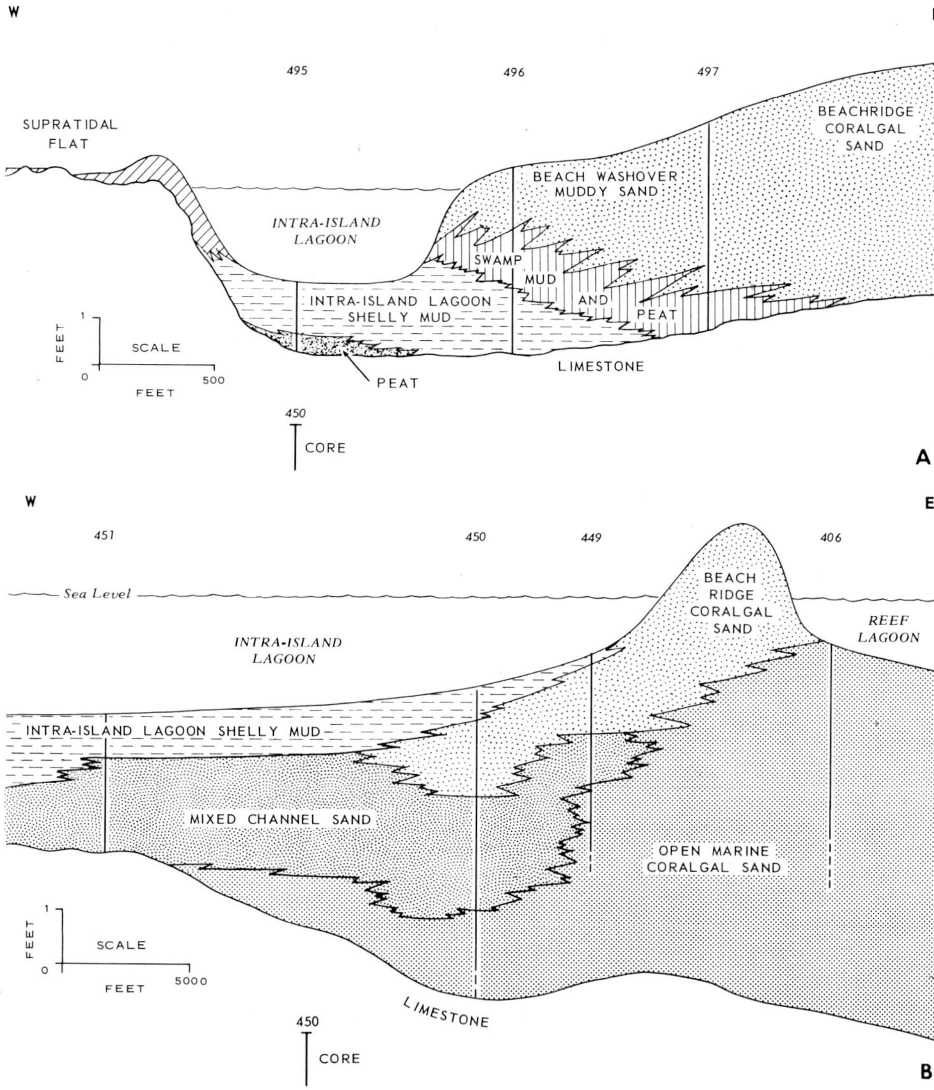

FIG. 24—Core traverses demonstrating transition between windward and leeward island deposits. Note difference in scale in two diagrams.

(80-200 ft or 24-60 m). These are analogous to other cays on isolated, steep-sided foundations, such as the atolls of the Pacific (Newell, 1956; McKee, 1959), the low islands of the Great Barrier Reef (Steers, 1929), or the islands of the Alacrán Reef (Folk and Robles, 1964). These "sand cays" have no great influence on the depositional environments around them; moreover, such cays are limited in areal extent to the size of the reef flat on which they form.

Shelf islands, which form by sediment accretion on a partly submerged topographic prominence, become a dominant and controlling feature in the sedimentation of an area. They provide shelter in which fine carbonate sediment can accumulate; they block the flow of bank water over the shelf-edge reef and encourage its growth; and they commonly form large bodies of distinctive supratidal and intertidal sediments. Ambergris Cay is one of these shelf islands, as are most of the Bahama Islands (Doran, 1955; Newell and Rigby, 1957), most of the other cays on the northern Belize shelf, and some of the Florida Keys (Ginsburg, 1956).

Mangrove cays have not been included in the shelf-island category because they are little more than mud mounds colonized by heavy vegetation. Few of the mounds have an appreciable subaerial surface or the varied environments of deposition that occur on shelf islands. Their sediments are more nearly a part of the submarine environment.

Also omitted here are islands that rise from fairly deep water to elevations of several hundred feet, with only fringing modern sediment; examples are the volcanic Pacific islands or the "continental islands" of the Great Barrier Reef lagoon (Steers, 1929). Such features may some day become shelf islands, but the chance is remote. These islands would be recognized in the geologic record as isolated areas of missing section (Schmidt, 1965), but they have little effect on surrounding deposits and develop no distinctive island sediments themselves.

Ambergris Cay

Ambergris Cay has formed as a result of the post-Wisconsin interglacial sea-level rise. Locally derived Holocene sediment has accumulated on and around a submerged topographic ridge developed on an older limestone unit. The configuration of this limestone surface is largely responsible for the present distribution of the island's sediments. Analogy with other modern islands suggests that Ambergris Cay is in a youthful stage of shelf-island development, and that the more mature conditions to be expected are predictable in terms of processes visible today.

Although there are locally incised stream channels and other irregularities, the main features of the limestone surface underlying the northern shelf are a wide, shallow basin under the present shelf lagoon and a raised rim, or ridge, near the seaward edge of the platform (Fig. 25). The ridge is broken at irregular intervals and islands have been built on each high segment. Figure 26 is a series of profiles of the Holocene sediment thickness across the island, demonstrating that the highs and lows of the present surface conform very well to the highs and lows of the limestone surface. The karstic nature of the weathered limestone surface is emphasized by the cratered appearance of the many shallow, round ponds which have formed over subcircular solution depressions.

During the post-Wisconsin sea-level rise and before approximately 6,000 years ago, the northern shelf of Belize was a coastal complex of marshes, shallow lakes, and sluggish rivers draining a low-relief Tertiary and Pleistocene, terrigenous and carbonate, inland terrain (Pusey, 1964; High, 1966; Wantland, 1967). Six thousand years before the present, the slowly rising sea level reached -16 ft (-4.9 m) and the first marine flooding of the shelf began. Coastal bays were formed and mangrove swamps developed over a large part of the area. The presence of a hilly limestone ridge near the east coast, or what is now the shelf edge, still prevented marine circulation across the shelf.

As the sea rose to -10 ft (-3.0 m), approximately 5,000 years ago (Pusey, 1964), more of the area was inundated, leaving the limestone ridge on the east coast as a long, narrow peninsula extending southward from the higher platform of the Yucatán mainland. Segments of this ridge became low, rocky islands seaward of the flooded, marshy bay to the west. Shallowly inundated areas of the peninsula, which is now Ambergris Cay, were covered by extensive mangrove swamps.

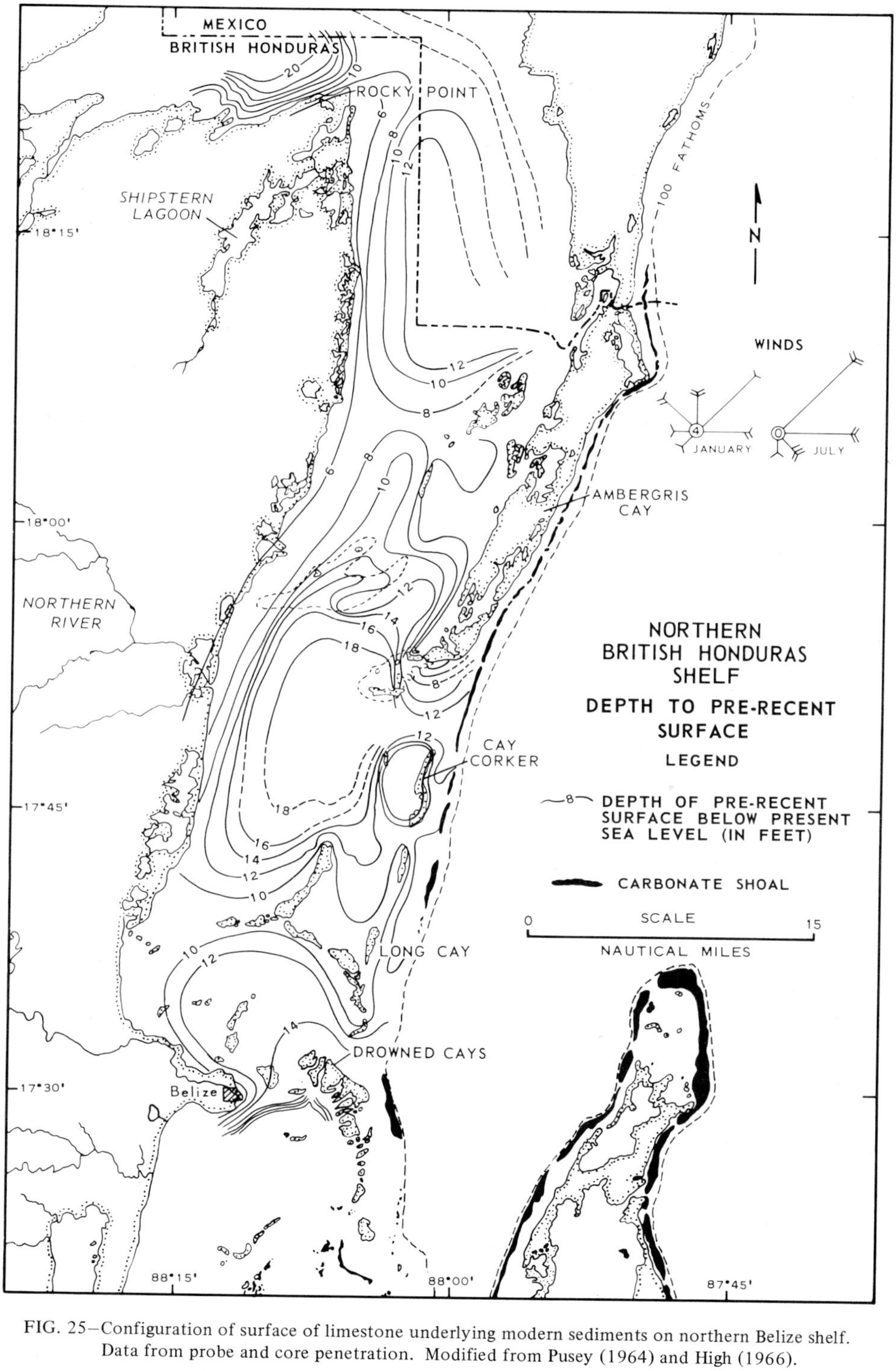

FIG. 25–Configuration of surface of limestone underlying modern sediments on northern Belize shelf. Data from probe and core penetration. Modified from Pusey (1964) and High (1966).

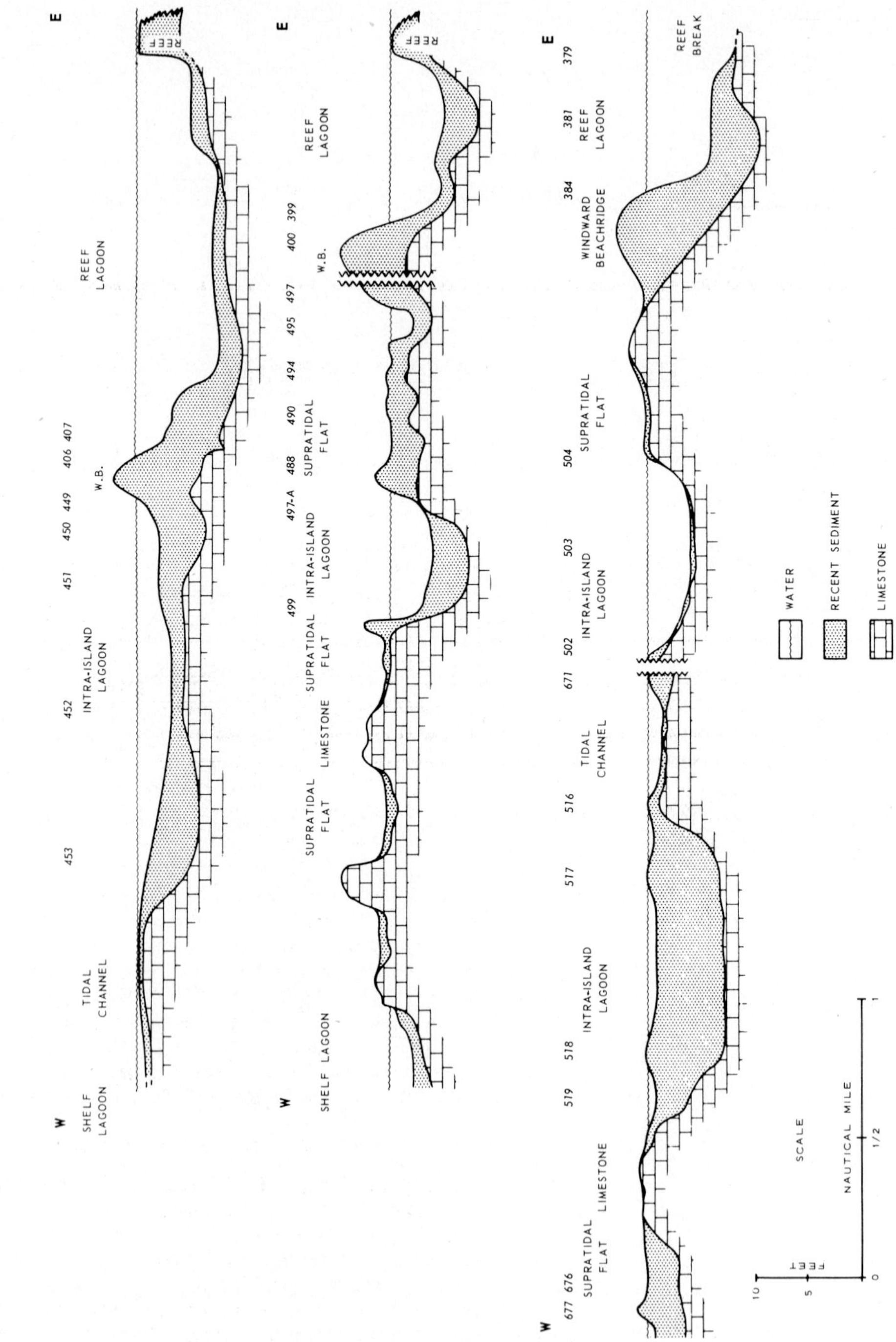

FIG. 26—Profiles of Holocene sediment thickness across Ambergris Cay. Numbers are sample localities where probe or core data provide control for cross sections.

Almost all deposition of terrigenous material in the coastal shelf lagoon ceased as the sea rose to -8 ft (-2.4 m) and coastal barrier bars trapped river effluent at the mainland shore. The mangrove swamps on the rocky peninsula spread, and occasional storms and increasing rainfall eroded most of the clay soil off the exposed areas of the limestone surface. A coral reef began to flourish on the continental-shelf edge at about the time that marine deposition became dominant on the inner shelf. Marine conditions seaward of the rocky peninsula promoted this coral growth as well as colonization of the reef lagoon by calcareous algae. Coralgal sediment from the growing reef and the reef lagoon accumulated on the shoaling rock bottom of the peninsula's eastern shore.

About 4,500 years before the present, continued flooding severed the surface connection of the rocky peninsula with the Yucatán mainland, and the Ambergris Cay area became a complex of rocky shoals with limited marine circulation. Reef-derived sediment continued to accumulate on the windward shore of the young island, impeding the circulation of water and enabling carbonate mud to accumulate in more sheltered leeward areas. The initial windward deposits grew and coalesced, forming a beach ridge with numerous tidal channels leading to tidal deltas on the leeward side. At the same time, intertidal and supratidal deposits were accumulating on the margins of rock shoals leeward of the windward beach ridge. Further accretion of muddy sediment around these shoals formed early intra-island lagoons.

As succeeding storms occluded and filled the remaining passages through the island, the early intra-island lagoons became increasingly separated from marine circulation and developed a characteristic fauna that contributed skeletal debris to the local environment. The island's sheltering effect on nearby submarine environments and on environments within the island itself resulted in the deposition of more and more muddy sediment in these areas and in the further growth of supratidal deposits at the expense of intra-island lagoons. The island's presence likewise prevented flow of brackish, less oxygenated bay water seaward and encouraged growth of the coral reef (Purdy and Imbrie, 1964).

In the present situation, Ambergris Cay strongly affects sedimentation in the surrounding environments and is itself the site of characteristic island sediments. The youth of the island is apparent from the character, variability, and thinness of the Holocene sediment. Its future development will probably result in the pattern exhibited in Andros Island, Bahamas (Purdy and Imbrie, 1964; Shinn et al, 1969). The land area of the island will continue to grow at the expense of that of the lagoons and ponds. Locally produced sediment from intra-island lagoons will become less available; and only seasonal storms will deliver sediment to the island surface. The isolation of intra-island lagoons will result in formation of smaller lakes that will derive most of their water from rain runoff. A few tidal channels will remain open but will develop natural levees. The leeward margin of the island will continue to grow westward by sediment accretion. Low, vegetated beach ridges will shelter the mud flats behind, and more fine sediment will accumulate in the lee of the large island.

Island Sedimentation

It is possible to construct a highly simplified model of island sediments, in relation to the marine deposits around them, by study of the Holocene sediment of Ambergris Cay. Figure 27 is an idealized and simplified cross section constructed from ideas expressed above. The cross section is somewhat artificial, as it is based on a composite of several

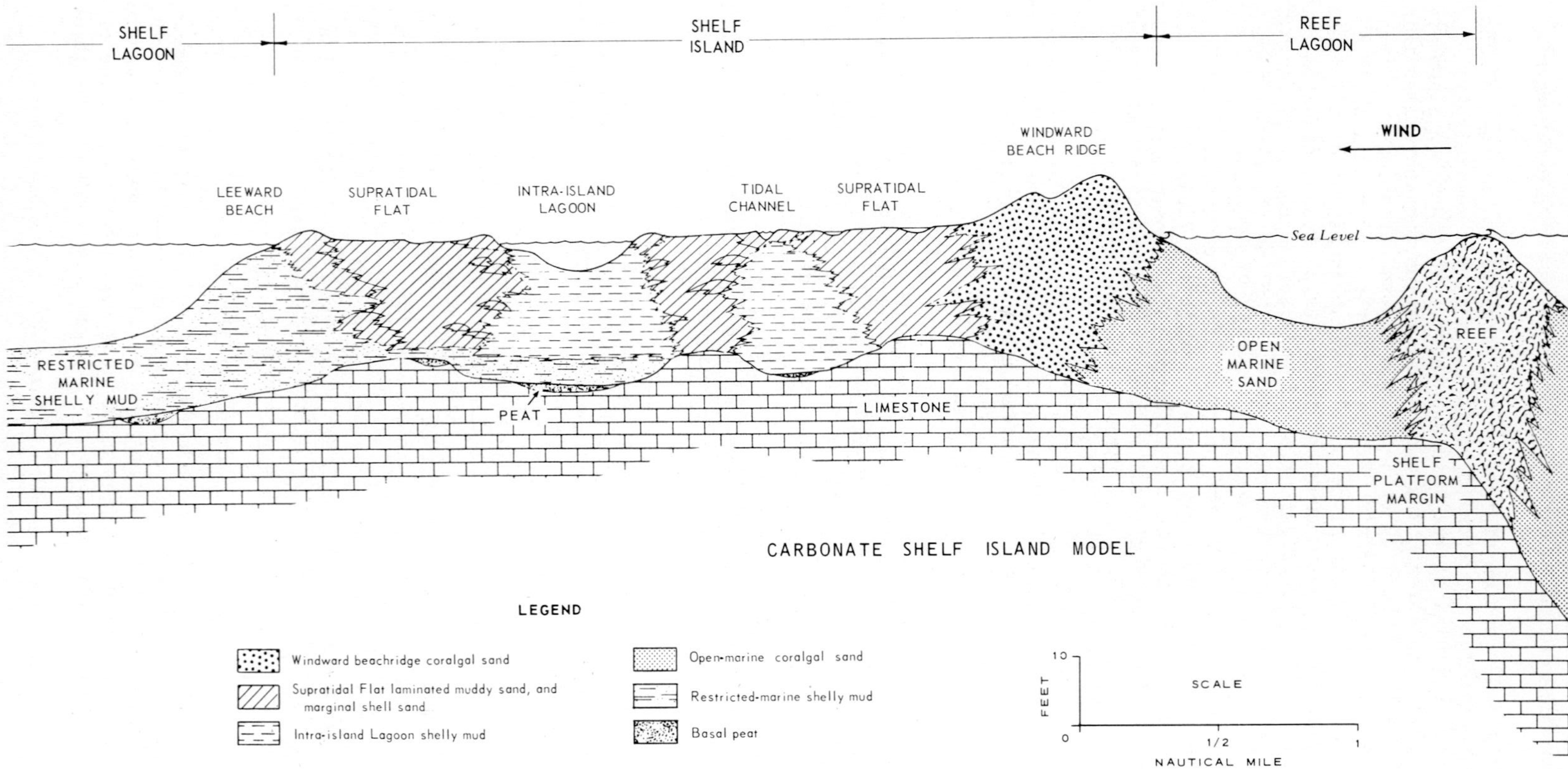

FIG. 27—Idealized cross section of shelf-island sediments and adjacent marine deposits. Cross section was constructed from composited core traverses and is somewhat artificial; also, sediment thickness is exaggerated (projected into future). See text for discussion.

core traverses across the island. The thickness of the Holocene sediment is exaggerated.

The schematic diagram emphasizes (1) localization of island deposits over a preexisting topographic high, on a rock surface or sediment buildup; (2) increasing restriction of circulation across the island site, with eventual supratidal deposition of sediment; (3) lateral accretion of island sediments and apparent "regressive" conditions in the midst of a marine transgression; and (4) variability of island deposits relative to associated submarine deposits.

The fate of island sediments during relative lowering of sea level is somewhat questionable. It might seem that, being subaerially exposed and no longer receiving accretionary sediment from surrounding areas, island deposits would be the first to be removed by erosion. This might be the case but, more likely, the supratidal deposits would continue to grow laterally over adjacent submarine deposits, provided that the rate of sea-level change were slow enough for sedimentation to keep pace. Continued progradation of island-margin deposits thus would be important in filling in the shallow shelf. Nevertheless, sediment bodies that are recognizable as shelf-island deposits are more likely to be preserved in transgressive sedimentary sequences than in regressive ones.

SUMMARY AND CONCLUSIONS

Islands are obvious and important features of modern carbonate-shelf seas. Their importance is threefold: (1) islands commonly influence the depositional environment of adjacent submarine areas two or three times their own size; that is, islands strongly affect the patterns of water circulation nearby and may cause the deposition of fine sediment that could not accumulate on a current-swept shelf; (2) islands comprise distinctive and potentially large bodies of carbonate sediment; that is, island sediments are unlike nearby submarine deposits, and they have the potential of increasing in areal extent with time; and (3) island sediments may be the key to the environmental reconstruction of large areas; that is, island deposits are worthy of recognition as a distinct facies because of their connotation of definite water depth and position near a shelf margin.

The great sequences of limestone in the geologic record contain many beds of recognizably shallow-water origin. If there were islands present in these ancient situations, as analogy with modern seas would suggest, recognition of the island sediments would be important to accurate paleoenvironmental analysis.

Ambergris Cay, a modern shelf island, has formed by accretion and supratidal deposition of sediment on and around a topographically prominent, karstic limestone surface under conditions of rising sea level. The configuration of the underlying limestone surface has strongly influenced the distribution of modern, unconsolidated sediments—supratidal deposits accumulate on the highs and lagoonal deposits form in depressions. In fact, the very position of the island near the shelf margin is controlled by the position of the underlying limestone ridge. The various environments of deposition on the island have developed in response to increasing restriction of wind and water circulation and consequent deposition of fine carbonate sediment.

The island's windward beach ridge is a long, narrow, coralgal sand body that has grown by storm deposition of sediment from the adjacent reef lagoon on a shoaling bottom. This ridge has provided a leeward, sheltered environment in which finer grained sediment could accumulate. It is distinguished from leeward deposits by the relative abundance of corals and calcified algae and by its elongate, lenticular geometry.

The supratidal-flat deposits are the island's most diagnostic sediments. These deposits are typified by their predominantly magnesium calcite mineralogy and by the relative importance of Foraminifera and mollusks among their constituents. These deposits are further characterized by their lateral and vertical variability, relatively high content of mud-size carbonate material, and assemblages of the following sedimentary structures: laminations, graded beds, stromatolites, desiccation cracks, intraformational conglomerates, and root casts.

Intra-island - lagoon sediments also contain a high percentage of mud-size carbonate material, but are nonlaminated and have very restricted faunal assemblages. Miliolids and peneroplids dominate the foraminiferal populations, and the specific composition of mollusk communities may differ from lagoon to lagoon as a result of the variable hydrography of these small water bodies.

Island sediments are also distinguished from submarine deposits by early diagenetic modifications consequent upon subaerial exposure. Particle dissolution, destruction, and weakening and early cementation are caused by meteoric water. No replacement of original skeletal mineralogy by calcite was noted in this study. In contrast, supratidal sediments associated with ephemeral, hypersaline ponds are dolomitized. The lithified crusts of dolomitic sediment form patchily distributed ledges in the sediment rimming these ponds.

Because the island's subaerial surface is expanding at the expense of its submarine surface, the sediments form a locally regressive sequence within a marine transgressive situation. This characteristic may be another key to their recognition in the geologic record.

REFERENCES CITED

Atwood, D. K., and J. N. Bubb, 1968, Distribution of dolomite in a tidal flat environment, Sugarloaf Key, Florida (abs.): Geol. Soc. America Spec. Paper 121, p. 10.

Berner, R. A., 1966, Diagenesis of carbonate sediments: interaction of magnesium in sea water with mineral grains: Science, v. 153, p. 188.

Billings, G. K., and R. C. Harriss, 1965, Cation analysis of marine waters by atomic absorption spectrometry, Gulf of Mexico coastal waters: Texas Jour. Sci., v. 16, p. 129-138.

Black, M., 1933, The algal sediments of Andros Island, Bahamas: Royal Soc. London Philos. Trans., ser. B, v. 222, p. 165-192.

Butler, G. P., 1969, Modern evaporite deposition and geochemistry of coexisting brines, the sabkha, Trucial Coast, Arabian Gulf: Jour. Sed. Petrology, v. 39, p. 70-89.

Butterlin, J., and F. Bonet, 1962, Les formations cénozoîques de la partie Mexicaine de la presqu'île du Yucatán, *in* 3d Caribbean Geol. Conf., 1962: Kingston, Jamaica, 45 p. (mimeographed).

Chave, K. E., 1962, The mineralogy of some Campeche Bank sediments: Texas A&M Univ., Dept. Oceanography and Meteorology, Rept. of A&M Projects 228, 287, and 286A.

Chayes, F., 1956, Petrographic modal analysis, an elementary statistical appraisal: New York, John Wiley & Sons, 113 p.

Dalrymple, D. W., 1964, Recent sedimentary facies of Baffin Bay, Texas: Houston, Texas, Rice Univ., PhD thesis, 192 p.

Deffeyes, K. S., F. J. Lucia, and P. K. Weyl, 1964, Dolomitization: observations on the island of Bonaire, Netherlands Antilles: Science, v. 143, no. 3607, p. 678-679.

——— ——— and ———1965, Dolomitization of Recent and Plio-Pleistocene sediments by marine evaporite waters on Bonaire, Netherlands Antilles, *in* Dolomitization and limestone diagenesis, a symposium: Soc. Econ. Paleontologists and Mineralogists Spec. Pub. 13, p. 71-88.

Doran, E., Jr., 1955, Landforms of the southeast Bahamas: Univ. Texas Pub. 5509, 38 p.

Dunham, R. J., 1962, Classification of carbonate rocks according to depositional texture, *in* W. E. Ham, ed., Classification of carbonate rocks: AAPG Mem. 1, p. 108-121.

——1969, Meniscus cement, *in* Carbonate cements: Bermuda Biol. Sta. Research Spec. Pub. 3., p. 314-318.

——1970, Keystone vugs in carbonate beach deposits (abs.): AAPG Bull., v. 54, p. 845.

Ebanks, W. J., Jr., 1967, Recent carbonate sedimentation and diagenesis, Ambergris Cay, British Honduras: Houston, Texas, Rice Univ., PhD thesis, 189 p.

Fischer, A. G., 1964, The Lofer cyclothems of the Alpine Triassic: Kansas Geol. Survey Bull. 169, v. 1, p. 107-148.

Flores, G., 1952, Geology of northern British Honduras: AAPG Bull., v. 36, p. 404-413.

Folk, R. L., and R. Robles, 1964, Carbonate sands of Isla Perez, Alacran Reef complex, Yucatan: Jour. Geology, v. 72, no. 3, p. 255-292.

Fuglister, F. C., 1947, Average monthly sea surface temperatures of the western North Atlantic Ocean: Massachusetts Inst. Technology and Woods Hole Oceanographic Inst., Papers Phys. Oceanography and Meteorology, v. 10, no. 2, 25 p.

Gebelein, C. D., 1968, Origin and growth rates of subtidal algal stromatolites, Bermuda (abs.): Geol. Soc. America Spec. Paper 115, p. 75-76.

Ginsburg, R. N., 1953, Beachrock in South Florida: Jour. Sed. Petrology, v. 23, p. 89-92.

——1956, Environmental relationships of grain size and constituent particles in some South Florida carbonate sediments: AAPG Bull., v. 40, p. 2384-2427.

——1957, Early diagenesis and lithification of shallow-water carbonate sediments in South Florida, *in* Regional aspects of carbonate sedimentation, a symposium: Soc. Econ. Paleontologists and Mineralogists Spec. Pub. 5, p. 80-100.

——et al, 1954, Laminated algal sediments of South Florida, and their recognition in the fossil record: Miami, Florida, Univ. Miami Marine Lab., Final report to the Natl. Sci. Found., 54-20, 33 p.

Goldsmith, J. R., and D. L. Graf, 1958, Structural and compositional variations in some natural dolomites: Jour. Geology, v. 66, p. 678-693.

Goreau, T. F., 1959, The ecology of Jamaican coral reefs; 1, Species composition and zonation: Ecology, v. 40, p. 67-90.

Graf, D. L., and J. R. Goldsmith, 1956, Some hydrothermal syntheses of dolomite and protodolomite: Jour. Geology, v. 64, p. 173-186.

High, L. R., Jr., 1966, Recent coastal sediments of British Honduras: Houston, Texas, Rice Univ., PhD thesis, 199 p.

Hoskin, C. M., 1963, Recent carbonate sedimentation on Alacran Reef, Yucatan, Mexico: Natl. Acad. Sci.-Natl. Research Council Pub. 1089, 160 p.

Hoyt, J. H., and V. J. Henry, Jr., 1964, Development and geologic significance of soft beach sand: Sedimentology, v. 3, p. 44-51.

Illing, L. V., 1954, Bahamian calcareous sands: AAPG Bull., v. 38, p. 1-94.

——A. J. Wells, and J. C. M. Taylor, 1965, Penecontemporary dolomite in the Persian Gulf, *in* Dolomitization and limestone diagenesis, a symposium: Soc. Econ. Paleontologists and Mineralogists Spec. Pub. 13, p. 89-111.

Kinsman, D. J. J., 1964, Recent carbonate sedimentation near Abu Dhabi, Trucial Coast, Persian Gulf: London, England, Univ. London, Imperial Coll., PhD thesis.

McKee, E. D., 1959, Storm sediments on a Pacific atoll: Jour. Sed. Petrology, v. 29, p. 354-364.

Moore, D. G., and P. C. Scruton, 1957, Minor internal structures of some Recent unconsolidated sediments: AAPG Bull., v. 41, p. 2723-2751.

Neumann, A. C., 1965, Processes of Recent carbonate sedimentation in Harrington Sound, Bermuda: Bull. Marine Sci., v. 15, p. 987-1035.

Newell, N. D., 1956, Geological reconnaissance of Raroia (Kon Tiki) atoll, Tuamotu Archipelago: Bull. Am. Mus. Nat. Hist., v. 109, art. 3, p. 311-372.

———and J. K. Rigby, 1957, Geological studies on the Great Bahama Bank, *in* Regional aspects of carbonate sedimentation: Soc. Econ. Paleontologists and Mineralogists Spec. Pub. 5, p. 15-79.

Ower, L., 1928, Geology of British Honduras: Jour. Geology, v. 36, p. 494-509.

Purdy, E. G., 1963, Recent calcium carbonate facies of the Great Bahama Bank; 1, Petrography and reaction groups; 2, Sedimentary facies: Jour. Geology, v. 71, p. 334-355, 472-497.

———1965, Diagenesis of Recent marine carbonate sediments (abs.), *in* Dolomitization and limestone diagenesis, a symposium: Soc. Econ. Paleontologists and Mineralogists Spec. Pub. 13, p. 169.

———1968, Carbonate diagenesis, an environmental survey: Estratto da Geologica Romana, v. 7, p. 183-228.

———and J. Imbrie, 1964, Carbonate sediments, Great Bahama Bank: Geol. Soc. America Guidebook for Field Trip No. 2, p. 37-43.

Pusey, W. C., 3d, 1964, Recent calcium carbonate sedimentation in northern British Honduras: Houston, Texas, Rice Univ., PhD thesis, 247 p.

Robertson, R., 1963, The mollusks of British Honduras: Proc. Philadelphia Shell Club, v. 1, p. 15-20.

Roehl, P. O., 1967, Stony Mountain (Ordovician) and Interlake (Silurian) facies analogs of recent low-energy marine and subaerial carbonates, Bahamas: AAPG Bull., v. 51, p. 1979-2032.

Romney, D. H., ed., 1959, Land in British Honduras, Report of the British Honduras land use survey team: Colonial Research Pub. 24, 327 p.

Russell, R. J., 1962, Beachrock: geological observations, *in* Guidebook, Field trip to Peninsula of Yucatan, 1962: New Orleans Geol. Soc., p. 64-72.

Sanders, J. E., 1965, Primary sedimentary structures formed by turbidity currents and related resedimentation mechanisms, *in* Primary sedimentary structures and their hydrodynamic interpretation: Soc. Econ. Paleontologists and Mineralogists Spec. Pub. 12, p. 192-219.

Schmidt, V., 1965, Facies, diagenesis, and related reservoir properties in the Gigas Beds (Upper Jurassic), northwestern Germany, *in* Dolomitization and limestone diagenesis, a symposium: Soc. Econ. Paleontologists and Mineralogists Spec. Pub. 13, p. 124-168.

Shinn, E. A., 1964, Recent dolomite, Sugarloaf Key, *in* Guidebook for Field Trip No. 1, South Florida carbonate sediments: Geol. Soc. America, p. 62-67.

———1968, Practical significance of birdseye structures in carbonate rocks: Jour. Sed. Petrology, v. 38, p. 215-223.

———R. N. Ginsburg, and R. M. Lloyd, 1965, Recent supratidal dolomite from Andros Island, Bahamas, *in* Dolomitization and limestone diagenesis, a symposium: Soc. Econ. Paleontologists and Mineralogists Spec. Pub. 13, p. 112-123.

———R. M. Lloyd, and R. N. Ginsburg, 1969, Anatomy of a modern carbonate tidal-flat, Andros Island, Bahamas: Jour. Sed. Petrology, v. 39, p. 1202-1228.

Steers, J. A., 1929, The Queensland coast and the Great Barrier Reefs: Geog. Jour., v. 74, p. 232-257.

Stoddart, D. R., 1962a, Three Caribbean atolls: Turneffe Islands, Lighthouse Reef, and Glover's Reef, British Honduras: Atoll Research Bull., no. 87, 151 p.

——1962b, Catastrophic storm effects on the British Honduras reefs and cays: Nature, v. 196, p. 512-515.

——1962c, Cambridge expedition to British Honduras, by J. E. Thorpe and D. R. Stoddart; 2, Physiographic studies on the British Honduras reefs and cays: Geog. Jour., v. 128, pt. 2, p. 161-173.

——1963, Effects of hurricane Hattie on the British Honduras reefs and cays, October 30-31, 1961: Atoll Research Bull., no. 95, 142 p.

Swinchatt, J. P., 1965, Significance of constituent composition, texture, and skeletal breakdown in some Recent carbonate sediments: Jour. Sed. Petrology, v. 35, p. 71-90.

Tebbutt, G. E., C. D. Conley, and D. W. Boyd, 1965, Lithogenesis of a distinctive carbonate rock fabric: Wyoming Univ. Contr. Geology, v. 4, p. 1-13.

Thorpe, J. E., and P. K. Bregazzi, 1960, Experiments and observations on the corals at Rendezvous Cay: Cambridge Exped. to British Honduras, 1959-60, Gen. Rept., p. 22-28.

U. S. Navy Hydrographic Office, 1952, Sailing directions for the east coasts of Central America and Mexico: H.O. Pub. 20, 5th ed., 274 p.

——1955, Atlas of pilot charts, Central American waters and South Atlantic Ocean: H.O. Pub. 160, 16 charts.

——1959, Punta Herrero to Cabo Gracias â Dios: Chart H.O. 394, 67th ed. (revised 1959), scale 1:936,430.

Vermeer, D. E., 1959, The cays of British Honduras: MA thesis, Univ. Calif., Berkeley, 127 p.

——1963, Effects of hurricane Hattie, 1961, on the cays of British Honduras: Zeitschr. Geomorphologie, N. F., v. 7, no. 4, p. 332-354.

Wantland, K. F., 1967, Recent benthonic Foraminifera in the sediments of the British Honduras shelf: Houston, Texas, Rice Univ., Phd thesis, 300 p.

APPENDIX A

Field Methods

The samples on which this study is based were collected during two summer field seasons, August and September 1964 and August 1965. Locations were determined from vertical air photos taken in 1945 by the U.S. Navy. The samples were taken along traverses across the island and in certain areas of detailed study, as shown in Appendix Figure 1. Stations occupied during field work numbered 193. From these, 170 bulk sediment samples and 110 sediment cores were collected. The bulk samples, which represent the upper 3-4 in. (7.6-10.2 cm) of sediment, were taken from a skiff, using a Petersen grab sampler, or by diving and scooping. On the surface of the island, sediment was collected by scooping it into a bottle. Bulk samples were stored wet in 1-quart glass jars until return to the laboratory. Cores, 2 in. (5.1 cm) in diameter, were taken and sealed in plastic tubing with a modified Ginsburg-Lloyd tripod-piston coring device.

Data on the thickness of the Holocene sediments were obtained from cores and by probing through soft sediment to hard rock with a metal rod. Rectangular box cores, 18-in. (45.7 cm) wide, were taken at several stations to study sedimentary structures. At 30 scattered localities, 2-oz (57 gm) water samples were taken for later analysis. In addition, measurements of water temperature, pH, and carbonate saturation were made at 12 localities.

Laboratory Methods

The parameters used in this study for characterizing the sediment are grain size, constituent particle composition, mineralogy, and sedimentary structures. The details of sample preparation and handling and analytical techniques are given by Ebanks (1967).

Grain size of the sediment is expressed as weight-percent of mud-size carbonate material, which was determined by wet sieving of subsamples. Constituent composition of the sediment fraction larger than 1/16 mm was determined by point counting of thin sections of plastic-impregnated, washed subsamples remaining after grain-size analysis. Purdy (1963) described this technique, based on the work of Chayes (1956), as applied to the study of modern carbonate sediments. A count length of 300-400 points was used in this study, giving an estimated accuracy within 8 percent of the reported value.

X-ray diffraction was used as a rapid means of analyzing a large number of samples for carbonate mineralogy to detect distinctive suites indicative of sedimentary or diagenetic environments (Chave, 1962; Neumann, 1965). Difficulty in obtaining reproducible results led to preparation of standard curves from Ambergris Cay carbonate sediment, relating weight-percent of the various carbonate minerals to peak area measured from a strip-chart record. On the basis of the scatter of points used to construct these curves and from reproducibility of duplicate analyses, the mineralogic compositions appear to be accurate to well within ±10 percent of the values reported.

Salinity of Ambergris Cay water samples was determined immediately upon return to the laboratory by a standard silver nitrate titration technique. Magnesium and calcium ionic concentrations of the natural waters of the island were determined by atomic-absorption flame photometry (Billings and Harriss, 1965).

Sediment cores were frozen in dry ice, sliced lengthwise on a rock saw, and cleaned for examination and photography (Geologic Research Center, Humble Oil & Refining Company; cited by Pusey, 1964, p. 50).

The map of environments of deposition of Ambergris Cay was constructed from high-level vertical air photos, low-level oblique, colored air photos, and field observations. The map is necessarily generalized and doubtless contains inaccuracies; nevertheless, it does portray the environmental complexity of this large island.

Petrography

Identification of the various sand-size particles was facilitated by reference to articles by Illing (1954), Ginsburg (1956), Hoskin (1963), Purdy (1963), and Pusey (1964). A more extensive discussion of Ambergris Cay sediment particles, with illustrations, was given by Ebanks (1967).

The skeletal particles in Ambergris Cay sediment comprise fragments of *Halimeda* (segmented green algae), encrusting and articulated coralline algae, corals, abundant soritid and miliolid foraminifers, various other less abundant foraminifers, pelecypod and gastropod mollusks, ostracods, echinoderms, bryozoans, sponge spicules, crustaceans, alcyonarians, and worm tubes.

Nonskeletal grains are common in island sediment. Ambergris Cay samples commonly contain the following widely recognized types: grapestone grains, which are present in only a few samples; mud aggregates, or fine carbonate bound into coarser particles, which are very common; fecal grains, which are not accurately represented because they survive wet sieving only if cemented; fragments of pre-Holocene limestone; fragments of dolomitized, cemented Holocene sediment; and scarce terrigenous sand grains, notably quartz.

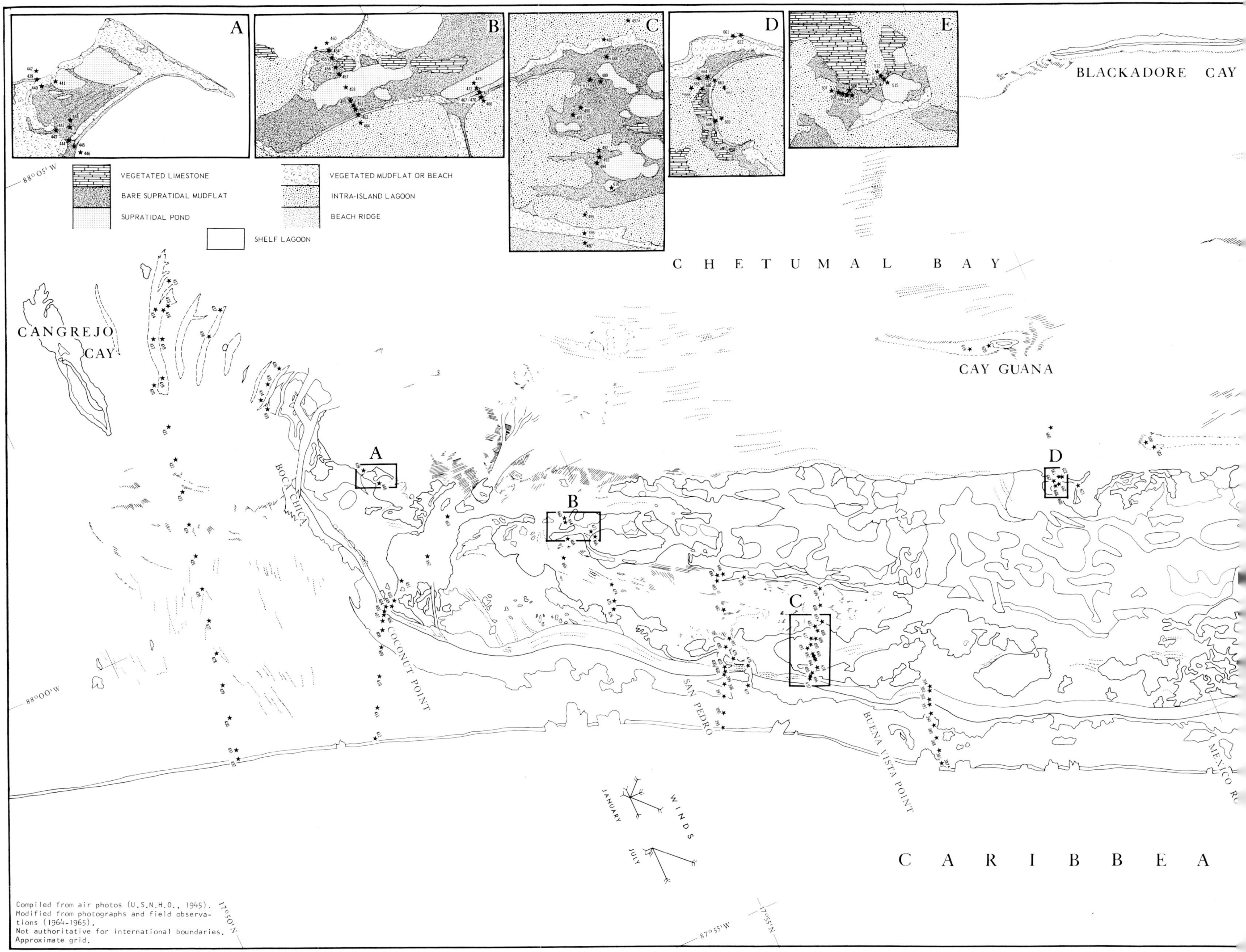

A
B
C
D
E
VEGETATED LIMESTONE
BARE SUPRATIDAL MUDFLAT
SUPRATIDAL POND
VEGETATED MUDFLAT OR BEACH
INTRA-ISLAND LAGOON
BEACH RIDGE
SHELF LAGOON
88°05' W
BLACKADORE CAY
C H E T U M A L B A Y
CANGREJO CAY
CAY GUANA
BOCA CHICA
COCONUT POINT
SAN PEDRO
BUENA VISTA POINT
MEXICO RO
88°00' W
JANUARY
JULY
WINDS
C A R I B B E A
Compiled from air photos (U.S.N.H.O., 1945).
Modified from photographs and field observations (1964-1965).
Not authoritative for international boundaries.
Approximate grid.
17°50' N
87°55' W
17°55' N

APPENDIX FIG. 1—Sample locations (overleaf).

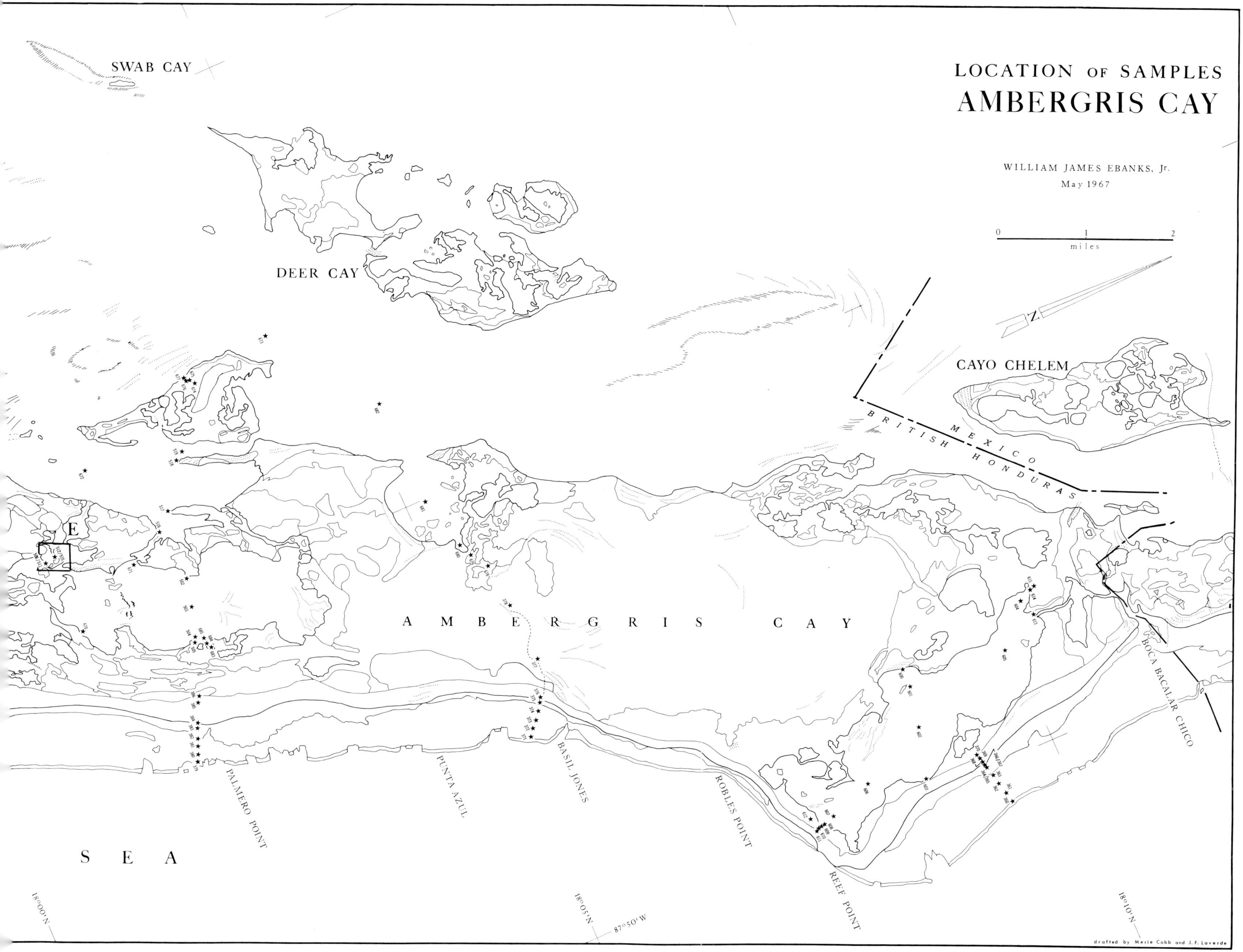
LOCATION OF SAMPLES
AMBERGRIS CAY
WILLIAM JAMES EBANKS, Jr.
May 1967
0
1
2
miles
N
SWAB CAY
DEER CAY
CAYO CHELEM
MEXICO
BRITISH HONDURAS
E
A M B E R G R I S C A Y
BOCA BACALAR CHICO
PALMERO POINT
PUNTA AZUL
BASIL JONES
ROBLES POINT
REEF POINT
S E A
18°00'N
18°05'N
87°50'W
18°10'N
drafted by Merle Cobb and J. F. Laverde

APPENDIX B

	Constituent Composition Volume Percentage of Fraction >1/16 mm																			Mineralogy Weight Percent of Carbonate				Calcite Comp.	
Colm.#	1	2	3	4	5	6	7	8	9	10	11	12	13	14	15	16	17	18	19	20	21	22	23	24	
Sample No.	Coral	Halimeda	Coralline Algae	Soritidae	Miliolidae	Other Foraminifera	Mollusks	Ostracodes	Miscellaneous Skeletal	Cryptocrystalline Skeletal	Total Skeletal	Composite Grains	Mud Aggregates	Limestone Fragments	Dolomitic Crust Fragments	Fecal Grains	Unidentified	Terrigenous Minerals	Wt. percent finer than 1/16 mm.	Aragonite	Mg-Calcite	Calcite	Dolomite	Mol percent $MgCO_3$ in Calcite	Sample No.
361	14.2	37.6	14.9	3.1	0	5.8	14.6	0	1.7	0	91.2	3.4	0	0	0	3.4	1.4	0	3.2	57	43	0	0	16	361
362	20.9	33.7	14.6	2.5	0	7.2	13.4	0.3	1.4	9.1	94.0	2.9	0	0	0	2.9	0.3	0	4.4	64	36	0	0	14	362
363	8.6	39.4	9.8	4.5	0	2.9	14.6	0	2.3	50.3	82.3	5.3	0	0	0	5.8	6.3	0	3.3	55	45	0	0	16	363
364	2.2	50.5	3.8	10.4	0	7.4	13.7	0.3	1.9	48.1	89.6	5.5	0	0	0	1.4	3.6	0	3.8	75	25	0	0	14	364
367	4.5	50.8	9.8	5.6	0	4.3	13.3	0	2.0	47.6	89.6	5.3	0	0	0	2.4	2.7	0	1.4	74	26	0	0	14	367
368	4.5	52.0	11.2	5.3	0	2.8	16.6	0	2.0	52.2	94.4	2.5	0	0	0	0.8	2.2	0	4.4	70	30	0	0	14,2	368
369	4.3	46.3	13.6	4.5	0	1.7	24.1	0	0.9	24.1	95.5	1.4	0	0	0	0.9	2.3	0	15.4	58	42	0	0	13,4	369
370	0	0.7	0	0	56.9	9.2	14.1	1.3	0	0.7	82.0	0	17.3	0	0	0	0.7	0	46.6	34	66	0	0	13	370
372	18.6	30.6	24.7	2.0	0	10.2	7.5	0.3	4.7	14.2	98.3	0	0	0	0	0.7	0.7	0	2.8	53	47	0	0	16	372
373	11.5	35.6	18.8	2.3	0.3	14.2	12.2	0.3	4.0	23.8	98.7	0.3	0	0	0	0	0.7	0	4.8	58	42	0	0	14	373
374	12.4	23.9	21.6	5.4	0	13.4	10.8	0.3	8.3	19.4	97.2	0	0	0	0	0.6	2.2	0	5.1	56	44	0	0	14	374
375	4.0	48.8	16.0	6.8	0	6.2	13.9	C.6	1.2	39.2	97.2	0	0	0	0	0	2.8	0	5.3	59	38	0	0	14	375
376	6.9	32.8	19.9	7.6	0.3	4.4	20.2	0	4.1	36.3	95.9	0	0	0	0	0.6	3.2	0.3	2.2	57	33	10	0	16,4	376
377	4.0	29.3	12.2	3.6	0.7	7.9	23.0	0	4.0	23.0	91.8	0	0	0	10.9	0	1.0	4.0	1.3	40	24	36	0	13,2	377
378	4.8	25.7	16.4	2.9	1.6	8.7	18.6	0	3.9	22.5	82.3	0	0	0	15.1	0	3.9	0	8.3	52	43	5	0	16,2	378
379	34.6	14.6	25.2	1.9	0	5.9	8.1	0	6.8	18.4	97.1	0	0	0	0	0.3	2.6	0	3.8	53	47	0	0	14	379
380	14.1	27.9	20.6	2.8	0	7.9	14.9	0.6	7.9	7.9	96.6	0	0	0	0	0.8	2.5	0	4.5	55	45	0	0	16	380
381	10.9	35.5	28.0	1.6	0	6.6	9.2	0	5.6	13.8	97.4	0	0	0	0	0	2.6	0	5.0	48	52	0	0	14	381
382	12.5	29.7	21.9	3.5	0	4.4	12.2	0	6.4	19.2	90.4	3.5	0	0	0	2.3	3.8	0	-	51	49	0	0	15	382
383	6.9	46.3	12.4	4.2	0	6.9	10.6	0	5.6	11.4	92.6	1.9	0	0	0	1.6	4.0	0	7.3	61	39	0	0	14	383
384	2.5	48.5	14.0	6.9	0	11.0	8.6	0	5.9	5.6	97.1	1.0	0	0	0	1.5	0.5	0	0	56	44	0	0	14	384
385	17.9	59.4	24.5	1.3	0	2.5	15.7	0	1.6	17.3	93.4	2.5	0	0	0	0.9	3.1	0	14.3	49	33	18	0	14,2	385
386	6.1	22.8	13.6	2.0	0	4.6	16.2	0	2.0	10.7	95.7	1.2	0	0	0	0.3	2.9	0	18.5	51	23	26	0	14,2	386
387-A	34.9	20.0	17.3	3.6	0	6.3	10.2	0.3	4.5	9.9	96.4	0.3	0	0	0	1.5	1.5	0	2.0	60	40	0	0	14	387-A
387	9.6	17.5	22.6	8.0	1.3	8.3	20.0	0.8	5.3	17.3	93.4	1.3	0	0	0	0.5	4.3	0	9.5	57	43	0	0	14	387
388	No sample																								388
389	12.9	43.1	18.9	5.3	0	3.6	14.2	0.3	1.0	29.1	98.0	1.0	0	0	0	0.3	1.0	0	1.7	54	46	0	0	14	389
390	2.6	63.0	5.0	2.3	0.9	7.9	6.4	0.9	9.6	10.5	98.5	0.3	0	0	0	0	1.2	0	19.6	58	42	0	0	16	390
391	5.2	48.8	9.2	6.1	0.3	1.5	19.0	0	3.7	31.6	93.5	1.2	0	0	0	2.1	2.8	0	7.7	58	38	0	0	14	391
392	16.2	40.8	16.2	0.6	0	0.9	12.0	0	1.2	43.8	88.4	1.2	0	0	0	5.8	4.6	0	0	67	33	0	0	14	392
393	4.3	19.8	15.6	2.4	8.5	3.6	14.9	1.8	10.0	22.2	81.2	1.5	9.5	0	0	1.5	6.4	0	44.0	45	55	0	0	13	393
394	6.4	31.4	18.1	3.8	0.6	1.6	29.2	0.3	3.2	35.9	94.6	2.5	0	0	0	0.3	0.3	0.3	19.6	54	29	17	0	13,5	394
395	22.6	15.0	25.8	1.9	0	5.7	15.9	0	9.6	6.7	96.5	0	0	0	0	2.2	1.3	0	0.7	48	49	0	0	14	395
396	23.0	22.4	20.2	3.8	0.3	6.6	15.1	0.3	1.9	14.5	93.7	0	0	0	0	4.7	1.6	0	0	55	44	0	0	13	396
397	17.1	29.5	19.9	4.7	0.6	7.2	12.1	0	3.4	22.7	94.4	0	0	0	0	4.0	1.6	0	0.4	67	33	0	0	14	397

APPENDIX B (Continued)

	1	2	3	4	5	6	7	8	9	10	11	12	13	14	15	16	17	18	19	20	21	22	23	24	
398	14.9	30.7	21.1	3.7	2.5	5.0	12.1	0	4.7	28.3	94.7	0.6	0	0	0	1.9	2.8	0	1.3	76	24	0	0	14	398
399	14.9	34.5	12.8	9.5	0.3	1.2	20.4	0	4.3	41.2	97.3	0.6	0	0	0	0.9	0.9	0	0	77	20	0	0	14	399
400	15.7	41.5	13.5	2.5	0	1.2	18.2	0	2.5	12.0	94.8	1.8	0	0	0	2.2	1.2	0	0	61	33	6	0	14,4	400
401	13.1	42.4	14.3	0.9	0.6	2.4	19.8	0.3	1.5	32.3	95.1	1.5	0	0	0	2.4	0.9	0	8.1	54	32	14	0	14,5	401
402	9.7	13.7	5.1	0.9	0	0	27.7	0	0.9	14.9	98.0	0	0	0	0	0.6	0.9	0.6	14.9	66	11	23	0	14,2	402
403	10.6	47.6	18.2	1.5	0.6	0.3	20.0	0	0.6	20.9	96.5	0.6	0	0	0	1.1	1.8	0	28.2	50	50	0	0	11	403
404	10.4	48.2	17.5	4.5	0	1.1	13.2	0	0.8	37.7	95.5	3.4	0	0	0	0.8	0.3	0	10.6	65	25	10	0	16,5	404
405	12.1	12.1	19.6	1.2	0	0.9	19.9	0	0.6	25.8	97.5	1.6	0	0	0	0.6	0.3	0	3.2	64	28	8	0	14,4	405
407	6.9	56.4	6.1	4.9	0	3.5	18.8	0.3	1.2	14.2	98.0	1.4	0	0	0	0.3	0.3	0	3.3	62	38	0	0	13	407
408	5.8	50.3	12.8	5.8	0	3.9	16.4	0	1.4	25.0	96.1	1.9	0	0	0	1.1	0.8	0	3.6	71	29	0	0	14	408
409	4.2	48.1	8.0	10.1	0.6	2.7	18.1	0	0.9	48.7	92.6	6.5	0	0	0	0.3	0.6	0	1.8	75	25	0	0	14	409
410	10.2	24.9	12.0	4.8	0	3.6	14.7	0.3	0.6	39.0	70.6	21.9	0	0	0	5.1	2.1	0	1.2	79	21	0	0	14	410
411	24.0	18.1	15.6	9.7	0	5.6	10.8	0	1.9	29.9	85.7	6.5	0	0	0	6.2	1.6	0	2.5	74	26	0	0	16	411
412	50.5	11.3	20.0	0.5	0	5.0	8.9	0	1.3	1.6	96.1	0.5	0	0	0	2.4	0	0	1.9	53	47	0	0	16	412
413	0	0	0	2.4	29.1	5.5	27.3	7.9	13.0	0.9	88.3	0	7.9	0.9	0	0	3.0	0	29.0	41	52	7	0	13,2	413
414	0	0.6	0.3	2.5	27.4	6.2	29.0	5.3	25.8	0.3	86.9	0	9.3	0	0	0	3.7	0	68.7	29	71	0	0	13	414
415	0	0	0	1.5	34.1	3.3	23.6	7.2	18.0	0.3	78.5	0	18.0	0.3	0	0	3.3	0	65.6	24	76	0	0	11	415
416	0	0.6	0	3.0	35.9	5.1	20.0	10.2	16.2	0.3	86.8	0	12.3	0	0	0	1.2	0	64.7	44	56	0	0	13	416
417	0	0	0	5.9	52.0	8.9	24.7	5.3	20.7	0.3	70.7	0	22.0	0	0	0	7.2	0	72.9	52	48	0	0	13	417
418	0	0	0	3.0	15.9	5.3	29.2	3.3	23.2	1.5	77.4	0	16.3	0	0	0	6.3	0	58.1	49	46	5	0	11	418
419	0	6.7	0.6	1.6	24.0	4.5	28.5	2.9	18.6	0.6	84.9	0	8.7	1.6	0	0.3	4.8	0	0.8	37	59	0	0	13	419
420	0	0.3	0.3	0	21.6	4.7	23.3	12.0	36.3	0	94.0	0	3.7	0	0	0	2.3	0	71.3	27	73	0	0	11	420
421	2.7	35.2	5.0	5.6	4.0	4.6	16.9	0	3.3	34.9	75.4	12.6	0	0.7	0	1.7	9.3	0.3	3.9	71	29	tr	0	14,2	421
422	1.5	47.0	2.7	12.0	2.1	1.5	16.2	0	2.1	52.4	82.9	12.0	0	0	0	0.3	4.8	0	2.8	76	20	tr	0	14,2	422
423	2.2	47.6	3.3	8.9	1.1	3.9	13.4	0	2.2	49.6	80.5	13.1	0	0	0	0	5.0	0.3	2.9	83	14	tr	0	14,2	423
424	3.1	43.8	5.2	4.0	2.4	3.6	5.5	0	2.4	45.3	67.6	22.9	0	0	0	0.6	7.0	0	1.4	87	11	tr	0	14,4	424
425	1.7	39.5	4.2	4.2	3.8	8.4	11.3	0	2.1	33.6	68.5	26.9	0	0	0	0	4.2	0	3.1	84	14	tr	0	14,4	425
426	1.7	44.9	5.1	2.6	0.4	2.9	14.9	0	0.4	49.2	65.4	29.1	0	0	0	1.3	3.4	0.9	1.0	86	12	0	0	11	426
427	6.1	34.2	4.6	4.9	1.5	3.9	15.2	0	1.9	47.9	69.2	19.4	0	0	0	2.3	8.0	0	0.5	85	12	tr	0	13,2	427
428	8.9	37.8	4.9	5.7	0.8	2.4	20.7	0	2.8	47.2	76.8	15.0	0	0	0	3.3	4.5	0	0.5	84	14	tr	0	13,2	428
429	11.6	31.0	5.8	6.6	1.2	3.3	25.3	0	1.7	39.0	80.2	13.7	0	0	0	1.7	3.7	0	0.5	84	14	tr	0	14,1	429
430	16.6	37.4	13.4	4.2	0	4.6	16.0	0	3.8	41.6	90.3	6.7	0	0	0	0.4	2.1	0	0.2	82	15	0	0	16	430
431	41.1	25.3	14.9	3.3	0	1.7	10.8	0	3.3	19.9	96.7	0	0	0	0	0	3.3	0	0.1	64	33	0	0	14	431
432	23.2	19.3	24.2	2.9	0.5	14.9	16.4	0	10.1	4.8	93.2	1.0	0	0	0	0	3.4	0	2.3	55	45	0	0	16	432
433	0	0	0	2.0	32.1	6.6	7.2	14.1	21.6	0.3	85.2	0	12.8	0	0	0.3	1.6	0	78.8	35	65	0	0	13	433
434	0.3	7.8	2.6	8.3	16.7	3.4	21.9	2.9	14.3	1.6	77.6	0	22.1	0	0	0	0.3	0	62.2	37	63	0	0	11	434
435	0	0.9	0.6	8.1	15.3	4.7	33.1	4.4	12.2	1.9	77.2	0	19.7	0	0	0.9	2.2	0	58.5	42	58	0	0	13	435
436	0	1.3	0	3.2	33.2	6.8	12.6	8.4	22.3	0.3	87.4	0	11.0	0	0	0.3	1.3	0	75.5	35	65	0	0	13	436
437	0	0	0	1.3	29.6	7.0	9.3	5.6	20.9	0	70.4	0	25.2	0	0	1.0	3.3	0	63.9	29	71	0	0	11	437
438	0	7.3	0	2.0	7.0	1.3	87.4	1.7	5.0	2.3	93.1	0	2.0	2.6	0	0	2.3	0	19.5	68	23	9	0	13,1	438
439	0	0.3	0	17.8	0	0.6	54.7	0	10.9	24.3	84.3	0	1.8	10.9	0	0.3	2.7	0	0.9	57	35	8	0	14,2	439
440	0	0.8	0	12.8	41.8	2.1	61.4	0	3.7	18.3	83.6	0	4.4	10.2	0	0	0.5	0	8.9	66	28	6	0	14,2	440
441	0	0	0	6.6	16.1	3.2	57.9	1.3	2.2	7.3	87.0	0.3	11.4	0.9	0.3	0	0	0	30.6	53	47	0	0	13	441
442	0	0	0	17.0	13.2	1.9	2.2	0.5	8.5	8.5	78.6	0	12.3	8.8	0	0	0	0.3	12.9	50	50	0	0	13	442
443	0	0	0	6.7	40.1	7.3	21.6	2.6	3.5	3.5	81.6	0	18.4	0	0	0	0	0	61.8	26	74	0	0	13	443

	1	2	3	4	5	6	7	8	9	10	11	12	13	14	15	16	17	18	19	20	21	22	23	24	
444	0	0	0	39.9	5.1	2.1	44.7	0.3	5.1	28.4	95.8	0	1.5	1.2	0	0.6	0.9	0	20.6	44	56	0	0	13	444
446	0	0	0	6.5	51.5	10.1	6.8	1.5	1.5	4.4	77.8	0	21.6	0	0	0	0.6	0	28.8	29	71	0	0	13	446
449	0.3	12.8	2.8	21.7	12.0	0.8	30.6	0	0.3	27.0	80.5	0.8	18.7	0	0	0	0	0	3.6	61	39	0	0	13	449
450	1.2	15.0	2.2	13.3	22.8	2.4	15.5	1.2	0.7	23.1	74.0	1.7	23.8	0	0	0	0.5	0	9.5	52	48	tr	0	14,4	450
452	0.5	18.7	0.8	7.9	14.4	3.3	33.6	0.5	8.7	10.8	87.3	0.8	10.6	0	0	0	1.1	0.3	17.5	54	46	0	0	13	452
453	0	2.7	0	33.0	18.0	2.3	17.0	0.7	8.0	18.3	81.3	0	18.0	0	0	0	0.7	0	20.7	30	70	0	0	13	453
454	0	0.3	0	71.1	2.7	4.9	10.4	0.3	3.8	58.6	93.2	0	6.5	0	0	0	0.3	0	0	7	93	0	0	13	454
455	0	0	0	69.5	2.7	3.0	14.0	0	3.7	53.4	93.9	0	4.6	0.6	0.6	0	0.3	0	1.3	16	84	0	0	13	455
456	0	0	0	13.1	27.0	2.2	15.8	1.4	1.4	10.4	60.7	0	38.3	0	0	0	1.1	0	32.0	15	85	0	0	11	456
457	0	0	0	8.4	32.2	3.0	7.2	1.8	3.0	6.9	56.1	0	43.6	0	0	0	0.3	0	43.4	19	81	0	0	14	457
458	0	0	0	3.2	35.7	7.4	11.3	1.3	1.6	2.6	59.8	0	39.2	0	0	0	0.3	0	46.3	6	94	0	0	13	458
459	0	0	0	5.3	41.5	8.5	6.0	1.6	0.6	5.0	63.5	0	36.5	0	0	0	0	0	67.8	4	96	0	tr	14	459
461	0	0	0	2.3	48.5	5.2	3.6	0.6	0.3	3.2	60.2	0	33.3	0.3	5.0	0	0.3	0	44.3	3	83	0	10	13	461
462	0	0	0	11.5	42.3	3.8	6.1	1.0	0.3	4.2	64.8	0	33.0	0.3	2.2	0	0.6	0	44.5	11	89	0	tr	13	462
463	0	0	0	17.3	2.8	1.9	28.7	0.3	2.8	29.0	84.3	0.3	13.6	1.2	0	0	0.6	0	8.0	29	71	0	0	14	463
464	0	0.3	0	3.1	25.8	2.0	11.0	0.8	0.8	11.0	43.6	0	56.1	0	0	0	0.3	0	8.9	16	84	0	0	14	464
465	0	0	0	20.5	31.1	2.9	5.9	2.6	2.1	6.2	65.1	0	34.3	0	0	0	0.6	0	53.6	26	74	0	0	13	465
474	0	7.0	0	18.5	32.4	2.1	15.2	3.9	6.1	5.2	85.2	0	13.3	0	0	0	1.5	0	39.5	33	58	9	0	13,2	474
475	0	15.6	3.5	8.1	13.3	4.0	32.3	2.3	9.2	17.9	87.9	0	11.5	0	0	0	1.2	0	21.5	42	53	5	0	14,2	475
476	2.5	41.8	3.9	8.6	3.6	0.6	34.8	0.3	1.7	39.6	96.9	0.3	1.4	0	0	0.8	0.6	0	4.8	53	47	0	0	14	476
477	11.5	44.2	8.8	7.4	0.3	3.9	18.0	0	1.5	26.3	94.7	0.6	0.3	0.3	0	2.7	0.6	0	3.8	52	48	0	0	14	477
478	7.1	47.4	12.0	3.4	0.6	4.6	16.3	0.3	4.6	16.6	95.4	0.6	0	0	0	1.8	1.5	0	12.8	56	44	0	0	14	478
479	9.3	51.8	8.2	0.8	1.1	2.3	19.4	0.3	5.9	7.6	99.2	0.3	0	0	0	0.6	0	0	16.2	67	33	0	0	14	479
480	0.3	16.9	1.4	4.4	8.3	6.1	37.6	1.4	23.8	7.5	99.0	0	0.3	0	0	0	0.3	0	41.4	42	37	21	0	13,2	480
481	0	1.0	0	1.3	34.9	5.1	9.3	2.2	44.9	1.0	98.4	0	1.3	0	0	0	0.3	0	63.9	25	75	0	0	11	481
482	0	7.7	0	21.4	5.5	1.1	48.4	0.3	0	5.5	94.0	0	0.6	4.7	0	0.6	0.3	0	15.6	51	49	0	0	13	482
483	0	0.9	0	25.9	4.9	2.0	35.6	0	20.1	35.3	89.1	0	5.5	4.9	0	0.3	0.3	0	7.3	42	58	0	0	14	483
484	0	0.3	0	53.6	3.4	1.5	24.1	0.9	10.9	38.4	93.4	0.3	2.6	2.0	0	0.3	0.6	0	3.2	36	64	0	0	16	484
485	0	0	0	32.2	12.8	3.0	38.7	0.6	9.3	24.8	95.9	0	1.6	1.9	0	0	0.3	0	19.8	51	49	0	0	14	485
486	0	0	0	63.1	1.3	0.3	19.8	0	8.2	49.6	91.0	0.6	4.8	2.6	0	0	0.8	0	13.5	29	71	0	0	14	486
487	0	0	0	49.4	3.3	1.2	23.5	0	13.6	51.2	89.2	0.6	7.2	1.8	0	0	0.3	0	20.5	18	82	0	0	13	487
488	0	0	0	2.8	49.4	7.2	5.3	0.6	2.2	0	67.5	0	23.4	0	5.6	0	0.3	0	50.1	2	74	0	24	14	488
489	0	0	0	1.7	22.7	2.9	9.6	0.6	3.5	0.6	41.3	0	13.4	0	45.3	0	0	0	13.6	tr	57	0	43	16	489
490	0	0	0	1.7	33.3	5.7	10.7	1.7	0	1.7	52.0	0	34.7	0.3	12.7	0	0.3	0	56.7	5	58	0	37	14	490
492	0	0.7	0	15.6	20.0	6.7	27.6	1.0	0.7	4.3	72.1	0	19.6	1.3	5.6	0	1.0	0.3	51.6	16	50	8	26	16,2	492
493	0	0	0	19.2	15.3	1.4	33.3	0.8	2.5	6.7	72.8	0.3	12.8	3.1	9.7	0.8	0.6	0	29.1	25	58	7	10	13,2	493
494	0	0	0	13.6	18.0	1.8	16.6	0.9	0.6	8.9	51.5	0	23.4	3.0	21.6	0.3	0	0.3	45.0	18	68	0	14	16	494
495	0	0.6	0.3	7.1	11.7	5.2	56.2	2.6	1.0	3.6	94.5	0	15.6	0	0	10.1	0	0	22.7	58	29	13	0	14,4	495
496	9.5	8.7	8.7	2.8	2.0	2.8	11.2	0.3	1.1	7.8	47.3	0	22.4	0	14.8	14.0	1.4	0	39.9	24	55	tr	21	16,4	496
498	0	0	0	23.2	14.1	3.1	48.0	0.7	7.9	2.0	93.8	0.3	3.9	1.6	0	0.3	0	0	39.0	59	36	5	0	13,2	498
499	0	0.3	0.3	37.9	7.9	3.6	25.1	0.5	15.4	41.8	91.0	3.6	3.6	0	0	0.3	1.5	0	7.8	20	80	0	0	13	499
500	0	0	0	58.5	2.2	0.6	12.1	0	0	54.2	73.1	12.7	13.3	0.3	0	0	0.6	0	5.0	5	95	0	0	13	500
501	0	0	0	60.8	0.9	0.6	6.6	0	0	61.1	69.0	13.2	13.2	0	0	0.9	3.8	0	0	4	96	0	0	13	501
502	0	0	0	11.7	0.5	3.3	48.1	0	0.3	15.0	86.5	0.3	3.3	9.2	0	0	0.5	0.3	2.3	61	25	14	0	13,2	502
503	0	0	0	12.5	18.6	7.3	42.8	0.5	1.0	10.5	82.6	0	4.2	13.2	0	0	0	0	29.3	56	36	8	0	2,2	503

APPENDIX B (Continued)

	1	2	3	4	5	6	7	8	9	10	11	12	13	14	15	16	17	18	19	20	21	22	23	24	
504	0	0	0	21.2	25.7	5.6	22.5	0.3	0.5	32.1	75.3	0.3	16.2	6.6	0	0	1.6	0	7.8	25	75	0	0	14	504
505	0	0	0.3	0.8	27.2	9.4	9.9	0.8	0.1	1.1	47.8	1.4	31.0	18.7	0	0	1.9	0.5	46.7	0	100	0	0	14	505
507	0	0	0	4.6	18.8	10.9	45.4	2.0	0	2.0	81.3	0	13.2	0.7	0	0	0.7	4.3	14.9	48	34	14	4	13,2	507
508	0	0	0	36.9	2.0	6.1	45.9	1.7	0	5.2	92.7	0	3.5	2.0	0	0.9	0	0.9	24.9	40	48	12	0	13,1	508
509	0	0	0	12.7	4.6	3.6	10.1	0	0.3	1.0	31.3	0	46.6	3.3	17.3	0.7	1.0	0	33.3	7	56	5	32	16,2	509
510	0	0	0	12.5	12.1	2.8	9.0	0	0	1.9	36.4	0	37.7	3.7	20.9	0.6	0	0.6	46.3	0	57	tr	43	17,1	510
511	0	0	0	2.7	2.3	1.0	0.3	0	0	0.7	6.3	0	8.7	1.0	83.7	0	0	0.3	21.9	0	2	0	97	13	511
512	0	0	0	3.1	40.7	17.4	18.9	3.4	0	0	83.2	0	10.9	2.2	3.7	0	0	0	59.3	18	42	17	23	14,2	512
513	0	0	0	10.8	12.7	7.3	3.2	1.0	0.3	2.5	36.0	0	5.7	1.3	36.3	19.1	0.6	1.0	46.9	0	54	9	37	13,1	513
514	0	0	0	7.8	28.8	10.5	34.6	1.3	0.3	0.7	83.3	0	3.6	3.3	6.5	1.6	1.3	0.3	54.1	5	67	13	15	14,2	514
515	0	0	0	5.8	41.8	14.1	22.8	5.1	0.3	0.3	90.0	0	3.9	0.3	4.5	0.6	0.6	0	44.9	18	50	15	17	13,2	515
517	0	0	0	60.5	3.0	2.2	13.7	0	0	41.0	79.0	0.5	14.0	5.1	0	0	1.3	0	3.5	22	78	0	0	13	517
518	0	0	0	38.9	17.3	4.0	11.6	2.8	1.3	24.4	75.9	0	23.4	0.5	0	0	0.5	0	27.8	0	100	0	0	11	518
519	0	0	0	41.1	9.7	1.9	6.3	0.3	0	23.8	58.9	9.4	31.0	0.6	0	0	0	0	16.0	0	100	0	0	11	519
600	0	0	0	23.7	33.4	7.3	26.6	0.5	0.7	27.1	92.0	0	6.8	0	0	0	1.2	0	9.8	25	75	0	0	13	600
601	0	0	0	43.6	16.9	1.3	33.5	0	0	44.7	95.3	0.8	1.6	0	0	0	1.6	0.3	8.4	44	56	0	0	14	601
602	0	1.7	0	37.0	31.4	5.9	8.1	0.8	0.8	11.8	95.5	0	4.5	0	0	0	0	0	34.3	37	63	0	0	13	602
603	0	0	0	33.5	5.6	0.6	55.5	0.6	1.5	36.8	97.3	0	1.2	0.3	0	0	1.2	0	62.9	29	71	tr	0	13,2	603
604	No Point Count																			30	58	12	0	13,2	604
605	No Point Count																			54	46	0	0	14	605
606	0	0	0	49.7	6.9	0.3	40.5	0	0.3	43.4	97.7	0	1.3	0.7	0	0	0.3	0	7.8	42	58	0	0	14	606
607	0	0	0	34.3	10.0	1.9	29.0	0.7	0	24.0	99.8	0	0.2	0	0	0	0	0	11.8	44	56	0	0	14	607
608	0	0.3	0	24.7	48.1	2.9	21.8	1.3	0	13.3	98.7	0	1.3	0	0	0	0	0	18.5	18	82	0	0	14	608
609	0	0	0	64.1	7.1	1.6	26.2	1.3	0	27.8	99.4	0	0.3	0	0	0	0.3	0	5.4	25	75	0	0	13	609
610	0	0	0	13.7	24.5	2.3	3.5	1.6	0.3	6.7	45.8	0	5.5	0	47.5	0.3	0.9	0	35.1	15	69	0	16	16	610
611	0	0.3	2.5	2.3	20.1	5.0	7.8	0.9	0.9	3.4	39.8	0.3	30.4	2.5	25.7	0	1.3	0	55.7	0	89	0	11	16	611
612	0	0.3	0.3	35.6	11.1	3.8	24.4	0.3	0.3	34.6	75.9	0	9.2	2.9	10.8	0	1.3	0	18.0	25	75	0	0	14	612
614	0	0	0	2.3	24.3	4.6	9.2	2.3	0.7	2.0	43.6	0	55.4	0	0	0	1.0	0	57.1	10	90	0	0	11	614
615	0	0	0	0	32.0	3.7	31.1	3.4	0.3	2.8	70.2	0	27.4	1.2	0	0	0.9	0.3	41.8	0	99	0	tr	11	615
616	0	0	0	2.1	39.3	8.8	31.7	0.9	0.6	2.1	83.3	0	8.2	0.6	7.0	0	0.9	0	13.0	20	61	5	14	14,2	616
660	0	0	0	20.2	12.6	3.0	7.8	0.8	0.3	23.1	44.4	18.3	36.8	0.3	0	0	0	0	14.1	10	90	0	0	14	660
661	0	0	0	30.7	0.3	0.9	19.0	0	0	33.3	50.9	25.0	10.4	13.4	0	0	0.3	0	0.1	18	82	0	0	13	661
662	0	0	0	3.4	25.1	21.2	2.9	1.5	0.2	10.5	39.5	7.6	52.0	0.2	0	0.2	0.5	0	20.5	4	96	0	0	13	662
663	0	0	0	23.3	6.3	0.5	7.7	0	0	17.2	37.8	23.8	36.2	1.6	0	0	0.5	0	10.2	9	91	0	0	14	663
664	0	0	0	19.3	11.2	1.4	5.3	0.8	0	14.6	38.1	12.3	40.6	0.3	0	0	0.3	0	30.6	3	96	0	tr	13	664
665A	0	0	0	10.3	7.5	1.1	10.3	0	0	11.5	29.3	5.0	10.9	0.3	54.5	0	0	0	1.8	15	77	0	8	13	665-A
666	0	0	0	19.2	12.2	3.2	13.4	0.2	0.2	17.4	48.5	11.9	36.3	1.2	2.0	0	0	0	20.8	24	76	0	tr	14	666
671	0	0	0	67.2	8.4	4.3	13.5	0.3	0	57.0	93.8	1.1	1.1	2.2	0	0	1.6	0.3	10.9	22	78	0	0	14	671
672	0	0	0	63.4	7.4	2.4	5.8	0	0	52.6	78.9	3.9	17.1	0	0	0	0	0	13.6	0	100	0	0	13	672
673	0	0	0	19.2	7.9	1.2	30.3	0	0.3	12.9	58.9	19.7	14.4	6.6	0.3	0	0	0.3	22.1	18	67	5	tr	11,1	673
680	0	0	0	23.5	21.9	1.9	29.6	1.3	0.3	12.8	78.1	3.2	17.6	0.8	0	0.3	0	0	35.4	23	77	0	0	14	680
681	0	0	0	0.3	22.9	0.9	65.1	2.1	0	0.3	91.3	0	7.8	0.9	0	0	0	0	12.6	77	23	0	0	10	681
682	0	0	0	11.3	6.1	1.8	11.6	1.2	0.3	9.1	32.2	14.9	52.3	0	0.6	0	0	0	17.2	11	89	0	0	11	682
683	0	0	0	2.7	12.1	4.8	73.0	0.3	0	11.2	93.3	0	5.8	0	0.9	0	0	0	18.0	54	46	tr	0	13,2	683

Paleoecology and Diagenesis of Pleistocene Limestone on Ambergris Cay, Belize [1]

GORDON E. TEBBUTT[2]

ABSTRACT The fauna in Pleistocene limestone of Ambergris Cay is homologous with the adjacent Holocene fauna; thus, ready comparison can be made between unaltered and diagenetic samples.

Four facies are recognizable in the Pleistocene limestone: (I) a reef-crest facies; (II) a backreef facies; (III) a shelf-lagoon facies composed of outer, middle, and inner shelf zones; and (IV) a mudbank facies. The first two facies and the outer shelf zone correspond closely to equivalent Holocene facies on the Belize shelf; however, unlike any nearby Holocene deposits, the Pleistocene middle shelf zone is oolitic. Facies I and II are biomicrites, III is a biopelmicrite, and IV is a dismicrite.

Aragonite persists in corals, mollusks, *Halimeda*, tunicate spicules, pellets, and ooliths, where not obviously replaced by calcite. Remarkably, magnesian calcite is retained in skeletons of encrusting algae and Foraminifera.

Skeletal materials show examples of four general categories of diagenetic alteration: (1) solution—evident in many skeletal and nonskeletal grains; (2) precipitation of calcium carbonate as drusy rims or coarse sparry mosaics of calcite, or as syntaxial overgrowths of the same mineral composition and optical orientation as the grains upon which the overgrowths are based; (3) replacement of aragonitic gastropods by calcite along a jagged "front," probably accompanied by solution and deposition on a minute scale; ghosts of primary structures typically remain, indicating absence of any major intermediate void stage; replacement of *Halimeda* and of corals such as *Montastrea annularis* occurs after occlusion of internal pores by sparry calcite; (4) recrystallization (i.e., alteration of crystals without change in mineralogy)—evident in a few pelecypods, where local patches of shell have been altered to coarse transverse blades of aragonite in which ghosts of primary structures may persist.

The limestone was cemented by a combination of interstitial precipitation of drusy and sparry calcite and recrystallization.

INTRODUCTION

Current usage of the term "diagenesis" includes all the alteration effects to which a sediment is subjected during or after deposition; it includes only processes considered to be of lower intensity than the lowest of metamorphic grades (Ginsburg, 1957, p. 80). Early diagenetic processes such as burrowing and boring may contribute important clues to the depositional environment; later diagenetic and lithification processes involve major changes in chemistry, mineralogy, or texture, and tend to obscure the primary sediment properties.

Diagenesis implies alterations of primary constituents of a sediment such that it is brought toward chemical and physical equilibrium with the local environment. The major minerals constituting modern carbonate sediments, aragonite and magnesian calcite, are metastable at or near normal earth-surface conditions; they tend to be removed or redistributed after deposition, in contrast to the preservation and/or emplacement of relatively stable calcite.

Where diagenesis has masked or obliterated primary textures or structures, recognition of its effects is essential in interpreting paleoenvironments on the basis of observed constituents. Throughout the last

[1]Manuscript received, June 20, 1972.

[2]Geologist, Aquitaine Company of Canada, Ltd., Calgary, Alberta. This paper is extracted from a doctoral thesis submitted to the Department of Geology, Rice University, Houston, Texas, in June 1967. The research was supported by the American Petroleum Institute (API Research Project 71-A) and by the Department of Geology, Rice University. Fellowships from the American Chemical Society, Rice University, and the National Research Council of Canada enabled completion of the study. The author is grateful to E. G. Purdy for suggesting the problem and for encouragement and aid during field work. Thanks are also extended to J. L. Wilson, W. J. Ebanks, L. R. High, and the many others who offered willing advice and useful criticism, and to Chevron Standard Ltd. and Banff Oil Ltd., Calgary, for aid in drafting and reproduction.

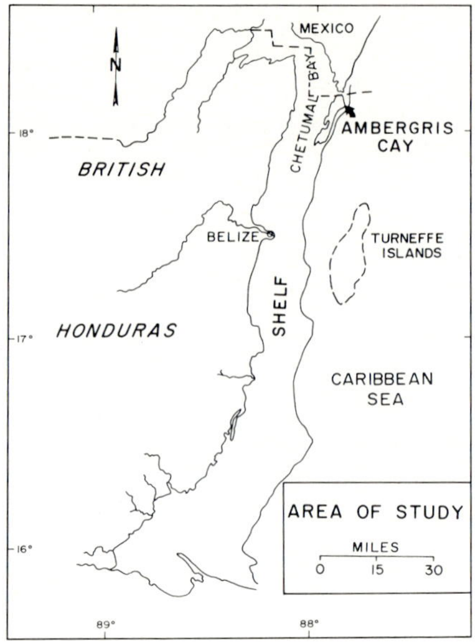

FIG. 1—Map of study area.

decade, increasingly detailed studies of modern carbonate environments have provided a basis for recognition of a great variety of ancient environments. In relating ancient deposits to modern sediments, and in rendering each more useful and translatable in terms of the other, one must determine not only what and how carbonate sediments are being formed and distributed today, but also which of these assemblages will be preserved in the rock record. The key to a valid interpretation is the recognition of the changes undergone by sediments before, during, and after lithification, and how these changes may be interpreted and anticipated.

Examples of intermediate stages illustrating partial alteration of carbonate sediments and limestones are scarce. The rocks involved in the present study not only show several varieties of diagenetic alterations observed in ancient limestones, but also are so young that many of the alteration processes have not reached completion. Furthermore, many of the facies and most of the faunal assemblages in the rock are homologous with those of the modern sedimentary regime that surround and overlap the limestone. These features provide an essential datum upon which a diagenetic model may be superposed—a starting point representing the prediagenesis state. If these comparisons are not available, a comprehensive study of alteration cannot be attempted.

Ambergris Cay is a sediment-mantled bedrock island on the seaward edge of the continental shelf in northernmost Belize (Fig. 1). It lies adjacent to the Mexican state of Quintana Roo, is approximately 20 by 5 mi (32 x 8 km) in size, shelters shallow-water Chetumal Bay to the west and, in turn, is sheltered by a narrow seaward lagoon and a virtually continuous barrier reef. Holocene sediments of the Cay and its immediate environs have been described by Ebanks (this volume).

In the text which follows, the geologic setting of Ambergris Cay is outlined and the petrographic character of diagenesis of the sediments is discussed.

NATURE OF INVESTIGATION

Most of the samples used in this study were collected by the writer during August 1965 from scattered localities on Ambergris Cay and a few nearby islands. Other rock samples collected at random from Ambergris Cay during previous summers proved useful in the study, as did selected samples of Holocene sediments and constituent grain types from the entire Belize shelf.

Laboratory research included petrographic study of approximately 130 large (3 x 2 in. or 7.6 x 5 cm) thin sections and corresponding slabs and hand specimens, in an attempt to develop criteria for identifying the kind, and tracing the sequence, of diagenetic alterations which the rocks have undergone. All constituent grains of frame and matrix were tabulated, and their interrelations were noted; any alteration products, and their mode of occurrence, were examined in detail.

To investigate mineralogic changes accompanying alterations in petrographic texture, X-ray diffractograms of strategic samples were produced, using copper Kα radiation and sample preparation as outlined by Matthews (1965, p. 23). Feigl's solution (Friedman, 1959) was useful for rapid discrimination of calcite from aragonite in thin section.

Nomenclature used in this paper follows that introduced by Folk (1959; e.g., biomicrite) and by Dunham (1962; e.g., packstone), except where otherwise specified.

GENERAL GEOLOGY OF AMBERGRIS CAY

A few regional geologic studies of the Yucatán Peninsula and of Belize include the northern shelf and Ambergris Cay, but none clearly defines the age of the limestone which constitutes the bedrock of this island. Flores (1952, p. 409) simply stated that "...the fossil coral reefs developed along the coast and on some of the major offshore cays [are]...Miocene to Pliocene [in age]." Early observations by Heilprin (1891) and Sapper (1896) noted the difficulty in separating Pliocene from post-Pliocene rocks in Yucatán. Post-Pliocene limestones appear in patches along the northern coast and in the interior of the peninsula (Sapper, 1896), and are "distinguished by the large numbers of *Venus cancellata* which fill the rock, making a true *V. cancellata* bed" (Heilprin, 1891). Sapper (1896) mapped Quaternary beds on Ambergris Cay and the peninsula directly to the north. *Chione cancellata* (formerly *Venus cancellata*) is common, and locally very abundant, in much of the outcrop on Ambergris Cay. However, this fact alone is insufficient to establish a precise age, because the authors cited recorded the presence of this pelecypod in Pliocene rocks, and it is certainly abundant in the Holocene biota. Butterlin and Bonet (1962) indicated that the Pliocene(?) Carrillo Puerto Limestone is well represented in Quintana Roo, Mexico, where it is characterized by peneroplids such as *Archaias*. Their geologic map does not include Ambergris Cay, but it shows the Carrillo Puerto extending to the Mexican border adjacent to the northern tip of the island, and shows undifferentiated Quaternary deposits fringing the mainland of Belize as far north as Chetumal Bay. Elsewhere in their paper, they stated in regard to the Quaternary deposits which overlie the Carrillo Puerto that the consolidated limestones are probably Pleistocene, and the "more marginal, softer" beds are Holocene.

The fauna in the limestones of Ambergris Cay shows close similarities with the Holocene biota of the adjacent sediments. In fact, all of the genera and most of the species identified by the writer from fossils collected on the island are found living in the Caribbean. The limestones, then, are probably not older than Pleistocene. As they are presumed to have been emergent at least since the Sangamon Interglacial Stage of high sea level, they may be considered to be of probable mid-Pleistocene age until more precise data are available. Reeflike coral banks constituting the Key Largo Limestone in the Florida Keys also are believed, on the basis of stratigraphy and Th-U age determinations, to have formed during the latest interglacial sea-level maximum (Stanley, 1966, p. 1928). Reef limestones of Pleistocene or Pliocene-Pleistocene age have been described from the Bahamas (Newell and Rigby, 1957, p. 58) and Barbados (Trechmann, 1933; Matthews and Mesolella, 1966).

OUTCROP AND SAMPLE DISTRIBUTION

The limestone of Ambergris Cay crops out at isolated localities; there are no large continuous exposures. The rock is usually very well lithified and produces a hackly, sometimes conchoidal fracture; it is typically light tan on fresh surfaces and weathers to medium to dark gray. Outcrop surfaces vary from smoothly rounded low mounds to extremely jagged pinnacles (Fig. 2A). Local relief rarely exceeds 1 ft. (0.3 m); consequently, no appreciable vertical sections are available, and the sampling is restricted generally to a random two-dimensional exposure. Systematic rock coring across the island and neighboring areas would yield much additional information; however, drilling attempts in this area have achieved little success.

Small outcrops appear throughout much of central and western Ambergris Cay and on a few nearby islands. In addition, bare rock underlies parts of shallow Chetumal Bay, west of Ambergris, and many supratidal ponds in the interior of the island.

All samples collected were exposed to the effects of air and fresh water; none were "submarine" with respect to present sea level. Of course, rocks now covered by only a few feet of seawater were exposed to subaerial conditions for an extended period of time, at least during the latest glacial stage, before being inundated by the slow postglacial rise in sea level during the last few thousand years (Shepard, 1964).

In the interior of Ambergris, outcrops are typically interspersed with supratidal mud flats (Fig. 2A), which may be covered by a living mat of blue-green algae but which more commonly are barren of flora. Ephemeral ponds, no more than a few inches deep, are present on these flats. Some of these ponds are rimmed or floored by crusts of Holocene sediments which have been dolomitized, penecontemporaneously with their initial deposition, by capillary waters of high salinity associated with the stagnant ponds (Ebanks, personal commun.). In a few localities, dolomitic crusts overlap the Pleistocene limestone. However, there is no evidence of any dolomite in the rock other than dolomitized mud which has seeped into pores.

The only rock exposure on the eastern shore is at Reef Point (Fig. 2B), where the outcrop is subjected to the full force of the surf at a gap in the modern reef. Most inland areas of high ground, reaching maximum elevations of a few tens of feet, are covered by a heavy mantle of soil and vegetation in which rock is rarely exposed.

PLEISTOCENE BIOFACIES AND LITHOFACIES

In general order of abundance, the grains recognized in the Pleistocene rocks of Ambergris Cay are: pellets, tunicate spicules, *Halimeda*,

FIG. 2–Pleistocene limestone outcrop.

A. Jagged pinnacles of limestone onlapped by supratidal mud flats, illustrating typical outcrop in interior of Ambergris Cay. Several *Strombus gigas* conchs weather in relief, in foreground, and mangroves colonize parts of mud flats.

B. Air view of Reef Point, looking southward. Here, Reef Point interrupts modern reef, which is marked by an irregular line of breakers bordering shallow-water shelf lagoon, left of photo. Patch of light at top is glare from window of airplane. Rows of cultivated coconut palms occupy seaward beach ridge of Ambergris Cay; supratidal ponds and mud flats cover interior.

ooliths, pelecypods, gastropods, corals, coralline algae, peneroplids, miliolids, echinoderms, *Homotrema*, other encrusting foraminifers, miscellaneous benthonic foraminifers, dasycladacean algae, ostracods, intraclasts, and alcyonarian spicules. Four distinct facies are recognizable, each distinguished by its fauna (or absence of fauna): (I) reef-crest facies, (II) backreef facies, (III) shelf-lagoon facies, and (IV) mudbank facies. The interpreted distribution of these facies is illustrated in Figure 3.

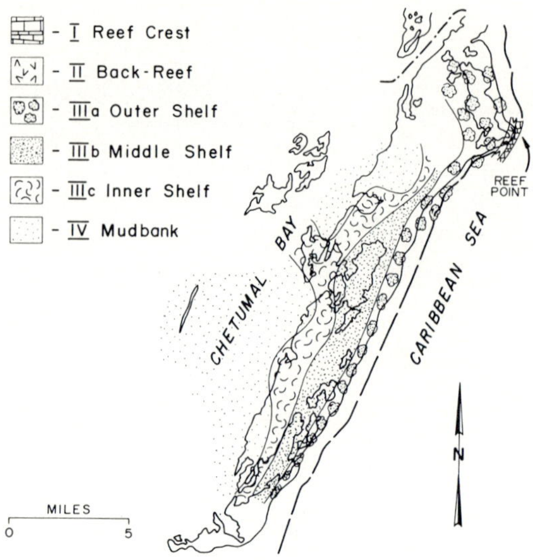

FIG. 3—Distribution of Pleistocene facies, Ambergris Cay.

Facies I—Reef Crest

The reef-crest facies is characterized particularly by a rigid framework of the "moosehorn" coral, *Acropora palmata*, with important biotic contributions by the encrusting coralline algae *Lithothamnium*, *Archaeolithothamnium*, and *Lithophyllum*, and by encrusting foraminifers including *Homotrema rubrum*. The fauna is more diverse in this facies than in the relatively protected environments of lower energy in the shelf-lagoon facies. Massive heads of *Montastrea* and *Diploria* are interspersed among the fronds of *Acropora palmata*, many of which have distinct lagoonward orientation (Fig. 4A), as would be expected near or at the reef crest. Shinn (1963) described and interpreted the significance of directional growth forms of *A. palmata* in modern reef environments. Fronds of this coral species are typically unoriented and palmate in forereef water depths exceeding 20 ft (6.1 m) and on the shallow, sheltered reef flat in the lee of the surf-battered crest. In the former situation, colonies are elongate vertically; in the latter, they tend to be broad and flat. On the true crest of the reef, which receives the brunt of the sea waves and swells, the fronds are directed sharply toward the backreef, in the direction of the wave surge (Shinn, 1963), and are more circular and massive in cross section than are the unoriented branches. This peculiar habit adapts the colonies for survival and vigorous growth in the most highly agitated zone of the reef. Pusey (1964, p. 114) noted similar orientations in the Holocene reefs of Belize. Whether spurs and grooves such as those reported in Holocene reefs of Florida (Shinn, 1963), British Honduras (Belize; Pusey, 1964, p. 113-114), Jamaica (Goreau, 1959), and the Pacific (Cloud, 1959, p. 406-410), and in Pleistocene reefs in Jamaica (Purdy, personal commun.), also are present in the rocks of Ambergris Cay is uncertain because of the small size of the reef-crest exposure; however, such structures would be anticipated.

In Holocene coral reefs near the coast of Jamaica, Goreau (1959) has emphasized the quantitative importance of massive *Montastrea annularis*. Large heads of this coral are the most important frame builders in that area. They grow to a large size, particularly in buttresses, on the seaward slope of the reef. In shallower water, *Acropora palmata*

FIG. 4—Reef and backreef facies, Pleistocene and Holocene.

A. Exhumed fronds of Pleistocene *Acropora palmata* (center) in reef facies, oriented away from open sea. Fronds are 1-1½ ft (0.3-0.5 m) long, enclosed in poorly sorted coralgal biomicrite.

B. Holocene *Acropora cervicornis* concentrations in backreef lagoon of Belize. *Diploria* head in right foreground, overturned *Montastrea* in center, *A. cervicornis* in middle background and foreground (arrow).

C. Pleistocene backreef exposure, a thicket of *A. cervicornis*, on Reef Point, in matrix of coralgal biomicrite. Bar length 1 ft (0.3 m).

becomes dominant, representing a climax phase of reef building. This rapidly growing coral provides great quantities of detritus to the reef because of its fragility relative to more massive heads of coral. The importance of *A. palmata* in Jamaican reefs lies more in its contribution of reef detritus than in its frame-building qualities (Goreau, 1959). Similarly, in the Alacrán reef complex, Kornicker and Boyd (1962) found that *Montastrea* and *Diploria* are the most abundant of the primary frame builders, and that *Acropora palmata* is typical of the reef front at shallow depths (<30 ft or 9.1 m). These observations are in accord with those of Newell (1958), who recognized in West Indian reefs an intermediate belt of *A. palmata* in turbulent waters of depths 6-30 ft (1.8-9.1 m), between an outer belt of *Montastrea annularis* and an inner rocky shoal.

A few mollusks and mollusk fragments are commonly found in the Pleistocene reef-crest facies; echinoid spines are more common in this facies than elsewhere. Large fragments of *Millepora* appear locally. Silt-size tunicate spicules, recognized by their characteristic three-dimensional stellate structure and aragonitic composition (Purdy, 1963, p. 342; Van Name, 1930, Fig. 14; Matthews, 1965, p. 59) are very abundant. Corals are important contributors to the sand-size fraction; benthonic foraminifers and worm tubes are commonly present. The debris which is interstitial to the frame-building corals and coralline algae is typically very poorly sorted, consisting of angular skeletal fragments of a wide range of sizes embedded in a micrite matrix. Small grains such as mollusk prisms, foraminifers, and tunicate spicules contribute to the high proportion of silt-size debris. In essence, then, this facies is characteristically a coralgal or coralgal-foram biomicrite filling a rigid frame composed of *Acropora palmata* and other massive corals.

The persistence of a matrix of calcilutite in such a turbulent environment is contrary both to an energy-index interpretation and to observations of many modern reefs, in which silt- and mud-size carbonate material is virtually absent (Ginsburg, 1956, p. 2411; Pusey, 1964, p. 117; Matthews, 1965, p. 38). However, several studies have shown appreciable proportions of micrite within reef frames in the stratigraphic record. Stanley (1966, p. 1934) reported abundant calcilutite in Pleistocene reefs of the Key Largo Limestone in the Florida Keys; Ladd and Schlanger (1960, p. 884) illustrated shallow cores of lithified reef rock from the reef flat of Eniwetok Atoll, in which "...lithified mud (fine sand to clay-size calcareous material) is abundant...and fills a large percentage of original voids."

Undoubtedly, much silt- and clay-size calcareous debris is being produced by attrition of skeletal materials (both by organisms and by wave action) in the reef environment (Emery et al, 1954, p. 87-88; Matthews, 1965, p. 124-133). The resulting fine-grained debris cannot be deposited interstitially in the coarse framework of the reef crest, which is exposed directly to surf turbulence. Stanley (1966) concluded that the presence of this material in the Key Largo reef suggests less turbulent conditions of deposition than in the modern regime. Fine detritus could filter downward into the deeper portions of the coarsely porous reef frame and thus be preserved in the record, even though it could not remain on the exposed upper few inches or feet of the reef crest. Numerous pockets trap this debris and enable it to accumulate, progressively enveloping the framework, at depth, in a blanket of carbonate silt and mud.

The Pleistocene reef-crest facies crops out only on Reef Point, where the living reef crest is directly on strike with its lithified ancient equivalent (Fig. 2B). For a short distance on each side of Reef Point, the modern reef becomes fringing, and it abuts on the point itself.

This is the only position along the entire island where the trend of the living reef is interrupted by land. The result is thus a superposition of similar ancient and modern facies in this area, with an intermediate unconformity. Such relations are evident in many shallow-water deposits in the geologic record. Unfortunately, this superposition has not been demonstrated beyond Reef Point, so that the length of the Pleistocene reef cannot be compared with that of the modern reef.

Facies II—Backreef

The backreef facies adjoins the Pleistocene reef and occupies a narrow zone directly lagoonward (westward) from the reef crest. Its most diagnostic feature is the presence of abundant "staghorn" coral, *Acropora cervicornis*, and the absence of *A. palmata*. Large heads of massive *Montastrea annularis* and *M. cavernosa*, as well as species of *Diploria*, are very common in this facies, generally constituting a higher proportion of the biota here than in any of the other facies except patch reefs. Encrusting *Lithothamnium*, *Lithophyllum*, and *Lithoporella* are common, as are branches of *Amphiroa*. Tunicate spicules, *Homotrema*, and several other encrusting foraminifers are very abundant in this facies. A few *Millepora* fragments are present also. Echinoid spines are common; a few peneroplids, other benthonic foraminifers, and a few *Halimeda* are typically present. Worm tubes encrust corals or other hard surfaces. A few small high-spired gastropods, some pelecypods, and molluscan debris of assorted sizes are scattered throughout the matrix. Rarely, the queen conch, *Strombus gigas*, is found here. A few *Lithophaga* appear in boreholes within coral heads, and one alcyonarian spicule was recognized in this facies.

Locally, branches of *Acropora cervicornis* have been concentrated into virtual thickets (Fig. 4B). Goreau (1959) noted the highest concentrations of *Acropora cervicornis* on the seaward slope of some Jamaican reefs, intermediate between the *A. palmata* buttresses and a deeper zone of *Montastrea annularis* (Goreau, 1959, p. 74). Newell and Rigby (1957, p. 42) indicated the presence of *A. cervicornis* in front of the *A. palmata* zone in the Andros reef, associated with other corals in water depths of 25 ft (7.6 m) or more. Lewis (1960) illustrated *A. cervicornis* at depths of 20 m in Barbados, associated with branching *M. annularis*, *M. cavernosa*, *Siderastrea*, and *Porites*. In contrast, Kornicker and Boyd (1962) described abundant *A. cervicornis* in shallow protected waters of the lagoon of the Alacrán reef complex. Similarly, in certain Pacific reefs, Cloud (1959, p. 381, 411) found species of *Acropora* which resemble *A. cervicornis* in morphology and hydrodynamic character occupying a zone in the backreef environment in the lee of the reef flat, in shallow lagoonal waters of relatively low turbulence. These corals are thriving in an area of coarse-grained, well-sorted carbonate sands, at water depths of 4-13 ft (1.2-4.0 m). In Belize as well, *A. cervicornis* thrives in protected lagoonal waters of the outer shelf in the lee of, and close to, the reef crest (Fig. 4C). This appears to be the environment in which the coral accumulated in the Pleistocene sediments, as it is found almost exclusively in the zone directly leeward of the reef-crest facies.

As in the reef facies, the matrix is primarily a silty micrite; therefore, this facies may also be classed as a coralgal-foram biomicrite. Poorly sorted angular grains are typical. Tunicate spicules and fragmented mollusk prisms constitute most of the silt-size fraction; sand grains are contributed by corals, mollusks, foraminifers, and echinoderms.

This backreef facies crops out only on Reef Point, bordering the leeward side of the *Acropora palmata* zone. Facies belts in the Pleistocene shelf-lagoon facies trend generally north-northeast, approximately parallel with both the elongation of Ambergris Cay and the modern barrier reef on the edge of the continental shelf. Consequently, the backreef and reef-crest facies, if they were as well represented in the Pleistocene as are the corresponding facies today, probably underlie the approximate positions of their modern equivalents. Unfortunately, no exposures other than that at Reef Point are located sufficiently far seaward to test this conjecture. Boreholes cored across the modern reef tract would contribute a great deal to this interpretation and, more generally, to the three-dimensional outline of the extent and to the depositional and diagenetic history of the Pleistocene limestone sequence in this area.

Facies III—Shelf Lagoon

The Pleistocene shelf-lagoon facies of Ambergris Cay incorporates a variety of faunal and lithologic assemblages. It encompasses most of the outcrop area, particularly in the interior and western parts of the island.

This lagoonal facies is typically a biopelmicrite with a normal-marine biota which is characteristic of shallow shelf environments. Pellets which are aggregates of silt- and clay-size calcium carbonate are very common. *Halimeda*, corals of several genera, gastropods, pelecypods, coralline algae, foraminifers, and ooliths are present in greatly diverse proportions. Locally high concentrations of one or another of these constituents are not exceptional.

Although a great variety of local environmental conditions is represented in shelf-lagoon facies, three zones or subfacies are definable. These zones are somewhat intergradational, but each can be distinguished by a few characteristic features. They have been designated as outer (IIIa), middle (IIIb), and inner (IIIc) shelf zones.

Facies IIIa—Outer Shelf, Including Patch Reefs

The outer shelf zone grades into the backreef *A. cervicornis* facies. It is characterized most particularly by the presence of corals, associated with abundant pellets and/or ooliths and an abundant fauna, predominantly of mollusks. *Montastrea annularis* is the most common coral species in the outer shelf zone. It appears in massive, rounded heads, up to 2 or 3 ft (0.6-0.9 m) in diameter; it is scattered throughout the zone, but the highest concentration is seaward. Massive heads of *Diploria* and scattered branches of *Acropora cervicornis* and *Millepora* are found in this zone as well. Another important coral in the outer-shelf biota is *Porites furcata*, a relatively fragile, branching colonial form known as the "finger coral." This coral is usually the last coral remaining in the zone as it passes westward into the adjacent area; toward the south of the island, it is the main indicator of the outer-shelf assemblage. It is also associated with larger, massive corals near the reef and in northern areas, but it is less obvious in their presence.

The most distinctive mollusk characterizing the outer-shelf fauna is the large and beautiful queen conch, *Strombus gigas*; conchs may be very abundant and they typically weather in relief above their matrix. Another gastropod common in this zone is a species of *Bulla*, probably *B. striata*. This smoothly rounded "bubble-shell" is most abundant in parts of the zone where corals are relatively scarce. It is usually found in a medium-grained pelmicrite matrix penetrated by abundant micrite-lined burrows. In some places, splintery flakes of calcilutite which are incorporated within the rock as intraclasts were obviously derived by reworking of these burrow walls.

Echinoderms, including whole irregular echinoids, are common in the outer-shelf fauna but diminish in abundance westward.

Halimeda flakes are more abundant here than in the seaward facies, and they may be, in part, indigenous. Other algae, particularly branching coralline types such as *Amphiroa*, and relatively few encrusting genera are also represented. A few *Homotrema* and other encrusting foraminifers appear, peneroplids are common but not abundant, and other benthonic foraminifers are usually present. Tunicate spicules are typically abundant, although not in such high proportions as in the reef-crest and backreef facies.

Many of the smaller pellets (0.15-0.20 mm) and fragments of mollusks and algae are coated with one to three layers of translucent aragonite, so that they may be classed as ooliths, or at least superficial ooliths, of the Bahamian type (Newell et al, 1960); a few are composite, with several nuclei enclosed by an outer envelope. Most of the associated larger grains are not coated.

The typical matrix for the outer shelf zone is a pelmicrite or oolitic pelmicrite, which produces a packstone, in contrast with the biomicrite-wackestone of the two more seaward facies.

A patch-reef area in northern Ambergris may be included in the outer-shelf biotic and lithologic zone. It comprises a dense concentration of large, massive heads of *Montastrea annularis* and of *Diploria*. Heads of *M. cavernosa* and branches of *Porites* are less abundant. The patch of corals reaches a diameter of about 200 ft (61 m) and is roughly circular in plan. Many *Strombus gigas* conchs lie nearby. This area also has a high proportion of recognizable ooliths in the characteristically pelletal (biopelmicrite) matrix.

Facies IIIb—Middle Shelf

A second zone or subfacies in the shelf-lagoon suite is that of the midshelf. Although not sharply delimited, it has sufficiently distinctive characters to warrant separation from adjacent zones. Typically, this zone is represented by sparsely fossiliferous, commonly oolitic, fine- to medium-grained calcarenites (pelmicrites or pellet grainstones). They are at least moderately well sorted, having a unimodal or bimodal distribution of size grades. The latter applies where irregular bedding or mottling has resulted from mixture of interbedded micrites and pelmicrites by burrowing organisms. Pusey (1964, p. 91-96) reported similar irregular bedding and mottling in Holocene carbonate sediments of the northern shelf.

Bulla cf. *B. striata* is locally common in this zone, but few other intact mollusks are evident. *Cerithium, Conus, Chione cancellata,* and thin-shelled pelecypods are represented locally but these mollusks do not aid in characterizing the zone. Tunicate spicules are common constituents of the silt-size fraction. A few *Halimeda,* echinoderm and coral fragments, miliolids, and peneroplids usually are present.

Ooliths are recognizable among the pellets. In the interior of the island, ooliths are predominant, approaching 60 percent of the total grains. Also, the fabric there is typically mottled, and the same fauna and flora are represented; the only difference appears to be in the abundance of ooliths relative to pellets of the same size (1/8-1/2 mm). Nuclei consist of mollusk and *Halimeda* fragments, pellets, tunicate spicules, dasycladacean algae, and echinoid spines. In this same area, a few branches of *Porites* and massive and encrusting red algae (cf. *Lithothamnium,* also *Lithoporella*) occur.

The abundance of ooliths suggests proximity to shoaling banks where agitation and calcium carbonate concentrations were great enough to cause physico-chemical precipitation of aragonite around moving sand and silt grains (Newell et al, 1960). The lack of abundant fossils is in accord with high mobility of the substrate and with carbonate saturation values of sufficient magnitude to permit the formation of ooliths; moderately good grain sorting indicates current or wave agitation. Micrite which persists as matrix in these rocks might indicate that the ooliths were generated elsewhere in a more agitated environment and introduced into a quieter, muddier area. Alternatively, the calcilutite might represent deposition between oolite bars. The interbedding of oomicrites (Folk, 1959) and micrites with fewer ooliths indicates a fluctuation of depositional conditions such as those associated with shifting, shallow-water bars. A few composite ooliths are evident, and all appear to be of typical Bahamian type. "Freshness" of the ooliths and the fact that tunicate spicules and other fine-grained skeletal materials found in adjacent environments serve as nuclei indicate that the grains were actively forming in the Pleistocene facies, and were not derived from an older oolite. The predominance of ooliths in north and central Ambergris, and their much diminished proportions relative to pellets southward, indicates that their main source lay in the more northerly areas. The southern parts of this midshelf zone probably represent deposition in deeper water than elsewhere on the shelf, and deposition was perhaps influenced by longshore currents.

Facies IIIc—Inner Shelf

The third zone of the shelf-lagoon facies, the inner shelf zone, bears some features in common with facies IIIb, but usually can be distinguished readily on the basis of its faunal content. In particular, it is characterized by a very high concentration of large, heavy-shelled pelecypods and abundant whole segments of *Halimeda*. *Chione cancellata* is the most common species; it is robust and very well preserved, commonly with pairs of valves still articulated. *Lucina pensylvanica*, another stout pelecypod, is nearly as abundant as *Chione*; and *Tellina interrupta*, *Laevicardium* spp., and *Anadara* spp. are also common. Among the gastropods represented, *Cerithium*, *Bulla*, and *Polinices* are most common; others include *Xancus angulatus*, *Olivella*, *Fasciolaria*, *Cerithidea costata*, and cf. *Calliostoma*. Scattered throughout the abundant biota are a few corals, including *Porites* and scarce *Acropora cervicornis* branches, *Siderastrea radians*, and fragments of *Diploria* and ?*Agaricia*. *Halimeda* flakes are usually present and, locally, a few echinoid spines and encrusting algal fragments have been found. The matrix typically is a pelletal packstone, a biopelmicrite of medium to coarse grain size. Foraminifera, including miliolids and peneroplids, in some areas contribute significantly to the sand fraction. A few burrows, up to 1/2 in. (1.3 cm) in diameter, are evident. Rocks of this zone show, on fresh surfaces, a coarsely and irregularly mottled fabric as a result of incomplete mixing of biopelmicrite and biomicrite by a burrowing infauna. Postlithification borings also are very common in this zone.

The faunal assemblage of facies IIIc includes mollusks characteristic of both *Thalassia* and carbonate-sand areas (Robertson, 1963); it is probably a mixture of fossils derived nearby from these two environments at least. The zone appears to represent the eastern edge of a shoaling bank, upon which a prolific flora and fauna grew and were concentrated by wave activity.

As in facies IIIb, rocks of the inner shelf zone are mottled and have abundant pellets; they are distinguished by their additional content of large numbers of intact pelecypods. The transitional boundary between these two zones appears to trend generally along the central axis of the island. Westward, zone IIIc is progressively more micritic and its transition into facies IV is broad and ill-defined. Pellets are less abundant, and the effects of burrowing are more distinct and are limited to a few definite burrows filled with pellets. Apparently, the more cohe-

TABLE 1. DISTRIBUTION OF MAJOR CONSTITUENTS, PLEISTOCENE SHELF-LAGOON FACIES

	Outer *IIIa*	*Middle* *IIIb*	*Inner* *IIIc*
Montastrea	4[1]	-	-
Diploria	3	-	-
Porites	3	-	1
Siderastrea	-	-	1
Halimeda	3	2	4
Strombus gigas	4	-	-
Bulla	2	3	2
Cerithium	2	1	3
Chione cancellata	3	1	4
Lucina pensylvanica	1	-	4
Tellina	2	-	3
Burrows	3	2	2
Pellets	3	4	3
Ooliths	3	4	2
	Coral-mollusk Biopelmicrite	Oolitic Pelmicrite	Mollusk Biopelmicrite

[1]Symbols are as follows: 4–abundant, 3–common, 2–few, 1–trace.

sive mud permitted better maintenance of burrows but was less preferred by burrowers than was the zone of abundant mottled pellets to the east. Alternatively, if deposition of the more micritic sequence was relatively rapid, the record of burrowing would be effectively diluted.

Associated with the increasing proportion of micrite and the diminution of fauna westward, a breccialike fabric is developed such as that which predominates in facies IV. Several samples from the transition area bear features characteristic of both facies IIIc and IV, certifying that these two facies are, indeed, time-equivalent.

Table 1 summarizes the distribution of major constituents in the three zones of facies III.

Near the patch reef in facies IIIa, a small area of black limestone rubble was found. This limestone differs from the surrounding fossil-rich biopelmicrite only by its color. Although a fetid odor is exuded when the rock is broken, the black material is not soluble in benzene and apparently is not a hydrocarbon. Its presence in one isolated outcrop is still unexplained.

Facies IV—Mudbank

Facies IV, tentatively classified as a mudbank facies, is distinctly different from the other facies. It is typically a light tan, very dense and tough, mottled limestone which is largely sublithographic. The permeative system of dark gray and brown wisps and blebs which give rise to its typical mottled aspect is due partly to the several sizes of borings and their respective fillings and, to a larger degree, to a dense and typically unsystematic pattern of breccialike partings which are now filled with calcite. By Folk's (1959; 1962, p. 69) terminology, these rocks would be classified as "disturbed micrites," or "dismicrites." Borings are of at least two sizes: small, tubular borings of about 1/4-

mm diameter which may or may not be filled with translucent, brown, microcrystalline calcite; and large, cylindrical borings filled with similar material. The larger borings, much less numerous than the smaller ones, resemble definite borings in other facies in which known hard substances such as shells have been penetrated. The breccialike partings are discontinous and usually have no definite systematic orientation.

Commonly, the only directly recognizable traces of a macrofauna are a few isolated shells or, more typically, molds of a small gastropod resembling *Bulla*. Molds of *Chione cancellata* are scarce. Benthonic foraminifers, including miliolids and peneroplids, are locally abundant.

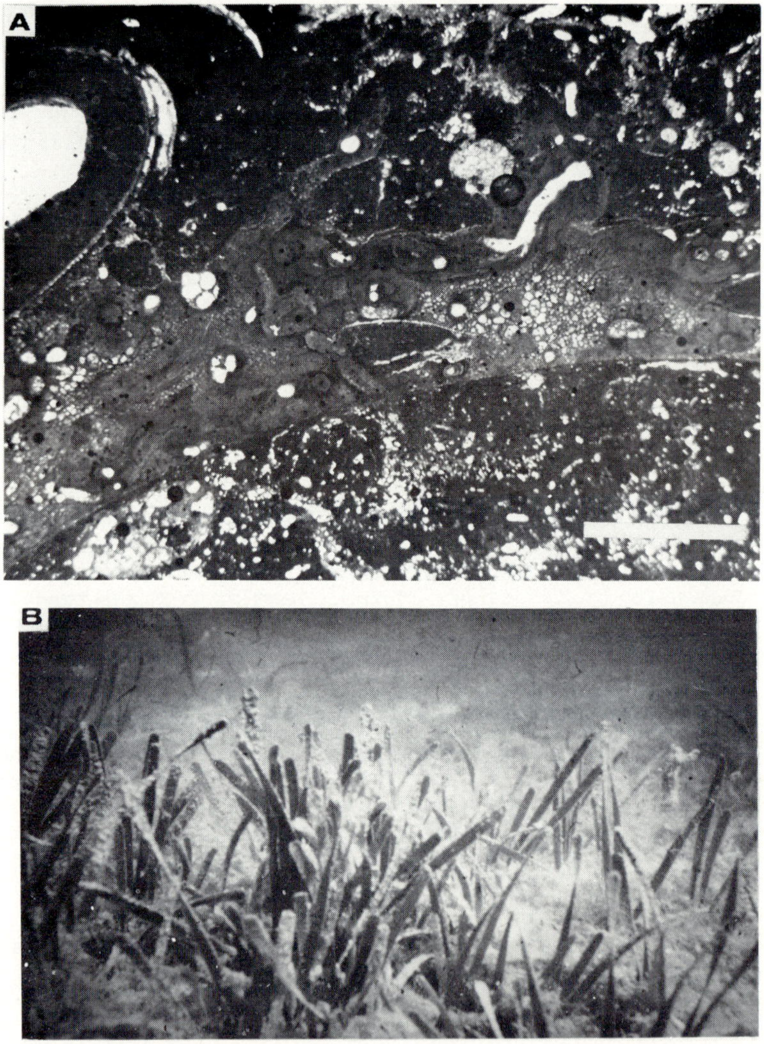

FIG. 5—**A.** Pleistocene "frothy" vesicular structures in band (center), with one flat side. Section of *Bulla* in upper left. Bar length 2 mm. Photomicrograph, plane-polarized light.

B. Holocene *Thalassia* in shallow, murky waters of Chetumal Bay; note organisms and sediment encrusted on grass blades, perhaps homologous with vesicular structures in **A.** Width of underwater photo represents about 3 ft (0.9 m).

Calcareous silt is incorporated in much of the micrite. Structures which appear to be rounded intraclasts of micrite may simply be steinkerns of gastropods.

A feature which is not restricted to facies IV but which is most strikingly evident and abundant in this facies is a puzzling "vesicular structure." It may fill large cylindrical "bore holes" or small fractures. The larger structures in a few samples display an array of vesicles resembling bryozoan zooecia; this affinity is most strongly suggested where the vesicles have a relatively systematic arrangement. One example illustrates, in thin section, a pattern of isolated circular "cells," up to 1/2 mm in diameter, in each of which are enclosed a few second-order vesicles. Rarely, however, is the structure so systematic; it commonly resembles, on a tiny scale, a piece of strained, flattened chicken wire. The vesicular structures constitute as much as 30 percent of the entire rock volume. In many places, the structure has a complexly variable, frothy appearance; where it occupies a somewhat linear band, one edge of the band may be smooth and regular (Fig. 5A). Vesicle sizes range from about 0.1 mm to 1.0 mm and vary abruptly and nonsystematically in all directions. A large mass of vesicles may exceed 2 cm in length and 3 or 4 mm in width; individual vesicles are generally globular, but many are elongate or flattened. Walls of vesicles nowhere show discrete microstructure such as oriented crystallites, lamination, or pores; they are always rather vaguely outlined, gray, and microcrystalline. The vesicles typically are lined or filled with medium-, fine-, or very fine-grained calcite spar, in a normal void-filling relation. Furthermore, some of this spar grades into the microcrystalline calcite of the vesicle walls, probably representing a complex process of recrystallization coupled with pore filling. Fauna associated with the vesicular structures show several grades of preservation. Peneroplids appear fresh and typically are filled with fine-grained spar. Small benthonic Foraminifera, including discorbids, commonly have been altered to a fine- or medium-grained spar mosaic after complete filling of chambers by similar spar; the resulting continuous mosaic is, in some places, difficult to recognize as an altered foraminifer. A similar masking of primary microstructure could have affected the vesicular structures, obliterating recognizable textures. Alternatively, as the primary texture is unknown and no gradational stages of alteration are evident, the present state of the vesicle walls could be primary.

Although the vesicular masses may occupy borelike or burrowlike positions, most are isolated. This would suggest that most of the structures were incorporated in the mud as primary constituents rather than introduced into burrows or borings at some later stage. Conceivably, any post-depositional shrinkage or compaction could initiate fractures contiguous with the spaces occupied by the vesicular masses. The result would be an overall breccialike fabric with fractures occupied in large part by vesicular structures. Alternatively, periodic subaerial exposure could result in such brecciation owing to desiccation. The complexity of the vesicular masses suggests to the writer that they are organic. A few of the structures have a degree of organization which suggests bryozoans, but most are much more irregular.

The closest resemblance to any organism described in the Holocene of Belize is to the coralline alga "*Melobesia*," described by Pusey (1964, p. 59-63, pl. 4). This alga has been shown to be epiphytic, commonly encrusting upon marine grasses such as *Thalassia* (Pusey, 1964). However, the vesicular structures found in the Pleistocene rock are considerably larger and less ordered than are the described thalli of *Melobesia*. They resemble more closely, in degree of variability and in approximate form and magnitude, some of the encrusting Foraminifera.

The disposition of the vesicular structures could represent primitive and very irregular Foraminifera which have encrusted upon marine grasses such as *Thalassia* that have left no direct record in the rock. This would explain the variety of orientations of the structures as the result of burial of encrusted grass in attitudes ranging from vertical to horizontal. In this way, the remains of the organisms could initiate minor fractures during compaction or dehydration. The resulting pseudobreccia fabric would display an irregular system of discontinuous partings, in most of which are incorporated the remains of organisms which once encrusted a sea-bottom flora. Numerous examples of such an epiphytic habit are evident in shallow areas of sea grasses today. The Belize shelf provides many such examples (Fig. 5B), as Pusey has demonstrated in *Thalassia* areas such as the mud shoals leeward of the southern tip of Ambergris Cay (Pusey, 1964, p. 154).

The dearth of fauna in facies IV and the general relations to other facies and to the present coastline suggest a very shallow-water near-shore, or perhaps even intertidal, depositional environment. If it were intertidal, the "brecciation" might have been caused by desiccation cracking. However, if the vesicular structures are remains of marine epiphytic organisms, and if some of the mottling is due to a burrowing infauna, depth of a few feet of water is suggested. Either of these interpretations places the environment within a few feet above or below sea level; the facies might then represent a shallow-marine, grass-covered mudbank or an intertidal mud flat. The few mollusks which are found could have been introduced from more seaward areas in either case, and are not dependable environmental indicators.

LAMINATED CRUSTS

Smooth, sheetlike crusts of brown cryptocrystalline calcite coat outcrops in many areas on the island. Laminations are vaguely visible in the crusts, but no fossils are recognizable. In outcrop, the crusts smooth over the irregular solution surface of the underlying limestone (Fig. 6A) and reach a maximum thickness of 1 or 2 in. (2.5-5.0 cm) on a horizontal surface and as great as 3 in. (7.6 cm) in small potholes. The crusts are extremely porous but have a "glazed" upper surface which probably inhibits weathering. The laminations are 1/2 mm or less in thickness, and are minutely crenulated, pinching and swelling irregularly; in small potholes, dips approach 90°. Most crusts contain scattered, very angular grains of fine quartz sand and silt. They typically appear near areas of heavy soil and vegetative cover.

Crusts which appear identical to those on Ambergris Cay have been reported from the Bahamas (Newell and Rigby, 1957, p. 50; Kornicker, 1958), the Florida Keys (Multer and Hoffmeister, 1968), and other tropical and subtropical areas of limestone outcrop. When first described, many were attributed to submarine or intertidal development of blue-green algal stromatolites. They now are generally interpreted as caliche deposits precipitated on limestone surfaces; if this interpretation is correct, they represent important records of subaerial exposure.

GROSS ASPECTS OF DIAGENESIS

Physical Aspects

Several physical processes are effective in changing sedimentary fabrics, structures, and textures after deposition. Some of these phenom-

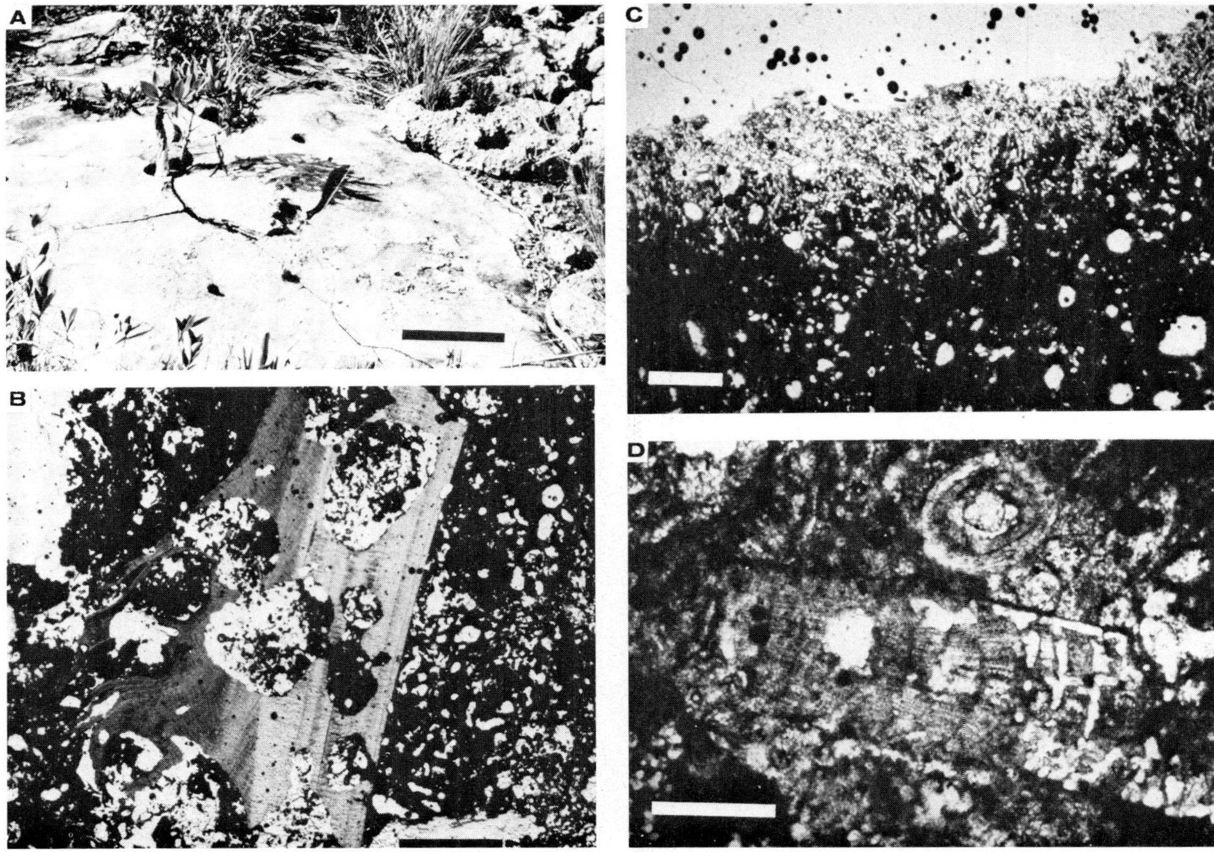

FIG. 6—**A**. Surficial crust of laminated, very porous microcrystalline calcite, coating Pleistocene limestone outcrop (right background) and interrupted by a living mangrove and several fractures. This appears to be a Holocene caliche crust. Bar length 1 ft (0.3 m).

B. Pelecypod in rocks of inner shelf zone, intensely bored, perhaps by a sponge. Bar length 2 mm. Photomicrograph, plane-polarized light.

C. Surface boring of rock by filamentous algae, resulting in progressive destruction of subaerially exposed surface. Bar length 0.5 mm. Photomicrograph, plane-polarized light.

D. Evidence of solution in branch of *Amphiroa* (nearly horizontal, in center of photo), corroded along boundaries of skeletal architecture. Red algae here retain original magnesian calcite composition. Bar length 0.25 mm. Photomicrograph, plane-polarized light.

ena occur before lithification, others after; most are related in some way to the life activity of organisms.

Where bottom conditions permit and an appropriate fauna exists, burrowing is a common modification of primary structures. Discrete burrows are best represented in zone IIIa, where some organism such as a marine worm or crab has constructed mud-lined cylindrical tubes averaging 13 mm in inside diameter. The outside of the micritic wall blends into the surrounding pelmicrite beyond a distance of 2 or 3 mm from the smooth interior. The burrows are filled with micrite and pellets and/or skeletal debris. Sections of burrows evident in the rock are typically nearly vertical, but some also pass through a gentle curve to a horizontal position. Distinct burrows appear also in the inner shelf zone and in the transitional area between inner and middle shelf. These burrows, although sharply defined, have no constructed walls; the infilling detritus consists of relatively clean-washed pellets, in contrast with the surrounding matrix of pelmicrite or micrite. Both types of burrows are represented, to a small degree, in facies IV. Burrows are one of the most common and most easily recognizable secondary biogenic structures observed in the geologic record. They are commonly preserved, and provide valuable clues to the presence and activities of many soft-bodied organisms which leave no other record. Perkins (1966) has emphasized the importance of distinguishing between burrows (in unconsolidated sediments) and borings (in rock). Burrows are common in shallow-water sediments of Belize (Pusey, 1964, p. 95).

In addition to contributing well-defined burrows to the record, burrowing organisms no doubt are responsible for the development of many mottled fabrics. The extent of this effect is not easily definable, but its intergradation with known burrows permits a roughly quantitative estimate of the contribution of burrowers. Mottling resulting from partial homogenization and intermixing of small-scale sedimentary units, and attendant loss of primary lamination, commonly can be attributed to burrowing (Pusey, 1964, p. 96).

In the rock samples studied, borings are even more common and more variable than are burrows. A few large borings in massive coral heads are the works of lithophagous pelecypods, which are found intact, imbedded within the coral skeleton. These borings are common in modern reef corals (Pusey, 1964, p. 115), where they probably represent the same genera which effected the Pleistocene bores. Many large mollusk and coral fragments reveal extensive systems of smaller borings which approach diameters of 1 or 2 mm (Fig. 6B); these suggest the activity of boring sponges, such as *Cliona*, which today are very important agents in the destruction and size reduction of skeletal grains (Ginsburg, 1957, p. 83-84).

On a microscopic scale, the minute tubular borings of filamentous blue-green algae are widespread, affecting both individual skeletal grains and large areas of outcrop surface. The effects of these hardy and virtually ubiquitous algae have been well documented (Ginsburg, 1957, p. 85; Pusey, 1964, p. 70; and many others). Skeletal grains in the Ambergris Cay limestones commonly bear peripheral networks of tubular borings, usually 2 or 4μ in diameter, which are probably caused by these algae. Their effect is even greater on limestone exposures, where they typically coat the surface with a gray or green film (Fig. 6C) that is perhaps associated symbiotically with boring fungi in a form of lichen. The rock surface beneath is riddled with their minute borings to a depth of a few millimeters, and the chemical and mechanical erosion of the limestone is thus expedited.

Chemical Aspects

Outcrops of all the Pleistocene facies locally are smoothly rounded, particularly in highlands associated with heavy soil cover and jungle growth. More typical, however, is the development of an extremely jagged surface with isolated pinnacles separated by irregular pits (Fig. 2A). Relief on these surfaces approaches a maximum of approximately 1 ft (0.3 m), as all exposures are not far above sea level. These surfaces are characteristic of solution-etched limestone in karst areas and, in the Alps, have been designated "lapiés" by Jovan Cvijics (Thornbury, 1954). Parker and Cooke (1944, p. 28) illustrated similar outcrops of Pleistocene limestones in Florida, and Illing (1954, p. 75) noted others from the Bahamas.

In facies containing massive colonial corals, large heads of *Montastrea* and *Diploria* are typically hollowed out by chemical weathering, leaving, in extreme cases, washtub-size pits in which water and sundry debris may collect. This selective destruction presumably results from a combination of mineralogic and porosity factors. The corals are composed of aragonite and hence would be expected to be more soluble, generally, than the enclosing matrix, which is predominantly calcite. Furthermore, the coral heads are much more porous, and thus more vulnerable to dissolving waters than is their fine-grained matrix.

Other aragonitic skeletons such as mollusks and dense corals such as *Acropora* (Fig. 4A, B) weather in relief above the calcitic matrix. This relation indicates the great importance of relative skeletal porosity, as reflected in taxonomic character—more important, indeed, than simply mineralogic differences in determining relative solubilities.

PETROGRAPHY AND MINERALOGY OF DIAGENESIS

The value of the study of limestone diagenesis lies primarily in enabling a worker to trace the kind, sequence, and perhaps the rate of diagenetic alteration which carbonate sediments and limestones have undergone, and to relate any chemical and mineralogic changes to observable changes in texture. The major goals of such studies are recognition and interpretation of diagenetic processes on the basis of criteria such as changes in petrographic texture, fabric, and mineralogy, which leave a direct record in the rocks. These criteria must be based upon knowledge of the primary condition of the materials being studied.

Many diagenetic effects are recorded in alterations of the skeletal constituents of a limestone. The present study has concentrated upon alteration of skeletal material, both because of its value as an indicator of diagenetic effects and because virtually all of the fauna is conspecific or at least congeneric with the organisms living nearby in environments resembling, if not truly homologous with, those represented in the Pleistocene limestone.

In terms of general preservation of skeletal materials in the rocks of Ambergris Cay, several points are notable. One of the most remarkable is the persistence of both aragonite and magnesian calcite (>4 percent $MgCO_3$) in many skeletons and constituent grains. Corals, mollusks, *Halimeda*, ooliths, pellets, and tunicate spicules remain as aragonite except where obviously replaced by calcite spar; furthermore, Foraminifera such as *Homotrema rubrum* and coralline algae remain as magnesian calcite. These two metastable mineral phases are so chemically reactive, to processes either of solution or ion exchange, that their retention in the geologic record is commonly limited to rocks of relatively recent origin. Aragonite has been described in other Tertiary sequences (Friedman, 1964; Stanley, 1966), but is not commonly preserved in older rocks. The persistence of magnesian calcite is even more exceptional (Stehli and Hower, 1961; Friedman, 1964). In Pleistocene rocks from the Caribbean region and elsewhere, Friedman (1964) and Stanley (1966) found that aragonitic fossils tended to be dissolved, and magnesian calcite fossils converted to calcite, when subaerially exposed. X-ray analyses of Ambergris Cay limestones indicate that magnesian calcite persists in coralline algae and encrusting foraminifers but is lost from echinoids. All samples of this Pleistocene limestone were collected from subaerial outcrops; the preservation of aragonite, and particularly of magnesian calcite, in the vadose zone is unusual for rocks of this age. Either the rocks are too young or too impermeable for alteration to have reached completion, or else Friedman's (1964) and Stanley's (1966) inferences are not applicable here. Friedman (1964) concluded that Pleistocene skeletal material sampled from the present spray zone or intertidal area retains primary aragonitic and magnesian calcitic compositions because of its prolonged exposure to seawater, whereas rocks entirely within the realm of freshwater influence lose aragonite and magnesium calcite through solution and alteration. However, the present position of the spray zone and sea level is a recent feature which has followed a long history of post-Pleistocene rise in sea level; therefore, this zone has not been exposed continuously to the influence of seawater, but has been subjected to subaerial influence throughout most of its postdepositional history. Similarly, the rocks which underlie shallow supratidal ponds on Ambergris Cay and the waters of Chetumal Bay have been subjected to a diagenetic regime which differs only superficially from that which affected the subaerial outcrops, in spite of present differences in environment of exposure.

Virtually all of the carbonate mudstone matrix in the Pleistocene samples is calcite, as indicated by both staining and X-ray analyses. The persistence of aragonite in microcrystalline pellets, in contrast to alteration of matrix carbonate mud to micrite, presents further interpretive problems. By analogy with Holocene sediments of the Belize shelf (Pusey, 1964; Matthews, 1965), it is assumed that the carbonate mudstone matrix was composed originally of aragonite, with variable proportions of magnesian calcite and minor amounts of calcite. Thus, the apparent conversion of much aragonite and magnesian calcite of clay size to calcite mudstone is a major problem in limestone diagenesis—an enigma for which data are virtually nonexistent. Folk (1965) presented a comprehensive analysis of the current philosophy dealing with this problem. Recognition of grain types, in all stages of alteration, is essential in order both to delineate significant facies within the Pleistocene rocks and to trace the sequence of diagenetic changes. Criteria for recognizing important grain types in Holocene sediments of Belize have been outlined and illustrated at length by Pusey (1964, p. 58-85) and Matthews (1965, p. 42-59). Most of the constituents of the Pleistocene rock are included in these categories and are distinguished in the same manner. Other literature by Bøggild (1930), Purdy (1963), Vaughan (1919), and Johnson (1961) also is applicable.

Skeletal Alteration

Alteration of skeletal materials may be approached from the viewpoints of four common categories: solution, precipitation in voids, replacement, and recrystallization. Each of these processes may be effective in altering primary aragonitic or magnesian calcitic skeletons; calcite containing less than about 4 percent $MgCO_3$ (i.e., the relatively stable polymorph) tends to be preserved in primary grains and introduced as a secondary constituent.

Solution

Solution effects are commonly evident in thin sections. Corrosion and pitting of grains and shredding of fossils are among the most common diagenetic alterations. Skeletal grains undergo solution both before

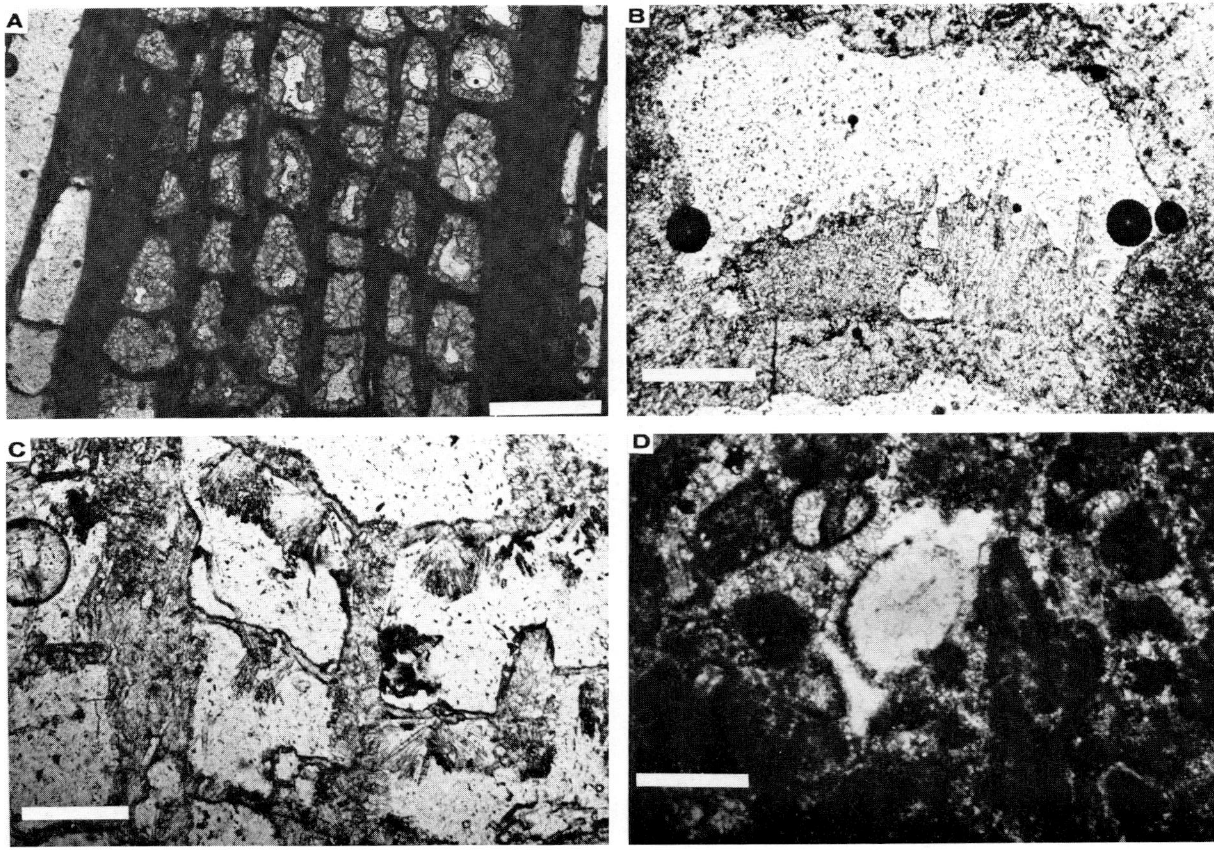

FIG. 7—Precipitated $CaCO_3$.

A. Longitudinal section of *Montastrea annularis*, showing filling of pores in aragonitic skeleton by calcite druse and spar mosaics. In such corals, void-filling spar crystals typically are equant throughout a pore, in contrast with a coarsening of crystal size away from void walls in general case. Bar length 0.5 mm. Plane-polarized light.

B. Prismatic needles of aragonite, based on "seed" crystals in wall of skeletal pore of *Montastrea annularis* and growing into empty pore. Bar length 0.05 mm. Plane-polarized light.

C. Fans of acicular aragonite growing into coral pores from points on pore walls, in *M. annularis.* Bar length 0.2 mm. Plane-polarized light.

D. Syntaxial overgrowth (white, in center of photomicrograph) on small, lentil-shaped echinoderm fragment within an oolitic biopelmicrite. Overgrowth appears to replace surrounding micrite. Bar length 0.5 mm. Plane-polarized light.

and after incorporation in a lithified sediment. When effective before deposition, solution may render outer morphology, internal structures, or wall microstructure unrecognizable, hindering proper interpretation of the kind and interrelation of constituent grains. When solution follows lithification, skeletal grains may be masked or removed, in which case the resulting voids may either remain as pores or be filled by sparry calcite. Much of the solution in the Pleistocene samples appears to be late, so that boundaries of solution pores are maintained rigidly by the enclosing matrix.

Halimeda flakes commonly are partly dissolved; internal porosity increases as utricles are enlarged. The severity of solution may vary locally; some *Halimeda* specimens may be almost completely dissolved, whereas others in the same thin section, isolated within a less permeable matrix, are untouched.

Coralline algae such as branching *Amphiroa* commonly show corrosion along boundaries of microstructure (Fig. 6D). Mollusks commonly are pitted and leached, and a micrite envelope which typically develops along the peripheries of mollusk grains is probably the result of corrosion accompanying microscopic borings (Bathurst, 1966; Winland, 1968).

Among the nonskeletal constituent grains, ooliths are most susceptible to solution, leaving a highly porous system of molds in some areas where the aragonitic ooliths have been leached from their calcitic matrix.

Precipitation in Voids

Much secondary crystalline calcite definitely is attributable to deposition in voids; some cases are less clearly evident. Bathurst (1958) prepared the first well-publicized set of criteria for distinguishing between void-filling and "recrystallization" calcite. Subsequently, other authors have attempted to verify or amplify these criteria (Folk, 1965; Dodd, 1966; Stanley, 1966). The problem is simplified where a surface now lined, for example, with sparry calcite can be determined definitely to have been a free surface at the time of deposition of the lining.

Precipitation in skeletal voids occurs in a variety of ways. Most typically, a rim of fine calcite druse lines the cavity and grades into an increasingly coarser spar mosaic in the interior; this is typical of pore fill in gastropods, Foraminifera, and paired valves of pelecypods or ostracods, and has been described many times in the literature (Bathurst, 1958, 1964; Dickson, 1966; Dodd, 1966). However, in some places, crystals of sparry calcite are coarse grained throughout the entire mosaic; this texture is more typical of pore fill in corals (Fig. 7A) of the present study.

Syntaxial overgrowths of calcite appear in several forms, depending upon the base or seed crystal on which each has developed. Each crystal is optically continuous with its host.

Extending from the periphery into the interior of pores of some corals are slender prisms or blades of aragonite which are syntaxial with the aragonitic crystals in the wall of the coral skeleton (Fig. 7B). These secondary crystals may continue growth until complete occlusion of the pore is achieved. A modified form of this type of overgrowth, resulting in crystals which are not strictly syntaxial, also is evident in the pores of some corals; radiating fans of fibrous or prismatic aragonite expand from a point of origin on the surface of the skeletal pore (Fig. 7C). Aragonite may completely fill a pore through continued growth of one or several systems of fans.

Fragments of echinoderms may or may not bear syntaxial overgrowths. Where an appreciable micrite envelope (Bathurst, 1964, 1966) coats such a grain, a composite drusy or bladed crust may develop. However, more typically, "rim cement" (Bathurst, 1958), known widely as (syntaxial) "overgrowth" (Lucia, 1962; Evamy and Shearman, 1965; and others), is the type of coating deposited (Fig. 7D). Enlargement of these syntaxial rims may replace parts of the surrounding micrite matrix.

Replacement

In this paper, the term "replacement" is limited to the introduction of a new mineral species in the place of a primary constituent. Any process, then, by which aragonite is removed and calcite is deposited within the same borders may be called "replacement"; the development of texture similar to this but involving no change in mineralogy must be termed "recrystallization" (Dodd, 1966; Tebbutt, in prep.).

Bathurst (1964) and several contemporary authors distinguish two basic means of replacement of aragonite by calcite: solution-deposition and *in situ* replacement. The solution-deposition process involves an intermediate void stage before introduction of the calcite, whereas *in situ* replacement has been effected with no large intermediate void. The *in situ* process is believed by some authors to involve a solid-state conversion; others believe a thin liquid film is necessary at the replacement interface (Folk, 1965, p. 42). The possibility of inversion in the solid state, without water, recently has been considered at length by several authors, and is generally regarded as inapplicable (Dodd, 1966, p. 733; Fyfe, 1964).

Most favorably regarded is a process of almost simultaneous solution of aragonite and deposition of calcite on a microscale (Friedman, 1964, p. 810; Bathurst, 1964, p. 373).

Where relics of primary structures are evident, the effective process may be termed "*in situ* replacement," involving the introduction of a different mineral (whether or not a polymorph of the original) in the presence of a thin aqueous film. Replacement, *sensu lato*, remains to be used where either relics cannot be recognized or solution has produced significant moldic voids prior to precipitation of calcite.

Gastropods—One example of *in situ* replacement in the Pleistocene limestones of Ambergris Cay is evident in gastropods, including *Strombus gigas*, *Bulla*, and *Cerithium*. Commonly, sparry calcite replaces primary aragonitic shells in which their characteristic crossed lamellar (Bøggild, 1930, p. 251) and other structures are well preserved. Typically, a replacement front advances from the interior surface of the shell outward, in some cases engulfing the entire skeleton. Stages of alteration are represented which are intermediate between fresh shells and shells completely altered to solid masses of calcite crystals (Tebbutt, in prep.). The area of replacement spar begins with broad-based crystals at the interior boundary of the shell and passes into a uniformly coarse mosaic; normally, there is no systematic increase or decrease in crystal size throughout the mosaic (Fig. 8). The contact between secondary calcite spar and primary aragonitic shell may be sharp, but is commonly ragged and interfingering. The most characteristic feature of the *in situ* replacement spar is the presence of inclusions, presumably remnants of the organic (conchiolin) matrix which permeates the microstructure of the shell (Fig. 8). These relic ghosts outline, in general, the primary structure of the aragonitic shell, and are continuous with structural features in adjacent parts of the unreplaced shell. Most significantly, they indicate that no large voids existed between the time of removal of the aragonite and introduction of the calcite. In thin section, the line along which replacement begins (e.g., the inner shell surface) is commonly sharp, marking a distinct contact with any void-fill calcite which has been precipitated on the inside of this free surface. This

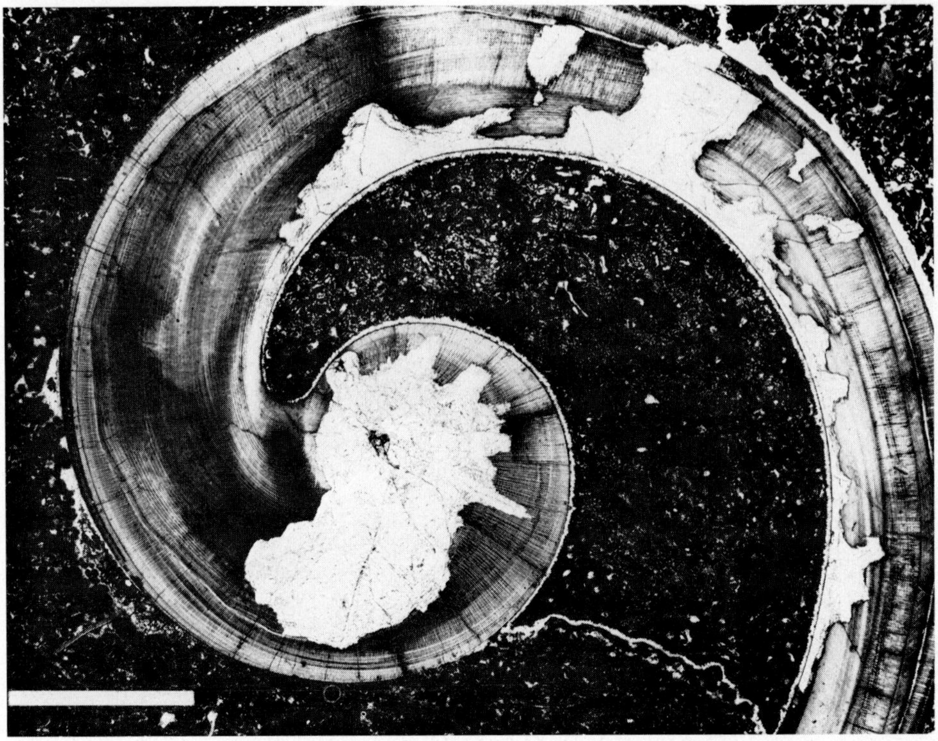

FIG. 8—Patches of coarse sparry calcite (white) replacing aragonitic shell of *Strombus gigas*, the queen conch. Replacement progresses along a jagged "front," typically from internal surface of shell, which is lined with fine drusy calcite, into shell. A few ghosts of primary structures are evident in patch to right of center, near upper edge of photomicrograph. Patch of coarse, elongate calcite crystals in columellar area (left of center) has grown radially from a central small cavity occupied by a few detrital pellets. A thin drusy rind of calcite coats interior of innermost whorl and covers top of sediment (near base of photo) which incompletely fills gastropod. Bar length 4 mm. Transmitted light.

latter drusy calcite is commonly much finer grained, particularly near its surface of initial growth, than is the coarse mosaic occupying the shell. Furthermore, crystals of replacement spar tend to be more anhedral than those of the drusy pore fill (Tebbutt, in prep.).

Crystals of replacement calcite reach a maximum length of more than 4 mm. Their optic axes usually appear to be randomly oriented and they are typically equant. However, in a few cases—such as in Figure 8, which illustrates replacement in the columellar area of *Strombus gigas*—definite elongation of spar crystals is evident. The optic axes coincide with the elongate *c*-axes in these bladed crystals, which are radially divergent from what appears to have been a secondary void—perhaps a boring or other gap in skeletal structure.

Very commonly, crossed lamellar structure has been replaced; other zones are also invaded, however, and no systematic preference is evident. The common replacement of crossed lamellar structure probably just reflects the relative abundance of this primary shell structure. In most examples of this type of replacement, the surface from which the front advances is coated with drusy calcite pore fill (Fig. 8), suggesting, perhaps, some interdependence of the two modes of occurrence of spar. Commonly, a gastropod is completely filled with coarse void-fill spar, and only small isolated salients (if any) extend into the shell as replacement crystals. The salient may be syntaxial with the adjacent pore-fill spar, and may incorporate a ghost outline of the shell boundary. A similar relation in corals will be described.

The gastropods described above retain visible relics of their primary structures in the form of vague ghosts of impurities which are continuous with discrete structural boundaries in adjacent unreplaced parts of a shell. However, where the shell has been completely replaced by a coarse mosaic, relics are less commonly evident. Otherwise, the mosaics are petrographically indistinguishable from those where ghosts are obvious; furthermore, all gradations between these extremes of preservation quality may be found. Therefore, probably only a single, variable phenomenon is involved. The process appears to be one of solution of aragonite and deposition of calcite, on a more or less minute scale, along an advancing replacement front. Where relics are preserved, the process may be called "*in situ* replacement" to signify the small scale of the replacement interface.

Corals—Several fossils in addition to gastropods in the Ambergris Cay limestones have also undergone *in situ* replacement. *Montastrea annularis* is the most common coral in these deposits; it illustrates many examples of the major diagenetic processes. All stages of pore filling of these aragonitic corals are evident. Almost invariably, the corallites are filled with either detrital or crustiform, saccharoidal brown micrite (i.e., calcite), and thus contain no crystalline spar. The coenosteal pores, however, may be partly or completely filled with coarse crystals of drusy calcite. Where incompletely filled with spar, the coenosteal skeleton typically remains aragonitic; however, where pores are occluded, the enclosing skeleton is replaced by sparry calcite (Fig. 9A). The resulting continuous mosaic of coarse sparry calcite retains relics of the original pore boundaries either as intracrystalline ghosts or as sharp contacts between fill and replacement spar (Tebbutt, in prep.). Many replacement crystals in the skeleton are syntaxial with adjacent crystals of pore-fill spar.

The progressive stages of pore filling of the aragonitic coenosteum, and replacement of the skeleton by calcite only in the case of complete pore occlusion, suggest that the pores were filled by calcite before replacement occurred. This is the reverse of the sequence interpreted by Matthews and Mesolella (1966) for alteration of Barbados corals of this same species and approximately the same age.

The fact that relics of skeletal boundaries are well preserved and not displaced implies that the alteration process was one of *in situ* replacement.

Halimeda—Probably the most common occurrence of *in situ* replacement is in the widespread and abundant flakes of the aragonitic green alga *Halimeda*. The manner of its replacement by calcite appears similar to that of the corals. Internal tubular pores (utricles) of *Halimeda* segments are commonly lined or filled with medium-grained drusy calcite. In many flakes in which pores are completely occluded, the enclosing skeleton has been replaced by a mosaic of sparry calcite in which crystals are comparable in size to those filling the pores. The resulting structure is a continuous mosaic of medium-grained calcite crystals (Fig. 9B), commonly with boundaries of primary pores indicated either by ghosts or by a sharp line dividing void fill from replacement spar. However, in the extreme case, relics are obliterated entirely, leaving a mass of equant spar which assumes only the outer form of the original fragment of *Halimeda*.

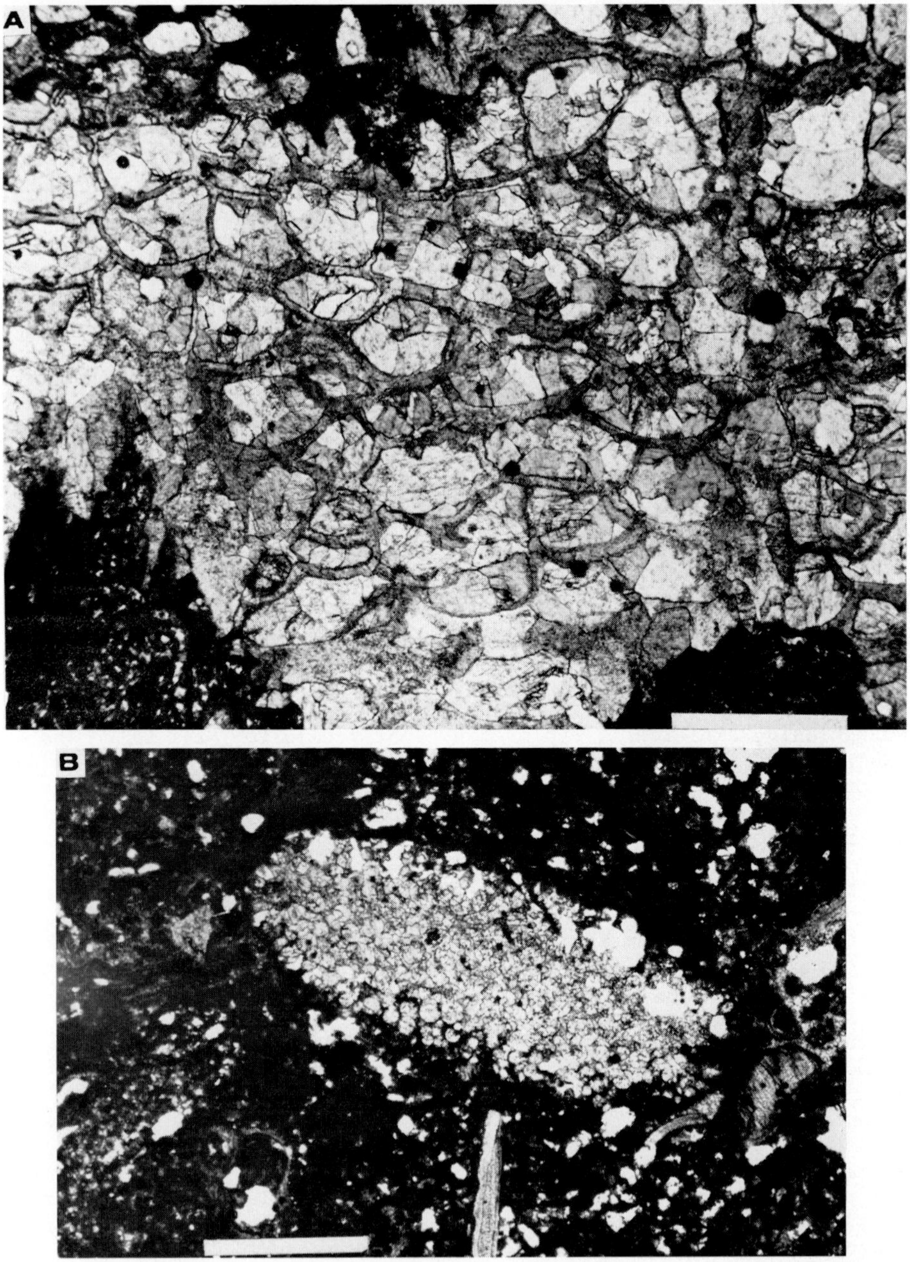

FIG. 9—**A.** Oblique section of *Montastrea annularis*, showing complete occlusion of coenosteal pores (parts of three micrite-filled corallites are evident, top and bottom) by sparry calcite. After filling of pores, enclosing skeleton was replaced by calcite; in specimens elsewhere, where pores are not completely filled by spar (Fig. 7A), skeleton remains aragonitic. Note, in center, that several crystals of replacement spar in skeleton are syntaxial with calcite crystals in adjacent pore fill. Bar length 0.5 mm. Plane-polarized light.

B. *Halimeda* flake in which utricles have been filled by medium and finely crystalline calcite spar, and enclosing skeleton has been replaced by a mosaic of calcite crystals of same size as those filling internal pores. Result is a continuous mosaic of calcite in which primary structures are obscured. Surrounding matrix is a fine-grained biopelmicrite. Bar length 1 mm. Plane-polarized light.

Recrystallization

This vague and commonly misused term is limited, in this text, to alteration involving change in size, shape, or arrangement of crystals but not in mineralogy. Thus aragonite does not, by this definition, recrystallize to calcite. In a very general sense, the term "recrystallization" may still find valid use where primary and/or secondary mineralogy are unknown and only a textural difference in crystallinity is evident (e.g., in field descriptions); however, the term should be used more precisely in petrographic studies, in which staining and other rapid analytical techniques are used.

Pelecypods—A few heavy-shelled pelecypods such as *Lucina pensylvanica* have been partly altered to coarsely crystalline mosaics which superficially resemble the replacement-spar mosaics found in the gastropods. Thin sections show transparent, altered patches which consist of coarse, bladed crystals trending perpendicular to the surfaces of the shell (Fig. 10A). Blades approach 1 mm in maximum length and 1/3 mm in width at the broadest base. Individual crystals are not sharply outlined, even in cross-polarized light; progressively smaller bladed crystals are intercalated between the larger ones as the latter wedge out. The secondary, bladed crystals are aragonite, as is the original shell, so that the phenomenon must be classed as recrystallization. The mass of secondary crystals may begin from some intermediate horizon within the shell, but it usually has a flat base, with crystals tapering toward the exterior; successive tiers of blades at several levels are typical. Some recrystallized areas are sharply defined and distinctly discordant with primary shell structures. Usually, at least a few ghosts of original structures are incorporated in the secondary crystals (Fig. 10A). Primary features appear to be virtually intact where the transverse blades are vaguely discernible and small. There appears to be every gradation between subtle, incipient development of secondary crystals, where structural outlines are well preserved, and extensive recrystallization to masses of large, distinct blades in which few or no relics are evident. A remarkable aspect of the secondary mosaics is their apparent distortional effect on primary structures. Isolated ghosts within the mosaic commonly appear to be displaced or rotated from their initial positions (Fig. 10A), and traces of shell lamination, both in the recrystallized blades and in the unaltered shell adjacent to the contact, are buckled considerably. Where the coarse blades reach the outer surface of a shell, they typically extend beyond the level of the adjacent unreplaced shell. The scarce examples of these partly recrystallized pelecypods are found in the same facies, and even in the same thin section, as are gastropods which have been replaced by sparry calcite. Recrystallization of this type might precede ultimate replacement by calcite, but no evidence was found in the Ambergris Cay limestone for such a sequence.

Cementation of Limestones

The mobility and instability inherent in the polymorphs of $CaCO_3$ which constitute limestones are responsible for the development of textures characteristic of both sedimentary and igneous rocks. We have already examined the role of replacement and recrystallization—two processes which are accelerated in limestones relative to their rate in more refractory sedimentary rocks.

For the purposes of the following discussion, granular cement (between grains) and drusy mosaics (within grains or fossils) are not differentiated, as each originates in the same manner and intermediate examples are common. The term "druse" will be used for finely crystalline crusts (usually calcite) which coat free surfaces.

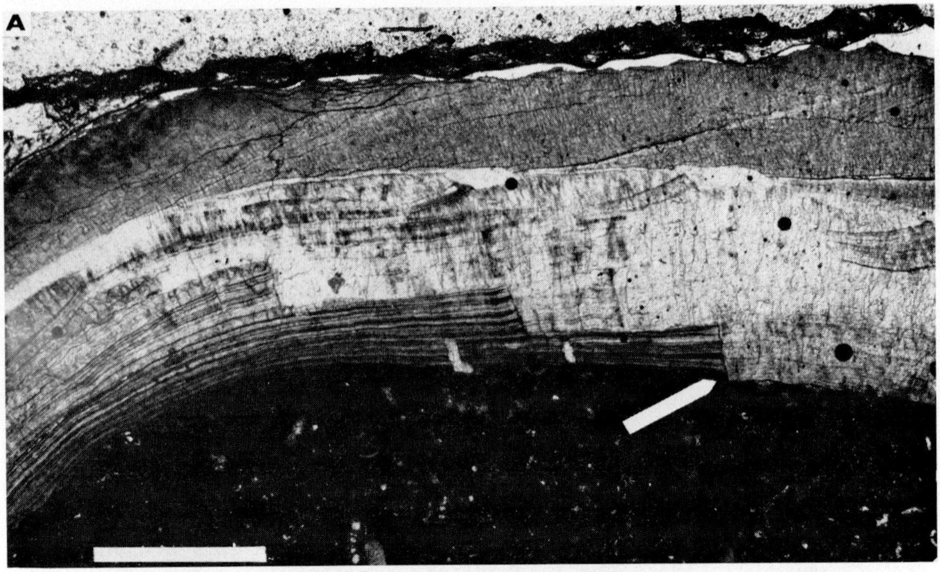

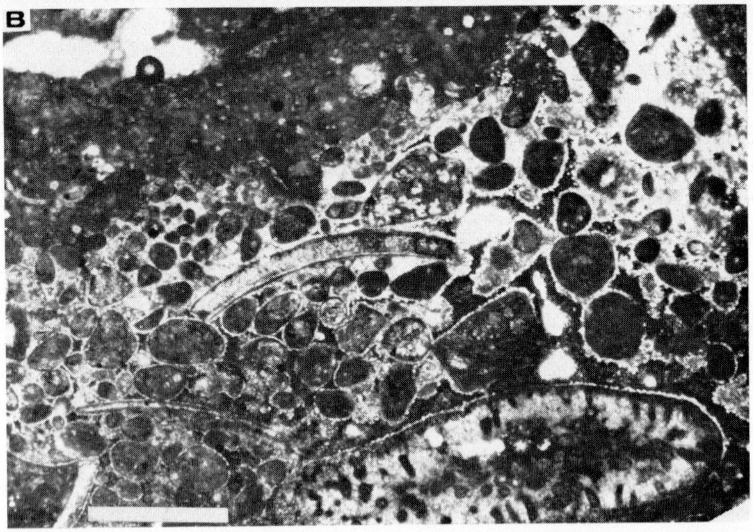

FIG. 10–A. Partly recrystallized shell of *Lucina pensylvanica.* Zone of recrystallization occupies lower half of shelf at right of photomicrograph, diminishing in height by a series of tiers which rise, to left, into center of shell. Edge of thin section is marked by heavy, rough line near top. Note apparent expansion in length of crystals in zone at right beyond level of unaltered shell (arrow). Ghosts are evident in much of recrystallized zone. Bar length 2 mm. Plane-polarized light.

B. Finely drusy rinds of calcite coating skeletal grains, pellets, and ooliths, and cementing grains together at and near grain contacts. Note large *Halimeda* segment in lower right, curved mollusk fragment in center. Bar length 0.5 mm. Plane-polarized light.

Most calcarenites, particularly where a calcilutite matrix is minor or absent, are cemented by the deposition of drusy calcite in interstitial pores. Initially, finely drusy rims commonly are precipitated upon the constituent grains, binding particles together by the encrusting process, at and near grain contacts (Fig. 10B). Continuing precipitation of calcite results in interlocking mosaics of spar which commonly increase progressively in size from the fine drusy rims toward the center of the interstitial voids (Bathurst, 1958, p. 15; 1964, p. 362; Friedman, 1964, p. 783; Dickson, 1966, p. 502). However, it is not uncommon, particularly in corals, for initial void-fill spar to be medium or coarse grained, and for the rest of the mosaic spar to be of about equal size (Fig. 7A).

Where a micrite matrix predominates, whether in a packstone or wackestone, consolidation is effected by an enigmatic process of recrystallization and/or replacement which might be called "micrite fusion." This confusing phenomenon involves an apparent, fine granular encrustation of constituent particles and some sort of suturing of microcrystalline grains of carbonate. The problem involves the lithification of carbonate mud, a subject which has been avoided by most carbonate petrologists. Folk (1965) presented a thoughtful and concise view of this development of micrite from carbonate mud. He proposed several alternative hypotheses by which unconsolidated aragonite needles are converted to mosaics of microcrystalline calcite. A major factor to consider is the introduction of large volumes of carbonate in solution, to fill the high initial porosity without permitting significant compaction; this factor is in addition to the chemical problem of converting aragonite to calcite at some intermediate stage of alteration. Progress in the study of micrite genesis and diagenesis must await further refinements in chemical and petrographic analytical techniques.

The environment of lithification of carbonate sediments is another major problem about which very little is known. Most workers agree that subaerial exposure is necessary for deposition of calcite and cementation of carbonate sediments (Purdy, 1965; Friedman, 1964), but a few cite examples of apparent submarine lithification (Milliman, 1966; Gevirtz and Friedman, 1966; Thompson et al, 1968). The limestones collected on Ambergris Cay all have been subjected to subaerial effects for such a long time that any previous submarine cementation would be indistinguishable. A study of O^{18}/O^{16} or C^{13}/C^{12} ratios in these rocks perhaps could determine where lithification took place (Gross, 1964).

SUMMARY AND CONCLUSIONS

Four facies are recognizable in the Pleistocene limestone exposures: reef crest, backreef, shelf lagoon, and mudbank. The reef-crest facies interrupts the trend of the modern reef, on Reef Point, and is characterized by the same fauna, including *Acropora palmata*, as found in the living reef. Thickets of *Acropora cervicornis* characterize the backreef facies, which lies directly lagoonward from the Pleistocene reef crest. Poorly sorted carbonate silt and mud with abundant tunicate spicules and larger, angular fragments of corals and a few mollusks constitute the matrix of these two seaward facies. The fine-grained sediments probably accumulated in these environments of high wave energy by filtering progressively down into the coarsely porous reef frame as growth continued above.

The shelf-lagoon facies encompasses most of the outcrop on Ambergris Cay and exhibits a wide variety of subfacies, among which are distinguished outer, middle, and inner shelf zones. All rocks of this facies are pelmicrites or biopelmicrites. Corals are abundant in outer shelf and patch-reef areas, diminishing away from the reef, and *Strombus gigas* also characterizes the outer shelf zone. The rock of the middle shelf consists principally of moderately well-sorted pelmicrite with mottled admixtures of micrite and ooliths; the latter increase in abundance toward the interior of Ambergris Cay. *Bulla* is the most common representative of the meager fauna in this zone. Rocks of the inner shelf consist of burrowed pelmicrites and micrites with high concentrations of *Halimeda* and of heavy-shelled pelecypods, chiefly *Chione cancellata, Lucina pensylvanica,* and *Tellina interrupta.*

The mudbank facies is relatively unfossiliferous; only a few molds of a small *Bulla*(?) typically are present. The characteristic rock type may be described as a dismicrite with many breccialike partings. Within these partings and in many possible bore holes are abundant vesicular structures which might represent epiphytic foraminifers or other "primitive" organisms which encrusted marine grasses and subsequently were buried in carbonate mud.

A reconstruction of Pleistocene depositional environments is illustrated in Figure 11. The reef-crest, backreef, and outer-shelf assemblages are virtually identical with their counterparts in the Holocene sediments of the Belize continental shelf. However, the middle shelf zone of the Pleistocene has no homologue in the Holocene; no ooliths have been reported from the Holocene sediments, and pellets are uncommon (Pusey, 1964, p. 80). The ooliths indicate that, in some respects, the Pleistocene depositional regime must have differed markedly from that of the Holocene. Carbonate saturation values and agitation must have reached levels higher than any attained on the shelf today in order for deposition of the ooliths to have occurred. In the classic area of oolith formation on Andros platform in the Bahamas (Purdy, 1963), the closest similarity to the Pleistocene reef- and shelf-sediment distribution of Ambergris Cay appears to be in the Joulters Cay area, north of Andros Island (Purdy and Imbrie, 1964). In this area, a windward coral reef fringes the platform, passing lagoonward through a narrow zone of coralgal carbonate sand into an area of oolitic sand. As oceanic waters pass over the rim of the platform, they become warmed, agitated, and depleted in CO_2, eventually reaching a saturation level at which precipitation can take place. A similar situation must have existed in the Ambergris Cay oolitic sands. The Pleistocene facies seem to be continuous and intergradational, and indicate no island or sheltered mud area between the coralgal sands and oolitic sands. This suggests that currents passing over the shelf edge were unidirectional; otherwise, the return flow of saline waters would have been detrimental to coral-reef growth. Perhaps currents were effective in directing these waters southward along the coast. The diminishing proportion of ooliths southward suggests increasing water depths where ooliths did not form and where pellets predominated. The middle shelf zone thus is interpreted as having been deposited as a series of shifting oolitic shoals to the north, grading into deeper waters of pellet-mud accumulation toward the south of the present island.

Stanley (1966) described a somewhat similar association of coral reefs and lagoonal shoals of oolite in Pleistocene limestones of the Florida Keys. Other near-Holocene oolites appear as dune deposits of presumed Pleistocene age on Isla Mujeres (Folk et al, 1962), about 200 mi (320 km) north of Ambergris Cay. Ooliths are not believed to be forming either in the Florida reef tract or near Isla Mujeres; thus, these are two more instances of abundant oolite in Pleistocene rocks with no counterpart in the adjacent Holocene sediments. If this observation has any broad implications, they are not yet evident; however, tidal effects might have been greater during the interglacial times of higher sea level (i.e., more water in all the oceans), causing more diurnal agitation of sediments on continental shelves—agitation which, in combination with other factors, might effect precipitation of carbonate as ooliths.

The inner shelf zone (IIIc) must have been more conducive to the growth of a vigorous biota. To some extent, ooliths are important in

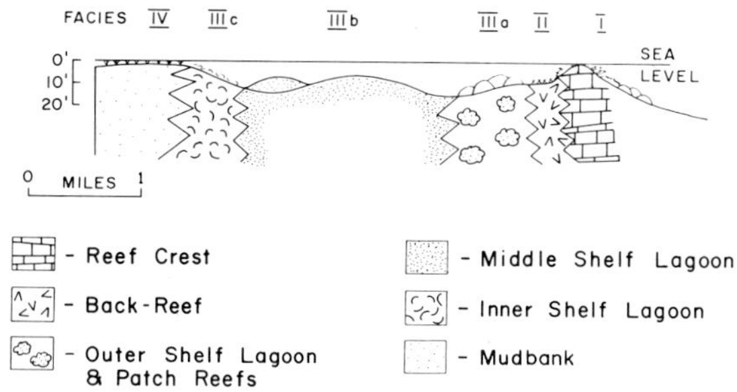

FIG. 11—Cross section showing interpreted Pleistocene facies distribution.

this zone as well, but the presence of abundant fossils distinguishes it from zone IIIb. The major mollusks of this facies are found today as burrowing infauna in shoaling mud bars and sandbars. Their lack of fragmentation indicates only short transport from their biocoenoses, and the locally very high concentrations of mollusks and *Halimeda* suggest a shallow-water bar or a beach deposit.

Facies IV appears to represent either a shoaling mudbank covered with *Thalassia* or a mud flat at or near sea level. The paucity of macrofauna suggests severe conditions of perhaps periodic subaerial exposure, and partings within the mud could represent shrinkage due to desiccation. However, no direct evidence of exposure was observed. If the facies represented deeper water deposition, mollusks would be expected to have been introduced from zone IIIc; such is not the case.

Laminated brown crusts of porous cryptocrystalline calcite which coat the Pleistocene outcrops in some areas are believed to represent Holocene caliche.

Burrows and borings are the most significant aspects of the physical diagenesis of the Ambergris Cay limestone. Borings include the work of mollusks and filamentous algae. Chemical effects include solution of coral heads in outcrop, from within their dense matrix of biomicrite. Density and porosity appear to be more important factors than mineralogy in selective solution of fossils.

The diagenetic history of the Pleistocene limestones is manifested, to a large extent, in alteration of skeletal materials. Both aragonite and magnesian calcite are preserved in these fossils where they are not obviously replaced by sparry calcite. Aragonite persists in mollusks, corals, pellets, and ooliths, and magnesian calcite in coralline algae and Foraminifera. In attempting to relate observed textural changes to changes in mineralogy of the fossils, the four common categories—solution, precipitation in voids, replacement, and recrystallization—may be considered.

Solution is evident on a microscale as well as on a larger scale in outcrops; many skeletal grains are corroded, leached, or shredded.

Precipitation in voids is evident as drusy calcite linings or fillings of cavities and as a variety of syntaxial overgrowths. These overgrowths appear as bladed or acicular aragonite extending into pores of aragonitic corals, and as enlargements of echinoderm fragments; in each case, the syntaxial crystals are of the same mineralogy and optical orientation as the host crystals upon which they are based.

Replacement of aragonitic shells by advancing "fronts" of coarsely crystalline calcite spar is evident in gastropods such as *Strombus gigas*, *Bulla*, and *Cerithium*. The replacement is of an *in situ* variety, probably effected by solution and nearly simultaneous precipitation on a minute scale, at the ragged boundary between fresh shell and invading spar. The presence of ghosts of primary microstructures indicates the absence of an intermediate stage of large voids. *In situ* replacement also occurs in corals and in *Halimeda* flakes; in both cases, pores first are filled with sparry calcite, then the surrounding skeletons are replaced, resulting in solid mosaics of calcite with outlines of original structures evident as ghosts.

Recrystallization, defined here as alteration in crystallinity without change in mineralogy, is evident in several skeletal types. Patches within a few pelecypod shells are recrystallized to coarse transverse blades of aragonite in which vague ghosts remain to outline primary structures. Most of the recrystallized patches show an apparent volumetric expansion of between 4 and 20 percent.

Cementation of limestones is effected primarily by recrystallization and by precipitation of drusy calcite in pores. All samples in this study have been subaerially exposed for a long time, so that any possible effects of submarine lithification processes are no longer evident.

The major purpose of this study has been to illustrate and interpret intermediate stages of skeletal alteration which are rarely evident in ancient limestones. Certain types of alteration are characteristic of specific groups of fossils and may be used to indicate their presence in an altered rock. Recognition of the types of diagenetic alteration which sediments have undergone is necessary for reconstructing conditions of deposition on the basis of constituent fossils.

SELECTED REFERENCES

Bathurst, R. G. C., 1958, Diagenetic fabrics in some British Dinantian limestones: Liverpool and Manchester Geol. Jour., v. 2, p. 11-36.

——1964, The replacement of aragonite by calcite in molluscan shell, *in* Approaches to paleoecology: New York, John Wiley & Sons, p. 357-376.

——1966, Boring algae, micrite envelopes and lithification of molluscan biosparites: Geol. Jour., v. 5, pt. 1, p. 15-32.

Bøggild, O. B., 1930, The shell structure of the mollusks: Koninkl Danske Vidensk. Selskab Skr. Naturvidensk. Math-Afd., raek 9, v. 2, p. 231-326.

Butterlin, J., and F. Bonet, 1962, Les formations cénozoîques de la partie Mexicaine de la presqu'île du Yucatán, *in* 3d Caribbean Geol. Conf., 1962: Kingston, Jamaica, 45 p. (mimeographed).

Chave, K. E., 1964, Skeletal durability and preservation, *in* Approaches to paleoecology: New York, John Wiley & Sons, p. 377-387.

Cloud, P. E., Jr., 1959, Geology of Saipan Mariana Islands, Part 4—Submarine topography and shoal-water ecology: U.S. Geol. Survey Prof. Paper 280-K, p. 361-445.

Cullis, C. G., 1904, The mineralogical changes observed in the cores of the Funafuti borings, *in* The atoll on Funafuti: Royal Soc. London, p. 392-420.

Dickson, J. A. D., 1966, Carbonate identification and genesis as revealed by staining: Jour. Sed. Petrology, v. 36, no. 2, p. 491-505.

Dodd, J. R., 1966, Processes of conversion of aragonite to calcite with examples from the Cretaceous of Texas: Jour. Sed. Petrology, v. 36, p. 733-741.

Dunham, R. J., 1962, Classification of carbonate rocks according to depositional texture, *in* W. E. Ham, ed., Classification of carbonate rocks: AAPG Mem. 1, p. 108-121.

Emery, K. O., J. I. Tracy, and H. S. Ladd, 1954, Geology of Bikini and nearby atolls: U.S. Geol. Survey Prof. Paper 260-A, p. 1-265.

Evamy, B. D., and D. J. Shearman, 1965, The development of overgrowths from echinoderm fragments: Sedimentology, v. 5, p. 211-233.

Fairbridge, R. W., 1950, Recent and Pleistocene coral reefs of Australia: Jour. Geology, v. 58, p. 330-401.

Fischer, A. G., and R. F. Garrison, 1967, Carbonate lithification on the sea floor: Jour. Geology, v. 75, no. 4, 488-496.

Folk, R. L., 1959, Practical petrographic classification of limestones: AAPG Bull., v. 43, p. 1-38.

——1962, Spectral subdivision of limestone types, *in* W. E. Ham, ed., Classification of carbonate rocks: AAPG Mem. 1, p. 62-84.

——1965, Some aspects of recrystallization in ancient limestones, *in* Dolomitization and limestone diagenesis: Soc. Econ. Paleontologists and Mineralogists Spec. Pub. 13, p. 14-48.

——M. O. Hayes, and R. Shoji, 1962, Carbonate sediments of Isla Mujeres, Quintana Roo, Mexico and vicinity, *in* Guidebook, Field trip to Peninsula of Yucatan, 1962: New Orleans Geol. Soc., p. 85-100.

Flores, G., 1952, Geology of northern British Honduras: AAPG Bull., v. 36, no. 2, p. 404-409.

Friedman, G. M., 1959, Identification of carbonate minerals by staining methods: Jour. Sed. Petrology, v. 29, p. 87-97.

——1964, Early diagenesis and lithification in carbonate sediments: Jour. Sed. Petrology, v. 34, no. 4, p. 777-813.

Fyfe, W. S., 1964, Calcite aragonite problem (abs.): AAPG Bull., v. 48, no. 4, p. 526.

Gevirtz, J. L., and G. M. Friedman, 1966, Deep-sea carbonate sediments of the Red Sea and their implications on marine lithification: Jour. Sed. Petrology, v. 36, no. 1, p. 143-151.

Ginsburg, R. N., 1956, Environmental relationships of grain size and constituent particles in some South Florida carbonate sediments: AAPG Bull., v. 40, p. 2384-2427.

——1957, Early diagenesis and lithification of shallow water carbonate sediments in South Florida, *in* Regional aspects of carbonate deposition: Soc. Econ. Paleontologists and Mineralogists Spec. Pub. 5, p. 80-99.

Goreau, T. F., 1959, The ecology of Jamaica coral reef, 1—Species composition and zonation: Ecology, v. 40, p. 67-90.

Gross, M. G., 1964, Variations in O^{18}/O^{16} and C^{13}/C^{12} ratios of diagenetically altered limestones in the Bermuda Islands: Jour. Geology, v. 72, no. 2, p. 170-194.

Heilprin, A., 1891, Geological researches in Yucatan: Acad. Natl. Sci. Philadelphia, Proc., p. 136-158.

Illing, L. V., 1954, Bahaman calcareous sands: AAPG Bull., v. 38, p. 1-95.

Johnson, J. H., 1961, Limestone-building algae and algal limestones: Golden, Colorado School of Mines, 197 p.

——1964, Ecology and oceanography of the coral-reef tract, Abaco Island, Bahamas: Geol. Soc. America Spec. Paper 79, 98 p.

Kornicker, L. S., 1958, Bahaman limestone crusts: Gulf Coast Assoc. Geol. Socs. Trans., v. 8, p. 167-170.

——and D. W. Boyd, 1962, Shallow-water geology and environments of Alacran reef complex, Campeche Bank, Mexico: AAPG Bull., v. 46, p. 640-673.

Ladd, H. S., and S. O. Schlanger, 1960, Drilling operations on Eniwetok Atoll: U.S. Geol. Survey Prof. Paper 260-Y, p. 863-905.

Lewis, J. B., 1960, The coral reefs and coral communities of Barbados, W.I.: Canadian Jour. Zoology, v. 38, p. 1133-1145.

Lucia, F. J., 1962, Diagenesis of a crinoidal sediment: Jour. Sed. Petrology, v. 32, p. 848-865.

Matthews, R. K., 1965, Genesis of lime mud in southern British Honduras: Rice Univ., PhD thesis, 139 p.

——and K. J. Mesolella, 1966, Subaerial diagenesis in reef-building corals from the Pliocene-Pleistocene of Barbados, West Indies (abs.): Geol. Soc. America Spec. Paper 87, p. 103.

Milliman, J. D., 1966, Submarine lithification of carbonate sediments: Science, v. 153, no. 3739, p. 994-997.

Multer, H. G., and J. E. Hoffmeister, 1968, Subaerial laminated crusts of the Florida Keys: Geol. Soc. America Bull., v. 79, no. 2, p. 183-192.

Newell, N. D., 1958, American coral seas (West Indies): New York Acad. Sci. Trans., ser. 2, v. 21, no. 2, p. 125-127.

——E. G. Purdy, and J. Imbrie, 1960, Bahamian oölitic sand: Jour. Geology, v. 68, p. 481-497.

——and J. K. Rigby, 1957, Geological studies of the Great Bahama Bank, *in* Regional aspects of carbonate deposition: Soc. Econ. Paleontologists and Mineralogists Spec. Pub. 5, p. 15-79.

Ower, L. N., 1928, Geology of British Honduras: Jour. Geology, v. 36, p. 494-509.

Parker, G. E., and C. W. Cooke, 1944, Late Cenozoic geology of southern Florida, with a discussion of the ground water: Florida Geol. Survey Geol. Bull., no. 27, p. 27-29.

Perkins, B. F., 1966, Rock-boring organisms as markers of stratigraphic breaks (abs.): AAPG Bull., v. 50, no. 3, pt. 1, p. 631.

Purdy, E. G., 1963, Recent calcium carbonate facies of the Great Bahama Bank—1, Petrography and reaction groups; 2, Sedimentary facies: Jour. Geology, v. 71, no. 3, p. 334-355; no. 4, p. 472-497.

——1964, Diagenesis of recent marine carbonate sediments (abs.): AAPG Bull., v. 48, no. 4, p. 542-543.

——1965, Diagenesis of Recent marine carbonate sediments (abs.), *in* Dolomitization and limestone diagenesis: Soc. Econ. Paleontologists and Mineralogists Spec. Pub. 13, p. 169.

——and J. Imbrie, 1964, Carbonate sediments, Great Bahama Bank: Geol. Soc. America, Guidebook, Field Trip no. 2, p. 31-37.

——and R. K. Matthews, 1965, Structural control of recent calcium carbonate deposition in British Honduras (abs.): Geol. Soc. America Spec. Paper 82, p. 152.

Pusey, W. C., 1964, Recent calcium carbonate sedimentation in northern British Honduras: Rice Univ., PhD thesis, 247 p.

Robertson, R., 1963, The mollusks of British Honduras: Philadelphia Shell Club Proc., v. 1., no. 7, p. 15-20.

Sapper, C., 1896, Geology of Chiapas, Tabasco, and Yucatan: Jour. Geology, v. 4, p. 938-947.

Schuchert, C., 1935, Historical geology of the Antillean-Caribbean region, or the lands bordering the Gulf of Mexico and the Caribbean Sea: New York, John Wiley & Sons, 811 p.

Shepard, F. P., 1964, Sea level changes in the past 6000 years: possible archeological significance: Science, v. 143, no. 3606, p. 574-576.

Shinn, E., 1963, Spur and groove formation on the Florida reef tract: Jour. Sed. Petrology, v. 33, no. 2, p. 291-303.

Sippel, R. F., and E. D. Glover, 1965, Structures in carbonate rocks made visible by luminescence petrography: Science, v. 150, p. 1283-1287.

Sorby, H. C., 1879, Anniversary address of the president: Geol. Soc. London Proc., v. 35, p. 56-74.

Stanley, S. M., 1966, Paleoecology and diagenesis of Key Largo limestone, Florida: AAPG Bull., v. 50, no. 9, p. 1927-1947.

Stehli, F. G., and J. Hower, 1961, Mineralogy and early diagenesis of carbonate sediments: Jour. Sed. Petrology, v. 31, p. 358-371.

Stoddart, D. R., 1963, Effects of Hurricane Hattie on the British Honduras reefs and cays, Oct. 30-31, 1961: Atoll Research Bull., no. 95, 142 p.

Taft, W. H., et al, 1968, Lithification of modern marine carbonate sediments at Yellow Bank, Bahamas: Bull. Marine Sci., v. 18, no. 4, p. 762-828.

Tebbutt, G. E., in preparation, Skeletal alteration in Pleistocene limestones, British Honduras.

Thompson, Geoffrey, et al, 1968, Lithified carbonates from the deep-sea of the equatorial Atlantic: Jour. Sed. Petrology, v. 38, no. 4, p. 1305-1312.

Thornbury, W. D., 1954, Principles of geomorphology: New York, John Wiley & Sons, p. 319-320.

Trechmann, C. T., 1933, The uplift of Barbados: Geol. Mag., v. 70, no. 823, p. 19-47.

Vaughan, T. W., 1919, Contributions to the geology and paleontology of the Canal Zone, Panama, and geologically related areas in Central America and the West Indies: Fossil corals from Central America, Cuba, and Puerto Rico, with an account of the American Tertiary, Pleistocene, and recent coral reefs: U.S. Natl. Mus. Bull. 103, p. 189-524.

Van Name, W. G., 1930, The ascidians of Porto Rico and Virgin Islands, *in* Scientific survey of Porto Rico and the Virgin Islands: New York Acad. Sci., v. 10, pt. 4, p. 405-535.

West, R. C., 1962, Physical geology of the Yucatan Platform, *in* Guidebook, Field trip to Peninsula of Yucatan, 1962: New Orleans Geol. Soc., p. 58-63.

Wilbur, K. M., 1960, Shell structure and mineralization in molluscs, *in* Calcification in biological systems: Am. Assoc. Adv. Sci. Pub. No. 64, p. 15-40.

Winland, H. D., 1968, The role of high Mg calcite in the preservation of micrite envelopes and textural features of aragonite sediments: Jour. Sed. Petrology, v. 38, no. 4, p. 1320-1325.

Wolf, K. H., 1965, Grain-diminution of algal colonies to micrite: Jour. Sed. Petrology, v. 35, no. 2, p. 420-427.

Distribution of Holocene Benthonic Foraminifera on the Belize Shelf[1]

KENNETH F. WANTLAND[2]

ABSTRACT The tropical, reef-bounded continental shelf of Belize encompasses a wide variety of marine and marginal-marine environments. The distribution patterns of foraminiferal assemblages in this area reflect this environmental diversity. The distribution is strongly influenced by the effect of both the seaward barrier and the adjacent landmass on the hydrography and sedimentation of the narrow shelf.

Nine foraminiferal assemblages are defined on the basis of taxonomic composition, faunal dominance, faunal diversity, and faunal density of total (living plus dead) foraminiferal populations.

A high-diversity miliolid-dominant assemblage occurs throughout the stenohaline, shallow-water areas of the barrier and outer platforms—areas characterized by calcium carbonate deposition and coral-reef growth. Areas of high turbulence on the windward edge of these platforms contain a unique *Archaias-Asterigerina*-dominant assemblage. A low-diversity miliolid-dominant assemblage is present in the northern shelf lagoon and Chetumal Bay, which are areas of variable salinity, shallow water, partially restricted circulation, and carbonate mud deposition. A unique low-diversity miliolid assemblage dominated by *Archaias* occurs in a part of Chetumal Bay behind a large carbonate island.

In the relatively deep southern shelf lagoon, a low-diversity *Cribroelphidium*-dominant assemblage occurs in two areas: (1) the landward part of the embayed Gulf of Honduras, an area influenced by high stream discharge, and (2) the lagoonal area that is adjacent to the barrier platform and is characterized by restricted circulation and mixed carbonate-terrigenous sedimentation. A high-diversity *Cribroelphidium*-dominant assemblage is present in deep lagoonal, stenohaline areas of uninhibited circulation. A *Cribroelphidium-Quinqueloculina*-dominant, mixed faunal assemblage marks the zone of influx of shoal-derived material into the lagoon. The deepest part of the continental shelf is the site of a *Globocassidulina*-dominant assemblage. The mainland coast of Belize is characterized by a *Quinqueloculina-Cribroelphidium*-dominant assemblage associated with nearshore quartzose sand and silt and a suite of marginal-marine assemblages including delta, marsh, and mainland-lagoon populations.

Data on living populations provide information concerning foraminiferal ecology but do not produce consistent, mappable faunal assemblages. Marked differences between the quantitative parameters of living and total populations in some areas are thought to result from recent shifting of environmental conditions and relict faunas, faunal displacement, seasonal and/or inadequate sampling of habitats, and unknown factors perhaps reflecting seasonal reproduction or differential rates of reproduction among species.

INTRODUCTION

An understanding of the structure and ecology of modern animal populations has proved invaluable in unraveling the paleoecology of their fossil counterparts. The objectives of this study are to document the distribution of modern foraminiferal populations in an environmentally diverse tropical area—the Belize shelf—and to relate this distribution to gross, measurable environmental characteristics.

The continental shelf of Belize encompasses an array of tropical marine environments including coral reefs and reef-related areas; shal-

[1]Manuscript received, October 3, 1974.

[2]Exploration and Production Research Department, Cities Service Oil Company, Tulsa, Oklahoma 74150.

The writer thanks E. G. Purdy and the Rice University Department of Geology for the opportunity to participate in the Belize shelf study. The writer is grateful to the Texas Producing Company, the American Chemical Society, and Rice University for financial support of his graduate studies, from which this report developed. Special thanks go to Cities Service Oil Company, Exploration and Production Research Department, for continued interest in tropical shelf studies, for the opportunity to revisit the Belize study area often, and for financial and material support in the preparation of the manuscript, including the use of their scanning electron microscope, operated by Allen White, for illustrating the Foraminifera. The writer acknowledges help and advice concerning the ecology and taxonomy of the Foraminifera from R. R. Lankford, formerly of Rice University; R. Cifelli, M. Buzas, and R. Todd, of the U.S. National Museum; and S. Streeter, Columbia University. However, the writer accepts all responsibility for errors in concept or identification in this report.

low tropical lagoons; relatively deep backreef lagoons; islands; coastal lagoons; and nearshore, marsh, estuarine, and deltaic settings.

Foraminifera that are abundant in all of the shelf sediments exhibit subtle variation of attributes reflecting sensitive adaptation to environmental conditions. Thus, this study of modern Foraminifera provides insight into principles of population structure and distribution patterns useful in understanding ancient populations and the environments in which they existed.

PREVIOUS WORK

The distribution and ecology of tropical Foraminifera have received relatively little attention, although the abundance and diversity of tropical faunas have intrigued taxonomists since the time of d'Orbigny (1839).

The only prior foraminiferal studies on the Belize shelf are reports by Cebulski (1969) and Wantland (1969a). Cebulski studied living and total populations in 89 samples from an area bounded by Belize City on the north and Placentia on the south. He recognized one assemblage characteristic of the barrier reef and one typical of the lagoon. In addition, he subdivided the barrier-reef assemblage into main-reef, reef-channel, and reef-margin subassemblages. He did not study nearshore marginal-marine environments. The foraminiferal assemblages of the coastal lagoons of Belize (including the northern shelf lagoon and Chetumal Bay) were described by Wantland.

Three other West Indian, tropical to subtropical areas dominated by calcium carbonate sedimentary regimes have been the subject of semiquantitative distribution studies of Foraminifera. They are the Bahama Banks (Streeter, 1963), the Gulf of Batabanó, Cuba (Bandy, 1964), and the southern Florida shelf (Bandy, 1956; Moore, 1957; Bush, 1958; Lynts, 1962; Scholz, 1962; Wright, 1964; Bock, 1967; Bock et al, 1971). Studies by Cushman (1918-1931, 1921, 1922, 1926, 1932-1942, 1959), Acosta (1940a, b), Norton (1930), Illing (1950, 1952), Drooger and Kaasschieter (1958), Boltovskoy (1958, 1963), Barker (1960), Sandberg (1961), Todd and Low (*in* Cloud, 1962), Cole (1965), Ayala-Castañares (1963), Seiglie and Bermudez (1963), Bermudez and Seiglie (1963), Seiglie (1964a, b, 1965a, b, 1968), Barnhart (1963), Davis (1964), Rivero (1964), Todd (1965), Howard (1965), Howard et al (1970), Hofker (1950-1953, 1964), Lidz and Lidz (1966), and Lynts (1965) have added in varying degrees to the knowledge of West Indian-Gulf of Mexico faunal composition and distribution. Some information concerning foraminiferal distributions in these areas can be derived from studies devoted to sediment distributions, specifically those by Creager (1958), Hoskin (1963), Pusey (1964), and Matthews (1966).

Because the Belize study area includes marginal-marine features and the shallow continental shelf in addition to the reef and reef-related environments, a large and familiar body of literature from the Gulf of Mexico is pertinent to this report. Faunal patterns and taxonomic identifications in that area have been established in papers by Kornfeld (1931), Lowman (1949, 1951), Treadwell (1955), Post (1951), Phleger (1951, 1954a, b, 1955, 1956, 1960a, b, c, 1964), Phleger and Lankford (1957), Phleger and Parker (1951), Parker et al (1953), Parker (1954), Bandy (1954, 1956), Lankford (1959), Rusnak (1960), Andersen (1961), Shifflett (1961), Waldron (1963), Walton (1955, 1964a, b), Upshaw et al (1966), and Wantland (1969b).

Studies devoted wholly or in part to the faunas of mangrove marshes include those by Benda and Puri (1962), Saunders (1957, 1958), and Todd and Bronnimann (1957). Studies of other marsh environments include those by Phleger and Walton (1950), Phleger (1965), Phleger and Bradshaw (1966), Parker and Athearn (1959), Andersen (1951, 1953), and Warren (1956, 1957). Murray (1973) has compiled a comprehensive review of studies of living benthic foraminifers.

METHODS OF STUDY

Field Methods

The samples for this study were collected during five summer field seasons from 1961 through 1965. The basic pattern of sample distribution is shown in Figure 1. The writer spent the summer of 1962 in the field. During that season approximately 200 of the 500 preserved sediment samples were collected, and the writer had the opportunity to visit nearly all of the shelf area.

Two sampling techniques were used to obtain equal-volume (15 cm^3) samples of surface sediment. In areas less than 20 ft (6.1 m) deep, samples were taken by hand. A short plastic tube (1 5/8-in. inside diameter) was inserted vertically into the sediment and stoppered. At the surface the core was extruded, and the top centimeter of sediment was sliced off and preserved in buffered formalin and seawater (see Aves, 1958, and Phleger, 1960c, for discussion of preservation techniques). In deeper areas, bottom samples were taken by a Petersen grab sampler. The sampler was opened carefully on deck in order to retain an undisturbed sediment surface. The surface was commonly characterized by a brown organic scum; occasionally, plants were retained in growth orientations. The plastic tube was inserted into the sediment surface, and the sample was collected and treated as above. A few plants and other vertical elements on which Foraminifera attach during life were collected and preserved. In addition, data were collected on temperature and salinity of surface and bottom waters, and bulk sediment samples were taken.

Laboratory Methods

Sample preparation—The preserved samples were stained with Rose Bengal (Walton, 1952), washed over a 230-mesh sieve, dried, and stored. All samples were examined for their foraminifer content. Approximately 120 samples were selected for quantitative analysis (Fig. 1). These samples were rewetted, and the stained (herein termed "living") individuals were counted.

These samples also were dried and split into a fraction containing 300-500 tests. These tests were counted dry; they constitute the total (living plus dead) population. In some samples the Foraminifera were concentrated by carbon tetrachloride flotation; but this was not possible in all samples because certain forms (e.g., *Archaias* and *Amphistegina*) do not always float.

Count data and photography—The data from living and total population counts were tabulated as percentages of each taxon for each quantified sample. These raw data are available elsewhere.[3] Representative specimens were photographed on a JEOLCO JSM II model scanning electron microscope. The illustrated specimens are stored in the writer's collection at the Cities Service Oil Company Exploration and Production Research Laboratory.

[3]ASIS/NAPS c/o Microfiche Publications, 305 East 46th Street, New York, New York 10017.

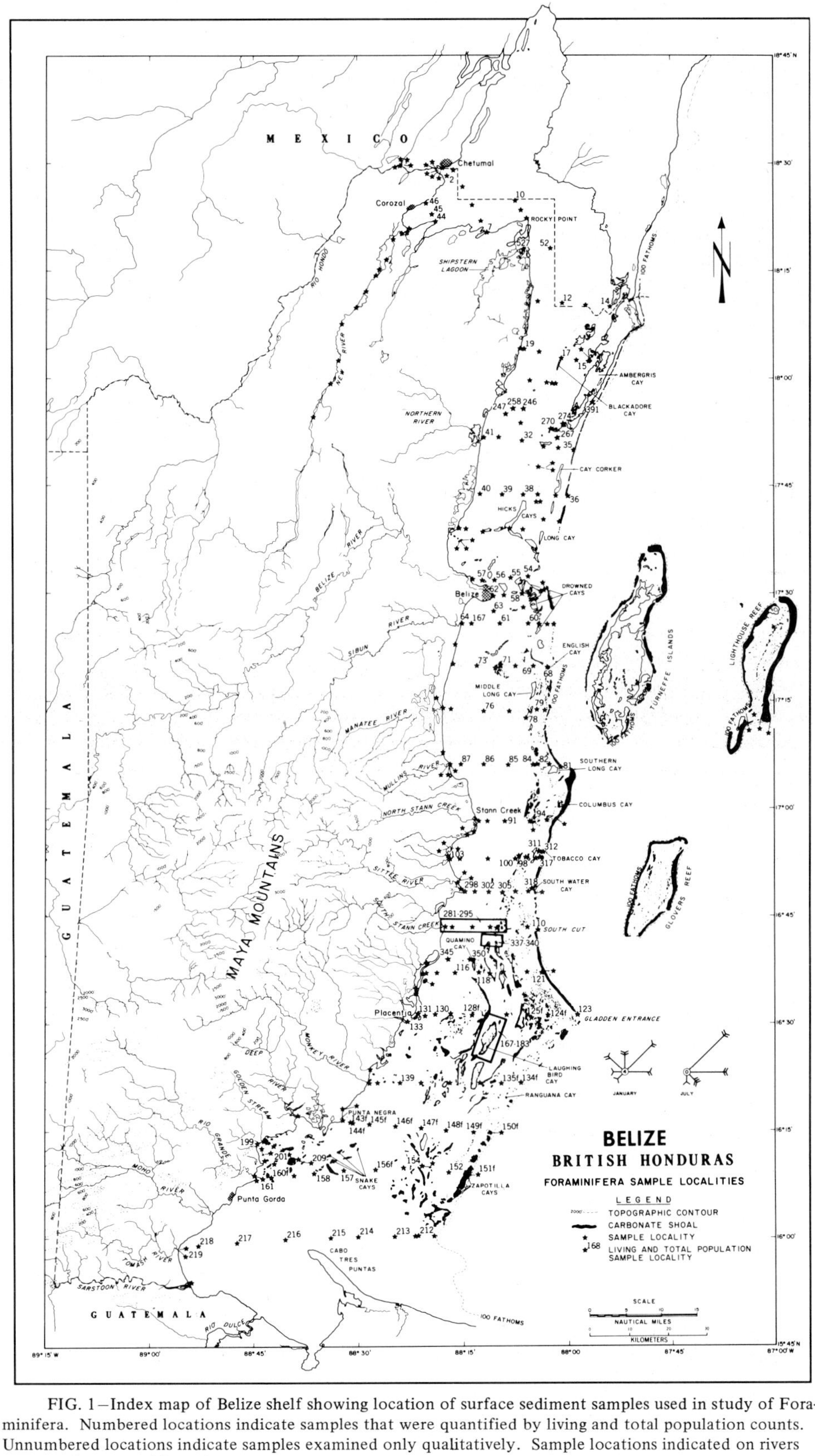

FIG. 1—Index map of Belize shelf showing location of surface sediment samples used in study of Foraminifera. Numbered locations indicate samples that were quantified by living and total population counts. Unnumbered locations indicate samples examined only qualitatively. Sample locations indicated on rivers are not relevant to this study.

DESCRIPTION OF AREA

Physiography and Bathymetry

The Belize shelf is divided into an inner lagoon and an outer barrier reef-cay complex. The lagoon is divisible into northern and southern segments at the latitude of Belize City. For descriptive purposes, these areas have been subdivided into bathymetric provinces by Purdy et al (this volume, Fig. 4). Northern Belize is divided into three provinces: the northern shelf lagoon, which extends parallel with the mainland from the Belize delta north to the Bulkhead mud shoal; Chetumal Bay, the largely enclosed water body extending north from the Bulkhead into Mexico landward of Ambergris Cay; and the outer platform, which parallels the mainland along the margin of the shelf seaward of Ambergris Cay. The barrier reef occupies the seaward edge of the outer platform. Throughout the northern shelf area, water depths are shallow, generally less than 3 fm.

The shelf south of Belize City is also divisible into three provinces: the southern shelf lagoon, the barrier platform, and the Gulf of Honduras. The southern shelf lagoon forms a trough that deepens gradually from the latitude of Belize City (depths of less than 10 fm) toward the south (depths as great as 35 fm). Seaward, the elongate, scarp-bounded reef-cay complex, termed the "barrier platform," borders the edge of the continental shelf. This feature ranges in width from 1 to 7 mi (1.6 to 11.3 km). Its upper surface is less than 5 fm below sea level and is the site of extensive reef development. The seaward margin of the platform is marked by almost continuous barrier-reef growth. The barrier platform is broadest from the area off Belize City to approximately the latitude of the Sittee River delta. South of this point the platform narrows markedly. Concurrently, the southern shelf lagoon widens and deepens and becomes the site of development of numerous steep-sided carbonate shoals. At the southernmost extent of the shelf in the Gulf of Honduras province, no barrier is present at the shelf margin, and deep parts of the shelf lagoon and the open Caribbean Sea are juxtaposed. The eastern edge of the Belize shelf is a scarp that drops steeply into depths exceeding 100 fm.

High (this volume) has studied and described the mainland coast of Belize as characterized by narrow beaches, coastal marshes, and many inshore lagoons and estuaries. Carbonate shoals are also encountered along the mainland margin of the southern shelf lagoon, particularly in the Port of Honduras area.

The contrasting features of the northern and southern shelf bathymetry are paralleled by striking differences in mainland topography and geology. Northern Belize is a low karst surface underlain by Cretaceous and Tertiary limestones, and southern Belize is a mountainous terrain underlain by Mesozoic and Tertiary clastic and carbonate rocks, Paleozoic metasedimentary rocks, and igneous intrusive rocks. Annual rainfall is high in the south (as much as 150 in. or 380 cm near Punta Gorda), and the mountains are drained by numerous streams. In the north, annual rainfall is less (50 in. or 127 cm near Chetumal), and only two major streams, the Rio Hondo and the New River, discharge onto the northern shelf.

Water Characteristics

Measurements of both surface- and bottom-water salinity were compiled (Purdy et al, this volume). These data show that bottom salinities are normal marine (36‰ S) over most of the southern shelf lagoon and the barrier and outer platforms. Along the southern mainland coast, dilution of marine waters by freshwater stream discharge creates a narrow zone of brackish-water bottom conditions. Surface salinities are reduced over a large part of the southernmost shelf, producing a marked "wedge effect" in the southern shelf lagoon where the less dense, fresher water rides out over the more dense, saline waters.

Waters of the northern shelf lagoon show a broad gradient from normal marine at the shelf edge to brackish along the northern coast. Very little evidence of a saltwater "wedge" is present, presumably because of thorough mixing of the shallow water column by wind action.

Chetumal Bay is an area of restricted circulation. The southern limit of the bay is bounded by a mud-shoal complex (the Bulkhead), and Ambergris Cay forms an effective barrier to the east. These features inhibit the circulation of marine waters into the bay. The Rio Hondo and the New River discharge fresh water at the head of the bay and produce a salinity gradient approximately along the axis of the bay. The mainland margin of Chetumal Bay is lined with shallow, irregularly shaped, pondlike lagoons. These enclosed ponds undoubtedly have marked variation in salinity and temperature. They are shallow enough to be considerably freshened by seasonal rainwater or, alternatively, to act as hypersaline evaporating pans. Only Northern River Lagoon receives a fairly constant supply of fresh water. Similar shallow lagoons with presumably varying water characteristics exist on the west side of Ambergris Cay.

In the shallow areas of the north, current patterns are complex owing to the interaction of runoff, prevailing winds, and slight tidal effect (Pusey, this volume). In the southern shelf lagoon, a surface drift current directed generally southward is indicated on nautical charts (U.S. Navy Hydrographic Office, 1931a, b; 1938). Tidal effects are not marked; the tidal range is 1.5 ft (0.5 m) under normal conditions (U.S. Navy Hydrographic Office, 1952). Strong northerly winds and hurricanes may produce short-term tidal effects of a much more extreme character.

Turbidity is variable. Throughout the barrier and outer platform areas, underwater visibility is maximum for the study area and has been estimated at 50 ft (15.3 m; Pusey, this volume). In the shallow water of the northern shelf lagoon and Chetumal Bay, wind-induced currents frequently suspend mud-size carbonate material, and underwater visibility may be zero.

Sediments

The sediments of the Belize shelf are diverse in composition and texture and reflect a wide array of depositional environments. Pusey (this volume) describes and delineates the sedimentary facies of northern Belize; the facies of southern Belize are described by Purdy et al in the initial paper of this volume.

The sediments of the barrier platform are pure calcium carbonate, almost wholly skeletal in origin but of contrasting texture. Sediments from agitated areas of the outer barrier margin are rounded, well-sorted carbonate sands. In areas protected from wave and current action, particularly those stabilized by marine-grass (*Thalassia*) meadows, the sediments are poorly sorted muds containing angular, fragmental skeletal grains. In some areas currents sweep the coral rock clean, and sediment accumulates only in localized depressions and other protected areas. The texture and grain size of sediments throughout this area are strongly influenced by *in situ* production of carbonate constituents.

Skeletal carbonate sands accumulate at the base of carbonate shoals in the southern shelf lagoon and grade outward into fine-grained carbonate muds away from the loci of shoal sediment influx. The lagoonal muds of the southern shelf grade from almost pure carbonate muds in the deeper areas surrounding the carbonate shoals to predominantly terrigenous muds

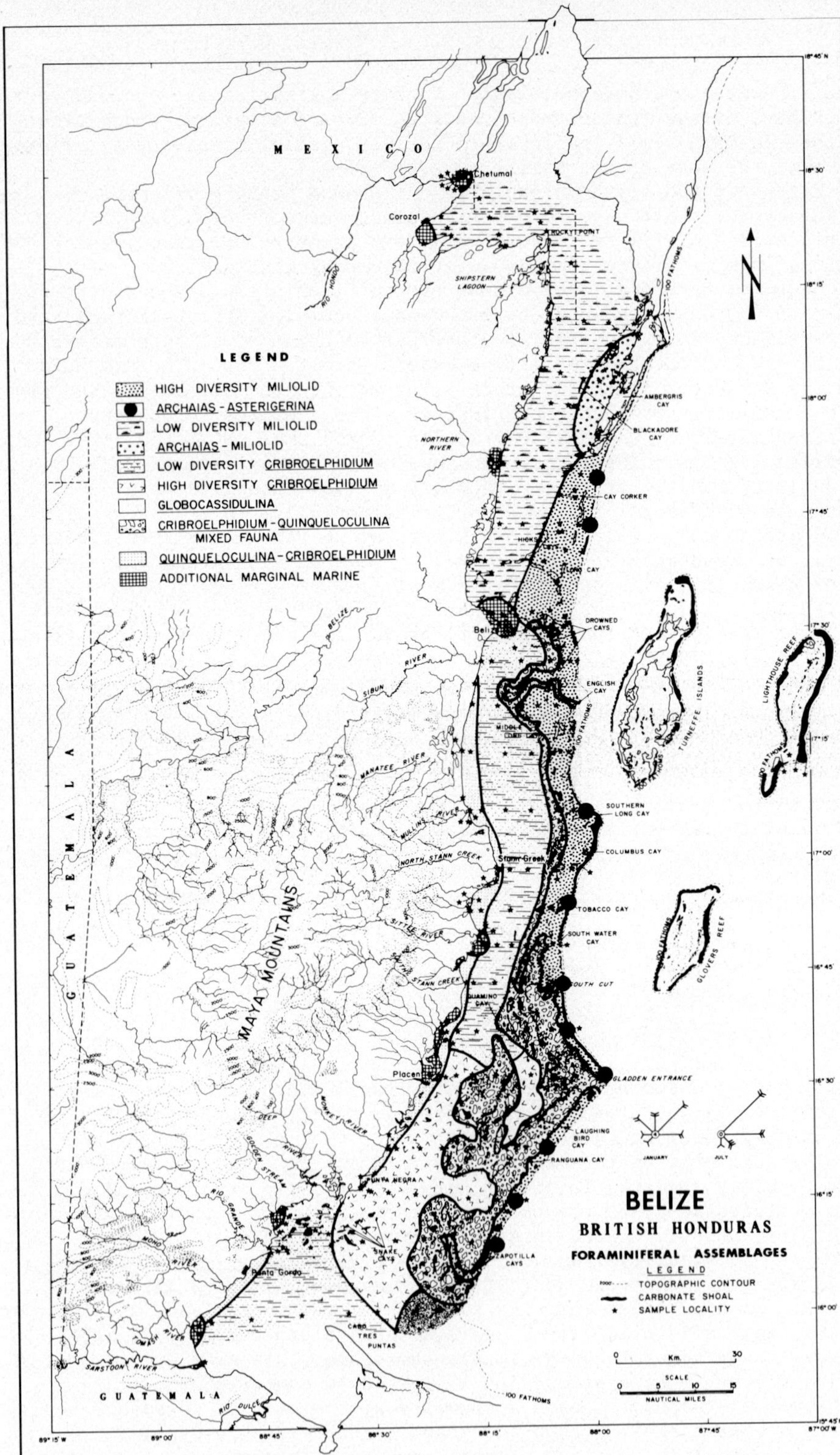

FIG. 2—Distribution of benthonic foraminiferal assemblages in surface sediments of Belize shelf.

toward the mainland. Various aspects of the fine-grained fraction of the lagoonal sediments have been described by Matthews (1966), Scholle and Kling (1972), and Scott (this volume). The sediments of the elongate inner part of the southern shelf lagoon are chiefly mixed carbonate-terrigenous sediments. The nearshore deposits bordering the mainland are almost wholly terrigenous in origin.

FORAMINIFERAL ASSEMBLAGES

The foraminiferal assemblages described in this study are defined on the basis of taxonomic composition, faunal diversity, faunal dominance, and faunal density of total (living plus dead) populations.

Taxonomic composition is a qualitative parameter consisting simply of the list of foraminiferal taxa in each sample. *Faunal diversity*, as used in this study, refers to the number of species or genera recorded in each volumetrically equal sample of sediment. This is the simplest measure of faunal diversity and does not take into account the proportions of species within populations. More sophisticated methods of expressing faunal diveristy in foraminiferal populations are discussed by various writers including Gibson and Buzas (1973) and Murray (1973). Faunal diversity is a distinctive aspect of population structure even though the factors affecting this parameter are poorly understood and somewhat speculative.

The concept of dominant taxa is used in several ways in this report. In the strictest sense it is the number, expressed in percent, of the most abundant single species in a sample; this is *faunal dominance* as defined by Walton (1964a). However, generic dominance is also used to characterize individual populations by the most common genera (e.g., *Heterillina-Triloculina-Miliolinella* association) and to define assemblages of populations (e.g., *Quinqueloculina-Cribroelphidium*-dominant assemblage).

Faunal density is also recorded and refers simply to the number of foraminifer tests (living or total) counted or extrapolated from a count of a fraction in a constant volume of sediment. Faunal density varies widely in certain areas and did not contribute significantly to the definition of all assemblages.

Foraminiferal assemblages based on the above criteria are found to be areally extensive, mappable faunal units. In most cases the boundaries separating these assemblages are gradational, but where they correspond to abrupt discontinuities in the physical environment they may be very sharp.

The assemblages are based on total populations rather than living populations for several reasons. Total populations are geologically significant because they reflect all biologic and sedimentologic effects influencing distribution at the time of burial. Further, total populations represent a degree of homogenization in highly variable environments, whereas living populations are subject to innumerable local and/or short-term, sometimes seasonal, effects. The living populations provide a wealth of detailed ecologic information concerning the habitats of individual species and populations; however, in a complex environmental setting such as a reef, the variation among samples is unmappable and the data are too detailed for the delineation of gross distributional patterns related to the relatively coarse environmental measurements available. Sampling was confined to the summer months, and no stations were revisited systematically. Therefore, no control exists on seasonal or other temporal variations of the living populations.

The distribution of assemblages is shown in Figure 2. The basic parameters that distinguish the assemblages are outlined in Table 1.

TABLE 1. MAJOR ATTRIBUTES OF FORAMINIFERAL ASSEMBLAGES OF THE BELIZE SHELF

Assemblage	*Dominant Genera*	*Dominant Species*	*Faunal Dominance (%)*	*Faunal Diversity (species/sample)*	*Faunal Density (individuals/sample)*
Quinqueloculina-Cribroelphidium-Dominant	*Quinqueloculina, Cribroelphidium*	*Cribroelphidium poeyanum*	20-25	75-115	10,000
Low-Diversity *Cribroelphidium*-Dominant	*Cribroelphidium, Ammonia, Florilus*	*Cribroelphidium poeyanum*	20-30	<50	7,000- 10,000
High-Diversity *Cribroelphidium*-Dominant	*Cribroelphidium, Fursenkoina, Florilus*	*Cribroelphidium poeyanum*	10-15	90-110	30,000- 40,000
Globocassidulina-Dominant	*Globocassidulina, Trifarina, Reussella*	*Globocassidulina subglobosa*	10-12	100	20,000- 90,000
High-Diversity *Miliolid*-Dominant	*Quinqueloculina, Triloculina*	Variable	5-10	100-150	Variable
Archaias-Asterigerina-Dominant	*Archaias, Asterigerina*	*Archaias angulatus*	25-30	50-100	1,000- 3,000
Cribroelphidium-Quinqueloculina-Dominant Mixed Fauna	*Cribroelphidium, Quinqueloculina*	*Cribroelphidium poeyanum*	10	100-140	30,000- 40,000
Low-Diversity *Miliolid*-Dominant	*Heterillina, Triloculina*	*Heterillina cribrostoma*	20-50	20-50	20,000-100,000
Archaias-Dominant	*Archaias, Heterillina*	*Archaias angulatus*	20-90	20-30	20,000- 60,000

Figures 3-16 illustrate representative suites of foraminiferal species that compose the assemblages. The figures are constructed to show the more common species and to illustrate the variety of foraminiferal morphologies that typify the assemblages.

Quinqueloculina-Cribroelphidium Assemblage

Total populations—A small number of samples contains distinct faunas dominated by certain species of the genera *Quinqueloculina* and *Cribroelphidium* and composed largely of the following species: *Cribroelphidium poeyanum* (d'Orbigny), *Quinqueloculina compta* Cushman, *Q. biocostata* d'Orgbigny, *Q. lamarckiana* d'Orbigny, and *Ammonia beccarii* (Linné) (Fig. 3).

Such faunas have relatively high total faunal diversity (75-115 species per sample) and high faunal dominance (20-25 percent), and the dominant species is always *Cribroelphidium poeyanum.* Total population density is approximately 10,000 tests per sample. None of the species is restricted to the area of assemblage occurrence; however, the conspicuous species of *Quinqueloculina* listed above characterize this assemblage.

Living populations—The associated living populations contain 25-65 species per sample, and none are dominated by *C. poeyanum.* The taxonomic compositions of the populations vary from sample to sample, but the following taxa are generally the most common: *Quinqueloculina* spp., *Discorbis* spp., *Buliminella elegantissima* (d'Orbigny), *Sagrina pulchella* d'Orbigny, *Brizalina striatula* (Cushman), and *Ammonia beccarii* (Linné).

Areal extent and ecology—Populations of this type are distributed along the mainland margin of southern Belize, except for the Port of Honduras embayment and areas of delta growth, everywhere in water less than 5 fm deep. Bottom-water salinity is less than normal marine, but these populations are not associated with areas of strong river influence and deltaic sedimentation. The distribution of this assemblage is wholly within and corresponds to the distribution of coastal, terrigenous, quartzose muddy sands. The mud content of the substrate is generally less than 25 percent, and carbonate content never exceeds 30 percent. The assemblage is essentially confined to nearshore conditions marked by wave action. In some areas the substrate is stabilized by the presence of nearshore marine-grass meadows. The grass offers the benthic microfauna many ecologic opportunities, which are reflected in the relatively high faunal diversity of this foraminiferal assemblage. South from the Port of Honduras to the mouth of the Tomash River there are no samples to substantiate the presence of this assemblage along the mainland coast. However, it is inferred to be present on the basis of the following environmental conditions: (1) salinities less than normal marine (less than 26°/₀₀ S), (2) low percentage of mud in the sediments (less than 25 percent), and (3) low percentage of carbonate in the sediments (less than 30 percent).

This assemblage is associated with a "continental shore" mollusk fauna (Robertson, 1963; Purdy et al, this volume). It grades seaward into the low-diversity and high-diversity *Cribroelphidium*-dominant assemblages. Shoreward it is bordered by the inshore lagoon and marginal marsh assemblages of the mainland coast. It is distinguished from the mixed *Cribroelphidium-Quinqueloculina*-dominant faunas of the shoal-influenced lagoon by (1) the relatively high frequencies of *Quinqueloculina compta, Q. bicostata,* and *Q. lamarckiana*; (2) low generic diversity; (3) the absence of most typical carbonate-shoal taxa (e.g., *Peneroplis, Hauerina, Articulina, Cymbaloporretta,* etc.); and (4) the absence or very low frequency of genera characteristic of the deeper lagoon (e.g., *Reussella, Fursenkoina, Nouria, Globocassidulina,* etc.).

FIG. 3—Species typical of *Quinqueloculina-Cribroelphidium*- dominant assemblage which is distributed along mainland coast of southern Belize. Although dominant species is *Cribroelphidium poeyanum*, assemblage is characterized by common quinqueloculine miliolids and relatively high faunal diversity. High diversity probably reflects presence of marine-grass meadows in nearshore coastal area.

Additional Marginal-Marine Assemblages

Several discontinuous assemblages are associated with nearshore, marginal-marine conditions other than the preceding. These assemblages have not been adequately sampled in this area; indeed, very few quantitative, preserved samples were collected in marginal environments. Examination of bulk sediment collections, however, demonstrates the existence of these marginal faunas; and they are noted here for completeness, although their areal extent and population structure have not been determined. These populations are not unlike marginal-marine faunas that have been described in considerable detail elsewhere. They include (1) an *Ammonia-Ammobaculites* fauna characteristic of the effluents of some major rivers (Belize, Sittee, Rio Grande, Tomash); (2) *Cribroelphidium-Ammonia* faunas prevalent in the low-salinity, mainland lagoons of southern Belize; (3) *Palmerinella* faunas near the present mouth of the Belize River; and (4) *Miliammina* and *Arenoparrella* faunas in the Belize River delta marshes and the marshes bordering the southern shelf lagoon. Figure 4 illustrates species typical of this assemblage.

Low-Diversity *Cribroelphidium* Assemblage

Total populations—Total foraminifer populations dominated by the genus *Cribroelphidium* characterize the sediments throughout much of the southern shelf lagoon. The first group of *Cribroelphidium*-dominant total populations is characterized by low faunal diversity and a rather uniform population structure. These are almost invariably *Cribroelphidium-Ammonia-Florilus* faunas composed of less than 50 species. The dominant species is *Cribroelphidium poeyanum* (d'Orbigny), occurring in frequencies of 20-30 percent. The major portion of the fauna in every case is composed of the following species: *Cribroelphidium poeyanum* (d'Orbigny), *Cellanthus gunteri* (Cole), *Cellanthus discoidale* (d'Orbigny), *Ammonia beccarii* (Linné), *Florilus atlantica* (Cushman), and *Florilus grateloupi* (d'Orbigny). Commonly associated species occurring in lower frequencies are *Buliminella elegantissima* (d'Orbigny), *Sagrina pulchella* d'Orbigny, *Fursenkoina pontoni* (Cushman), *Reussella atlantica* (Cushman), *Lagenammina atlantica* (Cushman), *Nouria polymorphinoides* Heron-Allen and Earland, and *Eggerella advena* (Cushman). No species are restricted to the areas of assemblage occurrence. Nearly all species noted in this assemblage occur also in the high-diversity *Cribroelphidium*-dominant assemblage. Those that do not occur in this latter assemblage do range

---

⟵

Explanation of Figure 3.

a. *Quinqueloculina bicostata* d'Orbigny, side view, x50.
b. *Quinqueloculina bicostata* d'Orbigny, side view, x75.
c. *Quinqueloculina compta* Cushman, side view, x75.
d. *Quinqueloculina funafutiensis* (Chapman), side view, x75.
e. *Quinqueloculina bidentata* d'Orbigny, side view, x50.
f. *Quinqueloculina seminulum* (Linné), side view, x100.
g. *Quinqueloculina seminulum* (Linné), side view, x75.
h. *Textularia agglutinans* d'Orbigny, side view, x50.
i. *Buliminella elegantissima* (d'Orbigny), side view, x150.
j. *Loxostomum mayori* (Cushman), side view, x100.
k. *Brizalina striatula* (Cushman), side view, x150.
l. *Ammonia beccarii* (Linné), ventral view, x150.
m. *Cribroelphidium poeyanum* (d'Orbigny), side view, x75.
n. *Cellanthus gunteri* (Cole), side view, x100.

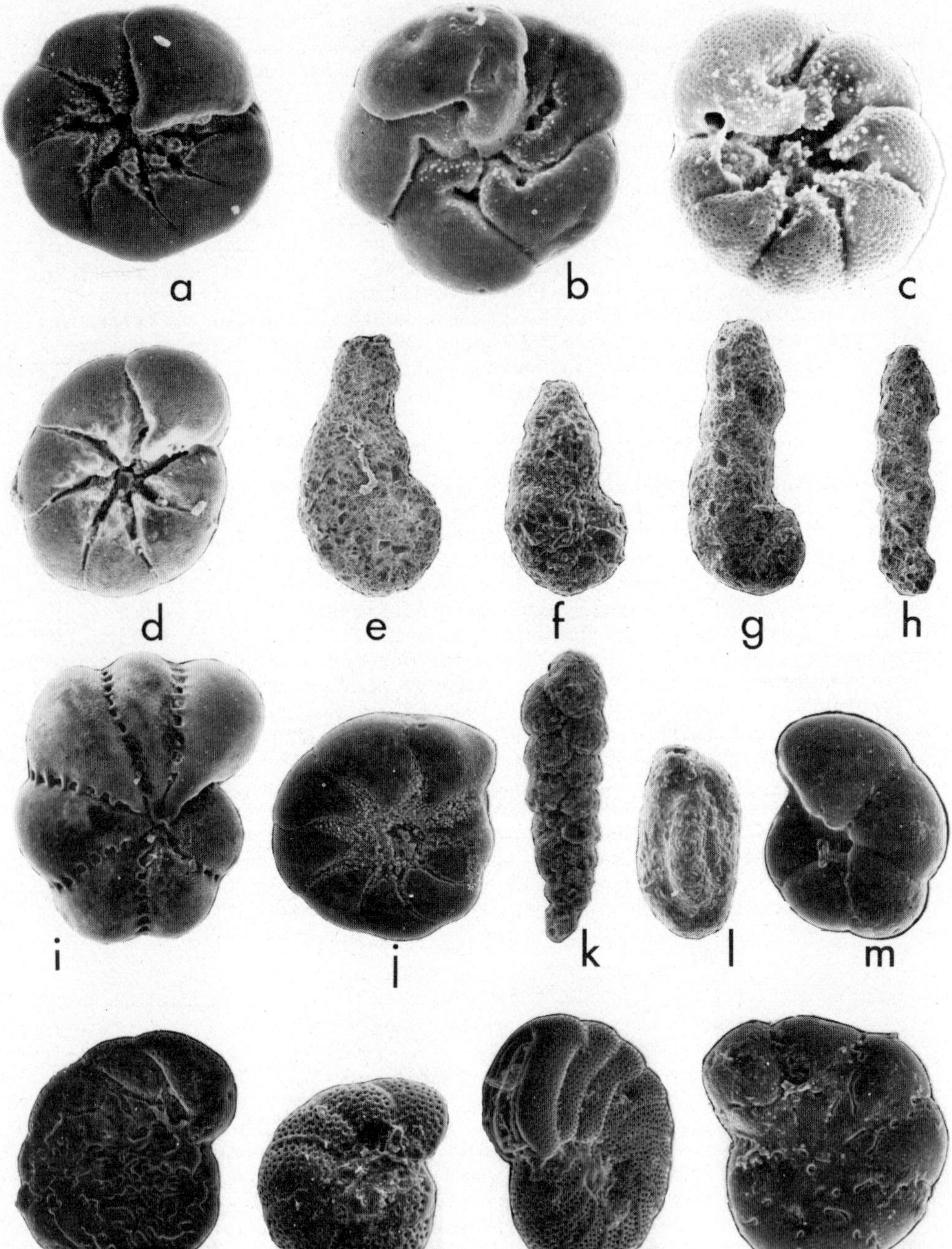

FIG. 4—Species typical of additional marginal-marine assemblages. These species are representative of foraminiferal faunas inhabiting a variety of different and disjunctly distributed marginal-marine environments of Belize shelf including coastal lagoons, deltaic margins, and a suite of coastal, island, and deltaic marshes.

into one or more of the marginal-marine assemblages (e.g., *Ammonia beccarii*). *Buliminella, Eggerella,* and *Nouria* attain their highest frequencies in the low-diversity *Cribroelphidium*-dominant assemblage and are scarce elsewhere on the shelf. Total-population density is a relatively constant parameter with nearly all samples in this assemblage containing 7,000-10,000 tests. Although population density ranges from 3,000 to 41,000 tests per sample, these extremes are represented by single samples. Representative species of this assemblage are illustrated in Figure 5.

Living populations—The living populations associated with this assemblage are somewhat variable in their taxonomic composition. They are low-diversity populations composed of 10-20 species, and they are nowhere dominated by *Cribroelphidium*. The dominant genera in the living populations vary from *Florilus-Fursenkoina-Ammonia* to *Buliminella-Florilus-Cribroelphidium* to *Fursenkoina-Eggerella-Cribroelphidium*. Faunal dominance ranges from 20 to 45 percent. Living-population density ranges from 20 to 200 individuals, but most samples contain an average of 100-150 stained tests. The distribution of these living populations is discussed in more detail in a subsequent section.

Areal extent and ecology—Samples assigned to the low-diversity *Cribroelphidium*-dominant assemblage occur in two separate areas in the southern shelf lagoon. The larger area extends from Belize City south approximately to the point where the lagoon narrows markedly, near Placentia. The assemblage is replaced seaward by a zone of faunal mixing at the base of the barrier platform. Toward the mainland the assemblage grades into a variety of coastal assemblages including the *Quinqueloculina-Cribroelphidium*-dominant assemblage. Southward it grades into the high-diversity *Cribroelphidium*-dominant assemblage. Water depths in the area

---

⟵

Explanation of Figure 4.

a. *Ammonia beccarii* (Linné), ventral view, x100.

b. *Ammonia beccarii* (Linné), ventral view, x200.

c. *Ammonia beccarii* (Linné), ventral view, x250.

d. *Ammonia beccarii* (Linné), ventral view, x100. A highly variable form that is common in all low-salinity, river-influenced environments of Belize shelf.

e. *Ammobaculites exilis* Cushman and Bronnimann, side view, x50.

f. *Ammobaculites exilis* Cushman and Bronnimann, side view, x70.

g. *Ammobaculites exilis* Cushman and Bronnimann, side view, x50.

h. *Ammotium salsum* (Cushman and Bronnimann), side view, x50.

Ammobaculites and *Ammotium* are common genera in sediments of low-salinity deltaic margins.

i. *Cribroelphidium poeyanum* (d'Orbigny), side view, x140. This species is common in all low-salinity coastal lagoons.

j. *Protelphidium delicatulum* (Bermudez), side view, x75. A distinctive form, this species is characteristic of delta-margin sediments and upper reaches of river estuaries (e.g., effluent of Rio Hondo).

k. *Gaudryina exilis* Cushman and Bronnimann, side view, x150. This small, delicate form is associated with *Ammonia-Ammobaculites* assemblage.

l. *Miliammina fusca* (H. B. Brady), side view, x70. Common in low-salinity marshes of southern Belize coast.

m. *Haplophragmoides* sp. One of several species of this genus that occur in low frequencies in low-salinity marsh faunas.

n. *Trichohyalus aguayoi* (Bermudez), vertical view, x70.

o. *Trichohyalus aguayoi* (Bermudez), dorsal view, x75. Species found in coastal and island marshes of northern Belize.

p. *Palmerinella palmerae* Bermudez, oblique view, x75. Species common in deltaic sediments, especially present effluent of Belize River.

q. *Bisaccium imbricatum* Andersen, side view, x75. Minor species associated most commonly with *Palmerinella*.

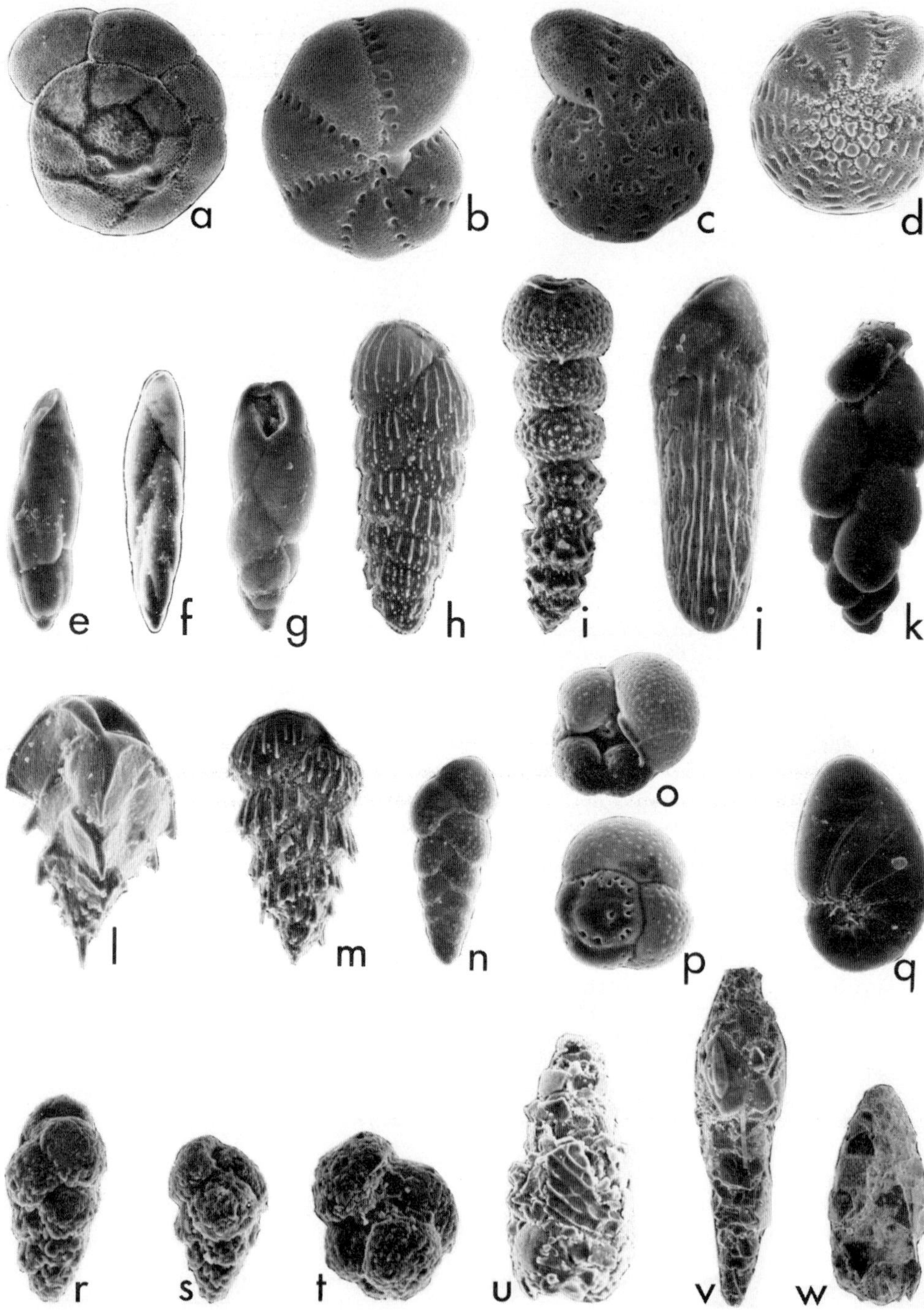

FIG. 5–Species typical of low-diversity *Cribroelphidium*-dominant assemblage. This assemblage is found in southern shelf lagoon from Belize City to Placentia and in landward parts of Gulf of Honduras.

of the assemblage range from 5 to 15 fm. Bottom salinities approximate normal-marine conditions—34-36°/₀₀ S. Little is known of the water circulation; however, the limited river influence along the mainland coast and the strong physical barrier of the shallow platform between the lagoon and the open ocean suggest that a somewhat restricted hydrographic regime is developed in the area. The sediments are largely mixed carbonate-terrigenous muds. The carbonate content of the lagoon sediments varies from 30 to 90 percent. The bottom is relatively level, and it supports virtually no macroscopic flora. The associated mollusk fauna is largely an infaunal mud assemblage (Robertson, 1963).

The second area of occurrence for this assemblage is the Gulf of Honduras in the extreme southwestern part of the Belize shelf. Environmental conditions are similar to those described in preceding paragraphs, except that the carbonate content of the sediments is less than 30 percent, periods of reduced bottom salinity result from the greater river influx, and bottom topography is somewhat irregular. Water depth and associated fauna and flora are similar.

In summary, this assemblage is limited ecologically by factors associated with depth, salinity, and perhaps circulation. Neither composition nor texture of the substrate is a significant factor affecting species distribution. It is not evident what depth-related factors exclude this assemblage from the shallowest parts of the lagoon. Some of the dominant taxa (*Cribroelphidium*, *Ammonia*) are abundant in extremely shallow and brackish waters, but the total assemblage does not persist to the lagoon margin. Apparently, most species of the assemblage are best adapted to conditions associated with deeper water, and these forms are replaced shoreward by forms adapted to nearshore conditions.

It is possible that the two areas of occurrence of this assemblage reflect slightly different factors affecting faunal diversity. In the

---

←

Explanation of Figure 5.

a. *Ammonia beccarii* (Linné), dorsal view, x100.
b. *Cribroelphidium poeyanum* (d'Orbigny), side view, x100.
c. *Cribroelphidium poeyanum* (d'Orbigny), side view, x100.
d. *Cellanthus gunteri* (Cole), side view, x75.
e. *Buliminella elegantissima* (d'Orbigny), side view, x150.
f. *Fursenkoina pontoni* (Cushman), side view, x100.
g. *Fursenkoina complanata* (Egger), side view, x150.
h. *Fursenkoina spinicostata* (Phleger and Parker), side view, x150.
i. *Rectobolivina advena* (Cushman), side view, x150.
j. *Brizalina striatula* (Cushman), side view, x200.
k. *Hopkinsina pacifica* Cushman, side view, x250.
l. *Reussella atlantica* Cushman, side view, x150.
m. *Sagrina pulchella* d'Orbigny, side view, x100.
n. *Brizalina lowmani* (Phleger and Parker), side view, x275.
o. *Rosalina bulbosa* (Parker), ventral view, x150.
p. *Rosalina bulbosa* (Parker), dorsal view, x150.
q. *Florilus atlanticus* (Cushman), side view, x75.
r. *Eggerella advena* Cushman emend. Loeblich and Tappan, side view of slender form, x150.
s. *Eggerella advena* Cushman emend. Loeblich and Tappan, side view of slender form, x150.
t. *Eggerella advena* Cushman emend. Loeblich and Tappan, oblique apertural view of inflated form, x150.
u. *Lagenammina atlantica* (Cushman), side view, x50.
v. *Reophax* sp. cf. *R. arayensis* Bermudez and Seiglie, side view, x50.
w. *Nouria polymorphinoides* Heron-Allen and Earland, side view, x35.

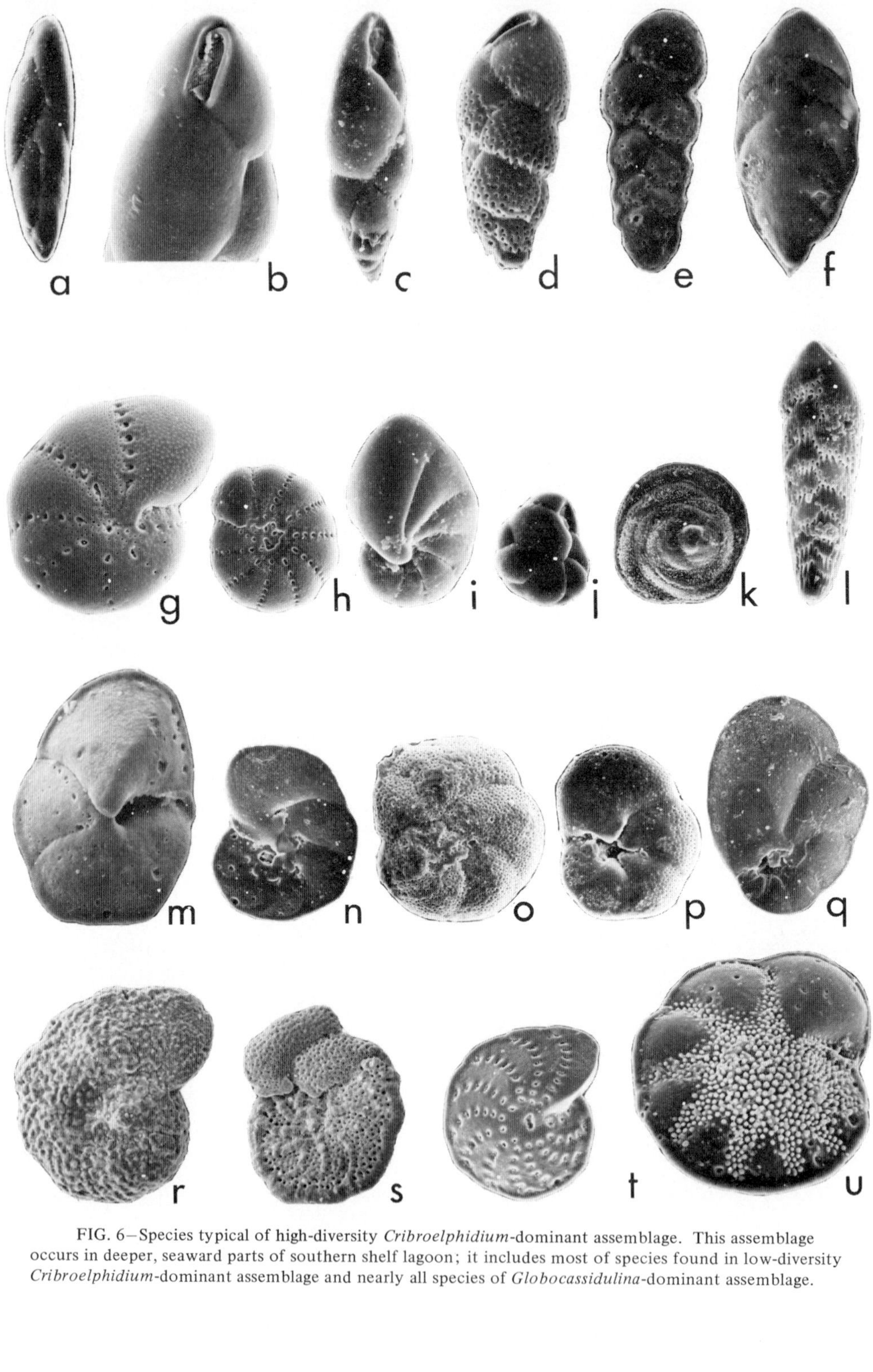

FIG. 6—Species typical of high-diversity *Cribroelphidium*-dominant assemblage. This assemblage occurs in deeper, seaward parts of southern shelf lagoon; it includes most of species found in low-diversity *Cribroelphidium*-dominant assemblage and nearly all species of *Globocassidulina*-dominant assemblage.

Gulf of Honduras the dominant factor limiting diversity may be the somewhat variable bottom salinity, whereas in the southern shelf lagoon the dominant factor may be restricted circulation.

High-Diversity *Cribroelphidium* Assemblage

Total populations—A second widespread group of *Cribroelphidium*-dominant total populations is characterized by relatively high faunal diversity and a distinct, though variable, taxonomic composition. Individual samples contain 90-110 species. Faunal dominance is only 10-15 percent, but the dominant species of the total populations is still *Cribroelphidium poeyanum*. The fauna is composed of a variety of taxa of which the most abundant and most commonly occurring are *Cribroelphidium poeyanum* (d'Orbigny), *Fursenkoina pontoni* (Cushman), *Florilus* spp., *Sagrina pulchella* d'Orbigny, *Globocassidulina subglobosa* (Brady), *Textularia* spp., *Reusella atlantica* Cushman, *Quinqueloculina* spp., *Lagenammina atlantica* (Cushman), and *Brizalina* spp. (Fig. 6). Commonly associated genera present in lower frequencies are *Discorbis*, *Cibicides*, *Hanzawaia*, *Neoconorbina*, *Triloculina*, and *Miliolinella*. The genus *Ammonia* is virtually absent from this assemblage, as are *Buliminella*, *Eggerella*, and *Nouria*.

Very few species are restricted to this high-diversity assemblage. Nearly all species, even rare forms, occur in at least one other assemblage on the Belize shelf. Thus it seems that high diversity in this case is not the result of restriction of certain species to an area by environmental variables, but is due to the existence of environmental conditions that permit the coexistence of a large number of taxa in a relatively small area.

Total-population density among these samples is extremely variable but high, ranging from 10,000 to 160,000 tests per sample. Most samples contain 30,000-40,000 specimens.

Living populations—The living populations in these samples are, again, *never* dominated by *Cribroelphidium*. The dominant genus is either

---

Explanation of Figure 6.

a. *Fursenkoina pontoni* (Cushman), side view, x100.
b. *Fursenkoina pontoni* (Cushman), close-up of aperture, x250.
c. *Fursenkoina complanata* (Egger), side view, x150.
d. *Fursenkoina spinicostata* (Phleger and Parker), side view, x150.
e. *Brizalina spinescens* (Cushman), side view, x150.
f. *Brizalina goesii*? (Cushman), side view, x150.
g. *Cribroelphidium poeyanum* (d'Orbigny), side view, x150.
h. *Cribroelphidium poeyanum* (d'Orbigny), side view, x150.
i. *Florilus atlanticus* (Cushman), side view, x150.
j. *Globocassidulina subglobosa* (Brady), side view, x150.
k. *Neoconorbina terquemi* (Rzehak), dorsal view, x75.
l. *Brizalina striatula* (Cushman), side view, x150.
m. *Eponides repandus* (Fichtell and Moll), ventral view, x100.
n. *Hanzawaia concentrica* (Cushman), ventral view, x75.
o. *Rosalina subauracana* (Cushman), dorsal view, x100.
p. *Rosalina subauracana* (Cushman), ventral view, x100.
q. *Cancris sagra* (d'Orbigny), ventral view, x75.
r. *Planulina exorna* Phleger and Parker, dorsal view, x75.
s. *Planulina exorna* Phleger and Parker, ventral view, x75.
t. *Elphidium advenum* (Cushman), side view, x70.
u. *Buccella frigida* (Cushman), ventral view, x150.

Florilus or *Fursenkoina*, and the most dominant generic associations are *Fursenkoina-Florilus-Cribroelphidium* and *Florilus-Fursenkoina-Hanzawaia*. Similar to the total populations, living populations in this assemblage are more diverse than those of the low-diversity *Cribroelphidium*-dominant assemblage. Twenty to 50 living species may be present, and uncommonly as many as 80 living species are recorded from a single sample. Faunal dominance varies from 20 to 30 percent, and the most commonly dominant living species is *Fursenkoina pontoni*. Living-population density is most commonly 300-500 specimens, although it ranges from less than 100 to 1,000 tests per sample.

Areal extent and ecology—This assemblage occurs over a large area of the southernmost shelf in water depths ranging from 5 to 25 fm. Bottom salinities are normal marine (36°/₀₀ S), and the sediments are muds or bioclastic muddy sands and sandy muds. Calcium carbonate content of the sediments ranges from 60 to 90 percent. There are few data on water movement, but one can speculate that in this area communication between the lagoon and the open ocean is increased because of the less continuous nature of the southern part of the barrier platform (see Purdy et al, this volume). This presumed decrease in restriction together with stability of water characteristics may be important to the development of a diverse, abundant benthonic foraminiferal fauna.

The distribution of this assemblage is limited by the marginal-marine influence toward the mainland, by the distribution of carbonate shoals, by factors associated with depth, and perhaps by circulation in the shelf lagoon. This assemblage does not persist in areas where bottom salinity is less than normal marine, and it loses identity in areas of shoal-influenced lagoonal sediments. Among the many shoals separated by relatively deep water in southern Belize, the high-diversity *Cribroelphidium*-dominant assemblage seems to be restricted to areas that are not shoal influenced, at least in terms of containing no shoal-derived carbonate sands. Toward the mainland this assemblage grades into the low-diversity

---

Explanation of Figure 7.

a. *Globocassidulina subglobosa* (Brady), oblique view, x250.
b. *Globocassidulina subglobosa* (Brady), side view, x375.
c. *Trifarina bella* (Phleger and Parker), side view, x150.
d. *Trifarina bella* (Phleger and Parker), side view, x150.
e. *Trifarina bradyi* Cushman, side view, x150.
f. *Bulimina tenuis* Phleger and Parker, side view, x105.
g. *Reussella atlantica* Cushman, side view, x150.
h. *Cribroelphidium poeyanum* (d'Orbigny), side view, x70.
i. *Pavonina atlantica* Cushman, side view, x70.
j. *Bigenerina irregularis* Phleger and Parker, side view, x70.
k. *Textularia mayori* Cushman, side view, x75.
l. *Textularia conica* d'Orbigny, side view, x70.
m. *Gaudryina compressa* Cushman, side view, x70.
n. *Liebussella soldanii* (Jones and Parker), side view, x50.
o. *Reophax* sp. cf. *R. arayensis* Bermudez and Seiglie, side view, x35.
p. *Reophax spiculifera* Brady, side view, x100.
q. *Lagenammina atlantica* (Cushman), side view, x70.
r. *Trochammina advena* Cushman, dorsal view, x100.
s. *Articulina sagra* d'Orbigny, side view, x70.
t. *Articulina sagra* d'Orbigny, side view, x70.
u. *Globigerinoides ruber* (d'Orbigny), side view, x70.
v. *Orbulina universa* d'Orbigny, side view, x70.

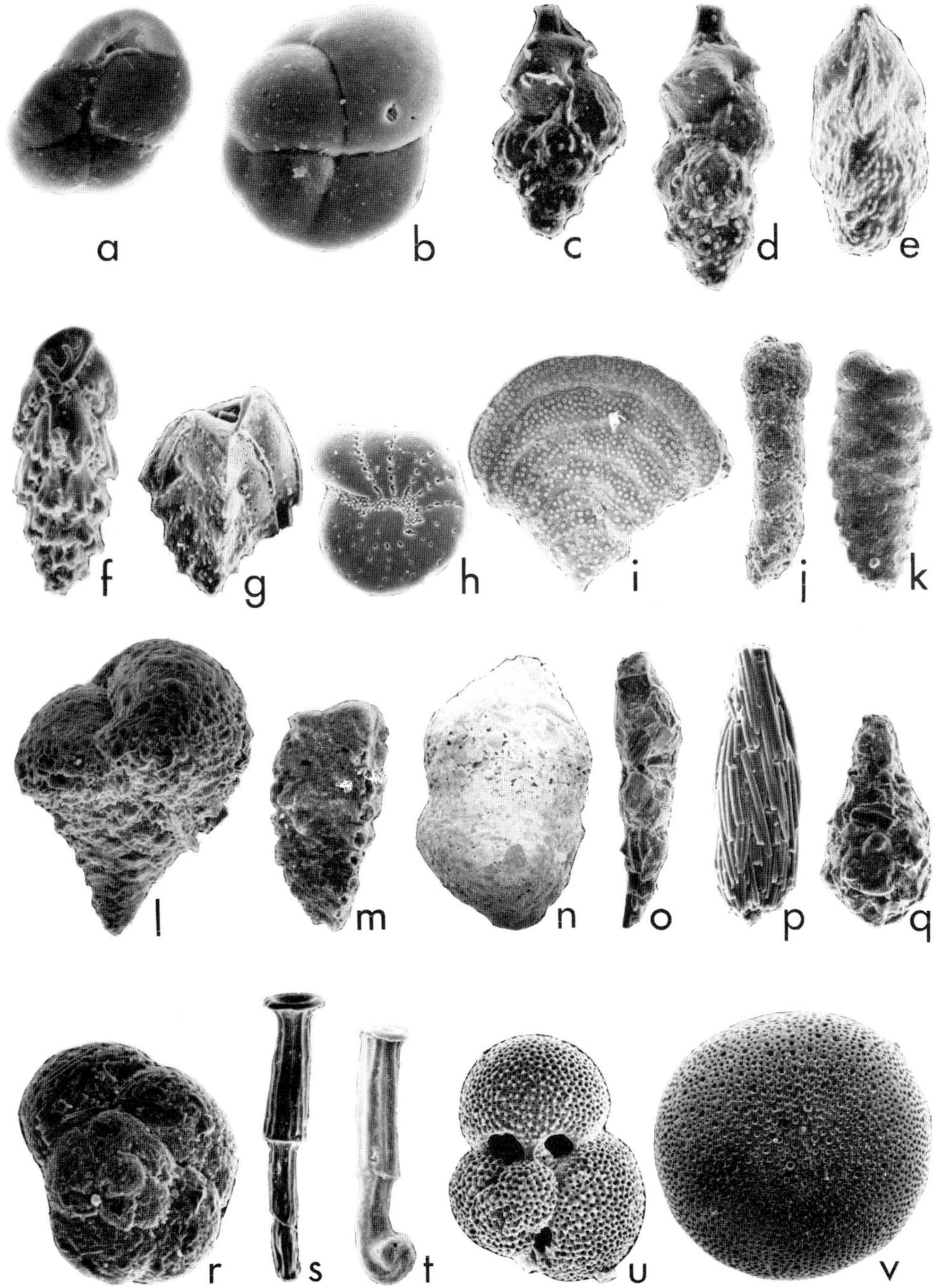

FIG. 7—Species typical of *Globocassidulina*-dominant assemblage. This assemblage includes forms indigenous to deepest part of southern shelf lagoon together with species derived from surrounding shoals (Fig. 7s, t) and planktonic species (Fig. 7u, v) indicative of open-ocean circulation.

Cribroelphidium-dominant assemblage in the Gulf of Honduras and into marginal-marine assemblages in other coastal areas. To the north, in the southern shelf lagoon, it grades into the other low-diversity *Cribroelphidium*-dominant assemblage. Several groups of calcium carbonate shoals which contribute a small amount of sediment to the surrounding bottom are developed within the distribution pattern of this assemblage (e.g., Snake Cays). However, the area of shoal influence is so slight that a mixed-fauna assemblage could not be shown.

Globocassidulina Assemblage

Total populations—Two samples (212, 213) on the Belize shelf contain a diverse fauna dominated numerically by *Globocassidulina* associated with *Cribroelphidium, Fursenkoina, Reophax, Reussella, Brizalina, Trifarina,* reef-derived miliolids, and the only significant occurrence of planktonic Foraminifera found in the shelf sediments (Fig. 7). *Globocassidulina subglobosa* (Brady) dominates the total populations and constitutes 12 percent of both samples. Approximately 100 species are present in each sample.

Living populations—Although *Globocassidulina* is the most abundant faunal element in the total populations, it was scarce in the living population. The living population in sample 212 is a *Bigenerina-Textularia-Quinqueloculina* fauna; in sample 213 it is a *Bigenerina-Trifarina-Fursenkoina* fauna. These populations are more typical of the channel- or shoal-influenced faunas of the lagoon.

Areal extent and ecology—This assemblage occurs in the deepest part of the Belize shelf (30-35 fm) at its southernmost extremity. In this area the barrier reef is absent, and the lagoonal waters are open to oceanic influence. There is no effect of river influx, and the bottom waters have normal-marine salinity. Sediments are sandy, bioclastic muds. Reef debris is common but not dominant in the sand-size fraction of the sediments. Except for the relatively high abundance of *Globocassidulina* and the presence of planktonic forms, this assemblage closely resembles the high-diversity *Cribroelphidium*-dominant assemblage into which it grades shoreward. This assemblage extends seaward to the limit

---

Explanation of Figure 8.

a. *Quinqueloculina poeyana* d'Orbigny, side view, x50.
b. *Quinqueloculina laevigata* d'Orbigny, side view, x75.
c. *Quinqueloculina candeiana* d'Orbigny, side view, x50.
d. *Quinqueloculina bosciana* d'Orbigny, side view, x100.
e. *Quinqueloculina bicornis* (Walker and Jacob), side view, x150.
f. *Quinqueloculina lamarckiana* d'Orbigny, side view, x50.
g. *Quinqueloculina bradyana* Cushman, side view, x50.
h. *Quinqueloculina polygona* d'Orbigny, side view, x70.
i. *Quinqueloculina tricarinata* d'Orbigny, side view, x25.
j. *Quinqueloculina berthelotiana* d'Orbigny, side view, x75.
k. *Quinqueloculina exsculpta* (Heron-Allen and Earland), side view, x75.
l. *Triloculina bicarinata* d'Orbigny, side view, x40.
m. *Triloculina quadrilateralis* d'Orbigny, side view, x70.
n. *Triloculina linneiana* d'Orbigny, side view, x50.
o. *Dentostomina bermudezi* Perez Farfante, side view, x25.
p. *Ammomassilina alveoliniformis* (Millett), side view, x150 (?).
q. *Hauerina occidentalis* Cushman, side view, x100.
r. *Heterillina cribrostoma* (Heron-Allen and Earland), side view, x100.

FIG. 8—Species typical of high-diversity miliolid-dominant assemblage. Figures 8-11 illustrate spectrum of forms representative of faunas inhabiting shallow carbonate platform and shoals of Belize shelf, especially in marine-grass meadows that cover much of platform area.

of sampling near the continental-shelf escarpment. The planktonic species are evidence of oceanic influence. Sample 212 contains 12 percent planktonic forms in the total population, and sample 213 contains 3-4 percent planktonic species. Planktonic tests are extremely scarce in the sediments throughout the rest of the Belize shelf.

The distribution of this assemblage falls within the *Gypsina*-marl sedimentary facies of Purdy et al (this volume). *Gypsina* is sometimes found as a free-living form, but commonly it has an encrusting habit; in this area it is found attached to *Halimeda* grains in sufficient volume to characterize the sediment petrographically. Thus the potential danger in attempting ecologic analysis with only one tool or one approach is apparent. Whereas the writer overlooked a volumetrically significant allochthonous foraminifer by studying only free-living forms, Purdy missed the ecologic significance of the small, deeper water benthonic foraminifer (*Globocassidulina*) and the open-ocean influence signified by the planktonic forms.

High-Diversity Miliolid Assemblage

Total populations—The family Miliolidae strongly dominates the foraminiferal assemblages of the shallow, marine areas of calcium carbonate deposition on the Belize shelf. The total populations grouped in this assemblage are characterized by high faunal diversity, and nearly all are dominated by the genera *Quinqueloculina* and *Triloculina*, although the populations vary widely with respect to taxonomic composition, faunal diversity, faunal dominance, and faunal density. The dominant generic associations are *Quinqueloculina-Triloculina-Planorbulina* and *Quinqueloculina-Triloculina-Heterillina*.

Faunal diversity of total populations for most samples of this assemblage ranges from 100 to 150 species per sample, and the average is approximately 120 species. The density of total populations varies widely from 3,000-5,000 to over 100,000 tests per sample. The greatest total-population diversity and density occur in areas of moderate to dense stands of turtle grass (*Thalassia*) growing in poorly sorted, skeletally derived sandy mud. For example, in sample 68, 148 species are recorded

---

⟶

Explanation of Figure 9.

a. *Miliolinella labiosa* (d'Orbigny), oblique apertural view, x250.
b. *Miliolinella labiosa* (d'Orbigny), side view, x100.
c. *Miliolinella labiosa* (d'Orbigny), oblique view, x100.
d. *Miliolinella labiosa* (d'Orbigny), oblique view, x70.
e. *Miliolinella labiosa* (d'Orbigny), apertural view, x100.
f. *Miliolinella labiosa* (d'Orbigny), side view, x70.
g. *Miliolinella labiosa* (d'Orbigny), side view, x100.
h. *Vertebralina atlantica* (Cushman and Hanzawa), side view, x35.
i. *Vertebralina cassis* d'Orbigny, side view, x40.
j. *Articulina lineata* Brady, side view, x40.
k. *Articulina mayori* Cushman, side view, x50.
l. *Peneroplis pertusus* (Forskal), side view, x50.
m. *Peneroplis bradyi* Cushman, side view, x22.
n. *Wiesnerella auriculata* (Egger), side view, x70.
o. *Archaias angulatus* (Fichtell and Moll), side view, x35.
p. *Archaias angulatus* (Fichtell and Moll), oblique apertural view, x150.
q. *Archaias angulatus* (Fichtell and Moll), side view, x50.
r. *Archaias angulatus* (Fichtell and Moll), side view, x22.

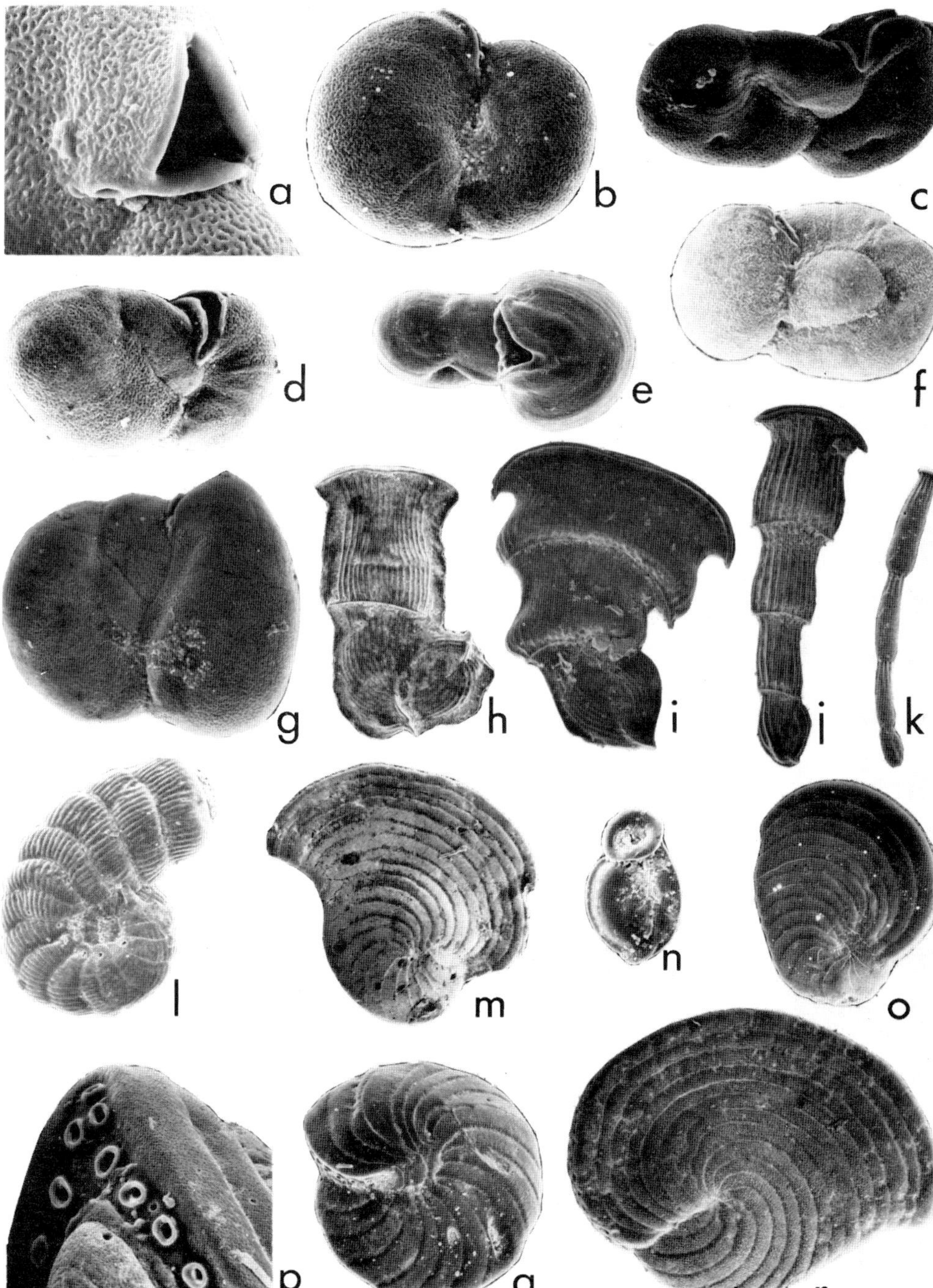

FIG. 9–Species typical of high-diversity miliolid-dominant assemblage.

in a population of over 30,000 tests. *Planorbulina acervalis* Brady, having a frequency of only 5 percent, is the dominant species in this sample. The dominant genera of the total population are *Triloculina, Quinqueloculina,* and *Planorbulina.* Similar populations are found in other areas characterized by *Thalassia* growth on the barrier and outer platforms and on the carbonate shoals of the southern shelf lagoon.

In contrast, several populations dominated by *Quinqueloculina* and *Triloculina* have lower total-population diversity and smaller size. In sample 118, 117 species are identified, but only about 3,000 tests are in the measured volume of sediment. This is a *Quinqueloculina-Planorbulina-Dentostomina* fauna from a rocky area near Pelican Cay. Unconsolidated sediments in this locality are sparse and coarse; this sample is characteristic of several shallow, current-swept areas in which little sediment accumulates.

Most species encountered in the high-diversity miliolid-dominant assemblage are restricted to the marine carbonate platforms and the nearby lagoonal areas subject to shoal-derived sediment influx. Approximately 100 species, more than two thirds of the species in this assemblage, are not found elsewhere on the Belize shelf. Several species that are not restricted to this assemblage occur elsewhere in much higher frequencies in both living and total populations. For example, *Heterillina cribrostoma* occurs in low frequencies in these populations but is the most abundant species in the populations of the low-diversity miliolid-dominant assemblage. Some of the more characteristic species of this high-diversity miliolid-dominant assemblage are shown in Figures 8-11.

Living populations—The living populations found associated with the total populations in the high-diversity miliolid-dominant assemblage are similar to the total populations with respect to taxonomic composition and dominant genera. Most living populations are dominated by *Quinqueloculina* and *Triloculina,* and nearly all of the taxa identified in the total populations are present.

Faunal density of living populations varies from approximately 100 to 1,000 individuals per sample and most samples contain 300-500 stained tests.

⟶

Explanation of Figure 10.

a. *Discorbis mira* Cushman, ventral view, x70.
b. *Discorbis mira* Cushman, dorsal view, x70.
c. *Rosalina candeiana* d'Orbigny, ventral view, x70.
d. *Rosalina candeiana* d'Orbigny, ventral view, x70.
e. *Rosalina floridana* (Cushman), ventral view, x150.
f. *Rosalina* sp., ventral view, x100.
g. *Rosalina* sp., dorsal view, x150.
h. *Discorbis* sp. cf. *D. australis* Parr, ventral view, x100.
i. *Hemidiscella palabunda* Bock, dorsal view, x150.
j. *Hemidiscella palabunda* Bock, ventral view (attached surface), x70.
k. *Bronnimannia palmerae* (Bermudez), dorsal view, x70.
l. *Bronnimannia palmerae* (Bermudez), ventral view, x70.
m. *Cymbaloporetta squamosa* (d'Orbigny), ventral view, x70.
n. *Cymbaloporetta squamosa* (d'Orbigny), dorsal view, x70.
o. *Cibicides mayori* (Cushman), dorsal view, x50.
p. *Cribroelphidium translucens* (Natland), side view, x75.
q. *Elphidium lanieri* (d'Orbigny), side view, x50.
r. *Brizalina variablis* (Williamson), side view, x150.
s. *Bolivinita rhomboidalis* (Millett), side view, x150.

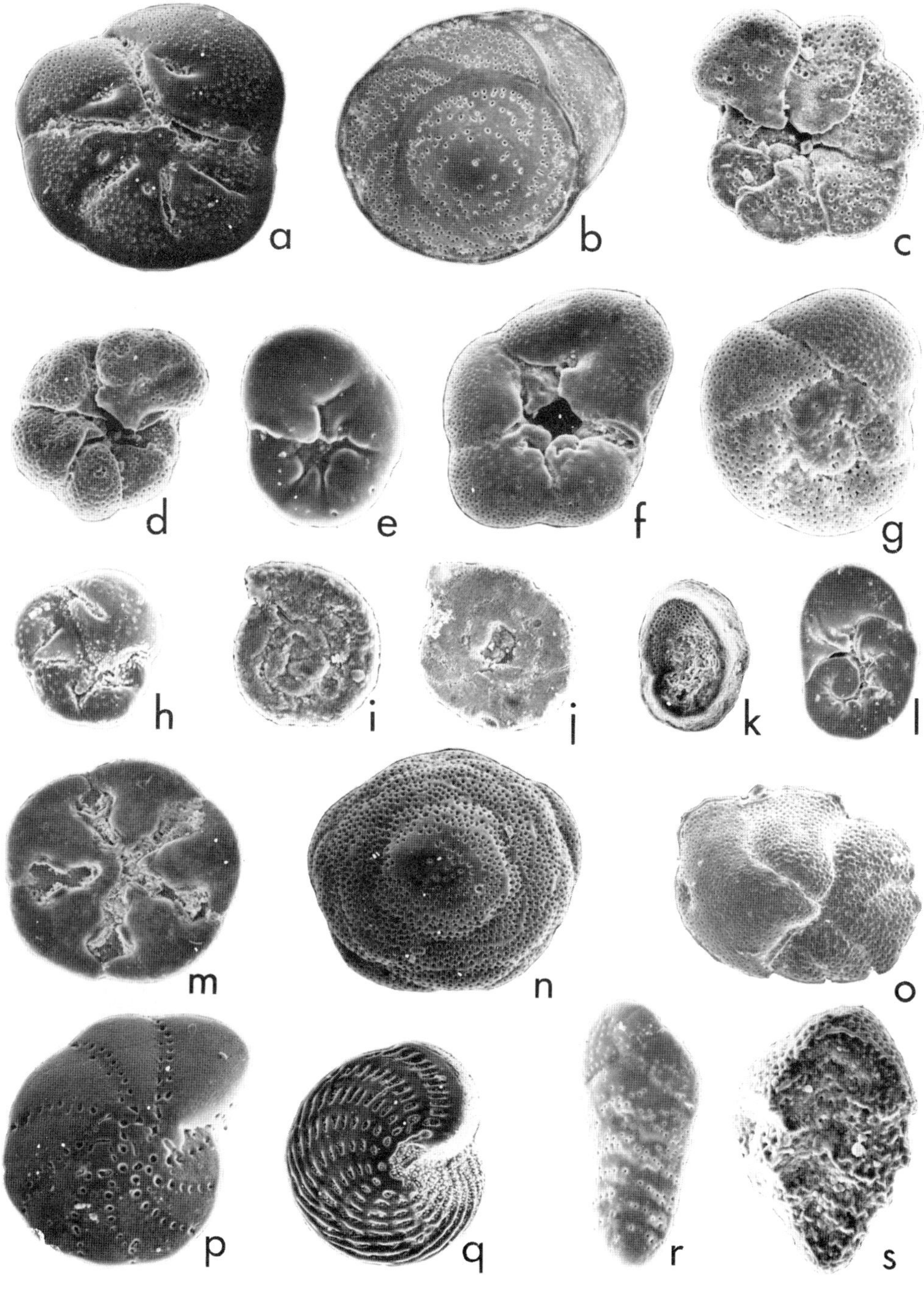

FIG. 10–Species typical of high-diversity miliolid-dominant assemblage.

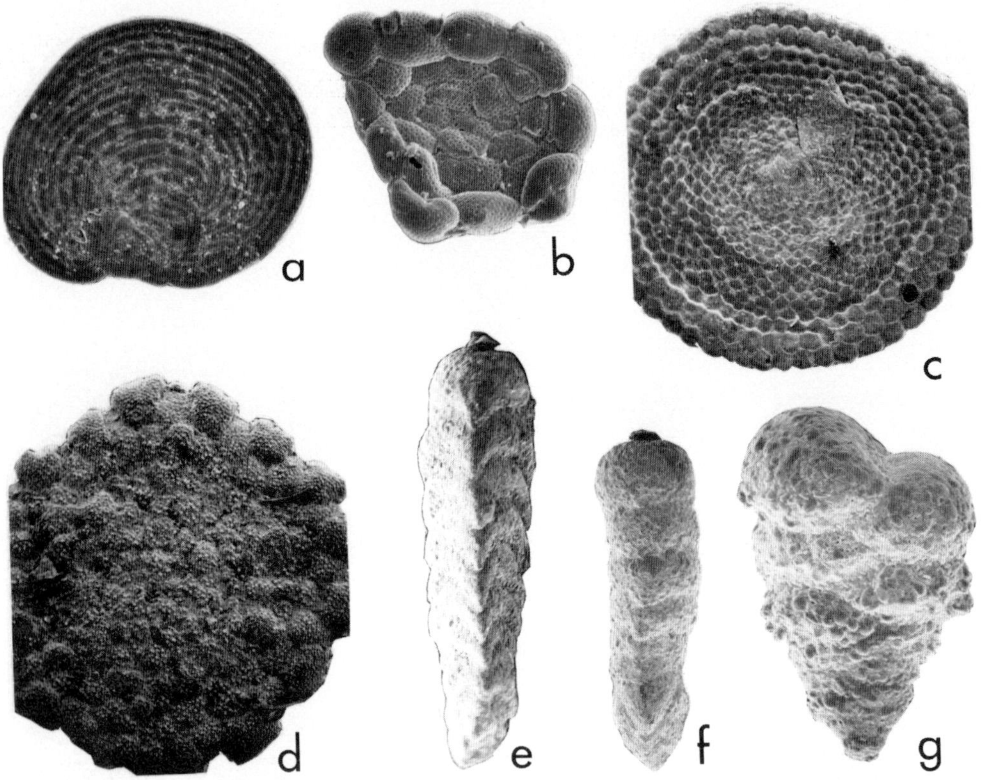

FIG. 11—Species typical of high-diversity miliolid-dominant assemblage.
a. *Sorities marginalis* (Lamarck), side view, x50.
b. *Planorbulina mediterranensis* d'Orbigny, dorsal view, x50.
c. *Amphisorus hemprichii* Ehrenberg, side view, x22.
d. *Planorbulina acervalis* Brady, dorsal view, x22.
e. *Clavulina tricarinata* d'Orbigny, side view, x30.
f. *Clavulina nodosaria* d'Orbigny, side view, x30.
g. *Textularia agglutinans* d'Orbigny, side view, x75.

Areal extent and ecology—The high-diversity miliolid-dominant assemblage is characteristic of areas of the Belize shelf in which these major environmental parameters are present: (1) shallow, clear water (less than 3-4 fm); (2) normal-marine salinity; (3) pure calcium carbonate sedimentary regime; and (4) high diversity of benthic habitats. Within this assemblage, population density, and to an extent diversity, is affected by the distribution of marine vegetation, current action, and sediment deposition. This assemblage is not present in the highly agitated environments of the windward reef and shoal areas.

Wright (1964), Murray (1965), and Bock (1967), among others, have pointed out that many, if not most, of the benthonic Foraminifera inhabiting shallow backreef environments live attached to plants and other floral and faunal elements above the sediment surface. Thus, the frequency distribution of living Foraminifera recorded from this study does not present an accurate picture of the true living-population structure.

As might be anticipated, therefore, the composition of living populations differs strikingly from that of the total populations in many samples. In several samples (e.g., samples 78, 79), *?Remaneica* sp. cf.

R. kelletae (Thalman) is a common living form—constituting as much as 6 percent of the living population. This tiny, attaching foraminifer is an extremely fragile, agglutinated form which would not likely be preserved in the sediment after death; certainly it is scarce in total populations. *Bolivinita rhomboidalis* (Millett) constitutes 6 percent of the living population in sample 82, but represents less than 1 percent of the total population; in sample 98, *Buliminella elegantissima* (d'Orbigny) represents 12 percent, and *Fursenkoina pontoni* (Cushman) 7 percent, of the living population, but each of these two species composes less than 1 percent of the total population.

It is evident that microenvironmental conditions are extremely complex and variable in the shallow backreef environment. Local and temporal conditions are reflected in the living populations, the effects of which are diluted by the composition of the total population.

A few species recorded in the total populations were never seen containing stained protoplasm. For example, *Amphistegina gibbosa* (d'Orbigny) is present in many and in one sample (134f) accounts for 50 percent of the total population, but it was not seen alive in the course of this study. Evidently the habitat of this species was never sampled.

Archaias angulatus is another foraminifer that is rarely encountered in living populations in this area. It was found in abundance living in the crowns of *Penicillus* plants collected in the Pelican Cay area, and empty tests of this species commonly accumulate in the sediment surrounding the plants. Surprisingly, several attaching species were found living in relatively high frequencies in the surface sediments of the *Thalassia* meadows. *Planorbulina*, a large, plano-convex form adapted to attachment on flat surfaces (such as marine-grass leaves) was found in consistently higher frequencies in the living population than in the total populations of bottom-sediment samples. Whether this represents artificial displacement during the sampling process is not known.

The diversity of the living populations attests to the complexity of the microenvironments sampled. Approximately 80-100 living species were identified in most individual samples. When it is considered that this represents the possible existence of 80-100 niches for the Foraminifera alone in the small volume of substrate collected, one begins to realize the degree of complexity of the inter-specific relation within the benthic community.

Areally this assemblage occurs throughout the barrier platform and carbonate shoals of southern Belize and the outer shelf lagoon of northern Belize (Fig. 2). On the northern shelf this assemblage grades westward into the low-diversity miliolid-dominant assemblage coincident with the salinity gradient. In southern Belize the boundary between the high-diversity miliolid-dominant assemblage and the *Cribroelphidium-Quinqueloculina* mixed-fauna assemblages is a sharp faunal discontinuity coincident with the sharp bathymetric boundary between the steep-sided shoal areas and the lagoon.

Archaias-Asterigerina Assemblage

Total populations—A small number of samples on the seaward edge of the continental shelf contain unique foraminifer populations. The total populations are characterized by low faunal diversity, relatively low population density, and high faunal dominance. The taxonomic composition is markedly different from other populations on the Belize shelf. The total populations of these samples are dominated chiefly by a combination of the genera *Archaias*, *Asterigerina*, and rarely *Amphistegina*, and to a lesser extent by *Glabratella*, *Neoconorbina*, *Trifarina*, and *Discorbis* (Fig. 12). A few samples included in this assemblage are dominated by

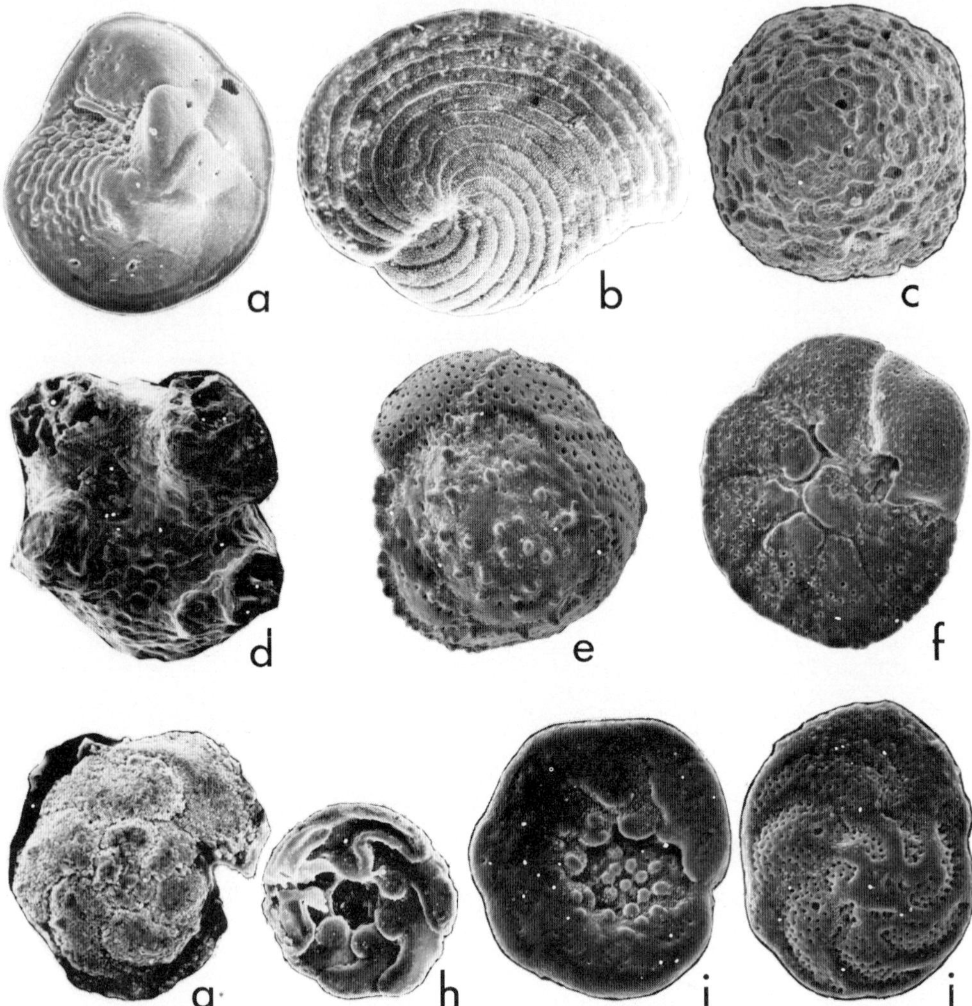

FIG. 12—Species typical of *Archaias-Asterigerina*- dominant assemblage which is found in mobile substrates along seaward margin of shelf-edge barrier platform. This assemblage is composed almost entirely of large, robust forms, either free-living or attached, and a suite of very small, attaching species.

Quinqueloculina and *Triloculina* but have significant frequencies of the preceding genera plus the distinctive population characteristics stated above. Diversity of the total populations varies from 50 to 100 species; faunal dominance is high—generally 25-30 percent; total population density is generally less than 3,000 tests; the lowest density is less than 1,000. The actual figures for the frequency of *Archaias angulatus* in this area are somewhat variable. The tests of this form are usually abraded and broken, and considerable subjective interpretation was necessary in counting test fragments and in identifying worn and recrystallized individuals. In this regard, considerable operator error may be involved, and the percentage figures may not be reproducible within large limits.

Living populations—The living populations generally contain less than 50 species, having a range of 23-87 species per sample. Faunal dominance is approximately 30 percent—the frequency of either *Glabratella opercularis?* (Brady), *Asterigerina carinata* d'Orbigny, or *Trifarina bella* (Phleger and Parker). *Archaias angulatus* is extremely scarce in living populations. Living-population density ranges from less than 100 to 350 individuals per sample.

Areal extent and ecology—Populations of this type are found in the well-sorted, commonly current-rippled carbonate sand areas near the outer shelf edge. These areas presumably receive maximum current agitation and also are virtually devoid of vegetation. This assemblage has little lateral continuity, and populations of this type are developed spottily along the seaward reef margin where the unique combination of limiting factors, particularly a mobile substrate and lack of vegetation, inhibits the development of the high-diversity miliolid-dominant assemblage.

None of the species in this assemblage are limited to the area of assemblage occurrence. All are found in varying but generally low frequencies distributed throughout the extent of the shallow, marine carbonate environment. The obvious exception is *Archaias*, which occurs in high frequency associated with a very different fauna in the assemblages of the northern shelf and Chetumal Bay areas.

Adaptive Morphology—The dominant living species in this assemblage are obviously adapted to a mobile substrate. Most are small, trochospiral forms that attach ventrally to various grains in the sediment. Individual test development is variable and conforms to the shape of the attachment surface. Examples of this type are *Glabratella opercularis?*, *Neoconorbina* spp., *?Remaneica* sp. cf. *R. kelletae*, and *Discorbis* spp. *?R.*

←

Explanation of Figure 12.

a. *Asterigerina carinata* d'Orbigny, ventral view, x75.
b. *Archaias angulatus* (Fichtell and Moll), side view, x25.
c. *Gypsina vesicularis* (Parker and Jones), dorsal view, x25.
d. *Homotrema rubrum* (Lamarck), dorsal view of fragmented specimen, x35.
e. *Discorbis rosea* (d'Orbigny), dorsal view, x75.
f. *Discorbis rosea* (d'Orbigny), ventral view, x75.
g. *?Remaneica* sp. cf. *R. kelletae* (Thalmann), dorsal view, x150.
h. *?Remaneica* sp. cf. *R. kelletae* (Thalmann), ventral view, x200.
i. *Glabratella opercularis?* (Brady), ventral view, x150.
j. *Glabratella opercularis?* (Brady), dorsal view, x150.
k. *Neoconorbina terquemi* (Rzehak), dorsal view, x100.
l. *Trifarina bella* (Phleger and Parker), side view, x100.
m. *Triloculina linneiana* d'Orbigny, oblique view, x75.
n. *Rosalina candeiana* d'Orbigny, ventral view, x75.

sp. cf. *R. kelletae* is an agglutinated form that implants a few calcareous grains in a largely organic test. The flexible nature of the test may prevent breakage. In contrast, *Asterigerina carinata* is larger and thick shelled and displays minimum phenotypic variability. This more robust form probably lives free in the carbonate sands.

The relatively high ratios of living to total populations in the samples in this assemblage probably reflect removal and destruction of abandoned tests. The abundant small tests of most species in this assemblage presumably are not in hydraulic equilibrium as free grains in the sediment; when not firmly attached, they tend to be transported into less agitated areas. The scope of this study is limited to benthonic, free-living forms, and no attempt was made to assess the quantitative significance of fragments of wholly encrusting forms. Many pieces of *Homotrema* and *Gypsina* were found, but they were not recorded quantitatively in this study.

Cribroelphidium-Quinqueloculina Mixed-Fauna Assemblage

Total populations—Several total populations are mixtures of miliolid-dominant and *Cribroelphidium*-dominant faunas. These populations are generally diverse *Quinqueloculina-Cribroelphidium* or *Cribroelphidium-Quinqueloculina* total populations. These faunas generally contain 100-140 species per sample; rarely do they contain as few as 50 species. In most of these samples the dominant species is *Cribroelphidium poeyanum*, which occurs in frequencies of 8-12 percent. This assemblage differs from the miliolid-dominant assemblages of the carbonate shoals by the relatively high frequencies of *Cribroelphidium* and by the presence of several taxa commonly associated with the *Cribroelphidium*-dominant assemblages but rarely, if ever, present in the miliolid-dominant assemblage. Such taxa include species of *Elphidium*, *Fursenkoina*, *Bigenerina*, *Reophax*, *Nouria*, *Hanzawaia*, *Reussella*, and *Ammonia*. The assemblage differs from the high-diversity *Cribroelphidium* assemblage in the abundance and diversity of miliolids, peneroplids, and other shoal-derived genera (e.g., *Planorbulina*, *Clavulina*, *Asterigerina*, etc.). Total-population density

---

⟶

Explanation of Figure 13.

a. *Cribroelphidium poeyanum* (d'Orbigny), side view, x100.
b. *Florilus atlanticus* (Cushman), side view, x75.
c. *Fursenkoina pontoni* (Cushman), side view, x150.
d. *Nouria polymorphinoides* Heron-Allen and Earland, side view, x40.
e. *Reophax* sp. cf. *R. arayensis* Bermudez and Seiglie, side view, x75.
f. *Bigenerina irregularis* Phleger and Parker, side view, x50.
g. *Bigenerina* sp., side view, x30.
h. *Bulimina tenuis* Phleger and Parker, side view, x150.
i. *Reussella atlantica* Cushman, side view, x100.
j. *Fursenkoina spinicostata* (Phleger and Parker), side view, x100.
k. *Brizalina striatula* (Cushman), side view, x75.
l. *Quinqueloculina candeiana* d'Orbigny, side view, x75.
m. *Quinqueloculina laevigata* d'Orbigny, side view, x70.
n. *Quinqueloculina bicostata* d'Orbigny, end view, x50.
o. *Textularia conica* d'Orbigny, side view, x75.
p. *Globocassidulina subglobosa* (Brady), oblique view, x300.
q. *Globigerinoides* sp., side view, x70.
r. *Tretomphalus bulloides* (d'Orbigny), dorsal view, x100.
s. *Tretomphalus bulloides* (d'Orbigny), ventral view, x100.

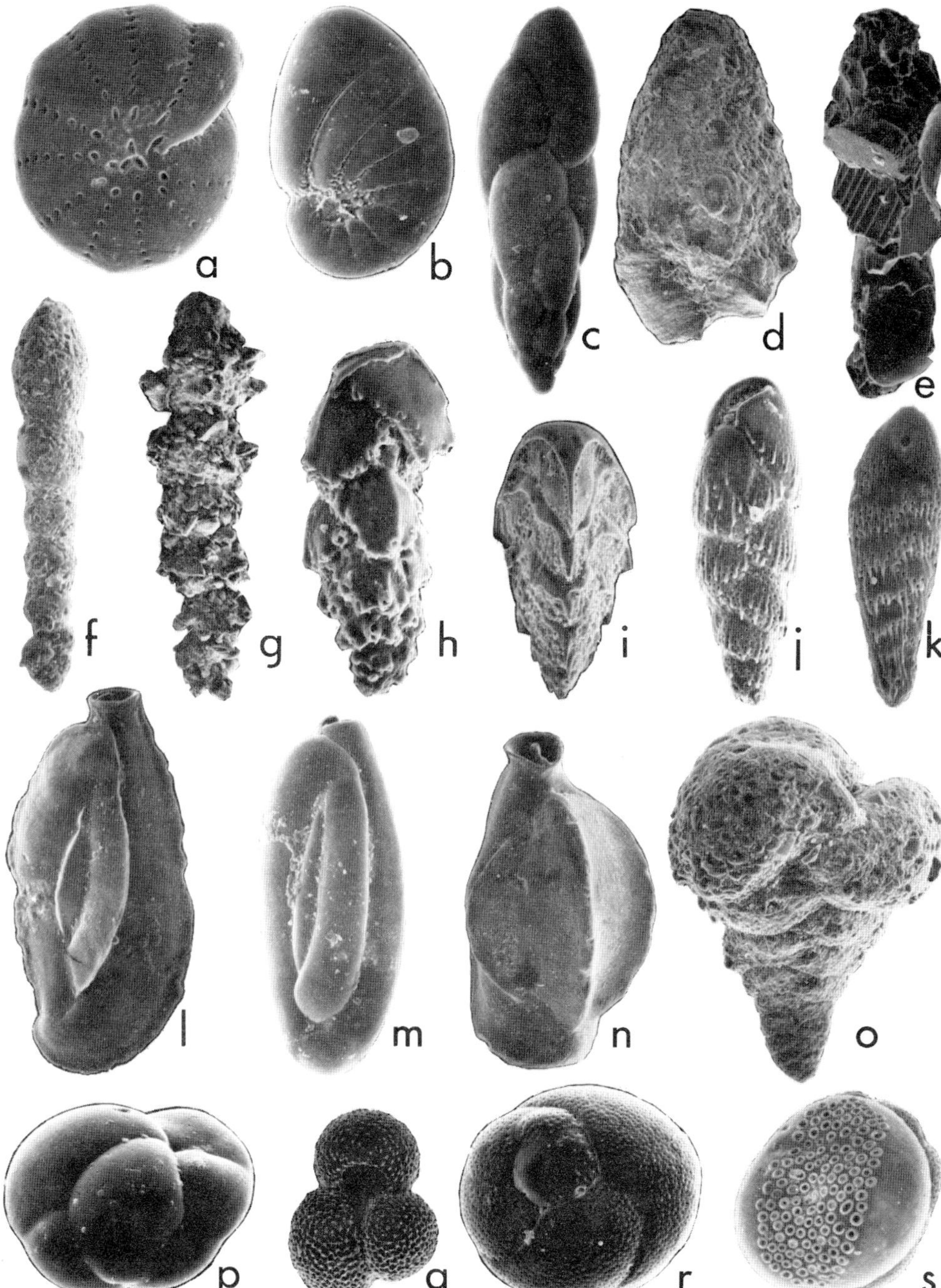

FIG. 13—Species typical of *Cribroelphidium-Quinqueloculina*-dominant mixed-fauna assemblage. Faunas of this type are found in southern shelf lagoon in close proximity to carbonate shoals; assemblage is a mixture of indigenous lagoonal and transported carbonate-platform species.

ranges from 7,000 to 100,000 tests per sample. Most samples contain large populations of about 30,000-40,000 tests. Figure 13 illustrates the species representative of this assemblage.

Living populations—As in most of the lagoonal assemblages described, the dominant taxa of the total populations rarely dominate the living populations. In the case of faunal displacement and mixing, this would be expected. The most abundant genus in most living populations in this assemblage is *Fursenkoina*. Commonly associated genera are *Quinqueloculina*, *Triloculina*, *Florilus*, *Bigenerina*, *Reophax*, *Cribroelphidium*, *Elphidium*, and, more rarely, *Textularia* and *Sagrina*. Living-population diversity varies widely; samples contain 3-90 species. Faunal dominance also is variable, ranging from 10 to 75 percent; however, in most samples faunal dominance is 25-30 percent. Among the living populations, the relation of high faunal dominance with lower faunal diversity was generally observed. The density of living populations ranges from 20 to 700 individuals; most samples contain 200-300 living specimens.

Areal extent and ecology—This assemblage represents a mixture of displaced shoal faunas and endemic deeper lagoon populations. It occurs in a narrow strip along the western margin of the well-developed barrier platform from Belize City south to the latitude of South Stann Creek. Throughout this area it grades abruptly shoreward into the low-diversity *Cribroelphidium*-dominant assemblage. Farther south it occurs throughout the complex shoals on the southern Belize shelf. It rims the shoals, lines the channels, and blankets the enclosed shelf-atoll lagoons. In each case it borders a development of the high-diversity miliolid-dominant assemblage, and away from the area of shoal influence it grades abruptly into *Cribroelphidium*-dominant assemblages. The area is generally one of deeper water (10-25 fm) on the shelf; hydrographic conditions are normal marine. The associated sediments, 60-100 percent carbonate in composition, are poorly sorted muddy sands and sandy muds that are almost wholly shoal derived. It is presumed that most of the area of assemblage occurrence is subjected to relatively rapid sediment accumulation. The zone of shoal influence and faunal mixing is very narrow. Because of sample density the boundary is generalized throughout; however, several detailed traverses provide some indication of absolute distance of transport off the shoals. For example, in the shelf lagoon a sampling traverse was made from the mouth of South Stann Creek to the barrier platform (Samples in sequence in the traverse are 281, 283, 285, 287, 289, 295, 291, 292, 294, and 98; Purdy et al, this volume, Fig. 3). Proceeding from the coast toward the barrier platform, marked reef influence (influx of miliolids) was not noted until sample 295, only 3/4 mi (1.2 km) west of the edge of the platform. Strong reef influence was evident in sample 291, 1/4 mi (0.4 km) west of the platform. Scarce, disc-shaped, juvenile *Planorbulina* spp. are found in mid-lagoon samples several miles shoreward of the carbonate shoals. They are an indication of shoal proximity, but the tests are irregularly distributed and uncommon in the sediments. This detailed traverse and others approaching individual atolls (e.g., Laughing Bird Cay atoll) suggest that most foraminiferal transport off the barrier platform and carbonate shoals is shoreward and terminates in a short distance.

Wantland and Pusey (1971) noted the marked compositional and textural similarity between these off-shoal sediments and the *Halimeda* marl of the shallow barrier platform. The presence of the indigenous lagoonal microfauna is unequivocal evidence of the deeper water depositional environment and would be an important tool in interpreting similar ancient sediments.

The strong anomalous dominance of *Fursenkoina pontoni* in the living populations in this assemblage may indicate that this species is best adapted to areas of relatively rapid sedimentation. In this particular area its importance is obscured in the total populations by displaced shallow-water foraminifers. In other areas, however, *Fursenkoina* dominance in an inorganic sedimentary regime might be a clue to relative sedimentary rates. Similarly, *Quinqueloculina candeiana*, *Bigenerina* spp., and *Reophax* spp. seem to proliferate under the same conditions that inhibit most other species.

Low-Diversity Miliolid Assemblage

Total populations—The total populations of many samples are characterized by (1) low faunal diversity, (2) overwhelming dominance by small miliolids, (3) high total-population density, (4) high generic dominance (*Heterillina-Triloculina-Miliolinella*), and (5) considerable intraspecific morphologic variation. The diversity of these populations ranges from 20 to 50 species per sample, although these figures are dubious because intraspecific variation is great and definition of species is more arbitrary than usual. Faunal dominance may be as high as 55 percent and is rarely less than 20 percent. Total-population density is high, ranging from 20,000 to over 100,000 tests per sample. This high density results in part from the small size of the tests of most species and the lack of any influx of clastic sediments.

The most common species in this assemblage are *Heterillina cribrostoma* (Heron-Allen and Earland) and *Triloculina bermudezi* Acosta. Accessory species include *Quinqueloculina* sp., *Triloculina fitterei* Acosta vars., *T.* sp., *Massilina protea* Parker, *Miliolinella circularis* (Bornemann), *Quinqueloculina laevigata* d'Orbigny, *Q. poeyana* d'Orbigny, and *Archaias angulatus* (Fichtell and Moll) (Figs. 14-16). Most of these taxa are characterized by small, smooth, unornamented tests. Exceptions are the costate species *Q. poeyana* and *M. protea*, costate varieties of *T. fitterei*, and the large, complex peneroplid *A. angulatus*.

Calcareous, perforate species occurring in significant frequencies in this assemblage include *Discorbis* sp. cf. *D. australis* Parr, *Cribroelphidium poeyanum* (d'Orbigny), *Cellanthus galvestonense* (Kornfeld), *C. gunteri* (Cole), and *Ammonia beccarii* (Linné).

The dominant generic associations found in this assemblage are *Heterillina-Triloculina-Quinqueloculina*, *Heterillina-Triloculina-Miliolinella*, and *Miliolinella-Triloculina-Cribroelphidium*.

Although most of the common species in this assemblage are found in other areas of the shelf (particularly in the high-diversity miliolid-dominant assemblage) and under different environmental conditions, these taxa reach their highest percentages in the low-diversity miliolid-dominant assemblage.

Locally, wave and current action on submarine shoals produces small patches of very different faunal character within the milieu of small-miliolid dominance. Samples belonging to the *Archaias*-miliolid assemblage, dominated by *Archaias angulatus*, are found on the large mud shoals as winnowed beach deposits associated with tiny cays formed by pioneer mangrove communities (e.g., sample 256). Intraspecific variability increases in areas of abundant marine-grass cover, where much of the phenotypic variability of test form results from the conformation of the shell to attaching surfaces (predominantly marine-grass leaves) during growth. The cell pattern of *Thalassia* leaves is clearly imprinted on the attaching surface of some tests of *Heterillina cribrostoma*, one of the more variable forms.

Living populations—Living populations are characterized by (1) low faunal diversity, (2) variable to low population density, (3) high faunal dominance, (4) anomalous high frequencies of certain nonmiliolid taxa, and (5) overall dominance of small miliolids.

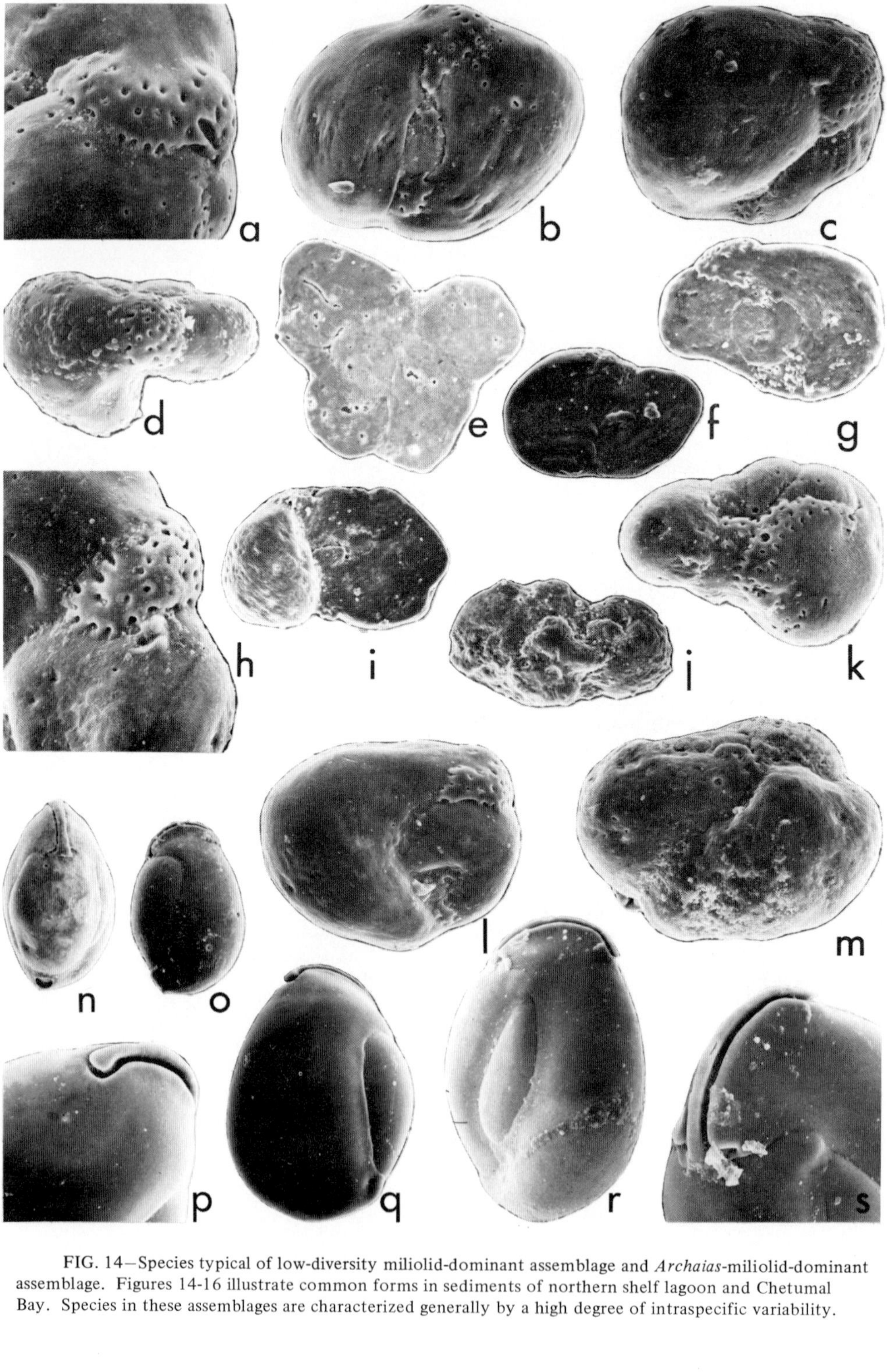

FIG. 14–Species typical of low-diversity miliolid-dominant assemblage and *Archaias*-miliolid-dominant assemblage. Figures 14-16 illustrate common forms in sediments of northern shelf lagoon and Chetumal Bay. Species in these assemblages are characterized generally by a high degree of intraspecific variability.

The diversity of living species ranges from 7 to 35 per sample, and in most samples a high percentage of the taxa in the total populations are found living. Living-population density ranges from 150-200 to approximately 1,000 individuals per sample, and most samples contain 200-500 stained tests. Faunal dominance is highly variable, ranging from less than 10 to 70 percent; faunal dominance generally is inversely related to faunal diversity.

In more than half of the samples, *Discorbis* is either the most abundant genus or one of the three or four most common genera. With this major exception, the taxonomic composition of the living populations corresponds rather well with that of the total populations. *Heterillina, Triloculina, Quinqueloculina,* and *Miliolinella* compose the bulk of the remainder of the faunas and dominate those samples in which *Discorbis* does not occur in high frequencies. *Elphidium, Cellanthus, Cribroelphidium, Ammonia, Valvulina,* and *Archaias* are locally significant.

The distribution of dominant living taxa in this area is vaguely consistent, though inexplicable. Living populations dominated by *Discorbis* sp. cf. *D. australis* occur in Chetumal Bay from the southern tip of Ambergris Cay north to a point opposite the northern tip of Ambergris Cay. The commonly associated fauna consists of *Triloculina bermudezi* and *Heterillina cribrostoma.* Both north and south of the *Discorbis*-dominant area, *Heterillina* and other miliolids common in the total populations dominate the living populations.

Areal extent and ecology—The distribution of this assemblage covers a large area of the northern shelf lagoon and most of Chetumal Bay. It extends from the latitude of Belize City north to the Mexican border (and probably beyond), and from the mainland margin seaward to a line corresponding approximately to north-south cay development in the lagoon that marks the boundary between the northern shelf lagoon and the outer platform.

The hydrography of this area is characterized by variable salinity that is generally less than normal marine. Hypersaline conditions may develop in the dry season throughout much of the shallow bay area and especially in the marginal, semirestricted lagoons. During the time of

---

⟵

Explanation of Figure 14.

a. *Heterillina cribrostoma* (Heron-Allen and Earland), oblique apertural view, x250.
b. *Heterillina cribrostoma* (Heron-Allen and Earland), side view, x150.
c. *Heterillina cribrostoma* (Heron-Allen and Earland), side view, x150.
d. *Heterillina cribrostoma* (Heron-Allen and Earland), apertural view, x200.
e. *Heterillina cribrostoma* (Heron-Allen and Earland), ventral attaching surface, x100.
f. *Heterillina cribrostoma* (Heron-Allen and Earland), ventral attaching surface, x70.
g. *Heterillina cribrostoma* (Heron-Allen and Earland), ventral attaching surface, x100.
h. *Heterillina cribrostoma* (Heron-Allen and Earland), oblique apertural view, x250.
i. *Heterillina cribrostoma* (Heron-Allen and Earland), ventral attaching surface, x70.
j. *Heterillina cribrostoma* (Heron-Allen and Earland), dorsal surface of attached form, x70.
k. *Heterillina cribrostoma* (Heron-Allen and Earland), apertural view, x150.
l. *Heterillina cribrostoma* (Heron-Allen and Earland), side view, x150.
m. *Heterillina cribrostoma* (Heron-Allen and Earland), side view, x150.
n. *Triloculina bermudezi* Acosta, edge view, x70.
o. *Triloculina bermudezi* Acosta, side view, x70.
p. *Triloculina bermudezi* Acosta, oblique apertural view, x250.
q. *Triloculina bermudezi* Acosta, side view, x150.
r. *Triloculina bermudezi* Acosta, side view, x150.
s. *Triloculina bermudezi* Acosta, oblique apertural view, x250.

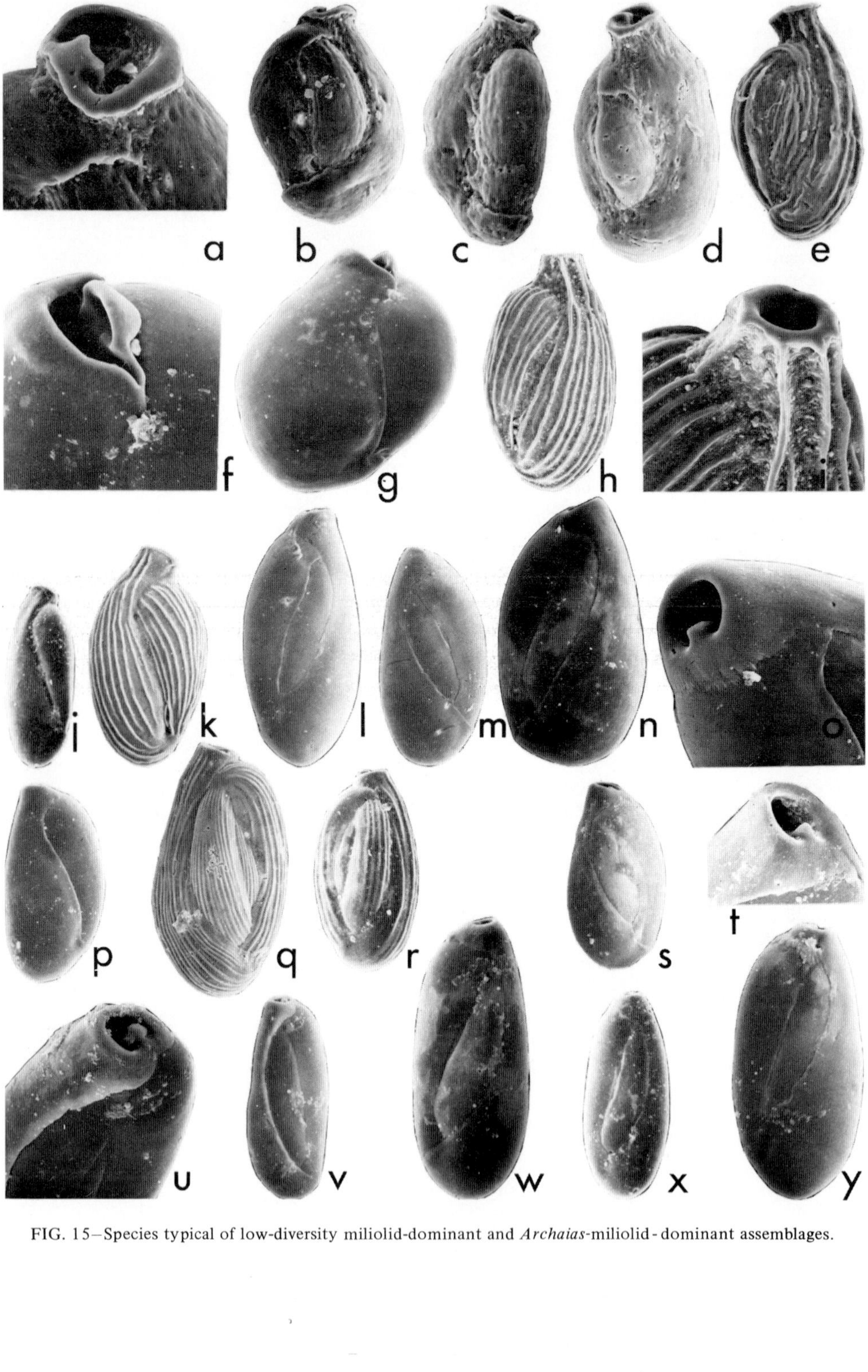

FIG. 15–Species typical of low-diversity miliolid-dominant and *Archaias*-miliolid-dominant assemblages.

sample collection, bottom salinities over the area ranged from 18°/₀₀ S near the mainland to 34-36°/₀₀ S toward the seaward margin of assemblage distribution. The salinity gradient developed along the axis of Chetumal Bay and from the mainland coast seaward corresponds to gradation in faunal composition. In upper Chetumal Bay the low-diversity miliolid-dominant assemblage grades over a wide area west of Rocky Point into an *Ammonia - small-miliolid* marginal-marine assemblage containing *Protelphidium delicatum* (Bermudez) (Fig. 4) near the effluents of the New River and the Rio Hondo. South of Chetumal Bay the low-diversity miliolid-dominant assemblage grades seaward into the high-diversity miliolid-dominant assemblage. This gradation parallels the salinity transition and coincides with the sedimentary transition from the "miliolid-mud" facies to the "*Halimeda*-sand" facies. The assemblage includes the areal extent of most of the "miliolid-mud," "cryptocrystalline-grain," and "terrigenous-sand" sedimentary facies (Pusey, this volume). To the south, this assemblage is replaced by an *Ammonia-Ammobaculites* marginal-marine assemblage at the mouth of the present Belize River and by the low-diversity *Cribroelphidium*-dominant assemblage of the southern shelf lagoon.

To the west, this assemblage is found in most of the inshore lagoons of the Belize mainland. A significant exception is Northern River Lagoon, which contains an *Ammonia-Elphidium* fauna (Wantland, 1969a). This is the only lagoon on the northern Belize coast that is subject to relatively constant freshwater influx.

The unique living *Discorbis-Triloculina-Heterillina* fauna seems to be associated with variable hydrography but does not persist in river-influenced areas such as upper Chetumal Bay or in the areas of more open circulation such as the northern shelf lagoon. It is significant that Streeter (1963) found a very similar fauna inhabiting the Inner Bank area

←

Explanation of Figure 15.

a. *Massilina protea* Parker, oblique apertural view, x250.
b. *Massilina protea* Parker, side view, x100.
c. *Massilina protea* Parker, end view, x125.
d. *Massilina protea* Parker, side view, x125.
e. *Masslina protea* Parker, side view, x125.
f. *Miliolinella circularis* (Bornemann), oblique apertural view, x250.
g. *Miliolinella circularis* (Bornemann), side view, x125.
h. *Triloculina fitterei* Acosta, side view of costate form, x100.
i. *Triloculina fitterei* Acosta, oblique apertural view, x250
j. *Triloculina fitterei* Acosta, side view of smooth form, x70.
k. *Triloculina fitterei* Acosta, side view of costate form, x75.
l. *Triloculina* sp., side view, x100.
m. *Triloculina* sp., side view, x100.
n. *Triloculina* sp., side view, x125.
o. *Triloculina* sp., oblique apertural view, x250.
p. *Triloculina* sp., side view, x100.
q. *Quinqueloculina poeyana* d'Orbigny, side view, x75.
r. *Quinqueloculina poeyana* d'Orbigny, side view, x70.
s. *Quinqueloculina* sp., side view, x100.
t. *Quinqueloculina* sp., oblique apertural view, x250.
u. *Quinqueloculina* sp., oblique apertural view, x250.
v. *Quinqueloculina* sp., side view, x100.
w. *Quinqueloculina* sp., side view, x70.
x. *Quinqueloculina* sp., side view, x150.
y. *Quinqueloculina* sp., side view, x125.

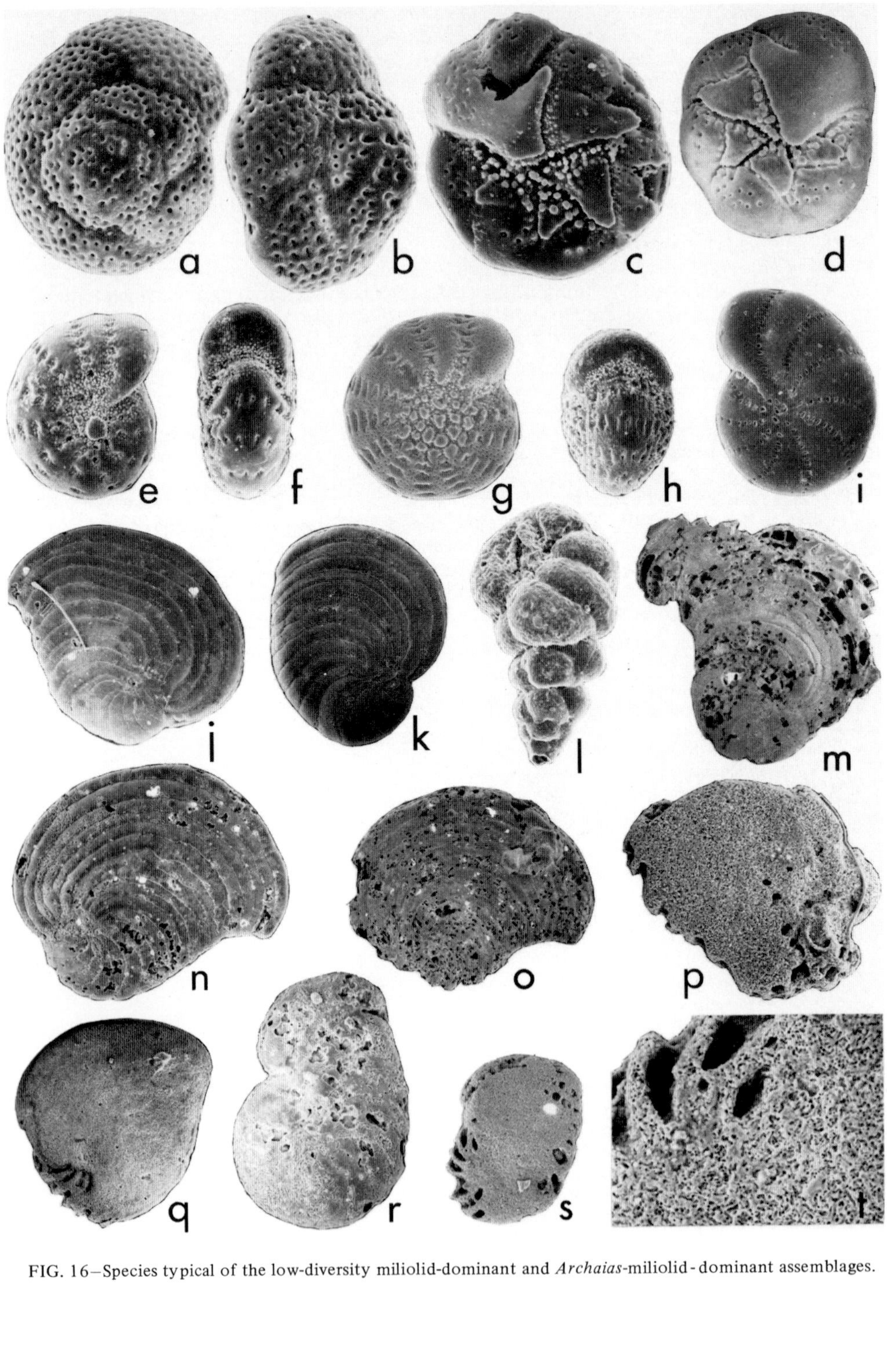

FIG. 16–Species typical of the low-diversity miliolid-dominant and *Archaias*-miliolid-dominant assemblages.

of the Bahamas west of Andros Island. He reported a low-diversity fauna characterized by *Triloculina eburnea* d'Orbigny, an undescribed species of *Triloculinella*, and an undescribed species of *Discorbis*. (These forms are thought to be conspecific with *Triloculina bermudezi*, *Heterillina cribrostoma*, and *Discorbis* sp. cf. *D. australis* of Belize—see Appendix.) This fauna (Assemblage I of Streeter) is found in calcium carbonate muds associated with restricted circulation, high salinity (46‰ S), and high temperature (Streeter, 1963, p. 21-22). These conditions perhaps exist at certain times in Chetumal Bay, although high temperature and salinity were not prevalent at the time of sample collection. It is suggested that the dominant factor for this fauna may be the fluctuating nature of the water characteristics rather than the specific extremes encountered.

In brief, the area in which this assemblage predominates is characterized by (1) very shallow (less than 5 fm), often turbid water; (2) variable salinity and water temperature; (3) deposition of soft carbonate mud to sandy carbonate mud; and (4) varied amounts of marine-grass cover, providing local increase in habitat diversity.

Archaias-Miliolid Assemblage

Total populations—The total populations of a few samples are dominated by *Archaias angulatus* associated with small, smooth miliolids. Faunal diversity is low—less than 30 species. Faunal dominance is high and may exceed 90 percent of the total population, although it is more commonly less. The associated genera are those common in the low-diversity miliolid-dominant assemblage. They include *Heterillina*, *Triloculina*, *Quinqueloculina*, *Massilina*, and more rarely *Cellanthus* and *Valvulina*.

Living populations—Living-population density in this assemblage is exceedingly low; in some samples no living individuals were seen. *Archaias* is extremely scarce in living populations. The most common living forms are *Heterillina cribrostoma*, *Massilina protea*, *Quinqueloculina* spp.,

---

←

Explanation of Figure 16.

a. *Discorbis* sp. cf. *D. australis* Parr, dorsal view, x150.
b. *Discorbis* sp. cf. *D. australis* Parr, edge view, x150.
c. *Discorbis* sp. cf. *D. australis* Parr, ventral view, x200.
d. *Discorbis* sp. cf. *D. australis* Parr, ventral view, x150.
e. *Cellanthus galvestonense* (Kornfeld), side view, x70.
f. *Cellanthus galvestonense* (Kornfeld), apertural view, x70.
g. *Cellanthus gunteri* (Cole), side view, x75.
h. *Cellanthus gunteri* (Cole), apertural view, x70.
i. *Cribroelphidium poeyanum* (d'Orbigny), side view, x75.
j. *Archaias angulatus* (Fichtell and Moll), side view, x35.
k. *Archaias angulatus* (Fichtell and Moll), side view, x35.
l. *Valvulina oviedoiana* d'Orbigny, side view of slender, megalospheric form, x40.
m-t. Bored, abraded, and recrystallized specimens of *Archaias angulatus* found in surface sediments and island beaches of Chetumal Bay.
m. *Archaias angulatus* (Fichtell and Moll), side view, x35.
n. *Archaias angulatus* (Fichtell and Moll), side view, x35.
o. *Archaias angulatus* (Fichtell and Moll), side view, x25.
p. *Archaias angulatus* (Fichtell and Moll), side view, x60.
q. *Archaias angulatus* (Fichtell and Moll), side view, x35.
r. *Archaias angulatus* (Fichtell and Moll), side view, x60.
s. *Archaias angulatus* (Fichtell and Moll), side view, x35.
t. *Archaias angulatus* (Fichtell and Moll), close-up of surface of Figure 16s, x150.

and *Discorbis* sp. cf. *D. australis*. Living-population diversity is low—generally less than 20 species.

Areal extent and ecology—Samples of this assemblage are located in the eastern area of lower Chetumal Bay bordering the western side of Ambergris Cay (Fig. 2). This distribution coincides with the "peneroplid-sand" sedimentary facies of Pusey. The concentration of *Archaias* tests in this area is considered to be more of a sedimentary phenomenon than a biologic one. The sands are commonly current rippled, and the component tests display a complete spectrum of preservation; all degrees of recrystallization and abrasion are evident (Fig. 16m-t). Purdy (1968) used material from this area to demonstrate penecontemporaneous recrystallization of surficial Holocene sediments.

The *Archaias* sand appears to be a lag deposit. Purdy (personal commun.) has observed cut banks and evidence of active erosion along the margin of the cays paralleling Ambergris Cay in Chetumal Bay. The material being eroded is a small-miliolid-*Archaias* mud. Marginal-island conditions similar to those in which *Archaias* is now living probably persisted over a much larger area of Chetumal Bay earlier in the Holocene. As sea level rose to its present position, this area was eroded and winnowed, leaving the *Archaias* facies as a lag deposit. The *Archaias*-miliolid-dominant assemblage grades rather abruptly in all directions into the low-diversity miliolid-dominant assemblage. As noted, *Archaias* occurs in variable, though generally low percentages throughout the low-diversity miliolid-dominant assemblage and is concentrated locally as a winnowed beach sand around small, newly formed cays on the mud shoals in Chetumal Bay.

The assemblage is bounded seaward by the little-studied populations of the saline lake and marsh complex of Ambergris Cay. Samples from this area are nonquantitative, but they demonstrate that *Archaias* is common in both living and total populations in several areas bordering the Chetumal Bay side of the large island. These samples are all low-diversity small-miliolid-dominant associations, but qualitative observation indicates that the contribution of *Archaias* to these foraminiferal populations can be significant.

Areas of *Archaias* dominance are not uncommon on carbonate platforms elsewhere. Streeter (1963) found a discontinuous distribution of such populations on the Bahama Banks and concluded that they were polygenetic. He thought that deposits of relatively fresh tests in muds of the Inner Bank represented living populations (Streeter did not have the opportunity to study stained samples) and that worn, abraded populations associated with carbonate sands on the bank margin were secondarily concentrated. Observations in Belize support these hypotheses. Bandy (1964) reported that a very large part of the marine carbonate shelf of Batabanó Bay, Cuba, is characterized by *Archaias angulatus* assemblages in which that species comprises up to 80 percent of the foraminiferal populations. These populations are associated with carbonate sands (ovoid grains of Daetwyler and Kidwell, 1959) in a current-swept area. Bandy decided that the total populations represented living populations and, considering their distribution, concluded that *Archaias angulatus* is a stenohaline species.

The distribution of *Archaias angulatus* in Belize and elsewhere indicates that this species is adapted not only to constant marine conditions, but also to variable marginal-marine conditions not influenced by terrigenous sediment influx.

FAUNAL DIVERSITY

Faunal diversity is a population statistic independent of taxonomic identification. If the ecologic significance of diversity gradients can be better understood, this parameter can be a valuable tool in paleoecologic interpretation, particularly among ancient populations in which analogues of adaptive morphology are absent or poorly understood.

The relative values and gradients of total-population foraminifer diversity in the Belize shelf area (Fig. 17) illustrate many of the distribution features of the foraminiferal populations. In northern Belize, gradients from low diversity (<25 species per sample) to high diversity (>100 species per sample) are evident. Minimum diversity is found at the effluents of the New River, the Rio Hondo, and the Belize River and along parts of the northern coast. Diversity increases southward along the axis of Chetumal Bay and eastward across the northern shelf lagoon. These gradients parallel the hydrographic trend from low salinity near rivers, to variable salinity in shallow bays, to stable marine conditions on the outer platform. Also, the sediment surface of most of the northern shelf is largely featureless and homogeneous, providing limited benthic habitat diversity and contributing to lower faunal diversity. In contrast, the adjacent outer platform supports extensive marine-grass (*Thalassia*) meadows and coral reefs. The increase of faunal diversity is gradual over most of the northern shelf to the "cay line" that separates the northern shelf lagoon from the outer platform. At that point, diversity increases abruptly. The strong barrier formed by Ambergris Cay sharply separates the low-diversity bay populations from the high-diversity platform populations. Clearly, the combination of stable marine conditions and high habitat diversity produces these diverse foraminiferal populations. The influence of associated flora on foraminiferal diversity can also be seen locally in the northern shelf area. The fringing marshes of Ambergris Cay and the Belize delta support more diverse faunas than the adjacent bay substrates.

In the southern Belize shelf area the diversity gradients are more complex. High diversity (>100 species per sample) exists on the seaward barrier platform and the complex of carbonate shoals. The environmental conditions are the same as on the outer platform—constant marine salinity and concomitant reefs and extensive marine-grass meadows. The *Thalassia* meadows provide a variety of potential habitats for microorganisms, and maximum faunal diversity (±150 species per sample) occurs in the skeletal carbonate muds associated with the grass areas. On the barrier platform, faunal diversity is reduced locally (50-100 species per sample) in areas of unstable, shifting carbonate sands, particularly at the windward shelf margin.

In the southern shelf lagoon, diversity increases from low in the north to high in the south along the axis of the lagoon; in the Gulf of Honduras, diversity increases from west to east, from the mainland coast to the open shelf margin.

Faunal diversity is relatively low (<50 species per sample) in the northern part of the southern shelf lagoon behind the barrier platform. The diversity gradient between platform and lagoonal populations is abrupt coincident with the sharp bathymetric discontinuity between the shallow platform and the relatively deep lagoon. The bottom of the southern shelf lagoon in this area is a featureless mud surface with little or no flora, and thus limited habitat diversity. Although little direct evidence is available, it is likely that the continuous seaward barrier strongly constrains the flow of oceanic waters into the lagoon, perhaps creating somewhat restricted benthic conditions. These factors of restricted circulation and low habitat diversity may combine to reduce the diversity of the benthic foraminiferal populations. Interestingly, faunal diversity increases from the shelf lagoon toward the mainland coast. This increase is coincident with the appearance of marine grass in shallow waters near the coast. The more diverse faunas are similar to those of the grass flats of the barrier platform, although less diverse. An

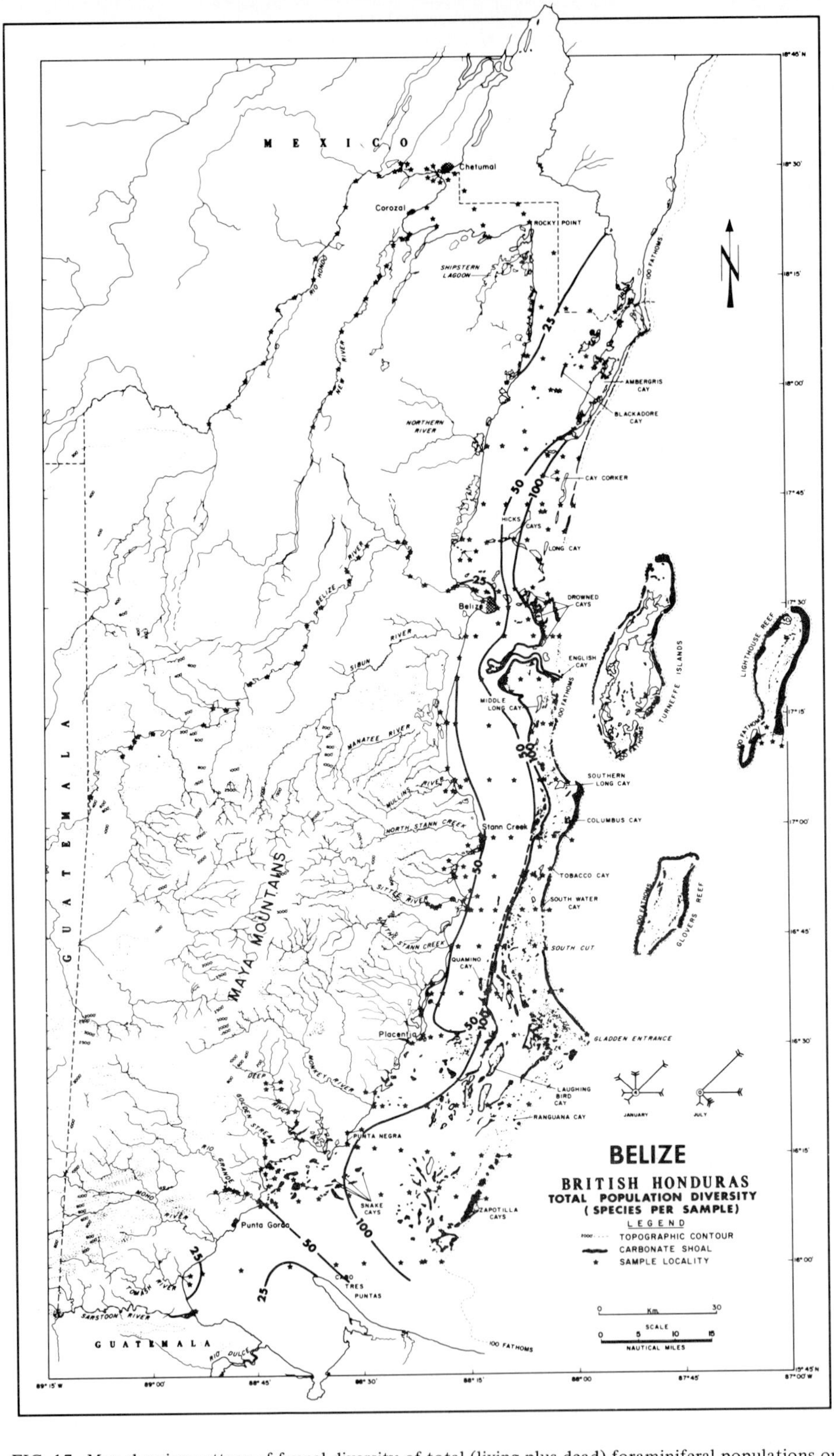

FIG. 17–Map showing pattern of faunal diversity of total (living plus dead) foraminiferal populations on Belize shelf. Data are plotted as species per sample (approximately 15 cm^3) of surface sediment.

increase in diversity also was observed to take place among the mollusk populations (data from Robertson *in* Purdy et al, this volume), where the epiphytic gastropods diversified in the grass areas both on the platform and along the coast.

The low-diversity populations of the southern shelf lagoon grade toward the south and southeast into more diverse populations (±100 species per sample). In this area the lagoon is deeper and has more bottom relief but, more importantly, the lagoon widens and the seaward barrier is narrow and discontinuous. Evidence of oceanic circulation is seen in the minor influx of planktonic species in the outer-shelf sediments. This more extensive area of stable marine benthic conditions with open circulation is perhaps conducive to a diverse foraminiferal assemblage. Although both lagoonal and barrier-platform populations reflect high diversity (>100 species per sample), their population structure is not the same. Faunal dominance in the diverse platform populations is consistently lower (Table 1) than that of the high-diversity lagoonal or deeper shelf populations. The proportional distribution of species within the high-diversity miliolid-dominant populations of the platform is much more even, whereas a relatively few species predominate and many others are very scarce in the high-diversity *Cribroelphidium*-dominant populations of the adjacent lagoon. A more sophisticated diversity statistic doubtless would reflect this difference.

In summary, the pattern of faunal diversity in the Belize shelf area is complex, reflecting the complex interaction of bathymetry and hydrography that governs the distribution of foraminiferal populations. The influence of barriers (cays, shoals, barrier platform) is evident. It has been demonstrated in open shelf areas (e.g., northeast Gulf of Mexico; Walton, 1964a) that there is a general trend of increased faunal diversity with increased distance from shore. This trend is interpreted to reflect increased environmental stability from the mainland coast toward the shelf edge, and this principle has been a useful factor in interpreting the paleoecology of Tertiary foraminiferal populations in the subsurface. The relation of faunal diversity to environmental stability, however, is a broad generalization; such a "rule of thumb" should not be applied without a knowledge of the more specific processes that affect population structure.

The evidence from this study indicates that the overall trend from unstable to stable marine conditions is only one factor that causes faunal diversity. In the Belize shelf area, the influence of barriers on shelf circulation and the local development of reefs and grass meadows create a more complex pattern of faunal diversity. It is clear, however, that the diversity pattern is closely related to the complex shelf environment, and that diversity, when linked with knowledge of the dominant species, defines distinct, ecologically significant foraminiferal assemblages.

PATTERN AND SIGNIFICANCE OF LIVING FORAMINIFERAL POPULATIONS

Comparison of data on total populations (living plus dead) with that on living populations demonstrates that the living and total foraminiferal populations were essentially in equilibrium in parts of the study area at the time of collection, whereas marked differences existed in other places.

Throughout most of the northern shelf lagoon and Chetumal Bay, the taxonomic composition and faunal dominance of living and total populations are similar. A significant exception is the group of *Discorbis*-

dominant living populations in lower Chetumal Bay. This anomaly is inexplicable, but one can speculate that the reason may lie in seasonal and/or differential reproduction of these species. A second exception in this area is the virtual absence of living populations associated with the *Archaias*-miliolid - dominant assemblage. This absence is thought to result from the essentially hostile environment in which the assemblage occurs and the sedimentary rather than biologic factors governing the distribution of the assemblage.

On the barrier and outer platforms, significant differences between living and total populations probably result from an inadequate sampling of living habitats. Collection was largely confined to sediment surfaces, although many Foraminifera live on plants, animal colonies, and rock surfaces and contribute only their empty tests to the sediments. Furthermore, certain forms can become adapted to living in areas in which their tests cannot remain after death. This is particularly true of the highly turbulent areas of the barrier reef and accounts for striking differences between living and total populations of the *Archaias*-*Asterigerina* - dominant assemblage. Similar arguments can be applied to data from populations of the shallow, nearshore turbulent environments of the southern Belize coast.

It is a consistent observation that the taxonomic composition and faunal dominance of the living populations differ from those of the total populations throughout the southern shelf lagoon. Nearly all of the total populations are dominated by *Cribroelphidium*, and this genus, although common, is rarely the most abundant taxon in living populations.

The living populations of assemblages in the southern shelf lagoon have a complex distribution. However, a semblance of pattern can be recognized in the shelf lagoon as a whole, and the pattern of generic dominance in the southern shelf lagoon is shown in Figure 18. In general, living species of *Florilus* dominate the populations of the shallower parts of the lagoon (exclusive of most coastal areas), particularly in the vicinity of the Belize delta in the north and the Gulf of Honduras in the south. *Fursenkoina* tends to dominate living populations in the deeper lagoon areas, particularly where influenced by carbonate shoal sedimentation. Locally, *Buliminella* and the arenaceous genera *Eggerella*, *Reophax*, and *Bigenerina* are the dominant taxa. This pattern is grossly oversimplified, however, and there are numerous exceptions. Moreover, evidence contrary to the living-population distribution outlined in this study was presented by Cebulski (1969). He stated that living populations in the part of the shelf he studied were in accord with total populations and that both were strongly dominated by *Elphidium* (=*Cribroelphidium*, *Cellanthus*, *Elphidium*); he found *Nonion* (=*Nonionella*) and *Streblus* (=*Ammonia*) to be consistent associated elements. This paradox cannot be reconciled at present. If it is assumed that the living populations recorded in the present study are representative of the true distribution of the prevalent foraminiferal faunas, it is imperative to discuss the difference between the living and total populations and the ecologic significance of the pattern of living Foraminifera.

Some factors that might result in differences between living and total populations cannot be evaluated adequately at present. For example, differential reproductive rates among foraminiferal species could give rise to higher contribution to total populations than living populations by some species, but very little is known about reproductive rates and life cycles of most Foraminifera. Differences between living and total populations could also result from seasonal changes in reproduction rates. The sampling for this study was, of necessity, conducted only in the summer season, and it is not known if the population statistics are repre-

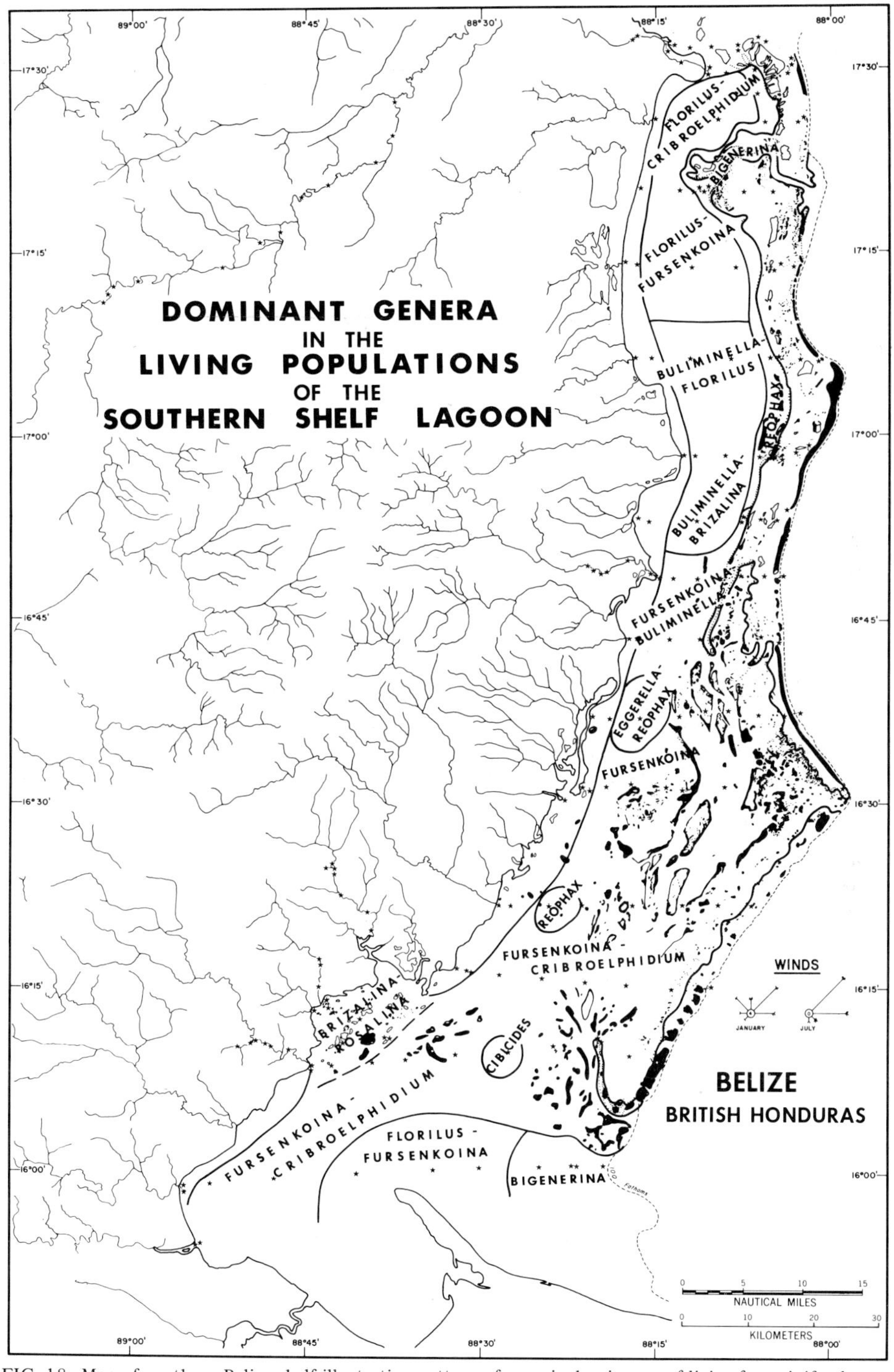

FIG. 18—Map of southern Belize shelf illustrating pattern of generic dominance of living foraminiferal populations in southern shelf lagoon.

sentative of all seasons. Another contributing factor may be changing environmental conditions concurrent with the development of the Belize shelf during the Holocene. Walton (1964a) concluded that recent ecologic changes account for faunas in which living populations are dominated by species not conspicuous in the total populations. Environmental conditions shift, living populations change, and the sediment contains numerous empty tests of the former dominant species. In short, there is a lag effect. Such mixed populations would be expected in marginal-marine and shallow reef-related environments that are subject to shifting conditions.

The dominant living foraminiferal taxa of the shelf lagoon of Belize are common elements in shallow shelf environments elsewhere. In the northeastern Gulf of Mexico, *Florilus*- and *Buliminella*-dominant faunas are common in 10-30 fm of water near the eastern margin of the Mississippi River delta (Walton, 1964a). Data presented by Phleger and Parker (1951) show local areas of dominance of arenaceous taxa and *Buliminella* on the inner continental shelf of the northwestern Gulf of Mexico. In a study of the Holocene Foraminifera of the Gulf of Cariaco, Seiglie and Bermudez (1963) described benthonic faunas dominated by *Buliminella silviae* Bermudez and Seiglie and *Virgulina pontoni* (=*Fursenkoina pontoni*). The *Buliminella*-dominant fauna described by Seiglie and Bermudez is characterized by low diversity (approximately four species) and high faunal dominance. It occurs in the deeper parts of the Gulf of Cariaco, which is a silled basin almost totally enclosed by land. Restricted circulation appears to be a significant limiting factor in the distribution of this biofacies. In the Belize shelf lagoon, living *Buliminella*-dominant faunas are grouped in an area coincident with the area of most continuous barrier-platform development, and perhaps minimum circulation, in the southern shelf lagoon. Although the living *Buliminella* faunas of Belize are not as limited in faunal diversity as those of the Gulf of Cariaco, they may reflect similar though less extreme conditions. The area of *Virgulina pontoni* (=*Fursenkoina pontoni*) dominance in the Gulf of Cariaco is a narrow band bordering the mainland coast in water depths of 18-41 m. The steep topographic gradient and poorly sorted land-derived sediments in this area may be indicative of relatively rapid sedimentation. In the Belize shelf area, *Fursenkoina pontoni* consistently dominates the living populations of near-shoal lagoonal areas, which presumably are subject to rapid sedimentation, but the high percentage of shoal-derived foraminiferal debris in these areas masks the significance of this species in total populations.

In summary, the present distribution of living and total foraminiferal populations in the southern shelf lagoon reflects, at least in part, the historical development of the shelf and the influence of the barrier platform. The barrier platform affects the distribution directly through its sediment contribution to the adjacent lagoon and, indirectly, through its effect on lagoonal water circulation.

CONCLUSIONS

1. Holocene benthonic foraminiferal assemblages in the Belize shelf area are well defined, largely on the basis of faunal diversity and faunal dominance of total (living plus dead) populations.

2. In this small but environmentally complex area, the distribution of assemblages closely parallels the interacting patterns of hydrography, bathymetry, and sedimentation. These patterns are controlled by the shelf-edge barrier platform, barrier reefs, and islands, as well as by the climate and topography of the adjacent landmass.

3. The dominant foraminiferal taxa in the shallow shelf areas (less than 5 fm deep) are miliolids. Major exceptions to this dominance are found in river effluents and in coastal lagoons that receive a constant supply of fresh water, as well as in the coastal zone of southern Belize characterized by quartzose sediments. The dominant miliolid genera in stenohaline marine areas are *Quinqueloculina* and *Triloculina*, whereas *Heterillina* and *Triloculina* dominate the miliolid populations of areas subject to variable salinity.

4. In deeper shelf-lagoon and open-shelf areas (greater than 5 fm deep), calcareous perforate forms dominate. The most abundant genus is *Cribroelphidium*; the most significant exception to this is the presence of *Globocassidulina*-dominant populations in the deepest part of the shelf.

5. The diversity of the benthic populations is strongly influenced by salinity, circulation, and benthic habitat diversity. Minimum diversity occurs with low or variable salinity, restricted circulation, and low habitat diversity (especially the absence of marine benthic flora). Conversely, maximum diversity results in areas of normal, constant marine salinity, free circulation, and high diversity of benthic habitat. The gradients of faunal diversity closely parallel the major ecologic gradients throughout the shelf.

6. The linear distribution of seaward barriers creates a complex pattern of faunal diversity; diversity gradients parallel the coast or increase seaward. In a few places, diversity increases toward the coast. The occurrence of anomalously low faunal diversity in ancient shelf populations may be a clue to partially restricted circulation and the presence of a seaward barrier.

7. The combination of faunal and sediment studies is clearly a stronger approach to paleoecologic interpretation than either discipline alone. Subtle variations in benthic populations in otherwise homogeneous sediments may provide important clues to water characteristics; presence of anomalously diverse populations may be indicative of benthic flora that is not preserved; and minor faunal constituents representing indigenous populations may be the only clues to depositional environment in areas of rapid deposition and faunal mixing.

REFERENCES CITED

Acosta, J. T., 1940a, *Triloculina bermudezi*, un nuevo Foraminifero de las Islas Bahamas: Mem. Soc. Cubana Hist. Nat., 14, p. 37.

——1940b, Algunos foraminiferos nuevos de las costas sur Cubanas: Torreia, Museo Poey, Universidad de la Habana, Cuba, no. 5, p. 36.

Andersen, H. V., 1951, Two new genera of Foraminifera from Recent deposits in Louisiana: Jour. Paleontology, v. 25, p. 31-34.

——1953, Two new species of *Haplophragmoides* from the Louisiana coast: Cushman Found. Foram. Research Contr., v. 4, p. 21-22.

——1961, Foraminifera of the mudlumps, lower Mississippi River delta: Louisiana Geol. Survey Geol. Bull. 35, pt. 2, 208 p.

Aves, Charles, 1958, Foraminiferal fixatives and preservatives, *in* Sedimentology of South Texas, Annual Field Trip, May 15, 16, 17: Corpus Christi Geol. Soc., p. 48-56.

Ayala-Castañares, Agustín, 1963, Sistematica y distribucion de los foraminiferos recientes de la Laguna de Terminos, Campeche, Mexico: Mexico Univ. Nac. Autonoma Inst. Geologia Bol. 67, pt. 3, 130 p.

Bandy, O. L., 1954, Distribution of some shallow-water Foraminifera in the Gulf of Mexico: U.S. Geol. Survey Prof. Paper 254-F, p. 125-140.

——1956, Ecology of Foraminifera in northeastern Gulf of Mexico: U.S. Geol. Survey Prof. Paper 274-G, p. 179-204.

——1964, Foraminiferal biofacies in sediments of Gulf of Batabanó, Cuba, and their geologic significance: AAPG Bull., v. 48, no. 10, p. 1666-1679.

Barker, R. W., 1960, Taxonomic notes on the species figured by H. B. Brady in his report on the Foraminifera dredged by H.M.S. Challenger during the year 1873-1876: Soc. Econ. Paleontologists and Mineralogists Spec. Pub. 29, 238 p.

Barnhart, J. T., 1963, Distribution of Recent sediments and Foraminifera across the Bermuda platform: Univ. Houston, MS thesis.

Benda, W. K., and H. S. Puri, 1962, The distribution of Foraminifera and Ostracoda off the Gulf Coast of the Cape Romano area, Florida: Gulf Coast Assoc. Geol. Socs. Trans., v. 12, p. 303-341.

Bermudez, P. J., 1935, Foraminiferos de la coasta norte de Cuba: Soc. Cubana Hist. Nat. Mem., v. 9, no. 3, p. 129-224.

——and G. A. Seiglie, 1963, Estudio sistematico de los foraminiferos del Golfo de Cariaco: Bol. del Inst. Oceanogr. de la Univers. de Oriente, v. 2, no. 2, p. 1-267.

Bock, W. D., 1967, Monthly variation in the foraminiferal biofacies on *Thalassia* and sediment in the Big Pine Cay area, Florida: Univ. Miami, PhD thesis, 291 p.

——et al, 1971, A symposium of Recent South Florida Foraminifera: Miami Geol. Soc. Mem. 1, 245 p.

Boltovskoy, E., 1958, The foraminiferal fauna of the Rio de la Plata and its relation to the Caribbean fauna: Cushman Found. Foram. Research Contr., v. 9, p. 18-21.

——1963, The littoral foraminiferal biocoenoses of Puerto Deseado (Patagonia, Argentina): Cushman Found. Foram. Research Contr., v. 14, p. 58-70.

Bush, James, 1958, The Foraminifera and sediments of Biscayne Bay, Florida, and their ecology: Univ. Washington, PhD thesis.

Cebulski, D. E., 1969, Foraminiferal populations and faunas in barrier-reef tract and lagoon, British Honduras: AAPG Mem. 11, p. 311-328.

Cloud, P. E., Jr., 1962, Environment of calcium carbonate deposition west of Andros Island, Bahamas: U.S. Geol. Survey Prof. Paper 350 (section on Foraminiferal Assemblages by R. Todd and D. Low, p. 34-35).

Cole, W. S., 1965, Structure and classification of some Recent and fossil peneroplids: Bulls. Am. Paleontology, v. 49, no. 219, p. 1-37.

Creager, J. S., 1958, Bathymetry and sediments of the Bay of Campeche: Texas A&M Research Found., Project 173, Ref. 58-12F, 188 p.

Cushman, J. A., 1918-1931, The Foraminifera of the Atlantic Ocean: U.S. Natl. Mus. Bull. 104, pts. 1-8, 1041 p.

——1921, Foraminifera from the north coast of Jamaica: U.S. Natl. Mus. Proc., v. 59, p. 156-172.

——1922, Shallow-water Foraminifera of the Tortugas region: Carnegie Inst. Washington Pub. 311, p. 1-82.

——1926, Recent Foraminifera from Puerto Rico: Carnegie Inst. Washington Pub. 344, p. 1-49.

——1932-42, The Foraminifera of the tropical Pacific collections of the "Albatross," 1899-1900: U.S. Natl. Mus. Bull. 161, pts. 1-3.

——1959, Foraminifera, their classification and economic use: Cambridge, Mass., Harvard Univ. Press, 4th ed.

——and P. Bronniman, 1948a, Some new genera and species of Foraminifera from brackish water of Trinidad: Cushman Found. Foram. Research Contr., v. 24, p. 15-21.

Daetwyler, C. C., and A. L. Kidwell, 1959, The Gulf of Batabano, a modern carbonate basin: 5th World Petroleum Cong., New York, 1959, Proc., Sec. 1, p. 1-21.

Davis, R. A., Jr., 1964, Foraminiferal assemblages of Alacran Reef, Campeche Bank, Mexico: Jour. Paleontology, v. 38, no. 2, p. 417-421.

d'Orbigny, Alcide, 1839, Foraminifères, *in* R. de la Sagra, Histoire physique, politique et naturelle de l'île de Cuba: Paris, 224 p.

Drooger, C. W., and J. P. H. Kaasschieter, 1958, Foraminifera of the Orinoco-Trinidad-Paria Shelf, v. 4 *of* Reports of the Orinoco Shelf Expedition: Koninkl. Nederlandse Akad. Wetensch., Afd. Natuurk. Verh., 1st Reeks, Deel 22, 108 p.

Gibson, T. G., and M. A. Buzas, 1973, Species diversity: patterns in modern and Miocene Foraminifera of the eastern margin of North America: Geol. Soc. America Bull., v. 84, p. 217-298.

Henson, F. R. S., 1950, Cretaceous and Tertiary reef formations and associated sediments in Middle East: AAPG Bull., v. 34, no. 2, p. 215-238.

High, L. R., 1967, Recent coastal sediments of British Honduras: Rice Univ., PhD thesis, 199 p.

——1975, Geomorphology and sedimentology of Holocene coastal deposits, Belize: this volume.

Hofker, J., 1950-1953, Recent Peneroplidae: Royal Microscop. Soc. Jour., pt. 1, v. 70, p. 388-396; v. 71, p. 223-239; pt. 2, v. 71, p. 342-356; pt. 3, v. 71, p. 450-463; pt. 4, v. 72, p. 102-122; pt. 5, v. 72, p. 40-46.

——1964, Foraminifera from the tidal zone in the Netherlands Antilles and other West Indian Islands, *in* P. W. Hummelinck, ed., Studies on the fauna of Curacao and other Caribbean islands: v. 21, 119 p.

Hoskin, C. M., 1963, Recent carbonate sedimentation on Alacran Reef, Yucatan, Mexico: Natl. Acad. Sci.-Natl. Research Council Pub. No. 1089, 160 p.

Howard, J. F., 1965, Shallow-water foraminiferal distribution near Big Pine Key, southern Florida Keys: Compass, v. 42, no. 4, p. 265-280.

——D. L. Kissling, and J. A. Lineback, 1970, Sedimentary facies and distribution of biota in Coupon Bight, lower Florida Keys: Geol. Soc. America Bull., v. 8, no. 7, p. 1929-1945.

Illing, M. A., 1950, The mechanical distribution of Recent Foraminifera in Bahama Banks sediments: Annals and Mag. Nat. Hist., ser. 12, v. 3, p. 757-761.

——1952, Distribution of certain Foraminifera within the littoral zone on the Bahama Banks: Annals and Mag. Nat. Hist., v. 5, p. 275-285.

Kornfeld, M. M., 1931, Recent littoral Foraminifera from Texas and Louisiana: Stanford Univ. Dept. Geology Contr., v. 1, no. 3, p. 77-101.

Lankford, R. R., 1959, Distribution and ecology of Foraminifera from east Mississippi delta margin: AAPG Bull., v. 43, p. 2068-2099.

Lidz, L., and B. Lidz, 1966, Foraminiferal biofacies of Veracruz reefs: AAPG Bull., v. 50, p. 1515-1517.

Loeblich, A. R., Jr., et al, 1964, Treatise on invertebrate paleontology—Pt. C, Protista 2, Sarcodina, chiefly "Theocamoebians" and Foraminiferida, v. 1-2: New York, Geol. Soc. America and Univ. Kansas Press, v. 1, p. C1-C510a; v. 2, p. C511-C900.

Lowman, S. W., 1949, Sedimentary facies in Gulf Coast: AAPG Bull., v. 33, p. 1939-1997.

——1951, The relationship of the biotic and lithic facies in Recent Gulf Coast sedimentation: Jour. Sed. Petrology, v. 21, no. 4, p. 233-237.

Lynts, G. W., 1962, Distribution of Recent Foraminifera in upper Florida Bay and associated sounds: Cushman Found. Foram. Research Contr., v. 13, pt. 4, p. 127-144.

——1965, Observations on some Recent Florida Bay Foraminifera: Cushman Found. Foram. Research Contr., v. 16, pt. 2, p. 67-69.

Matthews, R. K., 1966, Genesis of Recent lime mud in southern British

Honduras: Jour. Sed. Petrology, v. 36, no. 2, p. 428-454.
Moore, W. E., 1957, Ecology of Recent Foraminifera in northern Florida Keys: AAPG Bull., v. 41, p. 727-741.
Murray, J. W., 1965, The Foraminiferida of the Persian Gulf. 2. The Abu Dhabi region: Palaeogeography, Palaeoclimatology, Palaeoecology, v. 1, p. 307-332.
——1973, Distribution and ecology of living benthic foraminiferids: New York, Crane, Russak & Co., 274 p.
Norton, R. D., 1930, Ecologic relations of some Foraminifera: Scripps Inst. Oceanography Tech. Services Bull., v. 2, p. 331-388.
Parker, F. L., 1954, Distribution of the Foraminifera in the northeastern Gulf of Mexico: Mus. Comp. Zoology, v. 111, pt. 10, p. 453-588.
——and W. D. Athearn, 1959, Ecology of marsh Foraminifera in Poponesset Bay, Massachusetts: Jour. Paleontology, v. 33, p. 333-343.
——F. B. Phleger, and J. L. Peirson, 1953, Ecology of Foraminifera from San Antonio Bay and environs, southwest Texas: Cushman Found. Foram. Research Spec. Pub. 2, 75 p.
Phleger, F. B., 1951, Ecology of Foraminifera, northwest Gulf of Mexico, Part I—Foraminifera distribution: Geol. Soc. America Mem. 46, pt. 1, p. 1-88.
——1954a, Ecology of Foraminifera and associated micro-organisms from Mississippi Sound and environs: AAPG Bull., v. 38, p. 584-647.
——1954b, Gulf of Mexico Foraminifera: U.S. Fish and Wildlife Service Fishery Bull. 89, v. 55, p. 235-241.
——1955, Ecology of Foraminifera in southeastern Mississippi delta area: AAPG Bull., v. 39, p. 712-752.
——1956, Significance of living foraminiferal populations along the central Texas coast: Cushman Found. Foram. Research Contr., v. 7, pt. 4, p. 106-151.
——1960a, Ecology and distribution of Recent Foraminifera: Baltimore, Md., Johns Hopkins Press, 297 p.
——1960b, Sedimentary patterns of Foraminifera, northern Gulf of Mexico, *in* F. P. Shepard, F. B. Phleger, and T. H. van Andel, eds., Recent sediments, northwest Gulf of Mexico: Tulsa, Okla., Am. Assoc. Petroleum Geologists, p. 267-301.
——1960c, Foraminiferal populations in Laguna Madre, Texas: Tohoku Univ. Sci. Repts., 2d ser., Geology, no. 4, p. 83-91.
——1964, Foraminiferal ecology and marine geology: Marine Geology, v. 1, no. 1, p. 16-43.
——1965, Patterns of marsh Foraminifera, Galveston Bay, Texas: Limnology and Oceanography, v. 10, supp., p. R169-R184.
——and J. S. Bradshaw, 1966, Sedimentary environments in a marine marsh: Science, v. 157, p. 1551-1553.
——and R. R. Lankford, 1957, Seasonal occurrences of living benthonic Foraminifera in some Texas bays: Cushman Found. Foram. Research Contr., v. 8, pt. 3, p. 93-105.
——and F. L. Parker, 1951, Ecology of Foraminifera, northwest Gulf of Mexico, Pt. 2, Foraminifera species: Geol. Soc. America Mem. 46, 64 p.
——and W. R. Walton, 1950, Ecology of marsh and bay Foraminifera, Barnstable, Massachusetts: Am. Jour. Sci., v. 248, p. 274-294.
Post, R. J., 1951, Foraminifera of the South Texas coast: Inst. Marine Sci. Pub., v. 2, no. 1, p. 165-176.
Purdy, E. G., 1968, Carbonate diagenesis, an environmental survey: Geol. Romana, v. 7, p. 183-209.
——W. C. Pusey III, and K. F. Wantland, 1975, Continental shelf of Belize—regional shelf attributes: this volume.

Pusey, W. C., 1964, Recent calcium carbonate sedimentation in northern British Honduras: Rice Univ., PhD thesis, 247 p.

——1975, Holocene carbonate sedimentation on northern Belize shelf: this volume.

Rivero, F. C. de, 1964, Ecologia, paleoecologia y distribucion estratigrafica de los arrecifes organicos: GEOS, no. 11, p. 41-122 (Caracas).

Robertson, R., 1963, The mollusks of British Honduras: Philadelphia Shell Club Proc., v. 1, no. 7, p. 15-20.

Rusnak, G. A., 1960, Sediments of Laguna Madre of Texas, *in* F. P. Shepard, F. B. Phleger, and T. H. van Andel, eds., Recent sediments, northwestern Gulf of Mexico: Tulsa, Okla., Am. Assoc. Petroleum Geologists, p. 153-196.

Sandberg, P. A., 1961, A microfaunal study of Tamiahua Lagoon, Veracruz, Mexico: Louisiana State Univ., MS thesis.

Saunders, J. B., 1957, Trachamminidae and certain Lituolidae (Foraminifera) from the Recent sediments of Trinidad, British West Indies: Smithsonian Misc. Colln., v. 134, no. 5, p. 16.

——1958, Recent Foraminifera of mangrove swamps and river estuaries and their fossil counterparts in Trinidad: Micropaleontology, v. 4, no. 1, p. 79-92.

Scholle, P. A., and S. A. Kling, 1972, Southern British Honduras: lagoonal coccolith ooze: Jour. Sed. Petrology, v. 42, no. 1, p. 195-204.

Scholz, S., 1962, The distribution of Recent Foraminifera in Hawk Channel, Florida: Univ. Wisconsin, MS thesis, 123 p.

Scott, M. R., 1966, Distribution of clay minerals on the British Honduras shelf: Rice Univ., PhD thesis, 61 p.

——1975, Distribution of clay minerals on Belize shelf: this volume.

Seiglie, G. A., 1964a, Algunos foraminiferos arenaceos recientes de Venezuela: Univ. Oriente, Inst. Oceanog. Bol., v. 3, nos. 1-2, p. 5-12.

——1964b, New and rare foraminifers from Los Testigas reefs, Venezuela: Caribbean Jour. Sci., v. 4, no. 4, p. 497-511.

——1965a, Some observations on Recent Foraminifera from Venezuela, Part I: Cushman Found. Foram. Research Contr., v. 16, pt. 2, p. 70-75.

——1965b, Cuadro preliminar sobre la distribucion estratigrafica de los foraminiferos grandes de Cuba: Lagena, no. 7, p. 23-30, Cumana.

——1968, Bibliografia sobre los foraminiferos bentonicos recientes de las provincia Caribe-Antillana, con observaciones (Bibliography of Recent benthonic Foraminifera of the Caribbean-Antilles region, with annotations): Bol. Inf., v. 11, no. 1, p. 5-15.

——and P. H. Bermudez, 1963, Distribucion de los foraminiferos del Golfo de Cariaco: Univ. Oriente, Inst. Oceanog. Bol., v. 2, no. 1, p. 7-87 (with English summary).

Shifflett, E., 1961, Living, dead, and total foraminiferal faunas: Heald Bank, Gulf of Mexico: Micropaleontology, v. 7, no. 1, p. 45-54.

Smout, A. H., and F. E. Eames, 1958, The genus *Archaias* (Foraminifera) and its stratigraphical distribution: Palaeontology, v. 1, pt. 3, p. 207-225.

Streeter, S., 1963, Foraminifera in the sediments of the northwestern Bahama banks: Columbia Univ., PhD thesis.

Todd, R., 1965, The Foraminifera of the tropical Pacific collections of the "Albatross," 1899-1900: U.S. Natl. Mus. Bull. 161, pt. 4, 139 p.

——and P. Bronnimann, 1957, Recent Foraminifera and Thecamoebina from the eastern Gulf of Paria, Trinidad: Cushman Found. Foram. Research Spec. Pub. 3, 43 p.

Treadwell, R. C., 1955, Sedimentology and ecology of southeast coastal Louisiana: Louisiana State Univ. and Agr. and Mech. Coll. Tech. Rept., no. 6, 78 p.

U.S. Navy Hydrographic Office, 1931a, Approaches to Belize: H.O. Chart 1498, 1st ed., (revised 1965), scale 1:125,000.

——1931b, Ranguana Cay to Columbus Cay including inner channel and Glover reef: H.O. Chart 1497, 18th ed., scale 1:125,000.

——1938, Gulf of Honduras with Zapotilla Cays: H.O. Chart 1496, 24th ed. (revised 1964), scale 1:125,640.

——1949, Gulf of Mexico and Caribbean Sea including the West Indies: H.O. Chart 1290, 108th ed., scale 1:3,332,500.

——1955, Atlas of pilot charts, Central American waters and South Atlantic Ocean: H.O. Pub. 160, 16 charts.

——1952, Sailing directions for the east coasts of Central America and Mexico: H.O. Pub. 20, 5th ed., 274 p.

Upshaw, C. F., W. B. Creath, and F. L. Brooks, 1966, Sediments and microfauna off the coasts of Mississippi and adjacent states: Mississippi Geol. Econ. and Topog. Survey Bull. 106, 127 p.

Waldron, R. P., 1963, A seasonal ecological study of Foraminifera from Timbalier Bay, Louisiana—Office Naval Research Geography Br., Nonr 1575(03), Tech. Rept. 16, Pt. B: Louisiana State Univ. Coastal Studies Inst. Contr. 63-3, p. 132-188.

Walton, W. R., 1952, Technique for the recognition of living Foraminifera: Cushman Found. Foram. Research Contr., v. 3, p. 56-60.

——1955, Ecology of living benthonic Foraminifera, Todos Santos Bay, Baja Calif.: Jour. Paleontology, v. 29, p. 952-1018.

——1964a, Recent foraminiferal ecology and paleoecology, *in* J. Imbrie and N. Newell, eds., Approaches to paleoecology: New York, Wiley, p. 151-237.

——1964b, Ecology of benthonic Foraminifera in the Tampa-Sarasota Bay area, Florida: *in* Papers in marine geology—Shepard commemorative volume: New York, MacMillan Co., p. 429-454.

Wantland, K. F., 1969a, Foraminiferal assemblages of the coastal lagoons of British Honduras, *in* A. Ayala-Castañares and F. B. Phleger, eds., Lagunas Costeras, Un Simposio, Mem. Simp. Intern. Lagunas Costeras, UNAM-UNESCO, Nov. 28-30, 1967: Mexico, D.F., p. 621-644.

——1969b, Distribution of modern brackish-water Foraminifera in Trinity Bay, *in* R. R. Lankford and J. J. W. Rogers, eds., Galveston Bay geology: Houston Geol. Soc., chap. 6, p. 93-117.

Wantland, K. F., and W. C. Pusey III, 1971, A guidebook for the field trip to the southern shelf of British Honduras: Gulf Coast Assoc. Geol. Socs., 87 p.

Warren, A. D., 1956, Ecology of the Foraminifera of the Buras-Scofield Bayou region, Southeast Louisiana: Gulf Coast Assoc. Geol. Socs. Trans., v. 6, p. 131-151.

——1957, Foraminifera of the Buras-Scofield Bayou region, southeast Louisiana: Cushman Found. Foram. Research Contr., v. 8, pt. 1, p. 29-40.

Wright, R. C., 1964, Foraminiferal ecology in a back-reef environment, Molasses Reef, Florida: Univ. Illinois, PhD thesis.

APPENDIX

Faunal Reference List

Approximately 200 species of Foraminifera were identified in the shelf sediments of Belize. With very few exceptions the species were identifiable from published West Indies and Gulf of Mexico literature (see section on Previous Work).

Superfamily AMMODISCACEA Reuss, 1862

Lagenammina atlantica (Cushman) (Figs. 5u, 7q); *Proteonina atlantica* Cushman, 1949, Cushman Found. Foram. Research, Spec. Pub. 12, p. 5; *Lagenammina atlantica* (Cushman), Lankford, 1959, AAPG Bull., v. 43, p. 2098, Pl. 1, Fig. 1.

Ammodiscus gullmarensis Höglund; ?*Ammodiscus incertus* (d'Orbigny), Goess, 1894 (in part), K. Sv. Akad. Handl., v. 25, no. 9, p. 31; *Ammodiscus gullmarensis* Höglund, 1948, Cushman Found. Foram. Research Contr., v. 24, p. 45.

Glomospira gordialis (Jones and Parker); *Trochammina squamata* Jones and Parker var. *gordialis* Jones and Parker, 1860, Geol. Soc. London Quart. Jour., v. 16, p. 304; *Glomospira* cf. *gordialis* (Jones and Parker), Parker, 1954, Bull. Mus. Comp. Zoology, v. 111, no. 10, p. 485, Pl. 1, Fig. 13.

Hemidiscella palabunda Bock (Fig. 10i, j), 1968, Cushman Found. Foram. Research Contr., v. 19, pt. 1, p. 27, Pl. 4, Figs. 3-9.

Superfamily LITUOLACEA deBlainville, 1825

Reophax sp. cf. *R. arayensis* Bermudez and Seiglie (Figs. 5v, 7o, 13e); *Reophax arayensis* Bermudez and Seiglie, 1963, Bol. Inst. Oceanog. Univ. Oriente, Cumana, p. 146, Pl. 1, Figs. 1-3, Pl. 2, Fig. 1.

Reophax nana Rhumbler, 1911, Foram. Plankton Exped., pt. 1, Pl. 8, Figs. 6-12.

Reophax spiculifer Brady (Fig. 7p), 1884, Report Voyage *Challenger*, Zoology, v. 9, Pl. 31, Figs. 16, 17.

Ammoscalaria pseudospiralis (Williamson); *Proteonina pseudospirale* Williamson, 1858, Recent Foraminifera Great Britain, p. 2, Pl. 1, Figs. 2, 3; *Ammoscalaria pseudospiralis* (Williamson), Hoglund, 1947, Zoologiska Bidrag Från Uppsala, v. 26, p. 159-162.

Ammotium salsum (Cushman and Bronnimann) (Fig. 4h); *Ammobaculites salsus* Cushman and Bronnimann, 1948, Cushman Found. Foram. Research Contr., v. 24, pt. 1, p. 16, Pl. 3, Figs. 7-9.

Textularia agglutinans d'Orbigny (Figs. 3h, 11g), 1839, *in* De la Sagra, Histoire physique, politique, et naturelle de l'île de Cuba, Foraminifères, v. 8, p. 144, Pl. 1, Figs. 17, 18, 32-34.

Textularia candeiana d'Orbigny, 1839, *in* De la Sagra, Histoire physique, politique, et naturelle de l'île de Cuba, Foraminifères, v. 8, p. 143, Pl. 1, Figs. 25-27.

Textularia conica d'Orbigny (Figs. 7l, 13o), 1839, *in* De la Sagra, Histoire physique, politique, et naturelle de l'île de Cuba, Foraminiferes, v. 8, p. 143, Pl. 1, Figs. 19, 20.

Textularia mayori Cushman (Fig. 7k), 1922, Carnegie Inst. Washington, Pub. No. 311, p. 23, Pl. 2, Fig. 3.

Nouria polymorphinoides Heron-Allen and Earland (Figs. 5w, 13d), 1914, London Zool. Soc. Trans., v. 20, p. 376, Pl. 37, Figs. 1-15.

Miliammina fusca (H. B. Brady) (Fig. 4*l*); *Quinqueloculina fusca* H. B. Brady, 1870, Annals and Mag. Nat. History, Ser. 4, v. 6, p. 47 (286), Pl. 11, Figs. 2a-c, 3; *Miliammina fusca* (H. B. Brady), Hada, 1936, Zool. Mag., v. 48, p. 852, Text-fig. 5.

Haplophragmoides sp. (Fig. 4m).

Ammoastuta inepta (Cushman and McCulloch); *Ammobaculites ineptus* Cushman and McCulloch, 1939, Southern California Univ. Pub., Allan Hancock Pacific Exped., v. 6, no. 1, p. 89, Pl. 7, Fig. 6; *Ammoastuta inepta* (Cushman and McCulloch), Saunders, 1958, Micropaleontology, v. 4, p. 83, Pl. 1, Figs. 4-7.

Ammobaculites dilatatus Cushman and Bronnimann, 1948, Cushman Found. Foram. Research Contr., v. 24, pt. 2, p. 39, Pl. 7, Figs. 10-11.

Ammobaculites directus Cushman and Bronnimann, 1948, Cushman Found. Foram. Research Contr., v. 24, pt. 2, p. 38, Pl. 7, Figs. 3, 4.

Ammobaculites exiguus Cushman and Bronnimann, 1948, Cushman Found. Foram. Research Contr., v. 24, pt. 2, p. 38, Pl. 2, Figs. 7, 8.

Ammobaculites exilis Cushman and Bronnimann (Fig. 4e-g), 1948, Cushman Found. Foram. Research Contr., v. 24, pt. 2, p. 39, Pl. 7, Fig. 9.

Bigenerina irregularis Phleger and Parker (Figs. 7j, 13f), 1951, Geol. Soc. America Mem. 46, pt. 2, p. 4, Pl. 1, Figs. 16-21.

Bigenerina sp. (Fig. 13g). Test coarsely agglutinated, small initial biserial portion and elongate, straight uniserial portion; this form is similar to *B. irregularis* in size and shape, but distinct with respect to texture of the test wall.

Haeuslerella hoeglundi (Uchio); *Textularia bigenerinoides* Hoglund (not Lacroix) 1947, Zoologiska Bidrag från Uppsala, v. 26, p. 181, Pl. 13, Fig. 6, Text-fig. 159; *Haeuslerella hoeglundi* (Uchio), Lankford, 1973, Jour. Foram. Research, v. 3, no. 3, p. 122, Pl. 1, Fig. 9.

Siphotextularia sp. This species is characterized by an obliquely angled, lipped, areal aperture, the small size and lobulate shape of the test chambers, and the smoothly finished texture of the test wall. This form was found in only one sample (no. 212).

Trochammina advena Cushman (Fig. 7r), 1922, Carnegie Inst. Washington Pub. No. 311, v. 17, p. 261, Pl. 1, Figs. 2-4.

Trochammina inflata (Montagu); *Nautilus inflata* Montagu, 1808, Testacea Britannica, p. 81, Pl. 18, Fig. 3; *Trochammina inflata* (Montagu), Carpenter, Parker and Jones, 1862, p. 141, Pl. 11, Fig. 5.

Arenoparrella mexicana (Kornfeld) emend. Anderson; *Trochammina inflata* (Montagu) var. *mexicana* Kornfeld, 1931, Stanford Univ. Dept. Geology Contr., v. 1, p. 86, Pl. 13, Fig. 5a-c; *Arenoparrella mexicana* (Kornfeld) emend. Andersen, Saunders, 1957, Smithsonian Misc. Colln., v. 134, no. 5, p. 12, Pl. 4, Fig. 5a-x.

?*Remaneica* sp. cf. *R. kelletae* (Thalmann) (Fig. 12g, h); *Trochammina peruviana* Cushman and Kellett (not W. Berry, 1928), 1929, U.S. Natl. Mus. Proc., v. 75, art. 25, p. 4, Pl. 1, Fig. 8; *Trochammina kelletae* Thalmann, 1932, Eclogae Geol. Helvetiae, v. 25, no. 2, p. 313. The small, flattened test, the relatively small ratio of agglutinated material to organic cement, and the unique zigzag trace of the ventral sutures distinguish this species. The test is extremely fragile when dry. This species is well adapted to wave- and current-agitated environments, but it is doubtful if many specimens survive destruction after death, owing to the nature of test construction. The species is scarce in total populations and would be extremely scarce in the geologic record.

Gaudryina compressa Cushman (Fig. 7m), 1935, Smithsonian Misc. Colln., v. 91, no. 21, p. 3, Pl. 1, Fig. 9.

Gaudryina exilis Cushman and Bronnimann (Fig. 4k), 1948, Cushman Found. Foram. Research Contr., v. 24, pt. 2, p. 40, Pl. 7, Figs. 15-16.

Eggerella advena (Cushman) emend. Loeblich and Tappan (Fig. 5r-t); *Verneuillina advena* Cushman, 1922, Canadian Biol. Contr. (1921), no. 9, p. 141; *Eggerella advena* (Cushman) emend. Loeblich and Tappan, 1953, Smithsonian Misc. Colln., v. 121, no. 7, p. 36. This form is characterized by its extremely small size and by the triserial chamber arrangement of all but the oldest part of the test. It was originally counted as two separate categories, an obese form with highly inflated chambers (Fig. 5t) and a thin, elongate form (Fig. 5r, s). It was concluded that the two forms are probably distinct generations of the same species in that they occur in a relatively constant ratio and never occur separately.

Valvulina oviedoiana d'Orbigny (Fig. 16*l*), 1839, *in* De la Sagra, Histoire physique, politique, et naturelle de l'île de Cuba, Foraminifères, v. 8, p. 103, Pl. 2, Figs. 21, 22. This form is characterized in general by the large size, the triserial chamber arrangement, and the large tooth that almost completely fills the aperture. This species occurs in two distinct forms. One is characterized by a broad test with inflated chambers that increase rapidly in size (not illustrated); the other form has a thin, stemlike early portion and only the final whorl of chambers is inflated (Fig. 16*l*). These distinct forms have been considered to be distinct generations (Cushman, 1931; Lynts, 1965).

Clavulina tricarinata d'Orbigny (Fig. 11e), 1839, *in* De la Sagra, Histoire physique, politique, et naturelle de l'île de Cuba, Foraminifères, v. 8, p. 111, Pl. 2, Figs. 16-18.

Clavulina nodosaria d'Orbigny (Fig. 11f), 1839, *in* De la Sagra, Histoire physique, politique, et naturelle de l'île de Cuba, Foraminifères, v. 8, p. 110, Pl. 2, Figs. 19, 20.

Liebusella soldanii (Jones and Parker) (Fig. 7n); *Lituola soldanii* Jones and Parker, 1860, Quart. Jour. Geol. Sci., v. 16, no. 184, p. 307; *Liebusella soldanii* (Jones and Parker), Andersen, 1961, Louisiana Geol. Survey Bull. No. 35, p. 29, Pl. 3, Fig. 1.

Superfamily MILIOLACEA Ehrenberg, 1839

Cornuspira planorbis Schultze, 1854, Organismus Polythal., p. 40, Pl. 2, Fig. 21.

Fischerina helix Heron-Allen and Earland, 1915, Zool. Soc. London Trans., v. 20, pt. 2, p. 547, Pl. 46, Figs. 10-14.

Wiesnerella auriculata (Egger) (Fig. 9n); *Planispirina auriculata* Egger, 1893, Abh. K. bay. Akad. Wissy München (I, II), v. 18, p. 245, Pl. 9, Figs. 13-15; *Wiesnerella auriculata* (Egger), Cushman, 1933, Cushman Found. Foram. Research Contr., v. 9, p. 33.

Spiroloculina anderseni Todd and Bronnimann, 1957, Cushman Found. Foram. Research Spec. Pub. No. 3, p. 28, Pl. 4, Figs. 10-12.

Spiroloculina antillarum d'Orbigny, 1839, *in* De la Sagra, Histoire physique, politique, et naturelle de l'île de Cuba, Foraminifères, v. 8, p. 166, Pl. 9, Figs. 3, 4.

Spiroloculina attenuata Cushman; *Massilina inaequalis* Cushman, 1921, U.S. Natl. Mus. Proc., v. 59, p. 72, no. 2360, Pl. 17, Figs. 12, 13; *Spiroloculina attenuata* Cushman and Todd, 1944, Cushman Found. Foram. Research Spec. Pub. No. 11, p. 67, Pl. 9, Figs. 23-24.

Spiroloculina caduca Cushman, 1922, Carnegie Inst. Washington Pub. 311, p. 61, Pl. 11, Figs. 3, 4.

Spiroloculina communis Cushman and Todd, 1944, Cushman Found. Foram. Research Spec. Pub. No. 11, p. 63, Pl. 9, Figs. 4-8.

Spiroloculina guppyi Todd and Bronnimann, 1957, Cushman Found. Foram.

Research Spec. Pub. No. 3, p. 29, Pl. 4, Figs. 1, 2.

Vertebralina atlantica (Cushman and Hanzawa) (Fig. 9h); *Nodobaculariella atlantica* Cushman and Hanzawa, 1937, Cushman Found. Foram. Research Contr., v. 13, pt. 2, p. 42, Pl. 5, Figs. 7, 8. In the quantitative population counts this form was not differentiated from *V. cassis.*

Vertebralina cassis d'Orbigny (Fig. 9i), 1839, *in* De la Sagra, Histoire physique, politique, et naturelle de l'île de Cuba, Foraminifères, v. 8, p. 51, Pl. 7, Figs. 14, 15.

Quinqueloculina angulata (Williamson); *Miliolina bicornis* (Walker and Jacob) var. *angulata* Williamson, 1853, Recent Foraminifera Great Britain, Ray Soc. London, p. 88, Pl. 7, Fig. 196; *Quinqueloculina angulata* (Williamson), Bermudez, 1963, Bol. Inst. Oceanog. Univ. Oriente, Cumana, p. 127, Pl. 9, Fig. 5. This species is distinguished by angular, truncated chambers, relatively small size, and absence or small number of longitudinal costae. It tends to grade morphologically into a more densely costate form with rounded chambers herein assigned to *Q. bicornis* (Walker and Jacob). The end members are certainly distinct; the discontinuity between the forms appears constant and justifies their separation.

Quinqueloculina berthelotiana d'Orbigny (Fig. 8j), 1839, *in* Barker, Webb, and Berthelot, Histoire Naturelle Iles Canaries, v. 2, pt. 2, Foraminifères, p. 142, Pl. 3, Figs. 25-27.

Quinqueloculina bicornis (Walker and Jacob) (Fig. 8e); *Serpula bicornis* Walker and Jacob, 1798, *in* Kanmachers, ed., Adams Essays Micro., p. 633, Pl. 14, Fig. 2; *Quinqueloculina bicornis* (Walker and Jacob), Cushman, 1929, U.S. Natl. Mus. Bull. 104, pt. 6, p. 32, Pl. 5, Figs. 5-7, Pl. 6, Figs. 1-2.

Quinqueloculina bicostata d'Orbigny (Figs. 3a, b, 13n), 1839, *in* De la Sagra, Histoire physique, politique, et naturelle de l'île de Cuba, Foraminifères, v. 8, p. 195, Pl. 12, Figs. 8-10.

Quinqueloculina bidentata d'Orbigny (Fig. 3e), 1839, *in* De la Sagra, Histoire physique, politique, et naturelle de l'île de Cuba, Foraminifères, v. 8, p. 197, Pl. 12, Figs. 18-20.

Quinqueloculina bosciana d'Orbigny (Fig. 8d), 1839, *in* De la Sagra, Histoire physique, politique, et naturelle de l'île de Cuba, Foraminifères, v. 8, p. 34, Pl. 11, Figs. 22-24.

Quinqueloculina bradyana Cushman (Fig. 8g); *Miliolina undosa* Brady (not *Quinqueloculina undosa* Karrer), 1884, Report Voyage *Challenger*, Zoology, v. 9, p. 176, Pl. 6; *Quinqueloculina bradyana* Cushman, 1917, U.S. Natl. Mus. Bull. 71, pt. 6, p. 52, Pl. 18, Fig. 2.

Quinqueloculina candieana d'Orbigny (Figs. 8c, 13*l*), 1839, *in* De la Sagra, Histoire physique, politique, et naturelle de l'île de Cuba, Foraminifères, v. 8, p. 199, Pl. 12, Figs. 24-26.

Quinqueloculina collumnosa Cushman; *Miliolina cuvieriana* Heron-Allen and Earland (not d'Orbigny), 1915, Zool. Soc. London Trans., v. 20, p. 571, Pl. 4, Figs. 33-36; *Quinqueloculina collumnosa* Cushman, 1922, Carnegie Inst. Washington Pub. 311, p. 65, Pl. 10, Fig. 10.

Quinqueloculina compta Cushman (Fig. 3c), 1947, Cushman Found. Foram. Research Contr., v. 23, pt. 4, p. 87, Pl. 19, Fig. 2.

Quinqueloculina exsculpta (Heron-Allen and Earland) (Fig. 8k); *Miliolina exsculpta* Heron-Allen and Earland, 1915, Zool. Soc. London Trans., v. 20, pt. 17, p. 567, Pl. 42, Figs. 23-26.

Quinqueloculina funafutiensis (Chapman) (Fig. 3d); *Miliolina funafutiensis* Chapman, 1900, Linnean Soc. London Jour., Zoology, v. 28, p. 178, Pl. 19, Fig. 6; *Quinqueloculina funafutiensis* (Chapman), Cushman, 1922, Carnegie Inst. Washington Pub. 311, p. 67, Pl. 13, Fig. 3.

Quinqueloculina laevigata d'Orbigny (Fig. 8b, 13m), 1826, Ann. Sci. Nat., p. 301, no. 6.

Quinqueloculina lamarckiana d'Orbigny (Fig. 8f), 1839, *in* De la Sagra, Histoire physique, politique, et naturelle de l'île de Cuba, Foraminifères, v. 8, p. 189, Pl. 11, Figs. 14, 15.

Quinqueloculina poeyana d'Orbigny, (Figs. 8a, 15q, r), 1839, *in* De la Sagra, Histoire physique, politique, et naturelle de l'île de Cuba, Foraminifères, v. 8, p. 191, Pl. 11, Figs. 25-27.

Quinqueloculina polygona d'Orbigny (Fig. 8h), 1839, *in* De la Sagra, Histoire physique, politique, et naturelle de l'île de Cuba, Foraminifères, v. 8, p. 198, Pl. 12, Figs. 21-23.

Quinqueloculina sabulosa Cushman, 1947, Cushman Found. Foram. Research Contr., v. 23, p. 87, Pl. 18, Fig. 22.

Quinqueloculina seminulum (Linné) (Fig. 3f, g); *Sepula seminulum* Linné, 1753, Systemae Naturae, 10th ed., p. 786; *Quinqueloculina seminulum* (Linné), Cushman, 1929, Cushman Found. Foram. Research Contr., v. 5, pt. 3, p. 59, Pl. 9.

Quinqueloculina tricarinata d'Orbigny (Fig. 8i), 1839, *in* De la Sagra, Histoire physique, politique, et naturelle de l'île de Cuba, Foraminifères, v. 8, p. 187, Pl. 11, Figs. 7, 9, 11.

Quinqueloculina wiesneri Parr; *Quinqueloculina anguina* Terquem var. *wiesneri* Parr, 1950, B.A.N.Z. Antarctic Research Exped., 1929-1931, Repts. Ser. B, v. 5, pt. 6, p. 290, Pl. 6, Figs. 9-10; *Quinqueloculina wiesneri* Parr: Parker, Phleger, and Pierson, 1953, Cushman Found. Foram. Research Spec. Pub. No. 2, p. 13, pt. 2, Pl. 2, Figs. 20-22.

Quinqueloculina sp. (Fig. 15s-y). This quinqueloculine species is characterized by the very slender, elongate, rounded test, smooth chamber wall, and absence of neck. Streeter (1963) described a species from the Bahama Banks which is probably conspecific with *Q.* sp. of this study. He named it *Q. orbignyana* Streeter, but it is as yet unpublished.

Dentostomina bermudezi Perez Farfante (Fig. 8o), 1939, Soc. Cubana Hist. Nat. Mem., v. 13, p. 319, Pl. 45, Figs. 7-9.

Massilina protea Parker (Fig. 15a-e), 1953, Cushman Found. Foram. Research Spec. Pub. No. 2, p. 10, Pl. 2, Figs. 1-4, Text-fig. 2.

Pyrgo comata (Brady); *Biloculina comata* Brady, 1884, Report Voyage *Challenger*, Zoology, v. 9, p. 144, Pl. 3, Fig. 9; *Pyrgo comata* (Brady), Cushman, 1929, U.S. Natl. Mus. Bull. 104, pt. 6, p. 73, Pl. 19, Fig. 8.

Pyrgo denticulata (Brady); *Biloculina ringens* (Lamarck), var. *denticulata* Brady, 1884, Report Voyage *Challenger*, Zoology, v. 9, p. 143, Pl. 3, Figs. 4, 5; *Pyrgo denticulata* (Brady), Cushman, 1929, U.S. Natl. Mus. Bull. 104, pt. 6, p. 69, Pl. 18, Figs. 3, 4.

Pyrgo eburnea (d'Orbigny); *Triloculina eburnea* d'Orbigny, 1839, *in* De la Sagra, Histoire physique, politique, et naturelle de l'île de Cuba, Foraminifères, p. 180, Pl. 10, Figs. 21-23; *Biloculina eburnea* (d'Orbigny), Low *in* Cloud, 1962, U.S. Geol. Survey Prof. Paper 350, p. 34, 35, Table 12.

Pyrgo nasutus Cushman, 1935, Smithsonian Misc. Colln., v. 91, no. 21, p. 7, Pl. 3, Figs. 1-4.

Pyrgo sp. cf. *P. subsphaerica* (d'Orbigny); *Biloculina subsphaerica* d'Orbigny, 1839, *in* De la Sagra, Histoire physique, politique, et naturelle de l'île de Cuba, Foraminifères, v. 8, p. 162, Pl. 8, Figs. 25-27; *Pyrgo subsphaerica* (d'Orbigny), Cushman, 1929, U.S. Natl. Mus. Bull. 104, pt. 6, p. 68, Pl. 18, Figs. 1, 2.

Pyrgo triloculina (Wiesner); *Biloculina elongata* d'Orbigny var. *triloculina* Wiesner, 1923, Die miliolideon der ostlichen Adria, p. 87, Pl. 17, Fig. 249.

Triloculina bermudezi Acosta (Fig. 14n-s), 1940, Soc. Cubana Hist. Nat. Mem. La Habana, v. 14, no. 1, p. 37, Pl. 4, Figs. 1-5.

Triloculina bicarinata d'Orbigny (Fig. 8*l*), 1839, *in* De la Sagra, Histoire physique, politique, et naturelle de l'île de Cuba, Foraminifères, v. 8, p. 158, Pl. 10, Figs. 18-20.

Triloculina fitterei Acosta (Fig. 15h-k), 1940, Torreia, no. 3, p. 25, Pl. 4, Figs. 6-8. *T. fitterei* as defined in this study includes a plexus of forms including those that might be identified as *T. fitterei* var. *meningoi* Acosta. The species varies in length-width ratio and in number, coarseness, and regularity of costae. Scarce forms that are very similar to *Triloculina littoralis* Collins (1958, p. 369, Pl. 3, Fig. 12a-c) from Great Barrier Reef material were also included in this category.

Triloculina gracilis d'Orbigny, 1839, *in* De la Sagra, Histoire physique, politique, et naturelle de l'île de Cuba, Foraminifères, v. 8, p. 181, Pl. 11, Figs. 10-12.

Triloculina linneiana d'Orbigny (Figs. 8n, 12m), 1839, *in* De la Sagra, Histoire physique, politique, et naturelle de l'île de Cuba, Foraminifères, v. 8, p. 172, Pl. 9, Figs. 11-13.

Triloculina planciana d'Orbigny, 1839, *in* De la Sagra, Histoire physique, politique, et naturelle de l'île de Cuba, Foraminifères, p. 173, Pl. 9, Figs. 17-19.

Triloculina quadrilateralis d'Orbigny (Fig. 8m), 1839, *in* De la Sagra, Histoire physique, politique, et naturelle de l'île de Cuba, Foraminifères, v. 8, p. 173, Pl. 9, Figs. 14-16.

Triloculina rotunda d'Orbigny, 1826, Ann. Sci. Nat., v. 7, no. 4, p. 299, (nomen nudum); *Triloculina rotunda* d'Orbigny, Cushman, 1929, U.S. Natl. Mus. Bull. 104, pt. 6, p. 59, Pl. 14, Fig. 3.

Triloculina tortuosa Cushman; *Quinqueloculina agglutinans* Cushman (not d'Orbigny), 1971, U.S. Natl. Mus. Bull. 71, pt. 6, p. 42, Pl. 9, Fig. 1; *Triloculina tortuosa* (Cushman), 1921, U.S. Natl. Mus. Bull. 100, v. 4, p. 464, Pl. 9, Fig. 1.

Triloculina transversestriata (Brady); *Miliolina transversestriata* Brady, 1881, Jour. Micros. Sci., v. 21, p. 45; *Triloculina transversestriata* (Brady), Cushman, U.S. Natl. Mus. Bull. 104, pt. 6, p. 62, Pl. 16, Fig. 3.

Triloculina tricarinata d'Orbigny, 1826, Ann. Sci. Nat., ser. 1, v. 7, p. 299, mod. 94.

Triloculina trigonula (Lamarck); *Miliolites trigonula* Lamarck, 1804, Ann. Mus., v. 5, no. 3, p. 351; *Triloculina trigonula* (Lamarck), Cushman, 1929, U.S. Natl. Mus. Bull. 104, pt. 6, p. 56, Pl. 12, Figs. 10, 11, Pl. 13, Figs. 1, 2.

Triloculina sp. (Fig. 15*l*-p). This triloculine species is distinguished from other forms by the small size, elongate test, smooth chamber walls, and characteristic angled truncation of the aperture, which gives the aperture a "hooded" appearance. The rounded aperture contains a single simple tooth.

Miliolinella circularis (Bornemann) (Fig. 15f, g); *Triloculina circularis* Bornemann, Cushman, and Valentine, 1930, Stanford Univ. Dept. Geol. Contr., v. 1, p. 15, Pl. 4, Fig. 4a-c.

Miliolinella labiosa (d'Orbigny) (Fig. 9a-g); *Triloculina labiosa* d'Orbigny, 1839, *in* De la Sagra, Histoire physique, politique, et naturelle de l'île de Cuba, Foraminifères, v. 8, p. 178, Pl. 10, Figs. 12-14; *Miliolinella labiosa* Said, 1950, Cushman Found. Foram. Research Contr., v. 1, pts. 1, 2, p. 5, Pl. 1, Fig. 10.

Miliolinella fichtelliana d'Orbigny, 1839, Histoire physique, politique, et naturelle de l'île de Cuba, Foraminifères, v. 8, p. 171, Pl. 9, Figs. 8-10; *Miliolinella fichtelliana* (d'Orbigny), Drooger and Kaaschieter, 1958, Repts. Orinoco Shelf Exped., v. 4, p. 54, Pl. 3, Fig. 6.

Miliolinella oblonga (Montagu); *Vermiculum oblongum* Montagu, 1803, Testacea Britannica, p. 522, Pl. 14, Fig. 9; *Miliolinella oblonga* (Montagu),

Bermudez, 1963, Bol. Inst. Oceanog. Univ. Oriente, Cumana, p. 101.

Ammomassilina alveoliniformis (Millett) (Fig. 8p); *Massilina alveoliniformis* Millett, 1898, Royal Micros. Soc. Jour., art. 15, p. 609; *Ammomassilina alveoliniformis* (Millett), Cushman, 1933, Cushman Found. Foram. Research Contr., v. 9, pt. 2, p. 32.

Hauerina bradyi Cushman; *Hauerina compressa* d'Orbigny, Brady, 1884, Report Voyage *Challenger*, Zoology, v. 9, p. 190, Pl. 11, Figs. 12, 13; *Hauerina bradyi* Cushman, 1917, U.S. Natl. Mus. Bull. 71, pt. 6, p. 62, Pl. 23, Fig. 2.

Hauerina occidentalis Cushman (Fig. 8q); *Hauerina ornatissima* Karrer, Cushman, 1929, U.S. Natl. Mus. Bull. 104, pt. 6, p. 47, Pl. 10, Figs. 10-12; *Hauerina occidentalis* Cushman, 1946, Cushman Found. Foram. Research Contr., v. 22, p. 9, Pl. 1, Figs. 22-24.

Hauerina speciosa (Karrer); *Spiroloculina speciosa* Karrer, 1968, K. Akad. Wien, Math-Naturn. Cl., Sotzber, Wien, Ostereich, Bd. 58, Abt. 1, p. 135, Pl. 1, Fig. 8; *Hauerina speciosa* (Karrer), Cushman, 1946, Cushman Found. Foram. Research Contr., v. 22, pt. 1, p. 5.

Heterillina cribrostoma (Heron-Allen and Earland) (Figs. 8r, 14a-m); *Miliolina circularis* (Bornemann) var. *cribrostoma* Heron-Allen and Earland, 1915, Zool. Soc. London Trans., v. 20, pt. 17, p. 558, Pl. 41, Figs. 12-16. This species is characterized by irregular chamber arrangement; small size; smooth, polished test wall; and low, arched, cribrate aperture. As defined in this study it is highly variable in form. It varies from compressed attaching forms with one side strongly flattened (Fig. 14e, f, g, j) to nearly spherical forms characterized by rounded chambers (Figs. 14b, c, d, k, *l*, m). All gradations between flattened and rounded forms are present (Fig. 14i), including many forms with faceted chambers. The cell pattern of *Thalassia* leaves is visibly impressed on the flattened surface of a few specimens of compressed forms—evidence of an attaching habitat. This species is conspecific with the form described by Streeter (1963) as ?*Triloculinella bahamensis* from the Bahama Banks. The ecology of the forms in Belize and the Bahama Banks is very similar. Areas of maximum occurrence are characterized by restricted hydrography, fine-grained carbonate-mud deposition, and extremely shallow water. B. W. Logan (personal commun.) found a form he referred to *Triloculina cribrostoma* to be abundant in an extensive hypersaline area of Shark Bay, Australia. He noted that it is variable in form and has a cribrate aperture. In that area the species was part of a low-diversity fauna containing peneroplids.

Tubinella funalis (Brady); *Articulina funalis* Brady, 1884, Report Voyage *Challenger*, Zoology, v. 9, p. 185, Pl. 13, Figs. 6-11; *Tubinella funalis* (Brady), Cushman, 1929, U.S. Natl. Mus. Bull. 104, pt. 6, p. 54, Pl. 12, Fig. 8.

Articulina lineata Brady (Fig. 9j), 1884, Report Voyage *Challenger*, Zoology, v. 9, p. 183, Pl. 12, Figs. 19-21.

Articulina mayori Cushman (Fig. 9k), 1922, Carnegie Inst. Washington Pub. 311, p. 71, Pl. 13, Fig. 5.

Articulina sagra d'Orbigny (Fig. 7s, t), 1839, *in* De la Sagra, Histoire physique, politique, et naturelle de l'île de Cuba, Foraminifères, v. 8, p. 183, Pl. 9, Figs. 23-26.

Peneroplis bradyi Cushman (Fig. 9m); *Peneroplis planatus* Brady var. *laevigata* Brady, 1884, Report Voyage *Challenger*, Zoology, v. 9, Pl. 13, Figs. 12, 13; *Peneroplis bradyi* Cushman, 1930, U.S. Natl. Mus. Bull. 104, pt. 7, p. 40, Pl. 14, Figs. 8-10.

Peneroplis pertusus (Forskal) (Fig. 9*l*); *Nautilus pertusus* Forskal, 1775, Descr. Anim., p. 125, no. 65; *Peneroplis pertusus* (Forskal), Cushman, 1930, U.S. Natl. Mus. Bull. 104, pt. 7, p. 35, Pl. 12, Figs. 3-6.

Peneroplis proteus d'Orbigny; *Peneroplis protea* d'Orbigny, 1839, *in* De la Sagra, Histoire physique, politique, et naturelle de l'île de Cuba, Foraminifères, v. 8, p. 60, Pl. 7, Figs. 7-11; *Peneroplis proteus* d'Orbigny, Cushman, 1930, U.S. Natl. Mus. Bull. 104, pt. 7, p. 37, Pl. 13, Figs. 1-17.

Monalysidium politum Chapman; *Peneroplis lituus* H. B. Brady, 1884, Report Voyage *Challenger*, Zoology, v. 9, p. 205, Pl. 13, Figs. 24, 25; *Monalysidium politum* Chapman, Cushman, 1930, U.S. Natl. Mus. Bull. 104, pt. 7, p. 44, Pl. 15, Figs. 11, 12.

Spirolina acicularis (Batsch); *Nautilus (Lituus) acicularis* Batsch, 1791, Conch. Seesandes, p. 4, Pl. 6, Figs. 16a, b; *Spirolina acicularis* (Batsch), Cushman, 1930, U.S. Natl. Mus. Bull. 104, pt. 7, p. 42, Pl. 15, Figs. 1-3.

Spirolina arietinus (Batsch); *Nautilus (Lituus) arietinus* Batsch (part), 1791, Conch. Seesandes, p. 4, Pl. 6, Fig. 15c; *Spirolina arietinus* (Batsch), Cushman, 1930, U.S. Natl. Mus. Bull. 104, pt. 7, p. 43, Pl. 15, Figs. 4, 5.

Archaias angulatus (Fichtell and Moll) (Figs. 9o-r, 12b, 16j, k, m-t); *Nautilus angulatus* Fichtell and Moll, 1803, Test. Micro., p. 112, Pl. 21; *Orbiculina compressa* d'Orbigny, 1839, *in* De la Sagra, Histoire physique, politique, et naturelle de l'île de Cuba, Foraminifères, v. 8, p. 73, Pl. 8, Figs. 4-7; *Archaias angulatus* (Fichtell and Moll), Cushman, 1930, U.S. Natl. Mus. Bull. 104, p. 46, Pl. 16, Figs. 1-3, Pl. 17, Figs. 3-5. *A. angulatus* is distinguished by variability of external form, lack of obvious external ornamentation, and presence of chambers separated into numerous chamberlets. The taxonomy of this species has been treated at length by Cole (1965), Hofker (1950-53, 1964), Henson (1950), and Smout and Eames (1958). The classification used herein generally follows that of Cole (1965), modified where necessary because of the dependence on external characters for population study in this report. As defined in this study, *A. angulatus* includes the form *Cyclorbiculina compressa* (d'Orbigny) as described by Smout and Eames (1958) and Loeblich and Tappan (1964). This form of *A. angulatus* was very scarce in the collections from Belize. It was found living abundantly in the branching crown of a *Penicillus* plant and a single living specimen was imbedded in the stalk of the same plant (sample no. 359).

Amphisorus hemprichii Ehrenberg (Fig. 11c), 1838, Abh. k. bay. Akad. Wissy München, p. 134, Pl. 3, Fig. 3.

Sorites marginalis (Lamarck) (Fig. 11a); *Orbulites marginalis* Lamarck, 1816, Hist. Nat. Anim. sans Vert., v. 2, no. 1, p. 196; *Sorites marginalis* (Lamarck), Cushman, 1930, U.S. Natl. Mus. Bull. 104, pt. 7, p. 49, Pl. 18, Figs. 1-4.

Borelis pulchra (d'Orbigny); *Alveolina pulchra* d'Orbigny, 1839, *in* De la Sagra, Histoire physique, politique, et naturelle de l'île de Cuba, Foraminifères, v. 8, p. 70, Pl. 8, Figs. 19, 20; *Borelis pulchra* (d'Orbigny), Cushman, 1930, U.S. Natl. Mus. Bull. 104, pt. 7, p. 55, Pl. 15, Figs. 9, 10.

Superfamily BULIMINACEA Jones, 1875

Buliminella elegantissima (d'Orbigny) (Figs. 3i, 5e); *Bulimina elegantissima* d'Orbigny, 1839, Voy. Amér. Mérid., v. 5, pt. 5, Foraminifères, p. 51, Pl. 7, Figs. 13, 14; *Buliminella elegantissima* (d'Orbigny), Phleger, 1951, Geol. Soc. America Mem. 46, pt. 1, p. 7, Pl. 8, Figs. 3, 4.

Buliminella milleti Cushman, 1933, Cushman Found. Foram. Research Contr., v. 9, pt. 4, no. 137, p. 78.

Brizalina goesii? (Cushman) (Fig. 6f); *Bolivina goesii* Cushman, 1922, U.S. Natl. Mus. Bull., no. 104, p. 34, Pl. 6, Fig. 5. This is one of several scarce species that were included in the category *Brizalina* spp. in the quantification of populations for this study.

Brizalina lowmani (Phleger and Parker) (Fig. 5n); *Bolivina lowmani* Phleger and Parker, 1951, Geol. Soc. America Mem. 46, pt. 2, p. 13, Pl. 6, Figs. 20, 21.

Brizalina pacifica (Cushman and McCulloch); *Bolivina acerosa* Cushman var. *pacifica* Cushman and McCulloch, 1942, Univ. Southern Calif. Pub., Allan Hancock Pacific Exped., v. 6, no. 4, p. 185; *Bolivina pacifica* Cushman and McCulloch, Uchio, Cushman Found. Foram. Research Spec. Pub. No. 5, Pl. 7, Fig. 2.

Brizalina paula (Cushman and Cahill); *Bolivina paula* Cushman and Cahill, 1932, *in* Cushman and Ponton, 1932, Florida Geol. Survey Bull. No. 9, p. 84, Pl. 12, Fig. 6.

Brizalina spinescens (Cushman) (Fig. 6e); *Bolivina spinescens* Cushman, 1911, U.S. Natl. Mus. Bull. 71, p. 46, Text-fig. 76.

Brizalina striatula (Cushman) (Figs. 3k, 5j, 6l, 13k); *Bolivina striatula* Cushman, 1922, Carnegie Inst. Washington Pub. 311, p. 27, Pl. 3, Fig. 10.

Brizalina torqueata (Cushman and McCulloch); *Bolivina torqueata* Cushman and McCulloch, 1942, Univ. Southern Calif. Pub., Allan Hancock Pacific Exped., v. 6, no. 4, p. 215.

Brizalina variablis (Williamson) (Fig. 10r); *Textularia variablis* Williamson, 1858, Ray Soc. London, p. 76, Pl. 6, Figs. 162, 183; *Bolivina variablis* (Williamson), Parker, 1952, Bull. Mus. Comp. Zoology, v. 106, no. 10, p. 445, Pl. 4, Fig. 12.

Bolivinita rhomboidalis (Millett) (Fig. 10s); *Textularia rhomboidalis* Millet, 1899, p. 559, Pl. 7, Fig. 4; *Bolivinita rhomboidalis* (Millett), Lynts, 1965, Cushman Found. Foram. Research Contr., v. 16, pt. 2, p. 69, Pl. 7, Figs. 5, 6.

Rectobolivina advena (Cushman) (Fig. 5i); *Siphogenerina advena* Cushman, 1922, Carnegie Inst. Washington Pub. 311, p. 35, Pl. 5, Fig. 2; *Bifarina decorata* Phleger and Parker, 1951, Geol. Soc. America Mem. 46, pt. 2, p. 12, Pl. 6, Figs. 9a, b; *Rectobolivina advena* (Cushman), Parker, Phleger and Pierson, 1953, Cushman Found. Foram. Research Spec. Pub. 2, p. 13, Pl. 4, Figs. 10, 11.

Rectobolivina raphanus (Parker and Jones); *Uvigerina (Sagrina) raphanus* Parker and Jones, 1865, Philos. Trans., v. 155, p. 364, Pl. 18, Figs. 16, 17; *Siphogenerina raphanus* (Parker and Jones), Cushman, 1923, U.S. Natl. Mus. Bull. 104, pt. 4, p. 174, Pl. 42, Fig. 14.

Bulimina marginata d'Orbigny, 1826, Ann. Sci. Nat., ser. 1, tome 7, p. 269.

Bulimina tenuis Phleger and Parker (Fig. 7f, 13h), 1951, Geol. Soc. America Mem. 46, pt. 2, p. 16, Pl. 7, Figs. 33, 34.

Reussella atlantica Cushman (Figs. 5l, 7g, 13i); *Reussella spinulosa* (Reuss) var. *atlantica* Cushman, 1947, Cushman Found. Foram. Research Contr., v. 23, pt. 4, p. 91, Pl. 20, Figs. 6, 7; *Reussella atlantica* Cushman, Phleger and Parker, 1951, Geol. Soc. America Mem. 46, p. 18, Pl. 8, Figs. 8, 9.

Pavonina atlantica Cushman (Fig. 7i), 1922, U.S. Natl. Mus. Bull. 104, pt. 3, p. 51, Pl. 19, Fig. 1.

Chrysalidinella dimorpha (Brady); *Chrysalidina dimorpha* Brady, 1881, Quart. Jour. Micros. Sci., London, n.s., v. 21, p. 54; *Chrysalidinella dimorpha* (Brady), Schubert, 1907, Neues Jahrb. Mineralogie, Geologie u. Paläontologie, Beil.-Band, Abt., v. 25, p. 242.

Hopkinsina pacifica Cushman (Fig. 5k), 1933, Cushman Found. Foram. Research Contr., v. 9, pt. 4, p. 86, Pl. 8, Fig. 16.

Sagrina pulchella d'Orbigny (Fig. 5m), 1839, *in* De la Sagra, Histoire physique, politique, et naturelle de l'île de Cuba, Foraminifères, v. 8, p. 150, Pl. 1, Figs. 23, 24; *Bolivina pulchella* (d'Orbigny), Cushman, 1922, U.S. Natl. Mus. Bull. 104, pt. 3, p. 41, Pl. 7, Fig. 4.

Trifarina bella (Phleger and Parker) (Figs. 7c, d, 12*l*); *Angulogerina bella* Phleger and Parker, 1951, Geol. Soc. America Mem. 46, pt. 2, p. 12, Pl. 6, Figs. 7, 8.

Trifarina bradyi Cushman (Fig. 7e), 1923, U.S. Natl. Mus. Bull., no. 104, pt. 4, p. 99, Pl. 22, Figs. 3-9.

Superfamily DISCORBACEA Ehrenberg, 1838

Discorbis sp. cf. *D. australis* Parr (Figs. 10h, 16a-d); *Discorbis valvulata* Brady (not *Rosalina valvulata* d'Orbigny), 1884, Report Voyage *Challenger*, Zoology, v. 9, p. 644, Pl. 37, Figs. 5-7; *Discorbis australis* Parr, 1932, Royal Soc. Victoria Proc., v. 44, n.s., p. 227, Pl. 22, Fig. 31. This species is distinguished by the small, biconvex test; coarsely perforate, peaked dorsal side; dark brown color; white ventral portion of ultimate chamber; and apertural slit extending into ventral face of chamber.

Discorbis mira Cushman (Fig. 10a, b), 1922, Carnegie Inst. Washington Pub. 311, v. 17, p. 39, Pl. 6, Figs. 10, 11.

Discorbis reniformis (Heron-Allen and Earland); *Discorbina reniformis* Heron-Allen and Earland, 1915, Zool. Soc. London Trans., v. 20, pt. 17, p. 646, Pl. 52, Figs. 7-14.

Discorbis rosea (d'Orbigny) (Fig. 12e, f); *Rotalia rosea* d'Orbigny, 1826, Ann. Sci. Nat., v. 7, p. 272, no. 7, mod. no. 36; *Rotorbinella rosea* (d'Orbigny), Bandy, 1944, Jour. Paleontology, v. 18, p. 372.

Bronnimannia palmerae (Bermudez) (Fig. 10k, *l*); *Discorbis palmerae* Bermudez, 1935, Soc. Cubana Hist. Nat. Mem., v. 9, no. 3, p. 207; *Bronnimannia palmerae* (Bermudez), Bermudez, 1952, Venezuela Minist. Minas and Hydrocarb. Bull. Geol., v. 2, no. 4, p. 39.

Buccella frigida (Cushman) (Fig. 6u); *Fulvinulina frigida* Cushman, 1922, Contr. Canadian Biology (1921), no. 4, p. 12 (144); *Buccella frigida* (Cushman) emend. Anderson, Anderson, 1952, Washington Acad. Sci. Jour., v. 42, p. 144.

Neoconorbina parkerae (Natland); *Discorbis parkeri* Natland, 1950, Geol. Soc. America Mem. 43, pt. 4, p. 27, Pl. 6, Fig. 11; *Rosalina parkerae* (Natland), Parker, 1954, Harvard Mus. Comp. Zoology Bull., v. 111, no. 10, p. 525, Pl. 8, Figs. 24, 25.

Neoconorbina terquemi (Rzehak) (Fig. 6k, 12k); *Rosalina orbicularis* Terquem, 1876, Anim. sur. la Plage de Dunkerque, p. 75, Pl. 9, Fig. 4; *Discorbina terquemi* Rzehak, 1888, Verb. Geol. Reichanst., Wien, p. 227, (nam. subst.); *Neoconorbina terquemi* (Rzehak), Hofker, 1951, Siboga Exped., Foraminifera, pt. III, p. 435.

Rosalina advena (Cushman); *Discorbis advena* Cushman, 1931, U.S. Natl. Mus. Bull. 104, pt. 8, p. 13, Pl. 2, Fig. 8; *Conorboides advena* (Cushman), Barker, 1960, Soc. Econ. Paleontologists and Mineralogists Spec. Pub. No. 9, p. 180, Pl. 87, Fig. 1.

Rosalina bulbosa (Parker) (Fig. 5o, p); *"Discorbis" bulbosa* Parker, 1954, Harvard Mus. Comp. Zoology Bull., v. 111, no. 10, p. 523.

Rosalina candeiana d'Orbigny (Figs. 10c, d, 12n), 1839, *in* De la Sagra, Histoire physique, politique, et naturelle de l'île de Cuba, Foraminifères, v. 8, p. 47, Pl. 4, Figs. 2-4; *Discorbis candeiana* (d'Orbigny), Cushman, 1931, U.S. Natl. Mus. Bull. 104, pt. 8, p. 19, Pl. 7, Fig. 4. This species is identified by the coarsely perforate test, lobulate periphery, and overall variability of form. This variability is particularly expressed in degree of ventral flattening and test height. The species appears to be well adapted to an attaching habitat.

Rosalina floridana (Cushman) (Fig. 10e); *Discorbis globularis* Heron-Allen and Earland (not d'Orbigny), 1915, Trans. Zool. Soc. London, p. 694, Pl. 51, Figs. 36-39; *Discorbis floridana* Cushman, 1922, Carnegie Inst. Washington Pub. 311, v. 17, p. 39, Pl. 5, Figs. 11, 12.

Rosalina subauracana (Cushman) (Fig. 6o, p); *Discorbis subauracana* Cushman, 1922, Carnegie Inst. Washington Pub. 311, v. 17, p. 41, Pl. 7, Figs. 1, 2.

Rosalina sp. (Fig. 10f, g); *Discorbis floridana* Cushman, Drooger and Kaasschieter, 1958, Koninkl. Nederlandse Akad. Wetensch. Proc., afd. Natuurk, Reeks 1, Deel 22, p. 42, Pl. 2, Fig. 6. This species is characterized by finely perforate test, both dorsally and ventrally, and by rounded, slightly inflated chambers. *Rosalina* sp. demonstrates variability in test form, degree of chamber inflation, and degree of ventral flattening. This species is commonly associated with marine vegetation and appears to be suited to an attaching habitat. *Rosalina* sp. is identical to illustrations of *D. floridana* Cushman in Drooger and Kaasschieter. *Rosalina* sp. is distinct, however, being perforate on the ventral surface. Streeter (1963) noted a similar species in sediments of the Bahama Banks which he referred to *D. floridana* Drooger and Kaasschieter (not Cushman).

Tretomphalus bulloides (d'Orbigny) (Fig. 13r, s); *Rosalina bulloides* d'Orbigny, 1839, *in* De la Sagra, Histoire physique, politique, et naturelle de l'île de Cuba, Foraminifères, v. 8, p. 98. This encysted form with a globular float chamber was found in a few places in the shelf sediments; the benthonic stage was not recognized.

Cancris sagra (d'Orbigny) (Fig. 6q); *Rotalina sagra* d'Orbigny, 1839, *in* De la Sagra, Histoire physique, politique, et naturelle de l'île de Cuba, Foraminifères, v. 8, p. 77, Pl. 5, Figs. 13-15; *Cancris sagra* (d'Orbigny), Cushman, 1931, U.S. Natl. Mus. Bull. 104, pt. 8, p. 74, Pl. 15, Fig. 2.

Glabratella opercularis? (Brady) (Fig. 12i, j); *Discorbina opercularis* Brady (not d'Orbigny), 1884, Report Voyage *Challenger*, Pl. 89, Figs. 8, 9; ?*Pileolina opercularis* (d'Orbigny), Barker, 1960, Soc. Econ. Paleontologists and Mineralogists Spec. Pub. No. 9, p. 184, Pl. LXXXIX, Figs. 8, 9.

Siphonina pulchra Cushman, 1919, Carnegie Inst. Washington Pub. 291, p. 42, Pl. 14, Fig. 7.

Asterigerina carinata d'Orbigny (Fig. 12a), 1839, *in* De la Sagra, Histoire physique, politique, et naturelle de l'île de Cuba, Foraminifères, v. 8, p. 118, Pl. 5, Fig. 23, Pl. 6, Figs. 1, 2.

Palmerinella palmerae Bermudez (Fig. 4p), 1934, Soc. Cubana Hist. Nat. Mem., v. 8, no. 2, p. 84, Text-figs. 1-3.

Superfamily ROTALIACEA Ehrenberg, 1839

Ammonia beccarii (Linné), (Figs. 3*l*, 4a-d, 5a); *Nautilus beccarii* Linné, 1758, Systemae Naturae, 10th ed., p. 710; *Ammonia beccarii* (Linné), Brunnich, 1772, Zoologiae Fundamenta, p. 246.

Elphidium advenum (Cushman) (Fig. 6t); *Polystomella advena* Cushman, 1922, Carnegie Inst. Washington, v. 17, p. 56, Pl. 9, Figs. 11, 12; *Elphidium advena* (Cushman), Cushman, 1930, U.S. Natl. Mus. Bull. 104, pt. 7, p. 25, Pl. 10, Figs. 1, 2.

Elphidium lanieri (d'Orbigny) (Fig. 10q); *Polystomella lanieri* d'Orbigny,

1839, *in* De la Sagra, Histoire physique, politique, et naturelle de l'île de Cuba, Foraminifères, v. 8, p. 54, Pl. 7, Figs. 12, 13; *Elphidium lanieri* (d'Orbigny), Hofker, 1964, Natuur. Studiekring voor Suriname en de Nederl. Antil. No. 39, p. 113, Text-fig. 271.

Cellanthus discoidale (d'Orbigny); *Polystomella discoidale* d'Orbigny, 1839, *in* De la Sagra, Histoire physique, politique, et naturelle de l'île de Cuba, Foraminifères, v. 8, p. 56, Pl. 6, Figs. 23, 24; *Elphidium discoidale* (d'Orbigny), Cushman, 1930, U.S. Natl. Mus. Bull. 104, pt. 7, p. 22, Pl. 8, Figs. 8, 9. This species is distinct and easily identifiable in the large, adult form. In juveniles, however, the angular periphery and large umbilical boss are not evident; thus they are difficult to separate from *Cribroelphidium poeyanum* (d'Orbigny) and even *Cellanthus gunteri* (Cole). Samples in which the adult end members are distinct also contain numerous smaller forms for which specific assignment is difficult. Bandy (1964) noted the coexistence of *Elphidium discoidale* and *Elphidium poeyanum* in almost equal numbers in the deeper backreef areas of the Gulf of Batabanó. Streeter (1963, p. 137) noted that the two species appear to grade into one another in samples from the Bahama Banks. This is thought to be true also in the southern shelf lagoon of Belize; however, most of the smaller tests were assigned to *C. poeyanum.*

Cellanthus galvestonense (Kornfeld) (Fig. 16e, f); *Elphidium gunteri* Cole var. *galvestonensis* Kornfeld (part), 1931, Stanford Univ. Dept. Geol. Contr., v. 1, no. 3, p. 87, Pl. 15, Figs. 1a, b; *Elphidium galvestonense* Kornfeld; Parker, Phleger, and Pierson, 1953, Cushman Found. Foram. Research Spec. Pub. 2, p. 7, Pl. 3, Figs. 15, 16.

Cellanthus gunteri (Cole) (Figs. 3n, 5d, 16g, h); *Elphidium gunteri* Cole, 1931, Florida State Geol. Survey Bull. 6, p. 34, Pl. 4, Figs. 9, 10; *Elphidium gunteri* Cole var. *galvestonense* Kornfeld (part), 1931, Stanford Univ. Dept. Geol. Contr., v. 1, no. 3, p. 87, Pl. 15, Figs. 2a, b, 3a, b.

Cribroelphidium koebeense (LeRoy); *Elphidium koebeense* LeRoy, 1939, Natuurk Tijdschr. Nederl.-Indie, dl. 99, afl. 6, p. 240, Pl. 9, Figs. 6, 7; *Elphidium* sp. cf. *E. koebeense* LeRoy; Parker, Phleger, and Pierson, 1953, Cushman Found. Foram. Research Spec. Pub. 2, p. 8, Pl. 3, Figs. 22, 23.

Cribroelphidium poeyanum (d'Orbigny) (Figs. 3m, 4i, 5b, c, 6g, h, 7h, 13a, 16i); *Polystomella poeyana* d'Orbigny, 1839, *in* De la Sagra, Histoire physique, politique, et naturelle de l'île de Cuba, Foraminifères, p. 55, Pl. 6, Figs. 25, 26; *Elphidium poeyanum* (d'Orbigny), Cushman, 1930, U.S. Natl. Mus. Bull. 104, pt. 7, p. 25, Pl. 10, Figs. 4, 5.

Cribroelphidium translucens (Natland) (Fig. 10p); *Elphidium translucens* Natland, 1933, Scripps Inst. Oceanography Tech. Ser. Bull., v. 4, p. 144, Pl. 5, Figs. 3, 4.

Cribroelphidium spp. A number of small, indeterminate tests of this genus are found in shelf samples. Many are thought to be juveniles of described species, but this group probably also contains small forms belonging to ?*Cribroelphidium matagordanum* (Kornfeld) and varieties of *Cribroelphidium incertum* (Williamson).

Protelphidium delicatulum (Bermudez) (Fig. 4j); *Elphidium delicatulum* Bermudez, 1949, Cushman Found. Foram. Research Spec. Pub. 25, p. 168, Pl. 11, Figs. 22, 23.

Superfamily GLOBIGERINACEA Carpenter, Parker, and Jones, 1862

Globigerinoides ruber (d'Orbigny) (Fig. 7u); *Globigerina rubra* d'Orbigny, 1839, *in* De la Sagra, Histoire physique, politique, et naturelle de

l'île de Cuba, Foraminifères, v. 8, p. 82, Pl. 4, Figs. 12-14.
Orbulina universa d'Orbigny (Fig. 7v), 1839, *in* De la Sagra, Histoire physique, politique, et naturelle de l'île de Cuba, Foraminifères, v. 8, p. 3, Pl. 1, Fig. 1.
Globorotalia menardii (d'Orbigny); *Rotalia menardii* d'Orbigny, 1826, Ann. Sci. Nat., v. 7, p. 273, no. 26.

Superfamily ORBITOIDACEA Schwager, 1876

Eponides antillarum (d'Orbigny); *Rotalina antillarum* d'Orbigny, 1839, *in* De la Sagra, Histoire physique, politique, et naturelle de l'île de Cuba, Foraminifères, v. 8, p. 75, Pl. 5, Figs. 4-6; *Eponides antillarum* (d'Orbigny), Cushman, 1931, U.S. Natl. Mus. Bull. 104, pt. 8, p. 42, Pl. 9, Fig. 2.
Eponides repandus (Fichtell and Moll) (Fig. 6m); *Nautilus repandus* Fichtell and Moll, 1803, Test. Micro., p. 35, Pl. 3, Figs. a-d; *Eponides repandus* (Fichtell and Moll), Phleger and Parker, 1951, Geol. Soc. America Mem. 46, pt. 2, p. 21, Pl. 11, Figs. 5, 6.
Amphistegina gibbosa d'Orbigny, 1839, *in* De la Sagra, Histoire physique, politique, et naturelle de l'île de Cuba, Foraminifères, v. 8, p. 120, Pl. 8, Figs. 1-3; *Amphistegina lessonii* d'Orbigny, Cushman, 1931, U.S. Natl. Mus. Bull. 104, pt. 8, p. 79, Pl. 16, Figs. 1-3.
Planulina exorna Phleger and Parker (Fig. 6r, s), 1951, Geol. Soc. America Mem. 46, pt. 2, p. 32, Pl. 18, Figs. 5-8.
Cibicides lobatulus (Walker and Jacob); *Nautilus lobatulus* Walker and Jacob, 1798, Adams Essays, Kanmacher's ed., p. 642, Pl. 14, Fig. 36; *Cibicides lobatulus* (Walker and Jacob), Cushman, 1931, U.S. Natl. Mus. Bull. 104, pt. 8, p. 118, Pl. 21, Fig. 3.
Cibicides mayori (Cushman) (Fig. 10o); *Truncatulina mayori* Cushman, 1924, Carnegie Inst. Washington Pub. 342, p. 39, Pl. 12, Figs. 3, 4.
Cibicides pseudoungerianus (Cushman); *Truncatulina ungeriana* Brady (not *Rotalia ungeriana* d'Orbigny, 1826, 1846), 1884, Report Voyage *Challenger*, Zoology, v. 9, p. 664, Pl. 94, Fig. 9; *Truncatulina pseudoungeriana* Cushman, 1922, U.S. Geol. Survey Prof. Paper 129-E, p. 97, Pl. 20, Fig. 9; *Cibicides pseudoungeriana* (Cushman), Cushman, 1931, U.S. Natl. Mus. Bull. 104, pt. 8, p. 123, Pl. 22, Figs. 3-7.
Planorbulina acervalis Brady (Fig. 11d), 1884, Report Voyage *Challenger*, Zoology, v. 9, p. 657, Pl. 92, Fig. 4. This species is characterized by large, flat, coarsely perforate, plano-convex test and multiple apertures. This form is well adapted to an attaching habitat; however, living forms were commonly found in surface sediments as well as on the leaves of *Thalassia*. Through error, a few specimens probably belonging to the species *Acervalis inhaerens* Schultze, a related form without apertures, were counted in this group. In addition, no distinction was made between *Planorbulina acervalis* and *P. mediterranensis*. Many juvenile specimens were assigned to *Planorbulina* juvenile and counted separately because it was not possible to distinguish between juveniles of *Planorbulina* and *Cibicides*. These forms composed significant percentages of some high-diversity miliolid-dominant living populations. Some small juvenile forms were evidently transported some distance into the southern shelf lagoon and were recorded several miles from the landward edge of the barrier platform.
Planorbulina mediterranensis d'Orbigny (Fig. 11b), 1826, Ann. Sci. Nat., v. 7, p. 280, no. 2, Pl. 14, Figs. 4-6, Mod. no. 79.
Gypsina vesicularis (Parker and Jones) (Fig. 12c); *Orbitolina vesicularis* Parker and Jones, 1860, Annals and Mag. Nat. History, ser. 3, v. 6, no. 5, p. 31.

Cymbaloporetta bradyi Cushman; *Cymbalopora poeyi* (d'Orbigny) var. *bradyi* Cushman, 1915, U.S. Natl. Mus. Bull. 71, pt. 5, p. 25, Pl. 10, Fig. 2, Pl. 14, Fig. 2; *Cymbaloporetta bradyi* (Cushman), Todd and Bronnimann, 1957, Cushman Found. Foram. Research Spec. Pub. No. 3, p. 37, Pl. 11, Fig. 9.

Cymbaloporetta squamosa (d'Orbigny) (Fig. 10m, n); *Rotalia squammosa* d'Orbigny, 1826, Ann. Sci. Nat., v. 7, p. 272, no. 8; *Cymbaloporetta squammosa* (d'Orbigny), Cushman, 1931, U.S. Natl. Mus. Bull. 104, pt. 8, p. 83, Pl. 16, Fig. 4.

Homotrema rubrum (Lamarck) (Fig. 12d); *Millepora rubra* Lamarck, 1816, Hist. Nat. Anim. sans Vert., v. 2, p. 202. *Homotrema rubrum* (Lamarck), Hickson, 1911, Linnean Soc. London Jour., Zoology, Ser. 2, v. 14, p. 445-447, Pl. 30, Fig. 2, p. 31, Fig. 9, Pl. 32, Figs. 19, 22, 28.

Superfamily CASSIDULINACEA d'Orbigny, 1839

Fursenkoina complanata (Egger) (Figs. 5g, 6c); *Virgulina schriebersiana* Czjzek var. *complanata* Egger, Abh. k. bay. Akad. Wissy München, CI, II, v. 18, p. 292, Pl. 8, Figs. 91, 92; *Virgulina complanata* Egger, Phleger and Parker, 1951, Geol. Soc. America Mem. 46, pt. 2, p. 19, Pl. 9, Figs. 1-3; *Fursenkoina complanata* (Egger), Lynts, 1962, Cushman Found. Foram. Research Contr., v. 13, pt. 4, p. 144.

Fursenkoina pontoni (Cushman) (Figs. 5f, 6a, b, 13c); *Virgulina pontoni* Cushman, 1932, Cushman Found. Foram. Research Contr., v. 8, pt. 1, p. 17, Pl. 3, Fig. 7; *Fursenkoina pontoni* (Cushman), Lynts, 1962, Cushman Found. Foram. Research Contr., v. 8, pt. 4, p. 144.

Fursenkoina spinicostata (Phleger and Parker) (Figs. 5h, 6d, 13j); *Virgulina spinicostata* Phleger and Parker, 1951, Geol. Soc. America Mem. 46, p. 19, Pl. 9, Figs. 11-14.

Sigmavirgulina tortuosa (Brady); *Bolivina tortuosa* Brady, 1884, Rept. Voyage *Challenger*, Zoology, v. 9, Pl. 52, Figs. 31, 32; *Sigmavirgulina tortuosa* (Brady), Barker, 1960, Soc. Econ. Paleontologists and Mineralogists Spec. Pub. No. 9, p. 108, Pl. 52, Figs. 31, 32.

Loxostomum limbatum (Brady); *Bolivina limbata* H. B. Brady, 1884, Report Voyage *Challenger*, Zoology, v. 9, p. 414, Pl. 52, Figs. 26-28; *Loxostomum limbatum* (Brady), Cushman, 1937, Cushman Found. Foram. Research Spec. Pub. No. 9, p. 186, Pl. 21, Figs. 28, 29.

Loxostomum mayori (Cushman) (Fig. 3j); *Bolivina mayori* Cushman, 1922, Carnegie Inst. Washington Pub. 311, v. 17, p. 27, Pl. 3, Figs. 5, 6; *Loxostomum mayori* (Cushman), Bandy, 1956, U.S. Geol. Survey Prof. Paper 274-G, p. 195, Pl. 31, Fig. 11.

Globocassidulina subglobosa (Brady) (Figs. 6j, 7a, b, 13p); *Cassidulina subglobosa* Brady, 1881, Micros. Soc. London Quart. Jour., n.s., v. 21, p. 60.

Melonis barleeanus (Williamson); *Nonionia barleeana* Williamson, 1858, Recent Foraminifera of Great Britain, p. 32, Pl. 3, Figs. 63, 69; *"Nonion" barleeanus* (Williamson), Andersen, 1961, Louisiana Geol. Survey Bull. No. 35, pt. II, p. 82, Pl. 18, Figs. 6a, b.

Bisaccium imbricatum Andersen (Fig. 4q), 1951, Jour. Paleontology, v. 25, no. 1, p. 32, Text-fig. 2.

Florilus astricta (McCulloch); *Nonionella japonica* (Asano) var. *mexicana* Cushman and McCulloch, 1940, Southern Calif. Univ. Pub., Allan Hancock Pacific Exped., v. 6, no. 3, p. 160, Pl. 17, Fig. 10; *Nonionella astricta* McCulloch, 1965, Cushman Found. Foram. Research Contr., v. 16, pt. 4, p. 153.

Florilus atlanticus (Cushman) (Figs. 5q, 6i, 13b); *Nonionella atlantica* Cushman, 1947, Cushman Found. Foram. Research Contr., v. 23, p. 90, Figs. 4, 5.

Florilus grateloupi (d'Orbigny); *Nonionina grateloupi* d'Orbigny, 1826, Ann. Sci. Nat., v. 7, p. 294, no. 19; *Nonion grateloupi* (d'Orbigny), Cushman, 1939, U.S. Geol. Survey Prof. Paper 191, p. 21, Pl. 6, Figs. 1-7.

Trichohyalus aguayoi (Bermudez) (Fig. 4n, o); *Discorbis aguayoi* Bermudez, 1935, Soc. Cubana Hist. Nat. Mem., v. 9, p. 204, Pl. 15, Figs. 10-14; *Discorinopsis aguayoi* (Bermudez), Parker, Phleger, and Pierson, 1953, Cushman Found. Foram. Research Spec. Pub. No. 2, p. 7, Pl. 4, Figs. 23, 24; *Trichohyalus aguayoi* (Bermudez), Bermudez, 1963, Bol. Inst. Oceanog. Univ. Oriente, Cumana, p. 176, Pl. 26, Fig. 4.

Hanzawaia concentrica (Cushman) (Fig. 6n); *Truncatulina concentrica* Cushman, 1918, U.S. Geol. Survey Bull. 676, p. 64, Pl. 21, Fig. 3; *Cibicides concentrica* (Cushman), Renz, 1948, Geol. Soc. America Mem. 32, p. 127, Pl. 10, Fig. 8; *Hanzawaia concentrica* (Cushman), Drooger and Kaasschieter, 1958, Rept. Orinoco Shelf Exped., v. 4, p. 49, figure in Map 17.

Superfamily ROBERTINACEA Reuss, 1850

Cushmanella brownii (d'Orbigny); *Nonionina brownii* d'Orbigny, 1839, *in* De la Sagra, Histoire physique, politique, et naturelle de l'île de Cuba, Foraminifères, v. 8, p. 45, Pl. 7, Figs. 22, 23; *Cushmanella brownii* (d'Orbigny), Hofker, 1956, Skr. Univ. Zool. Mus. Copenhagen Pub. 15, p. 144, Pl. 22, Figs. 1-3, Pl. 23, Figs. 1-5, 21.

Distribution of Holocene Marine Ostracoda from Belize [1]

JAMES W. TEETER[2]

ABSTRACT Holocene marine ostracod species from Belize define four major biofacies. The *Chetumal Bay biofacies*, which extends southward along the mainland coast from Chetumal Bay to Belize City, is typified by low, variable salinities and water shallower than 20 ft (6.1 m).

The *southern nearshore biofacies* is restricted to a narrow strip along the coast south of Belize City. The faunal difference between the first two biofacies probably is controlled by differences in water chemistry of the influent streams; the rivers of northern Belize drain a low, flat-lying carbonate terrain, whereas the streams of southern Belize flow across the igneous and metamorphic rocks of the Maya Mountains.

The *main-lagoon biofacies* is characterized by a depth range from 20 to over 200 ft (6.1-61 m), normal-marine salinity, and a very fine-grained, commonly soupy substrate. This biofacies extends southward from the Chetumal Bay biofacies and is bounded by the southern nearshore biofacies on the west and by the carbonate-platform biofacies on the east.

The *carbonate-platform biofacies* faces the Caribbean Sea and is characterized by depths shallower than 30 ft (9.1 m) and water of normal-marine salinity, which supports abundant carbonate-secreting organisms, including reef-building corals. All biofacies, with the exception of the southern nearshore, may be further divided on the basis of their ostracod faunas.

The 105 species examined in this study include 4 new genera, 32 new species, and 1 new subspecies. Of the entire fauna, 29 species have been found as fossils, some dating back to the Eocene. Knowledge of the Holocene distribution of such long-ranging species offers a powerful tool for the paleoecologic interpretation of past sedimentary environments. At the generic level, assemblages from the various biofacies exhibit close similarity to ostracod associations found in Tertiary sedimentary formations of the Caribbean area.

INTRODUCTION

The impressive variety of ostracods already known from the Tertiary of the Caribbean area, and from the Gulf and Atlantic coastal plains, must certainly indicate that numerous environments existed at the time of deposition. Yet, few paleoecologic syntheses of these faunas have been attempted. This fact is particularly striking inasmuch as ostracods, which are almost totally benthonic, should be most useful in paleoecologic interpretation of facies. Perhaps the most significant reason for the failure to employ the few ecologic studies available in this area is their lack of scope. Most studies are restricted to nearshore areas where the depth of water is less than 5 or 10 ft (1.5 or 3.0 m; e.g., Benda and Puri, 1962; Kornicker, 1958, 1961, 1963; Hulings, 1958; and others). Those few studies which have sampled to depths of several tens

[1] Manuscript received, June 20, 1972.

[2] Associate Professor of Geology, University of Akron, Akron, Ohio. This research was carried out under the supervision of C. G. Croneis as part of a larger program directed by E. G. Purdy. The purpose of the research was to study the sediments and faunas of the Belize shelf. Financial support for the overall project was provided by the Humble Oil and Refining Company, Jersey Production and Research Company, Imperial Oil Limited, Petroleum Research Fund of the American Chemical Society, and the American Petroleum Institute.

The writer thanks all personnel who have participated in collecting the material on which this study is based. Appreciation is also extended to Reginald Barker, T. E. Pulley, and R. R. Lankford, who are among the many who contributed samples for comparative purposes. W. A. van den Bold and H. V. Howe very kindly provided not only comparative samples, but also considerable assistance with taxonomic problems and the use of the Howe type collection. The manuscript has benefited considerably from a review by W. A. van den Bold. Esso Production Research kindly offered their excellent photographic facilities. Margaret Geib drafted the maps.

The writer particularly thanks Frank van Morkhoven, not only for stimulating discussions and abundant comparative material, but also for his gracious hospitality.

For her constant encouragement, enthusiasm, perserverance in typing the manuscript, and editing, my wife, Gladys, is in large measure responsible for any merit this paper may possess.

of feet have been hampered by scarcity of Ostracoda (Keij, 1954b) and apparent faunal mixing (Benson and Coleman, 1963; Baker and Hulings, 1966).

The present work is unique because, to date, no Holocene studies have involved such a diverse sequence of environments as is present in Belize. Relatively little work has been done on modern carbonate ostracod assemblages in the Caribbean, and essentially none on modern assemblages in deeper waters.

TECHNIQUES

Field Methods

The approximately 225 samples used in this study were collected during the summers of 1960-64. The writer participated in collecting these samples during 1964. In addition to the piston core and bulk sediment samples, a constant volume, approximately 15 cm^3, of the upper centimeter of the sediment surface was also taken at each station in order to determine the abundance and variety of Foraminifera (Wantland, this volume) and Ostracoda. Where the water was shallower than 20 ft (6.1 m), these quantitative samples were obtained by diving. In deeper areas samples were obtained by using a Petersen grab sampler. In both cases, a short length of plastic tubing, 1 5/8 in. (4.1 cm) in diameter, was forced into the sediment and the upper centimeter extruded into a small sample jar containing buffered formalin. Sample stations and selected geographic locations are shown in Figure 1.

Laboratory Methods

A solution of Rose Bengal was added to the samples and preservative. In this technique, described by Walton (1952), the Rose Bengal stains the protoplasm of organisms living at the time of collection, enabling easy distinction from dead animals during counting.

Although several workers have experienced difficulty with this staining method, the writer suspects the reason to be the relatively low number of ostracods living in most samples at the time of collection. Also, apparently only the softer tissues of the organism are stained, whereas the harder, "chitinous" exoskeleton is not.

Samples were stained for 2-3 days with occasional agitation, after which they were freed of silt and clay by washing over a 230-mesh screen and then allowed to dry. The dried samples were divided several times using a microsplitter in order to obtain a fraction of convenient size for specimen counting. The count was then recalculated, depending on the number of times split, to yield the total number of ostracods in the whole sample. Counting categories included living specimens and dead specimens (both complete carapaces and single valves) of adults and juveniles. These data are recorded in Table 1. Many additional stations were examined and the species noted, although no accurate count was made.

The presence of appendages as a criterion to indicate organisms living at the time of collection is misleading, because appendages have been found preserved in fossil ostracods at least as old as Cretaceous (Martin, 1957).

Ideally, the possession of a bivalve carapace should make ostracods good indicators of autochthonous and allochthonous populations. Numerous articulated, empty carapaces suggest that the ostracods lived at that particular sampling station at some time during the past, although the environment may have differed from that encountered during sampling. Valve disarticulation is dependent upon hinge strength, current agitation,

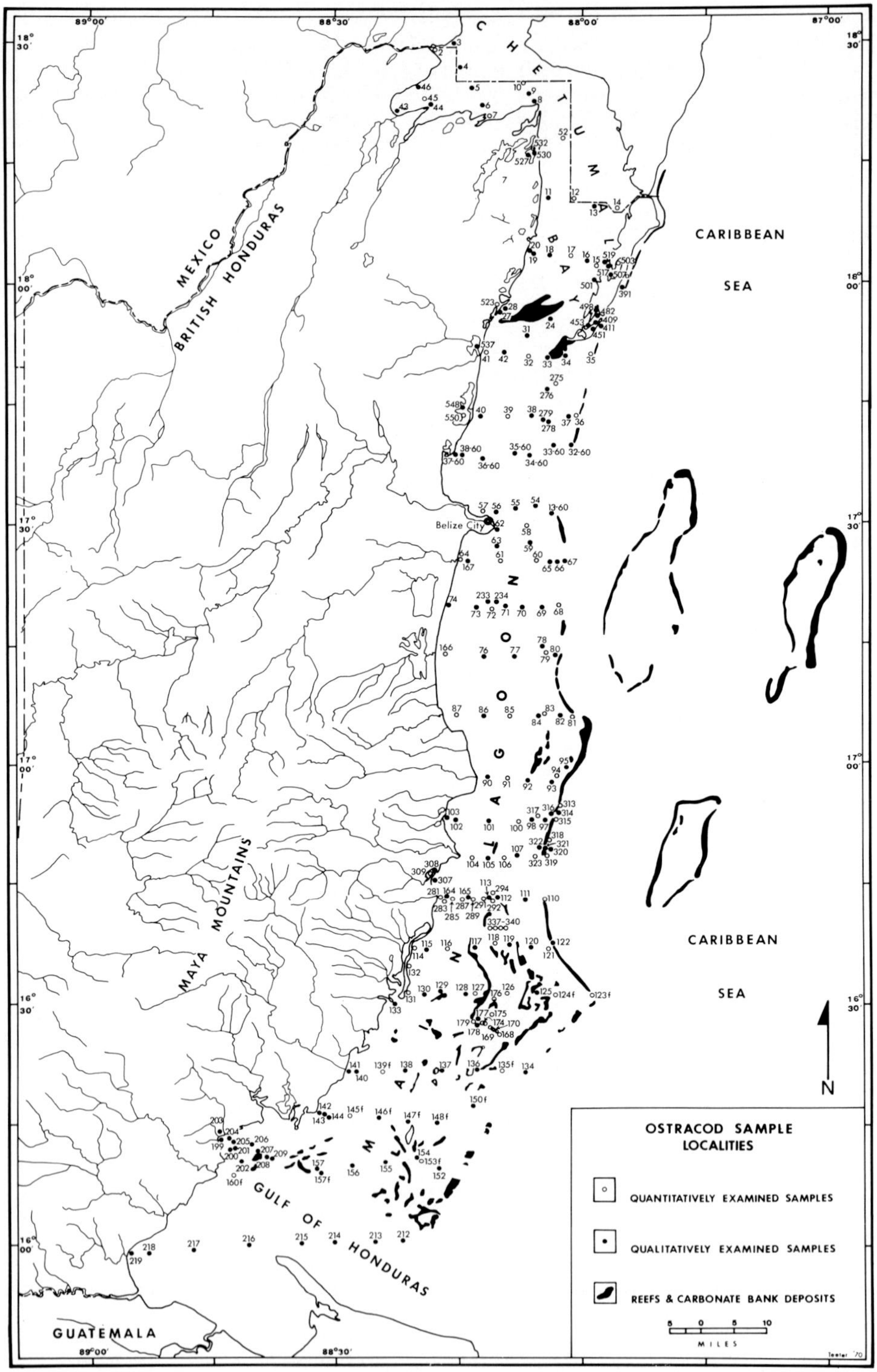

FIG. 1—Ostracod sample localities, Belize.

and the presence of an active benthonic fauna. The intense laboratory preparation of samples, however, may disarticulate many carapaces. Current agitation tends to winnow the earlier instars and smaller species, both of which may be present in areas disturbed by an active benthos. In areas where current agitation is strong enough, allochthonous elements may be introduced, producing faunal mixing. Such mixed samples should be well sorted and, perhaps, exhibit high faunal diversity.

PREVIOUS INVESTIGATIONS

There have been no previous investigations dealing with the ostracod fauna of Belize. However, many studies—both taxonomic and ecologic—concerning the ostracod faunas of the Caribbean, Gulf of Mexico, and Atlantic coastal areas proved useful in this research.

The earliest work in the area was carried out by Brady (1866, 1867-1871, 1880, 1902) and dealt with the description of many new Holocene Caribbean species. Subsequent investigations, primarily of a taxonomic nature, have been made by Tressler (1949, 1954) on Holocene faunas from Tortugas, Florida, and the Gulf of Mexico, and by Sandberg (1964a, 1964b) on the distribution and systematics of the brackish-water genera *Haplocytheridea* and *Cyprideis*.

The upper Tertiary ostracod fauna of the Atlantic and Gulf coastal plains and Florida has been well described in a series of papers by Mincher (1941), Edwards (1944), Malkin (1953), Puri (1953d), McLean (1957), Sabol (1960), Pooser (1965), and Swain (1951, 1968). Coryell and Fields (1937) described the middle Miocene ostracod fauna from the Gatun Formation near the Panama Canal. The greatest contributions to Caribbean ostracod study were made by van den Bold in an extensive series of papers written from 1946 to 1968 and covering a span ranging from Cretaceous to Holocene. Hall (1965), in a thorough study, summarized the results of ecologic studies on nearshore environments from the Gulf of Mexico and Atlantic Coast, and applied these data in a paleoecologic interpretation of Pleistocene ostracods from Sapelo Island, Georgia.

For a long period the classic works on ostracod ecology by Mueller (1894) and Elofson (1941) were the only investigations available in this field. The first paper on ostracod ecology in the United States was one by Tressler and Smith (1948), but it did little to stimulate interest in this subject. In recent years, however, probably as a result of increasing oil-company interest in paleoecology, several papers have appeared, particularly on Gulf Coast and Caribbean faunas.

Brackish-water areas in the northern Gulf Coast received particular attention in studies by Swain (1955), Curtis (1960), Byrne et al, (1959), and Engel and Swain (1967). These investigations indicated the presence of distinctive upper-bay and fluvial freshwater genera, and middle- to lower-bay, brackish-water genera. Typical ostracod genera indicating freshwater influence are *Potamocypris*, *Physocypria*, *Candona*, *Darwinula* and *Limnocythere*. With increasing salinity these genera are replaced mainly by *Cyprideis*, *Haplocytheridea*, *Perissocytheridea*, *Cytheromorpha*, and *Cytherura*. Taxonomic diversity generally increases with increasing salinity and the abundance of each taxon apparently decreases. Although these ostracod facies are controlled by salinity, Swain suggests that available food supply and, to a lesser extent, bottom type also may be important. In Swain's and Curtis' investigations, both living and dead specimens were used in defining facies.

The western and southern coasts of Florida were extensively discussed in a series of papers by Puri and Hulings (1957), Puri (1960), Benda and

TABLE 1. DISTRIBUTION OF HOLOCENE

	Chetumal Bay																	Southern Nearshore				
SPECIES / STATION	2	45	7	10	532	52	12	523	41	550	57	14	17	15	503	32	39	64	166	307	132	131
Cytherura sp. ex. gr. C. johnsoni	/56	/x	/4		2/102			/8		/8	/3								/6	/4		
(?) Dolerocypria inopinata	/8	/2		/x	x4/18		/64	/40		/2	x1/1											
Perissocytheridea subrugosa	/44	/3	/8		x/68		/12	/124	/18	x/118	/x							x/	/3	/x		/2
Aurila chetumalensis			/3		/6			/44	/13	2/2			/x	2/2	/2							
Loxoconcha levis	/8		/17		/98		/24	/76		/24	/2				/2							
Cytheromorpha apheles	1/352	/6	/5					/120		/8								/1				
Megacythere stephensoni			/2		x2/28			/28		/2			/x			/x	/8					
Quadracythere producta													/x		/18		1/					
Perissocytheridea bisulcata												/10		/14								
Actinocythereis bahamensis																						
Haplocytheridea kirkbii	1/4	1/x	/x		20/38		/16	/x20	/7	4/26	/1	x/14	/x	x/30	/8							
Cytherella arostrata																	x3/x					
Haplocytheridea placentiaensis											x1/2								/3	2/6	1/2	
Neomonoceratina mediterranea																			/1			
Megacythere johnsoni (?)											/3								/4	/x	/4	
Ruggieria dictyon																						
Pumilocytheridea sandbergi																			/3			/2
Triangulocypris laeva									/5								/x		/x		/x	
Costa variabilocostata recticostata																	/4					
Cytherura maya																						/2
Loxoconcha avellana																			/1			/1
Munseyella sp. cf. M. minuta																						
Cativella pulleyi																						
Puriana matthewsi																						
Basslerites parvissima																						
Caribbella puseyi																						/x
Pterygocythereis hondurasensis																						
Copytus baculoides																						
Parakrithella reniformis																						
P. hanaii																						
Cytherura swaini																						
C. subwardensis																						
? C. sp.																						
Kangarina ancycla																						
Cytheropteron barkeri																						
C. subreticulatum																						
Pellucistoma howei																			/x	/x	/2	/1
Cytherelloidea praecipua																						
C. sp. cf. C. leonensis																			/x			
Orionina serrulata																						/1
Reussicythere reussi																			/4	1/4	/x	1/x
Microcythere sp. cf. M. inflexa			/1			/x		/16	x/2			x8/12		x2/18	/4		1/x					/1
Eucytherura spp.																						
Cytherella polita																						
Paradoxostoma spp.									1/1						/2	1/1	/32					
Neonesidea longisetosa																						
Paranesidea bensoni																						
P. gigacantha																						
Bairdoppilata (Bairdoppilata) cushmani																						
B. (B.) fasciata																						
B. (Glyptobairdia) coronata																			/1			
B. (G.) trinodosa																						
Triebelina rugosa																						
Saipanetta mckenziei																						
Propontocypris multiporifera																			/1			
Macrocyprina propinqua																						
Aglaiocypris croneisi																						
Callistocythere cranekeyensis																						
Morkhovenia inconspicua																			x/4			/1
Keijia demissa																						
Hemicythere (?) auricula																						
Quadracythere lankfordi																						
Mutilus confragosus																						
Aurila wantlandi											/10											
Caudites nipeensis																						
C. angulata																						
C. howei																						
Orionina bradyi																						
O. ebanksi																						
Jugosocythereis pannosa																						
Hermanites hornibrooki																						/1
H. transoceanica																						
Occultocythereis angusta																						
Neocaudites pulchra																						
N. scottae																						
Puriana convoluta																						
Basslerites twincayensis																						
B. yoni																						
Protocytheretta pumicosa																						
Cushmanidea cristifera																						
Cytherura sablensis																						
C. sp. aff. C. cornuta																						
Hemicytherura cranekeyensis																						/2
H. bradyi																						/1
Paracytheridea tschoppi																						
P. edwardsi																						
Loxoconcha wilberti																						
L. fischeri																						
L. magna																						
L. fischeri minima																						
L. dorsotuberculata																						
L. oculocrista																						
L. suboculocrista																						
L. postdorsolata group																						
Xestoleberis punctata	/12			/x	/2		/16	/36	/4	/2	x/4	2/2	/x		/8	/5	/12				/6	/2
X. curassavica																						
X. sp. aff. X. aurantia															/2							
X. sp. 1							/x					x2/4		/x	/2							
X. sp. 2																						
X. sp. 3																/3						
X. sp. 4			/42	/16	/106		/42	/112	/4	/2		x/26	/64									
Cytherelloidea umbonata																						

EXPLANATION

Numerator – indicates living specimens
– "x" indicates juveniles
– number indicates adult abundance

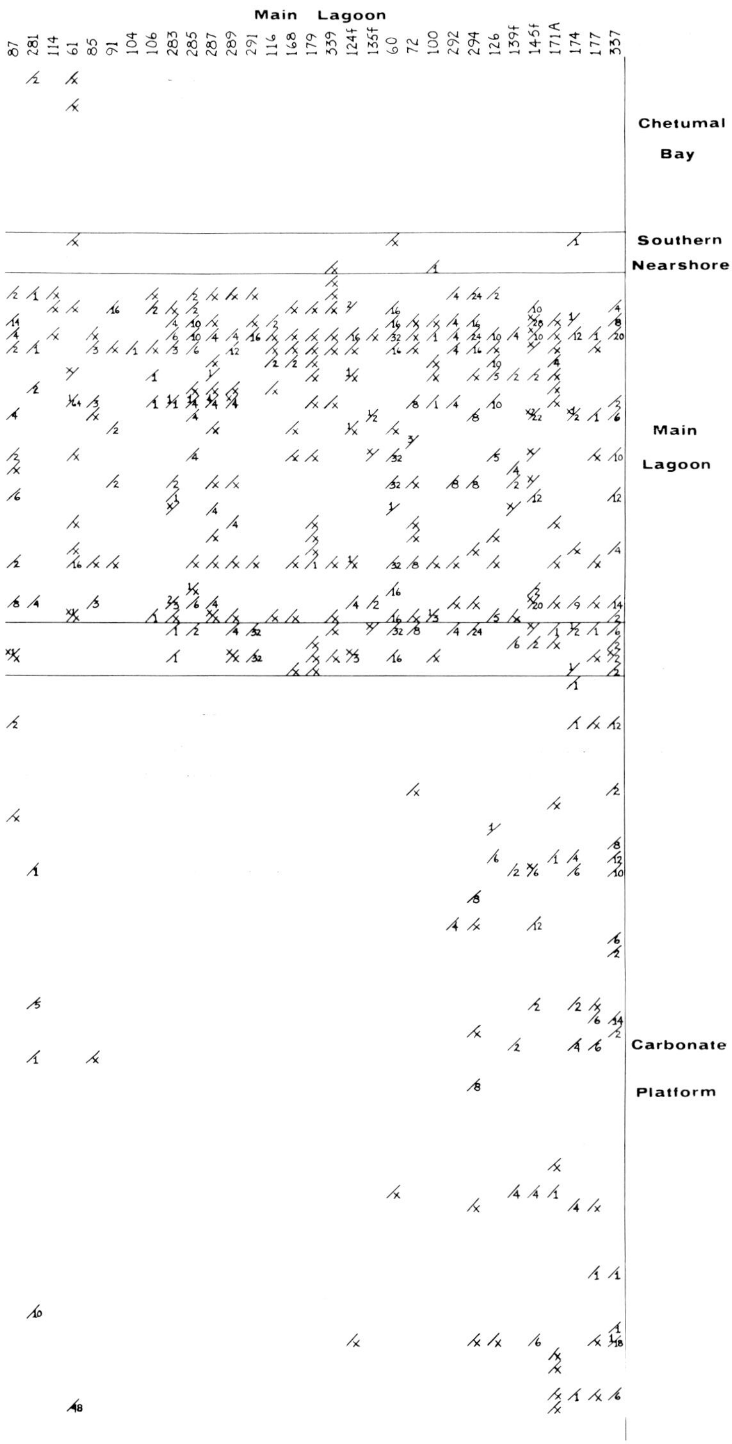

Denominator – indicates dead specimens
– "X" indicates presence of more than three single valves
– number indicates abundance of complete carapaces

TABLE 1 (Continued)

Carbonate Platform

SPECIES / STATION	35	36	81	313	315	318	319	110	123f	151f	79	83	121	118	172	173	175	340	338	275	58	68	94	317	323	176	127f	153f	160f	
Cytherura sp. ex. gr. C. johnsoni		/8																			/6	x/	/6			/4					
(?)Dolerocypria inopinata						2/	x1/2													/2										
Perissocytheridea subrugosa																															
Aurila chetumalensis																													Chetumal Bay
Loxoconcha levis																							/1						
Cytheromorpha apheles																															
Megacythere stephensoni																															
Quadracythere producta																			/2										
Perissocytheridea bisulcata								/3	/4		/x		/1	/x			/7	/1								/7			
Actinocythereis bahamensis																															
Haplocytheridea kirkbii												/1							/1			/2							
Cytherella arostrata																		/1	/1										
Haplocytheridea placentiaensis												/1							/1	x/		x/2		x/					Southern Nearshore
Neomonoceratina mediterranea																															
Megacythere johnsoni (?)	/1																													
Ruggieria dictyon	/x																													
Pumilocytheridea sandbergi	/1			x/			/4													/1	/3		1/12	/10	/2					
Triangulocypris laeva	x/2	/2							/1	/2						1/					/4	/2		/5	/8			/2		
Costa variabilocostata recticostata																															
Cytherura maya																					/x					/1			
Loxoconcha avellana																															
Munseyella sp. cf. M. minuta																															
Cativella pulleyi																				/2									
Puriana matthewsi																															
Basslerites parvissima																					/2	/2	/1						
Caribbella puseyi																															
Pterygocythereis hondurasensis																													Main Lagoon
Copytus baculoides																															
Parakrithella reniformis				/1																	1/	x3/8							
P. hanaii																															
Cytherura swaini																															
C. subwardensis																															
? C. sp.																															
Kangarina ancycla						/1																								
Cytheropteron barkeri																															
C. subreticulatum									/1																					
Pellucistoma howei							/x														/4		/2	/5						
Cytherelloidea praecipua																							/1	/x	/11						
C. sp. cf. C. leonensis																															
Orionina serrulata										/1											/4									
Reussicythere reussi																						/x		/8					
Microcythere sp. cf. M. inflexa	x/1		3/	x15/		x17/4	x5/8	2/	1/3	x1/	3/13	x/1	x2/	5/				x/	x2/8	/11		1/2	/8	/8	/2		/1	/2	x/	
Eucytherura spp.							/2	/1	/1		/2																		/1	/1	
Cytherella polita											/2										/4	x/								
Paradoxostoma spp.	/1				/3	x1/			/2	x3/1	2/2								/x	/1		/x	/4	/14				/x		
Neonesidea longisetosa				/2	/3		/4									/x		/14			/2	/2		/5	/14				/1	
Paranesidea bensoni	/1			/2	/5	/1	/4									/x		/1	/1				3/2		/2		/1			
P. gigacantha																					/1				/6					
Bairdoppilata (Bairdoppilata) cushmani					2/	1/			/1	/1				1/3				x/8	/1		x/4	/2		/3	/16		/1			
B. (B.) fasciata					/3		/2		/1											/14		/2							
B. (Glyptobairdia) coronata		/2					/2																						/2	
B. (G.) trinodosa				/2		/x	/2		/1	/2																		/2		
Triebelina rugosa					/5				/1																					
Saipanetta mckenziei	/x		1/	/2	/x	/10					/2					/x		/1							x3/					
Propontocypris multiporifera					x2/5	x/2	x/2							1/1				/x		/1	x4/		/x	/3				1/2		
Macrocyprina propinqua													/x		/x									/3						
Aglaiocypris croneisi				x/	x3/3		1/											x/						/8	x1/					
Callistocythere cronekeyensis												/1		/4				/2	/2		/4	/3		/5		/1	/9		/5	
Morkhovenia inconspicua	/1		x3/1	x1/8	x/13	x12/10	12/54	x2/15	/7	x/11				4/10	/x			x2/1	x5/4			4/42			/2	/2	/1	/14	x/10	
Keijia demissa												/2						1/											1/1	
Hemicythere (?) auricula				/2	/x		/2	/2		/2	/1																		/7		
Quadracythere lankfordi	/2			/1	/3		/2	/4		/4	/1																			
Mutilus confragosus			/2			/2	1/10			/2											/1		/2			/1	/1			
Aurila wantlandi					/2			/2						/1								1/4		/5		/3			
Caudites nipeensis		/16		/4		/1		/1	/2		/2					/x			/2			/28				/1	/1	/4		
C. angulata	/2			/4		/1	1/12	/9	/3	/7	/1									/2		/18			/6			/8		
C. howei	/2	/18		/2			/6	/2	/1					2/		/1						/2				/2			/4	
Orionina bradyi	/3	/4	/1	1/2	1/5	1/	3/8	/5		1/3	/4					1/		2/		/x		/34			/2			/6	/1	
O. ebanksi					2/					x/1															/8			1/18		
Jugosocythereis pannosa					/3					/6	x/	/1				/2		/2			/2		/2		/4					
Hermanites hornibrooki	/1	/2		/2	/10	/1	/10	/5	/1					/2								/8								
H. transoceanica				x/8	/13		1/10	/2		/1	/3				/x	/x										/1		/2		Carbonate Platform
Occultocythereis angusta							1/										/2												
Neocaudites pulchra																															
N. scottae		/4		/2	/5	/1	/2								/1	/1						/2						/2		
Puriana convoluta	/2	/2			/13	/4		/2	/2	/2	1/2		/1	/1								/4			/2			/10		
Basslerites twincayensis																									2/10					
B. yoni					/2				/1														/2							
Protocytheretta pumicosa		/2					1/2				x/2			/1			/1		/1							/1				
Cushmanidea cristifera		1/	2/	1/	/10	x5/	/2		1/1	x1/1	x/3									/1		/2						/2		
Cytherura sablensis										/1			/1										/4							
C. sp. aff. C. cornuta	x/	/2		4/6	2/3	x/1	3/4																		/2		/1	/4		
Hemicytherura cranekeyensis									/2														/3	/8		/1		/1	
H. bradyi	3/1	2/8		2/x	/10		3/6	1/2	/5	/2	/6		/1	/1				/1		/4		/12	/2	/5	/16		/5		/2	
Paracytheridea tschoppi					1/	/1	/x		/x				x/	x/x		/x	/x		/x			1/x								
P. edwardsi			x/2		/3			/1																					
Loxoconcha wilberti	1/x	/2		/4	/5						/1		/2							/1			1/16	/3						
L. fischeri												/2								/3									
L. magna	/6	/6			/1		/1	/11		/1	/2								/1			/2								
L. fischeri minima									/2										/1				/1	/1					
L. dorsotuberculata	/2			/8	/9	/2	/5	/4	/1	/2	/1		/3						/1				/3					/1		
L. oculocrista	/2	/4		/2	/4		/3	/4	/1	/5	/1											/1								
L. suboculocrista				/1						/6											/4		/1					/1	/6	
L. postdorsolata group		/4			/2	/1	/1	/1		/1	/5											/4						/2		
Xestoleberis punctata				1/2		/5	/6	/1		/2								/2	/2	/3	/16	/12	/16		/24		/4	/2	/7	
X. curassavica		/4			/5	/x		/2	/2		1/6		/2				/x			/1			/10							
X. sp. aff. X. aurantia	/2			x/2	/5		/2	/2		x4/1	1/1			1/1	/x					/1		/4	/4					/6		
X. sp. 1	/1			2/							1/1																			
X. sp. 2	/2		/2		/20	/1			/1		/10							/5	x/												
X. sp. 3	/3	/4		/4	/23	/2	/2	/1	/3	/2	/9		/1							/1				/3			/2			
X. sp. 4					/5					/x	/3	/1								/4		x/8		/13	/8		/1			
Cytherelloidea umbonata	/2																					1/2								

Numerator – indicates living specimens
– "x" indicates juveniles
– number indicates adult abundance

Denominator – indicates dead specimens
– "x" indicates presence of more than three single valves
– number indicates abundance of complete carapaces

Puri (1962), and Benson and Coleman (1963). Hulings and Puri (1964) summarized much of the foregoing literature and added new material. Various environmental factors have been suggested as controlling the ostracod distributions investigated in the above papers. Puri and Hulings, in the original study in this area, recognized two principal biofacies—a carbonate province and a clastic province. As a result, disproportionate emphasis has been placed on the substrate by these authors in their two papers. The writer believes that the carbonate substrate in the Florida Bay area is controlled by the virtual absence of terrigenous sediment and by the region's higher temperatures, permitting colonization by carbonate-secreting organisms characteristic of shallow waters in the lower latitudes. The ostracod fauna of the carbonate province also provides evidence of these higher temperatures. The Florida Bay fauna has many elements typical of the Caribbean province. Hulings and Puri (1964) have recognized several assemblages from four study areas along the Florida coast from Panama City to Cape Romano. Probably because these areas are widely separated, exactly similar faunas are not present at all locations. Hulings and Puri chose to recognize these differences by erecting different assemblages.

Of the environmental factors, salinity obviously is one of the major controls. Although there are few species in common, many of the brackish-water and marine genera of the northern Gulf Coast are also present in Florida. Substrate also plays a role, in that certain species are most commonly associated with marine grasses.

Although primarily a taxonomic study, Benson and Coleman's (1963) paper on the eastern Gulf of Mexico requires comment here. Preparation of the samples involved washing them on a 60-mesh screen. Thus, many of the smaller species, unfortunately, were lost. Although Benson and Coleman recorded no reefs or shoals in the vicinity, they reported several species of ostracods (in water from 60 to over 180 ft or 18-55 m) which are characteristically found on the reef and carbonate shoals of Belize. However, Lynch (1954, Fig. 16) indicated the presence of coral patches and heads near those stations with mixed faunas.

Morales (1966), in a study of the ostracods from the shallow Laguna de Terminos, Mexico, established three assemblages on the basis of living and dead adult specimens. His oyster-bank assemblage is low in diversity and is characterized by reduced salinity and a predominantly terrigenous clastic substrate. This assemblage is located in estuaries. The most diverse assemblage occurs throughout the lagoon and is characterized by higher salinities and finer substrates with a greater carbonate concentration. A washover-delta assemblage—influenced by salinities slightly below to slightly above normal-marine salinity, high carbonate substrate, and swift currents—occurs in one of the channels connecting the lagoon to the Gulf of Mexico. Only a few species are restricted totally to each assemblage. Morales suggests that submerged vegetation and the zone of mixing between clear Gulf and turbid lagoonal water may significantly control ostracod abundance.

Ecologic papers on ostracods from the Caribbean have been published by Klie (1933, 1939a) and Baker and Hulings (1966). Klie dealt largely in a taxonomic fashion with hyposaline to hypersaline faunas from Bonaire, Curaçao, and Aruba.

In a paper on Holocene marine ostracods from Puerto Rico, Baker and Hulings (1966) recognized five ecologic assemblages based mainly on ostracod species, sediment type, and water depth. Some of these assemblages range over depths of several hundred feet and contain species characteristic of much shallower water in Belize. The problem may lie in Baker and Hulings' sampling technique, which consisted of dragging a 10-in.-diameter (25.4 cm) steel casing over the bottom for variable lengths of time.

Such a casing may have dug deeply into the soft offshore muds, penetrating older sediments deposited in shallower water during the Holocene transgression.

In a primarily taxonomic study, Keij (1954b) recorded all ostracod species common to the Gulf of Paria. Analysis of population data indicates species distributions similar to those observed in Belize.

Several studies of an autecologic nature have been conducted on Gulf Coast and Caribbean species. Kornicker (1961, 1963) investigated the effects of water depth, salinity, substrate, temperature, and associated biota on Bahamian Bairdiidae and Cytherellidae. It is the opinion of the writer that the factors studied in the Bimini area exhibited too little variation to define the ecologic limits of species of these groups. In an excellent laboratory study, Kornicker and Wise (1960) subjected individuals of *Aurila conradi* (Howe and McGuirt) to various salinities, temperatures, and substrates in order to determine the limits and preferences of the species. Grossman (1965) studied *Aurila conradi* in its natural environment at Redfish Bay, Texas, and found it to live within the environmental limits established in the laboratory study.

OSTRACOD BIOFACIES

A glance at the ostracod distribution data in Belize (Table 1) suggests that the faunal assemblages represent three major biofacies. These biofacies include (1) a low- to variable-salinity, shallow-water assemblage which, on the basis of minor faunal differences, may be further subdivided into the *Chetumal Bay biofacies* and the *southern-nearshore biofacies*; (2) a deeper water assemblage of normal to near-normal salinity, the *main-lagoon biofacies*; and (3) a shallow-water assemblage of essentially normal-marine salinity, the *carbonate-platform biofacies*. The aerial extent of these biofacies is indicated in Figure 2.

Further examination of Table 1 reveals some scatter in the data. This may be explained by the fact that some species may have rather broad environmental limits. In view of the low living populations, this possibility is difficult to assess. A second explanation is that the environment in the various biofacies may have differed in the past, permitting colonization by species no longer capable of living there. In some instances ostracods undoubtedly have been washed into foreign environments. This has probably occurred in southern Belize, where shallow-water ostracods have been swept into the adjacent depths of the main lagoon.

Chetumal Bay Biofacies

Chetumal Bay and the shoreward part of the northern shelf lagoon contain a distinctive ostracod fauna whose distribution is closely related to salinity gradients. Euryhaline species diagnostic of this Chetumal Bay biofacies are *Aurila chetumalensis* Teeter, n. sp., *Perissocytheridea subrugosa* (Brady), *Haplocytheridea kirkbii* (Brady), and *Megacythere stephensoni* (Puri).

The most obvious factor controlling the distribution of ostracods in this biofacies is salinity. The water varies from fresh water at the mouths of New River and Rio Hondo to almost normal-marine salinity near the mouth of Chetumal Bay. A brackish-water tongue extending southward along the coast may be the result of southward-moving bay waters or the sluggish freshwater effluent from the coastal lagoons. Although observations have not been made during the winter dry season, salinities probably increase in Chetumal Bay. Thus, this area is characterized by low, fluctuating salinity.

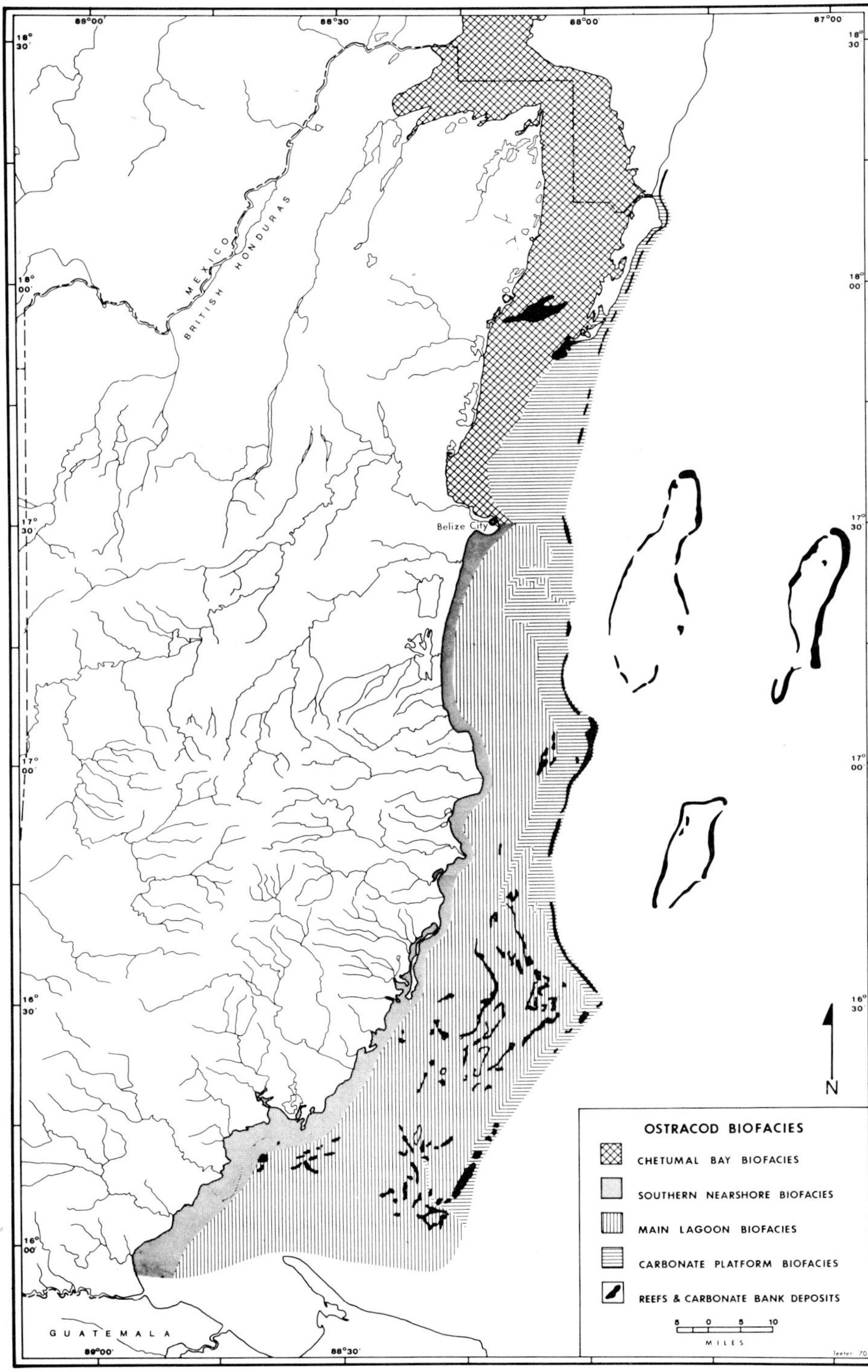

FIG. 2—Ostracod biofacies, Belize.

Toward the head of the bay where salinity is constantly low owing to freshwater river influx, a subfacies is present which is characterized by *Dolerocypria inopinata* Klie?, *Loxoconcha levis* Brady, *Cytheromorpha apheles* van den Bold, and *Cytherura* sp. ex. gr. *C. johnsoni* Mincher. *Cytheromorpha apheles* is the most restricted of these, occurring only in mainland coastal lagoons and near the mouths of rivers, where it is commonly associated with charophytes and a freshwater cyprid ostracod.

Increased salinity near the mouth of the bay and in the shallow lagoons on the west side of Ambergris Cay is accompanied by another change in the ostracod faunas. As the salinity increases and tends to become more stable, the number of species increases. Most of these species are also present in the carbonate-platform biofacies and commonly include *Perissocytheridea bisulcata* Teeter, n. sp., *Quadracythere producta* (Brady), *Actinocythereis bahamensis* (Brady), *Cytherella arostrata* Kornicker, *Xestoleberis* spp., *Loxoconcha* spp., and *Microcythere* sp. cf. *M. inflexa* Mueller. Of these species only *Perissocytheridea bisulcata* is restricted to the western bays of Ambergris Cay.

The effect of the pass through Ambergris Cay at San Pedro is apparent, for although the lower-bay fauna predominates, bairdiids, *Loxoconcha fischeri* (Brady), and *Callistocythere cranekeyensis* (Puri), which are characteristic of the reef, have at times migrated into the bay.

The boundary between the Chetumal Bay biofacies and the carbonate-platform biofacies, though gradational, is defined on the presence of bairdiids in the more diverse carbonate-platform faunas, and the boundary closely parallels the bottom-water isohalines.

Table 2 records the ostracod diversity and dominance of the various biofacies. Diversity—the number of species per station—exhibits a regular increase southward in Chetumal Bay, paralleling increasing salinity. In a stress environment, diversity is low, increasing with the stability of the environment. Dominance—the percentage abundance of the commonest species—is high in Chetumal Bay, although inexplicably variable. In a stress environment, one species commonly accounts for a disproportionately large percentage of the population. This dominance decreases as the environment becomes more stable or favorable for more species. Thus, diversity and dominance are inversely related.

Southern-Nearshore Biofacies

This biofacies extends, probably intermittently, southward along the coast from the northeast side of the Belize delta, and is similar to the upper Chetumal Bay biofacies in faunal constituents and low, variable salinity. The southern-nearshore biofacies grades eastward with increasing salinity and depth into the main-lagoon biofacies. The following species are characteristic of this biofacies: *Haplocytheridea placentiaensis* Teeter, n. sp., *Perissocytheridea subrugosa* (Brady), *Megacythere johnsoni* (Mincher)?, and *Cytherura* sp. ex. gr. *C. johnsoni* Mincher. *Cytheromorpha apheles* van den Bold is also present, although scarce, in areas of lowest salinity.

The diagnostic species of this biofacies is *Haplocytheridea placentiaensis*. Only in the vicinity of Belize have both this species and *H. kirkbii* been found associated. The writer suspects that the controlling factor in the distribution of these two species may be related to the water chemistry of the rivers draining northern and southern Belize. The rivers of northern Belize drain flat-lying carbonate sediments, thus introducing a superabundance of calcium ions to Chetumal Bay. The southern rivers drain Paleozoic noncarbonate rocks and contribute large amounts of coarse to fine siliceous clastic sediment to the southern coastal shelf. The Belize River drains both the northern carbonate lowlands and the

TABLE 2. OSTRACOD POPULATION STATISTICS[1]

Biofacies	*Salinity Range*	*Average Salinity*	*Depth Range (ft)*	*Average Depth*	*Diversity Range*	*Average Diversity*	*Average Dominance*
Upper Chetumal Bay (13)	1-21	15	3-10	7	1-10	7	70 (4)
Middle Chetumal Bay (27)	5-33.5	23	1-9	5	2-16	9	37 (7)
Lower Chetumal Bay (11)	27-36	32	4-17	10	1-24	13	71 (3)
Southern nearshore (10)	1.6-33	20	3-18	11	1-14	8	40 (7)
Main lagoon (32)	27.5-36	34	13-196	62	1-20	11	43 (12)
Barrier rim (9)	18.2-37	32.6	3-17	7	5-33	23	18 (9)
Reef apron (24)	29.3-37	34.4	3-25	11	1-37	15	25 (7)
Thalassia shoals (29)	18.2-37	31	2-23	9	3-36	19	19 (10)

[1]Figures in parentheses indicate the numbers of stations involved in calculations.

southern noncarbonate Maya Mountains. Interestingly, *H. kirkbii* (Brady) possesses a much more heavily calcified carapace than *H. placentiaensis*.

A similar explanation may account for the distribution of *Megacythere stephensoni* (Puri) and *M. johnsoni* (Mincher). The substrates in both areas differ (predominantly carbonate in the north; siliceous clastic in the south), although the texture, rather than the chemistry, of the substrate is probably of more significance to benthonic organisms.

Several other species, found most abundantly in the shallower parts of the main-lagoon biofacies, occur at most of the southern-nearshore stations. These include *Pumilocytheridea sandbergi* van den Bold, *Costa variabilocostata recticostata* van den Bold, *Cytherura maya* Teeter, n. sp., *Loxoconcha avellana* (Brady), *Caribbella puseyi* Teeter, n. sp., *Pellucistoma howei* Coryell and Fields, *Neomonoceratina mediterranea* (Ruggieri), *Orionina serrulata* (Brady), and *Reussicythere reussi* van den Bold.

Although the water depth of the southern-nearshore biofacies is variable, the maximum is probably 20 ft (6.1 m). As in Chetumal Bay, the diversity is low and the dominance high, reflecting the low, presumably variable salinity of this environment.

Main-Lagoon Biofacies

In the main lagoon, with only one narrow channel to the north and a wide connection with the Caribbean to the south, the bottom water gives no evidence of stagnation. However, the presence of pyrite within the uppermost centimeter of many samples from this biofacies suggests rapid depletion of oxygen within the sediment. Low oxygen content of these sediments almost certainly exerts some effect on the ostracod infauna, permitting only those burrowing species with reduced oxygen tolerance to exist in this environment. Species inhabiting the sediment surface may not be subject to oxygen limitation.

With increasing water depth and distance from land, the texture of the substrate becomes finer grained. At many of the deepest stations, a soupy gray-green mud is present. Several of the ostracod species exhibit adaptations to this plastic substrate. Some of the adaptations, presumably advantageous to sediment-surface-dwelling ostracods, are the development of ventrolateral alae as in *Pterygocythereis* and *Cytheropteron*, and spines as in *Cativella*. Small size, as in *Munseyella*, may also represent an adaptation to decrease sinkage into soft substrates.

The chemistry of the substrate has no apparent effect on ostracod distributions. Two stations, 124f and 135f, 94 and 102 ft (29 and 31 m) deep, respectively, are far enough from land to receive essentially no land-derived sediments. They are also far enough from neighboring shoals to receive only carbonate muds. Both stations support similar faunas characteristic of the main-lagoon biofacies.

The main-lagoon biofacies is subject to contamination from two other areas. Species typically restricted to the carbonate-platform biofacies have been swept by currents into the deeper waters of the main lagoon along its eastern boundary and throughout much of its southernmost extent, where numerous carbonate atolls are dispersed from the dissected platform almost to the mainland coast. With increasing distance from the platform, the proportion of disarticulated carapaces increases. Juvenile instars exhibit a wider distribution and greater disarticulation than adults. The presence of shallow-water carbonate-platform ostracods in the deeper water assemblages of the main lagoon has obvious paleoecologic applications.

Several samples from the main lagoon, notably near the Belize delta, contained traces of nearshore faunal elements. Most of these were disarticulated and many were blackened, indicating burial under reducing conditions. The writer suspects that they represent relict assemblages deposited during lower stands of sea level and subsequently disinterred by burrowing organisms or wave action.

Complete carapaces of several species occur quite consistently in depths shallower than approximately 80 ft (24 m). Most of these species occur consistently in very nearshore areas, as recorded above.

One species, *Munseyella* sp. cf. *M. minuta*, occurs consistently at depths in excess of 100 ft (30 m) and is scarce at shallower depths. The lower depth limit of this species is unknown, but it has been recorded as deep as 170 ft (52 m).

Ostracod diversity within the main-lagoon biofacies is surprisingly low. Although comparative information on ostracod diversity elsewhere is generally lacking, analysis of data provided by Puri et al (1964) from the Gulf of Naples suggests a maximum diversity of 80-100 species at depths comparable to those of the main lagoon. Maximum diversity—the total number of species—within the main lagoon is about 30 and the average, recorded in Table 2, is only 11. Dominance is correspondingly rather high. The reason for the anomalously low diversity is unknown, but it may be the result of poor circulation of water between the main lagoon and the Caribbean caused by the intervening shallow-water carbonate platform.

Carbonate-Platform Biofacies

This biofacies borders the continental shelf and extends from Ambergris Cay on the north to the Gulf of Honduras on the south. The width of the biofacies varies from less than a mile off Ambergris Cay to several miles throughout most of the area. From approximately the latitude of Placentia, south to the Gulf of Honduras, the western side of the platform is dissected and numerous shallow, carbonate shoals within the main lagoon complicate the areal distribution of the platform and main-lagoon biofacies. Thus, no attempt has been made to separate these in Figure 2.

Water depth ranges from a few inches to a maximum of approximately 30 ft (9.1 m). Because this biofacies is comparatively far removed from land, salinity is normal and exhibits little fluctuation. Only in the Port of Honduras, where carbonate shoals lie close to shore, must salinities be appreciably lowered by the influx of fresh water. Presumably, such influxes are of sufficiently short duration to prevent annihilation of organisms dwelling on carbonate shoals. Good water circulation reduces thermal fluctuation, although the fluctuation must be somewhat greater here than in the main-lagoon biofacies because of the shallower water.

The carbonate platform may be subdivided on the basis of sediment texture and geography. The barrier rim is the narrow seaward edge of growing reef, bordered on its landward side by the reef flat, an extremely shallow zone of rubble derived from the living coral and associated organisms and deposited during times of storm. Finer reef-derived material is swept over the reef flat and deposited in the gradually deepening water to the west as the reef apron. Small patch reefs dot the reef apron. In relatively quiet, nearshore areas, or in the deeper waters of the platform, grow dense stands of the marine grass *Thalassia*. This grass slows water motion sufficiently to permit deposition of the fine carbonate particles in suspension.

Water turbulence is greatest along the barrier rim and reef flat where waves driven by the northeast trade winds across the Caribbean Sea first encounter shallow water. This turbulence is sufficient to winnow the finest carbonate particles and deposit them farther to the west.

The carbonate-platform biofacies, with its variety of substrates, diverse faunal constitution, and constancy of environment, offers abundant niches. Thus, the ostracod diversity of this biofacies is not surprising. Of the more than 50 common species found in this biofacies, most are restricted to it and occur only as allochthonous elements. Especially indicative of this biofacies are the variety of bairdiid genera, including *Neonesidea, Paranesidea, Bairdoppilata,* and *Triebelina*; the abundance of coarsely ornamented *Loxoconcha*; and the variety of *Caudites* species. Other diagnostic species include *Aglaiocypris croneisi* Teeter, n. sp., *Callistocythere cranekeyensis* (Puri), *Morkhovenia inconspicua* (Brady), *Hemicythere* (?) *auricula* Teeter, n. sp., *Quadracythere lankfordi* Teeter, n. sp., *Orionina ebanksi* Teeter, n. sp., *O. bradyi* van den Bold, *Jugosocythereis pannosa* (Brady), *Hermanites* spp., *Neocaudites scottae* Teeter, n. sp., *Puriana convoluta* Teeter, n. sp., *Protocytheretta pumicosa* (Brady), *Cushmanidea cristifera* Teeter, n. sp., *Cytherura* sp. aff. *C. cornuta* Brady, *Hemicytherura bradyi* (Puri), *Paracytheridea* spp., and *Xestoleberis* spp.

Based on data recorded in Table 1 and on qualitative data, certain species appear to be restricted to some of the subdivisions of the platform. The following species are found consistently on the barrier rim, reef flat, and apron in areas generally devoid of vegetation, particularly *Thalassia*: *Bairdoppilata (Glyptobairdia) coronata* (Brady), *B. (G.) trinodosa* Teeter, n. sp., *Triebelina rugosa* Allison and Holden, *Hemicythere? auricula* Teeter, n. sp., *Quadracythere lankfordi* Teeter, n. sp., *Loxoconcha magna* Teeter, n. sp., *L. dorsotuberculata* (Brady), *L. oculocrista* Teeter, n. sp., *Xestoleberis* sp. 1, *X.* sp. 2, and *Saipanetta mckenziei* Teeter, n. sp. *Cytherura* sp. aff. *C. cornuta* Brady and *Paracytheridea* spp. are somewhat more questionably restricted to these areas lacking in vegetation. Of the named species, the first five typically are restricted to the barrier rim and reef flat.

Two taxa, *Loxoconcha fischeri minima* Teeter, n. ssp., and *L. suboculocrista* Teeter, n. sp., occur consistently with *Thalassia*. *Hemicytherura cranekeyensis* Puri may also be restricted to areas of this marine grass.

The ostracod diversity attains a maximum and, inversely, the dominance shows a marked decrease in the shallow, stable, carbonate-platform environment. Especially striking is the very high diversity of the barrier rim, attributable perhaps to the numerous niches available in complex organic reef associations. The decreased dominance reflects the rising competition between members of the association.

Cosmopolitan Species

Some species apparently may live consistently in more than one biofacies, although few do so in numbers great enough to confuse an environmental interpretation based on the whole fauna.

Thus, *Quadracythere producta* (Brady), *Actinocythereis bahamensis* (Brady), and *Cytherella arostrata* Kornicker, which are found in the more saline areas of the Chetumal Bay biofacies, appear periodically in samples from the platform. *Xestoleberis punctata* Tressler, *X.* sp. 1 and *X.* sp. 4, have a similar distribution. The only apparent difference between the environments of these two biofacies is the greater variability of salinity in Chetumal Bay.

Triangulocypris laeva (Puri) and *Pumilocytheridea sandbergi* van den Bold have a presumably greater tolerance for variable salinity, as these two species occur in the southern-nearshore, main-lagoon, and carbonate-platform biofacies.

Microcythere sp. cf. *M. inflexa* Mueller and *Paradoxostoma* spp. exhibit even wider distribution—they occur in all biofacies. The former species has been observed on *Thalassia* and on floating vegetation in Belize, a fact which may account for its wide distribution. *Paradoxostoma* is also known to be a plant dweller.

Several other species are found beyond their normal environmental range only rarely. Some of these are indicated in Table 1.

TABLE 3. STRATIGRAPHIC RANGES OF HOLOCENE BELIZE OSTRACODA WITH FOSSIL REPRESENTATIVES

Species	*Range*				
	Eoc.	*Olig.*	*Mio.*	*Plio.*	*Pleist.*
CHETUMAL BAY AND SOUTHERN-NEARSHORE BIOFACIES					
Perissocytheridea subrugosa				—	—
Cytherura johnsoni			—	—	—
Loxoconcha levis				—	—
Cytheromorpha apheles			—	—	—
Megacythere johnsoni			—	—	—
M. stephensoni			—	—	—
MAIN-LAGOON BIOFACIES					
Pumilocytheridea sandbergi			—	—	—
Orionina serrulata			—	—	—
Costa variabilocostata recticostata			—	—	—
Cativella pulleyi			—	—	—
Cytherura swaini			—	—	—
Kangarina ancycla			—	—	—
Cytheropteron subreticulatum			—	—	—
Loxoconcha avellana			—	—	—
Pellucistoma howei			—	—	—
Cytherella polita			—	—	—
CARBONATE-PLATFORM BIOFACIES					
Neonesidea longisetosa			—	—	—
Triebelina sertata				—	—
Keijia demissa	—	—	—	—	—
Mutilus confragosus			—	—	—
Quadracythere producta			—	—	—
Caudites nipeensis			—	—	—
C. angulata			—	—	—
Hermanites hornibrooki			—	—	—
Hemicytherura cranekeyensis			—	—	—
Paracytheridea tschoppi			—	—	—
Loxoconcha wilberti			—	—	—
L. fischeri			—	—	—
Cytherelloidea umbonata			—	—	—

STRATIGRAPHIC DISTRIBUTION

Several species of ostracods currently living in Belize range well back into the Tertiary (Table 3). Most species found in the Tertiary are restricted to the Caribbean or Atlantic and Gulf coastal plain; however, two species, *Triebelina sertata* and *Keijia demissa*, are present as fossils only in the Indo-Pacific region. Several other fossil species are strikingly similar to, and undoubtedly closely related to, several Holocene Belize species. These include *Quadracythere bichensis* (van den Bold) and *Jugosocythereis vicksburgensis* (Howe and Law). The modern relatives of these two species are *Quadracythere lankfordi* and *Jugosocythereis pannosa*, respectively.

Ecologic information on these long-ranging taxa provided by this study may aid in the paleoenvironmental interpretation of fossil assemblages containing these species.

CONCLUSIONS

The principal result of this study is the recognition of three major ostracod assemblages. These include:

1. A nearshore, shallow-water, low- to variable-salinity assemblage —Minor faunal differences, presumably related to the variable water chemistry of rivers draining northern and southern Belize permit the recognition of two subassemblages. In addition to their characteristic species, these two biofacies may be identified by low taxonomic diversity and high dominance.
2. An assemblage of species characteristic of essentially normal-marine salinity and inner to middle neritic depths—With increasing depth, the ostracod fauna exhibits a corresponding change. Although the diversity is greater than in the shallow-water, low-salinity assemblages, it is considerably less than might be anticipated on the basis of comparative studies of faunas elsewhere at similar depths—a fact perhaps explained by poor water circulation.
3. An offshore, shallow-water assemblage of essentially normal-marine salinity, characteristically associated with carbonate deposits— This assemblage, distinguished by its abundance of bairdiid and strongly ornamented *Loxoconcha* species, possesses remarkably high faunal diversity and low dominance. Different species and genera appear to be diagnostic of various substrates—such as reefs, sands, and marine grass—within this biofacies.

The fact that most species are very clearly restricted to a certain assemblage permits the recognition of transported faunas. Paleoecologically significant is the mixing of shallow-water species swept by currents from the steep-sided carbonate platforms into surrounding deeper water associations.

The fact that more than one fourth of the total species are present as fossils, many ranging back to the Miocene, enhances their value as paleoecologic indicators.

SYSTEMATIC DESCRIPTIONS

The following includes the taxonomic descriptions and discussions of 105 species and 1 subspecies of Ostracoda. These represent only the commonest taxa; at least 30 additional species are too scarce to warrant inclusion. Among these are included 33 new species, 1 new subspecies, and 4 new genera.

Holotypes and paratypes have been deposited in the Henry V. Howe collection at Louisiana State University, and additional paratypes and hypotypes in the micropaleontologic collection of Rice University.

All photographed specimens, except holotypes, were prepared by calcining, staining with Merck dye 1315 (light yellowish green), and immersing in glycerine. Holotypes were not calcined.

Subclass OSTRACODA Latreille, 1806
Order PODOCOPIDA Mueller, 1894
Suborder PODOCOPINA Sars, 1866
Superfamily BAIRDIACEA Sars, 1888
Family BAIRDIIDAE Sars, 1888
Subfamily BAIRDIINAE Sars, 1888

Genus NEONESIDEA Maddocks, 1969
Neonesidea longisetosa (Brady), 1902
Figures 3a, 4a

Bairdia longisetosa Brady, 1902, p. 197, Pl. 35 (Figs. 8, 9); van den Bold, 1966f, p. 45, Pl. 2 (Figs. 7a, b); 1968, p. 48, Pl. 2 (Figs. 3a, b).
Bairdia crosskeyiana Brady. Puri, 1960, p. 130, Pl. 6 (Figs. 11, 12).
Bairdia gerda Benson and Coleman, 1963, p. 19-20, Pl. 1 (Figs. 14-16); Fig. 8; Hulings, 1967, p. 637, Figs. 2a, 3a.
Neonesidea gerda (Benson and Coleman). Maddocks, 1969, p. 24-26, Figs. 7A-K.

Diagnosis—Characterized by fine spines along posteroventral margin of left valve, four elongate adductor muscle scars, and central, vertically elongate, opaque area.

Dimensions—Female: length 1.22 mm, height 0.72 mm. Male: length 1.14 mm, height 0.63 mm.

Distribution—Originally described by Brady from St. Thomas, West Indies, it has been found in southern Florida, the Bahamas, and Colón harbor, Panama. It has also been found in the upper Miocene of the northern Dominican Republic. In Belize it occurs in the carbonate-platform biofacies.

Genus PARANESIDEA Maddocks, 1969
Paranesidea bensoni Teeter, n. sp.
Figures 3b-d, 4b

Bairdia sp. cf. *B. bradyi* van den Bold (in part?). Benson and Coleman, 1963, p. 18-19, Pl. 2 (Figs. 1-3); Fig. 7.

Diagnosis—Characterized by the presence of a subhorizontal swelling at the anterior and posterior angulations of both valves.

Description—Dorsal margin with a straight, horizontal median part and equally inclined anterior and posterior parts, terminating abruptly at the anterior and posterior angulations. Anterior margin straight, vertically inclined dorsally, curving ventrally into straight ventral margin. Evenly rounded posteriorly. Antero- and posteroventral margins in left valve with numerous denticles, in right valve with striate frills. Surface of valves fairly coarsely punctate. Anterior and posterior angulations in right valve with subhorizontal swelling. In transmitted light, poorly defined opaque areas covering muscle-scar area.

Dimensions—Holotype: length 1.00 mm, height 0.64 mm.

Type locality—Station 314, in 4 ft (1.2 m) of water near Tobacco Cay.

Type specimens—Holotype: H.V.H. 8976. Paratype: H.V.H. 8977.

Remarks—This species is named in honor of R. H. Benson of the Smithsonian Institute.

In their 1963 paper, Benson and Coleman illustrated several views of a species of *Bairdia* they conferred to *B. bradyi* van den Bold. There appear to be two species illustrated, of which the one in their Figure 7 is probably conspecific with the species described above. The other species may be conspecific with *B. bradyi.*

An externally similar species, *Bairdia teeteri,* described by Allison and Holden (1971), differs from *P. bensoni* in the possession of hinge denticles.

Distribution—The writer has observed this species occasionally in samples from the Florida reef tract and the Bahamas. The Belize specimens typically are present in the carbonate-platform biofacies.

Paranesidea gigacantha (Kornicker), 1961
Figures 3e, 4c

Bairdia gigacantha Kornicker, 1961, p. 62-63, Pl. 1 (Figs. 2a-c); Figs. 6A-J, 10A, D.
Bairdia victrix Brady. Benson and Coleman, 1963, p. 17-18, Pl. 2 (Figs. 4-10); Fig. 6. (not) *Bairdia victrix* Brady, 1869, p. 152, Pl. 18 (Figs. 17, 18).
Paranesidea gigacantha (Kornicker). Maddocks, 1969, p. 45-46, Fig. 20.

Diagnosis—Characterized by variously shaped opaque areas.

Dimensions—Length 1.09 mm, height 0.72 mm.

Distribution—This species is present in the Bahamas and the Florida reef tract. In Belize it is found on the carbonate platform and is apparently commonest in *Thalassia* areas.

Genus BAIRDOPPILATA Coryell, Sample and Jennings, 1935
Bairdoppilata (Bairdoppilata) cushmani (Tressler), 1949
Figures 3f, 4d

Nesidea cushmani Tressler, 1949, p. 342, Figs. 4-8.
Bairdoppilata carinata Kornicker, 1961, p. 66-67, Pl. 1 (Figs. 5a-e); Text-figs. 9A-J, 10B, C, E.
Bairdoppilata triangulata Edwards. Benson and Coleman, 1963, p. 20-21, Pl. 3 (Figs. 1-3); Fig. 9.

⟶

FIG. 3

a. *Neonesidea longisetosa* (Brady). Male, complete carapace, right valve, station 340, x38.
b. *Paranesidea bensoni* Teeter, n. sp. Holotype (H.V.H. 8976), complete carapace, right valve, station 314, x43.
c. *P. bensoni* Teeter, n. sp. Paratype (H.V.H. 8977), complete carapace, right valve, station 314, x43.
d. *P. bensoni* Teeter, n. sp. Paratype, complete carapace, left valve, station 314, x43.
e. *Paranesidea gigacantha* (Kornicker). Complete carapace, right valve, station 323, x38.
f. *Bairdoppilata (Bairdoppilata) cushmani* (Tressler). Female, complete carapace, right valve, station 33, x38.
g. *Bairdoppilata (Bairdoppilata) fasciata* (Brady). Complete carapace, right valve, station 33, x43.
h. *Bairdoppilata (Glyptobairdia) coronata* (Brady). Complete carapace, right valve, station 325, x55.
i. *Bairdoppilata (Glyptobairdia) trinodosa* Teeter, n. sp. Holotype (H.V.H. 8978), complete carapace, right valve, station 320, x60.
j. *B. (G.) trinodosa* Teeter, n. sp. Paratype (H.V.H. 8979), complete carapace, right valve, station 320, x60.
k. *B. (G.) trinodosa* Teeter, n. sp. Paratype, complete left valve, station 320, x60.
l. *Triebelina sertata* Triebel. Complete carapace, right valve, station 325, x77.
m. *Triebelina rugosa* Allison and Holden. Complete carapace, right valve, station 151, x64.
n. *T. rugosa* Allison and Holden. Complete carapace, left valve, station 151, x64.
o. *T. rugosa* Allison and Holden. Complete carapace, right valve, station 151, x64.

Bairdoppilata (Bairdoppilata) cushmani (Tressler). Maddocks, 1969, p. 68-71, Figs. 34, 35h-m.

Diagnosis—Characterized by centrally located, ovoid, opaque area and marginal, lobate, opaque areas. Sexual dimorphism pronounced.

Dimensions—Female: length 1.15 mm, height 0.72 mm. Male: length 1.06 mm, height 0.64 mm.

Distribution—Originally described from Tortugas (Florida), this species was collected, living, by Kornicker, at depths of 1-5 m on sandy bottoms near Bimini. In Belize it occurs in the carbonate-platform biofacies.

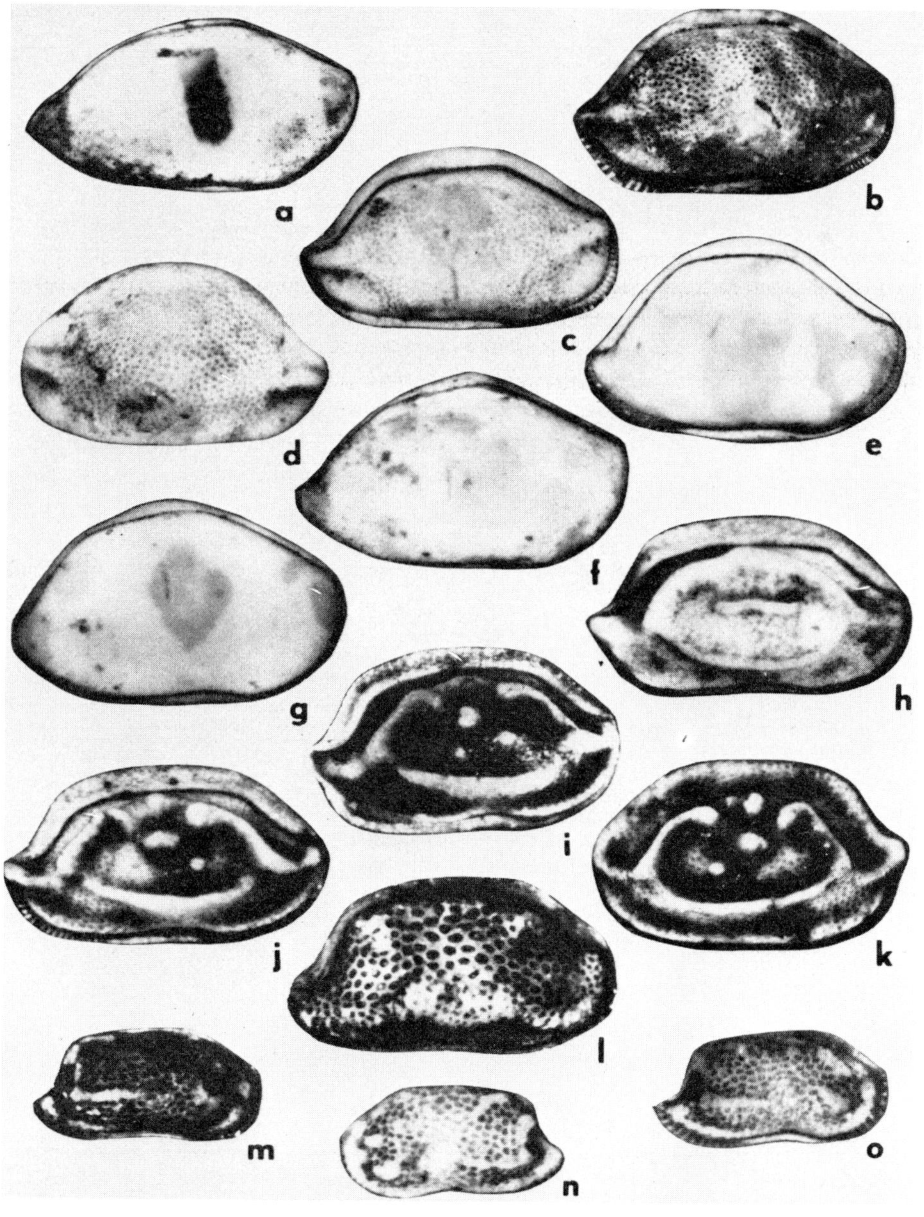

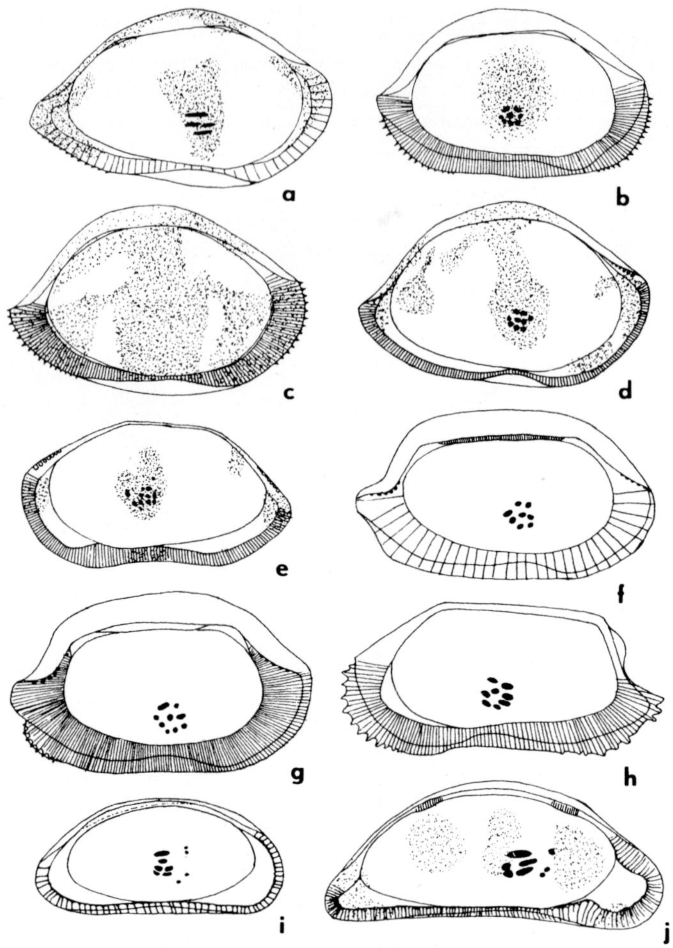

FIG. 4–Internal features.

a. *Neonesidea longisetosa* (Brady). Female, left valve, station 340, x31.
b. *Paranesidea bensoni* Teeter, n. sp. Paratype, left valve, station 314, x35.
c. *Paranesidea gigacantha* (Kornicker). Left valve, station 323, x35.
d. *Bairdoppilata (Bairdoppilata) cushmani* (Tressler). Female, left valve, station 33, x32.
e. *Bairdoppilata (Bairdoppilata) fasciata* (Brady). Male, right valve, station 33, x32.
f. *Bairdoppilata (Glyptobairdia) coronata* (Brady). Left valve, station 319, x51.
g. *Bairdoppilata (Glyptobairdia) trinodosa* Teeter, n. sp. Paratype, left valve, station 313, x54.
h. *Triebelina rugosa* Allison and Holden. Paratype, right valve, station 151, x90.
i. *Triangulocypris laeva* (Puri). Left valve, station 32-60, x38.
j. *Macrocyprina propinqua* Triebel. Left valve, station 97, x42.

Bairdoppilata (Bairdoppilata) fasciata (Brady), 1870
Figures 3g, 4e

Bairdia fasciata Brady, 1870, p. 193, Pl. 19 (Figs. 20, 21).

Diagnosis—Characterized by its evenly rounded anterior margin, low dorsal hump, and consistently arranged opaque areas. Sexual dimorphism pronounced.

Dimensions—Female: length 1.15 mm, height 0.72 mm. Male: length 1.07 mm, height 0.62 mm.

Remarks—This species possesses a remarkably consistent arrangement of opaque areas, including an ovate central spot and variously lobed, marginal patches. The opaque areas agree closely with Brady's original figure, as does the low dorsal hump.

Distribution—Originally described from St. Vincent in the Caribbean, this species is also found in Belize in the shallow water of carbonate shoals.

Bairdoppilata (Glyptobairdia) coronata (Brady), 1870
Figures 3h, 4f

Bairdia coronata Brady, 1870, p. 243, Pl. 32 (Fig. 9).
Glyptobairdia bermudezi Stephenson, 1946, p. 346, Pl. 42 (Figs. 1-3).
Triebelina coronata (Brady). Stephenson, 1947, p. 578; van Morkhoven, 1958, p. 366-368, Pl. 46 (Figs. 1-6); Puri, 1960, p. 131-132, Pl. 6 (Figs. 1, 2); Rome, 1960, p. 1-14, Figs. 1-4; Benda and Puri, 1962, p. 324, Pl. 5 (Figs. 1-3); van Morkhoven, 1963, pt. II, p. 36-37, Pl. 44a, b.
Bairdoppilata (Glyptobairdia) coronata (Brady). Maddocks, 1969, p. 84-86, Fig. 44.

Diagnosis—Both valves characterized by a coronalike rim, with ridge-like extensions toward the anterior and posterior angulations and surrounding a horizontal bar.

Dimensions—Length 0.75 mm, height 0.44 mm.

Distribution—This species occurs throughout the Caribbean and in the Florida Keys. In Belize, it is restricted to the barrier rim.

Bairdoppilata (Glyptobairdia) trinodosa Teeter, n. sp.
Figures 3i-k, 4g

Diagnosis—Both valves characterized by a coronalike rim, which is discontinuous dorsally and surrounds two prominent central nodes, with ridgelike extensions toward the anterior and posterior angulations.

Description—Carapace subhexagonal in lateral and dorsal view. Left valve strongly overlapping right, dorsally and ventrally. Both valves with prominent, submarginal, coronalike rim, dissected dorsally leaving a central, isolated node. Rim tending to be discontinuous antero- and posteroventrally in right valve. Prominent ridges extending horizontally from the rim to the anterior and posterior angulations. Two well-defined nodes located within submarginal rim. One directly below node produced by dissection of rim, second slightly anterior to and below the first. Surface with well-defined punctae.

Dimorphism apparent in left valve; posteroventral part of submarginal rim depressed in males, continuous in females.

Hinge characteristic of the subgenus, as defined by Maddocks (1969). Left valve with finely crenulate median bar and two terminal sockets; accommodation groove pronounced. Hinge in right valve with crenulate median groove and smooth terminal teeth; well-defined denticles which line the antero- and posteroventral slopes below the terminal teeth and fit into corresponding depressions in the left valve.

Central muscle scar with nine elongate secondary scars.

Dimensions—Holotype: length 0.70 mm, height 0.44 mm.

Type locality—Station 320, on reef apron near South Water Cay.

Type specimens—Holotype: H.V.H. 8978. Paratype: H.V.H. 8979.

Remarks—Specific name indicates the presence of three well-developed nodes.

This species shows affinities such as surface ornamentation and development of submarginal rim with *Bairdoppilata (Glyptobairdia) coronata* (Brady).

The presumed juveniles of either this species or *B. (G.) coronata* indicate a remarkable ontogeny. They have the bairdian outline, show pronounced dorsal overlap, and have a punctate surface; but they lack any indication of the prominent ridges common to the adult.

Distribution—This species is almost completely restricted to the barrier rim.

Genus TRIEBELINA Bold, 1946

Triebelina sertata Triebel, 1948
Figure 3*l*

Triebelina indopacifica van den Bold (in part), 1946, p. 74.
Triebelina sp. cf. *T. cubensis* van den Bold. Kingma, 1948, p. 69.
Triebelina sertata Triebel, 1948, p. 19-22, Figs. 1a, b, 2a, b; Keij, 1953, p. 158, Pl. 1 (Fig. 5); Puri, 1960, p. 132, Pl. 6 (Figs. 3, 4); Maddocks, 1969, p. 61-63, Pl. 2 (Figs. 9, 10); Figs. 31, 32b-e.

Diagnosis—Characterized by bituberculate ventral ridge in both valves and pronounced horizontal medial bar in left valve.

Dimensions—Length 0.56 mm, height 0.31 mm.

Remarks—Although only one specimen of this ostracod has been found in Belize, it is recorded here as further proof of its extremely widespread distribution.

Distribution—This species ranges from the lower Pliocene to the Holocene. It is found as a fossil, however, only in northern Sumatra (lower Pliocene) and on the island of Seram (Pleistocene). It was originally described from the Holocene of the Red Sea and has been reported subsequently from the Lesser Sunda Islands, Kai Island, and the Celebes. Puri records this species as common at Molasses Reef, Florida, and Maddocks has found it in northern Madagascar. The writer has observed it also in coral reef material from near Sorong, New Guinea. In Belize, the single specimen was found at station 325 in a coral sand at a depth of 5 ft (1.5 m).

Although usually relatively scarce, this species is an excellent reef indicator. Keij, however, recorded it from depths of 372-3,221 m but found only single valves; and, because all stations are close to land, it may be concluded that there has been some seaward transportation.

Triebelina rugosa Allison and Holden, 1971
Figures 3m-o, 4h

(not) *Triebelina bradyi* Triebel. Puri, 1960, p. 132, Pl. 6 (Figs. 7, 8).
Triebelina rugosa Allison and Holden, 1971, p. 179-180, Fig. 9.

Diagnosis—Characterized by diminutive size and distribution of numerous ridges and nodes.

Description—In lateral view, small, elongate, subhexagonal, left valve overlapping right slightly on all margins. Both valves with approximately 12 antero- and posteroventral denticles. In dorsal view, anterior rim prominent, truncate posteriorly.

Anterior marginal rim in right valve terminating at anterior angulation in a prominent node. Postdorsal from the node are three more nodes of which the first two commonly coalesce. Parallel with and ventral to the coalescing nodes, short prominent ridge inclined anteroventrally. Posteriorly, another ridge, C-shaped in outline, extending a short distance along the dorsal margin, approximately two thirds of the way ventrally along the posterior truncation and paralleling the ventral margin, to about midvalve. This ridge nodose in some specimens. Left valve exhibiting a similar arrangement of nodes and ridges. Surface of the valves reticulate. Apparently smooth hinge bar in left valve and corresponding groove in right.

Dimensions—Length 0.46 mm, height 0.22 mm.

Remarks—Although of approximately the same size, the numerous ridges and nodes of *T. rugosa* serve to distinguish it from the relatively unornamented *T. bradyi*.

Distribution—This species has been recorded from Molasses Reef in the Florida Keys. In Belize, it is found in the shallow waters of the carbonate platform.

Subfamily BYTHOCYPRIDINAE Maddocks, 1969

Genus TRIANGULOCYPRIS Teeter, n. gen.

Diagnosis—A bairdiid genus characterized by its triangular outline in lateral view.

Description—Carapace triangular in outline. Ventral margin straight to slightly concave; anterior broadly rounded, posterior bluntly pointed; antero- and posterodorsal margins approximately equally sloping. Greatest height attained about two fifths of the length from the anterior margin. Left valve larger, overlapping the right on all margins. Surface smooth.

Hinge in right valve with raised dorsal edge which engages groove in left valve. Inner lamella broad, exhibiting relatively small attachment area with outer lamella. Large anterior and posterior vestibula. Central muscle scar with six (more or less) secondary scars. Normal pores common, open. Radial pore canals common, straight, and simple.

Type species—*Triangulocypris laeva* (Puri) 1960.

Remarks—Generic name refers to the roughly triangular lateral outline.

This genus is similar to *Bythocypris*, differing from it mainly in having a triangulate rather than subquadrate lateral outline, and possessing a thicker carapace. The muscle scars differ only slightly in arrangement and number. Both genera possess fairly large vestibula.

Some other species possessing a very similar outline and presumably belonging to this genus are *T. elongata* (Brady), *T. gibonensis* (Howe and Chambers), *T. goodlandensis* (Alexander), and *T. howei* (Puri).

This genus ranges from Upper Cretaceous to Holocene and seems to be typically a shallow-water genus, in contrast to the deeper water genus *Bythocypris*.

Triangulocypris laeva (Puri), 1960
Figures 4i, 5a

Bythocypris laeva Puri, 1960, p. 132, Pl. 5 (Figs. 5, 6); Pl. 6 (Figs. 16, 17).

Paracypris? sablensis Benson and Coleman, 1963, p. 16-17, Pl. 1 (Figs. 11-13); Fig. 5.

?Bythocypris keiji van den Bold, 1963c, p. 374-375, Pl. 1 (Fig. 72b); Pl. 12 (Fig. 10).

Diagnosis—Characterized by the dorsal angulation and the slight postdorsal break in slope.

Dimensions—Length 0.84 mm, height 0.40 mm.

Remarks—Topotype material from Florida was compared with the specimens from Belize and found to be conspecific.

In their Figure 5, Benson and Coleman have apparently missed the two lowermost muscle scars in the muscle pattern. These seem to be particularly difficult to see in the Florida material.

B. keiji van den Bold is questionably included in this species. The only obvious difference is one of size: the Holocene material has slightly greater length.

Some variation in the development of the dorsal angulation was observed. Certain specimens (males?) seemed lower and had a considerably more pronounced angulation.

Distribution—This species is found mainly in southern Florida, but Puri has also found it in northern Florida at one locality. *B. keiji* was described from the upper Miocene Melajo clay of Trinidad. In Belize, this species is most commonly found in the deeper waters of the main lagoon, although it is also found commonly on the carbonate platform.

Family MACROCYPRIDIDAE Mueller, 1912

Genus MACROCYPRINA Triebel, 1960

Macrocyprina propinqua Triebel, 1960

Figures 4j, 5b

?Cytherideis maculata Brady (in part), 1866, p. 367, Pl. 57 (Figs. 12a, b).

Macrocyprina propinqua Triebel, 1960, p. 118-123, Pl. 14 (Figs. 7-10); Pl. 15 (Figs. 11-17); Pl. 16 (Figs. 18-24).

Diagnosis—Characterized by its large massive valves, each with three, vertically oriented, opaque areas. Ventral margin of right valve indented slightly, anteriorly.

Remarks—Brady, in 1866, described *Cytherideis maculata* from Australia, the West Indies, and Turks Island. On the basis of its wide distribution, length (1.16 mm), height/length index (0.37), and opaque, horizontally oriented areas, Triebel (1960) concluded that *Macrocyprina propinqua* differed significantly, in part at least, from *C. maculata*.

The Belize specimens are somewhat longer than Triebel's, ranging from 1.20 to 1.38 mm as compared to 0.95-1.08 mm; they also have lower height/length indices, ranging from 0.40 to 0.42 as compared to 0.44.

⟶

FIG. 5

a. *Triangulocypris laeva* (Puri). Complete carapace, right valve, station 317, x51.

b. *Macrocyprina propinqua* Triebel. Complete carapace, left valve, station 321, x47.

c. *?Dolerocypria inopinata* Klie. Right valve, station 6, x55.

d. *Propontocypris multiporifera* Teeter, n. sp. Holotype (H.V.H. 8982), complete carapace, left valve, station 315, x47.

e. *P. multiporifera* Teeter, n. sp. Paratype (H.V.H. 8983), complete carapace, right valve, station 315, x47.

f. *Aglaiocypris croneisi* Teeter, n. sp. Holotype (H.V.H. 8984), complete carapace, left valve, station 320, x60.

g. *A. croneisi* Teeter, n. sp. Paratype (H.V.H. 8985), complete carapace, left valve, station 320, x60.

h. *A. croneisi* Teeter, n. sp. Paratype, complete carapace, right valve, station 320, x60.

i. *Haplocytheridea kirkbii* (Brady). Male, left valve, station 8, x43.

j. *H. kirkbii* (Brady). Male, right valve, station 8, x43.

k. *Haplocytheridea placentiaensis* Teeter, n. sp. Holotype (H.V.H. 8986), female, complete carapace, right valve, station 132, x51.

l. *H. placentiaensis* Teeter, n. sp. Paratype (H.V.H. 8987), female, complete carapace, right valve, station 132, x55.

m. *H. placentiaensis* Teeter, n. sp. Paratype, female, complete carapace, right valve, station 132, x60.

n. *Cushmanidea cristifera* Teeter, n. sp. Paratype (H.V.H. 8989), male, complete carapace, left valve, station 314, x60.

o. *C. cristifera* Teeter, n. sp. Paratype, male, complete carapace, right valve, station 314, x60.

These differences, however, probably reflect only slight genotypic or ecotypic variation between the two areas.

Distribution—Triebel's material came from Punt Vierkant, Bonaire, in water 1-2 m deep. If *C. maculata* belongs in part to *M. propinqua*, this species is also present in the West Indies, at Turks Island, and possibly in Australia. In Belize this species occurs on the carbonate platform.

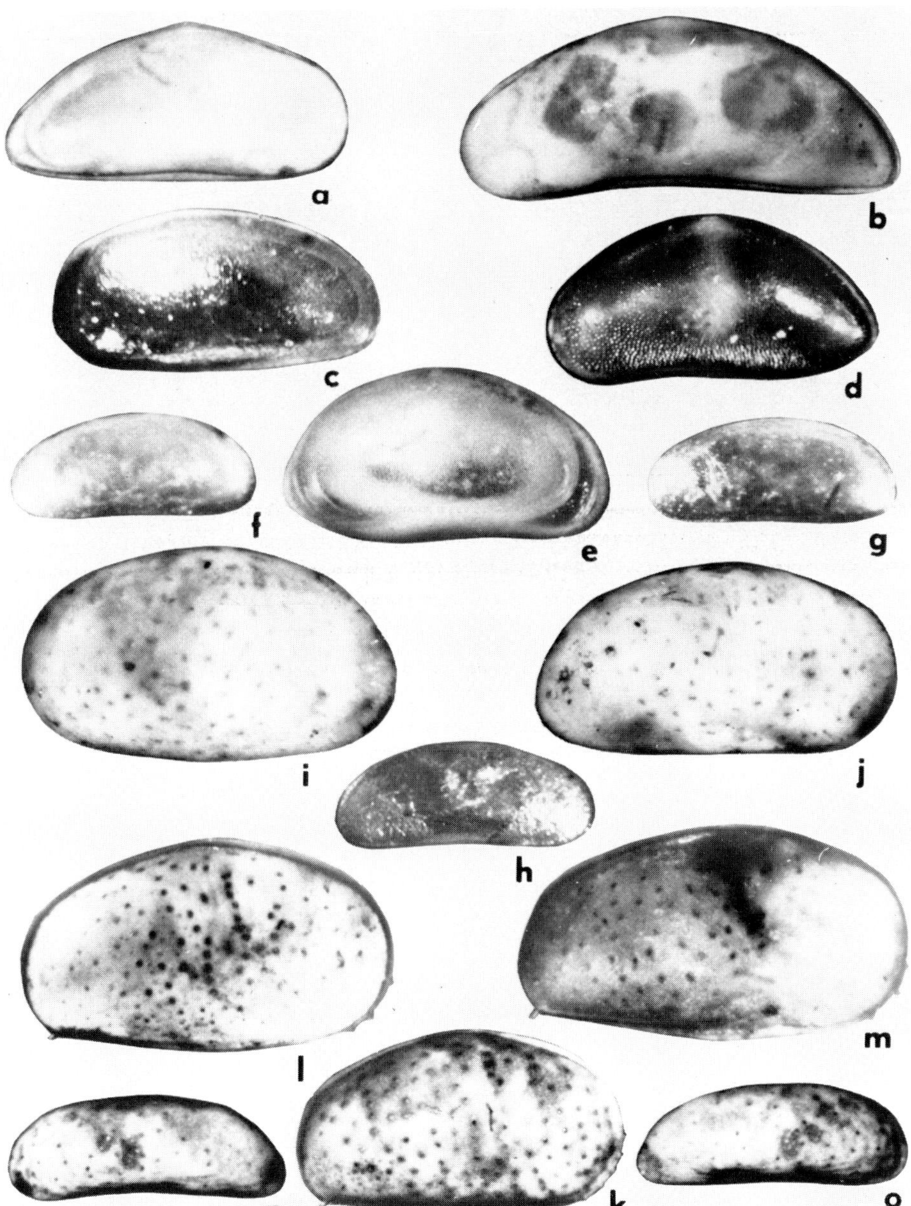

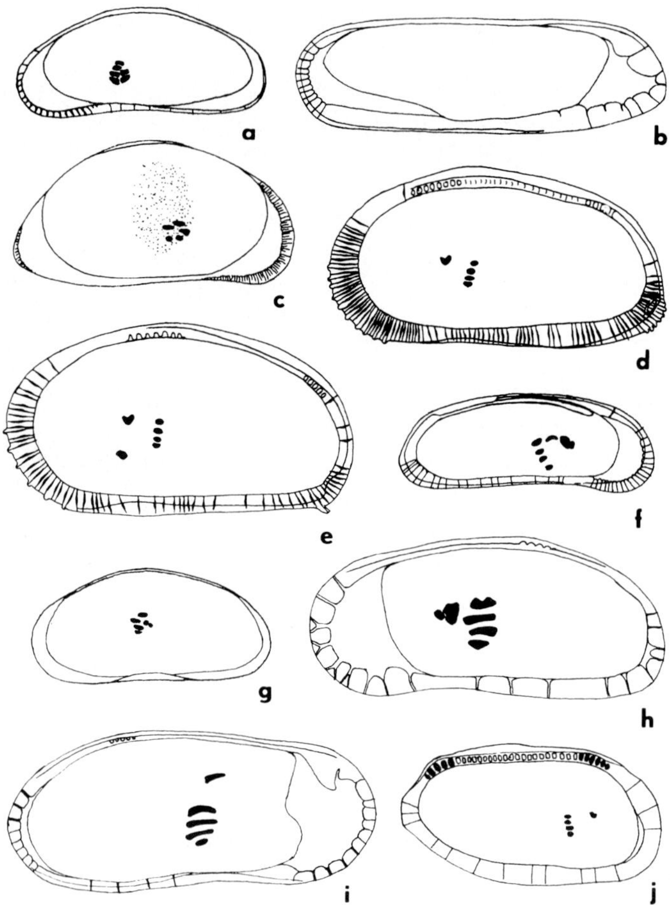

FIG. 6—Internal features.

a. *Aglaiocypris croneisi* Teeter, n. sp. Paratype, right valve, station 323, x64.
b. *Copytus baculoides* (Brady). Left valve, station 72, x99.
c. *Propontocypris multiporifera* Teeter, n. sp. Paratype, left valve, station 315, x38.
d. *Haplocytheridea kirkbii* (Brady). Male, right valve, station 8, x38.
e. *Haplocytheridea placentiaensis* Teeter, n. sp. Paratype, female, right valve, station 307, x54.
f. *Cushmanidea cristifera* Teeter, n. sp. Paratype, male, left valve, station 320, x54.
g. *?Dolerocypria inopinata* Klie. Right valve, station 6, x38.
h. *Parakrithella hanaii* Teeter, n. sp. Paratype, right valve, station 152, x96.
i. *Parakrithella reniformis* (Brady). Left valve, station 146f, x96.
j. *Perissocytheridea subrugosa* (Brady). Male, left valve, station 20, x54.

Superfamily CYPRIDACEA Baird, 1845
Family CYPRIDIDAE Baird, 1845
Subfamily CYPRIDINAE Baird, 1845

Genus DOLEROCYPRIA Tressler, 1937
?Dolerocypria inopinata Klie, 1939
Figures 5c, 6g

Dolerocypria inopinata Klie, 1939a, p. 6-10, Figs. 1-8.

Diagnosis—Thinly calcified shell, carapace bean-shaped in lateral view.

Dimensions—Length 0.80 mm, height 0.40 mm.

Remarks—The identification of many cyprid genera is based on the appendages; as this study did not involve investigation of the soft parts, other than presence or absence, some identifications are, of necessity, tenuous.

The Belize specimens resemble *D. inopinata* very closely in outline and configuration of the inner lamella.

Distribution—Originally described from hyposaline to hypersaline pools of the island of Bonaire. In Belize this species is most common in Chetumal Bay.

Family PONTOCYPRIDIDAE Mueller, 1894

Genus PROPONTOCYPRIS Sylvester-Bradley, 1947

Propontocypris multiporifera Teeter, n. sp.
Figures 5d, e; 6c

Diagnosis—Characterized by abundant fine pore canals and broadly rounded dorsal margin.

Description—Carapace bean-shaped; ventral margin concave, evenly rounded anteriorly, broadly rounded dorsally; bluntly pointed posteriorly. Surface smooth. Shell with extremely abundant, fine, normal pore canals.

Inner lamella with large anterior and posterior vestibules. Relatively narrow attachment area between inner and outer lamellae. Numerous fine, simple, straight, radial pore canals.

Muscle scar typical of genus, consisting of five secondary scars arranged in a subcircular grouping.

Dimensions—Holotype: length 0.93 mm, height 0.50 mm.

Type locality—Station 315, in 10 ft (3.0 m) of water on the south side of Tobacco Entrance.

Type specimens—Holotype: H.V.H. 8982. Paratype: H.V.H. 8983.

Remarks—The specific name reflects the extremely abundant, normal pore canals possessed by this species.

This species apparently has close resemblance to *P. simplex* Brady (1880), but lacks the rather pronounced antero-ventral angulation and has the posterior angulation displaced more dorsally. The writer has in his collection specimens from the Kerimba Archipelago, Mozambique, which are remarkably similar, if not conspecific.

Distribution—In addition to the Belize occurrence, this species has been observed by the writer in a sample donated by T. E. Pulley from Flower Garden Bank. In Belize, it occurs on the carbonate platform.

Family PARACYPRIDIDAE Sars, 1923

Genus AGLAIOCYPRIS Sylvester-Bradley, 1946

Aglaiocypris croneisi Teeter, n. sp.
Figures 5f-h, 6a

Diagnosis—Characterized by elongate shape in lateral view; low, gently convex, dorsal margin; and compact muscle area with few secondary scars.

Description—Carapace elongate, bean-shaped in lateral view. Almost perfectly symmetrical about median vertical line; posterior slightly more bluntly rounded. Compressed in dorsal view. Left valve slightly larger than right, overlapping ventrally, anterodorsally, and posterodorsally. Surface smooth.

Inner lamella wide anteriorly, slightly less so posteriorly. Area of attachment narrow, producing large anterior and posterior vestibula.

Radial pore canals straight, short, and simple; numerous anteriorly, fewer and longer ventrally, and very few, if any, posteriorly.

Normal pore canals few, simple, and widely scattered. Six secondary muscle scars arranged in a compact group.

Dimensions—Holotype: length 0.52 mm, height 0.22 mm.

Type locality—Station 320, in 5 ft (1.5 m) of water, on the reef apron of South Water Cay.

Type specimens—Holotype: H.V.H. 8984. Paratype: H.V.H. 8985.

Remarks—This species is named in honor of the late C. G. Croneis, Chancellor of Rice University.

This species closely resembles *A. complanata* (Brady and Robertson) described from the Gulf of Naples, but differs in being more elongate, less arched dorsally, and more concave centrally. The muscle scar in *A. croneisi* appears more ovate. *A. pulchella* (Brady), also from the Mediterranean, is similar, but more inflated in dorsal view.

Distribution—Occurs consistently on the carbonate platform.

Superfamily CYTHERACEA Baird, 1850
Family CYTHERIDEIDAE Sars, 1925
Subfamily CYTHERIDEINAE Sars, 1925

Genus HAPLOCYTHERIDEA Stephenson, 1936

Haplocytheridea kirkbii (Brady), 1866
Figures 5i, j; 6d

Cytheridea kirkbii Brady, 1866, p. 369, Pl. 58 (Figs. 14a-c).

Diagnosis—Characterized by large size; spinose anterior margins on both valves; short, thick, posteroventral spines on right valve; and numerous, shallow surface depressions, each marking a normal pore canal.

Dimensions—Female: length 1.13 mm, height 0.65 mm. Male: length 1.16 mm, height 0.64 mm.

Remarks—*H. kirkbii* is similar to *H. setipunctata* (Brady), but is less arched dorsally and much less truncate posteriorly; possesses few, more widely spaced, anterior spines, and a more weakly crenulate hinge.

Distribution—Brady records this species from Honduras in shallow water on calcareous sand substrates. This species typifies the Chetumal Bay biofacies.

Haplocytheridea placentiaensis Teeter, n. sp.
Figures 5k-m, 6e

Diagnosis—Characterized by numerous minute punctae, 5-6 short anterior marginal spines, and a prominent posteroventral spine in the right valve.

Description—Carapace subovate. Dorsal and ventral margins broadly rounded. Anterior margin evenly rounded with 5-6 short spines on each valve. Posterior margin obliquely truncate with prominent posteroventral spine in the right valve. Left valve overlapping right along all margins. Dimorphism prominent, males more elongate. Surface with numerous minute punctae tending toward arrangement in vertical rows.

Internal features characteristic for genus.

Dimensions—Holotype (female): length 0.85 mm, height 0.51 mm. Male: length 0.90 mm, height 0.47 mm.

Type locality—Station 132, in 12 ft (3.6 m) of water in Placentia Bay.

Type specimens—Holotype: H.V.H. 8986. Paratype: H.V.H. 8987.

Remarks—This species is named after the type locality, Placentia Bay, where it is very abundant.

The posteroventral spine is also present in *H. houghi* Butler, which possesses a more strongly developed ornamentation and numerous anterior marginal spines. *Cyprideis torosa* (Jones) is similar in shape and exhibits a prominent posteroventral spine, but differs obviously in hinge structure.

Distribution—This species is restricted to nearshore areas south of the Belize delta.

Subfamily NEOCYTHERIDEIDINAE Puri, 1957

Genus CUSHMANIDEA Blake, 1933

Cushmanidea cristifera Teeter, n. sp.
Figures 5n, o; 6f; 7a

Diagnosis—Characterized by its lateral outline, crescent-shaped vestibula, and curved surface ridges.

Description—Carapace elongate, bean-shaped. Dorsal margin broadly rounded. Anterior margin obliquely rounded, with several denticles anteroventrally on each valve. Ventral margin concave, more so in females. Posterior margin oblique dorsally, evenly rounded ventrally with a nipple-like extension. Surface with several faintly developed ridges paralleling the anterior and ventral margins. Dimorphism pronounced, males more elongate, less concave ventrally.

Hinge lophodont; selvage in right valve extremely strongly developed, terminating dorsally, anteriorly, and posteriorly in rather broad, flat, toothlike plates which fit into poorly defined sockets in the left valve; median bar in left valve appears faintly crenulate. Pouchlike depression isolated posteriorly by selvage in right valve. Inner lamella widest anteriorly with numerous, straight, simple radial pore canals. Crescent-shaped anterior vestibulum; smaller, similarly shaped posterior vestibulum. Adductor muscle scar with three closely spaced, obliquely oriented scars ventrally, and one isolated scar slightly dorsal to this group. Frontal scar large, subovate in shape. Prominent fulcral mark. Scattered, sieve-type, normal pores common.

Dimensions—Holotype (female): length 0.54 mm, height 0.25 mm. Male: length 0.60 mm, height 0.25 mm.

Type locality—Station 314, in 4 ft (1.2 m) of water on the reef flat on the south side of Tobacco Entrance.

Type specimens—Holotype: H.V.H. 8988. Paratype: H.V.H. 8989.

Remarks—The species name is derived from the Latin, meaning "ridge bearing."

C. cristifera resembles *C. miurensis* Hanai, differing from it in having a more concave ventral and lower posterodorsal margin. This species also somewhat resembles *C. sulcata* (Brady), but the anterior vestibulum is much more restricted and the posterior is lacking in the latter species.

Distribution—The writer has observed this species rarely in samples collected from the reef tract in the Florida Keys. In Belize it is restricted to the carbonate platform.

Genus COPYTUS Skogsberg, 1939

Copytus baculoides (Brady), 1890
Figure 6b

Cytherideis baculoides Brady, 1890, p. 512, Pl. 3 (Figs. 11, 12).

Diagnosis—Characterized by its cylindroidal shape and shallow, horizontal, anterior marginal depression.

Dimensions—Length 0.48 mm, height 0.16 mm.

Remarks—The writer has compared the specimens from Belize with some in his collection from Fiji, the type locality of this species, and has found them identical.

Distribution—Although scarce, its localized distribution in Belize, in conjunction with the only other recorded occurrence at Fiji, presents an interesting problem. In Belize, this species is restricted to the deeper parts of Eastern Channel, the only entrance to Belize Harbour available to ocean-going vessels, suggesting that these ostracods may be distributed by such vessels.

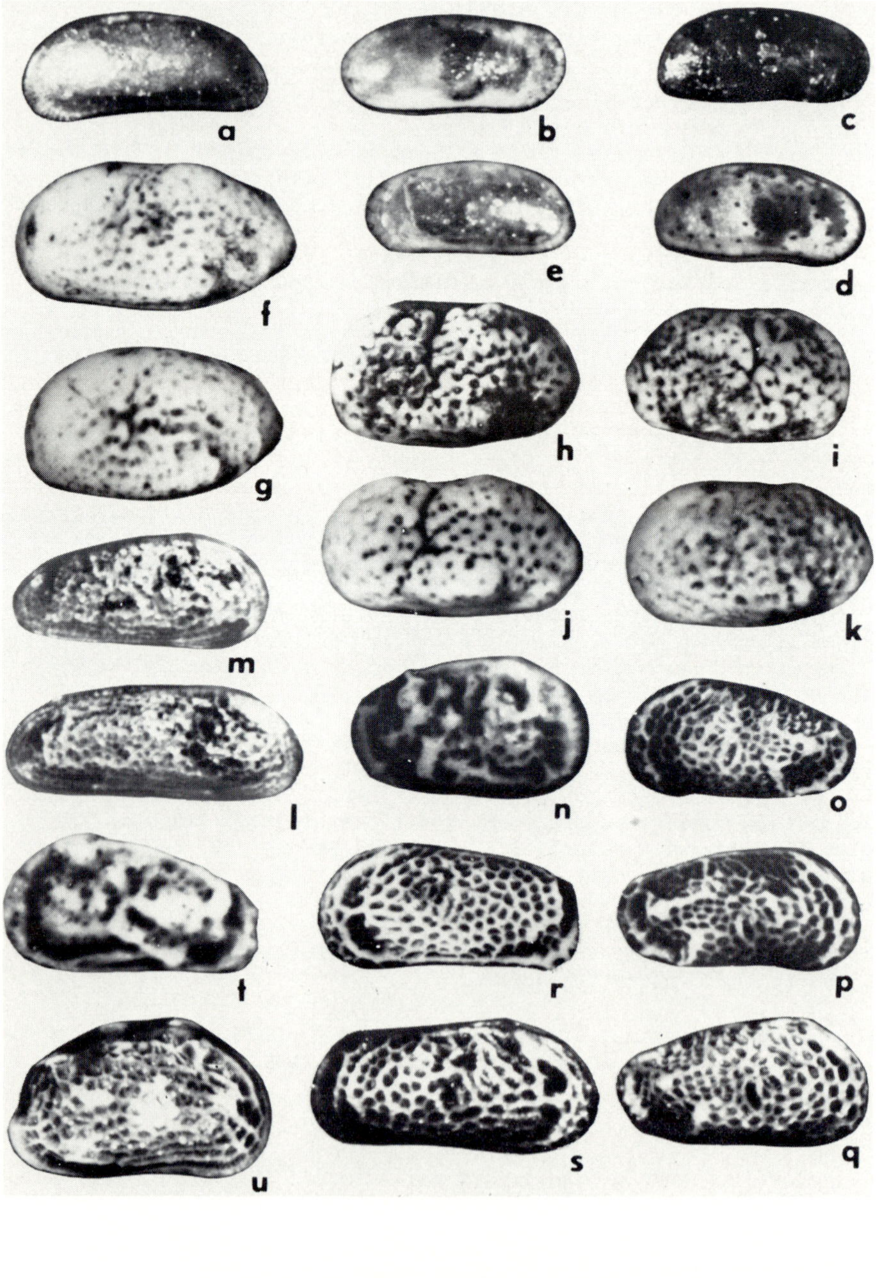

The specimens from Fiji were collected from the intertidal zone to depths of 4 fm. In Belize, living adults have been collected from stations 235, 237, and 72, in the Eastern Channel area at depths ranging from 64 to 110 ft (20-34 m).

Subfamily KRITHINAE Mandelstam, *in* Bubikan, 1958

Genus PARAKRITHELLA Hanai, 1959

Parakrithella reniformis (Brady), 1868

Figures 6i, 7b

Paradoxostoma reniforme Brady, 1868b, p. 224, Pl. 15 (Figs. 1, 2).
Krithe reniformis (Brady). Brady and Norman, 1889, p. 182, Pl. 21 (Figs. 23, 24); Mueller, 1894, p. 358-359, Pl. 28 (Fig. 24); Pl. 30 (Figs. 1, 3-16, 22, 23).
Cytherois? sp. Keij, 1954b, p. 229, Pl. 5 (Fig. 18).
Cushmanidea laevata (Brady)? van den Bold, 1963c, p. 378, Pl. 3 (Fig. 5).
(not) *Cytherideis laevata* Brady, 1880, p. 146, Pl. 6 (Figs. 5a-d); Pl. 35 (Figs. 5a-d).
Parakrithella sp. van den Bold, 1966f, Pl. 4 (Figs. 7a, b).

Diagnosis—Characterized by its very large, anteriorly somewhat expanded vestibulum and short, wide, radial pore canals.

Dimensions—Length 0.51 mm, height 0.22 mm.

Remarks—The Belize specimens are identical with the excellent illustrations of *P. reniformis* figured by Mueller from the Gulf of Naples.

This species differs from the original diagnosis (Hanai, 1959) in that the radial pore canals do not bifurcate and the ventral restriction of the vestibulum is lacking.

⟵

FIG. 7

a. *Cushmanidea cristifera* Teeter, n. sp. Holotype (H.V.H. 8988), female, complete carapace, left valve, station 314, x60.

b. *Parakrithella reniformis* (Brady). Complete carapace, right valve, station 146f, x55.

c. *Parakrithella hanaii* Teeter, n. sp. Holotype (H.V.H. 8990), complete carapace, right valve, station 152, x60.

d. *P. hanaii* Teeter, n. sp. Paratype (H.V.H. 8991), complete carapace, right valve, station 152, x60.

e. *P. hanaii* Teeter, n. sp. Paratype, complete carapace, left valve, station 152, x60.

f. *Perissocytheridea subrugosa* (Brady). Male, complete carapace, left valve, station 20, x60.

g. *P. subrugosa* (Brady). Female, complete carapace, left valve, station 20, x60.

h. *Perissocytheridea bisulcata* Teeter, n. sp. Holotype (H.V.H. 8992), female, complete carapace, left valve, station 15, x60.

i. *P. bisulcata* Teeter, n. sp. Paratype, female, complete carapace, right valve, station 15, x55.

j. *P. bisulcata* Teeter, n. sp. Paratype (H.V.H. 8993), male, complete carapace, left valve, station 15, x60.

k. *P. bisulcata* Teeter, n. sp. Paratype, female, complete carapace, left valve, station 15, x60.

l. *Pumilocytheridea sandbergi* van den Bold. Male, complete carapace, right valve, station 292, x102.

m. *P. sandbergi* van den Bold. Female, complete carapace, right valve, station 292, x72.

n. *Callistocythere cranekeyensis* (Puri). Complete carapace, right valve, station 13-60, x85.

o. *Morkhovenia inconspicua* (Brady). Complete carapace, left valve, station 314, x94.

p. *M. inconspicua* (Brady). Complete carapace, right valve, station 314, x94.

q. *M. inconspicua* (Brady). Complete carapace, right valve, Lanikai, Hawaii, x94.

r. *Keijia demissa* (Brady). Complete carapace, left valve, station 70, x72.

s. *K. demissa* (Brady). Complete carapace, right valve, Kerimba Archipelago, Mozambique, x72.

t. *Munseyella* sp. cf. *M. minuta* (van den Bold). Complete carapace, left valve, station 158, x115.

u. *Hemicythere? auricula* Teeter, n. sp. Holotype (H.V.H. 8994), complete carapace, right valve, station 315, x34.

Distribution—Brady and Norman (1889) listed it as present in (beach?) sand from Tenedos, at 180-200 fm in the Bay of Biscay, and in shallow water at Naples. Keij (1954b) records specimens assignable to this species from the Gulf of Paria. Analysis of his data indicates that the species never occurs above 6 fm and approximately 70 percent of the occurrences are below 10 fm. Van den Bold (1963c) figured a specimen externally identical with *P. reniformis* from the upper Miocene Melajo clay of Trinidad. In Belize this species is restricted to the main lagoon.

Parakrithella hanaii Teeter, n. sp.
Figures 6h, 7c-e

Diagnosis—Characterized by small size, deep crescent-shaped anterior vestibulum, and extremely small posterior vestibulum.

Description—Carapace small, bean-shaped. Dorsal and anterior margins evenly rounded. Ventral margin concave anteriorly, convex posteriorly. Posterior margin obliquely rounded, blunt posteroventrally. Left valve larger than right, overlapping dorsally and ventrally.

Narrow dorsal lip with posterior, finely crenulate part in right valve, hinges with corresponding submarginal groove in left valve. Inner lamella wide anteriorly, with deep crescent-shaped vestibulum, narrow ventrally and posteriorly. Very small posterior vestibulum. Seven anterior radial pore canals, of which the four median ones bifurcate. Scattered, large, simple, normal pores common. Vertical row of four large adductor muscle scars, uppermost one U-shaped. Frontal scar elongate, obliquely oriented.

Dimensions—Holotype: length 0.47 mm, height 0.22 mm.

Type locality—Station 152, in 156 ft (48 m) of water near Grass Cay.

Type specimens—Holotype: H.V.H. 8990. Paratype: H.V.H. 8991.

Remarks—This species is named in honor of T. Hanai of the University of Tokyo.

P. hanaii is considerably smaller than *P. pseudadonta* (Hanai). The Belize species lacks the ventral restriction of the anterior vestibulum and possesses a relatively smaller posterior vestibulum.

Distribution—This species is restricted to the main lagoon.

Subfamily PERISSOCYTHERIDEINAE van den Bold, 1963

Genus PERISSOCYTHERIDEA Stephenson, 1938

Perissocytheridea subrugosa (Brady), 1870
Figures 6j; 7f, g

Cythere subrugosa Brady, 1870, p. 238-239, Pl. 32 (Figs. 18, 19).
Ilyocythere meyer-abichi Hartmann? 1953, p. 310-316, Text-figs. 1-17.
Perissocytheridea subrugosa (Brady). Van den Bold, 1958a, p. 71; 1966f, Pl. 1 (Figs. 7a, b).
Perissocytheridea matsoni (Stephenson)? Curtis, 1960, p. 478, 484, 486, Text-fig. 5; Pl. 3 (bottom; Figs. 3, 5).
(not) *Cytheridea matsoni* Stephenson, 1938, p. 145, Pl. 23 (Fig. 10); Pl. 24 (Figs. 3, 4); Text-figs. 4, 7, 11.
Perissocytheridea subrugosa (Brady)? van den Bold, 1963c, p. 380, Pl. 4 (Figs. 2a-d).

Description—A reticulate species of *Perissocytheridea* characterized by slight ventral projection and bluntly pointed posterior.

Dimensions—Female: length 0.56 mm, height 0.33 mm. Male: length 0.60 mm, height 0.33 mm.

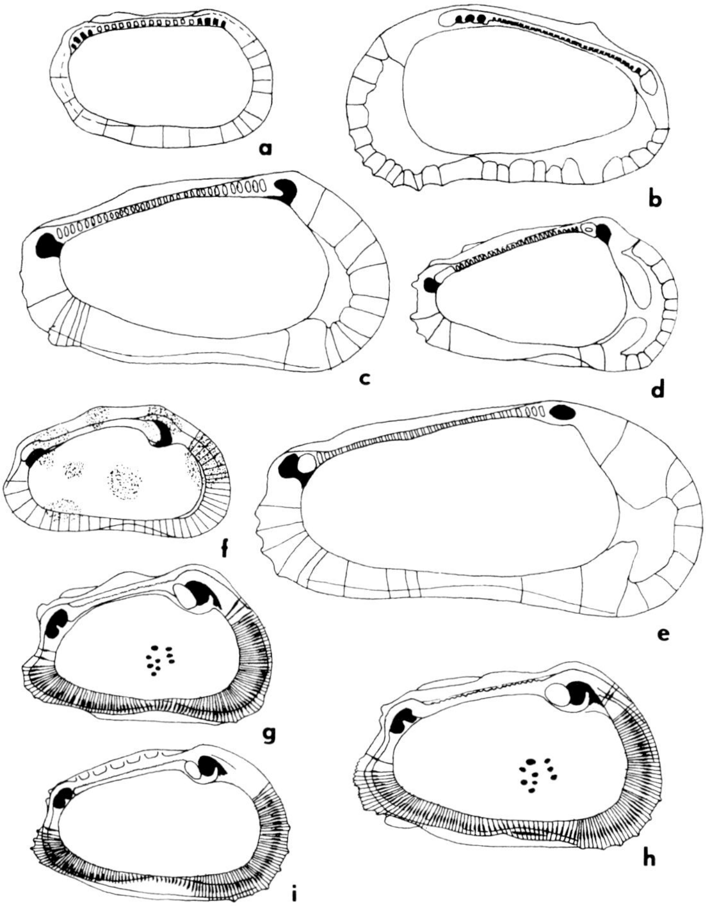

FIG. 8—Internal features.

a. *Perissocytheridea bisulcata* Teeter, n. sp. Paratype, female, left valve, station 15, x54.
b. *Callistocythere cranekeyensis* (Puri). Right valve, station 13-60, x118.
c. *Morkhovenia inconspicua* (Brady). Left valve, station 314, x118.
d. *Munseyella* sp. cf. *M. minuta* (van den Bold). Left valve, station 158, x115.
e. *Keijia demissa* (Brady). Left valve, station 70, x128.
f. *Hemicythere? auricula* Teeter, n. sp. Paratype, male, left valve, station 313, x29.
g. *Mutilus confragosus* (Edwards). Subadult, left valve, station 313, x54.
h. *M. confragosus* (Edwards). Adult, left valve, station 322, x54.
i. *Aurila wantlandi* Teeter, n. sp. Paratype, left valve, station 45, x48.

Remarks—This species is similar to *P. rugata* Swain but may be distinguished from it by its finer reticulate pattern and its more obvious ventral projection.

Distribution—Originally described from Santiago de Cuba by Brady in 1870, this species is present in Belize in the less saline parts of Chetumal Bay and in the southern-nearshore biofacies.

Perissocytheridea bisulcata Teeter, n. sp.
Figures 7h-k, 8a

Diagnosis—A reticulate species of *Perissocytheridea* characterized by a ventral knoblike extension and two well-developed sulci.

Description—Carapace subquadrate in lateral view, with anterior and posterior margins bluntly rounded. In dorsal view, males quadrate, females ovate. Male with postdorsal inflation, typical of genus, female with slight tendency to produce small postdorsal crest. Prominent knob-like extension ventrally, approximately one third the length from the posterior margin. Twin sulci located anterodorsally. Posterior one deeply incised, vertical, extending one third to one-half the height from the dorsal margin. Anterior sulcus less deeply incised, inclined at about 45° to dorsal margin, and slightly shorter than posterior sulcus. Surface fairly coarsely reticulate. Shell heavily calcified.

Hinge, inner lamella, normal and radial pore canals all typical for genus.

Dimensions—Holotype (female): length 0.53 mm, height 0.31 mm. Male: length 0.57 mm, height 0.30 mm.

Type locality—Station 15, near Deer Cay in Chetumal Bay.

Type specimens—Holotype: H.V.H. 8992. Paratype: H.V.H. 8993.

Remarks—The trivial name, bisulcata, reflects the presence of two well-developed sulci.

This species is similar to *P. subrugosa*, but differs in the possession of two sulci, subquadrate outline, and much more pronounced ventral extension.

Distribution—*P. bisulcata* is found in the more saline areas of Chetumal Bay, especially on the west side of Ambergris Cay.

Genus PUMILOCYTHERIDEA van den Bold, 1963
Pumilocytheridea sandbergi van den Bold, 1963
Figure 7*l*, m

Pumilocytheridea sandbergi van den Bold, 1963c, p. 382-383, Pl. 4 (Figs. 3a-d); Pl. 12 (Figs. 1a-d, 2a, b).

Diagnosis—Characterized by its vertical posterior ridge and compressed surface posterior to ridge.

Dimensions—Female: length 0.45 mm, height 0.19 mm. Male: length 0.38 mm, height 0.19 mm.

Distribution—This species was originally described from the Melajo clay, upper Miocene of Trinidad. It is also known to be present in the Pleistocene Morne l'Enfer of Trinidad, and the Miocene to possibly Pliocene Cubagua and Cumana Formations of Venezuela. In the Holocene, it is found in the Gulf Coast, both in nearshore areas and mudlump samples from the Mississippi, and in the waters around Trinidad.

Van den Bold (1963c) suggested that, environmentally, it is marine with some brackish-water influence. In Belize, it exhibits an amazing ecologic latitude, present as complete carapaces in nearshore areas, uncommonly in the deeper water of the main lagoon, and even on the carbonate shoal and reef platform. Therefore, van den Bold's suggested salinity range of 17-19‰ must be expanded to include normal-marine salinities.

Family LEPTOCYTHERIDAE Hanai, 1957

Genus CALLISTOCYTHERE Ruggieri, 1953
Callistocythere cranekeyensis (Puri), 1960
Figures 7n, 8b

Leptocythere n. sp. 2 Puri and Hulings, 1957, p. 174, Pl. 11 (right; Fig. 18).

Leptocythere cranekeyensis Puri 1960, p. 114, Pl. 4 (Figs. 12, 13).

Diagnosis—Characterized by its small size and surface of subdued ridges and depressions.

Dimensions—Length 0.35 mm, height 0.20 mm.

Remarks—The coarse ornamentation, heavily calcified carapace, and relatively short length, in comparison to *Leptocythere*, clearly place this species in the genus *Callistocythere*. Also, ecologically, *Leptocythere* is usually restricted to cold, commonly brackish water, and *Callistocythere* is restricted to predominantly warm water.

In accordance with Hanai's (1957a) paper on the Leptocytherinae, *C. cranekeyensis* would be placed morphologically in the *C. littoralis* group characterized by a comparatively thick carapace, subquadrangular lateral outline, strong ornamentation, and two or more anterior terminal teeth in the left valve. The Belize material closely resembles *C. undata* Hanai within this group, differing only in the development of the anterior and posterior marginal ridges.

Keij (1954a) illustrated "*Leptocythere* sp." (Pl. 1, Figs. 10a, b), which strikingly resembles *C. cranekeyensis*, but no definite conclusion can be reached because no description is given.

Distribution—This species occurs in Florida Bay near Crane Key in 6 ft (1.8 m) of water and on the shore at Bahia Honda. This species has also been observed in material in the writer's collection from Kawaikai, Hawaii. It is found rarely at the latter locality in a fine carbonate sand in 23 ft (7.0 m) of water. In Belize this species is restricted to the carbonate platform.

MORKHOVENIA Teeter, n. gen.

Diagnosis—A Pectocytherid genus characterized by a puncto-reticulate surface and two very distinct posteroventral angulations.

Description—*Morkhovenia* possibly a monotypic genus; thus the generic description essentially that of the type species presented below.

Type species—*Cythere inconspicua* (Brady) 1880.

Remarks—*Morkhovenia* is named in honor of F. P. C. M. van Morkhoven, paleontologist with Shell Oil Co., Houston, Texas.

This genus resembles particularly the females of *Leptocythere*, which have a well-developed postventral angulation, but differs in having coarser ornamentation, pentodont hinge, and particularly in the radial pore canals. *Morkhovenia* also resembles *Callistocythere*, but differs markedly in hinge and radial pore canals; ornamentation is also coarser in *Callistocythere*, and more inclined to development of ridges than punctae or reticulations. *Munseyella* is most closely related, for both hinge development and marginal features are almost identical. Surface ornamentation, however, differs radically, as *Munseyella* tends to development of a smooth carapace with only a few subdued ridges.

Only one species is known to belong to this genus at present. Brady (1890, Pl. 2, Figs. 6, 7), however, figured an ostracod, *Cythere cuneolus*, which strikingly resembles *M. inconspicua*. Only a few specimens were found and, according to Brady, they varied considerably in character.

Morkhovenia inconspicua (Brady), 1880
Figures 7o-q, 8c

Cythere inconspicua Brady, 1880, p. 70, Pl. 13 (Fig. 1).

Leptocythere inconspicua (Brady). Keij, 1954a, p. 354, Pl. 1 (Figs. 3a, b).

Leptocythere sp. cf. *L. inconspicua* (Brady). Van den Bold, 1963c, p. 365.

Description—Carapace small. Dorsal margin straight in anterior two thirds, flexed ventrally in posterior. Posterior margin subrounded, denticulate, with five or six denticles. Ventral margin concave in anterior half, convex posteriorly. Anterior margin broadly rounded. In dorsal view, subsagittate, blunt anteriorly, two steplike depressions posteriorly.

Two pronounced postventral depressions, the most ventral about one third to one fourth the length from the posterior margin, the second, dorsally, about one half this distance from the posterior margin.

Hinge pentodont with two terminal sockets in the left valve, and a median, finely crenulate bar with two low terminal crenulate teeth. Inner lamella and attachment area with outer lamella broad. Anterior vestibulum large, expanded somewhat distally. Posterior vestibulum small. About ten straight, simple, radial pore canals anteriorly, fewer ventrally and posteriorly.

Dimensions—Length 0.41 mm, height 0.22 mm.

Remarks—This species is extremely widespread. The writer has in his collection specimens from the Great Barrier Reef, Australia, the Philippines, Hawaii, Qatar, the Bahamas, and Florida.

Distribution—The Belize specimens are restricted to the carbonate shelf.

KEIJIA Teeter, n. gen.

Diagnosis—A reticulate pectocytherid genus characterized by its elongate, somewhat tubular shape, anterior submarginal depression, and posterior submarginal angulation.

Description—Carapace small to medium. Subquadrate in outline; dorsal margin straight; anterior broadly rounded; ventral, slightly concave anteriorly; posterior end blunt with five or six denticles. Subovate in dorsal view. Surface with reticulations tending to arrangement in rows parallel with the margins. A short distance posterior to and parallel with the anterior margin, a deep narrow depression divided variously into subdepressions by horizontal ridges. More or less well-defined subvertical angulation slightly anterior to posterior margin.

Internal features very similar to *Morkhovenia*, but with the median hinge bar more coarsely crenulate and terminated by low, smooth projections. The anterior vestibulum very similar to *Munseyella*.

Dimorphism variably accentuated, females shorter, higher, and may be more strongly ornamented.

Type species—*Cythere demissa* Brady 1868.

Remarks—This genus is quite similar to *Morkhovenia*, and for some time the writer considered the possibility that the two type species were, in fact, dimorphic forms of the same species. However, the differences in ornamentation as well as in internal features, the discovery of juveniles of both genera, and the apparent nonoccurrence of one species or the other throughout much of their distribution area, together lead the writer to conclude that these are not sexual dimorphs. Such differences in morphology as the presence of the two distinct posteroventral projections in *Morkhovenia* and the lack of this feature in *Keijia* are of the magnitude of generic characters.

In a sample from Tunganan Bay, Philippines, the writer observed another markedly dimorphic species clearly assignable to this genus (see *Remarks* under *K. demissa*).

Keijia demissa (Brady), 1868
Figures 7r, s; 8e

Cythere demissa Brady, 1868a, p. 180, Pl. 12 (Figs. 1, 2); 1880, p. 66, Pl. 12 (Fig. 7a-h); 1890, p. 497.

Leptocythere demissa (Brady). Hornibrook, 1952, p. 13, 17, 21; Keij, 1954a, p. 354, Pl. 1 (Fig. 3).

Diagnosis—A moderately reticulate species of *Keijia* with the anterior depression divided in two subdepressions, and the postero-submarginal angulation present only in the dorsal half of the carapace. Dimorphism not pronounced.

Dimensions—Length 0.45 mm, height 0.21 mm.

Remarks—The Belize and East African material differs from the Tunganan Bay, Philippines, species in being more finely reticulate, in having a shorter submarginal posterior angulation, and in the development of the anterior marginal depression. Other differences in shape, the prominence of posterior denticulations, and the pronounced dimorphism in the Tunganan Bay species may be used to separate these species.

In his paper of 1880, Brady figured 10 views of *K. demissa*, at least one of which (Pl. 12, Fig. 7E) is probably another species, and possibly even another genus.

Distribution—This species has a remarkable distribution. It has been observed by the writer in material from the Kerimba Archipelago, Mozambique, from Florida, and from Belize. In addition, this species has been recorded from Manila in the Philippines, and Hornibrook recorded it from Eocene beds to the Holocene in New Zealand. Belize specimens are restricted to the carbonate shelf.

Genus MUNSEYELLA van den Bold, 1957

Munseyella sp. cf. *M. minuta* (van den Bold), 1946

Figures 7t, 8d

Cytheromorpha minuta van den Bold, 1946, p. 105, Pl. 14 (Figs. 12a, b).
Munseyella minuta (van den Bold). Van den Bold, 1957b, p. 238; 1958b, p. 402, Pl. 4 (Figs. 1a-e); Pl. 5 (Figs. 2a-c); 1960, p. 162.

Diagnosis—Characterized by a submarginal ventral ridge which bifurcates anteriorly, and a dorsal ridge which thickens and extends ventrally at its posterior termination.

Dimensions—Length 0.29 mm, height 0.15 mm.

Remarks—This species tends to display much greater variability than do most ostracod species, as van den Bold (1958b) indicated. The Belize specimens resemble van den Bold's (1958b) Plate 4, Figure 1E, most closely, but some variants approach Figure 16, which shows a posterodorsal welt. The welt appears to develop on the site of a somewhat variable, subdued, vertical bar or V-shaped ridge. Forms intermediate between these developmental conditions exist.

The Holocene specimens from Belize, measuring about 0.29 by 0.15 mm, are smaller than the fossil ones described by van den Bold (0.33-0.40 by 0.18-0.24 mm).

The writer doubts that the Holocene material can be conspecific with the fossil material because some change may be expected during the transition from the late Eocene. Also, in spite of the diversity of the specimens from Belize, the material illustrated by van den Bold (1958b) shows far more variation, possibly representing more than one species.

Distribution—The species has a fossil range from uppermost Eocene to middle Miocene. In Belize, it is found in the deepest parts of the main lagoon.

Family HEMICYTHERIDEA Puri, 1953

Genus HEMICYTHERE Sars, 1925

Hemicythere? auricula Teeter, n. sp.

Figures 7u; 8f; 9h, i

Diagnosis—Characterized by auriculate outline, pronounced dorsal overlap by left valve, and prominent postdorsal "ear."

Description—Carapace auriculate in lateral view. Anterior margin broadly rounded; ventral margin concave centrally, upturned posteriorly, terminating in obtusely pointed posterior margin; dorsal margin horizontal centrally, posteriorly becoming more steeply inclined through two pronounced breaks in slope. Left valve with pronounced dorsal crest which overlaps right. Both valves with anterodorsal submarginal ridge accommadating eye tubercle dorsally. Subdued ventral submarginal ridge becoming more pronounced posteriorly. Strongly developed postdorsal "ear." Surface reticulate with horizontal ribbing accentuated.

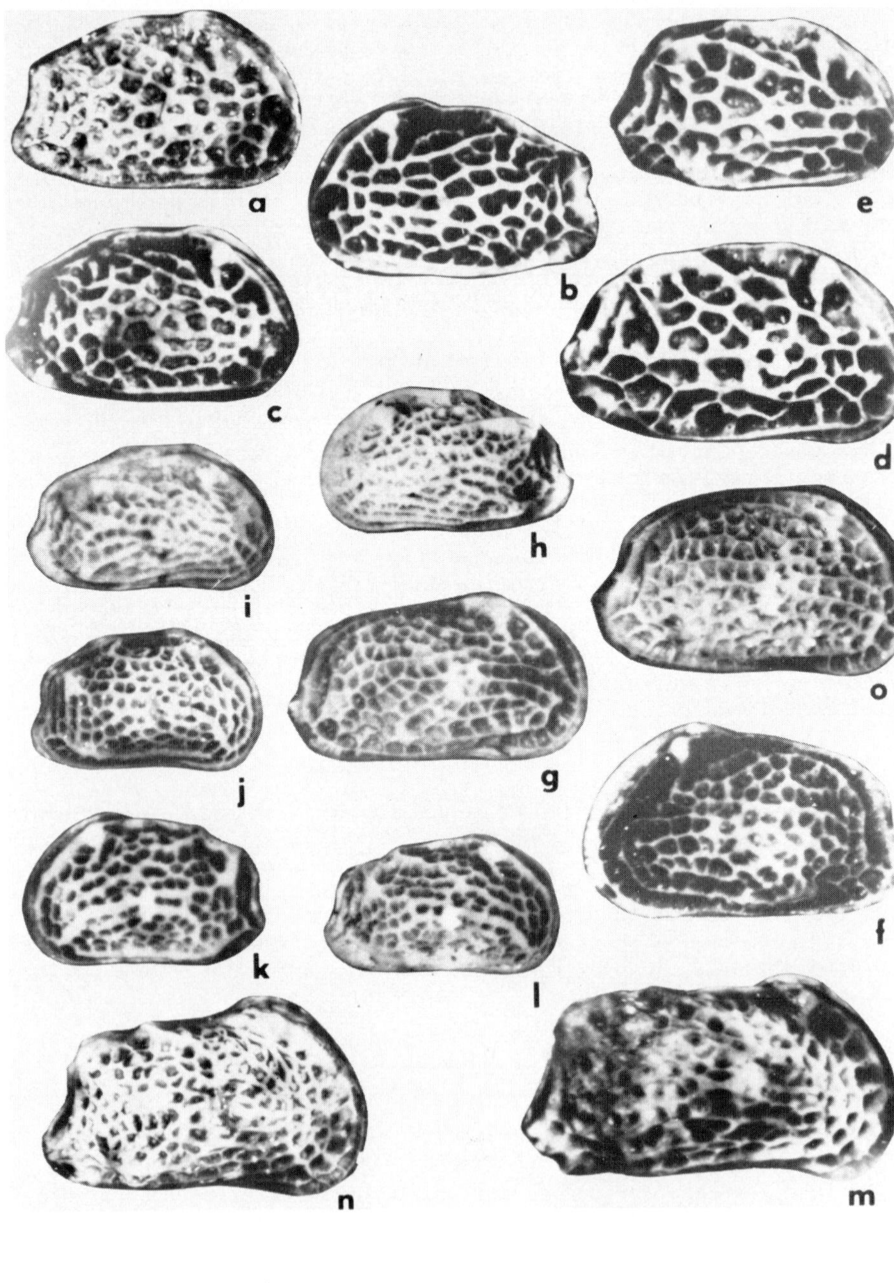

Hinge holamphidont, median bar of left valve noncrenulate; radial pore canals abundant; narrow anterior vestibulum, posterior more narrow. Small, scattered, sieve-type pore canals common. Small opaque area centrally, approximately six others marginally.

Dimensions—Holotype: length 0.55 mm, height 0.35 mm.

Type locality—Station 315, in 10 ft (3.0 m) of water on the south side of Tobacco Entrance.

Type specimens—Holotype: H.V.H. 8994. Paratype: H.V.H. 8995.

Remarks—The species name, *auricula,* refers to the ear-shaped appearance in lateral outline.

In the past, many species have been assigned to *Hemicythere*; however, more recent studies (e.g., Hazel, 1967; van Morkhoven, 1963; Pokorny, 1955, 1964) recognizing the complexity of the family Hemicytheridae have more precisely defined the genera within this family. Thus, this species is questionably included within the genus *Hemicythere.* It does not belong within any of the other genera of the Hemicytheridae.

This species is rather unlike any of the other species assigned to *Hemicythere.* It most closely resembles *H. palosensis* LeRoy (1943), but differs from it in having a more sloping posterodorsal margin and a stronger development of the "ear."

Distribution—This species appears to be restricted to the *Thalassia*-free areas of the barrier rim and reef flat.

Genus MUTILUS Neviani, 1928

Mutilus confragosus (Edwards), 1944

Figures 8g, h; 9d, e

Hemicythere confragosa Edwards, 1944, p. 518, Pl. 86 (Figs. 22-26); Swain, 1951, p. 43, Pl. 6 (Figs. 13-14); Puri, 1953c, p. 176, Pl. 1 (Figs. 4-6); 1953d, p. 266, Pl. 11 (Figs. 10-12); van den Bold, 1958a, p. 71; Brown, 1958, p. 66, Pl. 7 (Fig. 1).

Mutilus confragosa (Edwards). Puri, 1960, p. 130; Swain, 1968, p. 21-22, Pl. 4 (Figs. 8a-e); Pl. 5 (Figs. 5a-c); Pl. 7 (Figs. 3a-c).

⟵

FIG. 9

a. *Aurila wantlandi* Teeter, n. sp. Holotype (H.V.H. 8996), complete carapacr, right valve, station 45, x55.

b. *A. wantlandi* Teeter, n. sp. Paratype (H.V.H. 8997), complete carapace, left valve, station 13-60, x55.

c. *A. wantlandi* Teeter, n. sp. Paratype, complete carapace, right valve, station 13-60, x55.

d. *Mutilus confragosus* (Edwards). Adult, complete carapace, right valve, station 322, x60.

e. *M. confragosus* (Edwards). Subadult, complete carapace, right valve, station 313, x60.

f. *Aurila chetumalensis* Teeter, n. sp. Paratype, female, complete carapace, left valve, station 27, x60.

g. *A. chetumalensis* Teeter, n. sp. Paratype (H.V.H. 8999), male, complete carapace, right valve, station 27, x60.

h. *Hemicythere? auricula* Teeter, n. sp. Paratype (H.V.H. 8995), complete carapace, left valve, station 319, x34.

i. *H.? auricula* Teeter, n. sp. Paratype, complete carapace, right valve, station 319, x34.

j. *Quadracythere lankfordi* Teeter, n. sp. Holotype (H.V.H. 9000), female, complete carapace, right valve, station 315, x60.

k. *Q. lankfordi* Teeter, n. sp. Paratype, female, complete carapace, left valve, station 313, x60.

l. *Q. lankfordi* Teeter, n. sp. Paratype (H.V.H. 9001), male, complete carapace, right valve, station 313, x60.

m. *Quadracythere producta* (Brady). Male, complete carapace, right valve, station 451, x64.

n. *Q. producta* (Brady). Female, complete carapace, right valve, station 451, x64.

o. *Aurila chetumalensis* Teeter, n. sp. Holotype (H.V.H. 8998), female, complete carapace, right valve, station 27, x60.

Mutilus confragosa (Edwards)? Swain, 1967, p. 83, Pl. 6 (Figs. 1a, b); Fig. 52a.
Aurila confragosa (Edwards). Van den Bold, 1963c, p. 385, Pl. 8 (Fig. 1).
Mutilus confragosus (Edwards). Van den Bold, 1966f, Pl. 1 (Fig. 13).

Diagnosis—A species of *Mutilus* characterized by its coarse reticulation and pronounced postdorsal projection.

Dimensions—Adult: length 0.75 mm, height 0.47 mm. Subadult: length 0.60 mm, height 0.38 mm.

Remarks—Through the courtesy of R. A. Edwards of the University of Florida, the writer was able to secure some original material from the Duplin Marl of North Carolina. In comparing topotype specimens of *M. confragosus*, it was observed that the penultimate instar of material from Belize fell within the size range of the presumed adult instar from the Duplin Marl. Both Holocene subadults and fossil material exhibit every indication of having attained maturity. With respect to ornamentation, the fossil specimens are more similar to the adult Holocene material than to the subadults of comparable size. Thus, either this species has increased in size throughout time, or the fossil adult of *M. confragosus* has never been discovered. Differences between specimens from the Duplin Marl and the Holocene include the enlargement of the posterodorsal "ear" and coarser ornamentation of the latter.

The specimens from California, identified by Swain (1967), exhibit differences in the reticulate ornamentation and are questionably included in synonymy.

Distribution—In Belize this species is restricted to the carbonate platforms. It is widespread throughout the Caribbean, ranging back to the Miocene, and apparently is present in Holocene deposits in the Gulf of California.

Genus AURILA Pokorny, 1955

Aurila wantlandi Teeter, n. sp.

Figures 8i, 9a-c

?*Aurila conradi* (Howe and McGuirt). Puri, 1960, p. 129-130, Pl. 3 (Figs. 9, 10).

Diagnosis—Characterized by its fairly coarse reticulate pattern and short posteroventral "ear."

Description—Carapace subquadrate. Anterior and anterodorsal margin rounded. Dorsal margin horizontal centrally, inclined at about 30° posteriorly. Posterior margin vertical dorsally, upturned, with several small denticles, ventrally. Ventral margin horizontal anteriorly, upturned posteriorly. Anteroventral margin with numerous spines interspersed with fewer stronger denticles. Both the anteroventral and posterior denticles easily lost by abrasion.

Prominent submarginal ventral ridge, commencing near the anteroventral margin and terminating in a slightly expanded cusp, about one third the length from the posterior margin. Anterior marginal ridge commencing slightly below midvalve and passing into dorsal crest. Short posterodorsal "ear." Surface fairly heavily reticulate.

Dimensions—Holotype: length 0.65 mm, height 0.41 mm.

Type locality—Station 45. Uppermost Chetumal Bay, near the mouth of New River.

Type specimens—Holotype: H.V.H. 8996. Paratype: H.V.H. 8997.

Remarks—This species is named in honor of Frank Wantland, a coworker in the Belize project.

The quality of Puri's figures is rather poor but, after comparing specimens of *A. floridana* Benson and Coleman, 1963 (formerly assigned to *A. conradi*), and *A. wantlandi*, the writer believes Puri's specimens more likely belong to the latter species.

This species is larger than *A. floridana*, has a heavier ornamentation, and possesses a postdorsal "ear." *A. wantlandi* possesses a more closely meshed reticulation than *A. confragosa*, but is otherwise rather similar. *A. schmidtae* (Curtis), a Miocene species of the Atlantic coastal plain, differs in outline and lacks a postdorsal "ear." *Cythere compacta* Brady, 1866, resembles the Belize species, but is smaller and possesses a posterior ridge connecting the ventral and dorsal ridges.

Distribution—The writer has observed this species in samples from Florida Bay and the Bahamas. In Belize this species is restricted to the carbonate platform.

Aurila chetumalensis Teeter, n. sp.
Figures 9f, g, o; 10a

Diagnosis—Characterized by well-defined angulation at posterior cardinal angle and flaplike posterior extension of ventral ridge.

Description—Carapace subquadrate. Lateral outline characteristic of genus, but posterior cardinal angle more angular than most species of *Aurila*.

Surface reticulate with three submarginal ridges. Ventral ridge terminating posteriorly in a flaplike projection and gradually disappearing anteriorly. Anterior submarginal ridge beginning just below midpoint on anterior margin and angling dorsally, passing through and terminating just posterior to the eye tubercle. Subdued posterodorsal submarginal ridge parallel with free margin.

Dimensions—Holotype (female): length 0.70 mm, height 0.45 mm. Male: length 0.63 mm, height 0.38 mm.

Type locality—Chetumal Bay, just west of Bulkhead shoal.

Type specimens—Holotype: H.V.H. 8998. Paratype: H.V.H. 8999.

Remarks—The name "*chetumalensis*" reflects the restricted distribution of this ostracod, present only in Chetumal Bay.

A. chetumalensis is similar to *A. floridana* Benson and Coleman, but differs in several important features. *A. chetumalensis* exhibits a well-defined change in slope at the posterior cardinal angle, a well-defined flaplike termination of the ventral ridge, and larger size than *A. floridana*. There are minor differences in the development of the reticulate surface pattern and in the strength of the intervening ribs. In *A. chetumalensis* particularly, the horizontal ribbing is more pronounced.

Genus QUADRACYTHERE Hornibrook, 1952

Quadracythere producta (Brady), 1866
Figures 9m, n; 10b

Cythere producta Brady, 1866, p. 378, Pl. 59 (Figs. 7a-c).
?*Cythere compacta* Brady. Brady, 1870, p. 241, Pl. 30 (Figs. 18, 19).
(not) *Cythere compacta* Brady, 1866, p. 380, Pl. 62 (Figs. 3a-d).
Quadracythere producta (Brady). Van den Bold, 1968, p. 57-58, Pl. 5 (Figs. 1a, b).

Diagnosis—Characterized by its subquadrate lateral outline, pronounced posterodorsal ridge, slight, horizontal median ridge, well-defined posteroventral ridge, and reticulate surface.

Dimensions—Female: length 0.71 mm, height 0.45 mm. Male: length 0.76 mm, height 0.46 mm.

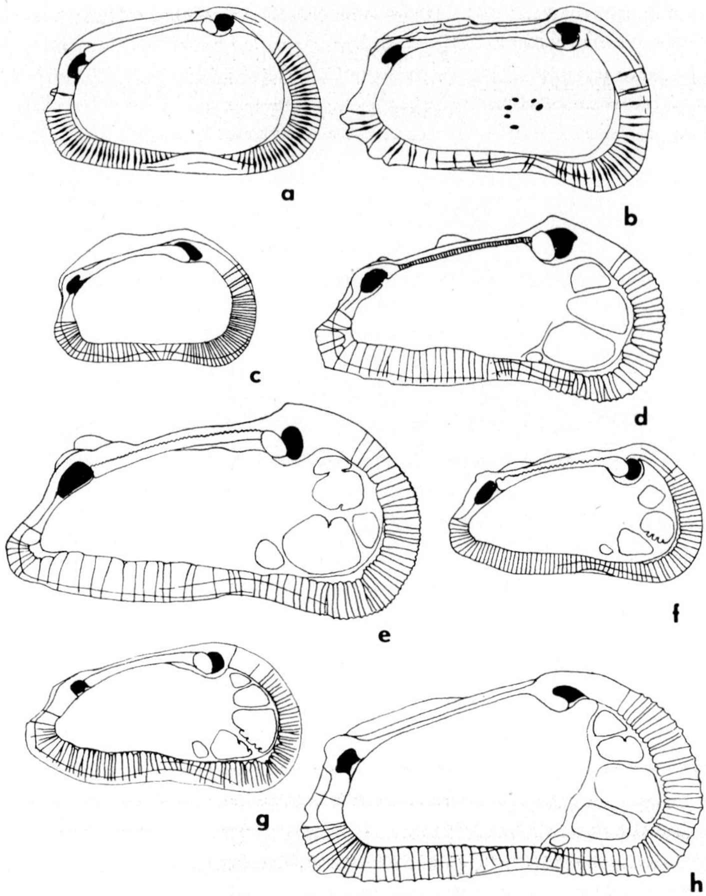

FIG. 10–Internal features.

a. *Aurila chetumalensis* Teeter, n. sp. Paratype, female, left valve, station 27, x51.
b. *Quadracythere producta* (Brady). Male, left valve, station 451, x51.
c. *Quadracythere lankfordi* Teeter, n. sp. Paratype, female, left valve, station 110, x51.
d. *Caudites nipeensis* van den Bold. Left valve, station 320, x96.
e. *Caudites angulata* Puri. Left valve, station 320, x90.
f. *Caudites howei* Puri. Left valve, station 320, x56.
g. *Caudites highi* Teeter, n. sp. Paratype, left valve, station 320, x58.
h. *Orionina bradyi* van den Bold. Left valve, station 320, x86.

Remarks—The two species herein assigned to *Quadracythere* resemble this genus in outline, development of ridges, muscle-scar pattern, possession of sieve-type pore canals, and, although rather weakly developed in comparison, reticulate ornamentation. They differ slightly from *Quadracythere* in lacking crenulations on the hinge bar and in the weak development of the ventral ridge.

In 1870 Brady figured a specimen which he assigned to *Cythere compacta* Brady. Although his illustrations are poor, this specimen is definitely not *C. compacta* but resembles closely his *C. producta*.

Distribution—This species was originally described from Turks Island in the Bahamas. Van den Bold found it ranging from upper Miocene to Holocene in the Dominican Republic. In Belize this species is largely restricted to lower Chetumal Bay, uncommonly appearing on the carbonate platform.

Quadracythere lankfordi Teeter, n. sp.
Figures 9j-l; 10c

Quadracythere compacta (Brady) 1866. Van den Bold, 1966f, p. 48, Pl. 1 (Figs. 9a, b).
(not) *Cythere compacta* Brady, 1866, p. 380, Pl. 62 (Figs. 3a-d).

Diagnosis—Characterized by small size, continuous anterior and dorsal ridges, and ventrally inclined termination of dorsal ridge.

Description—Carapace subquadrate in lateral view. Left valve overlapping right, midventrally, posterodorsally, and anterodorsally. Anterior margin broadly rounded; dorsal margin straight, slightly inclined posteroventrally; posterior margin with short, blunt, upturned projection. Subovate in dorsal view, depressed medially. Continuous anterior and dorsal crest, accomodating eye tubercle at anterior cardinal angle; dorsal ridge ventrally inclined posteriorly, also with short, anteriorly directed spur commencing about one third the length from the posterior margin at an angle of about 30° with the dorsal margin. Subdued ventral ridge becoming more prominent posteriorly. Fairly prominent dorsally flexed rib commencing medially and terminating near the anteroventral margin. Surface reticulate. Dimorphic, females shorter than males.

Hinge holamphidont, median hinge bar noncrenulate. Radial pore canals abundant. Vestibula exceedingly reduced. Normal pores common, scattered sieve-type.

Dimensions—Holotype (female): length 0.50 mm, height 0.35 mm. Male: length 0.53 mm, height 0.34 mm.

Type locality—Station 315, in 10 ft (3.0 m) of water near Tobacco Entrance.

Type specimens—Holotype: H.V.H. 9000. Paratype: H.V.H. 9001.

Remarks—This species is named in honor of R. R. Lankford.

The specimen (H.V.H. 8282) figured by van den Bold from Colón Harbour agrees well with this species, differing only in the strength of the ornamentation, the ridges being rather more exaggerated in the Belize specimens. R. H. Bate, of the British Museum of Natural History, has kindly compared specimens from Belize with the lectotype of *Cythere compacta* Brady, deposited in the Hancock Museum, and verifies that they are not conspecific with Brady's species.

Q. lankfordi closely resembles, and likely evolved from, *Q. bichensis* (van den Bold) 1963c. Van den Bold (1966f) also noted the similarity.

Distribution—This species is restricted to *Thalassia*-free areas of the carbonate platform.

Genus CAUDITES Coryell and Fields, 1937
Caudites nipeensis van den Bold, 1946
Figures 10d, 11a

Caudites nipeensis van den Bold, 1946, p. 103, Pl. 14 (Fig. 1); Keij, 1954b, p. 224, Pl. 4 (Fig. 14); Pl. 6 (Figs. 11a-d); van den Bold, 1957b, p. 239; 1963a, Text-fig. 6 (Fig. 10); 1963c, p. 386, Pl. 6 (Figs. 3a, b); 1964, p. 5, 7, Table 2; 1966c, p. 14, Tables 2, 5; 1966e, p. 361; 1968, p. 19.

Diagnosis—Characterized by emaciated appearance and two very prominent posterodorsal nodes.

Dimensions—Length 0.47 mm, height 0.24 mm.

Distribution—This species is reported from the Miocene to the Holocene in Trinidad, the Miocene of Cuba, and the Holocene of the Gulf of Paria and the north coast of Venezuela. The specimens in Belize are restricted to the carbonate platform.

Caudites angulata Puri, 1960
Figures 10e, 11b

Caudites angulata Puri, 1960, p. 130, Pl. 3 (Figs. 1, 2, 5, 6).
Caudites angulatus Puri. Van den Bold, 1963c, p. 386-387, Pl. 6 (Fig. 10).

Diagnosis—A species of *Caudites* characterized by a prominent median ridge.

Dimensions—Length 0.59 mm, height 0.32 mm.

Remarks—Puri, in the original description, figured a juvenile specimen (H.V.H. 4723). This specimen, however, has the dimensions of other

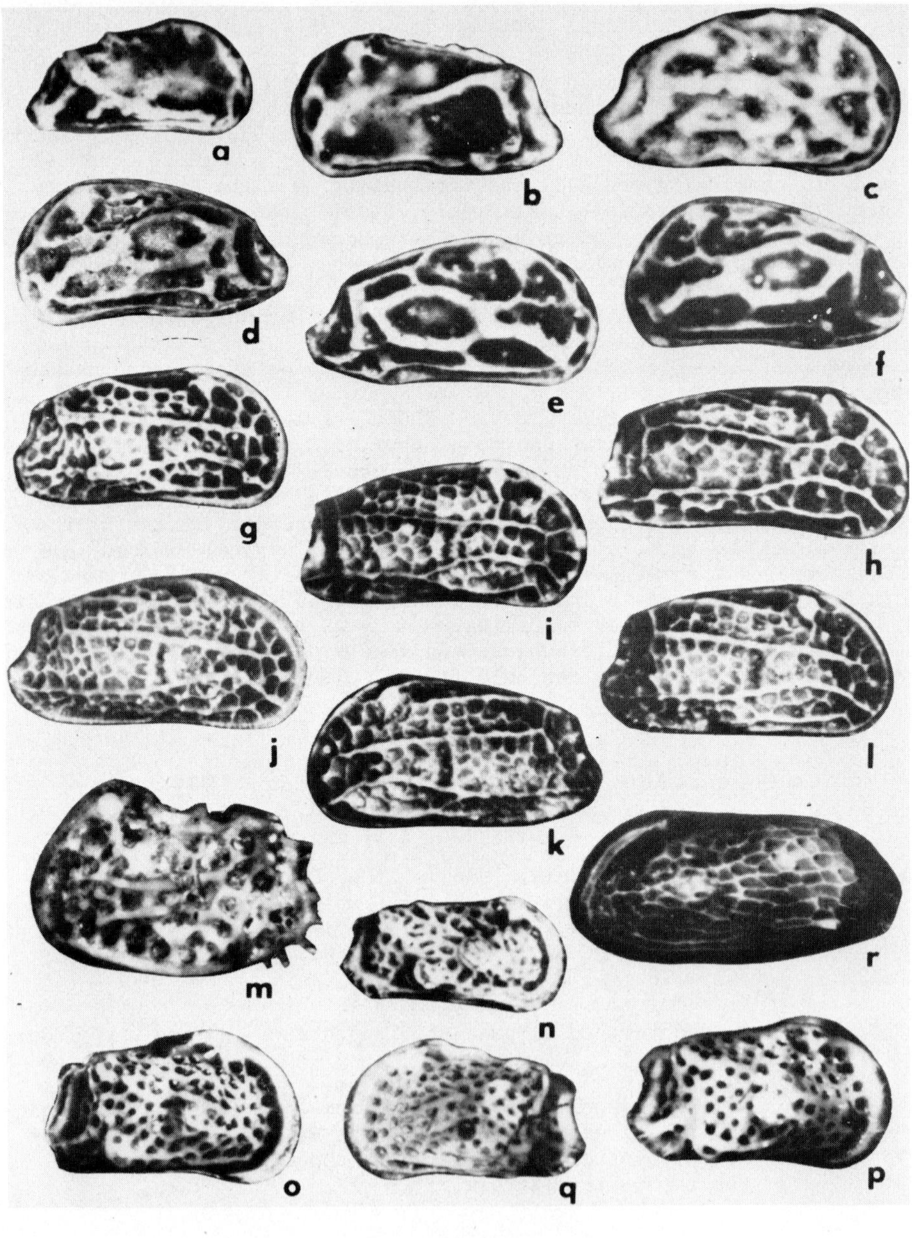

adults from the same sample, and thus probably represents the adult of a second species, possibly that of *C. rectangularis* (Brady) 1869.

C. angulata is similar to *C. rectangularis*, but differs in lacking the well-developed posteroventral ridge of the latter species. A closely related, perhaps ancestral, species is *C. medialis* Coryell and Fields, 1937, in which the transverse ridge crosses the valves obliquely, bifurcating anteriorly, whereas in *C. angulata* it terminates medially.

Distribution—Originally described by Puri from the Florida Keys area, *C. angulata* has also been found in the Miocene in Trinidad. The writer has observed this species in a sample from Bonaire, and in Belize it occurs on the carbonate platform.

Caudites howei Puri, 1960
Figures 10f, 11c

Caudites howei Puri, 1960, p. 130, Pl. 3 (Figs. 3, 4); van den Bold, 1964, p. 7, Table 2.

Diagnosis—A species of *Caudites* characterized by coarsely developed nodose ridges.

Dimensions—Length 0.58 mm, height 0.34 mm.

Remarks—This species resembles *C. rectangularis* (Brady) in the distribution of ridges, but those of *C. howei* are much more coarsely developed.

Distribution—This species was originally described by Puri from Florida. Van den Bold has observed it in the Holocene of Venezuela, and the writer has in his collection specimens from the Bahamas. This species occurs on the carbonate platform in Belize.

Caudites highi Teeter, n. sp.
Figures 10g, 11d-f

Diagnosis—Characterized by subtriangulate shape, expanded median ridge, and narrow submarginal ventral ridge.

⟵

FIG. 11

a. *Caudites nipeensis* van den Bold. Complete carapace, right valve, station 320, x60.
b. *Caudites angulata* Puri. Complete carapace, left valve, station 320, x55.
c. *Caudites howei* Puri. Complete carapace, right valve, station 320, x68.
d. *Caudites highi* Teeter, n. sp. Holotype (H.V.H. 9002), complete carapace, left valve, station 320, x55.
e. *C. highi* Teeter, n. sp. Paratype (H.V.H. 9003), complete carapace, right valve, station 320, x55.
f. *C. highi* Teeter, n. sp. Paratype, complete carapace, left valve, station 320, x55.
g. *Orionina bradyi* van den Bold. Complete carapace, right valve, station 320, x60.
h. *Orionina serrulata* (Brady). Male, complete carapace, right valve, station 73, x55.
i. *O. serrulata* (Brady). Female, complete carapace, right valve, station 73, x55.
j. *Orionina ebanksi* Teeter, n. sp. Holotype (H.V.H. 9004), complete carapace, right valve, station 153f, x64.
k. *O. ebanksi* Teeter, n. sp. Paratype (H.V.H. 9005), complete carapace, left valve, station 153f, x64.
l. *O. ebanksi* Teeter, n. sp. Paratype, complete carapace, right valve, station 153, x64.
m. *Jugosocythereis pannosa* (Brady). Female, complete carapace, left valve, station 54, x51.
n. *Hermanites hornibrooki* (Puri). Male, complete carapace, right valve, station 319, x60.
o. *Hermanites transoceanica* Teeter, n. sp. Holotype (H.V.H. 9006), complete carapace, right valve, station 313, x60.
p. *H. transoceanica* Teeter, n. sp. Complete carapace, right valve, Kawaikai, Hawaii, x60.
q. *H. transoceanica* Teeter, n. sp. Paratype (H.V.H. 9007), complete carapace, left valve, station 313, x60.
r. *Reussicythere reussi* (Brady). Male, complete carapace, left valve, station 302, x60.

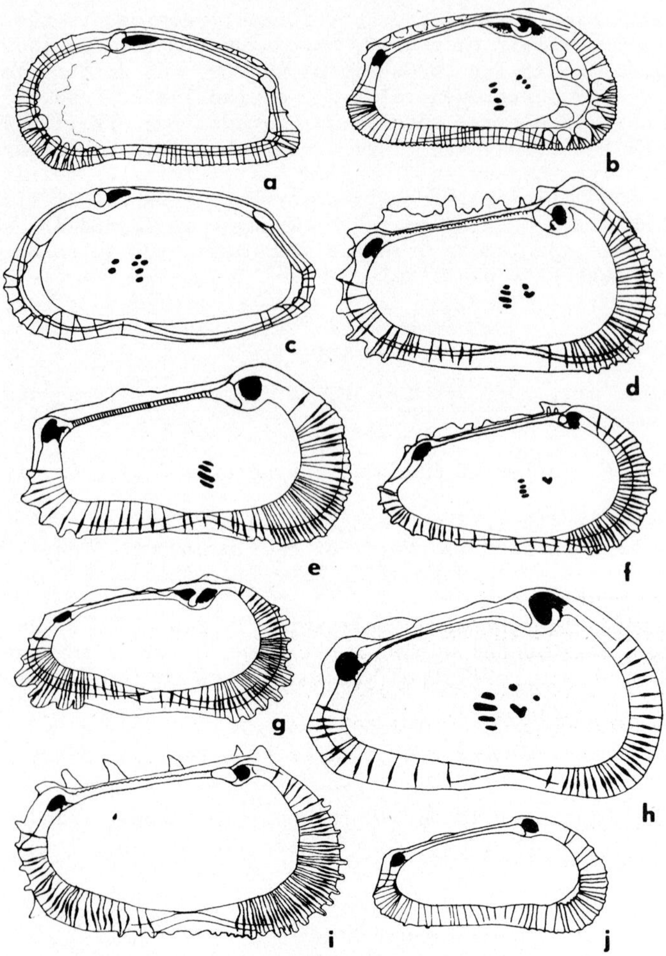

FIG. 12—Internal features.

a. *Orionina serrulata* (Brady). Female, right valve, station 278, x54.
b. *Orionina ebanksi* Teeter, n. sp. Paratype, left valve, station 153f, x58.
c. *Reussicythere reussi* (Brady). Female, right valve, station 302, x64.
d. *Jugosocythereis pannosa* (Brady). Male, left valve, station 54, x54.
e. *Hermanites hornibrooki* (Puri). Male, left valve, station 319, x80.
f. *Cativella pulleyi* Teeter, n. sp. Paratype, male, left valve, station 294, x58.
g. *Neocaudites pulchra* Teeter, n. sp. Paratype, female, left valve, station 85, x58.
h. *Hermanites transoceanica* Teeter, n. sp. Paratype, female, left valve, station 320, x83.
i. *Actinocythereis bahamensis* (Brady). Female, left valve, station 498, x54.
j. *Neocaudites scottae* Teeter, n. sp. Paratype, male, left valve, station 319, x61.

Description—Carapace subtriangular in lateral view; anterior margin broadly rounded; ventral margin sinuate; dorsal margin slightly convex with short concave part posterodorsally. Very short, blunt posterior margin. Surface of the carapace with several more or less prominent ridges. Vertical posterior ridge extending about two thirds the height from the posterior cardinal angle; obliquely connected ventrally to posteroventral margin by weakly developed ridge. Strong median ridge commencing near dorsal termination of vertical posterior ridge and extending obliquely, ventrally, to anteroventral margin; median ridge split in posterior half enclosing a horizontally oriented, ovate area. Submarginal horizontal ventral ridge beginning near anteroventral margin and termination of the posterior ridge. Several other short minor ridges present; surface between ridges very finely punctate.

Internal features characteristic for genus.

Dimensions—Holotype: length 0.63 mm, height 0.35 mm.

Type locality—Station 320, in 5 ft (1.5 m) of water, on reef apron near South Water Cay.

Type specimens—Holotype: H.V.H. 9002. Paratype: H.V.H. 9003.

Remarks—This species is named in honor of another of the writer's coworkers in the Belize project, Lee R. High.

C. highi resembles both *C. obliquecostatus* van den Bold and *C. rectangularis* (Brady), differing mainly in having subtriangulate rather than elongate shape, size, and disposition of minor ridges.

C. obliquecostatus is present but very scarce at station 209.

Distribution—This species is restricted to the carbonate platform.

Genus ORIONINA Puri, 1954

Orionina bradyi van den Bold, 1963
Figures 10h, 11g

Cythere bermudae Brady (in part), 1880, p. 90, Pl. 21 (Figs. 2a-d). New name for *Cythere serrulata* Brady 1869 (not Bosquet 1854). However, figured specimen differs from *Orionina serrulata* (Brady).

Orionina bermudae (Brady). Puri and Hulings, 1957, p. 188, Fig. 11; Hulings, 1958, p. 166; Puri and Vernon, 1959, p. 243; Puri, 1960, p. 123, 126, Pl. 1 (Figs. 15, 16).

not *Orionina bermudae* (Brady). Van den Bold, 1957b, p. 242, Table 1, *Orionina vaughani* (Ulrich and Bassler), Pl. 1 (Fig. 12; =*Orionina similis* n. sp.); van den Bold, 1958b, p. 403, (in part *Orionina vaughani*, in part *Orionina* sp. aff. *O. bradyi*), Pl. 5 (Fig. 9; =*Orionina serrulata*); Puri and Vernon, 1959, p. 151, 154, 156, 182, 197, 206 (=*Orionina vaughani*).

Orionina bradyi van den Bold, 1963a, p. 45-47, Pl. 3 (Figs. 7, 8); Text-fig. 6 (Figs. 5, 6, 7); Morales, 1966, p. 85, 87, Pl. 8 (Figs. 2a-c); Fig. 41.

Diagnosis—Characterized by relatively small carapace, subtriangular shape, and straight radial pore canals.

Dimensions—Length 0.59 mm, height 0.31 mm.

Remarks—This species is similar to *Orionina serrulata*, but may be easily distinguished by its smaller size and particularly by its straight radial pore canals.

Distribution—*O. bradyi* ranges throughout the Caribbean area, Florida, Bermuda, and the Bahamas. In Belize it is restricted to the carbonate platform.

Orionina serrulata (Brady), 1869
Figures 11h, i; 12a

Cythere serrulata Brady, 1869, p. 153, Pl. 18 (Figs. 11, 12). Not *Cythere serrulata* (Bosquet), Bosquet, 1854, p. 114, Pl. 9 (Figs. 9a-d) =*Cypridina serrulata* Bosquet, 1847, p. 370, Pl. 24 (Figs. 20a-e) =*Alatacythere serrulata* (Bosquet). Howe and Laurencich, 1958, p. 44.

Cythere bermudae Brady (in part), 1880, p. 90, not Pl. 21 (Figs. 2a-d).

Cythereis ramdohri Mueller, 1912, p. 358.

Cythereis vaughani (Ulrich and Bassler). Coryell and Fields, 1937, p. 9, Fig. 10.

Cythereis vaughani (Ulrich and Bassler) in part. Van den Bold, 1946, p. 88.

?*Cythereis (Elofsonella) reticulata* Hartmann, 1956a, p. 37, Figs. 45-52.

Orionina bermudae (Brady) in part. Van den Bold, 1957b, p. 242, Table 1, Pl. 1 (Figs. 12a, b); 1958, p. 403, Pl. 5 (Fig. 9).
Orionina reticulata (Hartmann). Van den Bold, 1958b, Pl. 5 (Fig. 8).
?Orionina reticulata (Hartmann). Hartmann, 1959a, p. 240, Pl. 48 (Figs. 197, 198).
Orionina serrulata (Brady). Van den Bold, 1963a, p. 44-45, Pl. 4 (Figs. 1-6); Text-Fig. 5 (Figs. 6, 7, 8a-d); 1963c, p. 386, Pl. 6 (Fig. 7; 1966f, p. 47, Pl. 1 (Fig. 8); 1967b, p. 311.

Diagnosis—Characterized by relatively large size, pronounced dimorphism, and peculiar bundling of anteroventral radial pore canals.

Dimensions—Female: length 0.68 mm, height 0.34 mm. Male: length 0.58 mm, height 0.35 mm.

Remarks—This species cannot be distinguished externally from *O. vaughani*. Internally, however, *O. vaughani* possesses straight, evenly distributed, radial pore canals, whereas those of *O. serrulata* are peculiarly bundled anteroventrally, rather as in *Leptocythere*.

Distribution—This species is widespread in the Caribbean area from upper Miocene to Holocene (van den Bold 1963a, p. 45). In Belize, *O. serrulata* is commonly present in the main lagoon and is extremely uncommon on the carbonate platform.

Orionina ebanksi Teeter, n. sp.
Figures 11j-1, 12b

Diagnosis—A species of *Orionina* characterized by an isolated anteroventral ridge, subparallel dorsal and ventral margins, and broad, bundled radial pore canals which tend to branch anteroventrally.

Description—Carapace subrectangulate. Ventral margin with slight concavity slightly anterior to middle, slightly upturned ventrally. Anterior margin broadly rounded. Dorsal margin straight with slight ventral inclination posteriorly, almost parallel with ventral margin. Posterior margin straight, slightly concave medially.

Dorsal ridge beginning approximately medially, slightly ventral to the dorsal margin, ascending almost to posterior cardinal angle, making a hairpin bend and extending to anteromedian position. Sinuate ventral ridge extending from posteroventral position to anteroventral. Short isolated anteroventral ridge extending from anteroventral to midventral position. Surface reticulate. Dimorphism not pronounced; females slightly higher posteriorly.

Inner features characteristic for genus. Anteroventral pore canals broad, bundled, as in *O. serrulata*; several bifurcate or trifurcate.

Dimensions—Holotype: length 0.62 mm, height 0.34 mm.

Type locality—Station 153f, in 2 ft (0.6 m) of water near Seal Cay.

Type specimens—Holotype: H.V.H. 9004. Paratype: H.V.H. 9005.

Remarks—This species is named in honor of James Ebanks, a coworker on the Belize project.

O. ebanksi is similar to the females of *O. serrulata* and *O. vaughani*. Externally, it differs from these species in being higher posteriorly and in possessing an isolated anteroventral ridge. Internally, *O. ebanksi* possesses bundled, thick, radial pore canals which exhibit a tendency to branch many times, differing from the bundled, thin, simple canals of *O. serrulata* and *O. reticulata*. *O. reticulata* is considerably smaller than the above three species.

Distribution—The writer has observed this species in samples from Klein Bonaire and Bermuda, in addition to Belize. In Belize it occurs on the carbonate platform.

Genus REUSSICYTHERE van den Bold, 1966

Reussicythere reussi (Brady), 1869

Figures 11r, 12c

Cythere reussi Brady, 1869, p. 153, Pl. 18 (Figs. 9, 10).
not *Cythere reussi* Brady. Brady, 1880, p. 74, Pl. 14 (Figs. 2a-d).
not *Cythere reussi* Prochazka, 1893, p. 79, Pl. 2 (Figs. 1a, b).
Reussicythere reussi (Brady). Van den Bold, 1966f, p. 46, Pl. 3 (Figs. 2a-d); Pl. 5 (Figs. 1a, b).

Diagnosis—A hemicytherid genus characterized by delicate reticulate ornamentation and prominent posteroventral spine.

Dimensions—Female: length 0.65 mm, height 0.32 mm. Male: length 0.69 mm, height 0.33 mm.

Distribution—This species was originally described from Colón, Panama. In Belize, it is almost completely restricted to the main lagoon.

Genus JUGOSOCYTHEREIS Puri, 1957

Jugosocythereis pannosa (Brady), 1869

Figures 11m, 12d

Cythere pannosa Brady, 1869, p. 154, Pl. 19 (Figs. 1, 2).
Cythereis vicksburgensis Howe and Law. Van den Bold, 1946, p. 89 (in part).
not *Jugosocythereis pannosa* (Brady). Van den Bold, 1957b, p. 240, Pl. 2 (Fig. 2).
Jugosocythereis pannosa (Brady). Van den Bold, 1966f, p. 47-48, Pl. 1 (Fig. 11).

Diagnosis—Characterized by heavy reticulate and spinose ornamentation.

Dimensions—Length 0.75 mm, height 0.46 mm.

Remarks—*J. pannosa* is remarkably similar to *J. vicksburgensis* (Howe and Law), differing only in its heavier ornamentation and increased spinosity. Whether such differences warrant specific separation is questionable.

Distribution—This species has been observed in samples in the writer's collection from Bermuda, the Bahamas, the Florida Keys, and the Caribbean. In Belize it occurs on the carbonate platform.

The only other extant species of *Jugosocythereis* known to the writer similarly inhabits shallow reefoidal areas near Sorong, New Guinea.

Genus HERMANITES Puri, 1955

Hermanites hornibrooki (Puri), 1960

Figures 11n, 12e

Trachyleberis sp. cf. *T. thoracophora* (van den Bold). Keij, 1954b, p. 223, Pl. 4 (Figs. 12a, b).
not *Cythereis thoracophora* van den Bold, 1946, p. 92, Pl. 10 (Figs. 25a, b).
Hermanites sp. aff. *H. hutchisoni* van den Bold. Van den Bold, 1957b, p. 240.
Hermanites sp. 2. Drooger and Kaasschieter, 1958, p. 90.
Bradleya hornibrooki Puri, 1960, p. 123, Pl. 1 (Figs. 9, 10); Text-figs. 27, 28.
Hermanites hornibrooki (Puri). Van den Bold, 1966b, p. 5; 1966f, Pl. 1 (Fig. 14); 1968, p. 66.

Diagnosis—A species of *Hermanites* characterized by its relatively small size, pronounced central tubercle, and nodes at termination of dorsal and ventral ridges.

Dimensions—Female: length 0.49 mm, height 0.27 mm. Male: length 0.48 mm, height 0.25 mm.

Distribution—First described from Florida, *H. hornibrooki* also is present in the shallow reefoidal areas near Trinidad. This species ranges back to the Miocene in the Caribbean. In Belize, it favors the shallow turbulent waters of the reef flat.

Hermanites transoceanica Teeter, n. sp.

Figures 11o-q, 12h

Diagnosis—Characterized by twin, oblique, weakly developed ridges on subcentral tubercle, and sieve-type pore canals.

Description—Carapace subquadrate in lateral view. Dorsal margin straight. Anterior margin bluntly rounded. Ventral margin straight, convex anteriorly. Posterior margin convex in ventral half, with several strong denticles, concave dorsally. Anterior marginal ridge commencing at anterior cardinal angle and terminating anteroventrally. Eye tubercle located on anterior ridge near cardinal angle. Dorsal ridge short, winglike extension confined to posterodorsal area. Fine ridge extending anteroventrally from this winglike extension across prominent subcentral tubercle. A second, more weakly defined, shorter ridge crossing the subcentral tubercle ventral to the first. Neither ridge reaches the anteroventral margin. Prominent posteroventral winglike extension. Surface reticulate.

Hinge holamphidont. Inner lamella of medium width. Radial pore canals abundant, straight, simple. Normal pores abundant, sieve-type. Three frontal muscle scars, one V-shaped. Adductor muscle scars elongate, second restricted medially.

Dimensions—Holotype: length 0.55 mm, height 0.34 mm.

Type locality—Station 313, in 2.5 ft (0.8 m) of water on the reef flat, south of Tobacco Entrance.

Type specimens—Holotype: H.V.H. 9006. Paratype: H.V.H. 9007.

Remarks—The trivial name "*transoceanica*" refers to the ostracod's widespread distribution, in the western Pacific (New Guinea), central Pacific (Hawaiian Islands), and the Caribbean. Hazel (1967) described a species of *Hermanites* from Hawaii, similar to *H. transoceanica* but lacking the twin ridges on the subcentral tubercle.

This is an unusual species of *Hermanites* in that it possesses sieve-type normal pores, a feature characteristic of the family Hemicytheridae, and an apparently V-shaped frontal scar with two subsidiary scars visible in fluoritized specimens. The V-shaped frontal scar is characteristic of the family Trachleberidae. Holden (1967) also noted the presence of a V-shaped frontal scar with four subsidiary scars in another species of *Hermanites* from Hawaii.

Distribution—This species is extremely widespread, occurring in the Pacific and Indo-Pacific areas, as well as in the Caribbean. In Belize it occurs on the carbonate platform.

Family TRACHYLEBERIDIDAE Sylvester-Bradley, 1948

Genus ACTINOCYTHEREIS Puri, 1953

Actinocythereis bahamensis (Brady), 1868

Figures 12i, 13a, b

Cythere bahamensis Brady, 1868, p. 241, Pl. 32 (Figs. 1, 2).
(not) *Actinocythereis bahamensis* (Brady). Curtis, 1960, Pl. 1 (Fig. 1).

?*Actinocythereis* sp. aff. *A. exanthemata* (Ulrich and Bassler). Benson and Coleman, 1963, p. 48, Pl. 6 (Figs. 12, 31).

Diagnosis—A species of *Actinocythereis* characterized by relatively delicately developed spines, particularly posteriorly.

Dimensions—Female: length 0.80 mm, height 0.46 mm. Male: length 0.83 mm, height 0.43 mm.

Remarks—Although Brady's original figures are of poor quality and apparently do not agree perfectly with the specimens from Belize, the writer believes, particularly on the basis of the fine posterior spines, that these Belize specimens are assignable to *A. bahamensis*.

The ostracod figured by Curtis is very clearly not *A. bahamensis*. The posterior spines are much too strongly developed.

Benson and Coleman figured several specimens which they concluded to have affinities with *A. exanthemata* (Ulrich and Bassler). This species also appears similar to *A. bahamensis*. The writer has observed another species of *Actinocythereis* in samples from Florida Bay which differs both from the Belize material and from *A. exanthemata*, and may be conspecific with species figured by Benson and Coleman.

Distribution—This species was originally described from New Providence in the Bahamas. In Belize it occurs uncommonly on the carbonate platform and in lower Chetumal Bay.

Genus CATIVELLA Coryell and Fields, 1937

Cativella pulleyi Teeter, n. sp.

Figures 12f, 13c-e

Bradleya? sp. Keij, 1954b, p. 223, Pl. 4 (Figs. 13a-c); Pl. 6 (Fig. 10).
?*Cythereis alata* Hartmann, 1956a, p. 39, Figs. 53-62.
Hermanites sp. 5 Drooger and Kaasschieter, 1958, p. 90.
Cativella semitranslucens (Crouch). Van den Bold, 1958a, p. 71.
Cativella sp. aff. *C. semitranslucens* (Crouch). Van den Bold, 1958b, p. 404, Pl. 3 (Fig. 3); 1963c, p. 389, Pl. 11 (Fig. 6); 1965b, p. 399.
?*Cativella* sp. cf. *C. runcinata* (Baird). Curtis, 1960, p. 481, Pl. 1 (Fig. 4).
(not) *Cythereis runcinata* Baird, 1850, p. 256, Pl. 18 (Figs. 7-9).

Diagnosis—A species of *Cativella* characterized by the presence of numerous interridge tubercles.

Description—Carapace subovate in lateral outline. Dorsal margin rounded, hinge line straight. Anterior and ventral margins evenly rounded. Anterior margin with numerous spines. Posterior margin convex below, with several prominent spines, and concave above. Anterior ridge commencing just posterior to and including prominent eye tubercle and extending marginally almost to posteroventral position. Prominent curved dorsal ridge. Well-developed, slightly oblique median ridge. Short, well-developed, ventral ridge terminating posteriorly by strong spine. All ridges supported by closely spaced spinelike extensions. Interridge surface with numerous tubercles. Third and earlier instars, and in a few specimens the preadult instar, lack the median ridge.

Hinge holamphidont. Inner lamella of medium width, nonvestibulate. Radial pore canals numerous. Normal pore canals simple, largely restricted to ridge "spines" and interridge nodes. Muscle scar trachyleberid.

Dimensions—Holotype: length 0.61 mm, height 0.38 mm.

Type locality—Station 283, at a depth of 34 ft (10.4 m) in the main lagoon, off South Stann Creek.

Type specimens—Holotype: H.V.H. 9008. Paratype: H.V.H. 9009.

Remarks—*C. pulleyi* is named in honor of Thomas E. Pulley of the Houston Museum of Natural Science.

This species is probably very closely related to *C. semitranslucens* (Crouch), but differs mainly in being very much smaller. Comparison with specimens of *C. semitranslucens* indicates that this species lacks the posterodorsal concavity of the margin and possesses far fewer interridge nodes than *C. pulleyi*. Van den Bold (1958b, Pl. 3 [Fig. 3]) illustrated a specimen that possesses fewer interridge tubercles than *C. pulleyi*, but agrees in all other characteristics, and is very likely conspecific. Curtis (1960) figured a specimen which would seem to be conspe-

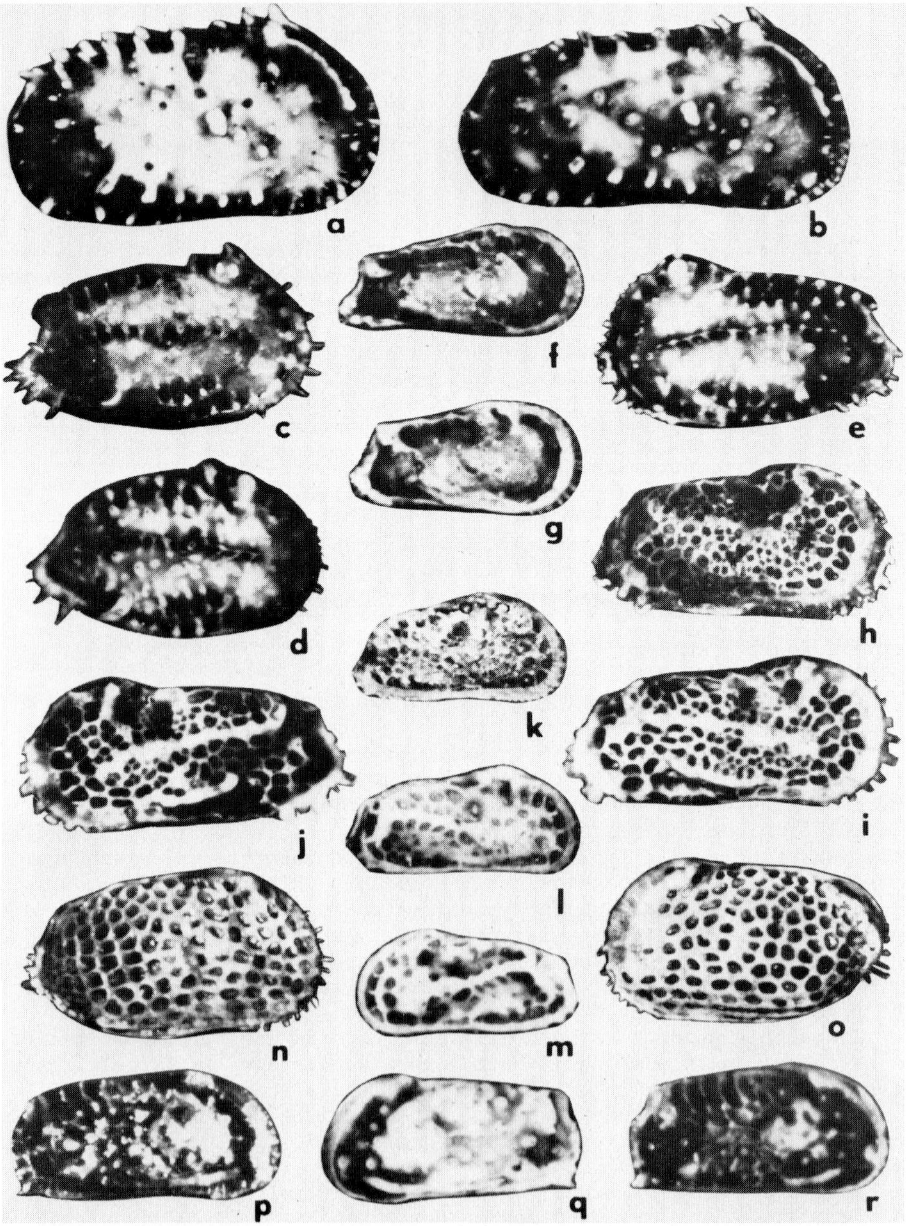

cific with the Belize species. However, specimens in the writer's collection, obtained from the northern Gulf of Mexico at depths similar to Curtis' specimens, belong to a species other than *C. pulleyi*.

Distribution—This species has been recorded from the Oligocene-Miocene to the Holocene of Trinidad (van den Bold, 1963c).

Analysis of Keij's (1954b) data indicated that half of the 50 occurrences of this species in the Gulf of Paria were in water 10 fm or deeper, whereas only 10 percent were found in water 6 fm or shallower. Curtis' (1960) species from the Mississippi delta region occurs in middle to outer neritic depths. In Belize, this species occurs in the main lagoon.

Genus OCCULTOCYTHEREIS Howe, 1951
Occultocythereis angusta van den Bold, 1963
Figures 13f, g; 14a

Cythereis deformis Brady, 1911, p. 597, Pl. 20 (Figs. 7, 8).
not *Cythereis deformis* Baird, 1850, p. 256, Pl. 18 (Figs. 4-6).
Occultocythereis angusta van den Bold, 1963c, p. 391-392, Pl. 9 (Figs. la-c); Pl. 12 (Fig. 6).

Diagnosis—Characterized by relatively small size, faint surface reticulation, and strongly developed node at posterior cardinal angle.

Dimensions—Female: length 0.50 mm, height 0.25 mm. Male: length 0.53 mm, height 0.24 mm.

Distribution—Originally described from Madeira, where it was collected at a depth of 70 fm, it subsequently has been recorded from the Holocene and upper Miocene of Trinidad. The writer has observed this species in other Holocene Caribbean samples. In Belize it is found on the carbonate platform.

←

FIG. 13

a. *Actinocythereis bahamensis* (Brady). Female, complete carapace, right valve, station 498, x60.

b. *A. bahamensis* (Brady). Male, complete carapace, right valve, station 498, x60.

c. *Cativella pulleyi* Teeter, n. sp. Holotype (H.V.H. 9008), female, complete carapace, right valve, station 283, x68.

d. *C. pulleyi* Teeter, n. sp. Paratype, female, complete carapace, right valve, station 292, x68.

e. *C. pulleyi* Teeter, n. sp. Paratype (H.V.H. 9009), male, complete carapace, left valve, station 292, x68.

f. *Occultocythereis angusta* van den Bold. Male, complete carapace, right valve, station 174, x60.

g. *O. angusta* van den Bold. Female, complete carapace, right valve, station 319, x60.

h. *Neocaudites pulchra* Teeter, n. sp. Holotype (H.V.H. 9010), female, complete carapace, right valve, station 33, x60.

i. *N. pulchra* Teeter, n. sp. Paratype (H.V.H. 9011), male, complete carapace, right valve, station 199, x60.

j. *N. pulchra* Teeter, n. sp. Paratype, male, complete carapace, left valve, station 199, x60.

k. *Neocaudites scottae* Teeter, n. sp. Holotype (H.V.H. 9012), female, complete carapace, right valve, station 320, x60.

l. *N. scottae* Teeter, n. sp. Paratype (H.V.H. 9013), male, complete carapace, right valve, station 320, x60.

m. *N. scottae* Teeter, n. sp. Paratype, male, complete carapace, left valve, station 313, x60.

n. *Ruggieria dictyon* van den Bold. Right valve, station 76, x60.

o. *R. dictyon* van den Bold. Left valve, station 76, x60.

p. *Puriana matthewsi* Teeter, n. sp. Holotype (H.V.H. 9014), complete carapace, right valve, station 289, x68.

q. *P. matthewsi* Teeter, n. sp. Paratype (H.V.H. 9015), complete carapace, left valve, station 171A, x68.

r. *P. matthewsi* Teeter, n. sp. Paratype, complete carapace, right valve, station 171A, x68.

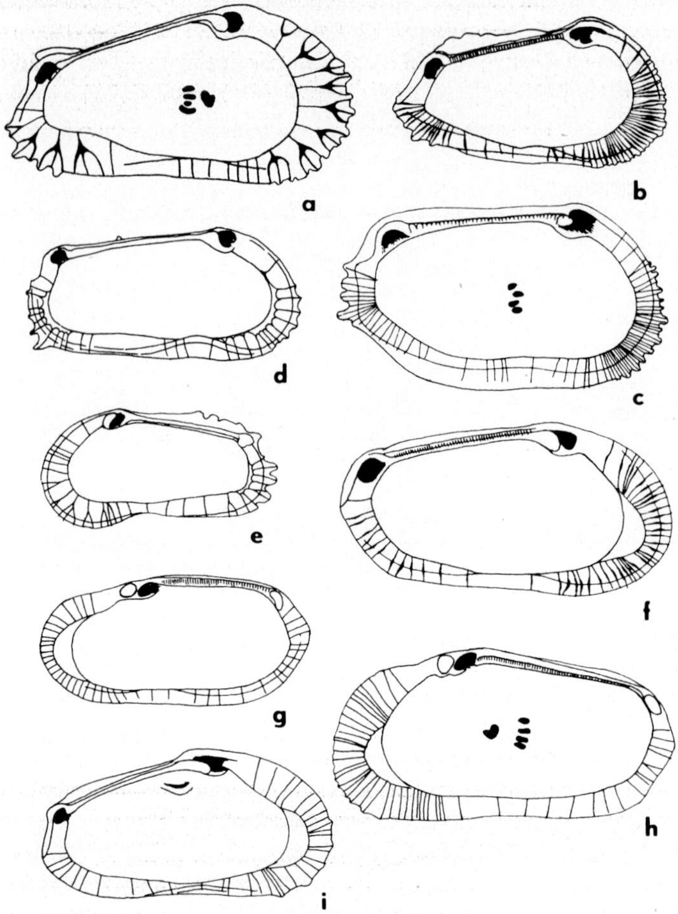

FIG. 14—Internal features.

a. *Occultocythereis angusta* van den Bold. Female, left valve, station 319, x90.
b. *Costa variabilocostata recticostata* van den Bold. Female, left valve, station 92, x51.
c. *Ruggieria dictyon* van den Bold. Left valve, station 219, x64.
d. *Puriana matthewsi* Teeter, n. sp. Paratype, left valve, station 171A, x70.
e. *Puriana convoluta* Teeter, n. sp. Paratype, right valve, station 315, x64.
f. *Basslerites twincayensis* Teeter, n. sp. Paratype, female, left valve, station 323, x96.
g. *Basslerites parvissima* Teeter, n. sp. Paratype, female, right valve, station 84, x96.
h. *Basslerites yoni* (Puri). Right valve, station 315, x102.
i. *Caribbella puseyi* Teeter, n. sp. Paratype, female, left valve, station 145f, x90.

Genus NEOCAUDITES Puri, 1960
Neocaudites pulchra Teeter, n. sp.
Figures 12g, 13h-j

Diagnosis—A species of *Neocaudites* characterized by reticulate ornamentation.

Description—Carapace subquadrate. Anterior margin evenly rounded with single row of large squamose denticles; posterior margin bluntly pointed with inner row of small, and outer row of large, squamose denticles. Dorsal and ventral margins slightly sinuate. Compressed in dorsal view. Dorsal ridge expanded posteriorly near posterior cardinal angle, and ventrally deflected. Poorly defined vertical sulcus just anterior to dorsal ridge. Ventral extension of dorsal ridge terminated about one fourth the height from the dorsal margin, with narrow ridge passing obliquely from termination to approximately the center of the valves. Thin, slightly obliquely inclined, submarginal, midventral ridge. This ridge slightly depressed in right valves of males; males much more elongate.

Hinge holamphidont. Inner lamella relatively wide. Radial pore canals numerous anteriorly. Anterior vestibulum relatively broad. Normal pore canals simple. Muscle scar normal for Trachyleberidinae.

Dimensions—Holotype (female): length 0.65 mm, height 0.36 mm. Male: length 0.75 mm, height 0.35 mm.

Type locality—Station 33, in 8 ft (2.4 m) of water just south of Congrejo Cay.

Type specimens—Holotype: H.V.H. 9010. Paratype: H.V.H. 9011.

Remarks—This genus is closely related to *Occultocythereis*, particularly in the position of the sulcus, and the similarity of the dorsal and, to some extent, the ventral ridge. Also, at least in some species, the males, in addition to attaining greater length, exhibit the same ventral dimorphic asymmetry. These genera differ markedly, however, in radial pore canals and presence of an oblique median ridge. Those species of *Occultocythereis* with unbranching radial pore canals, mentioned by van Morkhoven (1963, p. 197), may belong to *Neocaudites*.

Distribution—This species is present but scarce throughout Belize, on some of the coral patches nearshore in the Gulf of Honduras, and in lower Chetumal Bay.

Neocaudites scottae Teeter, n. sp.
Figures 12j, 13k-m

Diagnosis—Characterized by its relatively small size and prominent anterior reticulations.

Description—Carapace subtriangulate to subquadrate in outline, small to medium size. Dorsal margin straight, anterior margin broadly rounded, with numerous spines. Ventral margin slightly concave medially. Posterior margin convex below, with fewer spines, and concave above. Compressed in dorsal view. Dimorphism marked, males more elongate, subquadrate in lateral outline; females shorter, subtriangulate.

Anterior, ventral, and posterior margins with low, rounded marginal rim. Low, dorsal, arched ridge commencing middorsally and terminating at posterior cardinal angle in prominent node. Short ridge extends posteroventrally from this node. Larger ridge extends anteroventrally from the node to midvalve area and is accented by several paired punctations on each side. Surface smooth with scattered punctations and/or reticulations.

Hinge holamphidont, median bar crenulate. Inner lamella relatively wide. Numerous radial pore canals anteriorly, fewer ventrally and posteriorly. Well-developed anterior vestibulum of slightly irregular shape. Normal pore canals simple. Muscle scar trachyleberid.

Dimensions—Holotype (female): length 0.46 mm, height 0.25 mm. Male: length 0.49 mm, height 0.24 mm.

Type locality—Station 320, in 5 ft (1.5 m) of water on the reef apron near South Water Cay.

Type specimens—Holotype: H.V.H. 9012. Paratype: H.V.H. 9013.

Remarks—*N. scottae* is named to honor of another coworker on the Belize project, Martha Scott.

This species is very similar to, and likely evolved from, *N. macerta* (Stephenson). The Holocene species may be distinguished by its much smaller size and more widespread ornamentation.

Distribution—This species is restricted to the carbonate platform.

Genus RUGGIERIA Keij, 1957
Ruggieria dictyon van den Bold, 1966
Figures 13n, o; 14c

Trachyleberis goujoni (Brady). Keij, 1954b, p. 222, Pl. 4 (Fig. 10); Pl. 6 (Fig. 8).
Ruggieria dictyon van den Bold, 1966f, p. 49-51, Pl. 3 (Figs. 1a, b); Pl. 5 (Fig. 5).

Diagnosis—Characterized by posteroventral submarginal spine, numerous anterior marginal spines, five posterior marginal spines, and coarsely reticulate ornamentation with anterior marginal smooth area in left valve.

Dimensions—Length 0.68 mm, height 0.41 mm.

Distribution—This species has been recorded from the Gulf of Paria and Colón Harbour, Panama. In Belize it is located principally in near-shore areas, except Chetumal Bay.

Genus COSTA Neviani, 1928
Costa variabilocostata recticostata van den Bold, 1970
Figures 14b, 15d

Costa variabilocostata (van den Bold). Van den Bold, 1963c, p. 370, Table 6 (in part); 1964, p. 9-11 (in part).
Costa variabilocostata variabilocostata (van den Bold). Van den Bold, 1966c, p. 10, Table 1 (in part), p. 11, 27, Table 2; 1966a, p. 183, Pl. 22 (Fig. 5); 1966f, Pl. 1 (Fig. 10); Baker and Hulings, 1966, p. 116, Pl. 2 (Fig. 4).

⟶

FIG. 15

a. *Basslerites twincayensis* Teeter, n. sp. Holotype (H.V.H. 9016), female, complete carapace, right valve, station 323, x85.

b. *B. twincayensis* Teeter, n. sp. Paratype (H.V.H. 9017), male, complete carapace, left valve, station 323, x85.

c. *B. twincayensis* Teeter, n. sp. Paratype, male, complete carapace, right valve, station 323, x77.

d. *Costa variabilocostata recticostata* van den Bold. Female, complete carapace, right valve, station 337, x60.

e. *Basslerites parvissima* Teeter, n. sp. Holotype (H.V.H. 9018), female, complete carapace, right valve, station 85, x110.

f. *B. parvissima* Teeter, n. sp. Paratype (H.V.H. 9019), male, complete carapace, right valve, station 84, x110.

g. *Basslerites yoni* (Puri). Complete carapace, right valve, station 316, x94.

h. *Puriana convoluta* Teeter, n. sp. Holotype (H.V.H. 9020), complete carapace, left valve, station 315, x60.

i. *P. convoluta* Teeter, n. sp. Paratype, complete carapace, left valve, station 315, x60.

j. *P. convoluta* Teeter, n. sp. Paratype (H.V.H. 9021), complete carapace, right valve, station 315, x60.

k. *Caribbella puseyi* Teeter, n. sp. Holotype (H.V.H. 9022), female, complete carapace, right valve, station 145f, x85.

l. *C. puseyi* Teeter, n. sp. Paratype (H.V.H. 9023), male, complete carapace, left valve, station 145f, x85.

m. *C. puseyi* Teeter, n. sp. Paratype, male, right valve, station 145f, x85.

n. *Pterygocythereis hondurasensis* Teeter, n. sp. Holotype (H.V.H. 9024), female, complete carapace, right valve, station 146f, x51.

o. *P. hondurasensis* Teeter, n. sp. Paratype (H.V.H. 9025), male, complete carapace, left valve, station 180, x60.

p. *P. hondurasensis* Teeter, n. sp. Paratype, male, complete carapace, right valve, station 180, x60.

q. *Protocytheretta pumicosa* (Brady). Complete carapace, right valve, station 320, x55.

Costa variabilocostata recticostata van den Bold, 1970, p. 68, Text-fig. 1, Table 1, Pl. 1 (Fig. 4).

Diagnosis—Distinguished from other subspecies of *C. variabilocostata* by its smaller size and details of ornamentation.

Dimensions—Female: length 0.70 mm, height 0.40 mm. Male: length 0.63 mm, height 0.30 mm.

Distribution—Originally described from the middle Miocene of Venezuela, this species subsequently has been observed by van den Bold from the lower Miocene to Holocene in Trinidad. In Belize it is found in the main lagoon.

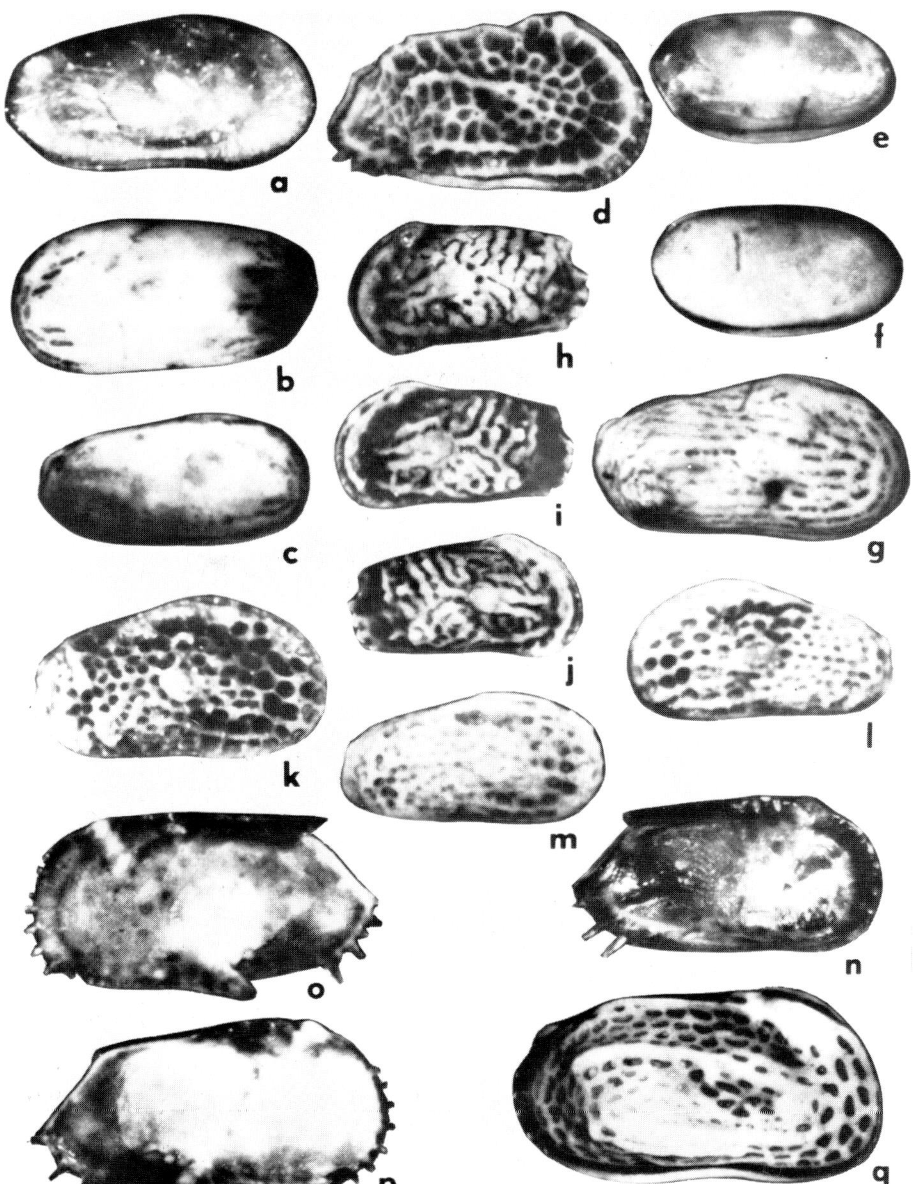

Genus PURIANA Coryell and Fields, 1953

Puriana matthewsi Teeter, n. sp.

Figures 13p-r, 14d

Diagnosis—A species of *Puriana* characterized by narrow ridges and tubercles, and weak development of posterior spines.

Description—Carapace subquadrate. Anterior margin broadly rounded with a few denticles. Ventral margin almost perfectly straight, very slightly concave anteriorly. Posterior margin vertical, with one prominent ventral spine and several subdued spines ventrally, slightly concave dorsally. Dorsal margin straight. Pronounced anterior marginal rim. Submarginal ventral ridge prominent anteriorly, reduced posteriorly. Subcentral tubercle enlarged, spreading. Surface with thin vertical ridges in posterodorsal quadrant; elsewhere with thin ridges and tubercles. Ornamentation variably developed. Dimorphism slight, males longer than females.

Internal features characteristic for the genus.

Dimensions—Holotype: length 0.51 mm, height 0.27 mm.

Type locality—Station 289, in the main lagoon at a depth of 79 ft (24 m).

Type specimens—Holotype: H.V.H. 9014. Paratype: H.V.H. 9015.

Remarks—This species, named in honor of R. K. Matthews of Brown University, is similar to *P. rugipunctata* (Ulrich and Bassler) but differs in the more finely developed nature of the posterodorsal ridges, in the possession of tubercles, and in the weakly spinose posterior margin. *P. matthewsi* also approaches *P. floridana* Puri, which is similarly tuberculate, but lacks the pronounced posterodorsal ridges and well-developed posterior spines.

Distribution—This species is restricted to the main lagoon.

Puriana convoluta Teeter, n. sp.

Figures 14e, 15h-j

Diagnosis—A species of *Puriana* characterized by its small size; four broad, short, posterior spines; and anteroventral and posterior, branching pore canals.

Description—Carapace small, subquadrate in lateral view. Anterior margin broadly rounded. Ventral margin concave medially. Posterior margin slightly convex in ventral two thirds with four broad, short spines; concave in dorsal third. Dorsal margin straight. Anterior marginal rim as in other species of *Puriana*. Short, submarginal, anteroventral ridge. Surface with numerous rounded ridges presenting the appearance of brain tissue; hence the name "*convoluta.*"

Internal features characteristic for genus with the exception of the peculiarly branching anteroventral and posterior pore canals.

Dimensions—Holotype: length 0.50 mm, height 0.28 mm.

Type locality—Station 315, in 10 ft (3.0 m) of water on the south side of Tobacco Entrance.

Type specimens—Holotype: H.V.H. 9020. Paratype: H.V.H. 9021.

Remarks—This species is remarkably similar externally to a Holocene species from Florida assigned by Puri to *P. rugipunctata* (Ulrich and Bassler). The writer doubts this identification, because the Florida specimens are considerably smaller than *P. rugipunctata* and differ slightly in ornamentation. The Belize specimens are slightly shorter than those from Florida, but differ only superficially in external ornamentation. Internally, however, *P. convoluta* possesses far fewer anterior marginal pore canals, and those in the anteroventral and posterior positions have multiple branches, a feature not observed in fossil or Holocene specimens of *P. rugipunctata*.

Distribution—This species is apparently widespread throughout the Caribbean, as the writer has observed it in samples from the Bahamas, Bonaire, Curaçao, the Virgin Islands, and probably Bermuda.

The very closely related Holocene Florida species inhabits the ecologically variable Florida Bay, whereas *P. convoluta* apparently prefers a much more constant environment on the carbonate platform.

Genus BASSLERITES Howe, 1937
Basslerites parvissima Teeter, n. sp.
Figures 14g; 15e, f

Diagnosis—Characterized by bluntly pointed posterior margin; depressed, fork-shaped posterior area; and relatively short, deep, anterior vestibulum.

Description—Carapace extremely small, subovate. Left valve larger than right, conspicuously so ventrally. Dorsal margin straight. Anterior margin very broadly rounded. Ventral margin convex. Posterior margin convex, bluntly pointed about one third the height from the dorsal margin. Surface smooth. Posterior surface with slightly depressed, fork-shaped area. Normal pores concentrated in the depressed area. Dimorphic, the females higher and slightly longer than the males.

Hinge holamphidont, median bar crenulate. Relatively wide attachment area between inner and outer lamella. Simple, radial pore canals common anteriorly and posteriorly. Anterior vestibulum relatively short and deep.

Dimensions—Holotype (female); length 0.36 mm, height 0.21 mm. Male: length 0.37 mm, height 0.20 mm.

Type locality—Station 85, in 61 ft (18.6 m) of water in the main lagoon off Mullins River.

Type specimens—Holotype: H.V.H. 9018. Paratype: H.V.H. 9019.

Remarks—The trivial name "*parvissima*" is from the Latin, meaning most small.

This species very closely resembles *B. minutus* van den Bold, 1958b, but differs consistently in several features. *B. parvissima* is slightly smaller than *B. minutus* and is more pointed posteriorly. Also, the depressed area and outline of the anterior vestibulum differ slightly in shape. *B. minutus* possesses several bifurcating radial pore canals.

Distribution—This species is restricted to the main lagoon in Belize.

Basslerites twincayensis Teeter, n. sp.
Figures 14f, 15a-c

Diagnosis—Characterized by anterior and posterior slitlike pits in an otherwise smooth carapace.

Description—Carapace subovate. Dorsal margin straight. Anterior margin broadly rounded. Ventral margin slightly concave medially. Posterior margin bluntly pointed. Left valve slightly larger, overlap greatest ventrally. Dimorphism apparent; females higher and wider posteriorly in dorsal view. Surface largely smooth. Posteriorly depressed as in certain other *Basslerites* species, but with numerous, short, slitlike pits in the depressed areas. Some pits also confined to and parallel with the anterior margin.

Internal features as in most species of *Basslerites*.

Dimensions—Holotype (female): length 0.45 mm, height 0.24 mm. Male: length 0.45 mm, height 0.23 mm.

Type locality—Station 323, in 3 ft (0.9 m) of water in *Thalassia* near South Water Cay.

Type specimens—Holotype: H.V.H. 9016. Paratype: H.V.H. 9017.

Remarks—The possession of slitlike pits is a feature not previously recorded in the genus *Basslerites*. However, all other features agree perfectly with the concept of this genus, and the presence of this new feature is not regarded as sufficient for the creation of a new genus.

Distribution—*Basslerites twincayensis* is restricted to the carbonate platform.

Basslerites yoni (Puri), 1960
Figures 14h, 15g

Leptocythere yoni Puri, 1960, p. 114, Pl. 4 (Figs. 8, 9).

Diagnosis—A species of *Basslerites* characterized by a reticulate arrangement of slitlike pits.

Dimensions—Length 0.44 mm, height 0.24 mm.

Remarks—Puri (1960) assigned this species to the genus *Leptocythere*, noting, "On shape and general outline of the carapace this species perhaps belongs to *Leptocythere*. Further material is needed before it could be definitely assigned to this genus." Puri further states that internal features were not observed as all the valves were articulated.

Several paired valves were separated in order to examine internal features, all of which identify this species as belonging to *Basslerites*. Anterior marginal features, particularly the numerous, straight, unbranching, radial pore canals and small, crescent-shaped vestibulum, are most diagnostic of this genus.

Van den Bold (1968) discussed this species in relation to a fossil species from the Yague Group of the Northern Dominican Republic, suggesting that it may be a new genus related to the Leptocytherinae. However, the lateral outline and internal features of the carapace agree well with *Basslerites*. This species lacks the irregular vestibulum, polyfurcate radial pore canals, and the complex hinge of the Leptocytherinae. The ornamentation of *B. yoni* seemingly represents the final stage in a progression from the irregularly depressed posterior surface of some species *Basslerites* (e.g., *B. parvissima*) through the anterior and posterior slitlike pits of *B. twincayensis*.

Brady (1869) described an ostracod from St. Vincent, *Cythere macra*, which agrees very closely with the above species in size and ornamentation. The lateral outline, particularly posteriorly, disagrees, but this may reflect the inaccuracies typical of many of the illustrations in *Les Fonds de la Mer*. *Cythere jurinei costata* Brady, 1866, resembles this species even more closely, but it is apparently larger and is known only from the Levant.

Distribution—This species has been recorded previously from Florida. In Belize it is restricted to the carbonate platform.

CARIBBELLA Teeter, n. gen.

Diagnosis—A trachyleberid genus characterized by its small size, rounded triangulate shape, and punctate to reticulate ornamentation.

Description—As no other species of *Caribbella* are known, the description of the genus is essentially that of the species below.

Remarks—The affinities of this genus are difficult to assess. Outwardly, it resembles *Triginglymus*, differing somewhat in shape. Internally, *Caribbella* possesses a so-called "anti-slip tooth" similar to *Triginglymus*, but has a V-shaped frontal scar rather than a double scar. *Triginglymus* is apparently restricted to the Eocene.

The shape and ornamentation agree closely with those of the genus *Proteoconcha* (Plusquellec and Sandberg, 1969), but the V-shaped frontal scar and simple, normal pore canals of the specimens from Belize preclude their inclusion in the subfamily Campylocytherinae.

Caribbella puseyi Teeter, n. sp.
Figures 14i, 15k-m

Description—Carapace small, subtriangulate. Dorsal margin straight. Anterior margin very broadly rounded with four or five short denticles. Ventral margin slightly concave medially. Posterior margin bluntly rounded. Left valve slightly larger, overlap conspicuous anterodorsally and ventrally. In dorsal view, anterior margin broad (widest near mid-valve), posterior truncate. Dimorphic, females higher in lateral view. Surface punctate, depressions arranged in more or less horizontal rows. Slightly raised subcentral tubercle.

Hinge holamphidont. So-called "anti-slip tooth" ventral and posterior to the anterior hinge elements in both valves. Radial pore canals common, straight, and simple. Normal pore canals simple, scarce. Anterior vestibulum crescent-shaped. Frontal muscle scar V-shaped. Other muscle scars not observed.

Dimensions—Holotype (female): length 0.43 mm, height 0.25 mm. Male: length 0.39 mm, height 0.20 mm.

Type locality—Station 145f, in 57 ft (17.4 m) of water in the main lagoon, north of Port of Honduras.

Type specimens—Holotype: H.V.H. 9022. Paratype: H.V.H. 9023.

Remarks—This species is named after Walter C. Pusey, a coworker on the Belize project.

Distribution—This species is restricted to the main lagoon.

Family BRACHYCYTHERIDAE Puri, 1954

Genus PTERYGOCYTHEREIS Blake, 1933

Pterygocythereis hondurasensis Teeter, n. sp.
Figures 15n-p, 16a

Pterygocythereis sp.? van den Bold, 1966f, p. 49, Pl. 3 (Fig. 5); Pl. 5 (Fig. 4).

Diagnosis—Characterized by shortened dorsal crest in right valve, confluence of dorsal and anterior crests, and double postalar spines.

Description—Carapace large, elongate, pellucid. Anterior margin evenly rounded. Dorsal margin straight. Ventral margin parallel dorsally, upturned posteriorly to posterior angulation. Posterodorsal margin slopes at about 45°. Dorsal margin in left valve with continuous fluted crest, housing prominent eye tubercle, and extending along anterior margin terminating anteroventrally in a series of flattened spines. Dorsal crest in right valve lacking posteriorly. Posteroventral margin with four or five elongate spines. Surface smooth with prominent fluted ventral ala. Two postalar spines continuing in plane of alae.

Internal features typical for genus.

Dimensions—Holotype (male): length 0.80 mm, height 0.41 mm. Female: length 0.78 mm, height 0.45 mm.

Type locality—Station 146f, in 95 ft (29 m) of water in the main lagoon, north of Port of Honduras.

Type specimens—Holotype: H.V.H. 9024. Paratype: H.V.H. 9025.

Remarks—The writer has compared this species with specimens in his collection from the northern Gulf of Mexico assigned to *Pterygocythereis* sp. aff. *P. americana* by Benson and Coleman (1963). The Gulf of Mexico specimens differ most in their strongly dimorphic dorsal crest and position of postalar spines. These also lack the continuous dorsal and ante-

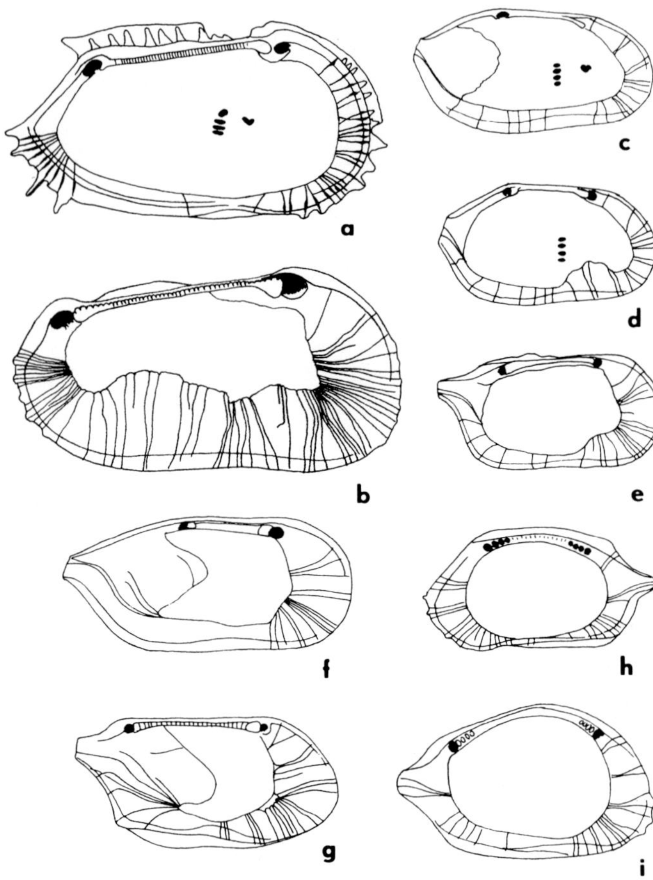

FIG. 16–Internal features.

a. *Pterygocythereis hondurasensis* Teeter, n. sp. Paratype, female, left valve, station 180, x64.
b. *Protocytheretta pumicosa* (Brady). Left valve, station 320, x61.
c. *Cytherura* sp. ex. gr. *C. johnsoni* Mincher. Male, left valve, station 2, x64.
d. *C.* sp. ex. gr. *C. johnsoni* Mincher. Female, left valve, station 2, x64.
e. *Cytherura sablensis* (Benson and Coleman). Female, left valve, station 278, x64.
f. *Cytherura subwardensis* Teeter, n. sp. Paratype, male, left valve, station 337, x102.
g. *Kangarina ancycla* van den Bold. Female, left valve, station 179, x96.
h. *Hemicytherura cranekeyensis* Puri. Right valve, station 323, x96.
i. *Hemicytherura bradyi* (Puri). Left valve, station 319, x96.

rior crests. This species resembles *P. americana* (Ulrich and Bassler) from the Atlantic and Gulf Coast Miocene, but differs in lacking the *en échelon* arrangement of the dorsal crest. There can be little doubt, however, that the latter species and *P. hondurasensis* are very closely related, for the juveniles of the Holocene species possess the *en échelon* dorsal crest in both valves.

The specimens from Belize agree rather closely with some described by van den Bold (1966f) from Colón Harbour, Panama, differing only in the lack of a crest.

Distribution—This species is restricted to the main lagoon.

Family CYTHERETTIDAE Triebel, 1952

Genus PROTOCYTHERETTA Puri, 1958

Protocytheretta pumicosa (Brady), 1866
Figures 15q, 16b

Cythere pumicosa Brady, 1866, p. 379, Pl. 61 (Figs. 3a-c).
Cythere danaiana Brady, 1869, p. 124, 241, Pl. 14 (Figs. 13, 14).
Paracytheretta daniana (Brady). Puri, 1952a, p. 210, Pl. 40 (Figs. 10, 11); Text-fig. 11.
Cytheretta daniana (Brady). Puri and Hulings, 1957, p. 174, 187, Fig. 11.
Cytheretta danaiana? (Brady). Hulings, 1966, p. 46, Pl. 6b; 1967, p. 642, Fig. 30.
(not) *Cytheretta daniana* (Brady). Curtis, 1960, p. 481, Pl. 2 (Fig. 2); Text-fig. 9.
Protocytheretta daniana (Brady). Puri, 1958b, p. 188, Pl. 3 (Figs. 7-11); 1960, p. 111, Pl. 1 (Figs. 1, 2); Text-figs. 18, 19; Benson and Coleman, 1963, p. 26-27, Pl. 5 (Figs. 5, 7, 9, 10); Fig. 13.

Diagnosis—Characterized by heavily reticulate ornamentation; three prominent, subhorizontal ridges; and five rounded, evenly spaced, posterior spines on each valve.

Dimensions—Length 0.84 mm, height 0.48 mm.

Remarks—Benson and Coleman (1963) noted the similarity of *Cythere pumicosa* and *Cythere daniana*, and suggested that the types be reexamined in order to establish whether these two species are synonymous. Previous workers apparently have not noted Brady's statement (1869, p. 241): "M. Brady croit . . . à l'identité du *Cythere danaiana* . . . avec le *Cythere pumicosa* . . . ." Also, the spelling of the junior synonym is not *daniana*, as recorded by Puri and accepted by succeeding workers, but *danaiana*.

The writer has examined material from Breton Island which contains specimens of the species incorrectly identified by Curtis as *C. danaiana*.

Distribution—Originally noted at Turk's Island in the Bahamas, this species was also subsequently noted by Brady from New Providence. Later workers have recorded it in Holocene sediments off Florida and from the Pliocene Caloosahatchee Formation of southern Florida. In Belize it is restricted to the carbonate platform.

Family CYTHERURIDAE Mueller, 1894

Genus CYTHERURA Sars, 1866

Cytherura sp. ex. gr. *C. johnsoni* Mincher, 1941
Figures 16c, d; 17a, b

Cytherura johnsoni Mincher, 1941, p. 343, Pl. 47 (Fig. 1); Swain, 1955, p. 627, Pl. 64 (Figs. 8a-c); Text-figs. 35b, 38 (Figs. 8a, b), 39 (Figs. 1a-c); Puri and Hulings, 1957, p. 187, Fig. 11; Puri, 1960, p. 114-115, Pl. 4 (Figs. 14, 15); Benson and Coleman, 1963, p. 31-32, Pl. 6 (Figs. 1-5); Fig. 18; Pooser, 1965, p. 45-46, Pl. 10 (Figs. 1-4, 6); Grossman, 1967, p. 70, Pl. 13 (Fig. 1); Pl. 19 (Figs. 1, 3).
Cytherura johnsoni Mincher? van den Bold, 1963c, p. 395, Pl. 9 (Fig. 3).
Cytherura sp. ex. gr. *C. johnsoni* van den Bold, 1966c, Table 2.
Cytherura ex. gr. *johnsoni* van den Bold, 1967b, p. 307, Table 1, Pl. 1 (Fig. 12).
Cytherura sp. cf. *lineata* (Brady). Van den Bold, 1946, p. 119, Pl. 9 (Fig. 17); Keij, 1954b, p. 227, Pl. 5 (Fig. 12).
Cytherura sp. 4. Drooger and Kaasschieter, 1958, p. 90.

Diagnosis—Characterized by its surface ornamentation of fairly strongly developed horizontal ribs and intervening, less prominent, vertical ribs producing a reticulate pattern, and by its pronounced variability.

Dimensions—Female: length 0.45 mm, height 0.26 mm. Male: length 0.46 mm, height 0.25 mm.

Remarks—Holocene specimens assignable to *C. johnsoni* have been compared with topotype material collected by the writer in 1967. The Holocene specimens have a much less pronounced, more dorsally oriented, caudal process, and less pronounced reticulate ornamentation. Specimens

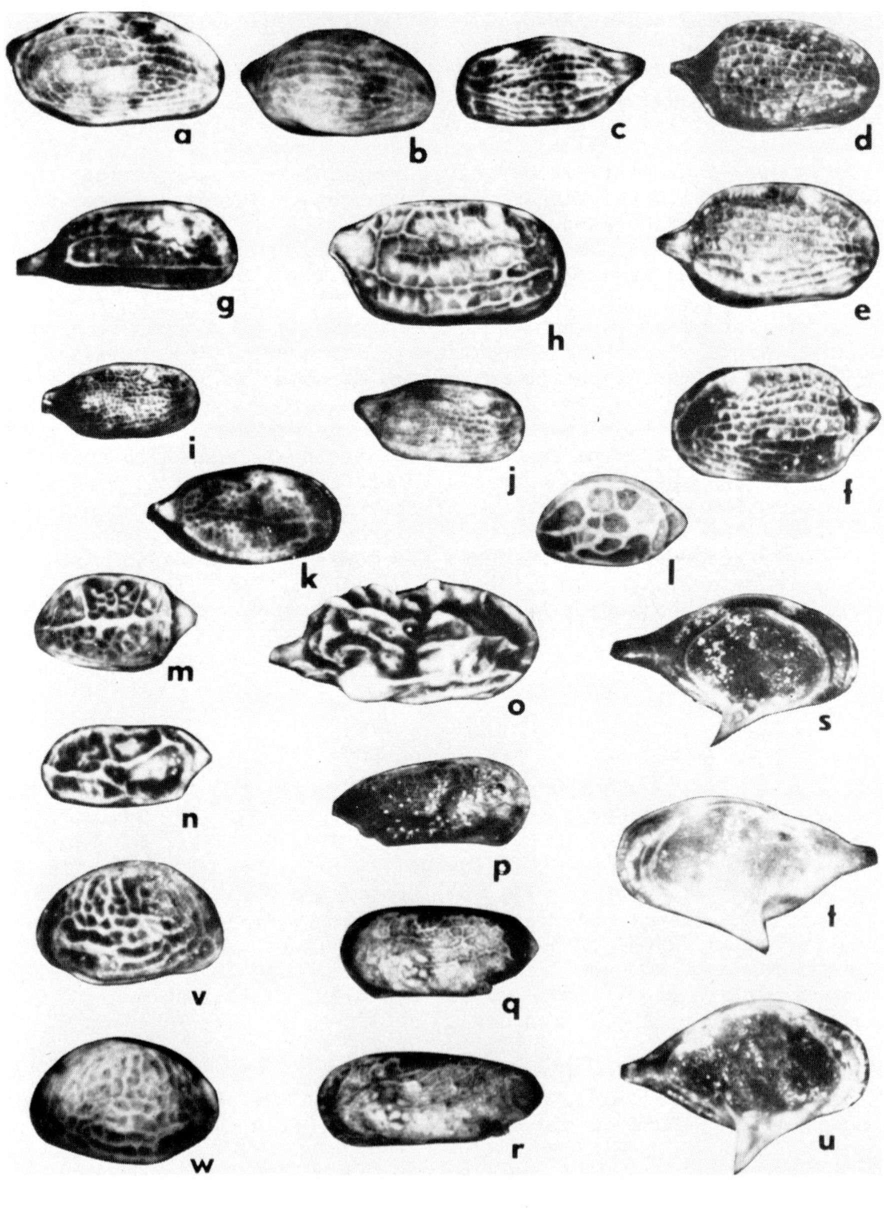

from Texas, in particular, are extremely variable both in ornamentation and size. Interestingly, many of the presumed males of this species possess the expanded posterior part of the inner lamella, a feature diagnostic of the genus *Semicytherura* Wagner 1957. This occurrence would seem to cast some question on the taxonomic validity of such a feature. Morales (1966) has also noted this.

Malkin (1953), in her work on the Miocene of the northern Atlantic coastal plain, noted that *C. forulata* Edwards and *C. elongata* Edwards were possibly conspecific. Swain (1955) distinguished these two species and *C. johnsoni* in Matagorda Bay. In her 1960 paper on the Mississippi delta area, Malkin noted the presence of *C. forulata*, which from her illustrations shows remarkable variation and which is probably in part conspecific with Holocene specimens of *C. johnsoni*. Benson and Coleman (1963) claim to have observed intergradation between all these species in their work in the Holocene of western Florida and have thus assigned to *C. johnsoni*, as junior synonyms, both *C. elongata* and *C. forulata*. Pooser (1965) has followed Benson and Coleman in his study of the Cenozoic Ostracoda of South Carolina.

Regardless of whether the Holocene specimens are conspecific with the fossil species mentioned above, there are, as Swain originally noted, three Holocene Gulf Coast species which are more or less closely related. Two of these species, *C. johnsoni* and *C. forulata* (sensu Swain), appear to possess extreme intraspecific variability but may be separated on the basis of lateral outline, particularly accentuation of the caudal process.

←—<

FIG. 17

a. *Cytherura* sp. ex. gr. *C. johnsoni* Mincher. Male, right valve, station 2, x60.

b. *C.* sp. ex. gr. *C. johnsoni* Mincher. Female, right valve, station 2, x55.

c. *Cytherura* sp. aff. *C. cornuta* Brady. Complete carapace, left valve, station 313, x68.

d. *Cytherura maya* Teeter, n. sp. Holotype (H.V.H. 9026), female, right valve, station 164, x60.

e. *C. maya* Teeter, n. sp. Paratype, female, right valve, complete carapace, station 164, x60.

f. *C. maya* Teeter, n. sp. Paratype (H.V.H. 9027), male, left valve, complete carapace, station 164, x60.

g. *Cytherura swaini* van den Bold. Complete carapace, right valve, station 291, x77.

h. *Cytherura sablensis* (Benson and Coleman). Male, complete carapace, right valve, station 278, x64.

i. *Cytherura subwardensis* Teeter, n. sp. Holotype (H.V.H. 9028), female, complete carapace, right valve, station 337, x60.

j. *C. subwardensis* Teeter, n. sp. Paratype (H.V.H. 9029), male, complete carapace, right valve, station 337, x60.

k. *Neomonoceratina mediterranea* (Ruggieri). Complete carapace, right valve, station 166, x64.

l. *Hemicytherura cranekeyensis* Puri. Female, complete carapace, left valve, station 315, x60.

m. *Hemicytherura bradyi* (Puri). Complete carapace, left valve, station 127f, x64.

n. *Kangarina ancycla* van den Bold. Male, complete carapace, left valve, station 179, x89.

o. *Paracytheridea tschoppi* van den Bold. Complete carapace, right valve, station 313, x60.

p. *Paracytheridea edwardsi* Teeter, n. sp. Holotype (H.V.H. 9030), female, complete carapace, right valve, station 122, x60.

q. *P. edwardsi* Teeter, n. sp. Paratype, female, complete carapace, left valve, station 122, x60.

r. *P. edwardsi* Teeter, n. sp. Paratype (H.V.H. 9031), male, complete carapace, left valve, station 122, x60.

s. *Cytheropteron barkeri* Teeter, n. sp. Holotype (H.V.H. 9032), complete carapace, right valve, station 178, x55.

t. *C. barkeri* Teeter, n. sp. Paratype (H.V.H. 9033), complete carapace, left valve, station 179, x55.

u. *C. barkeri* Teeter, n. sp. Paratype, complete carapace, right valve, station 179, x55.

v. *Cytheropteron subreticulatum* van den Bold. Complete carapace, right valve, station 215, x68.

w. *C. subreticulatum* van den Bold. Complete carapace, left valve, station 215, x68.

C. elongata (sensu Swain) may be easily distinguished from the other two species on the basis of its more elongate shape and more pronounced caudal process. The writer has observed all three species in a swash-mark sample from Surfside Beach, Freeport, Texas. A probable fourth species identical with Swain's (1955, Pl. 64 [Fig. 10c]) is also present. The writer has observed *C. johnsoni*, *C. elongata*, and *C. forulata* (sensu Swain) in two samples from Florida Bay.

In view of its variation and the difficulty of working with this species, the writer follows van den Bold in assigning his specimens to the *C. johnsoni* group until such time as these problems are resolved.

Distribution—*C. johnsoni* is present as a fossil in the middle Miocene Pascagoula Formation of Mississippi, the upper Miocene Springvale formation of Trinidad, the Miocene of Guatemala, and strata of similar age in Panama and Venezuela. In the Holocene it is found from the Gulf of Paria, north to the northern Gulf of Mexico.

In Belize, as elsewhere in the Holocene, *C. johnsoni* strongly prefers nearshore areas of lower salinity (less than 25°/₀₀). Extremely scarce specimens, seemingly lying within the variation of this species and *C. forulata* (sensu Swain), have been found in shallow carbonate areas of normal salinity (e.g., stations 36, 32-60, 317, 323, 409, 275).

Cytherura maya Teeter, n. sp.
Figures 17d-f, 18c

Diagnosis—A species of *Cytherura* characterized by a prominent caudal process medially located on the posterior margin; small, prominent eye tubercle; and a relatively heavy shell with deeply incised, constant ornamentation.

Description—Carapace subovate. Dorsal margin evenly arched. Anterior margin evenly rounded. Ventral margin concave anteriorly, convex posteriorly. Posterior margin with prominent median caudal process. Clear, glassy eye tubercle located on submarginal ridge extending from anteroventral to posterodorsal positions. Surface with deeply incised reticulate ornamentation. Dimorphism strong, males lower and more elongate.

Hinge apparently holamphidont, right valve with two terminal teeth, left valve with smooth median bar with small tooth at anterior end. Inner lamella broad anteriorly with numerous irregular radial pore canals. Inner lamella narrow ventrally and posteriorly in females, expanded posteriorly in males. False pore canals common, two consistently present in the posterior expansion of the males. Lacking vestibula. Central muscle scars not visible.

Dimensions—Holotype (female): length 0.46 mm, height 0.27 mm. Male: length 0.46 mm, height 0.25 mm.

Type locality—Station 164, at a water depth of 26 ft (7.9 m) in the main lagoon, off South Stann Creek.

Type specimens—Holotype: H.V.H. 9026. Paratype: H.V.H. 9027.

Remarks—This species is closely related to *C. johnsoni*. In degree of ornamentation and position and development of caudal process, *C. maya* is more akin to *C. johnsoni* than the Holocene specimens assigned to the latter. *C. maya*, however, differs in possessing an eye tubercle and is more heavily ornamented. It also differs ecologically; it lives in deeper waters and higher salinities than suspected of the original Miocene *C. johnsoni*.

C. maya very closely resembles *C. sandbergi* Morales, 1966, in development of ornamentation. However, the caudal process of *C. maya* is longer and located more ventrally than that of *C. sandbergi*. The two species appear to differ ecologically also: *C. maya* lives in deeper water.

Distribution—In Belize this species is restricted to the main lagoon.

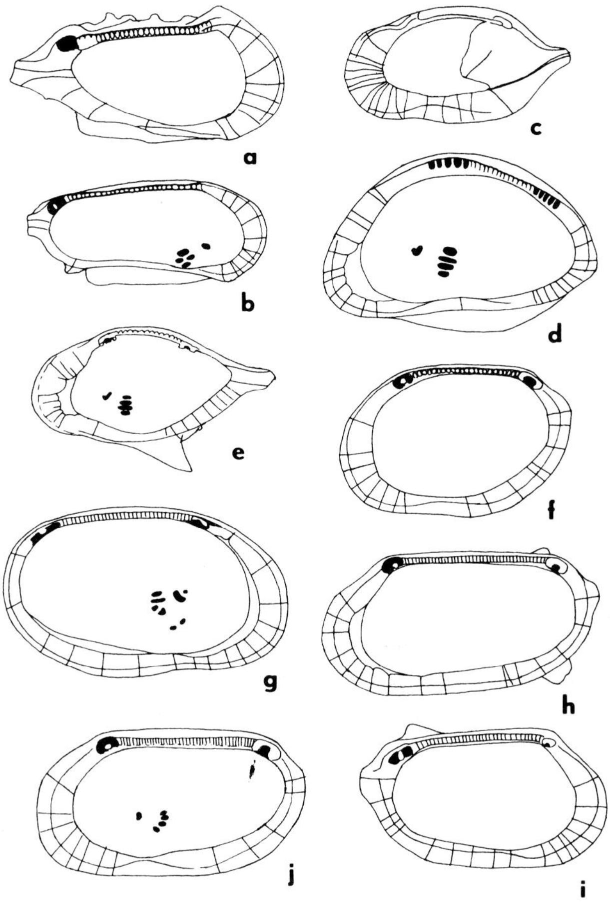

FIG. 18—Internal features.

a. *Paracytheridea tschoppi* van den Bold. Left valve, station 149f, x58.
b. *Paracytheridea edwardsi* Teeter, n. sp. Paratype, male, left valve, station 122, x64.
c. *Cytherura maya* Teeter, n. sp. Paratype, male, right valve, station 164, x64.
d. *Cytheropteron subreticulatum* van den Bold. Right valve, station 215, x96.
e. *Cytheropteron barkeri* Teeter, n. sp. Paratype, right valve, station 179, x58.
f. *Loxoconcha avellana* (Brady). Female, right valve, station 84, x64.
g. *Loxoconcha levis* Brady. Male, left valve, station 8, x64.
h. *Loxoconcha fischeri* (Brady). Male, right valve, station 58, x61.
i. *Loxoconcha fischeri minima* Teeter, n. sp. Paratype, female, left valve, station 201, x64.
j. *Loxoconcha wilberti* Puri. Male, right valve, station 313, x64.

Cytherura swaini van den Bold, 1963
Figure 17g

Cytherura swaini van den Bold, 1963c, p. 395, Pl. 9 (Figs. 4a, b); 1966c, Table 2; Morales, 1966, p. 53, Pl. 4 (Fig. 2); van den Bold, 1967b, p. 307, 312, Table 1.

Diagnosis—Characterized by its elongate shape, elongate subventral caudal process, and dorsoventrally flattened, ovate ridge.

Dimensions—Length 0.37 mm, height 0.14 mm.

Remarks—The specimens identified by Swain as *Cytherura costata* Mueller from San Antonio Bay and questionably assigned to this species by van den Bold, are not conspecific with *C. swaini*.

Distribution—This species is found in the main lagoon in Belize. It is also found in the upper Miocene Springvale formation of Trinidad.

Cytherura sablensis (Benson and Coleman), 1963
Figures 16e, 17h

Hemicytherura sablensis Benson and Coleman, 1963, p. 33, Pl. 6 (Figs. 6, 7); Fig. 19.
Hemicytherura? sablensis Benson and Coleman? Hall, 1965, p. 47, Pl. 16 (Figs. 24-30).

Diagnosis—Characterized by its sharp longitudinal ridges and subdorsal caudal process.

Dimensions—Female: length 0.46 mm, height 0.25 mm. Male: length 0.46 mm, height 0.24 mm.

Remarks—This species definitely does not belong to *Hemicytherura*. Among other differences, the dorsal margin is not highly arched as in the latter genus, nor is the hinge crenulate.

Specimens from Belize are slightly more heavily ridged than those in the writer's collection from Florida Bay. The specimens figured by Hall (1965) possess more horizontal ridges than those from Florida and are, therefore, questionably included in synonymy.

This species, scarce in Belize, is probably most closely related to *C. hermesi* van den Bold, 1946, an upper Eocene to middle Miocene species which possesses sharp longitudinal ridges oriented differently than those of *C. sablensis*.

Distribution—The writer has observed this species only in Florida Bay, its type locality, and in Belize. In the latter, it is restricted to the carbonate platform.

Cytherura sp. aff. *C. cornuta* Brady, 1868
Figure 17c

Cytherura cornuta Brady, 1868c, p. 445, Pl. 32, Figs. 12-15.
?*Cytherura* sp. cf. *cornuta* Brady. Benda and Puri, 1962, p. 324-328, Pl. 4. (Figs. 3, 4).

Diagnosis—Characterized by its small size, well-developed alae, and horizontal ribbing with weaker vertical ribbing.

Dimensions—Length 0.36 mm, height 0.18 mm.

Remarks—The Belize specimens are remarkably similar in ornamentation and shape to *C. cornuta*. However, the original specimens are more than double the size, and probably differ considerably in their ecology.

The specimens from Belize resemble *C. cresera* van den Bold, 1968, but are much less angular in dorsal view.

A second form, nonalate and much more expanded posterodorsally, occurs rarely with this species in Belize. The two forms agree in size and ornamentation and are assumed to be dimorphs. The rarer dimorph, with the posterodorsal expansion and posteriorly expanded inner lamella, is the male.

Distribution—Brady notes that this species was found in mud underlying 10 fm of water at Birterbuy Bay, Great Britain; off the southwest and north coasts of Scotland; at Galway in western Ireland; and in the Dardanelles, Turkey.

The writer has also observed this species in samples from Curaçao and the Bahamas.

In Belize this species is restricted to the carbonate platform.

Cytherura subwardensis Teeter, n. sp.
Figures 16f; 17i, j

Diagnosis—Characterized by its small size, reticulate ornamentation, and broad, shallow, dorsal sulcus.

Description—Carapace small, subquadrate. Dorsal margin arched in females, less strongly so in males. Anterior margin evenly, bluntly rounded. Ventral margin concave anteriorly, convex posteriorly. Posterior margin with prominent, dorsally positioned and inclined, caudal process. Dimorphism strong, males more elongate than females. Surface reticulate, horizontal ribbing pronounced. Prominent, small, glassy eye tubercle slightly posterior to anterior cardinal angle. Posterodorsal swelling well developed in both sexes. Broad, shallow sulcus anterior to swelling.

Internal features characteristic for genus. Males with flaplike projection of posterior part of inner lamella.

Dimensions—Holotype (female): length 0.36 mm, height 0.19 mm. Male: length 0.37 mm, height 0.18 mm.

Type locality—Station 337, in 70 ft (21 m) of water just east of Elbow Cay.

Type specimens—Holotype: H.V.H. 9028. Paratype: H.V.H. 9029.

Remarks—This species, as the name implies, is closely related to *C. wardensis* Howe and Brown, resembling the latter in ornamentation, posterior swelling, and possession of a broad, shallow sulcus. It differs from *C. wardensis* mainly in its much smaller size. Puri (1960) and Benda and Puri (1962) noted the presence of *C. wardensis* in the Holocene of Florida, but they neither illustrated nor gave dimensions of this ostracod. The writer has observed specimens in samples from Florida, probably assignable to *C. wardensis*, which are considerably larger in size than the Belize specimens and lack the dorsally oriented caudal process.

Distribution—This species is restricted to the main lagoon.

?*Cytherura* sp.

Remarks—Although these ostracods are extremely common in the main lagoon, the writer has been unable to assign them definitely. They are present only as juveniles, with extremely thin, slightly ribbed shells, possessing only very rudimentary inner lamellae or none at all. In outline, particularly posteriorly, they resemble both *Cytherura* and some species of *Basslerites*. The ribbed ornamentation is more suggestive of *Cytherura*.

Genus HEMICYTHERURA Elofson, 1941

Hemicytherura cranekeyensis Puri, 1960
Figures 16h, 17*l*

Cytherura clathrata Sars. Van den Bold, 1946, p. 118, Pl. 14 (Figs. 9, 10).
(not) *Cytherura clathrata* Sars, 1866, p. 77.
Hemicytherura videns (Mueller). Van den Bold, 1957b, p. 245, in part.
(not) *Cytheropteron videns* Mueller, 1894, p. 303, Pl. 20, Figs. 2, 8.
Hemicytherura cranekeyensis Puri, 1960, p. 115, 118, Pl. 4 (Figs. 4, 5); Morales 1966, p. 53-54, Pl. 4 (Figs. 8a, b); van den Bold, 1966c, Table 2; McKenzie and Swain, 1967, p. 291, Pl. 29 (Figs. 27a, b); Text-fig. 12; van den Bold, 1968, p. 72, Pl. 7 (Figs. 7a, b).
Hemicytherura sp. van den Bold, 1963c, p. 398, Pl. 9 (Fig. 5).

Diagnosis—Characterized by coarse pits and strongly developed ridges.

Dimensions—Male: length 0.33 mm, height 0.19 mm.

Distribution—Fossil representatives of this species are known from the upper Miocene Duplin Marl of North Carolina (observed in a sample in the writer's collection), the Miocene to Holocene of Trinidad, and the Miocene of the Dominican Republic and Jamaica. Originally described by Puri from Florida, it has also been observed in the Holocene in Laguna de Terminos, Mexico, and in Baja California. In Belize this species is found most commonly in *Thalassia*-rich areas of the carbonate platform.

Hemicytherura bradyi (Puri), 1960
Figures 16i, 17m

Kangarina bradyi Puri, 1960, p. 115, Pl. 4 (Figs. 6, 7).

Diagnosis—Characterized by prominent, horizontal median ridge and shorter, adjoining vertical ridges.

Dimensions—Female: length 0.33 mm, height 0.23 mm. Male: length 0.35 mm, height 0.19 mm.

Remarks—This species very closely resembles the preceding one in outline, lacking the subdorsal caudal process and the posteriorly expanded inner lamella of *Kangarina*, and is therefore assigned to *Hemicytherura*.

Distribution—This species is common throughout the Caribbean, Bahamas, and Florida. In Belize it is restricted to the carbonate platform.

Genus KANGARINA Coryell and Fields, 1937

Kangarina ancycla van den Bold, 1963
Figures 16g, 17n

Cytheropteron (Kangarina) quellita Coryell and Fields. Keij, 1954b, p. 226, Pl. 5 (Fig. 8).
Kangarina sp. Drooger and Kaasschieter, 1958, p. 90.
Kangarina ancycla van den Bold, 1963c, p. 396-397, Pl. 10 (Figs. 3a, b).

Diagnosis—Characterized by the distinctive pattern of the ridges, particularly the anterior projection of the "median" ridge to the anterior margin.

Dimensions—Female: length 0.35 mm, height 0.18 mm. Male: length 0.32 mm, height 0.15 mm.

Remarks—In their original description of the genotype *Kangarina quellita*, Coryell and Fields noted,

> ". . . broadly developed . . . interior lamella, leaving only an oval area in the posterior part of the anterior half uncovered . . . the inner extension of the lamellar platform rises sufficiently above the surface of the valve in the posterior ventral area to form a tunnel-like structure that opens anteriorly into the lower portion of the uncovered oval area."

Although this feature was poorly defined originally, the writer concludes from this quotation, the illustration in Figure 15C, and a careful examination of specimens of *K. ancycla* from Belize that Coryell and Fields are describing a posteriorly expanded inner lamella similar to that previously noted in the males of some species of *Cytherura*. This important taxonomic character regrettably has been neglected by subsequent workers with the result that another genus, lacking the internal feature described above and possessing an arched dorsal margin and thickened, slightly sinuate, ventral ridge, has become synonymous with *Kangarina* (e.g., *K. abyssicola*; van Morkhoven, 1963, v. 2, p. 353; van den Bold, 1963c, p. 397, Pl. 10 [Fig. 1]).

Kangarina is closely related to *Cytherura*, differing from it in that both sexes possess the posteriorly expanded inner lamella, ornamentation consists of a few well-defined ridges, and dorsal and ventral margins are straight and parallel. Another excellent Holocene example of this genus, probably representing more than one species, is *Cytherura costellata* Brady (Hornibrook, 1952, p. 50-51, Pl. 14 [Figs. 229, 230, 233, 234]). A second, slightly larger, undescribed species is very scarce in Belize.

Distribution—*K. ancycla* is found in the upper Miocene Springvale formation and in the Gulf of Paria at depths down to 60 fm. In Belize it is present in the main lagoon.

Genus EUCYTHERURA Mueller, 1894

Eucytherura spp.

Diagnosis—Characterized by small, highly ornamented carapaces.

Remarks—Although they are scarce, at least four species are included in this genus. Of these, two species are more common, one conspecific with that assigned questionably by van den Bold to *Eucytherura complexa* (Brady), the other closely resembling but not conspecific with *E. gibbera* Mueller. Of the two less common species, both resemble *E. alata* Mueller.

Distribution—Fossil specimens of *E. complexa*, according to van den Bold (1963c), are found from the upper Miocene to the Holocene. Van den Bold notes that Holocene specimens live at depths of from 40 to 200 ft (12-61 m). In Belize, specimens assignable to *E. complexa* are restricted to the main lagoon. The other species predominate in the shallow carbonate areas.

Genus PARACYTHERIDEA Mueller, 1894

Paracytheridea tschoppi van den Bold, 1946
Figures 17o, 18a

Paracytheridea tschoppi van den Bold, 1946, p. 85, Pl. 16 (Figs. 6, 7); Keij, 1954b, p. 220, Pl. 4 (Fig. 4); van den Bold, 1957b, p. 245, Pl. 4 (Fig. 7); Benson and Coleman, 1963, p. 33-34, Pl. 6 (Figs. 7, 9, 10); Fig. 20; van den Bold, 1968, p. 76, Pl. 4 (Figs. 8a-d).
Paracytheridea sp. 1 Drooger and Kaasschieter, 1958, p. 91.
Paracytheridea sp. van den Bold. Van den Bold, 1966f, Pl. 3 (Figs. 7a, b).

Diagnosis—Characterized by its long, pointed caudal process and the oblique, sharp ridges developed on the postdorsal projection.

Dimensions—Length 0.57 mm, height 0.30 mm.

Remarks—The specimens assigned to this species by Benson and Coleman (1963) are almost half again as large as the type and subsequently described Holocene specimens. Benson and Coleman suggested that the type specimen is a late instar. Holocene specimens from Belize and Colón Harbour, Panama, examined by the writer, are close in size to the type and are clearly adults. The writer has observed another larger species of *Paracytheridea*, possibly *P. vandenboldi*, in samples from Florida, and perhaps these have been included by Benson and Coleman in their measurements of *P. tschoppi*.

Distribution—This species has been identified in the Miocene of Cuba, Trinidad, and the northern Dominican Republic, and in the Holocene from the Gulf of Paria, Colón Harbour, and Florida. Belize specimens are restricted to the carbonate platform.

Paracytheridea edwardsi Teeter, n. sp.
Figures 17p-r, 18b

Paracytheridea sp? van den Bold, 1966f, Pl. 3 (Fig. 9).

Diagnosis—Characterized by a reduced ventral ala, subdorsal caudal process, and tumid outline in dorsal view.

Description—Carapace subquadrate, elongate in lateral view, ovate, tumid in dorsal view. Dorsal margin straight. Anterior margin evenly rounded. Ventral outline straight, margin obscured by ala in lateral view. Posteroventral margin uneven, inclined dorsally. Posterior margin with short subdorsal caudal process. Surface with subdued ventral ala, delicate tracery of ridges between which are numerous, very fine punctae.

Interior of valves deep. Hinge relatively weakly developed with crenulate teeth in right valve, and crenulate median bar in left. Inner lamella narrow, with few radial pore canals. Muscle scar peculiar, with three adductors in vertical row, the middle one split, and the frontal scar ovate, not V-shaped.

Dimensions—Holotype (female): length 0.45 mm, height 0.21 mm. Male: length 0.50 mm, height 0.21 mm.

Type locality—Station 122, in 4 ft (1.2 m) of water on the reef apron, northeast of Gladden Cay.

Type specimens—Holotype: H.V.H. 9030. Paratype: H.V.H. 9031.

Remarks—This species is named in honor of R. A. Edwards of the University of Florida, who kindly supplied ostracods from the Duplin Marl (Miocene) of North Carolina.

P. edwardsi is very similar to *P. mucra* Edwards from the Duplin Marl, but differs markedly in size. Also, in the Miocene material the caudal process is more ventrally located, and the sulcus and ventral alae are more markedly developed.

Distribution—This species occurs on the carbonate platform, most commonly in non-*Thalassia* areas.

Genus CYTHEROPTERON Sars, 1866
Cytheropteron barkeri Teeter, n. sp.
Figures 17s-u, 18e

Diagnosis—Characterized by small, very delicate shell, and narrow unornamented alae.

Description—Carapace small and delicate. Dorsal margin arched, slightly concave posteriorly. Anterior margin evenly rounded. Ventral margin convex, slightly concave posteriorly. Posterior margin with protracted caudal process. Outline sagittal in dorsal view. Ventral ala narrow, pointed, distal half of trailing edge at right angles to hinge line and unornamented. Trailing edge squared off. Thin, flaplike extension ventrally from proximal part of ala.

Hinge apparently with two crenulate terminal teeth in the right valve. Inner lamella wide anteriorly, narrower ventrally and posteriorly. Small crescent-shaped anterior vestibulum, with eight or nine radial pore canals. Muscle scar with four obliquely inclined adductors and V-shaped frontal scar.

Dimensions—Holotype: length 0.57 mm, height 0.28 mm.

Type locality—Station 178, in 119 ft (36 m) of water, west of Laughing Bird atoll.

Type specimens—Holotype: H.V.H. 9032. Paratype: H.V.H. 9033.

Remarks—This species is named in honor of R. W. Barker, who kindly assisted this research with comparative material from the Florida Bay area.

C. barkeri resembles several other species of *Cytheropteron*. It is considerably smaller than both *C. alatum* Sars and *C. pacificum* LeRoy, and lacks the alar ornamentation present in these species. Specimens assigned to *C. punctata* Lienenklaus by van den Bold from the lower Oligocene of Cuba are similar to the Belize specimens, but are considerably larger and possess longer, thinner alae. *C. nipeensis* van den Bold differs similarly with respect to the alae, as does *C.* sp. van den Bold, 1958b. *C. palton* van den Bold, 1966c, agrees closely with this species but is relatively higher and attains its maximum width at the tips of the alae, more posteriorly than *C. barkeri*.

Distribution—This species is confined to the main lagoon.

Cytheropteron subreticulatum van den Bold, 1946
Figures 17v, w; 18d

Cytheropteron subreticulatum van den Bold, 1946, p. 113, Pl. 14 (Figs. 6a-d); 1950a, p. 86; 1958b, p. 414, Pl. 4 (Figs. 4a, b); 1963c, p. 400; 1966e, p. 361; 1966f, Pl. 2 (Fig. 8).
Cytheropteron sp. 1 Drooger and Kaasschieter, 1958, p. 89, Pl. 4 (Fig. 10).

Diagnosis—Characterized by its semicircular, nonalate, noncaudate outline and subreticulate ornamentation.

Dimensions—Length 0.37 mm, height 0.24 mm.

Remarks—*C. ventrokurtosa* Swain, 1967, resembles this species closely in size and ornamentation, differing in having a more angular outline and fewer radial pore canals.

Distribution—This species occurs from Miocene to Holocene in the Caribbean. In Belize it is almost entirely restricted to the main lagoon.

Family SCHIZOCYTHERIDAE Howe, 1961

Genus NEOMONOCERATINA Kingma, 1948

Neomonoceratina mediterranea (Ruggieri), 1953
Figure 17k

Paijenborchella (Neomonoceratina) mediterranea Ruggieri, 1953, p. 4, Text-figs. 1-5; Morales, 1966, p. 80, Pl. 7 (Figs. 2a-c).
Payenborchella (Neomonoceratina) mediterranea Ruggieri. Keij, 1954a, p. 361, Pl. 3 (Figs. 12, 13); 1954b, p. 228, Pl. 5 (Fig. 15); Pl. 6 (Fig. 12).
Neomonoceratina sp. Swain, 1955, p. 643, Pl. 64 (Fig. 15).

Diagnosis—Characterized by typical development of ridges and finely punctate interridge surface.

Dimensions—Length 0.40 mm, height 0.22 mm.

Remarks—Although scarce in Belize, this species is included here because of its extremely wide distribution.

Distribution—This species is known from Port Said, Egypt (the type locality), the Bay of Djakarta, Java; Manilla, Philippines; Gulf of Paria, Venezuela; the Texas Gulf Coast; and Laguna de Terminos, Mexico. In Belize it is restricted to shallow nearshore areas south of Belize City.

Family LOXOCONCHIDAE Sars, 1925

Genus LOXOCONCHA Sars, 1866

Loxoconcha levis Brady, 1870
Figures 18g, 19a, b

Loxoconcha levis? Brady, 1870, p. 238, 240, 242, Pl. 32 (Figs. 7, 8).
not *Loxoconcha levis* Mueller, 1894; Mehes, 1941.
Loxoconcha levis Brady? Van den Bold, 1963c, p. 393, Pl. 9 (Figs. 10a, b).

Diagnosis—Characterized by smooth surface and consistently arranged normal pore canals.

Dimensions—Female: length 0.51 mm, height 0.34 mm. Male: length 0.56 mm, height 0.36 mm.

Distribution—This species was originally described by Brady from the Bahamas and Cuba. Van den Bold (1963c) subsequently recorded it from Holocene shallow-water deposits and Pliocene sediments of Trinidad. In Belize it occurs in the upper, less saline areas of Chetumal Bay.

Loxoconcha avellana (Brady), 1866
Figures 18f, 19c

Normania avellana Brady, 1866, p. 382, Pl. 61 (Figs. 15a-c).
?*Loxoconcha avellana* (Brady). Brady, 1890, p. 507; Tressler, 1954, p. 434.
Loxoconcha avellana (Brady). Van den Bold, 1964, p. 5, 7, Table 2; 1966c, p. 29, Pl. 2 (Figs. 5a, b); 1966f, p. 52, Pl. 3 (Figs. 6a, b).

Diagnosis—Characterized by dense surface pitting and lateral outline.
Dimensions—Female: length 0.48 mm, height 0.30 mm. Male: length 0.55 mm, height 0.35 mm.
Remarks—The writer observed, in a sample from the Fiji Islands, a species remarkably similar to *L. avellana* and probably identical with the specimens from Noumea assigned to this species by Brady (1890). Comparison with material from the West Indies indicates that the Pacific specimens are somewhat smaller, and possess minute posterodorsal projections. Larger, more similar, perhaps conspecific specimens have been observed by the writer from West Africa.

Van den Bold places those species from the Gulf Coast, identified as *L. australis* by other workers, in synonymy with *L. avellana*. These specimens represent, in fact, more than one species, and are definitely not *L. australis*, nor are they *L. avellana*.

Distribution—This species, originally described on the basis of a single carapace from the West Indies, has been subsequently reported by van den Bold from the Cubagua formation, and from the Holocene of Venezuela and Colón Harbour, Panama. In Belize it is restricted to the waters of the main lagoon.

⟶

FIG. 19

a. *Loxoconcha levis* Brady. Male, complete carapace, right valve, station 5, x64.
b. *L. levis* Brady. Female, complete carapace, right valve, station 5, x64.
c. *Loxoconcha avellana* (Brady). Female, complete carapace, right valve, station 76, x60.
d. *Loxoconcha wilberti* Puri. Female, complete carapace, right valve, station 313, x60.
e. *L. wilberti* Puri. Male, complete carapace, right valve, station 313, x60.
f. *Loxoconcha fischeri* (Brady). Male, complete carapace, left valve, station 58, x60.
g. *Loxoconcha fischeri minima* Teeter, n. ssp. Holotype (H.V.H. 9034), female, complete carapace, right valve, station 201, x60.
h. *L. fischeri minima* Teeter, n. ssp. Paratype (H.V.H. 9035), male, complete carapace, right valve, station 201, x60.
i. *L. fischeri minima* Teeter, n. ssp. Paratype, male, complete carapace, left valve, station 201, x60.
j. *Loxoconcha magna* Teeter, n. sp. Holotype (H.V.H. 9036), female, complete carapace, left valve, stations 35, x60.
k. *L. magna* Teeter, n. sp. Paratype (H.V.H. 9037), male, complete carapace, left valve, station 35, x60.
l. *L. magna* Teeter, n. sp. Paratype, female, complete carapace, right valve, station 35, x60.
m. *L. magna* Teeter, n. sp. Paratype, male, complete carapace, right valve, station 35, x60.
n. *Loxoconcha dorsotuberculata* (Brady). Female, complete carapace, right valve, station 315, x55.
o. *L. dorsotuberculata* (Brady). Male, complete carapace, right valve, station 315, x55.

Loxoconcha wilberti Puri, 1954
Figures 18j; 19d, e

Loxoconcha wilberti Puri, 1953d, p. 274-275, Pl. 10 (Fig. 1a); Text-figs. 11a, b; Puri and Hulings, 1957, p. 187, Fig. 11; Puri, 1960, p. 114, Pl. 4 (Figs. 18, 19); Swain, 1968, p. 24-25, Pl. 3 (Figs. 6a-c); Pl. 7 (Fig. 6); Text-figs. 21A, B.

Diagnosis—Characterized by extremely subdued posterodorsal projection, narrow posteroventral ridge, and reticulate surface ornamentation.

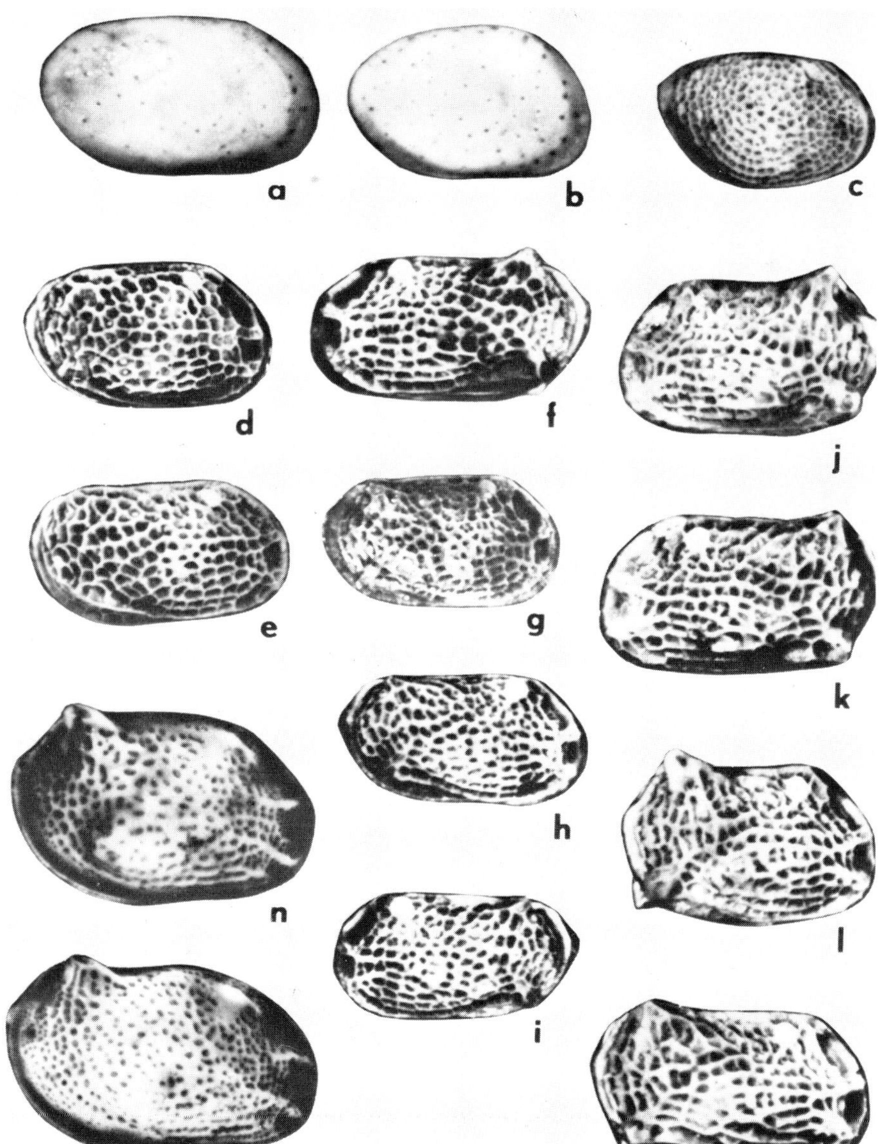

Dimensions—Female: length 0.52 mm, height 0.32 mm. Male: length 0.56 mm, height 0.32 mm.

Remarks—The writer compared the Belize specimens with the holotype in the Howe collection and found them to be identical. This species resembles *L. fischeri*, but may be distinguished from it by the much less pronounced nodes.

Distribution—Originally described by Puri from the middle Miocene of Florida, it subsequently has been recorded from the Holocene of northern and western Florida. In Belize this species occurs on the carbonate platform.

Loxoconcha fischeri (Brady), 1869
Figures 18h, 19f

Cythere? fischeri Brady, 1869, p. 152, Pl. 18 (Figs. 15, 16).
Loxoconcha antillea var. *oblongata* van den Bold, 1946, p. 110, Pl. 15 (Fig. 5).
Loxoconcha fischeri (Brady). Keij, 1954b, p. 225, Pl. 5 (Fig. 3); van den Bold, 1963c, p. 393-394, Pl. 8 (Figs. 8, 9); 1966f, p. 51, Pl. 3 (Fig. 7).
Loxoconcha fischeri (Brady; in part). Van den Bold, 1958b, p. 395, Table 1.
(not) *Loxoconcha fischeri* (Brady). Van den Bold, 1957b, p. 244, Pl. 2 (Fig. 7) (=*Loxoconcha antillea* van den Bold, 1946).
Loxocorniculum fischeri (Brady; in part?). Benson and Coleman, 1963, p. 39, Pl. 7 (Figs. 3, 4); Fig. 24.

Diagnosis—A species of *Loxoconcha* characterized by a prominent glassy eye tubercle from which extends a rather prominent, posteroventrally inclined ridge. Two prominent ventral projections and one posterodorsal projection.

Dimensions—Female: length 0.52 mm, height 0.31 mm. Male: length 0.59 mm, height 0.31 mm.

Remarks—Benson and Coleman (1963) erected the genus *Loxocorniculum* for reticulate species of *Loxoconcha* possessing posterodorsal hornlike protuberances. This would seem advantageous ecologically as such forms tend to be restricted to the tropics, but the degree of emphasis of the projection is extremely varied and, in several species, rather difficult to classify. Therefore, this new taxon is not adopted here.

Those species of *Loxoconcha* possessing prominent projections are extremely numerous and must be examined extensively in order to be speciated.

Van den Bold (1963c) has examined and figured Brady's type of *Cythere? fischeri* from Colón, Panama, and kindly provided topotype material, which is identical to the writer's specimens.

Benson and Coleman (1963, Fig. 24) illustrated a species of *Loxoconcha* which is *not* conspecific with *L. fischeri*. This is a new species, and may be differentiated easily from *L. fischeri* in that the former possesses only one posteroventral tubercle. Benson and Coleman's other illustrations (Pl. 7 [Figs. 3, 4]) are probably of specimens conspecific with *L. fischeri*. Brady's original illustrations agree very closely with the males of this new species.

The holotype, figured by van den Bold (1963c), with its pronounced horizontal ribbing and apparently prominent, anterior eye ridge, appears very similar to *L. magna* Teeter n. sp. The writer has not observed *L. magna* in the sample from Colón. Also, the specimen of *L. fischeri* figured by van den Bold (1963c) from the upper Miocene Manzanilla formation of Trinidad is remarkably similar to *L. wilberti* Puri. In conversation with the writer, van den Bold suggested that *L. wilberti* might be an underdeveloped *L. fischeri*. In the writer's opinion, the two species are not conspecific.

Distribution—*L. fischeri* ranges from Miocene to Holocene in the Caribbean area. In Belize it occurs on the carbonate platform.

Loxoconcha fischeri minima Teeter, n. ssp.
Figures 18i, 19g-i

Diagnosis—Characterized by its small size and prominent eye tubercle.

Description—Carapace small, subquadrate. Lateral outline as in *L. magna* Teeter, n. sp.; node development as in *L. fischeri*. Eye tubercle prominent, lacking associated ridges. Surface irregularly reticulate. Dimorphism apparent, males more elongate.

Internal features as in *L. fischeri*, with small crescent-shaped anterior vestibulum.

Dimensions—Holotype (female): length 0.51 mm, height 0.30 mm. Male: length 0.51 mm, height 0.26 mm.

Type locality—Station 201, in 2 ft (0.6 m) of water in Port of Honduras.

Type specimens—Holotype: H.V.H. 9034. Paratype: H.V.H. 9035.

Remarks—This species is smaller than both *L. fischeri* and *L. magna*, and possesses a somewhat more haphazard reticulate ornamentation. It lacks the rudimentary tubercle of *L. magna* and possesses a small anterior vestibulum, a feature not found in the latter species. In the males the anterior posteroventral tubercle is much reduced in comparison with that of *L. fischeri*. *L. fischeri minima* is remarkably similar to *L. antillea* in outline and development of nodes, differing only in size—*L. fischeri minima* is larger—and in details of the ornamentation.

Distribution—This species is restricted to carbonate shoal areas.

Loxoconcha magna Teeter, n. sp.
Figures 19j-m, 20e

Diagnosis—Characterized by its large size, small eye tubercle, prominent anterior eye ridge, and pronounced horizontal ribbing.

Description—Carapace large, subquadrate. Dorsal margin straight. Anterior margin bluntly, obliquely rounded. Ventral margin straight. Posterior margin upswept, slightly concave near dorsal margin. Surface with two prominent posteroventral tubercles and one posterodorsal tubercle. All tubercles more prominent in the female, a condition common to the genus. Subdued, fourth tubercle midway between most anterior of the ventral tubercles and the dorsal tubercle. Very prominent ridge anterior to reduced eye tubercle. Surface reticulate, with horizontal ribbing pronounced. Dimorphism striking, males more elongate, possessing more weakly developed tubercles. Hinge gongylodont. Inner lamella relatively wide. No anterior vestibulum. Posteroventral vestibulum small. Only three or four radial pore canals anteriorly.

Dimensions—Holotype (female): length 0.56 mm, height 0.35 mm. Male: length 0.59 mm, height 0.33 mm.

Type locality—Station 35, in 12 ft (3.6 m) of water, just east of Congrejo Cay.

Type specimens—Holotype: H.V.H. 9036. Paratype: H.V.H. 9037.

Remarks—This species differs from *L. fischeri* in being slightly larger and in having pronounced horizontal ribbing, a rudimentary fourth tubercle, and a smaller eye tubercle with prominent anterior ridge.

The writer still questions whether or not this may be conspecific with Brady's *Cythere? fischeri* (van den Bold, 1963c, Pl. 8 [Fig. 9]).

Distribution—This species is restricted to the carbonate platform in Belize.

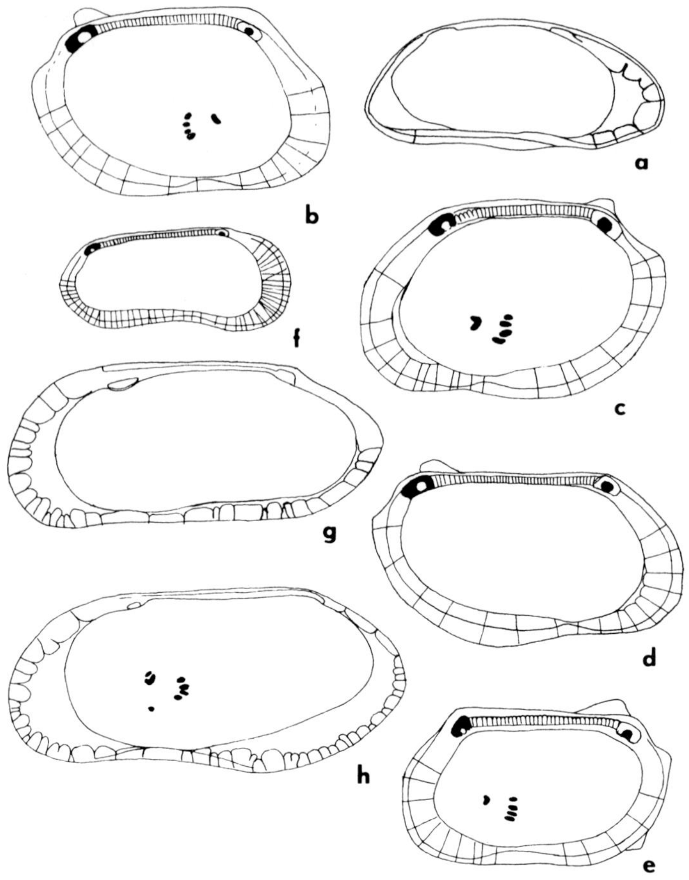

FIG. 20–Internal features.

a. *Microcythere* sp. cf. *M. inflexa* Mueller. Left valve, station 320, x154.
b. *Loxoconcha suboculocrista* Teeter, n. sp. Paratype, female, left valve, station 281, x64.
c. *Loxoconcha dorsotuberculata* (Brady). Female, right valve, station 315, x64.
d. *Loxoconcha oculocrista* Teeter, n. sp. Paratype, male, left valve, station 315, x61.
e. *Loxoconcha magna* Teeter, n. sp. Paratype, female, right valve, station 35, x61.
f. *Cytheromorpha apheles* van den Bold. Male, left valve, station 3, x58.
g. *Megacythere johnsoni* (Mincher)? Male, right valve, station 62, x96.
h. *Megacythere stephensoni* (Puri). Right valve, station 5, x99.

Loxoconcha dorsotuberculata (Brady), 1866
Figures 19n, o; 20c

Normania dorsotuberculata Brady, 1866, p. 383, Pl. 61 (Figs. 14a-g).
Loxoconcha dorsotuberculata (Brady). Brady, 1868, p. 56; 1890, p. 507; van den Bold, 1946, p. 111; 1966f, p. 51, Pl. 3 (Figs. 3a, b).
Loxoconcha dorso-tuberculata (Brady). Tressler, 1954, p. 435.

Diagnosis—Characterized by its subtrapezoidal outline, punctate ornamentation, and low, inconspicuous, posterodorsal projection.

Dimensions—Female: length 0.65 mm, height 0.44 mm. Male: length 0.71 mm, height 0.45 mm.

Remarks—This is one of several species, described subsequently, which are all remarkably similar. This group possesses a rather strongly developed posterodorsal projection and may, or may not, have a postmedian projection. In *L. dorsotuberculata*, the latter projection is more strongly developed in the female, and the male either lacks it or possesses a much reduced postmedian projection.

Van den Bold (1966f) questionably included *L. postdorsoalata* Puri in synonymy. However, this species is considerably smaller and lacks the postmedian projection completely.

Distribution—This species is restricted to the carbonate platform in Belize.

Loxoconcha oculocrista Teeter, n. sp.
Figures 20d, 21a-c

Loxocorniculum fischeri (Brady; in part). Benson and Coleman, 1963, p. 39, Fig. 24.

Diagnosis—Characterized by small eye tubercle with closely associated X-shaped ridge. Dorsal margin of each valve with ribbed crest.

Description—Carapace subtrapezoidal in lateral view; males subovate in dorsal view, females subsagittate. Dorsal margin straight. Anterior margin bluntly and obliquely rounded, straight and vertical from midpoint almost to ventral margin. Ventral margin slightly concave medially. Posterior margin upswept, slightly concave near dorsal margin. Prominent posterodorsal projection. Males with small swelling posterior and ventral to midvalve; swelling much enlarged in females. Males more elongate. Small eye tubercle with closely associated X-shaped ridge. Surface deeply and rather coarsely reticulate with emphasis on the horizontal ribs. Many of the reticulations subdivided, particularly anteriorly, into smaller squares. Two short, prominent, horizontal ridges terminate at anterior margin.

Internal features characteristic of genus.

Dimensions—Holotype (female): length 0.60 mm, height 0.38 mm. Male: length 0.64 mm, height 0.38 mm.

Type locality—Station 315, in 10 ft (3 m) of water on the south side of Tobacco Entrance.

Type specimens—Holotype: H.V.H. 9038. Paratype: H.V.H. 9039.

Remarks—The species name refers to the prominent eye (Latin "oculus") ridge (Latin "crista").

This is almost undoubtedly the species Benson and Coleman (1963, Fig. 24) assigned to *L. fischeri*. This species also closely resembles that originally figured by Brady for *Cythere? fischeri*.

Distribution—*L. oculocrista* occurs along the west coast of Florida. The writer has also observed it in samples from Florida Bay and the Bahamas. In Belize this species is found on the carbonate platform.

Loxoconcha suboculocrista Teeter, n. sp.
Figures 20b, 21d-f

Diagnosis—Characterized by V-shaped ridge associated with eye tubercle and thin low ridge along the dorsal margin of each valve.

Description—Carapace subtrapezoidal. Dorsal margin straight. Anterior margin with inclined dorsal half, vertical ventral half. Ventral margin prominently concave anteriorly. Posterior margin upswept; concave near dorsal margin. Large, pointed posterodorsal spine. Females with pronounced subcentral projection; males with low subcentral swelling. Males more elongate. Well-developed eye tubercle with prominent V-shaped ridge anteriorly. Surface reticulopunctate. Two prominent subhorizontal ridges terminate at anterior margin.

Internal features typical for genus.

Dimensions—Holotype (female): length 0.60 mm, height 0.40 mm. Male: length 0.69 mm, height 0.41 mm.

Type locality—Station 281, in 13 ft (4 m) of water near the mouth of South Stann Creek.

Type specimens—Holotype: H.V.H. 9040. Paratype: H.V.H. 9041.

Remarks—The species name refers to the reduced V-shaped eye ridge.

This species is very similar to the preceding one. However, *L. suboculocrista* lacks the ribbed dorsal crest of *L. oculocrista*, has a reduced development of the eye ridge, and has slightly different ornamentation.

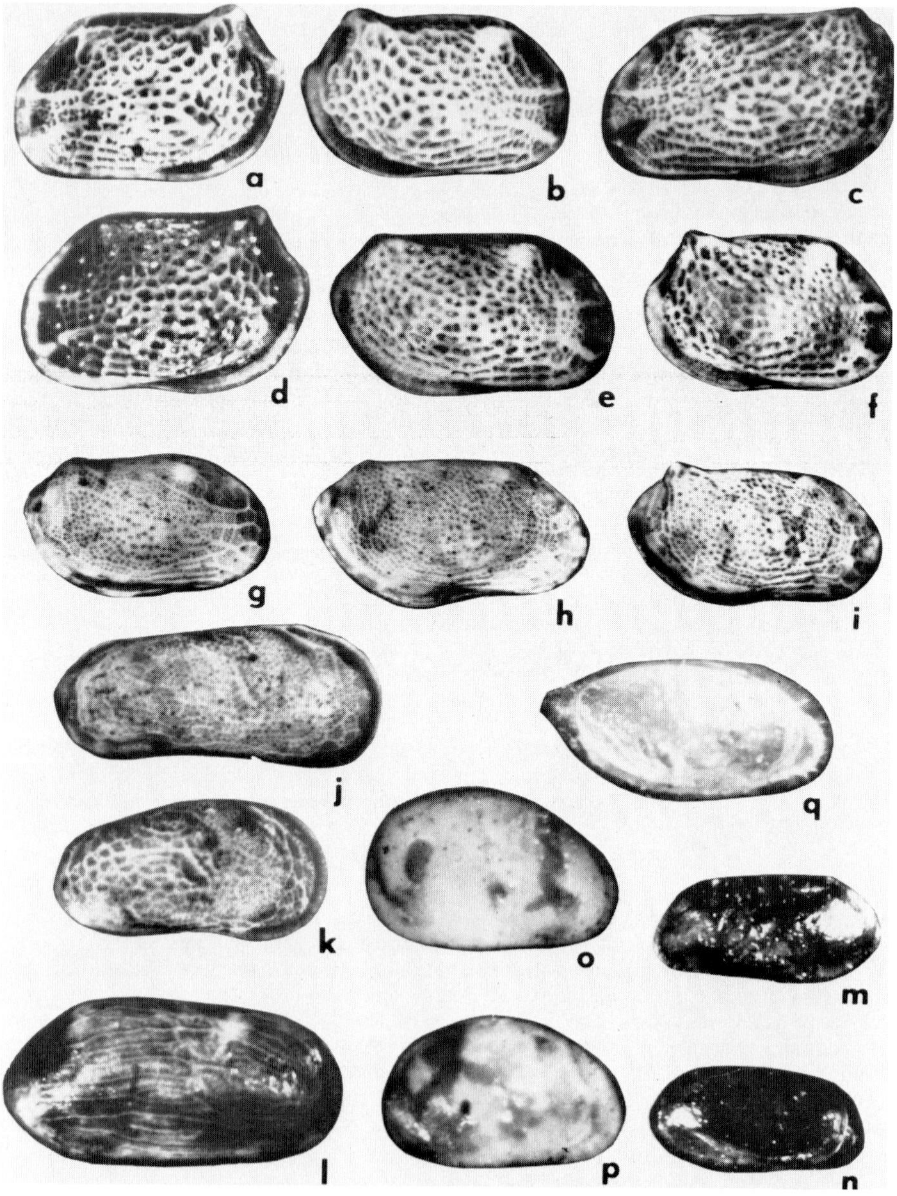

Scarce specimens of another closely related new species with very markedly developed postdorsal and ventral projections have been observed in Belize and the Bahamas. These would seem to be the binodal analogue of *L. rugosa*, which possesses three very prominent projections.

Distribution—Typically restricted to the carbonate platform.

Loxoconcha postdorsoalata group
Figure 21g-i

Loxoconcha anderseni Puri. Puri and Hulings, 1957, Fig. 11.
Loxoconcha postdorsoalata Puri, 1960, p. 111, Pl. 3 (Figs. 17, 18); Text-figs. 35, 37.
Loxocorniculum postdorsolatum (Puri). Benson and Coleman, 1963, p. 39, 40, Pl. 7 (Figs. 1, 2); Fig. 25.

Diagnosis—This group is characterized by the presence of a weak to pronounced posterodorsal projection, a finely punctate to reticulate ornamentation, and four fine, subhorizontal, anterior marginal ridges.

Remarks—This is an extremely difficult group with which to work. The writer is aware of at least five and very likely seven species, all very similar, which may be assigned here.

Loxoconcha matagordensis Swain, with its barely visible posterodorsal projection, illustrates the difficulty of separating the genera *Loxoconcha* and *Loxocorniculum*. This is the most northerly species belonging to the group.

Loxoconcha postdorsoalata Puri has a strongly developed posterodorsal projection and a well-defined reticulate ornamentation. Specimens of *L. postdorsoalata*, very similar to those examined by the writer in Florida samples, are common in Belize.

A second species which has probably been incorrectly assigned to *L. postdorsoalata* (Benson and Coleman, 1963, Fig. 25) is present in Florida Bay. This undescribed species possesses a low posterodorsal projection, fine reticulate ornamentation, and a more rounded anterior cardinal angle than *L. postdorsoalata*. A more finely ornamented species, possibly conspecific with the undescribed one from Florida Bay, occurs in Belize.

⟵

FIG. 21

a. *Loxoconcha oculocrista* Teeter, n. sp. Holotype (H.V.H. 9038), female, complete carapace, left valve, station 315, x60.

b. *L. oculocrista* Teeter, n. sp. Paratype, female, complete carapace, right valve, station 315, x60.

c. *L. oculocrista* Teeter, n. sp. Paratype (H.V.H. 9039), male, complete carapace, left valve, station 315, x60.

d. *Loxoconcha suboculocrista* Teeter, n. sp. Holotype (H.V.H. 9040), female, complete carapace, left valve, station 281, x60.

e. *L. suboculocrista* Teeter, n. sp. Paratype (H.V.H. 9041), male, complete carapace, right valve, station 281, x55.

f. *L. suboculocrista* Teeter, n. sp. Paratype, female, complete carapace, right valve, station 281, x55.

g-i. *Loxoconcha posdorsoalata* group. Complete carapaces, right valves, stations 33 and 68, x47; g, female, h and i, males.

j. *Cytheromorpha apheles* van den Bold. Male, complete carapace, right valve, station 3, x81.

k. *C. apheles* van den Bold. Female, complete carapace, right valve, station 3, x85.

l. *Megacythere johnsoni* (Mincher)? Male, complete carapace, right valve, station 62, x94.

m. *Megacythere stephensoni* (Puri). Complete carapace, left valve, station 532, x55.

n. *M. stephensoni* (Puri). Complete carapace, right valve, station 532, x55.

o. *Xestoleberis punctata* Tressler. Female, complete carapace, right valve, station 391, x64.

p. *X. punctata* Tressler. Male, complete carapace, right valve, station 391, x64.

q. *Pellucistoma howei* Coryell and Fields. Right valve, station 305, x85.

Probably at least three other species are present in Belize but, owing to the uncertainty involved in splitting these taxa, they are also included in this group.

In 1956a, Hartmann described *Loxoconcha bullata*, a species very similar to those mentioned above.

Somewhat similar to several other species of *Loxoconcha* from Belize (e.g., *L. dorsotuberculata*, *L. oculocrista*, and *L. suboculocrista*), members of the group may be distinguished by the presence of four fine, anterior ridges and the absence of the posteroventral projection.

Distribution—The group is found mainly on the carbonate platform although some of its members appear in lower Chetumal Bay samples.

Genus CYTHEROMORPHA Hirschmann, 1909
Cytheromorpha apheles van den Bold, 1963
Figures 20f; 21j, k

Cytheromorpha apheles van den Bold, 1963c, p. 384, Pl. 5 (Figs. 4a-d); Pl. 12 (Fig. 7).

Diagnosis—Characterized by fine, reticulate carapace, evenly rounded anterior margin, and carinalike rim.

Dimensions—Female: length 0.43 mm, height 0.22 mm. Male: length 0.48 mm, height 0.22 mm.

Distribution—Van den Bold described this originally from the upper Miocene Morne l'Enfer formation and Melajo clay of Trinidad. In Belize it is found in upper Chetumal Bay and, rarely, in nearshore areas south of Belize City.

Family PARADOXOSTOMATIDAE Brady and Norman, 1889
Subfamily PARADOXOSTOMATINAE Brady and Norman, 1889

Genus PARADOXOSTOMA Fischer, 1855
Paradoxostoma spp.

Remarks—Several, probably new, species of this genus occur throughout the area. However, they generally are present in such low numbers, are so widespread, and possess such delicate, easily disintegrated shells that they are of little ecologic value.

Subfamily MICROCYTHERINAE Klie, 1938

Genus MICROCYTHERE Mueller, 1894
Microcythere sp. cf. *M. inflexa* Mueller, 1894
Figure 20a

Microcythere inflexa Mueller, 1894, p. 328, Pl. 24 (Figs. 30-32, 40-42, 48, 50).
Microcythere sp. cf. *M. inflexa* Mueller. Puri, 1960, p. 119, Pl. 5 (Figs. 13, 14).

Diagnosis—Characterized by its extremely small size and slightly concave ventral margin.

Remarks—The specimens from Belize are almost certainly conspecific with those from Florida. Length, 0.25 mm, agrees perfectly, as does outline. The original material, however, measures 0.33 mm in length, and may not be conspecific with the material from the Caribbean and Florida.

Distribution—This species exhibits an extremely broad distribution; its carapace is found in all environments.

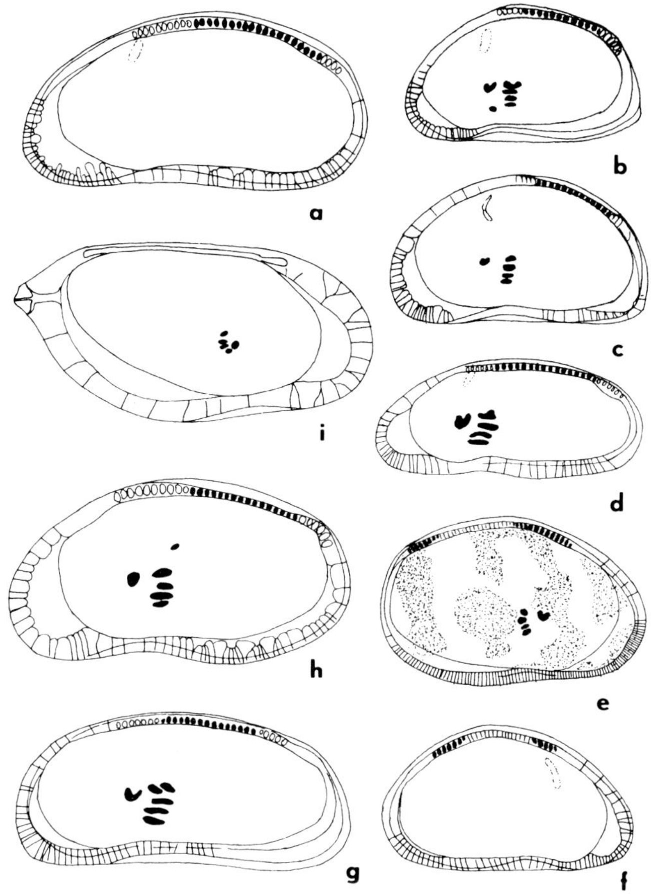

FIG. 22–Internal features.
a. *Xestoleberis* sp. 3. Right valve, station 391, x96.
b. *Xestoleberis* sp. 1. Male, right valve, station 482, x67.
c. *Xestoleberis curassavica* Klie. Female, right valve, station 409, x61.
d. *Xestoleberis* sp. 4. Female, right valve, station 317, x96.
e. *Xestoleberis punctata* Tressler. Female, left valve, station 391, x64.
f. *Xestoleberis* sp. aff. *X. aurantia* (Baird). Female, left valve, station 320, x61.
g. *Xestoleberis* sp. 2. Male, right valve, station 314, x106.
h. *Xestoleberis* sp. 5. Right valve, station 33, x96.
i. *Pellucistoma howei* Coryell and Fields. Left valve, station 305, x96.

Subfamily CYTHEROMATINAE Elofson, 1939

Genus PELLUCISTOMA Coryell and Fields, 1937
Pellucistoma howei Coryell and Fields, 1937
Figures 21q, 22i

Pellucistoma howei Coryell and Fields, 1937, p. 17, Text-figs. 18a-c; van den Bold, 1967b, p. 314, Pl. 1 (Fig. 17); Pl. 2 (Fig. 5).

Diagnosis—Characterized by straight dorsal margin and thin-shelled carapace.
Dimensions—Length 0.48 mm; height 0.25 mm.
Remarks—The present specimens agree closely with the original figures, are thin shelled, have the same dimensions, and possess identical marginal

features. However, there is much variation in the outline of the dorsal margin, which may be either long and straight or arched.

Distribution—This species was described originally from the middle Miocene Gatun Formation of Panama. In Belize it is almost completely restricted to the main lagoon.

Genus MEGACYTHERE Puri, 1960

Megacythere johnsoni (Mincher)? 1941

Figures 20g, 21*l*

Microcythere johnsoni Mincher, 1941, p. 344, Pl. 47 (Figs. 4a-d); Puri, 1953d, p. 299; Swain, 1955, p. 641, Pl. 63 (Figs. 2a-c); Pl. 64 (Fig. 7); Text-fig. 39/3; van den Bold, 1957b, p. 237, Pl. 4 (Fig. 1).

?*Paracytheroma costata* Hartmann, 1959a, p. 158, Figs. 124-130.

Megacythere johnsoni (Mincher). Puri, 1960, p. 122; Morales, 1966, p. 73-75, Pl. 7 (Figs. 3a, b).

Paracytheroma johnsoni (Mincher)? Van den Bold, 1963c, p. 412, 414, Pl. 10 (Fig. 10); Pl. 11 (Figs. 7a, b).

Paracytheroma sp. aff. *P. johnsoni* (Mincher). Van den Bold, 1966f, Pl. 4 (Fig. 6).

Diagnosis—Characterized by its horizontal, curving ribs, which fail to reach the anterior and posterior margins.

Dimensions—Female: length 0.54 mm, height 0.28 mm. Male: length 0.49 mm, height 0.23 mm.

Remarks—The writer has compared the Belize specimens with topotype material from the Miocene Pascagoula Formation. The Holocene specimens are larger and more quadrate, and possess narrow, more continuous ridges. Also, the shells are less heavily constructed. The writer has also examined specimens from the Gulf Coast assigned to this species by Swain (1955). These specimens possess wider ribs than both the fossil and the Belize material, but are similar in size to the fossil specimens. Dimorphism is obvious in the specimens from the Caribbean and the Texas coast. The males from Belize are more elongate and possess finer, much foreshortened ribs compared to the females. Males from the Gulf Coast are more elongate, but the ornamentation does not differ between the sexes. The writer feels that the two groups of Holocene specimens are probably not conspecific with *M. johnsoni* or with each other. However, as van den Bold (1963c) noted, there may be rather considerable variation in this species, and the writer has decided to use open nomenclature for the present.

Distribution—This species appears most commonly in nearshore areas south of Belize City.

Megacythere stephensoni (Puri), 1953

Figures 20h, 21m, n

Microcythere stephensoni Puri, 1953d, p. 291, Pl. 16 (Figs. 11, 12); Text-figs. 12g, h.

Microcythere cf. *M. stephensoni* Puri. Curtis, 1960, p. 478, Pl. 2 (Fig. 17 [bottom]).

Megacythere stephensoni (Puri). Puri, 1960, p. 122; Morales, 1966, p. 75-77, Pl. 7 (Figs. 5a-d).

Diagnosis—Characterized by its smooth carapace and deep anterior and posteroventral vestibula.

Dimensions—Length 0.52 mm, height 0.26 mm.

Distribution—Originally described from the Miocene of Florida, this species has been recorded from the Holocene of Florida and Laguna de Terminos, Mexico. In Belize it is restricted almost completely to upper Chetumal Bay.

Family XESTOLEBERIDIDAE Sars, 1928

Genus XESTOLEBERIS Sars, 1866

Xestoleberis punctata Tressler, 1949
Figures 21o, p; 22e

Xestoleberis punctata Tressler, 1949, p. 342-343, Fig. 12; 1954, p. 434; Puri, 1960, p. 122, Pl. 5 (Figs. 9, 10).
Xestoleberis margaritea (Brady)? Van den Bold, 1963c, p. 402, Pl. 10 (Fig. 13).
Xestoleberis sp. aff. *X. margaritea* (Brady)? Van den Bold, 1966f, p. 52, Pl. 4 (Fig. 11); Pl. 5 (Fig. 3).
Xestoleberis rigbyi Morales? 1966, p. 87-89, Pl. 8 (Figs. 4a-d).
Xestoleberis sp. 3 van den Bold? 1968, p. 79, Pl. 1 (Figs. 7a-c).

Diagnosis—Characterized by rather variable pellucid patches and, in the female, almost vertical posterior margin.

Dimensions—Female: length 0.51 mm, height 0.35 mm. Male: length 0.49 mm, height 0.30 mm.

Remarks—This is undoubtedly the same species as described by Tressler (1949) and figured subsequently by Puri. The writer has examined several samples from the Florida Bay area and observed specimens identical with those from Belize. Tressler did not mention the pellucid patches observed by the writer in specimens from the above areas. These patches apparently disappear soon after the death or molting of the organism, and the whole carapace assumes a milky appearance.

The pattern of the pellucid areas is rather similar to that of *X. variegata* Brady. In outline, however, *X. variegata* possesses a more evenly rounded posterior margin and a more obliquely rounded anterior margin.

Specimens figured by van den Bold (1963c, 1968) agree in size and shape, and are questionably included in synonymy. Another specimen figured by van den Bold (1966f) is similar in outline, but differs in detail in the anterior radial pore canals. Morales' (1966) *X. rigbyi* is similar in size and shape, but that writer reported that sexual dimorphism was not noticeable.

Distribution—This species is present throughout Chetumal Bay and the carbonate platform.

Xestoleberis curassavica Klie, 1939
Figures 22c; 23c, d

Xestoleberis curassavica Klie, 1939a, p. 16-19, Figs. 23-32.

Diagnosis—Characterized by somewhat angulate outline of right valve and pronounced overlap by posteroventral corner of the same valve.

Dimensions—Female: length 0.55 mm, height 0.34 mm. Male: length 0.52 mm, height 0.31 mm.

Distribution—Originally described by Klie from Curaçao. This writer has also observed this species in the Bahamas. In Belize it is restricted to the carbonate platform.

Xestoleberis sp. aff. *X. aurantia* (Baird), 1838
Figures 22f, 23e

Cythere aurantia Baird, 1838, p. 143.
Xestoleberis aurantia (Baird). Wagner, 1957, p. 94-95, Pl. 45.

Diagnosis—Characterized by fragile carapace, smooth surface, and W-shaped frontal muscle scar.

Dimensions—Female: length 0.55 mm, height 0.35 mm. Male: length 0.51 mm, height 0.31 mm.

Remarks—For complete synonymy, see Elofson (1941).

The specimens from Belize agree closely in lateral view and size. The carapace, however, is rather thick and the frontal muscle scar is V-shaped rather than W-shaped.

Whereas this and most subsequent taxa may represent new species, the writer has hesitated to erect them in view of the vast number of species assigned to this genus and the lack of comparative material.

Distribution—This species is restricted to the carbonate platform.

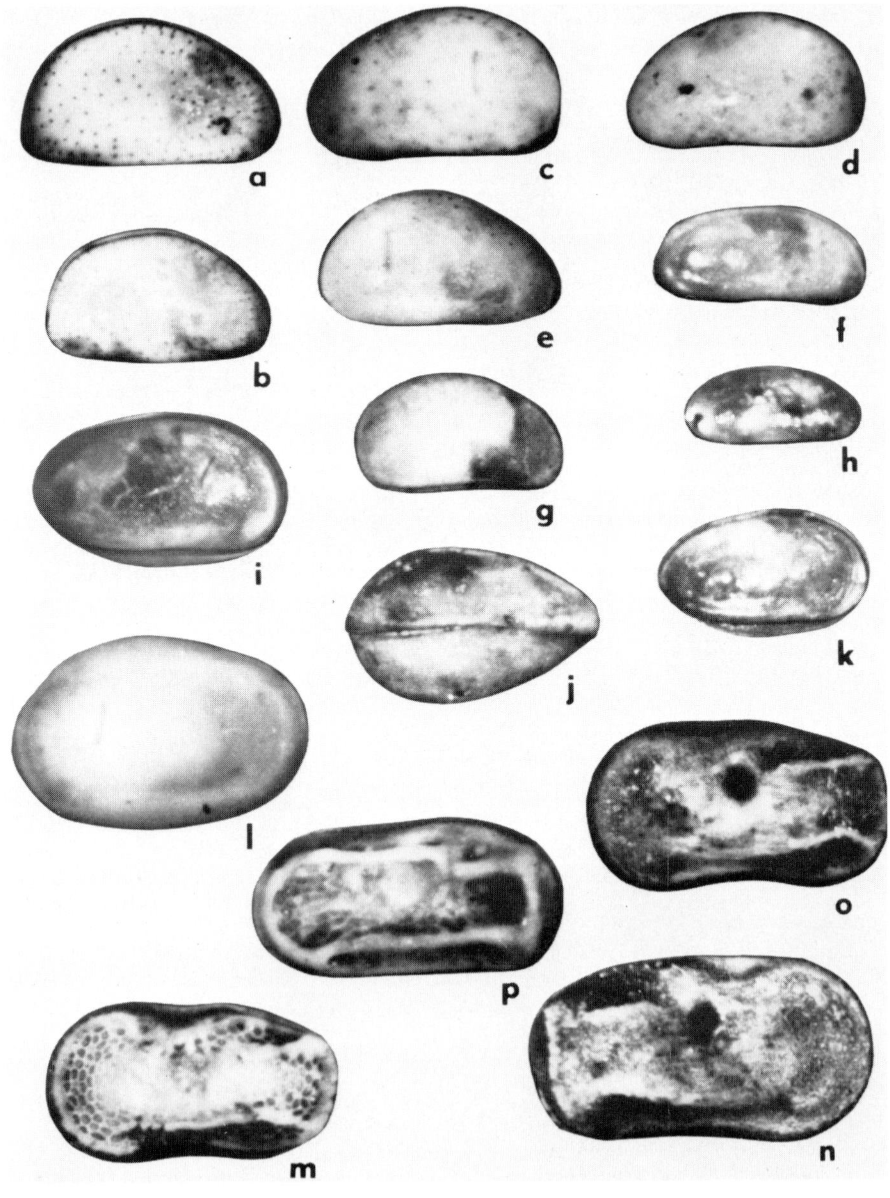

Xestoleberis sp. 1
Figures 22b; 23a, b

Diagnosis—Posteroventral margin with inflection. Papillae present or lacking on posteroventral angle. Ventral surface flattened.

Description—Dorsal margin evenly arched. Anterior margin obliquely rounded. Ventral outline almost perfectly straight. Carapace expanded laterally and ventrally, producing inflected ventral surface which is most pronounced in females. Posteroventral angle possessing or lacking papillae. Posterior margin straight, inclined anteriorly. Surface with very fine punctae. Dimorphism pronounced, females much larger. Submarginal ventral spine present or lacking.

Internal features characteristic of genus. Normal pore canals scattered laterally, arranged longitudinally ventrally.

Dimensions—(Spinose form) Female: length 0.56 mm, height 0.33 mm. Male: length 0.50 mm, height 0.29 mm. (Nonspinose form) Female: length 0.57 mm, height 0.34 mm. Male: length 0.46 mm, height 0.28 mm.

Remarks—This species is variable, particularly with respect to development of papillae, submarginal ventral spine, and size. The larger, papillate, spinose form is not present in Chetumal Bay. Smaller, nonpapillate, nonspinose specimens are almost completely restricted to Chetumal Bay. In other features the two forms are identical.

This species resembles *X? ventrostriata* Swain, particularly in the possession of a striate ventral surface. Swain's species, however, is lower, and the posterior, dorsal, and anterior margins are more confluent. *X. choctawhatcheensis* Puri is remarkably similar but differs, as in the former species, in dorsal outline. *X. angulata* Brady is similar in shape, but possesses two very strongly developed lateral spines.

Distribution—The writer has observed the smaller variation of this species in Florida Bay.

Xestoleberis sp. 2
Figures 22g, 23f

Diagnosis—Characterized by elongate shape and inflected ventral margin.

⟵

FIG. 23

a. *Xestoleberis* sp. 1. Female, complete carapace, right valve, station 482, x60.
b. *X.* sp. 1. Male, complete carapace, right valve, station 482, x64.
c. *Xestoleberis curassavica* Klie. Female, complete carapace, left valve, station 316, x60.
d. *X. curassavica* Klie. Male, complete carapace, left valve, station 316, x60.
e. *Xestoleberis* sp. aff. *X. aurantia* (Baird). Male, complete carapace, right valve, station 320, x60.
f. *Xestoleberis* sp. 2. Male, complete carapace, right valve, station 314, x68.
g. *Xestoleberis* sp. 3. Complete carapace, right valve, station 391, x60.
h. *Xestoleberis* sp. 4. Male, complete carapace, right valve, station 317, x68.
i. *Saipanetta mckenziei* Teeter, n. sp. Holotype (H.V.H. 9042), complete carapace, right valve, station 318, x72.
j. *S. mckenziei* Teeter, n. sp. Paratype (H.V.H. 9043), complete carapace, dorsal view, station 318, x72.
k. *S. mckenziei* Teeter, n. sp. Paratype, complete carapace, right valve, station 318, x60.
l. *Cytherella polita* Brady. Female, right valve, station 58, x60.
m. *Cytherelloidea praecipua* van den Bold. Male, complete carapace, left valve, station 199, x77.
n. *Cytherelloidea umbonata* Edwards. Female, complete carapace, right valve, station 68, x60.
o. *C. umbonata* Edwards. Male, complete carapace, left valve, station 68, x60.
p. *Cytherelloidea* sp. cf. *C. leonensis* Howe. Male, complete carapace, left valve, station 145f, x77.

Description—Carapace elongate. Dorsal margin straight. Anterior margin obliquely rounded. Ventral outline slightly concave anteriorly. Ventral surface similar to *X.* sp. 1 but lacking striae. Posterior margin straight, inclined. Dimorphism pronounced, females more elongate, expanded posteriorly. Surface smooth.

Internal features characteristic for genus.

Dimensions—Female: length 0.45 mm, height 0.21 mm. Male: length 0.41 mm, height 0.20 mm.

Distribution—This species is restricted to the carbonate platform.

Xestoleberis sp. 3
Figures 22a, 23g

Xestoleberis sp. van den Bold, 1963c, p. 402, Pl. 10 (Figs. 14a, b).
Xestoleberis sp. 4 van den Bold, 1968 (? in part), p. 79, Pl. 1 (Figs. 6a-d).

Diagnosis—Characterized by its elongate, bean-shaped outline and polyfurcating, anterior, radial pore canals.

Description—Carapace elongate, bean-shaped. Dorsal margin evenly arched. Anterior margin obliquely rounded. Ventral margin slightly concave anteriorly. Posterior margin evenly, bluntly rounded. Surface smooth, shells largely pellucid.

Internal features typical for genus. Anterior and several posterior radial pore canals somewhat irregular.

Dimensions—Length 0.45 mm, height 0.25 mm.

Remarks—Van den Bold (1963c) figured a specimen of *Xestoleberis* which is similar in shape and size to *X.* sp. 3, differing only in possession of an apparently more truncate anterior margin. In his 1968 paper, Figure 6c of Plate 1 is also similar in shape but smaller in size.

Distribution—The writer has observed scarce specimens of this species in samples from Florida Bay. In Belize it is restricted to the carbonate shelf.

Xestoleberis sp. 4
Figures 22d, 23h

Diagnosis—Characterized by its small size and thin, glassy carapace.

Description—Carapace small. Dorsal margin evenly arched. Anterior margin obliquely rounded, somewhat angular anteroventrally. Ventral outline very slightly concave centrally. Ventral margin inflexed posteriorly. Posterior margin evenly to slightly obliquely rounded. Surface smooth.

Internal features characteristic for genus.

Dimensions—Female: length 0.36 mm, height 0.17 mm. Male: length 0.33 mm, height 0.15 mm.

Distribution—The writer has observed this species in samples from Florida Bay and St. Barts in the Caribbean. In Belize it is present both in Chetumal Bay and on the carbonate platform.

Xestoleberis sp. 5
Figure 22h

Diagnosis—Characterized by small, bean-shaped carapace with large, evenly scattered normal pores and polyfurcating radial pore canals in the adult.

Description—Carapace small, bean-shaped. Dorsal margin evenly arched. Anterior margin very obliquely rounded. Posterior margin evenly rounded. Ventral margin markedly sinuate. Surface smooth.

Internal features typical for genus. Radial pore canals polyfurcate in adults, straight and simple in subadults. Numerous, evenly spaced, large, normal pores.

Dimensions—Female: length 0.48 mm, height 0.28 mm. Male: length 0.44 mm, height 0.25 mm.

Distribution—This species appears to be restricted to the main lagoon.

Suborder METACOPINA Sylvester-Bradley, 1961
Superfamily HEALDIACEA Harlton, 1933
Family SAIPANELLIDAE McKenzie, 1967

Genus SAIPANETTA McKenzie, 1967
Saipanetta mckenziei Teeter, n. sp.
Figure 23i-k

Diagnosis—Distinguished from other Holocene species by its slightly smaller size, ventrolateral overhang, and greater ventral overlap.

Description—Subovate in lateral view, posterior margin bluntly pointed, anteroventral margin bluntly rounded. Subsagittate in dorsal view, widest approximately two thirds the distance from the anterior. Left valve larger than right, overlapping markedly along the free margin, particularly ventrally. Both valves with ventrolateral overhang, especially posteriorly in right valve. Surface smooth; shell brown in fresh specimens.

Inner lamella delicate, narrow, apparently poorly developed. Right valve with finely crenulate terminal teeth with corresponding sockets in left valve. Muscle scar not observed.

Dimensions—Holotype: length 0.46 mm, height 0.28 mm.

Type locality—Station 318, on the reef flat near South Water Cay.

Type specimens—Holotype: H.V.H. 9042. Paratype: H.V.H. 9043.

Remarks—This species is named in honor of K. G. McKenzie, of the British Museum of Natural History, for his contribution to the understanding of this genus.

Distribution—This species has been observed in a sample from Flower Garden Bank, kindly donated by T. E. Pulley of the Houston Museum of Natural Science. In Belize it occurs on the shallow carbonate platform.

Suborder PLATYCOPINA Sars, 1866
Family CYTHERELLIDAE Sars, 1866

Genus CYTHERELLA Jones, 1849
Cytherella polita Brady, 1869
Figure 23*l*

Cytherella polita Brady, 1869, p. 161, Pl. 19 (Figs. 5-7); van den Bold, 1966c, Table 2; 1966e, Table 1; 1966f, Pl. 1 (Figs. 2a, b); 1967b, Table 1.

Diagnosis—Characterized by its small, ovate, thick carapace.

Dimensions—Female: length 0.66 mm, height 0.45 mm. Male: length 0.61 mm, height 0.40 mm.

Remarks—For the rather extensive synonymy, see van den Bold (1963c).

Distribution—This species was originally described from Port au Prince, and subsequently has been observed from the Miocene to the Holocene of the Caribbean. Other occurrences are in doubt. In Belize it is restricted to the main lagoon.

Cytherella arostrata Kornicker, 1963

Cytherella arostrata Kornicker, 1963, p. 66-67, Text-figs. 13-25, 35-38.

Diagnosis—Characterized by reticulate surface tracery and very shallow median sulcus.

Dimensions—Female: length 0.76 mm, height 0.43 mm. Male: length 0.69 mm, height 0.35 mm.

Remarks—Several relatively scarce species of *Cytherella* have been found in Belize, including *C. pandora* Kornicker and possibly *Platella muelleri* Puri, among others. Owing to their relative scarcity, these species have not been described or illustrated.

Distribution—Kornicker described this ostracod from Bimini, in the Bahamas, in water shallower than 15 ft (4.6 m); it occurs on sand bottoms commonly in association with marine grass and algae. The writer has also observed this species in samples from Florida. In Belize it is found in the more saline areas within Chetumal Bay and, rarely, on the carbonate platform.

Genus CYTHERELLOIDEA Alexander, 1929

Cytherelloidea umbonata Edwards, 1944
Figure 23n-o

Cytherelloidea umbonata Edwards, 1944, p. 506, Pl. 85 (Figs. 1, 2); van den Bold, 1946, p. 61, Pl. 5 (Figs. 2a, b).

Diagnosis—Characterized by numerous, short, blunt, anterior marginal denticles; and two rather well-defined, horizontal ridges—the shorter dorsal, the longer ventral—which terminate posteriorly in a vertical ridge.

Dimensions—Female: length 0.75 mm, height 0.41 mm. Male: length 0.67 mm, height 0.39 mm.

Remarks—This agrees almost perfectly with Edwards' original description, although the surface ornamentation in the Holocene specimens appears somewhat finer peripherally.

Distribution—Originally described by Edwards from the Miocene Duplin Marl from North Carolina, this species subsequently has been recorded by van den Bold (1946) from the Miocene of Cuba and questionably from the Quaternary of the island of Seram. The writer has observed one specimen in a sample from the Florida Bay area. In Belize this species is restricted to the carbonate platform.

Cytherelloidea praecipua van den Bold, 1963
Figure 23m

Cytherelloidea praecipua van den Bold, 1963b, p. 75-78, Pl. 1 (Figs. 1-7).

Diagnosis—Characterized by peripherally reticulate ornamentation and smooth central area.

Dimensions—Female: length 0.58 mm, height 0.33 mm. Male: length 0.51 mm, height 0.29 mm.

Remarks—This species is similar to *C. umbonata* Edwards but is much smaller. A second, similar species, perhaps conspecific with *C. sarsi* Puri, occurs rarely in Belize. This second species may be distinguished from *C. praecipua* by the presence of several anterior, marginal, elongate pits.

Distribution—Van den Bold has observed this species in coral sand from both Pigeon Point, Tobago, and Chaguaramus Bay, Trinidad. In Belize, *C. praecipua* apparently is restricted to the main lagoon.

Cytherelloidea sp. cf. *C. leonensis* Howe, 1934
Figure 23p

Cytherelloidea leonensis Howe, 1934, p. 34, Pl. 5 (Fig. 9); Coryell and Fields, 1937, p. 2, Text-figs. 1a-c; van den Bold, 1946, p. 62, Pl. 9 (Fig. 23); 1950a, p. 80; Puri, 1953d, p. 301, Pl. 17 (Fig. 3); Text-fig. 14c; van den Bold, 1958b, p. 396; 1963c, p. 372; 1967b, Table 1.
Cytherelloidea sp. A van den Bold, 1966f, Pl. 4, Fig. 9.

Diagnosis—Characterized by a continuous dorsal, anterior and ventral marginal ridge with a short subdorsal, posterior ridge and longer subventral ridge joining the ventral ridge anteriorly.

Dimensions—Female: length 0.56 mm, height 0.31 mm. Male: length 0.53 mm, height 0.29 mm.

Remarks—The Holocene specimens from Belize are remarkably similar to those originally described by Howe. Most Holocene specimens differ, however, in being slightly shorter and possessing a rather flexed dorsal ridge and extremely delicate reticulate surface.

Distribution—Howe (1934) described this species from specimens obtained from the Miocene "Choctawhatchee" formation of northern Florida. It subsequently has been described from the Miocene throughout the Caribbean. Van den Bold (1958b) noted its presence in the Pliocene of Trinidad. In Belize this species is restricted to the main lagoon.

SELECTED REFERENCES

Allison, E. C., and J. C. Holden, 1971, Recent ostracodes from Clipperton Island, eastern tropical Pacific: San Diego Soc. Nat. History Trans., v. 16, p. 165-214.

Baird, W., 1838, The natural history of British Entomostraca: Mag. Zoology and Botany, v. 2, p. 132-144.

——1850, Description of several new species of Entomostraca: Zool. Soc. London Proc., v. 18, p. 254-257.

Baker, J. H., and N. C. Hulings, 1966, Recent marine ostracod assemblages of Puerto Rico: Texas, Inst. Marine Sci., v. 11, p. 108-125.

Bate, R. H., 1963, The Ostracoda collected during the voyage of H.M.S. Challenger: Micropaleontology, v. 9, p. 79-84.

Benda, W. K., and H. S. Puri, 1962, The distribution of Foraminifera and Ostracoda of the Gulf Coast off the Cape Romano area, Florida: Gulf Coast Assoc. Geol. Socs. Trans., v. 12, p. 303-341.

Benson, R. H., 1959, Ecology of Recent ostracodes of the Todos Santos Bay region, Baja California, Mexico: Kansas Univ. Paleont. Contr., Arthropoda, art. 1, p. 1-80.

——and G. L. Coleman II, 1963, Recent marine Ostracoda from the eastern Gulf of Mexico: Kansas Univ. Paleont. Contr., Arthropoda, art. 2, p. 1-52.

——and R. L. Kaesler, 1963, Recent marine and lagoonal ostracodes from the Estero de Tastiota region, Sonora, Mexico (northeastern Gulf of California): Kansas Univ. Paleont. Contr., Arthropoda, art. 3, p. 1-34.

Blake, C. H., 1933, Order Ostracoda, *in* William Procter, Biological Survey of the Mount Desert Region, pt. 5, Marine fauna: Philadelphia, Wistar Inst. Anatomy and Biology, v. 5, pt. 3, p. 229-241.

Blake, D. B., 1950, Gosport Eocene Ostracoda from Little Stave Creek, Alabama: Jour. Paleontology, v. 24, p. 174-184.

Bosquet, J., 1847, Description des Entomostraces fossiles de la craie de Maestricht: Soc. Royale Sci. Leige Mem., v. 4, p. 353-378.

——1854, Monographie des crustaces fossiles du terrain cretace du Duche Limbourg: Mem. Comm. Descr. Carte Geol. Neerlande, v. 2, p. 53-126.

Brady, G. S., 1866, On new or imperfectly known species of marine Ostracoda: Trans. Zool. Soc. London, v. 5, p. 359-393.

——1867-1871, Ostracoda, *in* Les Fonds de la Mer: Paris, Folin & Perier, v. 1, (1-2), p. 54-247.

——1868a, Contributions to the study of the Entomostraca, II, Marine Ostracoda from the Mauritius: Annals and Mag. Nat. History, ser. 4, v. 2, p. 178-184.

——1868b, Contributions to the study of the Entomostraca, III, Marine Ostracoda from Tenedos: Annals and Mag. Nat. History, ser. 4, v. 2, p. 220-225.

——1868c, A monograph of the Recent British Ostracoda: Linnean Soc. London Trans., v. 26, p. 353-495.

——1880, Report of the Ostracoda dredged by the H.M.S. Challenger during the years 1873-1876: Report on the scientific results of the voyage of H.M.S. Challenger: Zoology, v. 1, p. 1-184.

——1886, Notes on Entomostraca collected by A. Haly in Ceylon: Linnean Soc. London Jour. Zoology, v. 19, p. 293-317.

——1890, On Ostracoda collected by H. B. Brady in the South Sea Islands: Royal Soc. Edinburgh Trans., v. 35, p. 489-525.

——1902, On new or imperfectly-known Ostracoda, chiefly from a collection in the Zoological Museum, Copenhagen: Zool. Soc. London Trans., v. 16, p. 179-210.

——1911, Notes on marine Ostracoda from Madeira: Zool. Soc. London Proc., v. 2, p. 595-601.

——and A. M. Norman, 1889, A monograph of the marine and freshwater Ostracoda of the north Atlantic and of northwestern Europe, Section I, Podocopa: Royal Dublin Soc. Sci. Proc., v. 4, ser. 2, p. 63-270.

——H. W. Crosskey, and D. Robertson, 1874, A monograph of the post-Tertiary Entomostraca of Scotland, including species from England and Ireland: Paleont. Soc., 274 p.

Brown, P. M., 1958, Well logs from the coastal plain of North Carolina: North Carolina Dept. Conserv. and Devel., Div. Mineral Resources Bull., no. 72, p. 1-99.

Butler, E. A., 1963, Ostracoda and correlation of the upper and middle Frio from Louisiana to Florida, with a discussion of the Gulf Coast Miocene-Oligocene boundary: Louisiana Geol. Survey Geol. Bull. 39, 100 p.

Byrne, J. V., D. O. LeRoy, and C. M. Riley, 1959, The chenier plain and its stratigraphy, southwestern Louisiana: Gulf Coast Assoc. Geol. Socs. Trans., v. 9, p. 237-270.

Chapman, F., 1926, The Cretaceous and Tertiary Foraminifera of New Zealand, with an appendix on the Ostracoda: New Zealand Geol. Survey, Paleont. Bull., no. 11, p. 1-22, 98-106.

Coryell, H. N., and S. Fields, 1937, A Gatun ostracode fauna from Cativa, Panama: Am. Mus. Novitates, no. 956, p. 1-18.

Crouch, R. W., 1949, Pliocene Ostracoda from southern California: Jour. Paleontology, v. 23, p. 594-599.

Curtis, D. M., 1960, Relation of environmental energy levels and ostracode biofacies in east Mississippi delta area: AAPG Bull., v. 44, p. 471-492.

Cushman, J. A., 1906, Marine Ostracoda of Vineyard Sound and adjacent waters: Boston Soc. Nat. History Proc., v. 32, p. 359-385.

Drooger, C. W., and J. P. H. Kaasschieter, 1958, Foraminifera of the Orinoco-Trinidad-Paria shelf: Report of the Orinoco Shelf Expedition,

v. 4: Koninkl. Nederlandse Akad. Wetensch. Afd. Natuurk., ser. 1, v. 22, p. 1-108.

——J. P. H. Kaasschieter, and A. J. Keij, 1955, The microfauna of the Aquitanian-Burdigalian of S. W. France: Koninkl. Nederlandse Akad. Wetensch. Proc., ser. 1, v. 11, no. 2, Ostracoda, pt. 4, p. 101-136.

DuBar, J. R., and J. F. Howard, 1964, Duplin formation (late Miocene), at the Muldrow place, Sumter County, South Carolina: South Carolina Devel. Board Div. Geology, Geol. Notes, v. 8, nos. 3-4, p. 25-34.

Edwards, R. A., 1944, Ostracoda from the Duplin Marl (upper Miocene) of North Carolina: Jour. Paleontology, v. 18, p. 505-528.

Elofson, O., 1941, Zur Kenntnis der marinen Ostracoden Schwedens mit besonderer Berucksichtigung des Skagerraks: Uppsala Univ. Zool. Bidr., v. 19, p. 215-534.

Engel, P. L., and F. M. Swain, 1967, Environmental relationships of Recent Ostracoda in Mesquite, Aransas, and Copano Bays, Texas Gulf Coast: Gulf Coast Assoc. Geol. Socs. Trans., v. 17, p. 408-427.

Goerlich, F., 1952, Über die Genotypen und den Begriff der Gattungen *Cyprideis* und *Cytheridea* (Ostracoden): Senckenbergiana Lethaea, v. 33, nos. 1-3, p. 185-192.

Grossman, S., 1965, Morphology and ecology of two podocopid ostracodes from Redfish Bay, Texas: Micropaleontology, v. 11, p. 141-150.

——1967, Ecology of Rhizopodea and Ostracoda of the southern Pamlico Sound region, North Carolina: Kansas Univ. Paleont. Contr., Ecology, art. 1, pt. 1, p. 1-82.

Hall, D. D., 1965, Paleoecology and taxonomy of fossil Ostracoda in the vicinity of Sapelo Island, Georgia, *in* Four reports of ostracod investigations: Michigan Univ. Mus. Paleontology, Rept. 4, 79 p.

Hanai, T., 1957a, Studies on the Ostracoda from Japan; I, Subfamily Leptocytherinae n. subfam.: Tokyo Univ. Fac. Sci. Jour., sec. 2, v. 10, pt. 3, p. 431-468.

——1957b, Studies on the Ostracoda from Japan; II, Subfamily Pectocytherinae n. subfam.: Tokyo Univ. Fac. Sci. Jour., sec. 2, v. 10, pt. 3, p. 469-482.

——1957c, Studies on the Ostracoda from Japan; III, Subfamilies Cytherurinae G. W. Müller (emend. G. O. SARS 1925) and Cytheropterinae n. subfam.: Tokyo Univ. Fac. Sci. Jour., sec. 2, v. 11, pt. 1, p. 11-34.

——1959, Studies on the Ostracoda from Japan; IV, Family Cytherideidae Sars 1925: Tokyo Univ. Fac. Sci. Jour., sec. 2, v. 11, pt. 3, p. 291-308.

Harding, J. P., and P. C. Sylvester-Bradley, 1953, The Ostracod genus *Trachyleberis*: British Mus. (Nat. History) Bull., Zoology, v. 2, no. 1, p. 3-15.

Hartmann, G., 1953, *Iliocythere meyer-abichi* n. sp., ein Ostracode des Schlickwattes von San Salvador: Zool. Anz., v. 151, nos. 11-12, p. 310-316.

——1956a, Weitere neue marine Ostracoden aus Brasilien, *in* E. Titchak and K. H. W. Koepeke, Beitrage zur neotropischen Fauna: Gustav Fischer Verlag, p. 19-62.

——1956b, Zur Kenntnis des Mangrove-Estero-Gebietes von El Salvador und seiner Ostracoden-Fauna: Kieler Meeresforsch., v. 12, no. 2, p. 219-248.

——1957a, Zur Kenntnis des Mangrove-Estero-Gebietes von El Salvador und seiner Ostracoden-Fauna: Kieler Meeresforsch., v. 13, no. 1, p. 134-159.

——1957b, Ostracoden aus den Mangroven Brasiliens: Zool. Anz., v. 158, nos. 9-10, p. 209-217.

——1959a, Zur Kenntnis der lotischen Lebensbereiche der pazifischen Kuste von El Salvador unter besonderer Berucksichtigung seiner Ostracodenfauna; III, Beitrag zur Fauna El Salvadors: Kieler Meeresforsch., v. 15, no. 2, p. 187-241.

——1959b, Neue Ostracoden von Teneriffa: Zool. Anz., v. 162, nos. 5-6, p. 160-171.

——1959c, Beitrag zur Kenntnis des Nicaragua-Sees unter besonderer Berucksichtigung seiner Ostracoden (mit Beschreibung von 5 neuen Arten): Zool. Anz., v. 162, nos. 9-10, p. 269-294.

——1963, Zur Morphologie und Okologie rezenter Ostracoden und deren Bedeutung bei der Unterscheidung mariner und nichtmariner Sedimente: Fortschr. Geologie Rheinland u. Westfalen, 10, p. 67-80.

Hazel, J. E., 1967, Classification and distribution of the Recent Hemicytheridae and Trachyleberidae (Ostracoda) off northeastern North America: U.S. Geol. Survey Prof. Paper 564, p. 1-45.

Hedgpeth, J. W., ed., 1957, Treatise on marine ecology and paleoecology: Geol. Soc. America Mem. 67, v. 1, 1296 p.

Hill, B. L., 1954, Reclassification of winged *Cythereis* and winged *Brachycythere*: Jour. Paleontology, v. 28, p. 804-826.

Holden, J. C., 1967, Late Cenozoic ostracodes from the drowned terraces in the Hawaiian Islands: Pacific Sci., v. 21, p. 1-50.

Hornibrook, N. de B., 1952, Tertiary and Recent marine Ostracoda of New Zealand, their origin, affinities, and distribution: New Zealand Geol. Survey Paleont. Bull. 18, p. 5-82.

Howe, H. V., 1934, The Ostracod genus *Cytherelloidea* in the Gulf Coast Tertiary: Jour. Paleontology, v. 8, p. 29-34.

——1951, New Tertiary ostracode fauna from Levy County, Florida: Florida Geol. Survey Geol. Bull. 34, p. 1-43.

——1955, Handbook of ostracod taxonomy: Louisiana State Univ. Studies, Phys. Sci. Ser., no. 1, 386 p.

——and J. Chambers, 1935, Louisiana Jackson Eocene Ostracoda: Louisiana Geol. Survey Geol. Bull. 5, p. 1-65.

——and J. B. Garret, Jr., 1934, Louisiana Sabine Eocene Ostracoda: Louisiana Geol. Survey Geol. Bull. 4, p. 1-64.

——and L. Laurencich, 1958, Introduction to the study of Cretaceous Ostracoda: Louisiana State Univ. Press, 536 p.

——and J. Law, 1936, Louisiana Vicksburg Oligocene Ostracoda: Louisiana Geol. Survey Geol. Bull. 7, p. 1-96.

——et al, 1935, Ostracoda of the Arca zone of the Choctawhatchee Miocene of Florida: Florida Geol. Survey Geol. Bull. 13, p. 1-37.

Howe, R. C., 1963, Type saline bayou Ostracoda of Louisiana: Louisiana Geol. Survey Geol. Bull. 40, 62 p.

Hulings, N. C., 1958, An ecological study of the Recent ostracods of the Gulf Coast of Florida (abs.): Gulf Coast Assoc. Geol. Socs. Trans., v. 8, p. 166.

——1966, Marine Ostracoda from the western North Atlantic Ocean off the Virginia Coast: Chesapeake Sci., v. 7, p. 40-56.

——1967, Marine Ostracoda from the western North Atlantic Ocean between Cape Hatteras, North Carolina and Jupiter Inlet, Florida: Bull. Marine Sci., v. 17, p. 629-659.

——and H. S. Puri, 1964, The ecology of shallow water ostracods of the west coast of Florida: Pubbli. Staz. Zool. Napoli, 33 suppl., p. 308-344.

Jones, T. R., 1856, A monograph of the Tertiary Entomostraca of England: London, Palaeont. Assoc., p. 1-68.

——and C. D. Sherborn, 1889, A supplemental monograph of the Tertiary Entomostraca of England: Palaeontology, v. 42, p. 1-55.

Keij, A. J., 1953, Preliminary note on the Recent Ostracoda of the Snellius Expedition: Koninkl. Nederlandse Akad. Wetensch. Proc., ser. B, v. 56, no. 2, p. 155-168.

——1954a, Some Recent Ostracoda of Manilla (Philippines): Koninkl. Nederlandse Akad. Wetensch. Proc., ser. B, v. 57, no. 3, p. 351-363.

——1954b, Identification and description of species, *in* T. van Andel and H. Postma, Recent sediments of the Gulf of Paria; Report of the Orinoco Shelf Expedition, v. 1: Koninkl. Nederlandse Akad. Wetensch., Natuurk., ser. 1, v. 20, no. 5, p. 218-245.

——1957, Eocene and Oligocene Ostracoda of Belgium: Inst. Royal Sci. Nat. Belgique Mém. 136, 210 p.

——1958, Note on the Lutetian Ostracoda of Damery (Marne), France: Koninkl. Nederlandse Akad. Wetensch. Proc., ser. B, v. 61, no. 1, p. 63-73.

Kingma, J. T., 1948, Contributions to the knowledge of the young-Caenozoic Ostracoda from the Malayan region: Utrecht Univ., thesis, 119 p.

Klie, W., 1933, Suss-und Brackwasser-Ostracoden von Bonaire, Curaçao und Aruba: Zool. Jahrbuch, v. 64, no. 5, p. 369-390.

——1939a, Ostracoden aus den marinen Salinen von Bonaire, Curaçao und Aruba: Capita Zoologica, v. 8, pt. 4, p. 1-19.

——1939b, Brackwasserostracoden von Nordostbrasilien: Zool. Jahrbuch Abt. Syst., v. 72, p. 359-372.

Kornicker, L. S., 1958, Ecology and taxonomy of Recent marine ostracodes in the Bimini area, Great Bahama Bank: Texas Univ. Inst. Marine Sci. Pub., v. 5, p. 194-300.

——1959, Distribution of the ostracode suborder Cladocopa and a new species from the Bahamas: Micropaleontology, v. 5, p. 69-75.

——1961, Ecology and taxonomy of Recent Bairdiinae (Ostracoda): Micropaleontology, v. 7, p. 55-70.

——1963, Ecology and classification of Bahamian Cytherellidae (Ostracoda): Micropaleontology, v. 9, p. 61-70.

——1964a, A seasonal study of living Ostracoda in a Texas bay (Redfish Bay) adjoining the Gulf of Mexico: Pubbli. Staz. Zool. Napoli, 33 suppl., p. 45-60.

——1964b, Ecology of Ostracoda in the northwestern part of the Great Bahama Bank: Pubbli. Staz. Zool. Napoli, 33 suppl., p. 345-360.

——and C. D. Wise, 1960, Some environmental boundaries of a marine ostracode: Micropaleontology, v. 6, p. 393-398.

Kruit, C., 1955, Sediments of the Rhone Delta: Koninkl. Geol.-Mijnb. Genoot., Verh., Geol. Ser., v. 15, p. 359-514 (ostracods, p. 475-489).

Krutak, P. R., 1961, Jackson Eocene Ostracoda from the Cocoa Sand of Alabama: Jour. Paleontology, v. 35, p. 769-788.

LeRoy, D. O., 1964, Two new species of Recent Ostracoda from southern Louisiana: Jour. Paleontology, v. 38, p. 1097-1099.

LeRoy, L. W., 1943, Pleistocene and Pliocene Ostracoda of the coastal region of southern California: Jour. Paleontology, v. 17, p. 354-373.

Lynch, S. A., 1954, Geology of the Gulf of Mexico, *in* Chap. 2 *of* Gulf of Mexico, its origins, waters, and marine life: U.S. Fish and Wildlife Service Fishery Bull. 89, p. 67-86.

Maddocks, R. F., 1969, Revision of Recent Bairdiidae (Ostracoda): U.S. Natl. Mus. Bull. 295, p. 1-126.

Malkin, D. C., 1953, Biostratigraphic study of Miocene Ostracoda of New Jersey, Maryland, and Virginia: Jour. Paleontology, v. 27, p. 761-799.

Martin, G. P. R., 1957, A new method of recovering remains of the chitinous integument of fossil Ostracoda: Micropaleontology, v. 3, p. 291-292.

Matthews, R. K., 1965, Genesis of Recent lime mud in southern British Honduras: Rice Univ., PhD thesis, 139 p.

McKenzie, K. G., 1967, Saipanellidae: A new family of podocopid Ostracoda: Crustaceana, v. 13, p. 103-113.

——1968, *Saipanetta*, a new name for *Saipanella* McKenzie, 1967 (Ostracoda, Podocopida) non *Saipanella* Chamberlin, 1945 (Myriapoda): Crustaceana, v. 14.

——and F. M. Swain, 1967, Recent Ostracoda from Scammon Lagoon, Baja California: Jour. Paleontology, v. 41, p. 281-305.

McLean, J. D., Jr., 1957, Ostracoda of the Yorktown formation in the York-James Peninsula of Virginia: Bulls. Am Paleontology, v. 38, no. 167, p. 57-103.

Mehes, G., 1941, Die ostracoden des oberoligozäns der umgebung von Budapest: Geol. Hungary Ser. Palaeontology, v. 16, p. 1-64.

Mincher, A. R., 1941, The fauna of the Pascagoula formation: Jour. Paleontology, v. 15, p. 337-348.

Moore, R. C., ed., 1961, Treatise on invertebrate paleontology, Part Q, Arthropoda 3, Crustacea, Ostracoda: Kansas Univ. Press, 442 p.

Moos, B., 1957, Die Ostracoden-Gattung *Triginglymus* D. B. Blake als Synonym von der Gattung *Leguminocythereis* H. V. Howe: Geol. Jahrb., v. 72, p. 501-502.

Morales, G. A., 1966, Ecology, distribution, and taxonomy of Recent Ostracoda of the Laguna de Terminos, Campeche, Mexico: México Univ. Nac. Autónoma Inst. Geología Bull. 81, 103 p.

Morkhoven, F. P. C. M. van, 1958, On the validity of the ostracod genera *Glyptobairdia* and *Bairdoppilata*: Jour. Paleontology, v. 32, p. 366-368.

——1962-1963, Post-Palaeozoic Ostracoda; their morphology, taxonomy, and economic use: Amsterdam, London, New York; Elsevier, v. 1, 204 p., v. 2, 478 p.

Mueller, G. W., 1894, Die Ostracoden des Golfes von Neapel und der angrenzenden Meeres-abschnitte: Naples Sta. Zool., Fauna und Flora des Golfes von Neapel, Monographie 21 (Berlin), p. 1-404.

——1912, Ostracoda: Tierreich, v. 31, p. 1-434.

Plusquellec, P. L., and P. A. Sandberg, 1969, Some genera of the ostracode subfamily Campylocytherinae: Micropaleontology, v. 15, p. 427-480.

Pokorný, V., 1955, Contribution to the morphology and taxonomy of the subfamily Hemicytherinae Puri 1953 (Crustacea, Ostracoda): Carolinae Univ. Acta, Geologica, no. 3, 35 p.

——1958, Grundzüge der zoologischen Mikropaläontologie; Bd. II: Berlin, Deutscher Verlag Wiss., 453 p.

——1964, The taxonomic delimitation of the subfamilies Trachyleberidinae and Hemicytherinae (Ostracoda, Crustacea): Carolinae Univ. Acta, Geologica, no. 3, p. 275-284.

Pooser, W. K., 1965, Biostratigraphy of Cenozoic Ostracoda from South Carolina: Kansas Univ. Paleont. Contr., Arthropoda, art. 8, p. 1-80.

Prochazka, V. J., 1893, Das Miocaen von Seelowitz in Mahren und dessen Fauna: Rozpr. Ceska. Akad. Praze, v. 2, no. 24, p. 1-90.

Puri, H. S., 1952a, Ostracode genera *Cytheretta* and *Paracytheretta* in America: Jour. Paleontology, v. 26, p. 199-212.

——1952b, Ostracode genus *Cytherideis* and its allies: Jour. Paleontology, v. 26, p. 902-914.

——1953a, Taxonomic comment on: "Ostracoda from the wells in North Carolina, Pt. I: Cenozoic Ostracoda" by F. M. Swain: Jour. Paleontology, v. 27, p. 750-752.

——1953b, The ostracode genus *Trachyleberis* and its ally *Actinocythereis*: Am. Midland Naturalist, v. 49, p. 171-187.

——1953c, The ostracode genus *Hemicythere* and its allies: Washington Acad. Sci. Jour., v. 43, no. 6, p. 169-179.

——1953d, Contribution to the study of the Miocene of the Florida Panhandle: Florida Geol. Survey Geol. Bull. 36, 345 p.

——1957, Notes on the ostracode family Cytherideidinae Puri 1952: Washington Acad. Sci. Jour., v. 47, p. 305-306.

——1958a, Ostracode genus *Cushmanidea*: Gulf Coast Assoc. Geol. Socs. Trans., v. 8, p. 171-181.

——1958b, Ostracode subfamily Cytherettinae: Gulf Coast Assoc. Geol. Socs. Trans., v. 8, p. 183-195.

——1960, Recent Ostracoda from the west coast of Florida: Gulf Coast Assoc. Geol. Socs. Trans., v. 10, p. 107-149.

——and N. C. Hulings, 1957, Recent ostracode facies from Panama City to Florida Bay area: Gulf Coast Assoc. Geol. Socs. Trans., v. 7, p. 167-190.

——and R. O. Vernon, 1959, Summary of the geology of Florida and a guidebook to the classic exposures: Florida Geol. Survey Spec. Pub. 5, 312 p.

——G. Bonaduce, and J. Malloy, 1964, Ecology of the Gulf of Naples: Pubbli. Staz. Zool. Napoli, v. 33, p. 87-199.

Pusey, W. C., 1964, Recent calcium carbonate sedimentation in northern British Honduras: Houston, Texas, Rice Univ., PhD thesis, 247 p.

Reyment, R., and E. Reyment, 1959, *Bairdia ilaroensis* sp. nov. aus dem Palaozan Nigeriens und die Gultigkeit der Gattung *Bairdoppilata* (Ostracoda, Crustacea): Stockholm Contr. Geology, v. 3, p. 59-69.

Rome, R., 1960, Note sur *Triebelina coronata* (Brady) et sa position systematique: Inst. Royal Sci. Nat. Belgique, v. 36, no. 3, p. 1-14.

Ruggieri, G., 1953, Ostracodi del genere *Paijenborchella* viventi nel Mediterraneo: Atti Soc. Ital. Sci. Nat., v. 92, p. 3-7.

Sabol, J. W., 1960, The microfauna of the Yorktown formation from James River, Surry County, Virginia: Bull. Am. Paleontology, v. 41, no. 191, p. 211-247.

Sandberg, P. A., 1964a, Larva-adult relationships in some species of the ostracode genus *Haplocytheridea*: Micropaleontology, v. 10, p. 357-368.

——1964b, The ostracode genus *Cyprideis* in the Americas: Stockholm Contr. Geology, v. 12, 178 p.

Sars, G. O., 1866, Oversigt af Norges marine ostracoder: Förh. Vidensk. Selsk. Christiania, v. 7, p. 1-130.

——1922-1928, An account of the Crustacea of Norway: Bergen Mus. (Norway), v. 9, p. 1-277.

Sexton, J. V., 1951, The ostracode genus *Cytherelloidea* in North America: Jour. Paleontology, v. 26, p. 808-816.

Skogsberg, T., 1928, Studies on marine ostracods; II, External morphology of the genus *Cythereis* with descriptions of 21 new species: California Acad. Sci. Occasional Paper 15, 154 p.

Sohn, I. G., 1951, Check list of salinity tolerance of post-Paleozoic fossil Ostracoda: Washington Acad. Sci. Jour., v. 41, no. 2, p. 64-66.

——1960, Paleozoic species of *Bairdia* and related genera: U.S. Geol. Survey Prof. Paper 330-A, p. 1-105.

Stephenson, M. B., 1935, Some microfossils of the *Potamides matsoni* zone of Louisiana: Louisiana Dept. Conserv. Geol. Bull. 6, p. 187-196.

——1938, Miocene and Pliocene Ostracoda of the genus *Cytheridea* from Florida: Jour. Paleontology, v. 12, p. 127-148.

——1944, New Ostracoda from subsurface middle Tertiary strata of Texas: Jour. Paleontology, v. 18, p. 156-161.

——1946, *Glyptobairdia*, a new genus of Ostracoda: Jour. Paleontology, v. 20, p. 345-347.

——1947, Notes on the ostracode genus *Triebelina*: Jour. Paleontology, v. 21, p. 577-579.

Swain, F. M., 1946, Ostracoda from the Tertiary of Florida: Jour. Paleontology, v. 10, p. 374-383.

——1951, Ostracoda from wells in North Carolina, Pt. I, Cenozoic Ostracoda: U.S. Geol. Survey Prof. Paper 234-A, p. 1-58.

——1955, Ostracoda of San Antonio Bay, Texas: Jour. Paleontology, v. 29, p. 561-646.

——1967, Ostracoda from the Gulf of California: Geol. Soc. America Mem. 101, 139 p.

——1968, Ostracoda from the upper Tertiary Waccamaw formation of North Carolina and South Carolina: U.S. Geol. Survey Prof. Paper 573-D.

Sylvester-Bradley, P. C., 1946, Some ostracod genotypes: Annals and Mag. Nat. History, ser. 11, no. 13, p. 192-199.

——1948, The ostracode genus *Cythereis*: Jour. Paleontology, v. 22, p. 792-797.

——1950, The shell of the ostracod genus *Bairdia*: Annals and Mag. Nat. History, ser. 12, no. 3, p. 751-756.

Tressler, W. L., 1941, Ostracoda, Pt. 4 *of* Geology and biology of North Atlantic deep-sea cores between Newfoundland and Ireland: U.S. Geol. Survey Prof. Paper 196, p. 95-106.

——1949, Marine Ostracoda from Tortugas, Florida: Washington Acad. Sci. Jour., v. 39, p. 335-343.

——1954, Marine Ostracoda, *in* Gulf of Mexico, its origin, waters, and marine life: U.S. Fish and Wildlife Service Fishery Bull. 89, p. 1-604 (ostracods, p. 429-437).

——and E. M. Smith, 1948, An ecological study of seasonal distribution of Ostracoda, Solomons Island, Maryland, region: Chesapeake Biol. Lab. Pub. 71, 61 p.

Triebel, E., 1939, Zur Ostracoden-Fauna von Brasilien: Senckenbergiana Lethaea, v. 21, nos. 5-6, p. 363-374.

——1941, Zur Morphologie und Oekologie des fossilen Ostracoden: Senckenbergiana Lethaea, v. 23, p. 294-400.

——1948, Zur Kenntnis der Ostracoden-Gattung *Triebelina*: Senckenbergiana Lethaea, v. 29, p. 17-22.

——1960, Die taxionomische Stellung und die Gattungen der Unterfamilie Macrocypridinae (Ostracoda): Senckenbergiana Lethaea, v. 41, nos. 1-2, p. 109-124.

——1961, Geschlects-Dimorphismus und Asymmetrie der Klappen bei der Ostracoden-Gattung *Occultocythereis*: Senckenbergiana Lethaea, v. 42, nos. 3-4, p. 205-225.

Ulrich, E. O., and R. S. Bassler, 1904, Systematic paleontology of the Miocene deposits of Maryland: Maryland Geol. Survey, Miocene Rept., Ostracoda, p. 98-130.

van den Bold, W. A., 1946, Contribution to the study of Ostracoda with special reference to the Tertiary and Cretaceous microfauna of the Caribbean region: Amsterdam, J. H. de Bussy, Utrecht Univ., thesis, 167 p.

——1950a, Miocene Ostracoda from Venezuela: Jour. Paleontology, v. 24, p. 76-88.

——1950b, A check list of Cuban Ostracoda: Jour. Paleontology, v. 24, p. 107-109.

——1957a, Ostracoda from the Paleocene of Trinidad: Micropaleontology, v. 3, p. 1-18.

——1957b, Oligo-Miocene Ostracoda from southern Trinidad: Micropaleontology, v. 3, p. 231-254.

——1958a, Distribution of fresh-water ostracodes in Trinidad: Micropaleontology, v. 4, p. 71-74.

——1958b, Ostracoda of the Brasso formation of Trinidad: Micropaleontology, v. 4, p. 391-418.

——1960, Eocene and Oligocene Ostracoda of Trinidad: Micropaleontology, v. 6, p. 145-196.

——1961, Some new Ostracoda of the Caribbean Tertiary: Koninkl. Nederlandse Akad. Wetensch. Proc., ser. B, v. 64, p. 627-638.

——1963a, The ostracode genus *Orionina* and its species: Jour. Paleontology, v. 37, p. 33-50.

——1963b, Anomalous hinge structure in a new species of *Cytherelloidea*: Micropaleontology, v. 9, p. 75-78.

——1963c, Upper Miocene and Pliocene Ostracoda of Trinidad: Micropaleontology, v. 9, p. 361-424.

——1964, Nota preliminar sobre los Ostracodos del Mioceno-Reciente de Venezuela: Separata de la Rev. GEOS., Caracas, no. 11, 9 p.
——1965a, *Pseudoceratina*, a new genus of Ostracoda from the Caribbean: Koninkl. Nederlandse Akad. Wetensch. Proc., ser. B, v. 68, p. 160-164.
——1965b, Middle Tertiary Ostracoda from northwestern Puerto Rico: Micropaleontology, v. 11, p. 381-414.
——1966a, Ostracoda of the Pozón section, Falcón, Venezuela: Jour. Paleontology, v. 40, p. 177-185.
——1966b, New species of the ostracod genus *Ambocythere*: Annals and Mag. Nat. History, ser. 13, v. 8, p. 1-18.
——1966c, Miocene and Pliocene Ostracoda from northeastern Venezuela: Koninkl. Nederlandse Akad. Wetensch. Proc., ser. 1, v. 23, p. 5-43.
——1966d, Ostracode zones in the Caribbean Miocene: AAPG Bull., v. 50, p. 1029-1031.
——1966e, Upper Miocene Ostracoda from the Tubará Formation (northern Colombia): Micropaleontology, v. 12, p. 360-364.
——1966f, Ostracoda from Colón Harbour, Panamá: Caribbean Jour. Sci., v. 6, p. 43-64.
——1967a, Miocene Ostracoda from Costa Rica: Micropaleontology, v. 13, p. 75-86.
——1967b, Ostracoda of the Gatún Formation, Panama: Micropaleontology, v. 13, p. 306-318.
——1968, Ostracoda of the Yague Group (Neogene) of the northern Dominican Republic: Bulls. Am. Paleontology, v. 54, no. 239, p. 1-106.
——1970, The genus *Costa* (Ostracoda) in the upper Cenozoic of the Caribbean region: Micropaleontology, v. 16, p. 61-75.
Wagner, C. W., 1957, Sur les Ostracodes du Quaternaire Récent des Pays-Bas et leur utilisation dans l'Étude géologique des Dépôts Holocènes: Paris Univ., thesis, 259 p.
Walton, W., 1952, Technique for the recognition of living Foraminifera: Cushman Found. Foram. Research Contr., v. 3, p. 56-60.
Wells, J. W., 1957, Coral Reefs, *in* Treatise on marine ecology and paleoecology: Geol. Soc. America Mem. 67, v. 1, p. 609-631.
Wise, C. D., 1960, Variations in size of ostracods cultured from dried soil in the laboratory: Ecology, v. 41, no. 2, 1 p.

Petrology and Diagenesis of Carbonate Eolianites of Northeastern Yucatán Peninsula, Mexico[1]

WILLIAM C. WARD[2]

ABSTRACT Recognition of features resulting from subaerial diagenesis of young carbonate rocks is important in defining times of subaerial exposure during the diagenetic history of ancient limestones. Meteoric-water cementation and other products of subaerial diagenesis of limestones can be characterized most unequivocally in carbonate units which were deposited in the subaerial environment and subsequently remained above marine waters. Such limestones were deposited as dune ridges along the northeast coast of the Yucatán Peninsula, Mexico. Late Pleistocene calcareous dunes accumulated along the seaward edge of a platform now 5 fm below sea level off the northeast coast of Quintana Roo, and Holocene eolianites are being deposited and lithified in the same area. Consolidated late Holocene dune rock on this coast provides a missing link in the study of progressive diagenesis from unconsolidated dune sand to lithified Pleistocene eolianite.

On the basis of grain composition and geomorphic relations, the Pleistocene eolianites can be divided into three different limestones, and the Holocene eolianites can be divided into two limestones. Each eolianite represents a different diagenetic stage. Youngest Holocene rocks have about the same composition as dune and beach sands: 75-85 percent aragonite, 15-20 percent Mg calcite, and less than 5 percent low-Mg calcite. Older ridges of the younger Holocene eolianite contain as much as 22 percent low-Mg calcite. The older Holocene eolianite contains 69-84 percent aragonite and less than 5 percent Mg calcite (this eolianite is low in Mg-calcite bioclasts).

The youngest Pleistocene eolianite originally contained as much as 45 percent Mg calcite, and there is high retention of Mg calcite in some beds. Several samples contain 20-32 percent Mg calcite (12-14 mole percent $MgCO_3$); some samples have no Mg calcite. Aragonite ranges from 45 to 65 percent. The second youngest Pleistocene eolianite has 48-75 percent aragonite and less than 5 percent Mg calcite. The oldest eolianite contains 40-60 percent aragonite and less than 5 percent Mg calcite. Although there is a general trend toward calcitization of these dune rocks, each eolianite has a different route and rate of progressive diagenesis.

Cement types are important clues to the environment of early lithification of limestones. The Holocene eolianites contain grain-contact cement, microstalactitic druse, and large syntaxial overgrowths on echinoderm fragments. Finer grained layers are preferentially cemented. "Micrite envelopes" may form around grains in the vadose zone, and microcrystalline inclusions are common in sparry cement of the eolianites. Cement in the Holocene rocks is derived from a local source.

Much of the Pleistocene eolianite contains grain-skin cement in pores which contain "root-hair sheaths" and blocky spar in pores where the sheaths are absent. This fact suggests that early cementation was influenced by transpiration of dune plants. "Needle-fiber" cement is present in Pleistocene eolianites near ancient weathered surfaces. Pleistocene eolianites now immersed in the intertidal-subtidal environment are enriched in Mg calcite as a result of precipitation of Mg-calcite cement in the pores.

Laminated crusts and conglomerates containing "oolitic" and "pisolitic" bodies are developed on the weathered surfaces of the Pleistocene eolianites. Blackened "caliche-algal micrite" crusts were laid down around hypersaline lakes on Isla Mujeres.

INTRODUCTION

Significance of Eolianites

It is the aim of this report to document the stratigraphy and progressive diagenesis of the wind-laid Quaternary limestones on the northeastern coast of the Yucatán Peninsula. Wind-deposited calcarenite, or eolianite, is a common facies in Pleistocene and Holocene carbonate sequences in

[1]Manuscript received by editor, January 1971; received by AAPG, June 20, 1972.

[2]Department of Earth Sciences, Louisiana State University of New Orleans, New Orleans, Louisiana 70122.

This paper is based on part of the writer's PhD thesis, Rice University, Houston, Texas. Petroleum Research Fund Grant No. 3028-A2 financed the research. I am grateful to James Lee Wilson, who supervised my thesis work, and to R. L. Folk, W. J. Ebanks, and J. E. Warme, who made helpful suggestions and editorial criticisms during preparation of the manuscript. G. P. Salas and D. A. Cordóba were cooperative in sponsoring my research in Yucatán. Many others who deserve thanks are acknowledged in the writer's PhD thesis.

many parts of the world. In spite of this, the vast literature on carbonate rocks includes relatively little on eolianite limestones, and calcareous dune rock rarely is reported in rocks older than Quaternary. This paucity in the literature exists either because geologists have not recognized eolianite, or because it is scarce in ancient sections. Even if wind-laid limestone were peculiar to younger sections, the study of Quaternary eolianite would have general significance for several reasons:

1. It seems reasonable to assume that meteoric-water cementation and other products of subaerial diagenesis of limestones can be characterized most unequivocally in carbonate units which were deposited in the subaerial environment and subsequently remained above marine waters. Once recognized, subaerial diagenetic features could be important in defining times of subaerial exposure and in interpreting diagenetic history of ancient limestones.

2. Establishment of criteria for characterizing Quaternary dune rock may lead to the recognition of eolianite in older carbonate sections. Identification of eolianite would facilitate location of ancient strandlines and islands—and this is important in stratigraphic analyses of carbonate provinces.

3. Ridges of Quaternary calcareous dunes have a profound effect on the distribution patterns of subsequent sediments (Ball, 1967, p. 582-583). Recognition of dune ridges in the ancient record, then, would be of significant value in predicting sedimentation patterns of younger, overlying deposits.

4. Eolianites are potential oil and gas reservoirs because they are cemented early, enhancing their chances for preservation. Porosity in eolianite may be high, and in some carbonate sequences eolianite is the most updip calcarenite. Whether or not any reservoirs exist in eolianitic limestones, the knowledge of subaerial diagenesis gained from the study of calcareous dune rock is applicable to petroleum exploration. Porosity and permeability in carbonate rock can be either enhanced or destroyed by subaerial weathering (e.g., Robinson, 1967; Roehl, 1967; Dunham, 1969b; Choquette and Pray, 1970).

Purpose

There is need for further documentation of the effects of subaerial diagenesis of eolian limestones in order to determine the general applicability of recent studies by other workers. Coastal northeast Yucatán is an excellent locality for the study of progressive diagenesis in wind-laid limestones, because it is possible to establish the relative ages of Holocene and late Pleistocene eolianites and to estimate their original compositions. Holocene eolianites that have accumulated on Isla Blanca, Isla Cancun, and the Puerto Juarez coast can be compared with their Pleistocene counterparts on Isla Contoy, Isla Mujeres, and Isla Cancun (Fig. 1). Furthermore, these eolianites exhibit a wide range of diagenetic effects—from slightly indurated crustal patches on sand dunes, to friable older dune rock, to still older and more indurated limestones.

Unlike previously published reports on progressive diagenesis, this paper deals extensively with subaerial cementation and diagenesis in late Holocene rocks. The observations made here may provide criteria for recognizing and interpreting subaerial deposition and diagenesis in more ancient carbonate sections and may aid in distinguishing between submarine and subaerial diagenetic features.

Previous Studies

Evans (1900) presented a comprehensive review of the nineteenth-century papers describing wind-deposited limestones in many parts of the world. He formulated criteria for recognizing calcareous dune rocks and

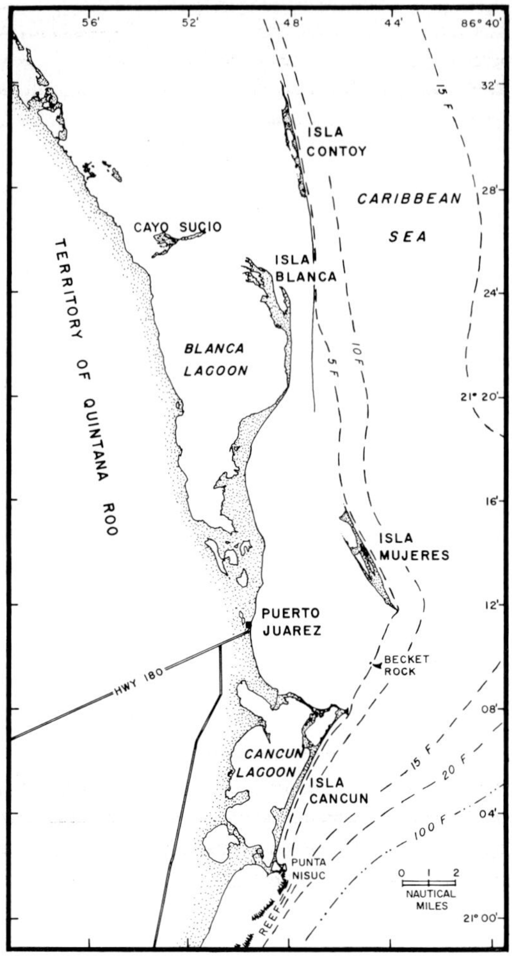

FIG. 1—General location map of northeast coast of Territory of Quintana Roo, Mexico, showing localities discussed in text.

applied his criteria to the identification of eolianite in the Jurassic Great Oolite Series of England. In an early work on the limestones of Bermuda, Sayles (1931) set the custom of using the term "eolianite" to mean calcarenitic dune rock, a usage followed in most later works and in this paper. Teichert (1947) and Fairbridge (1950) described eolianites on the west coast of Australia, where the calcareous dunes are the most extensive in the world.

Eolianites in the Caribbean have been described by Vaughan (1914), Illing (1954), Newell and Rigby (1957), Kaye (1959), and Ball (1967). Eolianitic limestones of northeast Quintana Roo were described briefly by Folk et al (1962), Folk (1967), Harding (1964), and Ward (1967).

Previous studies of diagenesis in eolianite include those of Schwarz and Rogers (1898; cited by Chapman, 1900), who noted cementation of calcareous dunes as a result of percolation of surface water; Friedman (1964), who studied mineralogic diagenesis and cementation of limestones; and Gross (1964), who described oxygen and carbon isotope variations in dune limestones of Bermuda. Recently, Robinson (1967), Land et al (1967), and Gavish and Friedman (1969) reported on the progression of subaerial diagenesis in calcarenitic limestones of Quaternary age.

NORTHEAST YUCATÁN SHELF

Since Early Cretaceous time the Yucatán region has been the site of carbonate sedimentation (Meyerhoff, 1967, Fig. 8). The resulting limestone platform underlies the low, flat Yucatán Peninsula and Yucatán shelf on the southeastern extremity of Mexico. On the north and west submerged part of the Yucatán platform is a broad shelf 150-250 km wide. In contrast, on the east the shelf is only a few kilometers wide. The area of this study is the northeastern tip of the peninsula where the broad Gulf of Mexico shelf passes southward into the narrow Caribbean shelf. Discussions of the regional geologic setting can be found in Murray (1961, p. 146-201), Murray and Weidie (1962), Meyerhoff (1967), Logan et al (1969, p. 23-24), Dengo and Bohnenberger (1969), and Bryant et al (1969).

Climate

Rainfall on Isla Mujeres averages about 100 cm (40 in.) annually, according to records of the Secretaria de Recursos Hidraulicas in Mérida, Yucatán. Evaporation averages about 160 cm (63 in.) per year. Precipitation exceeds evaporation 1-4 months a year, and heavy rainfall is not predictably seasonal. Daily maximum temperatures on Isla Mujeres are 95-102°F (rare) during July-September; minimum temperatures are 59-61°F in December-February.

Prevailing winds are from the northeast to east in the winter and from the southeast to east in the summer (H. Boilio, Centro Meteorológico in Mérida, personal commun.). The meteorological bureau does not keep records of wind velocities on this coast. Figure 14 includes wind-velocity data for the general vicinity of the northeastern coast (Hoskin, 1963).

Hydrography

Currents and Tides

The water regime is dominated by the Yucatán Current, which flows northward along the coast. In the strait between Isla Mujeres and the mainland the velocity of the current is 1-2 knots.

There is no station for measuring tides on the northeast coast of Quintana Roo. Observations made during the course of this field work suggest that fluctuation during spring tides is about 0.5 m.

Salinity and pH

Salinities recorded during the summers of 1967 and 1968 are shown in Figure 2. Open-water salinity is 35-36‰ (determined with an AO Instrument Company temperature-compensated refractometer). Salinities are higher in restricted lagoons and ponds. Figure 2 also shows pH values, which range between 8.2 and 8.8 (estimated with a Hellige pocket comparator).

Temperature

Surface temperature in open water was about 82°F during the summers of 1967 and 1968. Winter temperature of the sea surface in that latitude is about 76°F (Leipper, 1954). Midday temperatures as high as 89-93°F were recorded in shallow water on the island shores in June 1968. Water at the edge of hypersaline lakes on Isla Mujeres is hotter than 120°F (limit of thermometer) in the early afternoon on sunny summer days.

Shelf Morphology

A platform presently lying about 5 fm below sea level borders the eastern coast of Quintana Roo (Fig. 1). This terrace apparently is 3-4 km wide near the south end of Isla Cancun and widens northward to 10-15 km west of Isla Contoy. Seaward of the 5-fm terrace the bottom slopes rather steeply to an apparent shelf lying approximately 10 fm below sea level.

At the south end of the study area the edge of the continental shelf is within a few kilometers of the shore (Fig. 1). Northward, the slope of the bottom is more gentle and deep water is farther from the coast. At the point where the narrow continental shelf begins to broaden there is a dramatic change in the sedimentation regime on the 5-fm platform. South of Isla Cancun, coastal sediments accumulate in a barrier reef - reef lagoon environment. North of that island, shallow-shelf sediments are deposited in a strait - barrier island - lagoon setting.

Preexisting topography on this shallow shelf strongly influences the distribution of Holocene sediments by its effect on current strength and direction. Where the flow is funneled into the narrow strait between the mainland and Isla Mujeres, the current is accelerated, scouring the

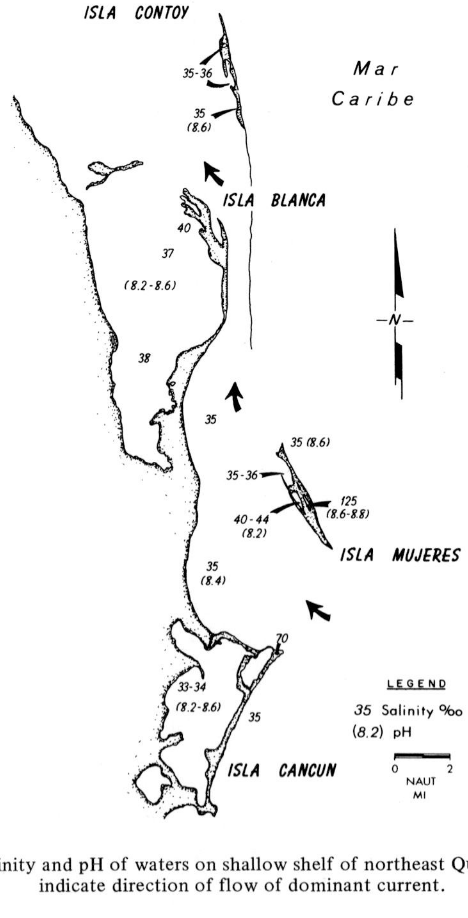

FIG. 2—Map showing salinity and pH of waters on shallow shelf of northeast Quintana Roo. Bold arrows indicate direction of flow of dominant current.

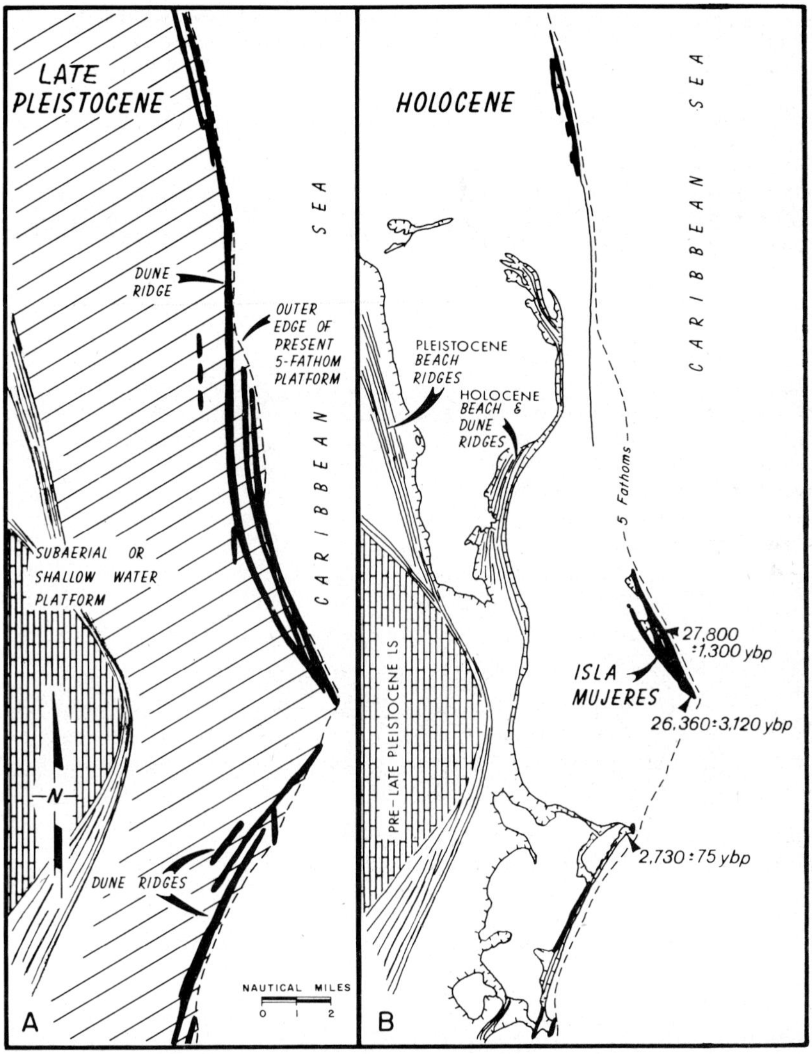

FIG. 3—Map of Pleistocene dune ridges and Holocene accretions.

A. Traces of late Pleistocene dune ridges deposited along seaward edge of a platform now 5 fm below sea level (from air-photo interpretation).

B. Present configuration of coastal areas resulting from partial inundation and erosion of Pleistocene dune ridges and accretion of land areas during Holocene. Ages of Holocene deposits (lower part of map) are based on C-14 age dates of marine gastropod shells, and ages of Pleistocene dune rocks (Isla Mujeres) are based on whole-rock age dates.

bottom in some places and piling up subaqueous dunes in other areas. Carbonate mud accumulates in spots protected from the current by vestigial topographic prominences. The location of Holocene spits, tombolos, and barrier islands is directly influenced by the position of Pleistocene dune ridges.

Pleistocene Eolian Ridges

The islands of Contoy, Mujeres, and Cancun are, in large part, remnants of eolian ridges that were deposited on the outer edge of the 5-fm terrace during a lower stand of sea level, probably during the early stages of the Wisconsin regression (Fig. 3A). These eolian ridges are progressively younger seaward. The islands have no intertidal terraces above the modern intertidal zone; therefore, they were not in existence during any stand of sea level higher than the present level. Radiocarbon ages of 26,360±3,120 and 27,800±1,300 years B.P. were obtained for whole-rock samples from Isla Mujeres.

Milliman and Emery (1968) postulated that the last interglacial rise of sea level reached the present level between 30,000 and 35,000 years ago. According to their data, the age dates for the Isla Mujeres samples correspond to a period of regression when sea level was 15-20 m below the present level.

Holocene Accretion of Land Area

Figure 3B shows the coastal land area, from Isla Blanca south to Isla Cancun, that has been built up by accretion of beach ridges and dune ridges during the Holocene transgression. Tombolos connected both ends of Isla Cancun to the mainland, and Holocene dunes accumulated seaward of the Pleistocene eolian ridges on Cancun. Relatively minor sand spits and beaches were added to Isla Mujeres and Isla Contoy. The development of mangrove swamps added to the land area in protected places around muddy lagoons and lakes.

Holocene Subaqueous Sediments

A variety of bioclastic and lithoclastic carbonate sediments has accumulated on the shallow shelf of northeast Quintana Roo. Brief descriptions of these submarine deposits are given here to call attention to the setting in which the eolianites have accumulated and the effect of the Pleistocene and Holocene eolian ridges on the distribution of Holocene sediments. Further details on the subaqueous deposits are given in Ward (1970, p. 20-51), Folk et al (1962), Folk (1967), and Wilson et al (1970).

The Holocene eolianites are the subaerial part of a belt of oolitically coated sand which parallels the coast from the Caribbean side of Isla Cancun northward beyond Isla Blanca (Fig. 4). Subaqueous oolitic sands are present in the western part of the strait between Isla Mujeres and the mainland, and farther northward they cover nearly the entire shallow shelf between Isla Blanca and the line of rocks extending south from Isla Contoy. In the strait the oolites are deposited in submarine dunes, ripple-marked spillover lobes, and megaripples. The coated-sand deposit also extends into the beach, which is the source of the eolian sediments. Nuclei of the ooids are bioclasts of calcareous plants and animals, microcrystalline peloids, and lithoclasts reworked from the limestone shelf.

The belt of Holocene oolites is thickest on its landward edge, where coastal dunes accumulate. For example, on Isla Cancun a ridge of oolitic calcarenite that stands as high as 17 m above the surf is based on Holocene sediments (sands?) that may be as thick as about 9 m (depth to 5-fm terrace). Maximum thickness of the subaqueous oolitic sands is about 8 m where the deposits are piled up in the constricted strait seaward of Isla Blanca.

Apparently the oolites are forming today. A radiocarbon age of 750±80 years B.P. was determined for a sample of oolitically coated sand from a depth of 75 m off the northern Caribbean shore of Isla Cancun. The seaward side of Cancun seems to be an ideal place for oolite coating to form. Here currents (upwelling?) bring water onto the warm, high-energy shallow shelf, where presumably loss of CO_2 induces precipitation of the aragonitic coatings. This area may be the source of supply of most of

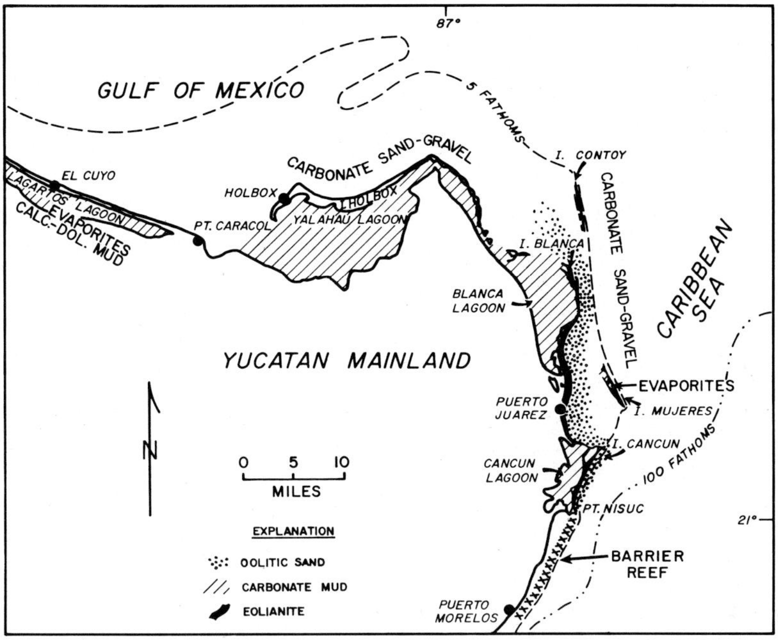

FIG. 4—Map of northeast Yucatán Peninsula showing sites of deposition of Holocene carbonate and evaporite sediments. Reprinted from Wilson et al (1970, p. 747).

the oolites deposited along the coast farther north, but oolite coatings also are precipitated in the narrow strait seaward of Isla Blanca. This fact is indicated by the presence of coated dolomite lithoclasts—a grain type that is brought uncoated into the northern part of the strait from the area seaward of Isla Contoy.

The oolitic sand passes into uncoated bioclastic sand and gravel in the eastern part of the strait near Isla Mujeres. West of Isla Contoy the oolite grades seaward into uncoated bioclastic and dolomite lithoclastic sands. The northwestern limit of the subaqueous oolites was not determined, but northwest of Isla Blanca the oolite probably is mixed with carbonate mud that is produced in the Blanca lagoon (Fig. 4).

Carbonate-mud deposits in northeastern Yucatán are extensive. A continuous belt of muddy sediments about 80 km long parallels the coast behind barrier islands from Isla Blanca to Isla Holbox on the Gulf of Mexico (Fig. 4). Another fairly large mud lagoon is present landward of Isla Cancun, and smaller areas of mud have accumulated in protected places on Isla Contoy and Isla Mujeres. In some places the lagoonal sediments are as much as 10 m thick (M. J. Brady, personal commun.). Patches of muddy sand occur in the *Thalassia* banks which border each side of the strait, and some muddy sediments accumulate behind the barrier reef south of Isla Cancun.

The northern limit of the world's second largest barrier reef is at Punta Nisuc on the southern tip of Isla Cancun (Fig. 4). Pleistocene dune ridges replace the barrier reefs at the outer edge of the 5-fm terrace north of Isla Cancun. The reason for the northern termination of the barrier reef is an unanswered question, but perhaps shelf morphology is the controlling influence. Deep water, which probably provides nutrients for coral growth, is much closer to shore south of Nisuc than it is north of there. From another point of view, the more gentle slope of the shelf north of Cancun probably facilitated supply of sediment for the deposition of Pleistocene dune ridges on the platform edge, eliminating a potential site of reef growth. Only small patches and single heads of coral grow on the rocky platform on the Caribbean side of Isla Mujeres. A small *Acropora palmata* - *Diploria* reef fringes the lee coast of Isla Mujeres 900 m northwest of the south point of the island, and an *Acropora palmata* patch reef is based on a 5-fm bottom in the strait 1 n. mi west of the south tip of Isla Mujeres.

Shelf Islands

Isla Contoy

The low, narrow island of Contoy lies about 7 n. mi off the mainland where the Caribbean Sea meets the Gulf of Mexico (Fig. 1). The island attracts thousands of birds and is a major rookery for sea turtles.

Isla Contoy (Fig. 5) is about 4.5 n. mi long and less than half a mile wide, and the main elements of the island are remnants of Pleistocene eolian ridges. The trend of rocky sea cliffs and modern dune ridges on the Caribbean side is unbroken and projects about 1,000 m farther northwest than the discontinuous rows of rocky hillocks and sandy beaches on the mainland side (Fig. 5). Between the ridges are shallow, muddy lagoons bordered by mangrove swamps, and the ridges are connected by sandy spits on the north and south ends. Ward (1970, p. 25-31) briefly described beach deposits and lagoonal sediments of Isla Contoy.

For about 8 n. mi south of Isla Contoy, waves break over a rocky, knife-edge continuation of the eastern ridge of the island. These eolian-

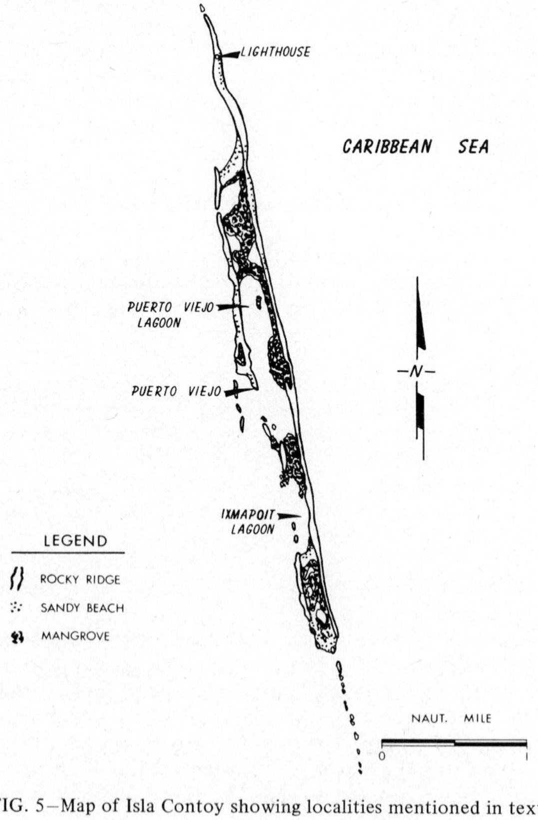

FIG. 5—Map of Isla Contoy showing localities mentioned in text.

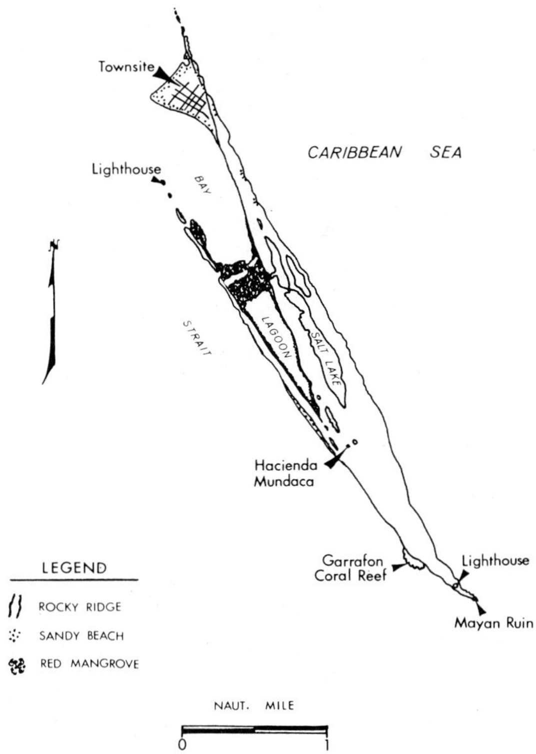

FIG. 6—Map of Isla Mujeres showing localities mentioned in text.

itic limestones emerge less than 1 m at low water, and they mark the seaward edge of the 5-fm terrace. The subaqueous continuation of this line of rocks can be traced into the westerly eolian ridge of Isla Mujeres (Fig. 3A).

About half the Caribbean coast of Isla Contoy is rocky with small pocket beaches. Low limestone sea cliffs as high as 3 m rise above the intertidal zone and are crested by vegetated sand dunes as high as 12 m above sea level. The other half of the windward shore, especially the northern part, is edged by stretches of steep beaches. Beach sands on the northern half of the eastern coast are highly polished, rounded, and medium to coarse grained. Bioclasts of mollusks, coralline algae, coral, and echinoids make up about half the grains, and lithoclasts of dolomite and limestone compose the other half. Dolomite grains give the Caribbean beaches of Isla Contoy a light brown color, contrasting with the typical white of other Quintana Roo beaches. A storm-ridge deposit of imbricated boulders and cobbles of limestone and dolomite is present 1-3 m above the high-tide mark on both the north and south quarters of the Caribbean shore (Ward, 1970, p. 28-29).

In the protection of the easterly windward ridge, the depositional topography of the older Pleistocene dune trends is more or less preserved on the leeward coast. A row of eolian hillocks, about 10 m high on its south end at Puerto Viejo, gradually decreases in elevation to below sea level on the north end. A string of lower knolls parallels the Caribbean ridge on the southern part of the island (Fig. 5). Intertidal nips are well developed on this rocky part of the quiet lee coast. These limestone hills are overgrown by low shrubs and cactus.

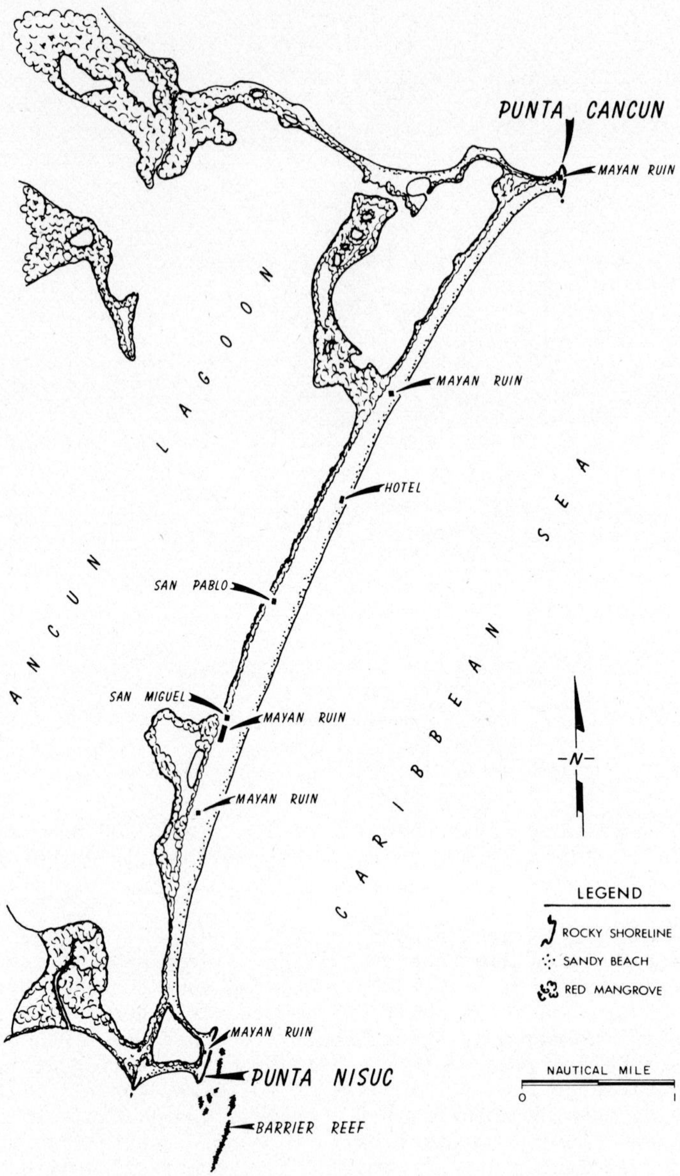

FIG. 7–Map of Isla Cancun showing localities mentioned in text.

Isla Mujeres

Isla Mujeres, owing to the tourist trade and the 7th Zone naval base, is one of the largest population centers of the Territory of Quintana Roo. About 2,000 people live on the island, and the continuing construction of tourist facilities, stores, restaurants, and houses insures an increasing population. In the past 10 years, Isla Mujeres has grown from a small fishing village with sandy streets and one truck to an internationally known tourist resort with one-way streets, some paved, and several taxis and automobiles. Isla Mujeres ("Women Island") probably was named either by Grijalva in 1518 or by Cortés in 1519. Tales that Amazon-type women inhabited the area, or perhaps the discovery of numerous female-like idols in stone temples, inspired the explorers to choose the name "Mujeres."

The backbone of Isla Mujeres is formed by three main ridges of Pleistocene eolian limestone. The oldest dune ridge, forming three fourths of the island's west side, is less than 6 m in elevation and strikes N30°W (Fig. 6). On the Caribbean side, two younger megaridges strike N22°W, intersecting the oldest dune line about 2 km northwest of the island's south point. The approximately 30-m elevation of the crest of the southern ridge is not matched anywhere else in the northeastern Yucatán Peninsula. From the south end the island ridge tapers northwestward to an elevation only a few meters above sea level. The extraordinary height of the south end of Isla Mujeres appears to have been achieved when the younger sands were blown atop the earlier dunes. Northward divergence of the oldest ridge from the two younger limestone ridges forms a protective crotch which partly encloses the bay and lagoon, where carbonate mud is accumulating (Ward, 1970, p. 36-38). Near the middle of Isla Mujeres, three shallow hypersaline lakes partly fill swales between the low limestone ridges (Folk et al, 1962; Ward et al, 1970; Ward, 1970, p. 38-42).

The island's shoreline is whittled by wave erosion on the Caribbean side and molded by sand accretion on the lee side. The Caribbean coast

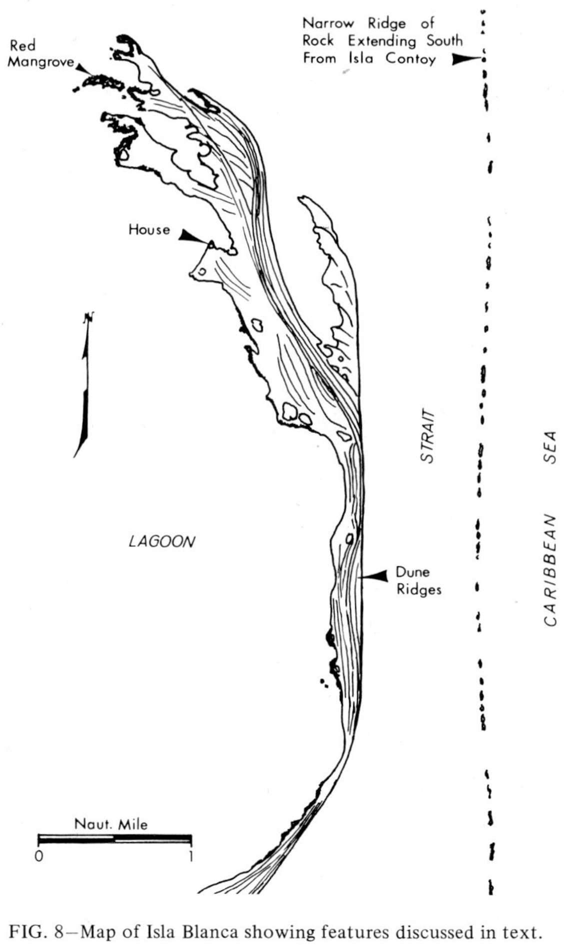

FIG. 8—Map of Isla Blanca showing features discussed in text.

is rocky and has sandy and gravelly pocket beaches; the village is built on a sandy spit accreted on the lee side of the north end of the middle limestone ridge. Most of the dune topography of the town area is being obliterated as encroaching civilization levels the sand. About 1.5 km of sandy beach has developed on the lee coast just south of mid-island. The rest of the lee coast is rocky.

The surface of the limestone island is a karst. Solution holes less than a meter in diameter pit the surface in many places, and a large solution well (cenote) is developed near the middle of the island. This cenote provided water for the hacienda of Mundaca, a Spanish pirate who lived on the island in the 1860s. Soils are thin and rocky, and a few small patches of terra rossa are developed at higher elevations. Most of the soil in well-vegetated lowlands is composed largely of land-snail fecal pellets.

Isla Cancun

Isla Cancun is about 7 n. mi long and less than half a mile wide; it is a complex of Pleistocene and Holocene deposits. Punta Nisuc on the south and Punta Cancun on the north are remnants of Pleistocene eolian ridges (Fig. 7). From these rocky points, Holocene tombolos extend toward the mainland. Complete connection with the mainland is prevented by the presence of tidal channels at each end of Cancun lagoon.

Only a few people inhabit the island, but the Mexican government has announced plans to make Isla Cancun a major tourist attraction. A hotel is being built there, as well as a bridge across the northern tidal channel. A complex of Mayan ruins is overgrown with vegetation on the jungle-like lagoonal side of the southern part of the island. The remains of other Mayan buildings are located on the crests of several of the highest hills.

Except at the two ends, the entire Caribbean shore is a brilliantly white beach of oolitic sand. Back of the beach, fine- to medium-grained oolitic sand is blown into dunes, veneering older Holocene dunes which have been built up as high as 17 m above the surf. Rapid lithification of the dune sands has produced a ridge of Holocene limestone.

Some of the strata exposed at the base of the Holocene eolian ridge contain lenticular layers of rounded pebbles and cobbles of corals, *Strombus gigas*, and pelecypods mixed with angular blocks of oolitic limestone (intraclasts). A *Strombus* shell from one of these layers has a radiocarbon age of 2,730±75 years B.P.

Unconsolidated dunes on the Caribbean side of Isla Cancun have rather steep windward slopes, many dipping more than 30° toward the sea. In this environment where dunes are being deposited uphill (climbing dunes), the inclination of the windward face is commonly equal to or greater than the dip of the lee slope. Low shrubs and grass vegetate the crest of each small dune.

The leeward side of the main island ridge is heavily vegetated with shrubs, trees, scrub palms, cactus, and orchids. Coconut palms have been planted in the low-lying area between the foot of the main ridge and the lagoon.

Two ridges of Pleistocene eolianite are exposed 2-3 m above sea level on the lagoonal side of the line of Holocene dunes. These rocks provide high ground on which several small houses are built along the east shore of Cancun lagoon.

Isla Blanca

Isla Blanca is a low barrier island about 5 n. mi long and half a mile wide (Fig. 8). The position and trend of the island suggest that

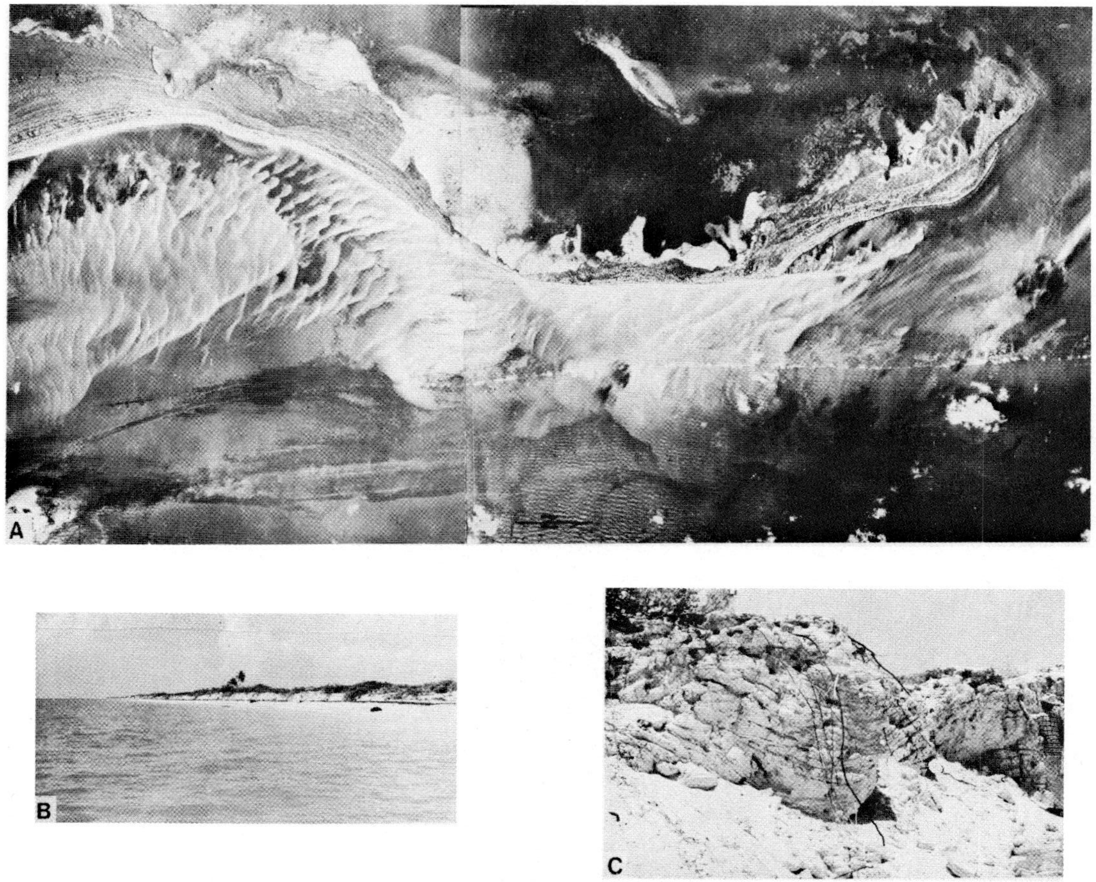

FIG. 9–**A**. Air photo of Isla Blanca and connected peninsula on south (refer to Fig. 8). Letter "A" is over northwest end of Isla Mujeres. Note series of coastal dune ridges on land area. Subaqueous dunes, spillover lobes, and megaripples are seen in strait. Muddy Blanca lagoon is at top (west) of photo. **B**. Wave-eroded Caribbean coast of Isla Blanca. **C**. Closeup of "sea cliffs" shown in **B**. Weakly lithified Holocene eolianite is exposed along seaward shore of Isla Blanca.

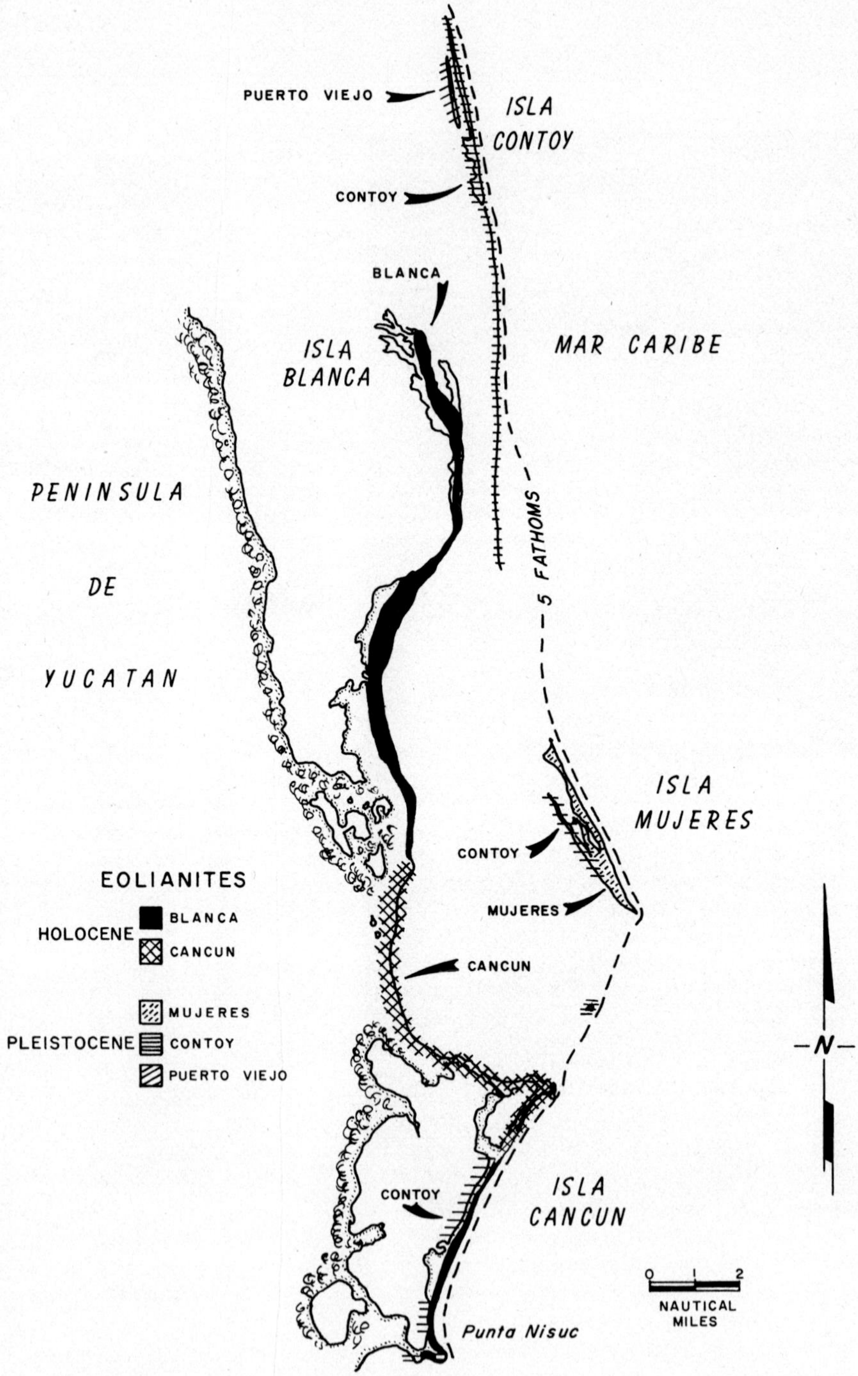

FIG. 10–Geologic map of Pleistocene and Holocene eolianites of northeastern Yucatán.

it may be underlain by a Pleistocene dune ridge paralleling the Contoy ridge.

On its south end, Isla Blanca is tied to a peninsula that projects northward from the mainland vicinity of Puerto Juarez (Fig. 9A). The west side of the peninsula and of the island is formed by a series of beach ridges, washover lobes, and tidal-channel deposits. This complex passes seaward into a series of dune ridges (Fig. 9). The third and fourth ridges landward from the modern beach contain the largest dunes, some as high as 3.5 m. These Holocene dune ridges are already in early stages of lithification. Swales between the dunes are heavily vegetated with shrubs such as uva del mar and icaco, low trees, scrub palms, and orchids. The leesides of dunes are more vegetated than the windward slopes. Red mangrove lines the lagoonal shore of the island. On the north end of Isla Blanca, fingerlike recurved spits and washover lobes project northwestward in the downcurrent direction. A recurved spit has formed north of the eroded coast, and comparison of 1959 and 1968 aerial photographs shows that this spit was elongated about 450 m in that 9-year period. On the leeside of this spit, a grassy supratidal flat with several shallow tidal channels has developed.

The southern part of the coast along the strait provides a high-energy environment. Coastal erosion at mid-island has truncated the four youngest dune ridges, and the interiors of the dunes are exposed above the beach (Fig. 9). This situation permits the sampling of internal beds of Holocene dune rocks of several different and known relative ages.

The thinly coated sand of the Isla Blanca dunes is very well sorted (σ_I = 0.25-0.50), ranging in mean size from 1.7ϕ (0.32 mm) to 2.4ϕ (0.19 mm). The grain-size frequency curves tend to be unimodal mesokurtic, and slightly negatively skewed (grain-size parameters are those of Folk and Ward, 1957). These data are based on sieve analysis of 28 samples, the details of which are not discussed further here.

PETROLOGY AND STRATIGRAPHY OF EOLIANITES

Eolianites of northeast Quintana Roo are essentially 100 percent carbonate, consisting of skeletal fragments of marine organisms, oolites, and lithoclasts of older carbonate rocks. The eolian ridges which form the backbones of the islands of Contoy, Mujeres, and Cancun are regarded as Pleistocene, and the dune rocks which crest the coastal beach ridges, spits, and tombolos are considered to be Holocene. The relative ages are based on both geomorphic relations and radiocarbon dates.

Differences in grain composition permit division of the Pleistocene rocks into at least three separate limestones and the Holocene eolianites into at least two types (Fig. 10). The various eolianites are given informal names to facilitate discussion, and the age relations established here are applied to interpretation of diagenetic changes.

General Properties

Color

Fresh surfaces of the eolianites are weak, extremely light orange to very weak, extremely light orange (10 YR 8/2 to 10 YR 8/1, following modification of National Research Council Rock-Color Chart by Folk, 1969). Weathered, algal-tinted surfaces are very light to medium gray (N-5, N-6, and N-7).

Induration

Cementation of the eolianites varies from layer to layer and from locality to locality. Generally both Pleistocene and Holocene eolianites

FIG. 11–Exposures of Holocene limestones. **A.** Blanca eolianite on Caribbean shore of Isla Blanca. **B.** Blanca limestone on Caribbean shore of Isla Cancun. **C.** Cancun eolianite on northern tombolo of Isla Cancun. Exposure face is parallel with dip direction of leeward crossbeds. **D.** Windward strata of same dune pictured in **C.** Exposure face is perpendicular to dominant wind direction (i.e., perpendicular to dip direction of leeward crossbeds).

FIG. 12–Exposures of Holocene and Pleistocene eolianites. **A.** Cancun eolianite (Holocene) on northern Isla Cancun. Boy on crest of dune is 1.6 m tall. **B.** Base of lithified dune shown in **A.** **C.** Mujeres eolianite (Pleistocene) on south end of Isla Mujeres; looking north with strait on left (west). Large-scale leeward cross strata dip steeply in a westerly direction. Sea cliffs are about 16 m high. **D.** Mujeres eolianite exposed in a sea cliff near middle of Isla Mujeres. Caribbean Sea on right (east).

are fairly friable, and greatest induration is not necessarily correlated with oldest age. Typically, fine layers are alternately friable and indurated; thus, in outcrop, bedding characteristics are accentuated by differential weathering of the laminae.

Stratification

Eolian limestones of northeast Yucatán are laminated, abundantly cross-stratified, and rarely ripple-marked. Individual layers are commonly no more than a few millimeters thick. Leeward crossbed dips are commonly 28-32°; the maximum dip is 39°. Surfaces of the leeward foreset beds are either planar, concave upward, or slightly convex upward (Figs. 11, 12). Sets of leeward crossbeds are truncated either in the windward or the leeward direction. Many leeward cross strata meet the underlying surface at acute angles; others are tangent to the base. Windward crossbeds, for the most part, dip less than 10°, but steeper windward strata developed where beds accumulated on acclivitous surfaces of preexisting topography. Festoon crossbedding of various scales is present within the core of the dunes, and contorted layers record overload slumping on the dune slopes. Many rather flat-lying beds also are present within both the Pleistocene and Holocene dune complexes. The near-horizontal strata may be windward topset beds, interdunal beds, or even lee-slope layers. Cooper (1958, p. 29) noted that "...in places the slipface is absent, and the lee slope is gentle from top to bottom..." in coastal dunes of Oregon. On Isla Cancun, where climbing dunes are being deposited, many lee slopes dip less than 10° and windward slopes dip as much as 30°.

High-angle crossbeds in the older limestones indicate that prevailing winds in the Pleistocene blew from the northeast, east, or southeast, as they do at present. Rose diagrams of crossbed directions in the Pleistocene dune rock of Isla Contoy and Isla Mujeres are plotted in Figure 13.

Rhizocretions

Pleistocene eolianites of northeast Quintana Roo contain root-cast structures, which are the result of caliche formation and cementation along the root systems of dune vegetation (Fig. 14A). Similar concretionary structures were named "rhizocretions" by Kindle (1923, p. 631), who modified the term "rhizomorphs" of Northrop (1910). Root casts are present in Quaternary eolianites in many parts of the world, and have been described by Darwin (1844, p. 260-263), Dolley (1889, p. 132), Teichert (1947), Fairbridge and Teichert (1953), Kaye (1959, p. 79), Johnson (1967), and others.

In the Pleistocene dune rock of northeast Quintana Roo, most of the branching and anastomosing rhizocretions have a brown microcrystalline (caliche) solid core, which grades outward into an encasement of indurated calcarenite. Cellular structures are preserved in some of the microcrystalline cores. Another type of root cast, which is less abundant, is a hard, hollow sheath similar to those of Johnson (1967, p. 154). Diameters of the rhizocretions range from a few millimeters to several centimeters. Typically, these hard root structures weather in relief at the outcrops of upper layers of Pleistocene dunes (Fig. 14B). Although hard-core rhizocretions are characteristic of the Pleistocene eolianites of northeast Yucatán, they are scarce in the Holocene limestones. Relatively inconspicuous horizontal and vertical "root casts" are found in the Holocene eolianites in several places. These structures apparently result from preferential cementation of carbonate sand that filled root holes after the plant material decayed.

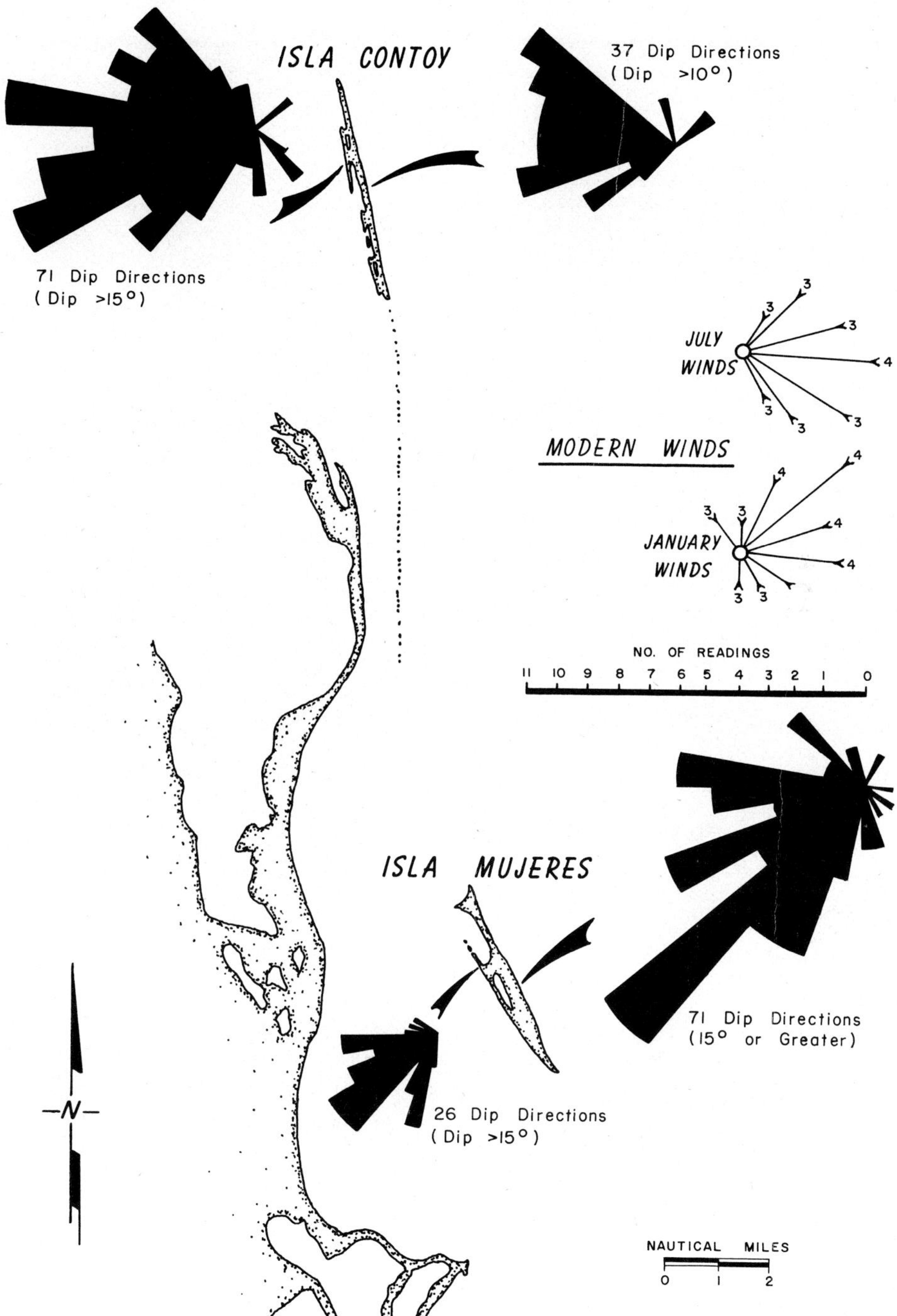

FIG. 13–Directions of steep crossbedding in Pleistocene eolianites on Isla Contoy and Isla Mujeres. Modern wind data are from Hoskins (1963). Figures at barbs of arrows show force of wind on Beaufort scale: 3 = 6.1-8.7 mi/hr; 4 = 9.6-14.0 mi/hr.

FIG. 14—Rhizocretions in Pleistocene eolianites. **A.** Rhizocretions exposed in small quarry northwest of airport on Isla Mujeres. **B.** Stereoscopic photo of rhizocretions weathering in relief on a Caribbean sea cliff of Isla Mujeres.

Root-hair Sheaths

Small tubules of microcrystalline calcite (Fig. 15A) are found in the intergranular areas of most of the Pleistocene eolianite samples examined petrographically. The hollow sheaths are 5-15μ in diameter with a wall thickness of 1-2μ. Some thin sections also show 75-100μ-diameter "main stalks" from which the small tubules branch out (Fig. 15B). Where they are abundant, the tubes form an interlaced network, enveloping grains and partly filling the interparticle pores. For the most part, the tiny tubes are confined to the interstitial areas, but there is some evidence that these filaments penetrate grains (Fig. 15C), and they do invade dissolution cavities (The interesting relation of these tubules to cementation of the eolianites is reported in the section on "Intergranular Diagenesis").

The microcrystalline sheaths apparently have no association with present weathered surfaces. For instance, tubules are present in samples of water-well diggings taken 15 m below the surface on Isla Mujeres. There is a relation between the abundance of the sheaths and the location of ancient zones of weathering. Tubes are abundant, for example, near rhizocretions and in highly micritized areas.

At least three sources for the small tubular structures may be suggested:

1. Algal filaments—In the present eolian environment, filamentous blue-green algae seem to be confined to the few millimeters at the surface of the calcarenite dunes where grains are highly bored and micritized. Particles in eolianites containing tubes, however, are relatively free of borings, and most show no evidence of having been at a surface of exposure for very long. The probability that many of the sheaths were precipitated in a lightless subterranean environment probably rules out an algal origin.

2. Root hairs of angiosperm dune plants—Root hairs flourish in the dark. Dittmer (1949) found that root hairs have diameters of 5-17μ and range in length from 80 to 1,500μ. Within any one species the size and length of the root hairs are about the same. Most species examined by Dittmer have relatively straight and uniform hairs which are never septate. Tubules in the eolianites are the same size as the root hairs measured by him, and in any one thin section most of the microcrystalline sheaths are approximately equal in diameter. Apparently the eolianite tubes are

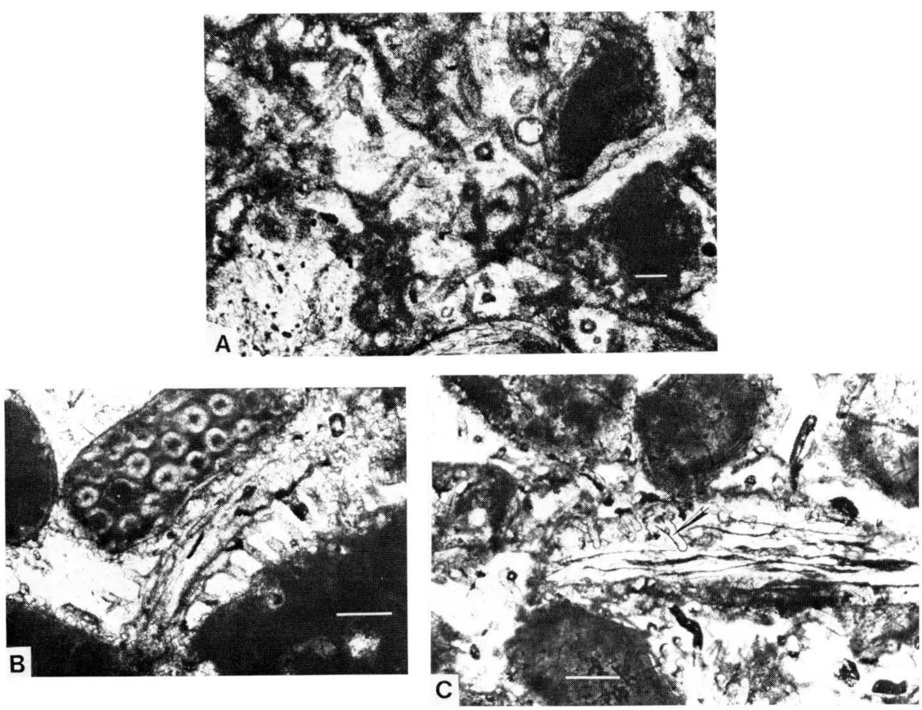

FIG. 15—Photomicrographs of root-hair sheaths.

A. Microcrystalline root-hair sheaths in intergranular pores of Pleistocene eolianite from Isla Mujeres. Bar scale = 50μ. Plane light.

B. Root-hair sheaths extending from "rootlet." Smaller tubules appear to be nonseptate extensions of epidermal cells of larger multicellular tube. Pleistocene eolianite from Isla Cancun. Bar scale = 100μ. Plane light.

C. Root-hair tubules which appear to penetrate mollusk fragment. Arrow points to boring in pelecypod bioclast. Borings have same diameter as root-hair tubules seen in intergranular pores surrounding bored bioclast. Bar scale = 100μ. Plane light.

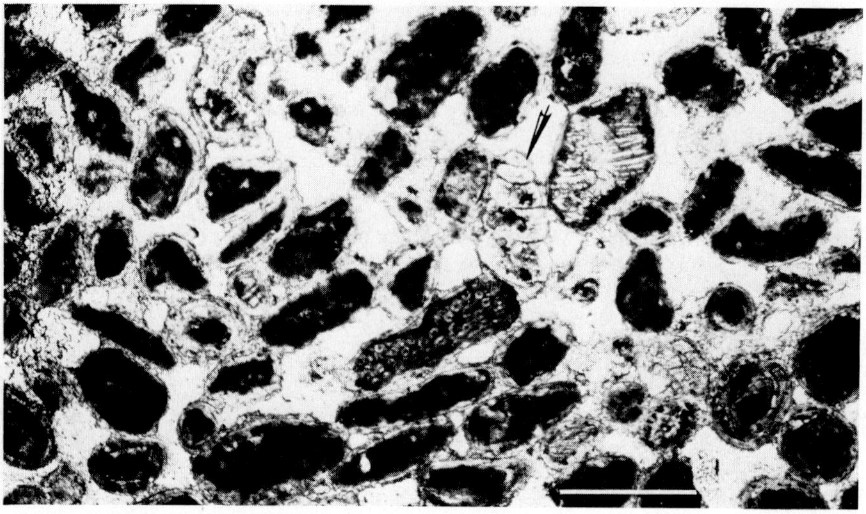

FIG. 16—Photomicrograph showing grain orientations in Holocene eolianite from Isla Blanca (section cut normal to bedding, which parallels bottom of photo). Arrow points to a vertically oriented microgastropod, apex of which points upward. Bar scale = 500μ. Plane light.

nonseptate and nonbranching. They appear to be single-celled tubular ramifications from larger cylindrical multicellular structures (Fig. 15B); this form is consistent with that of root hairs extending from epidermal cells of rootlets. The tubes are found only in Pleistocene eolianites, as are the rhizocretions, and in many rocks tubes are particularly abundant near rhizocretions. The only similar structures noted in the Holocene limestones are vague microcrystalline tubular forms within the upper 2 mm at weathered surfaces and in caliche deposits.

3. Fungal hyphae—Dittmer (1949) found a mantle of fungal mycelia growing in close association with root hairs of nearly all the plants he examined. Fungal hyphae, then, grow in the dark and are abundant near roots of dune plants (Webley et al, 1952)—two conditions which favor their existence in the eolianites. However, the size of the eolianite tubules is inconsistent with this origin; fungal hyphae are typically less than 5μ in diameter, on the basis of illustrations in several texts (e.g., Seaward, 1898, p. 207-222).

Most likely, then, the tubular structures in the eolianites were precipitated around root hairs of dune plants. In this report all small tubules of the size and structure described above are referred to as "root-hair sheaths."

Fabric

Quintana Roo eolianites are fine to medium grained (mean size 0.2-0.35 mm). Mean grain size increases upward in the individual layers, ranging as much as 1 phi unit from bottom to top of the beds, which are a few millimeters thick. Bagnold (1941, p. 238-240) explained the gradation into finer and coarser layers on the windward slope as the result of the initial deposition of finer grains during periods of accretion and the lag of coarse grains during periods of erosion. On the leeward slope, gradation is the result of shear. Coarsest grains rise to the top of avalanching layers, and "...the diameter change from outer to the inner surface of any sheared layer is always from large to small" (Bagnold, 1941).

Grain shapes are determined, to a great extent, by the source of the particles. Constituent grains in the Quintana Roo eolianites are (a) platy (predominantly *Halimeda* and mollusks), (b) more or less ellipsoidal (oolites, lithoclasts, etc), (c) rod-shaped (red algae, echinoid spines, etc), and (d) discoidal (peneroplid foraminifers, etc). In general, platy grains are the main constituent of coarser grained layers, and ellipsoidal clasts dominate finer grained laminae.

In vertical sections, orientation of the apparent long axes of the eolianite particles varies from parallel with bedding to perpendicular to stratification (Fig. 16). Microgastropods deposited in the vertical position are oriented with their apex upward. Imbricated grains and cross-lamination on a millimeter scale also are abundant in the eolianites. Orientations of apparent long axes in five thin sections chosen at random are noted in Table 1. At least 300 grains were counted in each of the sections, which are cut perpendicular to bedding. Samples, which include both Pleistocene and Holocene eolianites, are from the islands of Contoy, Blanca, Mujeres, and Cancun.

These "erratic" grain orientations may be normal in a deposit where particles on the windward accretion surface are "...fitted selectively into the most secure position by the gentle agitation of the surface caused by the bombardment of the saltation," and grains on the leeward foreslope are "...piled anyhow, as a result of avalanching" (Bagnold, 1941, p. 248). The original orientation of grains in calcareous dunes may be changed by roots, burrowing animals, blowing grass, and even raindrops.

Grain packing in the eolianites is normal to slightly overpacked compared to the illustrations of Dunham (1962, Pls. I, II, p. 109-110). An estimate of relative packing, based on counts of the different types of grain contacts (Griffiths, 1967, p. 173), indicates that the Pleistocene eolianites are only slightly more packed than are the Holocene dune rocks (these data are discussed under "Intergranular Diagenesis").

Mean intergranular porosity is 20-26 percent in the Holocene limestones and 5-18 percent in the Pleistocene dune rocks (estimated by point-counting interstitial voids in thin section—a method that probably gives lower than true values because very small, thin, or irregularly shaped crystals of intergranular cement, which may be present in various planes of the thin section, were counted as though the apparent maximum projection areas were in a single plane). Dissolution of bioclasts results in secondary

TABLE 1. ORIENTATIONS OF LONG GRAIN AXES IN FIVE THIN SECTIONS

	Percentage Parallel[1] *with Bedding*	*Percentage Perpendicular*[1] *to Bedding*	*Percentage Intermediate Orientations*
Holocene	34.7±2.6	13.7±1.8	51.7±2.9
	61.7±2.7	6.9±1.3	31.4±2.6
Pleistocene	44.3±2.8	20.7±2.4	35.0±2.6
	36.9±2.6	23.7±2.4	39.4±2.8
	42.8±2.8	20.9±2.3	36.3±2.6

[1]±10°.

TABLE 2. SUMMARY OF COMPOSITION OF FRAMEWORK GRAINS OF QUATERNARY EOLIANITES OF COASTAL NORTHEAST YUCATÁN PENINSULA

	Eolianite	*Mean % (±1 std. dev.) Ooids*	*Approximate Thickness of Oolitic Coatings*	*Estimated Mean Grain Size*	*Root-hair Sheaths*	*Predominant Ooid Nuclei*	*C-14 Age Dates*
HOLOCENE	BLANCA	91.7±5.0	Thin		Absent	*Halimeda*	
	I. Blanca facies		80%<10µ 2%>30µ	0.21 mm- 0.36 mm		Cryptocrystalline grains	
	I. Cancun facies		45%<10µ 10%>30µ	0.24 mm		Red Algae Mollusks Lithoclasts	
	CANCUN	82.0±8.5	Thick		Absent		
	P. Juarez facies		30%<10µ 40%>30µ	0.27 mm		Cryptocrystalline grains	
	I. Cancun facies		30%<10µ 20%>30µ	0.22 mm- 0.27 mm		*Halimeda*	*Strombus* shell at base of dune 2730±75 yr B.P.
PLEISTOCENE	MUJERES	9.0±7.2	Thin 75%<10µ 10%>30µ	0.35 mm- 0.20 mm	Abundant to absent	(uncoated) *Halimeda* Mollusks Foraminifers	Whole rocks 26,360±3,120 yr B.P. 27,800±1,300 yr B.P.
	CONTOY	85.8±6.0	Thin 80%<10µ 1%>30µ	0.30 mm- 0.18 mm	Abundant to present	*Halimeda* Cryptocrystalline grains	
	PUERTO VIEJO	92.0±5.2	Thick 10%<10µ 40%>30µ	0.25 mm	Abundant to absent	Cryptocrystalline grains	

intraparticle porosity, but in these rocks this type of porosity is not extensively developed. Additional information on porosity is presented in the section on "Intergranular Diagenesis."

Summary of Composition of Eolianites

Table 2 summarizes the composition of the Pleistocene and Holocene eolianites mapped in Figure 10. Pleistocene eolianites can be divided into two main types: (1) dune rock with a mean of over 85 percent oolitically coated grains and (2) a younger calcarenite which contains an average of less than 10 percent coated particles. The first kind is divided into the Puerto Viejo and Contoy eolianites, and the second type is the Mujeres eolianite. Holocene eolianites can be divided into two oolitic types: (1) an older Holocene dune rock, the Cancun eolianite, that has "thickly" coated particles, the nuclei of which average less than 5 percent red algal bioclasts, and (2) a younger limestone, the Blanca eolianite, that has "thinly" coated grains, the nuclei of which average more than 10 percent red algal fragments.

Pleistocene Eolianites

Several ridges of Pleistocene eolianites can be delineated on the shallow shelf of northeast Quintana Roo (Figs. 3A, 10). Each dune ridge has a unique composition, reflecting (1) changes with time in source of supply in the same geographic area and (2) differences in material supplied to different locations at approximately the same time.

Puerto Viejo Eolianite

Location—The Puerto Viejo eolianite crops out in the westerly ridge of Isla Contoy (Fig. 17) and is considered to be the oldest dune rock sampled for this study. "Puerto Viejo" is the local name for a mid-island site at the north edge of the entrance to the central lagoon (Fig. 5).

Composition—Puerto Viejo limestone contains 92 percent coated grains. Based on averages in eight thin sections, 40 percent of the coated particles are rimmed with oolite coatings over 30μ thick, and 9 percent have coatings less than 10μ thick (see Ward, 1970, Appendix, for point-count data). Maximum thickness of the oolite rims is about 100μ. The dominant nuclei of the ooliths are cryptocrystalline grains, presumably micritized bioclasts and lithoclasts. *Halimeda*, red algae, and mollusk fragments also are common nuclei, and the bulk of the uncoated particles is molluscan and red algal fragments. Mean grain size is estimated to be about 0.25 mm (2ϕ). The Puerto Viejo eolianite is supermature oosparite (Folk, 1962).

Rhizocretions are not abundant in the Puerto Viejo eolianite, and many of the root casts that are present are the hollow-sheath type. Root-hair sheaths are found in 7 of the 10 thin sections of this limestone.

Contoy Eolianite

Location—The name "Contoy eolianite" is used for the limestone of several ridges from Isla Contoy to Isla Cancun (Figs. 17, 18, 21). The bulk of the Contoy eolianite is found in the eastern dune ridge of Contoy and its extension into the western ridge of Mujeres. Also included in the Contoy eolianite are the dune rocks on the middle ridge of Contoy and most of the Pleistocene limestones on Cancun.

Composition—Oolitically coated grains make up about 85 percent of the Contoy eolianite (based on the mean composition of 32 thin sections). In the Contoy-Mujeres ridge and the lagoon-side eolian ridge of Cancun, about 78 percent of the grains are coated with rims less than 10μ thick (mean of 26 thin sections), and only about 1 percent of the grains have coatings thicker than 30μ. Eolianites on the south part of the ridge on the west side of Isla Contoy (samples Cty-62 and 63) and on the southwesternmost dune ridge of Isla Cancun (samples Can-60 and 61) contain a different coated-grain composition. In these two eolian ridges, 10-20 percent of the oolitic coatings are thicker than 30μ, and 25-50 percent are thinner than 10μ. From 50 to 75 percent of the Contoy ooid nuclei are *Halimeda* and microcrystalline clasts of undetermined origin. Coated mollusks, red algae, lithoclasts, and microcrystalline peloids, in that order of abundance, make up an average of 30 percent of the superficial ooids. Corals, foraminifers, and echinoderm fragments are minor nuclei. Noncoated bioclasts include *Halimeda*, mollusks, and red algae. The Contoy eolianite is fine to medium grained. Mean grain size is about 0.3 mm (1.75ϕ) in coarser laminae and about 0.18 mm (2.5ϕ) in finer layers. This limestone is a mature to supermature bioclastic oosparite.

Rhizocretions weather out in abundance on the north end of Isla Contoy and appear locally in other places in the Contoy eolianite. Root-hair sheaths are found in 31 of 32 thin sections of this dune rock.

Mujeres Eolianite

Location—The major limestone ridges of Isla Mujeres are remnants of the Mujeres eolianite (Fig. 18). Also tentatively included in this eolianite is the limestone of Punta Cancun on the rocky northeast tip of Isla Cancun (Fig. 20).

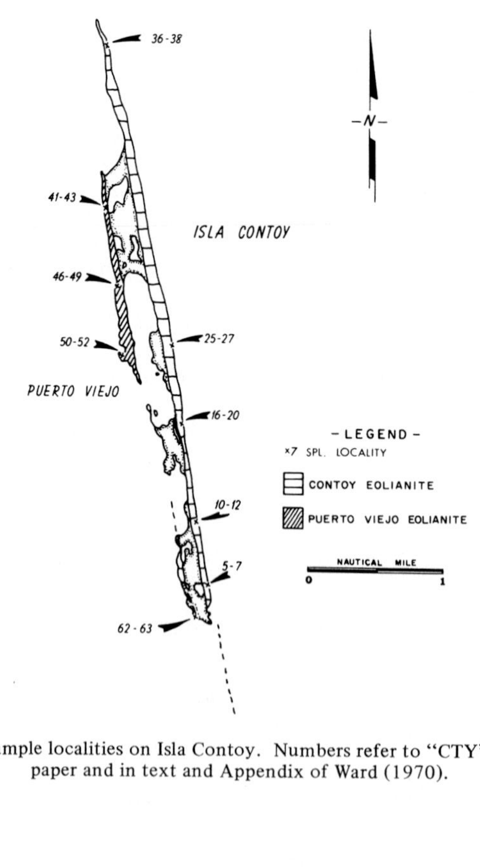

FIG. 17—Map showing sample localities on Isla Contoy. Numbers refer to "CTY" samples discussed in this paper and in text and Appendix of Ward (1970).

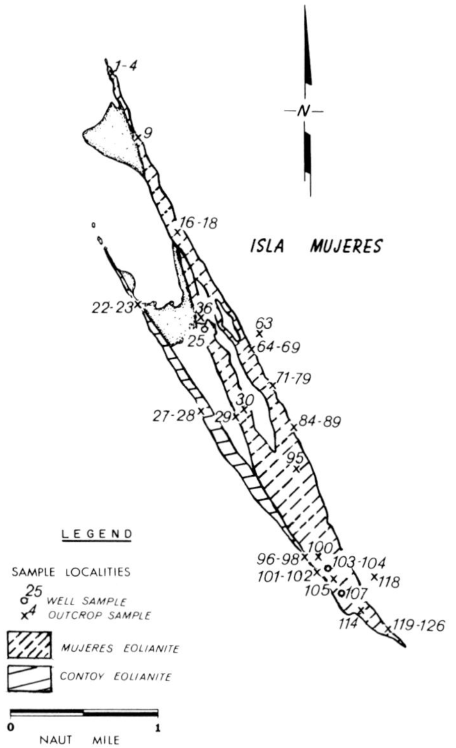

FIG. 18—Map showing sample localities on Isla Mujeres. Numbers refer to "IM" samples discussed in this paper and in text and Appendix of Ward (1970).

Composition—The Mujeres eolianite, which is considered to be the youngest of the Pleistocene dune rocks, contains a mean of only 9.0 percent coated grains (standard deviation is ±7.2 percent). *Halimeda* is the most abundant particle type of both coated and noncoated grains. About 20 percent of the total grains are micritized clasts of undetermined origin. Of the identifiable noncoated constituents, 25-45 percent is *Halimeda*; 10-20 percent, mollusks; and 5-20 percent, coralline algae and Foraminifera. Based on estimates from thin sections, the mean grain size of coarser laminae is 0.30-0.35 mm (1.5-1.75ϕ) and that of the finer grained layers is 0.2-0.25 mm (2-2.35ϕ). The Mujeres eolianite is a mature to supermature algae-mollusk-Foraminifera biosparite. In a few samples, "secondary" micrite associated with root-hair sheaths in the interstitial areas gives the rock the appearance of a poorly washed biosparite.

Rhizocretions are conspicuously well developed in the Mujeres eolianite, particularly within a 3-5-m interval just below the present soil zone. This root-cast zone generally follows the topography of the island rock and presumably was produced by cementation along the roots of plants which stabilized the dunes toward the end of their accumulation. Root-hair sheaths are common, being present in 60 percent of 38 thin sections.

A few massive, extremely light orange (10 YR 8/2) paleosols are found within the Isla Mujeres eolianite (Fig. 19A). These discontinuous soil zones are as much as half a meter thick in ancient topographic lows and thinner over old dune crests. The paleosols of Isla Mujeres are

FIG. 19—**A.** Paleosol in Mujeres eolianite. Hammer head rests on top of old soil zone. Exposure is in a sea cliff on Caribbean shore, mid-island on Isla Mujeres. **B.** Probably calcified cocoons of a soil-dwelling insect(?). Collected from paleosol shown in **A**.

azonal regosols (Thorp and Smith, 1949); they contain no contrasting color bands of red-brown laterite like those found in eolianite sections in other parts of the Caribbean (Kaye, 1959, p. 79; Ball, 1967, p. 572). Content of the Isla Mujeres paleosols is essentially the same as that of the eolianites, and the paleosols have not undergone extensive leaching. In fact, Mg calcite persists in some of the old soils. In thin section some grains appear micritized and grain boundaries are indistinct. Some areas of the thin sections show a lacy fabric of microcrystalline calcite that results from a myriad of tiny tubules (root-hair sheaths?) and a network of bored or rootlet-penetrated, partly dissolved bioclasts. The massive bedding of the paleosols suggests that they have been extensively burrowed and churned, probably by roots, ghost crabs, and snails.

Peculiar megafossils in the ancient soils are 3-cm-long egg-shaped cases of cemented calcarenite (Fig. 19B). Similar calcareous "eggs" are found in paleosols of Western Australia, which is almost the antipode of eastern Yucatán. Lea (1925) identifies the ovoid cases as calcified puparia of soil-dwelling insects, and he attributes many of the Australian cocoons to a weevil, *Leptops duponti*.

Age—One sample of Mujeres eolianite has a radiocarbon age of 26,360 ±3,120 years B.P. (Mobil Research Laboratory). This sample was collected at 4.5 m above sea level on an 18-m sea cliff at the southeastern end of Isla Mujeres (Fig. 18, sample locality IM 122). The rock contains little cement and approaches its original mineralogic composition (Fig. 26, sample 122). Another whole-rock sample from a water well in the middle ridge of Isla Mujeres (Fig. 18, sample IM 25) has an age of 27,800±1,300 years B.P. (LSUNO Radiocarbon Lab). These age dates give only a minimum age for this eolianite.

Holocene Eolianites

Most samples of both Holocene eolianites contain about 90 percent oolitically coated grains. The Cancun eolianite has a mean of about 13.5 percent cement, and the younger Blanca eolianite has about 5.5 percent cement. The Blanca eolianite is weak, extremely light orange (10 YR 8/2), and the Cancun dune rock is very weak, extremely light orange (10 YR 8/1), almost white.

Cancun Eolianite

Location—The Cancun eolianite was deposited atop the Puerto Juarez area beach ridge and the tombolo that connects Punta Cancun to the mainland, and it was piled against the Pleistocene dune ridges that run the length of Isla Cancun (Figs. 20, 21).

Composition—The Cancun eolianite has two facies. On the mainland coast from the Meco lighthouse to just south of Puerto Juarez, the limestone contains over 40 percent ooids with coatings more than 30μ thick, and the most abundant ooid nuclei are cryptocrystalline grains. On Isla Cancun the Cancun oolite has 15-20 percent grains with rims over 30μ thick, and the most abundant nuclei are *Halimeda*. Because of their geomorphic position, these two facies are considered here to be approximately the same age. In the Cancun eolianite many grains with oolite shells over 10μ thick (70 percent of the rock) show evidence of second-cycle coating. For example, clear, thin outer rims cover partly micritized inner coatings. In addition, some thinly rimmed multigranular lithoclasts contain ooids that are coated to the same diameter as most of the Cancun eolianite grains. The Cancun oolite is fine to medium grained, and the mean grain size is estimated to be 0.27 mm (1.9ϕ) in the Puerto Juarez facies and 0.22 mm (2.2ϕ) to 0.27 mm (1.9ϕ) in the Isla Cancun facies. This eolianite is a mature to supermature oosparite.

Age—A whole-rock sample taken near the western end of the north tombolo of Isla Cancun has a radiocarbon age of 13,600±200 years B.P. (Oak Ridge Associated Universities, Inc.). This age most likely is considerably older than the eolianite and is a reflection of the substantial content of second-cycle ooids. A presumably more accurate age was obtained for an unaltered *Strombus gigas* shell dug out of a conglomeratic layer that is overlain by about 7 m of eolianite on the Caribbean coast of northern Isla Cancun (sample locality Can-19). The C-14 date of the *Strombus* is 2,730±75 years B.P. (LSUNO Radiocarbon Lab).

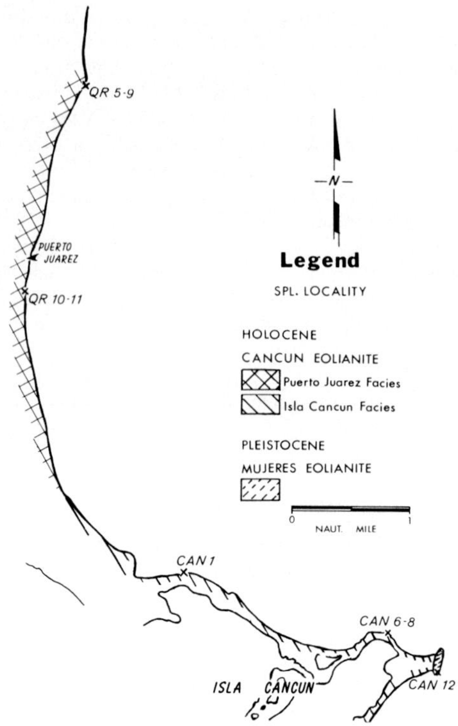

FIG. 20–Map showing sample localities from Puerto Juarez to northern Isla Cancun. "QR" and "CAN" numbers refer to samples discussed in text and Appendix of Ward (1970).

Blanca Eolianite

Location—Blanca eolianite is found in the lithified dune ridges of Isla Blanca and the peninsula that adjoins the southern tip of the island. It also includes the dune rock that veneers the Cancun eolianite on the southern two thirds of Isla Cancun (Figs. 21, 22).

Composition—Over 90 percent of the grains in the Blanca eolianite are thinly coated. On Isla Blanca, approximately 80 percent of the eolian particles have coatings less than 10μ thick. On Isla Cancun, about 45 percent of the grains in the youngest eolianite have oolitic rims less than 10μ thick. About 30-50 percent of the ooid nuclei are *Halimeda* fragments. Cryptocrystalline grains, red algae, mollusks, and lithoclasts also are important nuclei.

Mean grain size of the Blanca eolianite is estimated to range from 0.21 mm (2.3ϕ) to 0.36 mm (1.5ϕ) on Isla Blanca and is 0.24 mm (2.1ϕ) on Isla Cancun. The Blanca eolianite is a weakly cemented, mature to supermature oosparite.

DIAGENESIS

Introduction

There is no evidence that the Pleistocene and Holocene eolian ridges along this coast were ever more inundated than they are today. All of the various dune rocks presently above sea level have remained in a coastal subaerial environment, but for different periods of time. Inas-

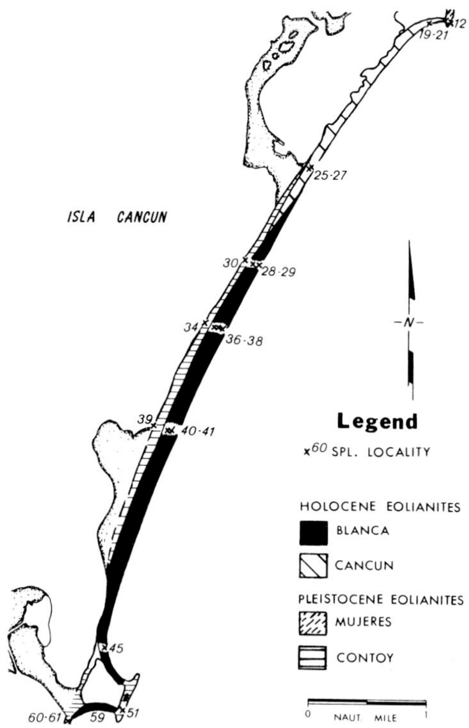

FIG. 21—Map showing sample localities on Isla Cancun. Numbers refer to "CAN" samples discussed in this paper and in text and Appendix of Ward (1970).

much as the relative ages of the eolianites have been determined, early diagenetic changes with time can be studied. Land et al (1967, p. 1002) found that Bermuda "...formations are not diagenetic units, even though there is a gross positive correlation between age and degree of alteration." Gavish and Friedman (1969) were able to relate advance in geologic age to progressive changes in texture, mineralogy, and geochemistry in late Tertiary to Holocene eolianites on the Mediterranean coast of Israel. Other than those two recent studies, there have been no published investigations of progressive diagenesis in calcareous dune rocks. Changes in mineralogy and texture with advancing age of the Quintana Roo eolianites are documented in this report.

As yet the question of the dominant environment of lithification of carbonate rocks is unresolved. Features which are diagnostic of cementation and diagenesis in the realm of meteoric water should be found in limestones that were deposited in the subaerial environment and that always have remained above marine water. Quaternary eolianites, therefore, may contain diagenetic textures which could help determine the environment of early cementation of older limestones. Cement types in the eolianites of northeast Quintana Roo are described in a following section. Other interstitial and intragranular alterations that may be indicative of subaerial diagenesis also are discussed. In addition, certain products of surficial weathering of eolianites that may be useful in interpreting ancient carbonate sections are described in this study.

Most rock samples for this study were taken from within a few inches of the weathered surface. Such a sampling procedure obviously is less than ideal for an investigation of diagenesis of limestones. Nevertheless,

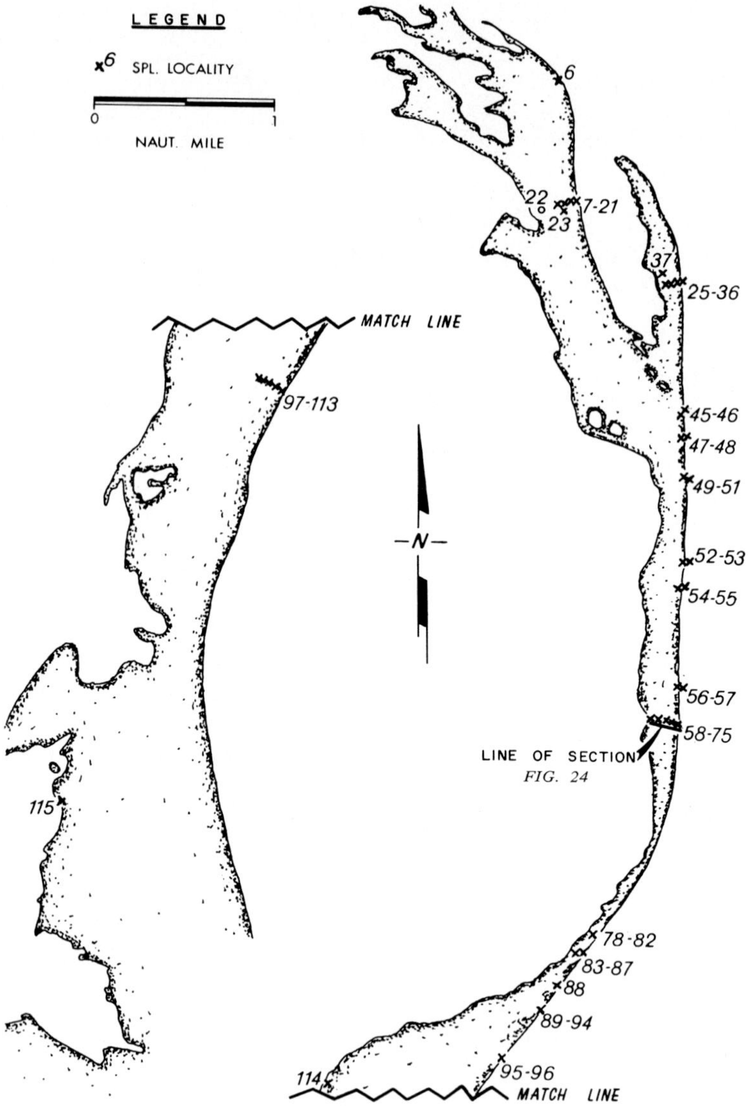

FIG. 22–Map showing sample localities on Isla Blanca. Numbers refer to "IB" samples discussed in this paper and in text and Appendix of Ward (1970).

Mujeres eolianite samples obtained from recently dug water wells and from fresh exposures on newly collapsed sea cliffs fit nicely into the model of diagenesis defined by other samples.

Mineralogy

Bulk mineralogy of the eolianite samples was determined by X-ray diffraction. The ratios of the areas under the diffraction peaks of aragonite (111), low-Mg calcite (104), and Mg calcite (104) were compared to curves constructed to give weight percentages of aragonite and Mg calcite (Ward, 1970, p. 83-85). Error in the determination of weight percentages is presumed to be less than 10 percent. For this study, Mg calcite is defined as calcite with over 10 mole percent $MgCO_3$, and low-Mg calcite is calcite with 5 mole percent $MgCO_3$ or less. Very few samples have intermediate values. $MgCO_3$ content of the Mg calcite was determined by entering the distance between the peaks of Mg calcite (104) and an internal standard, silicon (111), into a curve based on the relation between the position of the Mg-calcite peak and mole percent $MgCO_3$ (Ebanks, 1967, p. 34, from data by Goldsmith and Graf, 1958).

Blanca Eolianite

The bulk mineralogy of the Blanca eolianite (Holocene) is close to that of unconsolidated dune and beach sands on Isla Blanca (Fig. 23).

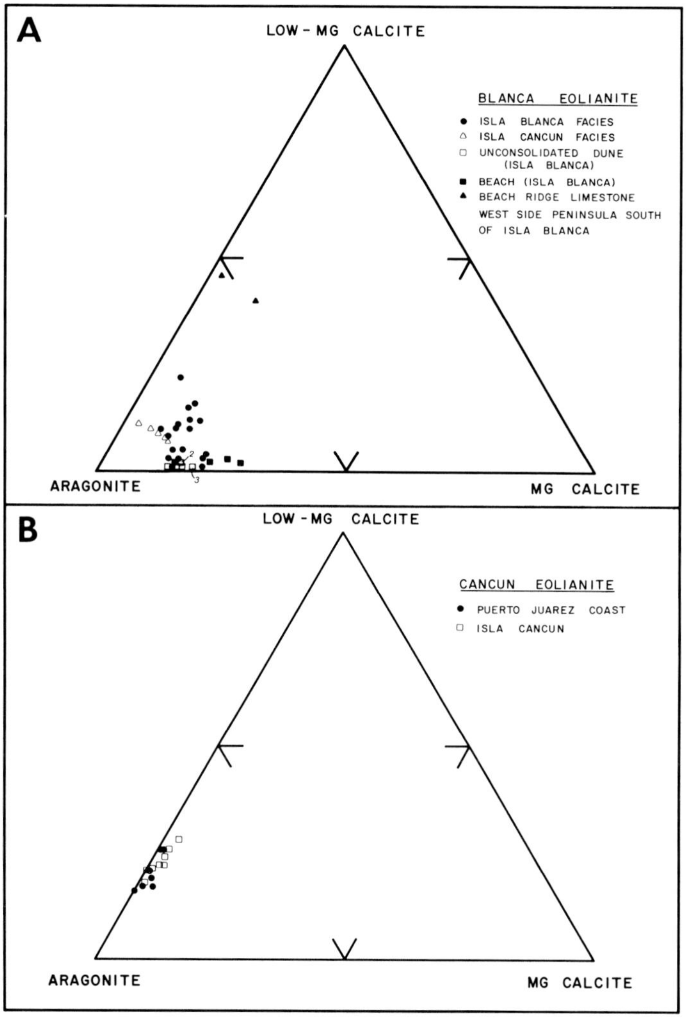

FIG. 23—Bulk mineralogy of Holocene eolianites of northeastern Yucatán.

A. Mineralogy of 23 samples of Blanca eolianite. Mineralogic composition of 8 unconsolidated dune sands and 4 beach samples from Isla Blanca are plotted for comparison. Also included are two samples of older Holocene beach ridges on west side of peninsula that adjoins southern tip of Isla Blanca.

B. Mineralogy of 15 samples of Cancun eolianite.

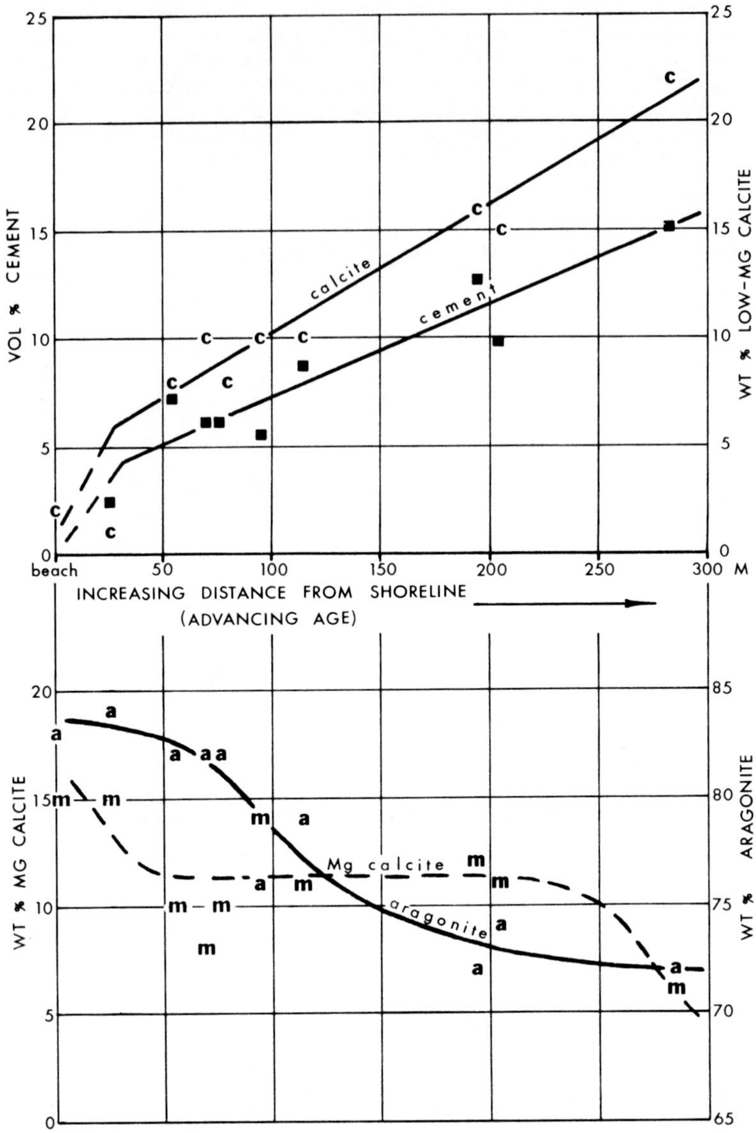

FIG. 24—Graph showing change in mineralogy and increase in intergranular cement with increasing distance from shore (increasing age) in Holocene eolianites on Isla Blanca (see line of section in Fig. 22). c = low-Mg calcite, a = aragonite, and **m** = Mg calcite. Solid squares = cement. Trend lines are fitted by inspection.

Much of the low-Mg calcite in this eolianite is in the form of cement; the mean weight percent of low-Mg calcite (X-ray data) is 8.7, and the mean volume percent of cement (grains + cement = 100 percent; point-count data) is 8.4. As a general rule, the content of low-Mg calcite and the amount of cement increase with advancing age (distance from shoreline) with no accompanying landward change in the constituent-grain content. The percentage of low-Mg calcite in the Blanca eolianite increases only slightly more rapidly than the volume of cement (Fig. 24). The limited data presented in Figure 24 indicate that an increase in very early vadose cement (and relative percent of low-Mg calcite) is accompanied by a reduction in the amount of aragonite. The relative content of Mg calcite in these late Holocene dune rocks apparently is not reciprocally related to an increase in the percent of low-Mg calcite with increase in age. Pingitore (1970) found that calcite cement within subaerially exposed coral is derived from an excess of $CaCO_3$ available after inversion of the aragonitic skeleton to calcite.

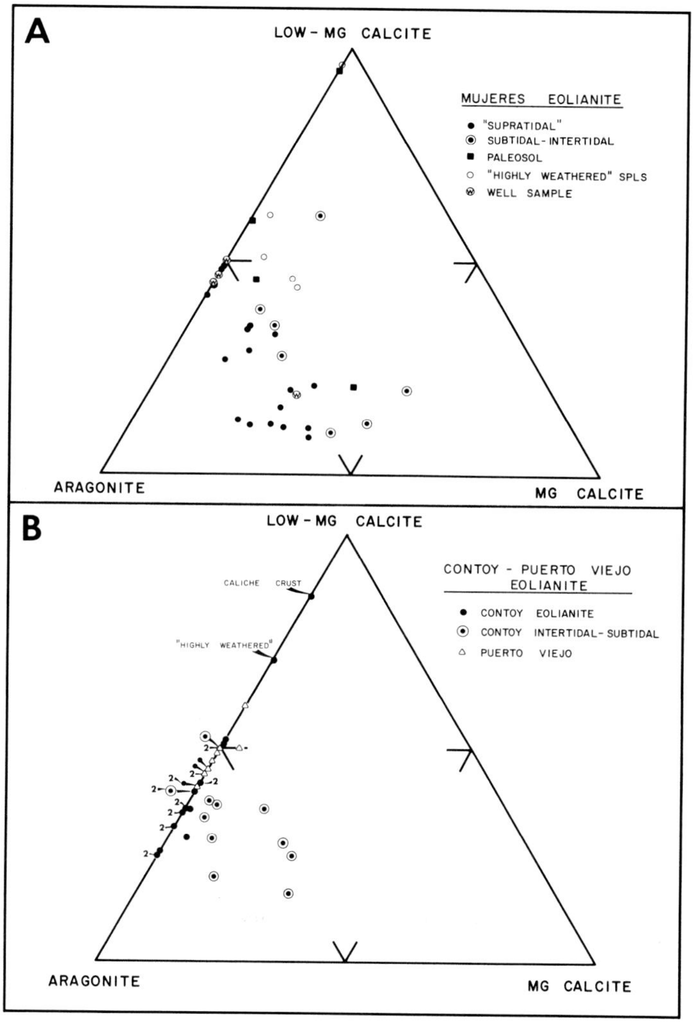

FIG. 25—Mineralogy of Pleistocene eolianites of northeastern Yucatán.

A. Mineralogic composition of 38 samples of Mujeres eolianite, including samples cropping out in splash zone ("supratidal") and in intertidal zone to 6 m below sea level. Well samples are from hand-dug water wells. "Highly weathered" is a subjective description of samples taken a few centimeters below "calichified" weathered surfaces.

B. Mineralogic composition of 33 samples of Contoy eolianite (including 12 samples taken above intertidal zone) and 10 samples of Puerto Viejo eolianite (intertidal and "supratidal" samples are not differentiated).

On the order of 40 percent low-Mg calcite was found in two samples from older beach ridges (sample localities IB 114 and 115, Fig. 22), in which cement content is less than 20 volume percent. The relatively high percentage of Mg calcite in one of these beach-ridge samples (Fig. 23A) reflects an originally high concentration of Mg-calcite bioclasts (25 percent of grains, estimated in thin section). From these analyses it appears that there is more or less increasing inversion of Mg calcite and aragonite to low-Mg calcite in the progressively older limestones of Isla Blanca.

Cancun Eolianite

The Cancun eolianite contains less Mg calcite than the other Holocene eolianite (Fig. 23B), mainly because this oolite contains few Mg-calcite bioclasts. The mean weight percent of low-Mg calcite is 21.6, and the mean volume percent of calcite cement is 16.8 (grains + cements = 100 percent). In addition, the calcite content of the Cancun eolianite probably is influenced by the content of second-cycle ooids, which would be expected to contain less Mg calcite than younger ooids. Thus, the apparent diagenetic stage of this eolianite may be partly inherited.

Mujeres Eolianite

Mineralogy of the Mujeres eolianite (Fig. 25A) is more difficult to interpret than that of the Holocene units, particularly in regard to the abundance of Mg calcite in these Pleistocene rocks. The original Mg-calcite content was fairly high (in most samples, between 10 and 20 percent of the recognizable bioclasts are red algae and Foraminifera), and there is high retention of the Mg calcite in many samples.

In the samples taken above the intertidal zone, the relation of the Mg-calcite composition to any other megascopic or microscopic characteristics was not discovered. Adjacent units with vastly different ratios of Mg calcite to low-Mg calcite may have similar grain compositions, equal volumes of cement, and equal quantities of aragonite, and may appear essentially the same in thin section and on the outcrop. In the samples which contain Mg calcite there seems to be no consistent correlation between Mg-calcite content and proximity to present or fossil weathered zones, except that Mg calcite is in low concentration or absent within a few centimeters of more weathered and "calichified" modern exposures. Surprisingly, a sample from a paleosol has one of the highest weight percents of Mg calcite. Figure 26 illustrates the erratic Mg-calcite composition of beds in the Caribbean sea cliffs at the south end of Isla Mujeres.

Of the 37 samples of Mujeres eolianite analyzed by X-ray (other than caliche and highly weathered surface samples), only 9 contained no detectable Mg calcite. Two of these are paleosol samples, which have been leached of Mg calcite. Another four of the samples were collected from the lower part of some of the highest Caribbean sea cliffs on the southeastern half of Isla Mujeres (sample localities IM 74, 86, 121, and 123; Fig. 18). The units devoid of Mg calcite are both underlain and overlain by beds which contain the mineral and which are essentially identical in thin section. The other three samples which lack detectable Mg calcite are diggings from two water wells recently hand-dug on the southeast end of Isla Mujeres (IM 103, 104, and 107). In one well, samples were from depths of 10 m (6 m above sea level) and 15 m (1 m above sea level and above the water table) and, in the other well, samples presumably were from 10 m (0.3 m below sea level and in the top of the present phreatic zone where salinity is 1°/₀₀). Aragonite in the samples lacking Mg calcite ranges from 50 to 58 percent, the same as the percentage found in

SAMPLE	% ARAGONITE	% MG CALCITE	MOLE % $MgCO_3$
126	42	7	12
125	52	28	13
124	54	12	12
123	52	0	
122	58	31	12
121	52	0	
120	53	12	13
119	29	51	14

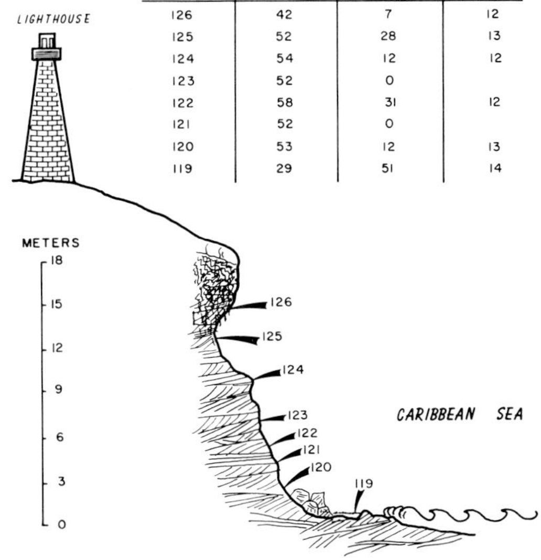

FIG. 26—Diagram showing variation in mineralogic composition within Mujeres eolianite cropping out in a sea cliff on southern end of Isla Mujeres (refer to Fig. 18 for sample localities).

most other samples of Mujeres eolianite. Cement consists of microcrystalline calcite or of less-than-25-μ blocky spikes of calcite coating most of the grains, and root-hair sheaths are absent to scarce. Most pelecypod fragments and peneroplid tests and some red algae appear well preserved.

Six of the samples in which the Mg calcite has been replaced completely are from the lower part of the present vadose environment. Purdy (1968, p. 198) speculates that on an oceanic island the zone of replacement of aragonite and Mg calcite by low-Mg calcite is in the lower part of the vadose environment. If this model can be applied to Isla Mujeres, perhaps beds which underlie the non-Mg-calcite layers have retained Mg calcite, because waters percolating downward became saturated with respect to Mg calcite before reaching the lower beds. As another explanation, Mg calcite could have been removed preferentially from beds which were preferred routes of vadose-water flow. The vuggy condition of the outcrop from which two of the non-Mg-calcite samples came may lend support to speculation that vadose water moved preferentially in these layers. Furthermore, removal of Mg calcite could have resulted from early leaching of these beds when they lay near the depositional surface, but there is no evidence that these units remained below weathered surfaces longer than Mg-calcitic layers.

Samples from another well in the middle of Isla Mujeres (sample locality IM 25, Fig. 18) add further inconsistencies to the Mg-calcite pattern. These rocks have 29 weight-percent Mg calcite and 52 percent aragonite. Taken from the top of a pile of diggings, these samples presumably are from the lower vadose or upper phreatic zone (the top of the water table in this well is 0.5 m above sea level, and the salinity of the water is 18°/₀₀). This limestone is weakly cemented and has been invaded by rootlets and numerous root hairs. No evidence for precipitation of Mg-calcite cement could be found in thin sections. The Mg-calcite content presumably may be attributed to the preservation of Mg calcite in foraminifer, echinoderm, and red algal bioclasts.

Although the content of Mg calcite is highly variable from layer to layer and from locality to locality, the aragonite composition of the Mujeres eolianite is more consistent. Most samples contain between 47 and 66 weight-percent aragonite. The samples that contain less than 47 percent aragonite fall into three categories: (a) paleosols, (b) samples taken near highly weathered, "calichified" surfaces, and (c) dune rocks presently in the marine environment. In the first two categories the relative aragonite composition is reduced by the addition of low-Mg calcite cement to the rock and by the replacement of aragonite by low-Mg calcite, which becomes more important as weathering and "calichification" progress. The relative percent of aragonite in the eolianite now in marine water is affected by the addition of Mg-calcite cement to the rock.

Much of the Mujeres eolianite presently immersed in the subtidal or intertidal environment contains a high percentage of Mg calcite (Fig. 25A). Mg calcite has been added to the rock as a pore-rimming cement precipitated in 5-25-μ-long blades oriented perpendicular to the grains. Mg-calcite cement of this type is found in beachrock (e.g., Ebanks, 1967; Taylor and Illing, 1969) and in lithified, modern, subtidal carbonate units (e.g., Ginsburg et al, 1967; Land, 1971; Glover and Pray, 1971). The mineralogy of pore cement in the Mujeres eolianite presently in the marine environment was inferred by comparison with Mg-calcite cement in intertidal-zone eolianite on Isla Contoy (discussed below) and confirmed by staining with a mixture of Clayton yellow, NaOH, and EDTA (Winland, 1971). This intertidal and subtidal addition of Mg calcite to the previously meteorically cemented Mujeres eolianite probably invalidates Friedman's (1964, p. 804-805) interpretation of the mineralogic differences between a sample of Mujeres limestone taken from the intertidal zone and a sample taken above the splash zone. In the intertidal rock he noted drusy fibrous calcite "...arranged normal to the periphery of the grains," but interpreted this as representing "...an early stage of (subaerial?) drusy calcite infilling." He suggested that the greater abundance of Mg calcite in the intertidal sample than in the supratidal eolianite is the result of the preservational effect of the marine environment. As already observed in this study, the retention of Mg calcite in the Mujeres eolianite is not necessarily related to elevation above sea level. Moreover, the Pleistocene dune rock that is in the present intertidal zone must have stood in the subaerial environment for a much greater period than it has been immersed in seawater.

Progressive diagenesis of the bulk of the Mujeres eolianite appears to follow a trend of preferential replacement of Mg calcite relative to aragonite, maintaining more or less the same composition of aragonite (about 45-65 percent) until the Mg calcite is depleted. Assuming an original composition of 60-70 percent aragonite and 30-40 percent Mg calcite with a minor amount of low-Mg calcite, there appears to be relatively little replacement of aragonite in the vadose zone before the Mg calcite is largely obliterated.

Contoy - Puerto Viejo Eolianites

Most of the limestone above the intertidal zone of the older Pleistocene eolian ridges has no detectable Mg calcite (Fig. 25B). Only three of the samples analyzed have minor amounts of Mg calcite (less than 5 percent). In part, this is a reflection of an original composition relatively low in Mg-calcite bioclasts. Also, there may be a sampling bias, because no samples from wells or recently made outcrops were available. The mean weight percentage of aragonite in the Contoy eolianite (above the intertidal zone) is 61.5 percent, and in the Puerto Viejo limestone it is 52 percent. A relatively highly weathered sample from the Contoy

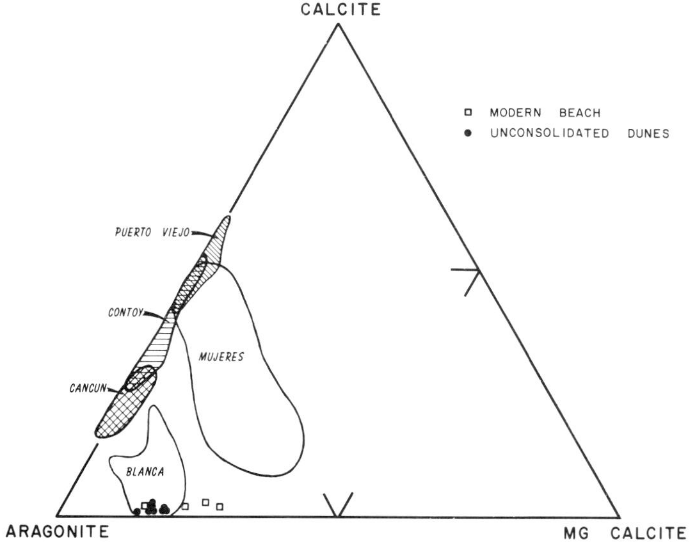

	NO. SAMPLES	MEAN WEIGHT PERCENT MG CALCITE	MEAN WEIGHT PERCENT ARAGONITE	RATIO VOL. % CEMENT / WT % LOW-MG CALCITE
HOLOCENE	I. BLANCA UNCONSOLIDATED BEACH (4)	23	75	
	DUNE (8)	17	82	
	EOLIANITES BLANCA (23)	12	79	0.83
	CANCUN (15)	1.5	77	0.79
PLEISTOCENE	MUJERES "SUPRATIDAL" (24)	15	52	0.29
	[SUBMARINE (6)]	[32]	[42]	[0.60]
	CONTOY "SUPRATIDAL" (23)	0.2 2 SAMPLES	62	0.42
	[INTERTIDAL W/ MG-CALCITE CEMENT (9)]	[15]	[56]	[0.56]
	PUERTO VIEJO (10)	0.4 1 SAMPLE	52	0.52

A

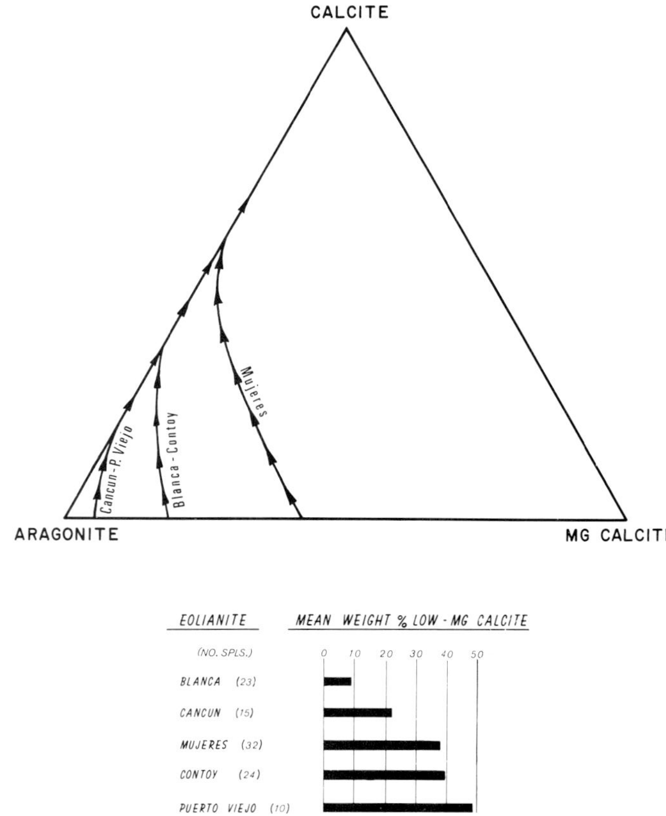

B

FIG. 27—Summary of mineralogic composition and diagenesis of Pleistocene and Holocene eolianites of northeastern Yucatán.

A. Summary of eolianite mineralogy. Samples included here are from present vadose zone. Eolianite cropping out in present intertidal-subtidal environment and "highly weathered" eolianite are excluded. Compositions of a few unconsolidated dune and beach sands from Isla Blanca are plotted for comparison.

B. Diagrams showing trends of progressive mineralogic diagenesis. Arrows indicate general directions of mineralogic change toward calcitization. The Puerto Viejo eolianite (Pleistocene) and Cancun eolianite (Holocene) probably follow approximately same route of progressive diagenesis. Another general trend is followed by Contoy (Pleistocene) and Blanca (Holocene) eolianites. Mujeres eolianite, originally relatively high in Mg-calcite content, follows a third trend of mineralogic change. Mean percent of low-Mg calcite increases with increasing relative age of eolianites.

eolianite ridge of Isla Mujeres has 29 percent aragonite, and caliche crust from Isla Contoy contains only 14 percent aragonite.

Most intertidal samples of the Contoy eolianite have some Mg calcite, but no more aragonite than the other samples. Not all Contoy dune rock now in the intertidal zone contains Mg calcite. In that which does, the grains are fringed with blades of Mg-calcite cement from 15 to 30μ long. The long axes of the cementing crystals are arranged normal to the surface of the particles. The mineralogy of the cement is inferred from the presence of a Mg-calcite peak on the X-ray diffractograms of all those samples which contain the bladed, "beachrock-type" cement. Staining with the Clayton yellow - NaOH - EDTA mixture corroborates the evidence. Spotts (1952, cited in Fairbridge, 1957, p. 143) postulated that Holocene cement is responsible for the magnesian carbonate enrichment in intertidal Pleistocene eolianites of Western Australia. Intertidal samples of the Puerto Viejo eolianite, on the calm-water side of Isla Contoy, have received no Mg-calcite cement. This fact suggests a relation between agitation of seawater and precipitation of intertidal cement.

Summary of Mineralogic Compositions

The bulk mineralogic content of all five eolianites is plotted on a triangular diagram and also stated in chart form in Figure 27A. For the northeast Quintana Roo eolianite, it would be difficult to trace a single trend of progressive diagenesis toward the low-Mg calcite pole, which has been reached by virtually all limestones older than Pleistocene, according to many workers (e.g., Stehli and Hower, 1961; Gavish and Friedman, 1969). Each eolianite, depending on its original composition, has its own diagenetic route. For example, the fact that the ratio of cement to low-Mg calcite for the Mujeres eolianite is low (Fig. 27A) probably results from the replacement of Mg calcite by low-Mg calcite without the production of abundant cement. The other two Pleistocene eolianites, originally high in aragonite, have substantially more cement than the Mujeres eolianite. Several workers (Gross, 1964; Stanley, 1966; Land et al, 1967; Gavish and Friedman, 1969) have found evidence that low-Mg-calcite cement in subaerially lithified calcarenite was derived from the aragonite bioclasts.

On the basis of grain compositions, the Blanca eolianite might be considered more or less a Holocene equivalent of the Contoy dune rock, and the Cancun eolianite, a Holocene counterpart of the Puerto Viejo oolite. The Mujeres eolianite has no Holocene equivalent. Under these assumptions it is possible to draw at least three trends of progressive diagenesis for the northeast Quintana Roo eolianites (Fig. 27B). These separate trends indicate that early subaerial diagenesis, even in one type of limestone in one limited geographic area, is complex. The mineralogic diagenetic stage is related to original composition, to proximity to weathering surfaces, and to position in the vadose zone, as well as to geologic age.

In the northeast Quintana Roo area, however, the five eolianites can be considered to represent progressive stages in mineralogic diagenesis. With increasing age of the eolianites, there is an increase in mean percent of low-Mg calcite, which is considered to have made up no more than a few percent of the dune rock at the time of its deposition (Fig. 27B).

Phreatic Water

Although most eolianite samples included in this study are from the present vadose zone, the composition of well-water samples allows some speculations about mineralogic diagenesis that can take place within the present phreatic zone. In Table 3 are listed various data concerning

TABLE 3. DATA ON PHREATIC WATERS FROM SHELF ISLANDS OF NORTHEAST QUINTANA ROO

Phreatic Water Containing Limestone or Sand	*Sample*	*Water Table (m Above Sea Level)*	*Salinity (°/oo)*	*pH*	*Ca++ (ppm)*	*Mg++ (ppm)*	*Sr++ (ppm)*	*Mg:Ca Molar Ratio*	*Sr:Ca Molar Ratio*
Holocene									
I. Mujeres	1		3.3	7.3	100	172	4	2.8	0.018
townsite	2		0		40	26	1	1.1	0.011
	3		3.6[1]	7.3[1]	145[1]	166[1]	3.3[1]	1.8	0.009
I. Blanca	1	0	0		50	74	1	2.4	0.009
Pleistocene									
Mujeres	1	0.6	17.6	8.1	410	530	10	2.1	0.011
Eolianite,	2	1.6	2.2	7.8	210	129	5	1.0	0.011
I. Mujeres	3		0		75	28	2	0.6	0.012
	4	0.3	1.1	7.9	55	107	4	3.2	0.033
Contoy Eolianite									
I. Cancun	1	0.7	3.3	7.6	192	133	10	1.1	0.024
	2		0	7.5	30	38	4	2.1	0.061
I. Contoy	1	0.3	2.2	8.0	110	82	3	1.2	0.012
Seawater, Caribbean coast									
I. Mujeres			35.2	8.6	400	1310	5	5.4	0.006

[1]From data in Back and Hanshaw (1967).

phreatic water that was dipped from hand-dug wells on the islands of Contoy, Blanca, Mujeres, and Cancun (the analysis of a sample of seawater is added for comparison). Thin lenses of fresh water, probably never more than a few meters thick, float on top of the brackish water that is present at shallow depths under Isla Mujeres (Back and Hanshaw, 1967), and probably under the other islands too. Near the mainland coast the freshwater lens may extend deeper, inasmuch as natural artesian wells discharge fresh(?) water into the marine strait 300 m offshore 1.2 n. mi south-southeast of Puerto Juarez. Back and Hanshaw (1967; 1970, p. 361-362) believe that the chemical character of the groundwater from Isla Mujeres results from the mixing of ocean water with meteoric water that has percolated down through carbonate rock. They found relatively high concentrations of Ca, Mg, HCO_3, and Sr in the groundwater, which presumably is enriched in these ions because of solution of carbonate minerals in the vadose zone.

Winland (1969) calculated that water with a Mg/Ca molar ratio of less than 2 is required, at near-surface conditions, for low-Mg calcite to be stable. Thus, data presented in Table 3 indicate that the metastable carbonate minerals, aragonite and Mg calcite, may be preserved even in the uppermost parts of the present phreatic zone in some places under Isla Blanca and Isla Mujeres. The Mg/Ca ratio is expected to be greater than 2 at shallow depth under the Quintana Roo islands, because the freshwater layer of the phreatic zone is so thin. Taft and Harbaugh (1964) suggested that a sufficient concentration of Mg ions in interstitial water will inhibit the transformation of aragonite and Mg calcite to low-Mg calcite. Back and Hanshaw (1967, p. 74) found that Isla Mujeres phreatic water is supersaturated with respect to calcite, aragonite, and dolomite. Apparently, then, under the present phreatic conditions, calcitization by inversion from aragonite and Mg calcite generally takes place in the vadose zone and uppermost phreatic zone of these islands.

Intergranular Diagenesis

Cementation

Eventually, the mineralogy of all the eolianites of northeast Quintana Roo will be monotonously low-Mg calcite. During the transformation of the mineralogic composition, what will mark these limestones as products of early lithification in a subaerial environment? Cement types may be the important clues. It is widely held that most cement in subaerially lithified limestones is precipitated as low-Mg calcite (e.g., Friedman, 1964). If this is true, then the stable mineralogy should give the subaerial cement a chance for survival into the ancient record. It is relevant, therefore, to examine the types of cement that are characteristic of early subaerial diagenesis.

Pore-filling cement deposited very early in the history of subaerial lithification can be studied in the Blanca eolianite, which presumably is no more than a few hundred years old. Somewhat more advanced stages of cementation are found in the earlier Holocene Cancun eolianite, and Pleistocene dune rocks of northeast Quintana Roo have types of cement not found in the Holocene eolianites.

"Drusy cement" and "blocky cement" are used here as defined by Dunham (1969a, p. 139-141). "Druse" refers to elongate or bladed crystals, commonly widening outward from the base and tending to rim voids. "Blocky cement" refers to mosaics of equidimensional crystals rimming or filling voids. "Microcrystalline" is used for crystals less than 4μ in diameter, and "cryptocrystalline" for crystals less than 1μ. "Spar" is used as a general term to refer to calcite cement with crystals larger than about 15μ occurring as blades or equant blocks. Other cement terminology is that of Folk (1965).

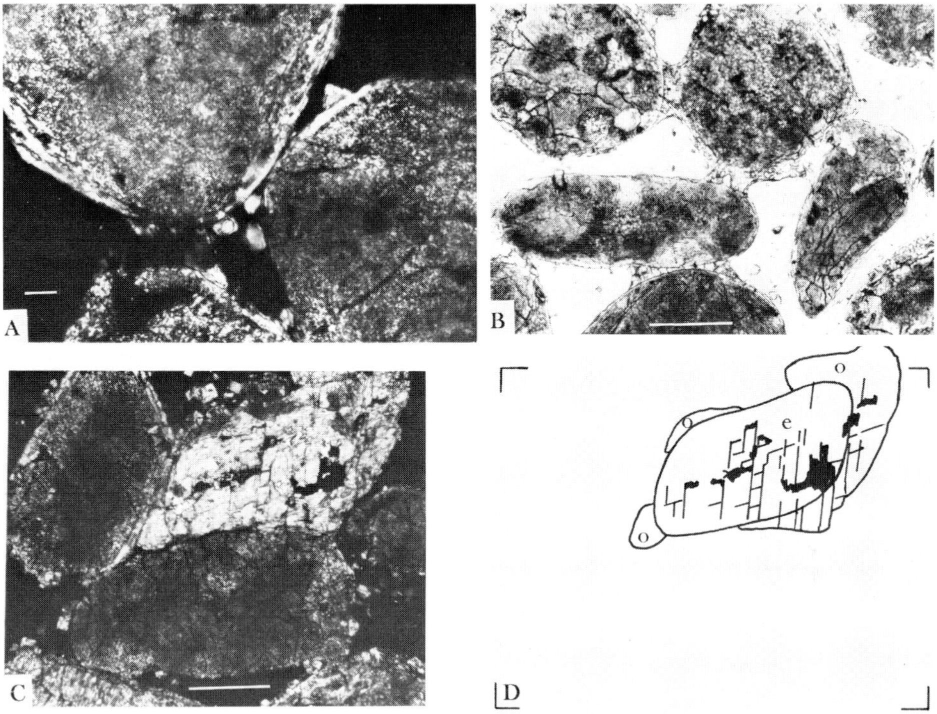

FIG. 28—Photomicrographs of early vadose cement in Blanca eolianite on Isla Blanca. **A.** Equant crystals at grain contacts. Bar scale = 20μ. Crossed nicols. **B.** Meniscus cement resulting in rounding of intergranular pores. Bar scale = 100μ. Plane light. **C, D.** Large syntaxial overgrowths (o) on echinoderm fragment (e). Bar scale = 100μ. Crossed nicols.

Cement in Holocene Eolianites

Cementation near grain contacts—In the vadose zone of the Blanca eolianite, early-lithification cement is deposited mostly near grain-to-grain boundaries. At grain contacts, surface tension holds droplets of percolating meteoric water, from which the cement is precipitated (Dunham, 1971). Cement in the youngest dune ridge of the Blanca eolianite is in the form of 5-15μ crystals of calcite (Fig. 28A). A few cement crystals as long as 45μ were found. The tiny equant cement, both rhombic blocks and dog-tooth crystals, is precipitated in optical continuity with the oolitic rims of the host grains. Many grains apparently have no cement but, even in the youngest eolianite on Isla Blanca, a few interparticle voids are completely filled by the enlargement of syntaxial cement of closely adjacent ooids. Most crystals of the pore-filling cement in this youngest eolianite are broadest at the base and may interlock with outwardly tapering cement from adjacent particles, fitting like cogs on gears. Other broad-base cement grains have irregular, curving boundaries and, under crossed nicols, show undulating, composite, or unit extinction. In addition to these types of early cement, single crystals of cement fill a few pores in the Isla Cancun facies of the Blanca eolianite (milled planks of driftwood firmly cemented within this rock attest to its young age). Single-crystal cement, which takes the shape of the void, is in masses as large as 80 by 130μ in the Cancun eolianite on Isla Cancun.

Staining with Feigel's solution (aragonite stain) and Clayton yellow - NaOH - EDTA mixture (Mg-calcite stain; Winland, 1971) indicates that most of the small grains of early cement are low-Mg calcite. Many blocks of

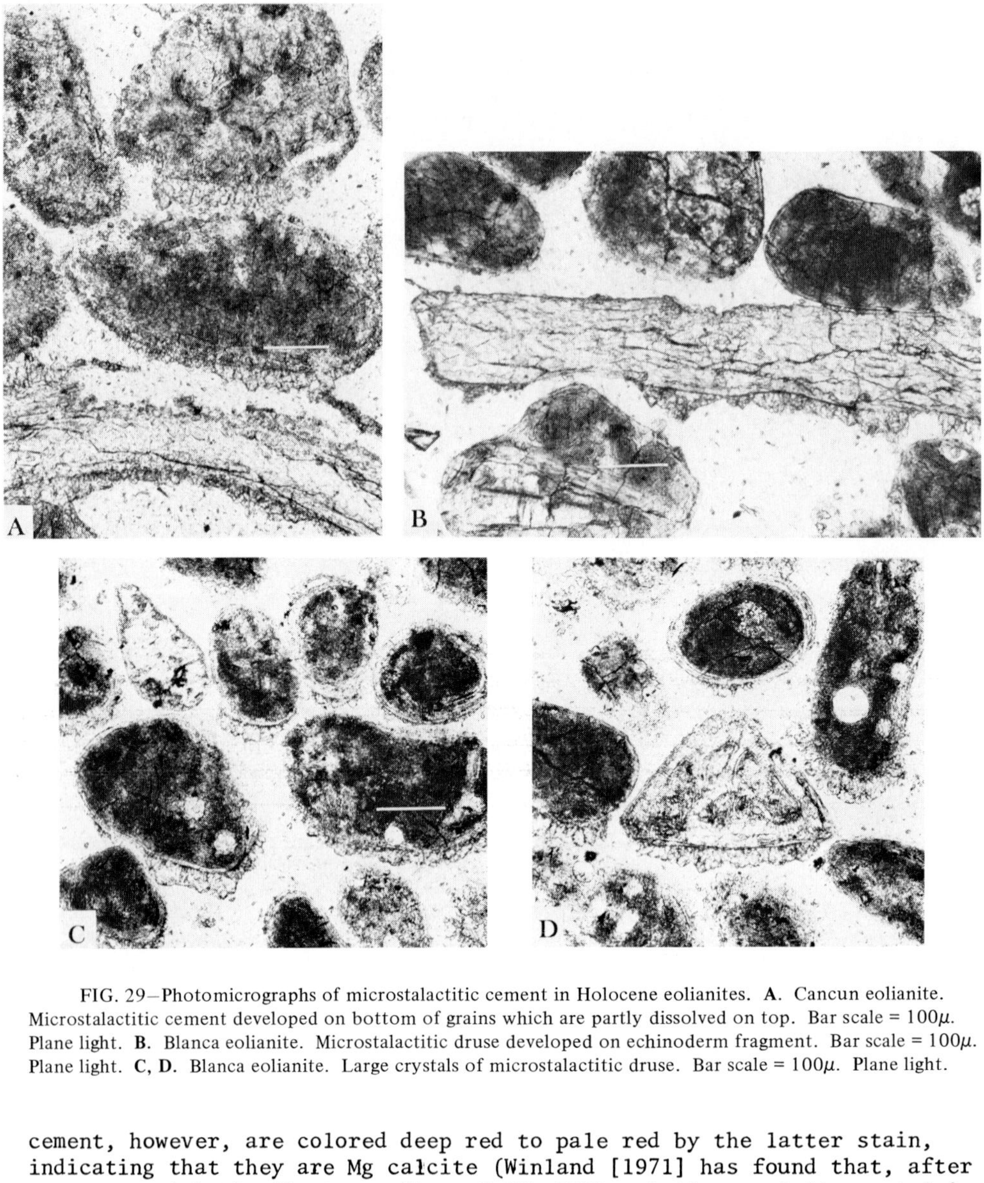

FIG. 29—Photomicrographs of microstalactitic cement in Holocene eolianites. **A**. Cancun eolianite. Microstalactitic cement developed on bottom of grains which are partly dissolved on top. Bar scale = 100μ. Plane light. **B**. Blanca eolianite. Microstalactitic druse developed on echinoderm fragment. Bar scale = 100μ. Plane light. **C, D**. Blanca eolianite. Large crystals of microstalactitic druse. Bar scale = 100μ. Plane light.

cement, however, are colored deep red to pale red by the latter stain, indicating that they are Mg calcite (Winland [1971] has found that, after treatment with the Clayton yellow - NaOH - EDTA solution, calcite containing over 9 mole percent $MgCO_3$ stains deep red, and calcite with 5-8 percent $MgCO_3$ takes on a pink to pale red color). Apparently, then, sparry calcite with more than 5 mole percent $MgCO_3$ is precipitated on both aragonite and Mg-calcite grains in the vadose zone of the Blanca eolianite.[3] This apparent Mg-calcite cement is most abundant near Mg-calcite bioclasts, suggesting that early vadose cement may include material dissolved from the Mg-calcitic grains as well as from the aragonitic particles. The patchy occurrence of pink- and red-stained spar in a single thin section suggests that various tiny pockets of groundwater in the same area of the vadose zone have different saturations with respect to the carbonate

[3]Microprobe analysis by R. K. Matthews shows about 3-5 weight percent $MgCO_3$ in the pink-stained cement.

minerals or that different cement minerals were precipitated at different times. Other sources of the Mg carbonate in the cement may be sea spray or Mg-rich water pumped up from the phreatic zone by intense evaporation.

As far as can be determined, this is the first report of subaerial cement that is apparently Mg calcite. Other authors have considered replacement of Mg calcite in bioclasts by low-Mg calcite as a result of either solution-deposition (Friedman, 1964; Purdy, 1968) or "incongruent dissolution" (Land, 1967; Schroeder, 1969), but no one has reported redeposition of Mg calcite in the vadose zone. Further detailed analysis of these cements is needed to confirm the mineralogy indicated by the staining technique.

In the older Holocene dune ridges on Isla Blanca, rocks are slightly more cemented than in the first ridge. A few interstitial areas are filled with cement, and blocky cement masses smaller than 25μ in diameter are scattered about the surface of many particles, not only at grain contacts. Pores are plugged both by expansion of syntaxial druse from neighboring grains and by the deposition of randomly oriented, blocky calcite in pores rimmed by syntaxial cement. Cements in the Cancun dune rock are the same types as those found in the Blanca eolianite, but they are more abundant and more extensive (Fig. 30A). Equant blocks of sparry cement in the Cancun limestone are as large as 80μ and blades may be as long as 170μ (Fig. 30B).

Meniscus cement—Dunham (1971) described the growth of grain-contact cement in the vadose zone, where crystal growth is restricted by the meniscus or outer limit of the water held in the corners of pores. Cement confined by the water-air boundary grows blunt-ended instead of dog-tooth terminations. As growth of the crystals proceeds, the pore face of the cement takes on the shape of the meniscus, resulting in rounding of the corners of the pores. Meniscus cement is found in the youngest dune rock on Isla Blanca and is well developed in most of the Holocene eolianite in northeast Yucatán (Fig. 28B).

Echinoderm syntaxial overgrowths—Although most particles in the young eolianite have received only tiny grains of cement, if any at all, some echinoderm clasts have taken on relatively large overgrowths (Fig. 28C, D). Apparently, then, the well-known, large syntaxial overgrowths on echinoderm fragments (Sorby, 1879, p. 66; Bathurst, 1958, p. 21-22) can form very rapidly in the vadose zone. Staining demonstrates that these overgrowths may contain more than 5 mole percent $MgCO_3$.

Microstalactitic cement—Drusy cement forms on the bottom side of a few grains in the youngest eolianite on Isla Blanca (Fig. 29). In older ridges and swales of the Blanca eolianite and in some beds of Cancun eolianite, the bottom druse is more abundant. A single grain may have both microstalactitic and grain-contact cement. Some of the underside cement has dog-tooth terminations, but much has rather blunt ends. In some places the whole mass of cement has a bulbous shape—the shape of a water droplet hanging on the underside of a particle. This cement type also has been recognized by Taylor and Illing (1969; and cited in Evamy, 1969, p. 791), Purser (1969), Müller (1971), and Nelson (1971). Most of these workers have called this cement "microstalactitic druse."

Drusy cement predominantly on one side of a grain is "...formed by slow evaporation of droplets which tend to form repeatedly at the same site as a result of local pore geometry...[and]...can only form in rock where pores are at least intermittently occupied by air" (Taylor and Illing, 1969, p. 95). Both microscopic stalactites and stalagmites were found by Taylor and Illing in intertidal beach rocks on the Qatar coast.

Staining indicates that the downward-growing druse in the Yucatán dune rock may be either low-Mg calcite or calcite with more than 5 mole percent $MgCO_3$. The nonaragonitic mineralogy of microstalactitic druse enhances the chance of its preservation in older limestones.

Figure 29A-B pictures microstalactitic cement on Blanca eolianite grains that appear to have been dissolved at their upper boundary and to have received preferential cementation on a sharp, undissolved lower boundary. This relation in this young limestone implies a very local source for early cement.

Micritic "inclusions" and "envelopes"—Light brown crypto-microcrystalline, organic-rich(?) material is included in much of the intergranular calcite in the Holocene eolianites, giving the cement a dirty-brown, cloudy appearance (Figs. 29B, 30A). Also, the bottoms of many grains are rimmed with cryptocrystalline Mg calcite (according to staining) which grades outward into apparently lower-Mg-calcite microstalactitic druse. Precipitation of both the fine-grained bottom rims and the clouding "inclusions" may have been induced by periodic intense evaporation. Another source of the fine material may be the micritized areas that rim dissolution cavities on the upper surfaces of some grains (Fig. 29A). If some of this micritic material is mechanically removed (not dissolved) and washed downward, then it could be deposited, because of evaporation, on lower grains or on preexisting cement. Regardless of its origin, this intergranular micrite could be an important indication of the environment of cementation if it were preserved following later cement growth. Dunham (1971) doubts that meniscus cement is common in ancient limestones, because when the cement is later "...placed in the phreatic zone [it] tends to grow euhedral faces on the previously blunted crystals and, with continued growth, to fill the rounded pore spaces, making all even." Micrite inclusions, such as those which occur in much of the vadose cement in Holocene eolianites of Quintana Roo, might preserve the outline of the meniscus cement even after euhedral faces were added.

In a few millimeters just below the weathered surface of the indurated dunes, grains are coated with low-Mg-calcite "micrite envelopes" (not micritization of the edges of particles as in the micrite envelopes of Bathurst [1966], but micrite added to the grain periphery). This crypto-microcrystalline calcite is filtered downward from the surface, where algae and lichen have created a bored and micritized crust 1-2 mm thick.

Preferential cementation of finer grained layers—In general, finer grained laminae and patches in the eolianites receive more sparry cement than the coarser layers. Bagnold (1938) observed that in the shallow beds of sand dunes water tends to seep along finer grained layers. Similarly, after a heavy rain on Isla Cancun, the finer grained layers near the surface of the unconsolidated dunes were damp and the coarser layers were dry (below about 8 cm, all the sand was damp). Laminae of fine particles and patches which are tightly packed have smaller pore spaces with greater surface tension. It should follow, then, that in calcareous dune rocks the layers and patches which preferentially draw or hold water are also the areas in which cementation takes place. Dunham (1971) noted "...preferential cementation of small-scale pore space, or, said another way, the inhibition of cementation in easily drained pore space." Preferential cementation of the finer grained layers accounts for the differential weathering of laminae, bringing about the characteristic appearance of exposed surfaces on which alternating laminae weather in relief.

Summary and discussion—Cement in these Holocene eolianites is highly variable. Within the same 1 or 2 mm of a thin section, some grains are rimmed with cement and others apparently lack cement; some pores are filled and some remain empty; some pores are rounded (meniscus cement) and others are lined with dog-tooth blades; some grains have microstalactitic druse and others are coated with radiating blades of cement; etc. Echinoderm fragments take on large syntaxial overgrowths in very early stages of cementation. The variability of the cements cannot be related to variability of host-grain mineralogy, because nearly all grains are

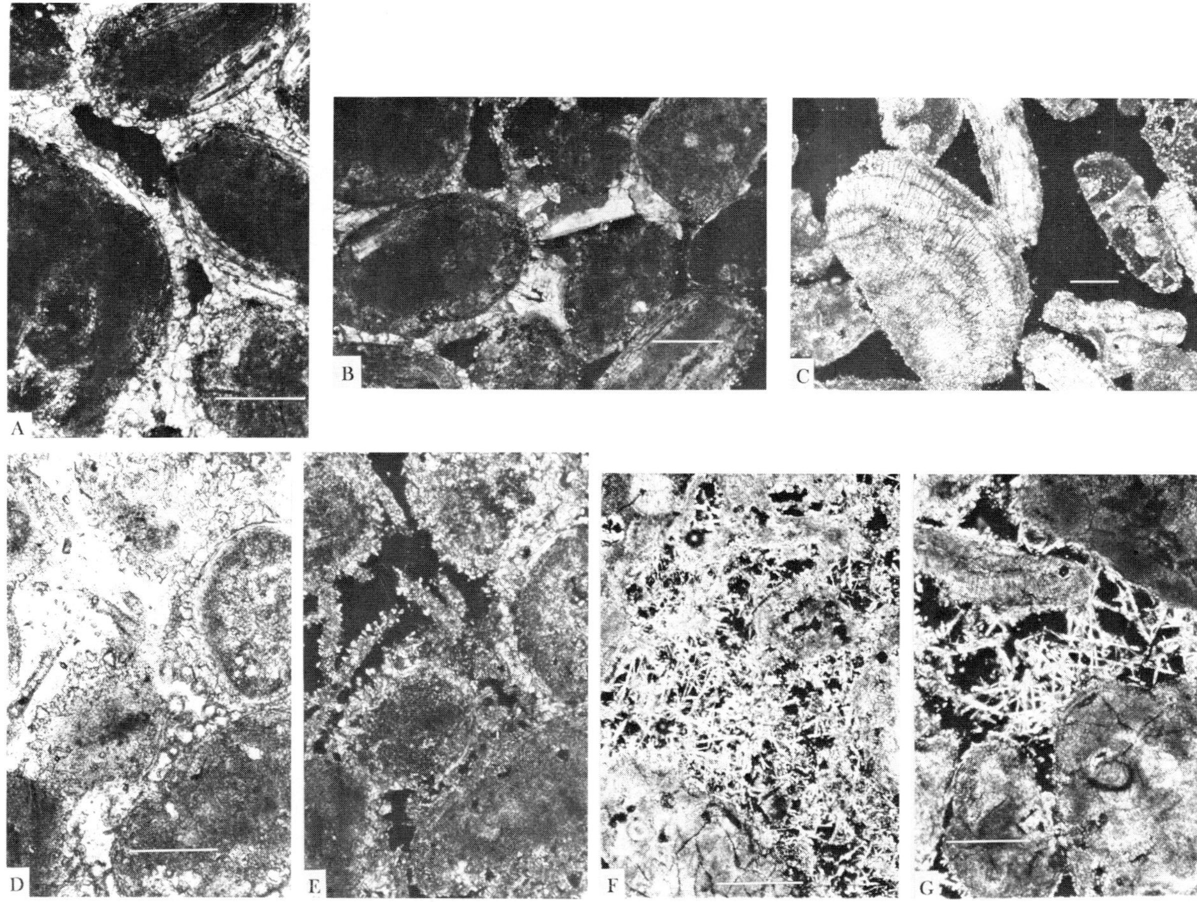

FIG. 30–Photomicrographs of intergranular cement in Holocene and Pleistocene eolianites.

A, B. Sparry cement in Cancun eolianite (Holocene). Rounded corners of pores and variation in amount and size of cement are characteristic of grain limestone lithified in vadose environment. Bar scale = 100μ. Crossed nicols.

C, D, E. Grain-skin cement in Pleistocene eolianites. **C**, Mujeres eolianite with fine-grained cement coating grains. Bar scale = 50μ; crossed nicols. **D**, grains and root-hair sheaths coated with microcrystalline cement which grades outward into fine-grained druse. Center of photo shows numerous end views of root-hair sheaths. Upper right-hand corner shows blocky spar precipitated in void which contains no root hairs. Bar scale = 100μ; plane light. **E**, same as **D** but with crossed nicols.

F, G. Needle-fiber cement in Mujeres eolianite (Pleistocene). Bar scales = 100μ; crossed nicols.

coated with aragonitic rims. Grain size, sorting, packing, amount of rainfall and rate of percolation, and intensity of evaporation are most likely the important controls on the highly variable pattern of cementation in the vadose zone. "Variability may well turn out to be the most diagnostic attribute of the shallow subaerial diagenetic environment" (Dunham, 1969a, p. 163).

Where mosaics of blocky, sparry calcite fill interstitial voids, the boundaries between cement crystals are both straight and curved. Single-crystal cement, which takes on the shape of the void, is developed in a few layers of the Blanca and the Cancun eolianites on Isla Cancun. Microcrystalline and cryptocrystalline calcite inclusions are common in much of the sparry cement of both the Blanca and the Cancun oolites.

Some of the sparry cement in the Holocene dune rocks is apparently Mg calcite. In the older Holocene limestone, the Cancun eolianite, peneroplid fragments take on a deep red when treated with the Mg-calcite-sensitive stain, but most red algae bioclasts are only slightly colored or remain unstained. Perhaps, then, Mg dissolved from the red algae is the source of this Mg in the cement. The first deposits of cement, particularly if microcrystalline, apparently have a higher $MgCO_3$ content than the later pore-filling calcite (preferential tinting of finer material may account for the differences in staining).

Cement in Pleistocene Eolianites

One might expect that cementation in the Pleistocene eolianites would be simply an amplification of that found in the Holocene dune rocks. In large part, however, this is not the case. Although the amount and types of cement in any given sample vary greatly as in the Holocene rocks, cement in the Pleistocene eolianites generally is more finely crystalline than in the Holocene limestones.

Sparry-calcite cement—Where root-hair sheaths are absent, equant blocks and drusy blades of sparry calcite are deposited in intergranular voids, preferentially in fine-grained layers. In sparry mosaics filling pores, the size of the various cement crystals increases toward the center of the voids. In Puerto Viejo oolite, blades are as large as 30 by 80μ, equant blocks are up to 100μ in diameter, and completely filled pores are fairly common. The younger, chiefly nonoolitic Mujeres eolianite, in contrast, is cemented largely with microcrystalline calcite and minute bladed and equant crystals of calcite less than 25μ in greatest dimension, and few intergranular voids are completely filled. Sparry cement in the Contoy eolianite is highly variable in size and abundance, but microcrystalline cement is abundant and most crystals are less than 25μ in diameter.

Sparry-calcite cement generally is absent where root-hair sheaths occur. In a single thin section, some intergranular areas may contain mosaics of sparry calcite and other areas may be occupied by root-hair sheaths, but a mixture of the two is uncommon.

The function of the root-hair system is to expand the surface through which plants soak up ground moisture. In the vadose zone and the uppermost phreatic zone, water taken in by plants would not be available to serve as the dissolution and reprecipitation medium that apparently is necessary for the accumulation of subaerial calcite cement. It has been established that significant amounts of sparry cement may be deposited soon after the calcareous dune sands are laid down, but it seems probable that rapid transpiration of ground moisture would induce nucleation of microcrystalline cement rather than sparry crystals. Indeed, as is discussed in the following section, the pores containing root-hair sheaths are lined with microcrystalline cement. Rapid transpiration of pore water has the same effect as rapid evaporation (e.g., precipitation of microcrystalline caliche), but extraction of groundwater by roots may be effective to a much greater depth than extraction by evaporation (Davis and DeWiest, 1966, p. 19). In the Pleistocene eolianites, pores which were not intruded by root hairs were able to hold water for a longer period; therefore, nuclei of larger crystals could have been formed in these interstitial areas. Even today, most of the voids which contain root-hair sheaths remain essentially free of sparry cement.

Grain-skin cement—In the vicinity of root-hair sheaths, grains are coated with an irregular rind of microcrystalline cement (1-2μ) that grades out into tiny spikes of druse (5-20μ) and minute equant rhombs of calcite (less than 10μ). Similar cement in limestones of Bermuda is called "grain-skin cement" by Land et al (1967). This same type of deposit forms the root-hair sheaths (Fig. 30D, E). Precipitation of the initial calcitic sheaths around root hairs may have been induced by the biochemical processes of heterotrophic bacteria or other microbes associated with the root system. Krumbein (1968) found evidence from laboratory experiments that bacteria from a caliche-crust extract played a part in the generation of aggregates of calcite crystals in a carbonate-enriched medium. Caumartin (1963) described a host of microorganisms which effect the crystallization of various minerals, including $CaCO_3$, in the subterranean environment.

In particularly "rooty" rocks, grain-skin cement is the only type present. Several Isla Mujeres samples that contain neither root-hair sheaths nor large crystals of spar are cemented by the fine-grained (rapidly precipitated?) rinds (Fig. 30C). Paleosols also have this type of cement. Pore rounding (meniscus cement) is not evident in eolianites lithified with grain-skin cement, although many apparently were cemented in the vadose zone. In some rocks grain-skin cement includes incipient microstalactitic druse.

Needle-fiber cement—A peculiar fibrous cement is found in many samples that have abundant root-hair sheaths (Fig. 30F, G). The needlelike fibers are as much as 200μ long and less than 10μ wide (mostly 2-4μ wide). This acicular cement is present both in intergranular pores and in intraparticle dissolution voids as randomly oriented, very straight, needlelike crystals having a large length/width ratio. Supko (1971) found similar cement in subsurface samples of calcarenite from the Bahamas. For this cement type he proposed the name "whisker crystal cement." The name "needle fiber" is preferred here, because Folk (1965, p. 25) already has defined as "fibers" any pore-filling cement with a length/width ratio of over 6:1.

Supko (1971) proposed that needle-fiber cement is indicative of the phreatic zone. In the Pleistocene eolianites of northeast Quintana Roo, this cement is common in the vadose zone. Needle fibers are abundant in paleosols and in eolianites that contain rootlet sheaths and rhizocretions, around which are dissolution voids and grains riddled with 10-15μ holes (root-hair size). In other words, needle fibers are found almost exclusively in eolianites that were near ancient weathered surfaces and soil zones. Samples which contain needle-fiber cement were collected from many different present-day elevations, ranging from 6 m below sea level to 15 m above sea level.

The needle fibers of this study apparently are low-Mg calcite. They are stained by Alizarin red ($CaCO_3$ stain), but show no coloration after treatment with Feigel's solution (aragonite stain) or the Mg-calcite stain. The "whisker crystals" of Supko (1971) are aragonitic.

The origin of the needle-fiber cement is unknown. It seems probable that the 2μ-wide needles were associated with fungal hyphae, which are about the same diameter and may occur as straight threads in the same random arrangement (Porter and Zebrowski, 1937, Fig. 1D; Ellis, 1918, Fig. 7; Kidston and Lang, 1921, Pl. 8). Apparent fungal borings are common in grains in close proximity to needle-fiber cement, and in a few places the calcite needles continue into hyphal borings. Some root-hair sheaths in the Pleistocene eolianites are partly made up of needle fibers, the long axes of which lie tangent to the tubules. Fungal hyphae grow in close association with root systems in modern dunes (Webley et al, 1952) and are one of the few filamentous plants that can exist in the shallow subsurface. If needle fibers are related to fungal hyphae, then this is further evidence that the acicular cement is indicative of precipitation near a weathered surface. Brown (1958) found that fungi are usually absent below about 18 in. (46 cm) in some British sand dunes.

Cement in dune rocks in the intertidal zone—Bladed and fibrous cement, unlike any found above the splash zone, occurs in Pleistocene eolianite cropping out in the intertidal and shallow subtidal zone of Isla Mujeres and Isla Contoy. This cement, typical of beachrock and subtidal cement described by many previous workers, is the product of submarine cementation following the last sea-level rise.

Mg calcite occurs as pore-rimming blades 4μ wide by 20μ long (Fig. 31A) in Pleistocene eolianites in the intertidal zones of Isla Mujeres and Isla Contoy. In one intertidal sample from Isla Mujeres, the grains are coated with a fringe of aragonite fibers 2μ wide and up to 30μ long (Fig. 31B). No submarine cement was found in samples of intertidal Puerto Viejo eolianite, which crops out on the calm-water side of Isla Contoy. Of four Contoy eolianite samples from the strait side of Isla Mujeres, two had minor amounts of Mg-calcite cement and two had none. In contrast, nearly every splash zone and intertidal sample from the wave-pounded Caribbean coasts of Mujeres and Contoy contained Mg-calcite cement.

Eolianites recemented by the more evenly distributed submarine cement are more highly indurated than the limestones cemented only by subaerial cement. In these intertidal rocks the vadose-zone cement is obscured by the more recent submarine cement. Blocky cement is confined to intraparticle pores such as *Halimeda* utricles and peneroplid chambers and to isolated intergranular pores, and it appears that subaerial cement has been replaced by submarine cement. Figure 31C shows the typical relation of subaerial cement and the later submarine cement. The pore is fringed with Mg-calcite blades that stain deep red when treated with the Mg-calcite stain. The edge of the vadose-type grain-contact cement that is juxatoposed to the pore-rimming druse takes on a red stain that fades to pink outward from the void.

In an intertidal sample from Isla Contoy, a few clear fibers 8μ wide and over 300μ long apparently are the Mg-calcite replacements of root-hair sheaths (Fig. 31D). Other fibers appear to represent replacement and enlargement of needle-fiber cement.

Summary of Eolianite Cementation Patterns

Grain-contact cement—Initial cement is in the form of tiny equant crystals, syntaxial to the host grain (Fig. 32). Cementation is confined mainly to corners of pores where water droplets are held by surface tension. As grain-contact cement expands, it takes on the shape of the air-water interface, and meniscus cement is formed. Meniscus cement results in rounding of the pore corners and is indicative of cementation where voids are partly filled with air. This contrasts with grain-edge cementation that takes place in pores uniformly filled with water in the phreatic or the submarine environment.

Also early in the history of cementation, crystals of druse develop in scattered pores, and large syntaxial overgrowths develop on echinoderm fragments (not shown in Fig. 32). Scattered pores are occluded by the

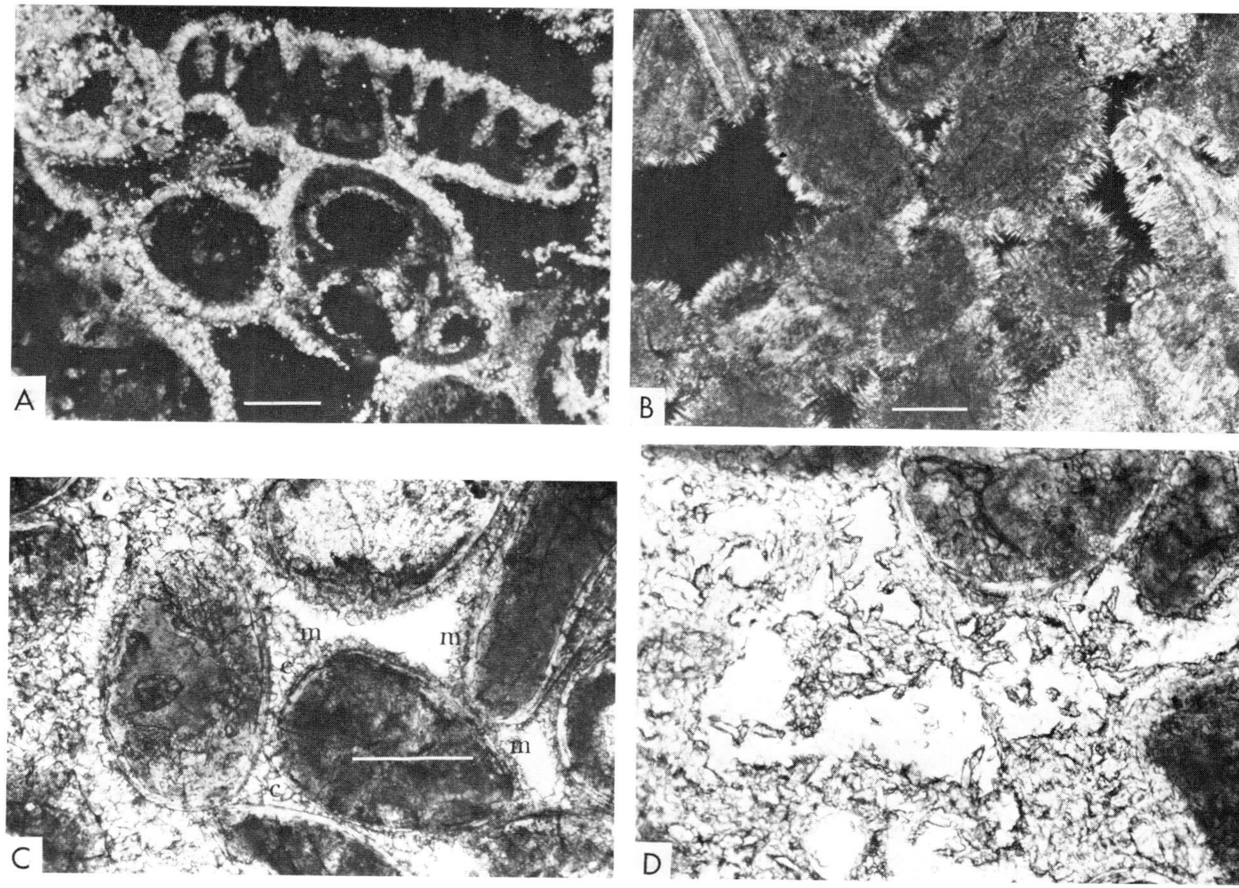

FIG. 31 – Intergranular cement in Pleistocene eolianite presently immersed in intertidal zone.

A. Mg-calcite blades precipitated in voids of Mujeres eolianite which crops out in intertidal zone on Caribbean shore of Isla Mujeres. Bar scale = 100μ. Crossed nicols.

B. Aragonite fibers coating grains of Mujeres eolianite which crops out in intertidal zone on Caribbean coast of Isla Mujeres. Bar scale = 100μ. Crossed nicols.

C. Mg-calcite cement in intertidal Contoy eolianite on Isla Contoy. Mg-calcite cement (m) lines voids; low-Mg-calcite cement (c) is at grain contacts. It appears that low-Mg calcite (vadose-zone cement) is being replaced by Mg calcite (present intertidal-zone cement). In this photo, Mg calcite is stained red (see text) and appears darker than low-Mg-calcite cement. Bar scale = 100μ. Plane light.

D. Same rock as in **C.** Elongate fibers of Mg calcite which presumably replaced root-hair sheaths. Same scale as **C.** Plane light.

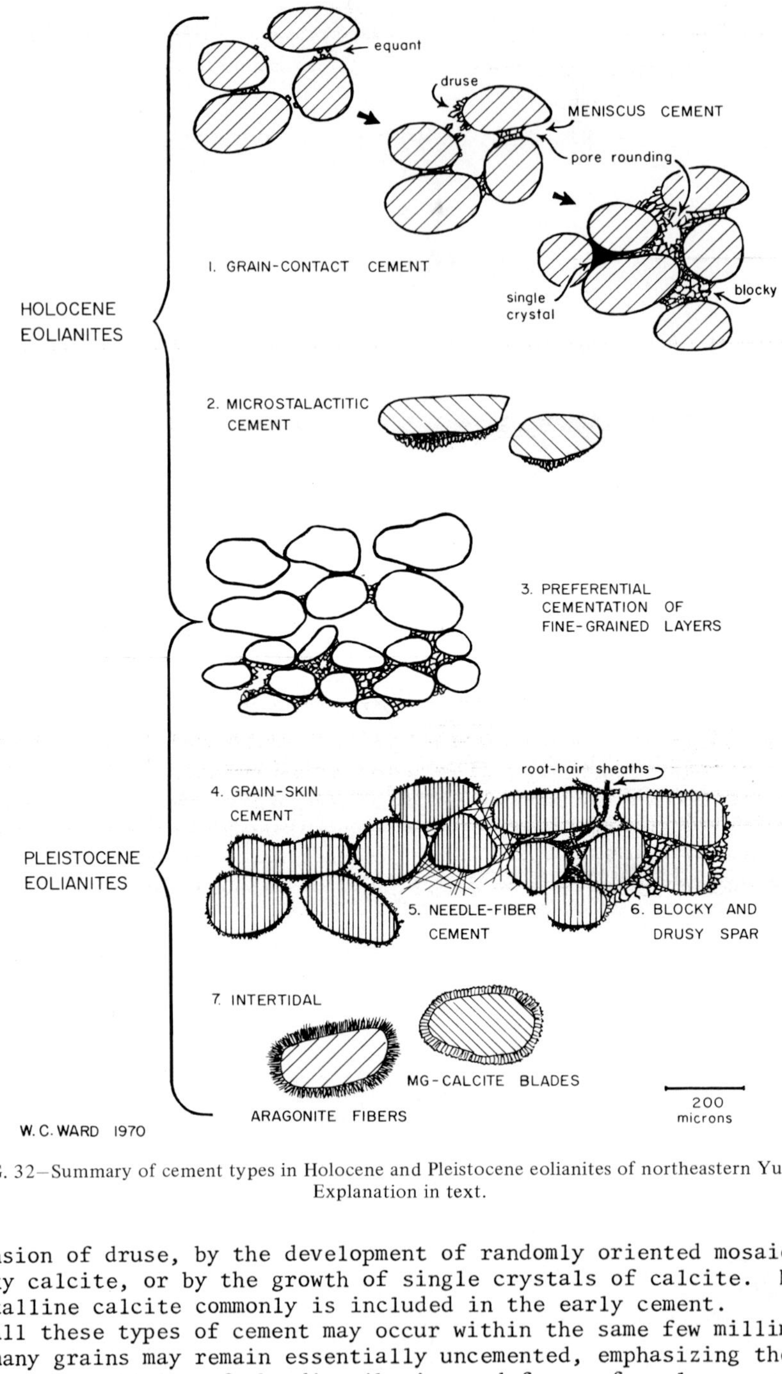

FIG. 32—Summary of cement types in Holocene and Pleistocene eolianites of northeastern Yucatán. Explanation in text.

expansion of druse, by the development of randomly oriented mosaics of blocky calcite, or by the growth of single crystals of calcite. Microcrystalline calcite commonly is included in the early cement.

All these types of cement may occur within the same few millimeters, and many grains may remain essentially uncemented, emphasizing the characteristic variability of the distribution and forms of early cementation in the vadose zone.

Microstalactitic cement—Druse preferentially developed on the bottom of the grains is common in the Holocene eolianites. This form also is indicative of cementation in the vadose zone.

Preferential cementation of fine-grained layers—Vadose water is held preferentially in the finer grained layers in sand dunes. These layers are more extensively cemented, and the coarser grained, more easily drained layers remain only lightly cemented. Patches of rock with finer grains or with tighter packing also are preferentially cemented. This phenomenon, which is common in both Holocene and Pleistocene eolianites, is particularly striking, because the finer, well-cemented laminae and coarser, less-cemented laminae weather in relief.

Grain-skin cement—In much of the Pleistocene eolianite, grains are irregularly coated with microcrystalline calcite that grades out into spikey crystals less than 20μ long. This type of cement develops most typically where microcrystalline root-hair sheaths are present. Fine-grained cement probably is associated with a microenvironment strongly influenced by evapotranspiration and consequent rapid precipitation.

Needle-fiber cement—Randomly oriented, straight, elongate fibers of calcite commonly are present near ancient weathered surfaces and where rhizocretions and root-hair sheaths are abundant. This unusual crystal habit of low-Mg calcite may be the product of organic activity associated with fungal hyphae.

Blocky and drusy spar in Pleistocene dune rock—Calcite druse and blocky calcite mosaics are developed in some pores where root-hair sheaths are absent. This relation suggests that the precipitation of sparry cement is permitted by longer retention of pore water where transpiration by root hairs does not occur. Sparry, grain-skin, and needle-fiber cements all may be present within the same few millimeters.

Intertidal cement—In the present intertidal zone, Pleistocene eolianites initially cemented with meteoric cement are recemented with fringes of Mg-calcite blades or aragonite fibers. These cements are similar to those found in modern beachrock. There is evidence that low-Mg calcite is replaced by Mg calcite in the intertidal zone.

Petrofabric of Eolianites

Packing—Relative grain packing in the eolianites was estimated by counting particle contacts in thin sections. On the basis of Griffiths' (1967, p. 173) classification, three types of contacts were recognized: (1) floating (no apparent contact), (2) tangent, and (3) long.

Little increase in packing with advance in age is indicated by the data in Table 4. Floating grains are slightly more abundant in the younger eolianites, but the number of long contacts is more dependent on grain shape than on an increase in packing. The Mujeres eolianite contains

TABLE 4. GRAIN CONTACTS IN YUCATÁN EOLIANITES

Eolianite	*No. Thin Sections*	*Counts/ Eolianite*	*Mean % Floating*	*Mean % Tangent*	*Mean % Long*
Holocene					
Blanca	31	3100	6±1.3	79±2.3	15±2
Cancun	14	1400	5±0.6	85±0.9	10±0.8
Pleistocene					
Mujeres	22	2200	3±0.3	63±1	34±1
Contoy	23	2300	3±0.3	71±0.9	26±0.9
P. Viejo	9	900	1.5±0.4	81±1.3	17.5±1.2

TABLE 5. POROSITY OF YUCATÁN EOLIANITES

Eolianite (no. samples)	*Intergranular Area*			*Estimated Original Mean % Porosity*
	Mean % Intergranular Voids ±1 std. dev.	*Mean % Intergranular Cement*	*Mean % Root-hair Sheaths Where Present (no. samples included)*	
Holocene				
Blanca (43)	25.8±4.5	5.6	0	31.4
Cancun (16)	20.4±2.6	13.5	0	33.9
Pleistocene				
Mujeres (25)	18.5±8.2	8.9	9.3 (15)	33.0
Contoy (32)	11.6±5.8	15.2	7.9 (31)	34.5
[Intertidal samples with Mg calcite]		[17.0]		
Puerto Viejo (8)	5.3±3.5	23.4	6.3 (5)	32.6

abundant platy grains and has the greatest number of long contacts. The Puerto Viejo eolianite, composed chiefly of subspherical oolites, has as many tangent contacts as the Holocene eolianites. Another indication of slight packing is that the estimated porosity before cementation is 30-35 percent, probably close to the porosity before consolidation. The spatial framework fixed by early cementation has been more or less maintained in these rocks. Overburden is too slight to have contributed significantly to packing.

Porosity—Mean intergranular porosity decreases with advancing age of the Quintana Roo eolianites. Relative porosity estimates, based on point counting of intergranular voids, are compared to percentages of cement and root-hair sheaths in Table 5. Original porosity is estimated by totaling the percentage of voids, cement, and root-hair sheaths (reduction of porosity by packing is assumed to be neglibible). True porosity values are difficult to obtain by point counting, because of the variability in pore geometry (Choquette and Pray, 1970, p. 210).

Porosity is variable in eolianites because of highly variable cementation. Finer grained layers and more tightly packed patches tend to have a higher percentage of occluded pores. In the Pleistocene dune rocks, root-hair sheaths occupy part of the void space, but the pores which they occupy rarely are plugged with sparry cement. Intertidal Contoy eolianite samples with Mg calcite have about the same percent porosity as do the total samples.

Intraparticle Diagenesis

Like the products of intergranular diagenesis, the effects of intragranular diagenesis are highly variable within any one eolianite sample. Grains of the same original composition in the same part of a thin section may be in various stages of alteration. Part of this variability is the result of the uneven distribution of interstitial water in the vadose zone, but much of the diversity is inherited. Grains that reach the beaches are in various stages of diagenetic alteration, apparently depending on their residence time in the marine environment. Some grains are highly bored and micritized; others are fresh. Besides this, some grains are reworked from older carbonate rocks and previously have undergone diagenesis under both subaerial and submarine conditions.

Inherited Intraparticle Diagenesis

Fibrous aragonite commonly occurs in the internal cavities of gastropod fragments and micropelecypods (Fig. 33A) in both Holocene and Pleistocene eolianites. In marine water, fibrous aragonite is deposited perpendicular to the walls of intraparticle voids in aragonitic bioclasts (e.g., Shinn, 1969; Hoskins, 1971; Pingitore, 1970; and Glover and Pray, 1971). In the eolianites of coastal Quintana Roo the aragonite persists into the subaerial environment, but the angular terminations of the crystals are rounded by dissolution. Utricles of *Halimeda* also may be lined or filled with fibrous or microcrystalline aragonite, which is preserved in both the Holocene and Pleistocene dune rocks.

In marine waters the chamber walls of peneroplid tests may be coated with a fringe of Mg-calcite crystals a few microns long and oriented perpendicular to the pore wall (Glover and Pray, 1971). In the Blanca eolianite, some peneroplid chambers are lined with Mg-calcite crystals 4-5µ long and may be further filled with aragonitic micrite (mineral determinations by staining). This internal lining of Mg-calcite cement was not found in the Pleistocene dune rocks, but microcrystalline aragonite does form a filling in *Halimeda* utricles and other intraparticle voids.

Carbonate particles of all types tend to be recrystallized to microcrystalline aragonite or Mg calcite in the submarine environment (see review by Purdy, 1968). Consequently, any one sample of eolianite, because of the diverse origin of its particles, contains grains in many different stages of micritization (Fig. 33B).

Many grains in the submarine and beach sands contain one to several holes, 20-50μ in diameter, that presumably were made by marine boring organisms. When carried into the subaerial environment, these bore holes may be filled with sparry calcite or may be enlarged by solution.

Intraparticle-Void Fill

Blocky cement mosaics develop in intraparticle pores under subaerial conditions. For example, in the youngest dune rock on Isla Blanca, intragranular cement in *Halimeda* utricles ranges from scattered 3μ equant crystals to mosaics of randomly oriented 15μ crystals that almost fill the cavities. In slightly older rocks, many internal pores in *Halimeda* (Fig. 33C), foraminifers, echinoderms, and mollusks contain blocky cement, but the amount and crystal size of sparry calcite are highly variable and are not necessarily proportional to the age of the eolianite.

Neomorphism

Sparry calcite—Minor replacement of grains and oolite coatings by sparry cement has taken place in the youngest Blanca dune rocks. This replacement occurs predominantly at grain contacts and at spots on the grains which have been micritized or otherwise "damaged." In general, replacement of the grains by spar is volumetrically minor and usually confined to the periphery of a particle.

There is no appreciable recrystallization of aragonitic bioclasts to blocky calcite other than in minor spotty patches. Mollusk fragments are entirely replaced by blocky spar only near surfaces of extreme weathering, such as directly below caliche crusts. Similarly, recrystallization of coralline-algae bioclasts to sparry calcite has taken place in small patches at the edges of grains and around bore holes; however, this recrystallization is volumetrically minor.

Microcrystalline calcite—It is difficult to determine micritization of bioclasts in the subaerial limestones because of the extensive recrystallization to microcrystalline calcite that takes place in the marine environment where the grains were produced. Subaerial micritization of particles apparently takes place in the development of caliche crusts and microcrystalline rhizocretions and at the weathered surface of many dune rocks where a layer only 1 or 2 mm thick is intensely bored by microscopic algae. Also, micritization apparently precedes dissolution of the upper surfaces of some grains (Fig. 29).

Dissolution of Grains

In general, total dissolution of grains is scarce in the Quintana Roo rocks. Moldic porosity is best developed in Pleistocene eolianites now in the intertidal or splash zone and in particularly "rooty" rocks. The bioclasts that most commonly dissolve to leave intraparticle porosity are microcrystalline allochems of unknown origin and Mg-calcitic red-algae fragments, recognized by remnants of their fine cellular structure (Fig. 33D). In contrast, Friedman (1964), Land et al (1967), and Gavish and Friedman (1969) found that, in subaerial diagenesis, the fabric of Mg-calcitic bioclasts tended to be preserved in detail and that aragonitic shells dissolved out, forming molds. Most of the coralline algae in the older Yucatán dune rocks have been converted to low-Mg calcite without

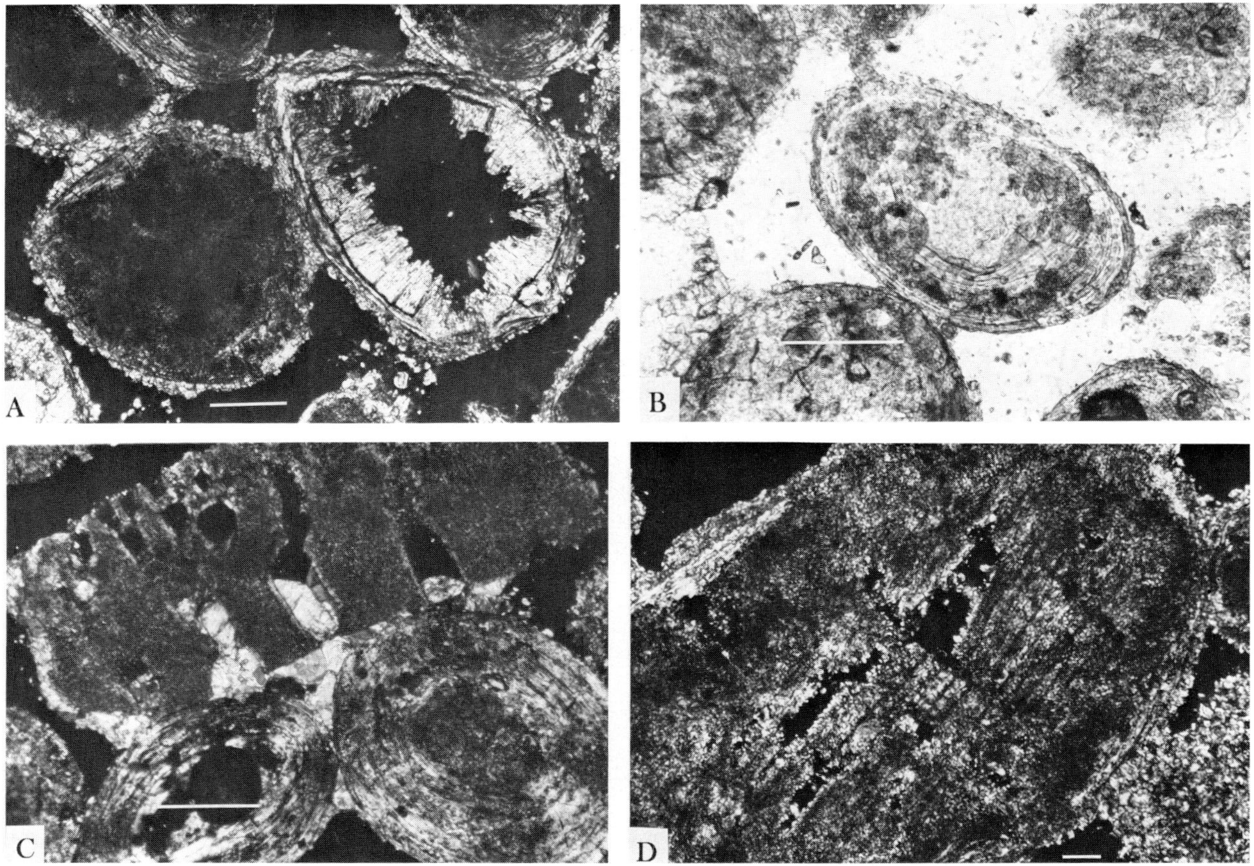

FIG. 33—Examples of intragranular diagenesis, both submarine and subaerial.

A. Photomicrograph of acicular aragonite cement lining cavity in pelecypod in Cancun eolianite (Holocene). Presumably this aragonite was precipitated in submarine water before grain was brought into eolian environment. Bar scale = 100μ. Crossed nicols.

B. Micritized areas in oolite grain (center) in Cancun eolianite. Presumably, micritized areas were bored by algae in marine water before grain was blown into dune environment. Bar scale = 100μ. Plane light.

C. *Halimeda* utricles (upper photo) partly filled, from bottom, with blocky low-Mg calcite that was precipitated in vadose zone. Blanca eolianite (Holocene). Bar scale = 100μ. Crossed nicols.

D. Partly dissolved red algae grain with fine-grained druse lining cavity. Contoy eolianite (Pleistocene), splash zone. Bar scale = 50μ. Crossed nicols.

a fabric change, but solution and removal of red algal material has taken place in all the eolianites.

Staining suggests that the removal of Mg calcite from the red algal fragments commonly takes place at several random sites within a single grain. In many algal and peneroplid bioclasts, however, the Mg calcite appears to have been preferentially removed from the center of grains and retained along the outer edges (i.e., the peripheral dark red stain fades toward the centers). Peneroplid grains in the Pleistocene eolianites take on a red coloration when stained, but most coralline algal particles remain uncolored or take on a pale tint. Hence, in the subaerial environment, Mg calcite apparently is retained more easily in peneroplid foraminiferal bioclasts than in coralline algal fragments. If this is true, the retention of high concentrations of Mg calcite in many Mujeres eolianite samples may be related to their relatively high content of peneroplid bioclasts.

Aragonitic grains generally are rather well preserved. Some moldic porosity is developed in microcrystalline aragonitic grains in the older and more weathered dune rocks. Microcrystalline grains or areas micritized by boring algae (e.g., see Bathurst, 1966) appear to be more readily dissolved than coarsely crystalline particles, presumably because of the exposure to pore water of greater surface areas in the microcrystalline material.

In some samples of Holocene eolianites, particularly those with abundant microstalactitic cement, the tops of grains (both aragonitic and Mg calcitic) are preferentially dissolved (Fig. 29). This dissolution appears to have been preceded by micritization of the upper sides of the grains. The edges of noncoated grains also are rather susceptible to solution. Oolitic coatings, which are among the most stable constituents of these limestones, appear to protect most nuclei from the effects of early subaerial diagenesis.

The slight dissolution of the periphery of grains presumably is important in early cementation. Together with that dissolved from the aforementioned algal-bored surficial crust, the $CaCO_3$ dissolved from the edges of some grains probably provides the material that is initially precipitated at grain contacts. Land et al (1967) and Gavish and Friedman (1969) consider the earliest cement to be "allochthonous" or "of distant origin." That distance can be only millimeters in the Holocene eolianites which are lithified at the surface and have no unconsolidated overburden. In this same regard, it is worth noting the relation of cemented calcarenite to the recrystallized areas of a *Strombus* shell taken from a late Pleistocene beach ridge on the mainland west of Isla Cancun. Most of the calcareous sand surrounding the *Strombus* was unconsolidated and easily cleaned away, revealing a well-preserved aragonitic shell. However, directly underlying several patches of calcarenite that are tightly cemented to the gastropod, the aragonite shell has been converted to calcite. Nearly every patch of lithified calcarenite adhering to the *Strombus* is adjacent to an area of recrystallized shell material, strongly implying that the cement is derived essentially at the site of deposition, from $CaCO_3$ left over after the conversion of the aragonitic shell to calcite.

Features of Uncertain Origin

Borings—Many grains are riddled with holes 15-25μ in diameter in Pleistocene paleosols and eolianites near large rootlet "borings." The holes are about the same size as the diameter of root-hair sheaths, some of which are preserved in the holes, and may be the result of "boring" by root hairs. Whatever their source, the patches of holes appear to be restricted to areas of rootlet penetration.

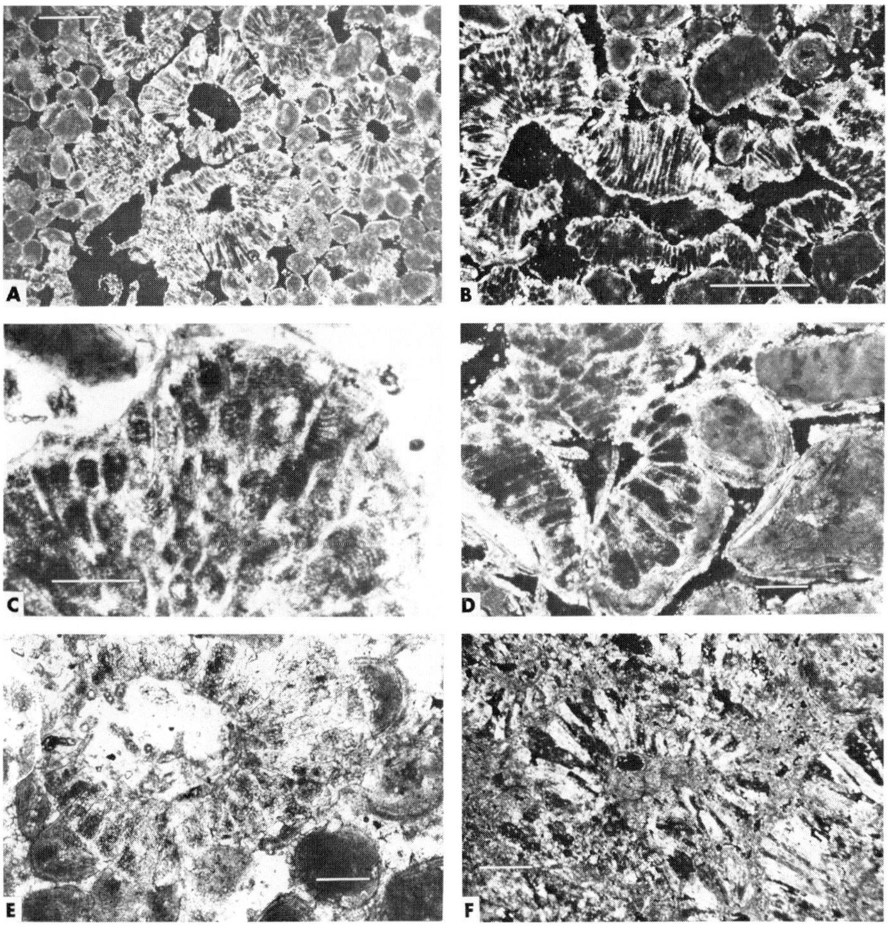

FIG. 34—Photomicrographs of radial structures of uncertain origin in Pleistocene eolianites.

A-E. Radial structures in Contoy eolianite, Isla Cancun.

A, B. Sheaths of low-Mg calcite radiate from central cavity. Structures are several times larger than eolianite grains. Bar scale = 500μ. Crossed nicols.

C. "Cellular" structures which are apparent in only a few of radial structures. Bar scale = 100μ. Plane light.

D, E. Apparent replacement of eolianite grains by low-Mg-calcite radial structures. Bar scale = 100μ. **D**, crossed nicols; **E**, plane light.

F. Radial crystals of low-Mg calcite in Puerto Viejo eolianite, Isla Contoy. Bar scale = 100μ. Crossed nicols.

Radial structures—Also common in "rooty" areas in the Pleistocene eolianites are unidentified radial structures, the properties of which suggest both an organic and inorganic origin (Fig. 34). The bodies characteristically are composed of radiating shafts which widen outward from a central cavity (Fig. 34A). The individual shafts are more or less hexagonal in cross section and contain varied amounts of cryptocrystalline calcite. Most of these structures appear to be subspherical, but others are elongate masses (Fig. 34B). In hand specimens viewed under a binocular microscope, it is evident that some of the shafts radiate out from tubular hollows that follow slightly sinuous paths such as rootlet holes.

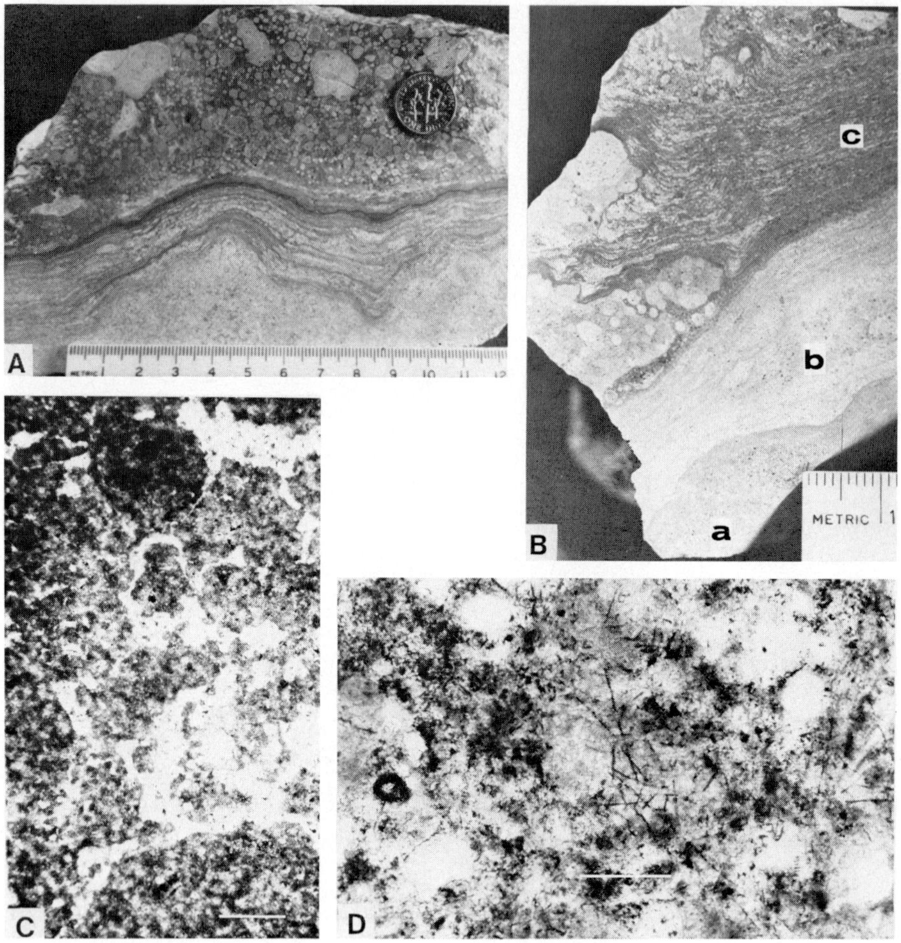

FIG. 35—Textures found in subaerial crusts (caliche) from Isla Mujeres.

A. Wavy, laminated crust overlain by conglomeratic layer, which shows reverse-graded bedding.

B. Mixed laminated and conglomeratic layers (c) above porous laminated layers (a and b).

C. Photomicrograph showing microscopic texture of conglomeratic layer shown in **A**, above. Generally clotty texture, rounded bodies of many different sizes, and fracturing around these rounded bodies are characteristic. Intergranular areas are filled with blocky spar. Bar scale = 100μ. Plane light.

D. Center of photomicrograph shows dark, thin, straight, randomly oriented borings that commonly penetrate caliche deposits. These may be fungal hyphae borings. Bar scale = 100μ. Plane light.

A few of the bodies show a microtexture that is suggestive of cellular structure (Fig. 34C). Their mineralogy, however, is different from that of the surrounding bioclasts. Eolianite grains derived from marine organisms are aragonitic or Mg-calcitic, but staining shows that these radiating structures are low-Mg calcite.

Whether or not these bodies are organically derived, they appear to have grown in place. Surrounding grains are replaced by the shafts, many of which appear to have grown outward at the same time as part of a dissolution-precipitation front (Fig. 34D, E). In more weathered rocks the shafts recrystallize into single-crystal prisms (Fig. 34F) resembling sinter crust on moss tufa (e.g., see Irion and Müller, 1968, Fig. 12, p. 167).

Subaerially produced structures such as these probably would be mistaken for marine organic structures if they were found in older limestones. In thin section the radial bodies described here bear resemblance to the alga(?) *Microcodium*, pictured by Johnson (1961, Pl. 35, Figs. 1-2) and tentatively classified by him in the family Codiaceae. *Microcodium* occurs in Mesozoic and Tertiary limestones in Western Europe and the Pacific. Cuvillier and Sacal (1956, Pl. LXIV, Fig. 1) noted that "...these algae(?) show the transition from an organized structure to crystallizations of calcite, the origin of which was still being discussed quite recently, being often attributed to a whim of nature."

Caliche

Calcitic surficial crusts a few millimeters to several centimeters thick have developed on some weathered surfaces of the Pleistocene eolianites. These crusts include massive and thinly laminated layers of cryptocrystalline to microsparitic calcite and commonly contain "oolitic" and "pisolitic" bodies (Fig. 35A-B). In some places the surficial deposits have been fragmented and recemented to form *in situ* conglomerates (Fig. 35A, B). The microfabric of the crusts is quite similar to that in the core of rhizocretions. All of these deposits are classed here under the broad category "caliche."

Crusts of subaerial origin have been noted in the Bahamas (Kornicker, 1958), in the Florida Keys (Multer and Hoffmeister, 1968), and on Ambergris Cay, Belize (Tebbutt, this volume). Caliche pisolites that are similar to structures in caliche of northeast Yucatán were described by Brown (1956), Swineford et al (1958), Thomas (1965), Dunham (1965, 1969b), and others.

Caliche is generally considered to be the product of inorganic accumulation of calcium carbonate as the result of soil-forming processes in semiarid regions where evaporation is substantial and leaching is moderate (Reeves, 1970, p. 353). It has been suspected that organic processes also may influence the accumulation of subaerial crusts (Kornicker, 1958; Multer and Hoffmeister, 1968, p. 186, 189). Krumbein (1968) found that crusts on certain Turonian limestone in Israel contain an abundant microbial flora that may take part in production of the caliche.

The presence of rhizocretions and a few thin crusts within the Pleistocene eolianites and paleosols and the absence of these within the Holocene dune rocks suggest that the late Pleistocene climate was different from the recent climate. Perhaps the late Pleistocene had more extreme seasonal variations with periods of torrential rainfall followed by periods of intense evaporation, the first causing leaching and the latter inducing deposition of root casts, root-hair sheaths, and caliche crusts. There is evidence, however, that caliche also has accumulated during the Holocene on the islands with Pleistocene bedrock. One sample of caliche conglomerate collected on Isla Mujeres (sample locality IM 36, Fig. 18) has a radiocarbon age of 1900±90 years B.P. (Oak Ridge Associated Universities, Inc.). In addition, a marine gastropod shell was found partly enveloped by the surficial crust on a bare rock slope several meters above sea level on Isla Mujeres. Apparently the caliche crust was precipitated after the shell was brought to that spot by a hermit crab or other animal, thus indicating that caliche is being built up on bare rock surfaces at the present time.

Laminated Crusts

Megascopic features—Surficial crusts generally less than 4 cm thick are made up of wavy thin laminae of porous and dense microcrystalline calcite (Fig. 35A-B). Color ranges from weak, extremely light orange to weak, dark brown (10 YR 8/2 to 10 YR 4/2, following modification of

National Research Council Rock-Color Chart by Folk, 1969). Porous layers are light colored, and dense layers are dark. Crusts unconformably overlie bedrock and tend to thicken in depressions, smoothing out underlying irregularities. Thin crusts are developed along vertical fractures and on the walls of solution cavities. Within the crusts are many unconformities, the result of alternating solution and deposition. Almost all the truncation of the laminae is upward, indicating that the crusts were built up by accretion of layers onto the exposed surface, not thickened downward by adding onto the bottom as a result of evaporation-deposition. The probable sources of the $CaCO_3$ in the surficial crusts on the Yucatán islands include: (1) Pleistocene eolianite bedrock, (2) Pleistocene and Holocene soils (probably most of the crusts were initiated in the Pleistocene), (3) older caliche crusts, (4) sea spray and mist, and (5) Holocene carbonate sand blown or washed on top of Pleistocene bedrock (locally significant at low elevations).

Multer and Hoffmeister (1968, p. 185-186) classed subaerial crusts into three types: (1) *porous laminated*, with 5-20 percent horizontal root tubes, organic debris, and various bioclastic and lithoclastic inclusions, and with common solution holes less than a millimeter in diameter; (2) *dense laminated*, with 0-5 percent organic debris and bioclastic and lithoclastic inclusions, and with scarce to common solution holes less than a half millimeter in diameter; and (3) *microcrystalline rind*, with scattered ghosts of pellets and ooids in a dense microcrystalline calcite mosaic, and with scarce pinpoint solution holes. The porous and laminated crusts occur together or separately, and the microcrystalline rind is generally at the bottom of the crust, directly overlying the bedrock. In general, these descriptions of crusts in the Florida Keys fit the caliche crusts on the Quintana Roo islands except that a dense microcrystalline rind is always the uppermost or outermost layer in the crustal caliche described here.

Microscopic features—In thin section the laminated crusts appear to consist of vaguely pelleted or clotted, dark brown cryptocrystalline and light-brown microcrystalline calcite (Fig. 35C). Most of the cryptocrystalline "pellets" are less than 20μ in diameter. This clotty texture may be the result of patchy recrystallization of cryptocrystalline calcite to microcrystalline-microsparry calcite. More likely, however, this microtexture is depositional, because no layers without the mottled appearance were found.

An equally characteristic feature of the caliche deposits is the "pelletoid texture." Subspherical peloids, both massive and "oolitic," form in a wide range of sizes (Fig. 35C) and are produced by a combination of desiccation fracturing, solution, deposition, and slight transportation.

Also scattered about in the crusts are tiny clear spherulites up to 40μ in diameter and composed of fibers about 20μ long and a fraction of a micron wide, radiating from a central point. The crusts also include varied amounts of bioclasts that apparently were derived from the eolianitic bedrock. Many aragonitic bioclasts and oolite rims withstand this inclusion without calcitization.

Another characteristic feature of these caliche deposits is the random array of elongate, very straight borings about 0.5-2μ wide (Fig. 35D). These borings, probably the work of fungal hyphae, commonly penetrate eolianite grains up to 1-2 mm below the caliche deposit.

In porous laminated crusts, vugs may be lined with microcrystalline calcite similar to grain-skin cement or may be plugged with blocky calcite mosaics. Some pores are fringed with fans of bluntly terminated blades of calcite as long as 40μ. The porous laminated layers appear to have included stems, rootlets, and the structures of lower plants. Abundant randomly oriented, cylindrical vugs 70-100μ in diameter suggest holes left by oxidized plant material. For the most part, these tubular voids do not extend into adjacent dense layers. This suggests that the porous layers owe their character to buildup during times of active plant growth (i.e., warm, wet seasons). The darker brown (more organic-rich), denser layers apparently were formed during periods when finely disseminated plant material was not flushed from the site of deposition (drier periods?).

In dense layers the peripheries of many pores are cracked into a mosaic of elongate rhombs and polyhedra. This "pore-cracking" (name suggested by R. L. Folk) probably is the result of the alternate wetting and drying to which surficial caliche deposits are constantly subjected. The effects of pore cracking appear similar to artifact fractures at the edge of thin sections, but the fact that they can be seen in acetate peels of polished rock surfaces indicates that the pore cracks are natural.

The lower boundary of the caliche crusts is rather sharp, grading into the underlying limestone within a transitional layer no more than a few centimeters thick. "Calichification" of the eolianite surface apparently was preliminary to the crustal buildup. By whatever process this was accomplished, it resulted in a fine-grained layer that probably acts as an aquiclude between the underlying eolianite and the overlying caliche deposits, thereby exerting a certain control over runoff, evaporation, and diagenesis. Eolianite beneath the capping crust would be somewhat protected from the effects of percolating meteoric water. At the same time, the caliche deposits above the initial layer would be subjected to the action of soil water that is forced to flow horizontally. The evidence of solution, deposition, and transportation of particles within the caliche layers may be explained by flow of water in the upper vadose zone (Dunham, 1969a, 1969b).

Caliche "oolites" in layered crusts—In some dense caliche layers, cracking has produced subspherical oolite-sized structures. For the most part, these are massive cryptocrystalline bodies, many of which contain one or more bioclasts and/or black caliche fragments. At certain spots within these layers there is a gradation from small, polygonally fractured blocks to well-rounded spheroids. These ball-shaped masses apparently are produced essentially *in situ* when the corners of small fractured blocks are progressively dissolved until subspherical bodies are formed. Some of the "oolites" have a thin outer shell of microcrystalline calcite, and a few samples contain massive oolite- and pisolite-sized bodies coated with fairly regular, clear calcite shells 15μ thick. In more porous layers, caliche oolites are best developed along fractures and just above impermeable(?), dense cryptocrystalline rinds. Thus their origin is related to the flow of vadose water.

Conglomeratic Layers

In some places laminated caliche crusts are overlain by conglomeratic layers containing rounded fragments of crustal material up to several centimeters in diameter. Small pieces of black caliche and scarce fragments of eolianite also are included. These clasts are set in a limonite-stained caliche matrix, which has the same clotty and "pelletoid" texture as the crusts. The conglomerate grains are in roughly reverse-graded layers (Fig. 35A).

Alternate wetting and drying have created globular masses of many sizes separated by nontectonic fractures within the matrix and within the crustal clasts. These fractures tend to be curved, and the fragments tend to become spherical masses. Fracturing around the grains produced by alternate shrinkage and expansion is called "circumgranular cracking" by R. L. Folk (personal commun., 1968). Circumgranular cracks commonly are filled with finely crystalline, blocky calcite.

The production of caliche "pisolites" and "oolites" in these caliche deposits appears to result from cracking, shrinkage, and dissolution perhaps more than from concretion. Where the spheroidal bodies are well defined and the surrounding cracks are filled with sparry calcite, the deposits might be mistaken for intrasparite or oncolite biosparite. The very poor sorting in the caliche "sparites" may help to distinguish them from the marine deposits which they mimic.

Blackening of Eolianite and Caliche Adjacent to Salt Lakes

Black "caliche-algal micrite" crusts formed on the periphery of hypersaline lakes on Isla Mujeres (Ward et al, 1970). The source of the dark color is organic matter, apparently derived from the organic (algal?) muck that is formed at the shoreline of the hypersaline lakes. In some places the finely disseminated black material also infiltrated the bedrock to a depth of a few centimeters. Desiccation fracturing of the surface layers at the edge of the lakes has produced an angular rubble of blackened and lighter colored caliche and eolianite fragments. Similar deposits are present on the periphery of hypersaline areas in other parts of Yucatán. These hypersaline environments apparently provide favorable conditions for the blackening of certain carbonate rocks. It follows, then, that the presence of dark-colored fragments in limestone sections may call attention to surfaces which are the product of subaerial exposure in the proximity of hypersaline water. Black carbonate-rock fragments may aid in locating ancient coastal areas or the landward edge of evaporite deposits.

At unconformities within Pennsylvanian beds of the Oro Grande basin, blackened limestone pebbles are present (Wilson, 1967, p. 808). Similarly, widespread disconformities in the Jurassic of the Middle East (e.g., top of Uwainat and Araej Formations) are marked by blackened sand-size particles (J. L. Wilson, personal commun.). The recognition of significant unconformities within carbonate sequences is difficult. Some types of blackened limestones apparently are indications of subaerial exposure and, therefore, should prove to be important in interpreting breaks in the record.

SUMMARY

1. Meteoric-water cementation and other products of subaerial diagenesis of limestones can be characterized most clearly in carbonate units which were deposited in the subaerial environment and subsequently remained above the influence of marine water. During the late Pleistocene and Holocene, such limestones were deposited and lithified in dunes along the northeast coast of the Yucatán Peninsula. Previously published works have dealt with unconsolidated late Holocene (recent) dune sands and lithified early Holocene and Pleistocene eolianites. The presence of consolidated late Holocene dune rock on the northeast Yucatán coast provides a missing link in the study of progressive diagenesis of eolianites.

2. On the basis of grain composition and geomorphic relations, the late Pleistocene eolianites can be divided into at least three different limestones. The oldest of these, the Puerto Viejo eolianite, forms the western ridge of Isla Contoy. A younger limestone, the Contoy eolianite, crops out along the eastern ridges of Isla Contoy, on the western ridge of Isla Mujeres, and along several ridges on Isla Cancun. The youngest Pleistocene dune rock, the Mujeres eolianite, forms the main part of Isla Mujeres. The two older eolianites are composed predominantly of

coated grains; the youngest Pleistocene eolianite is made up mostly of noncoated bioclasts.

Holocene eolianites can be separated into at least two limestones. The older Cancun eolianite was deposited along the Puerto Juarez coast and the northern and Caribbean coasts of Isla Cancun. A younger Holocene limestone, the Blanca eolianite, is accumulating on Isla Blanca and the southern two thirds of Isla Cancun.

The Pleistocene eolianites contain rhizocretions and root-hair sheaths, which are absent in the Holocene dune rocks. This fact probably indicates that evapotranspiration was seasonally more intense in the late Pleistocene than at present.

3. On the basis of the mineralogy of these eolianites, each of the limestones represents a different diagenetic stage. The youngest dune rock on Isla Blanca has the same composition as unconsolidated dune and beach sands on that island: 75-85 percent aragonite, 15-20 percent Mg calcite, and less than 5 percent low-Mg calcite. Progressively older dune ridges on Isla Blanca contain increasingly more low-Mg calcite. The Cancun eolianite contains 69-84 percent aragonite and less than 5 percent Mg calcite. Point-count data indicate that the composition of this Holocene oolite is low in Mg-calcite bioclasts. The composition of the Pleistocene Mujeres eolianite is highly variable. Originally the Mujeres dune rocks contained as much as about 40 percent Mg calcite, and there is high retention of Mg calcite in some beds. Several samples contain from 20 to 32 percent Mg calcite (12-14 mole percent $MgCO_3$). Aragonite content of most Mujeres samples ranges from 45 to 65 percent. The Contoy eolianite (Pleistocene) contains 48-75 percent aragonite and less than 5 percent Mg calcite. The oldest Pleistocene dune rock, the Puerto Viejo eolianite, is the most diagenetically mature, containing 40-60 percent aragonite and less than 5 percent Mg calcite.

4. There is no single trend of progressive diagenesis for the northeast Yucatán eolianites; each eolianite has its own route toward complete calcitization. It is apparent that early subaerial diagenesis in one type of limestone in one limited geographic area can be complex. The mineralogic diagenetic stage is related to original composition, climate, and position in the vadose zone as well as to geologic age.

5. An enrichment in Mg calcite was found in many samples of Contoy and Mujeres eolianites (Pleistocene) collected from present intertidal and subtidal zones. Beachrock-type Mg-calcite cement was precipitated in the interstices of these Pleistocene dune rocks now immersed in marine waters.

6. Cement types may be important clues in determining the environment of early lithification of limestones. In the vadose zone, initial cement is confined to corners of pores where water droplets are held by surface tension. As grain-contact cement expands, it takes on the shape of the air-water interface, and meniscus cement is formed. This type of cementation contrasts with more uniform grain-edge cementation which takes place in the phreatic or submarine environment where pores are filled with water. Also, early in the history of vadose cementation, crystals of druse grow in scattered pores, and large syntaxial overgrowths develop on echinoderm fragments. Microcrystalline calcite commonly is included in the early void-rimming spar in the Holocene eolianites. Furthermore, many grains remain essentially uncemented, emphasizing the characteristic variability of the distribution and forms of early cementation in the vadose zone. Druse preferentially precipitated on the bottom of grains is common in the Holocene eolianites. This form also is indicative of cementation in the vadose zone and may be preserved in ancient limestones.

In sand dunes water is preferentially held in finer grained layers and more tightly packed patches. In calcareous dunes these layers and patches which attract vadose water are preferentially cemented.

Apparently the source of the cement in the Holocene dune rock is local. The fact that surface layers of Holocene rocks are indurated indicates that the cement was not brought from a distant source. Stained tests indicate that some of the early cement in the vadose zone of the Holocene dunes may be Mg calcite.

7. In the Pleistocene eolianites, grain-skin cement is developed around pores that contain root-hair sheaths. Calcite druse and blocky calcite mosaics are developed in some of the pores where root-hair sheaths are absent. This relation suggests that the early influence of transpiration by root hairs inhibited the precipitation of sparry cement.

The Pleistocene eolianites commonly contain randomly oriented, straight elongate fibers of low-Mg calcite. This "needle-fiber" cement is present near ancient weathered surfaces where rhizocretions and root-hair sheaths are abundant.

In the present intertidal zone, Mg-calcite submarine cement apparently replaces low-Mg calcite meteoric cement with which the Pleistocene dune rocks were initially cemented.

8. The spatial framework fixed by early cementation has been more or less maintained in the older Yucatán eolianites. Original intergranular porosity of the Holocene eolianites is more than 20 percent, and the average porosity of the Pleistocene eolianites decreases with advanced age.

9. Certain intragranular diagenetic features, such as micritized areas in bioclasts and fibrous-aragonite fringes in voids, were inherited from the marine environment. Minor dissolution of grains has taken place in the subaerial environment. The tops of some bioclasts are preferentially dissolved relative to the bottoms. Micritized areas in grains are more readily dissolved than coarsely crystalline areas. Coralline algal bioclasts appear to be the most easily dissolved grains in the eolianites. Stained tests indicate that, in the vadose environment, peneroplid Foraminifera retain Mg calcite longer than coralline algae.

10. Laminated crusts and conglomerates are developed on weathered surfaces of the Pleistocene eolianites. These caliche deposits contain "oolitic" and "pisolitic" bodies, which apparently are formed as a result of alternate wetting and drying. Fracturing is produced by alternate shrinkage and expansion, and corners of the fracture blocks are dissolved. In some layers the subspherical bodies have taken on outer rims of microcrystalline or clear calcite. Black crusts of "caliche-algal micrite" were formed on the periphery of hypersaline lakes on Isla Mujeres. The source of the dark color is organic matter, apparently derived from the algal muck that is formed at the shoreline of the lakes (Ward et al, 1970).

SELECTED REFERENCES

Back, William, 1970, Comparison of chemical hydrogeology of the carbonate peninsulas of Florida and Yucatan: Jour. Hydrology, v. 10, p. 330-368.

——and B. B. Hanshaw, 1967, Hydrogeology of the northern Yucatan Peninsula, Mexico, *in* A. E. Weidie, ed., Yucatan field trip guide book, 2d edition: New Orleans Geol. Soc., p. 64-78.

Bagnold, R. A., 1938, Grain structure of sand dunes and its relation to their water content: Nature, v. 142, p. 403-404.

——1941, The physics of blown sand and desert dunes: London, Methuen, 265 p.

Ball, M. M., 1967, Carbonate sand bodies of Florida and the Bahamas: Jour. Sed. Petrology, v. 37, p. 556-591.

Bathurst, R. G. C., 1958, Diagenetic fabrics in some British Dinantian limestones: Liverpool and Manchester Geol. Jour., v. 2, p. 11-36.

——1966, Boring algae, micrite envelopes and lithification of molluscan biosparites: Geol. Jour., v. 5, p. 15-32.

Brown, C. N., 1956, The origin of caliche on the northeastern Llano Estacado, Texas: Jour. Geology, v. 64, p. 1-15.

Brown, J. C., 1958, Soil fungi of some British sand dunes in relation to soil type and succession: Jour. Ecology, v. 46, p. 641-664.

Bryant, W. R., et al, 1969, Escarpments, reef trends, and diapiric structures, eastern Gulf of Mexico: AAPG Bull., v. 53, p. 2506-2542.

Caumartin, V., 1963, Review of the microbiology of underground environments: Natl. Speleol. Soc. Bull., v. 25, p. 1-14.

Chapman, F., 1900, Notes on the consolidated aeolian sands of Kathiawai: Geol. Soc. London Quart. Jour., v. 56, p. 584-588.

Choquette, P. W., and L. C. Pray, 1970, Geologic nomenclature and classification of porosity in sedimentary carbonates: AAPG Bull., v. 54, p. 207-250.

Cooper, W. S., 1958, Coastal sand dunes of Oregon and Washington: Geol. Soc. America Mem. 72, 170 p.

Cuvillier, J., and V. Sacal, 1956, Stratigraphic correlations by microfacies in western Aquitaine: Internat. Sedimentary Petrogr. Ser., v. 2, Leiden, E. J. Brill, 33 p.

Darwin, Charles, 1844, Geological observations on the volcanic islands and parts of South America visited during the voyage of HMS Beagle: G. T. Bettany, ed., London, Ward and Lock, 549 p.; 1896, New York, D. Appleton, 648 p.

Davis, S. N., and R. J. M. DeWiest, 1966, Hydrogeology: New York, John Wiley and Sons, 463 p.

Dengo, G., and O. Bohnenberger, 1969, Structural development of northern Central America, *in* A. R. McBirney, ed., Relations of northern Central America and the western Caribbean—the Bonacca Expedition: AAPG Mem. 11, p. 203-220.

Dittmer, H. J., 1949, Root hair variations in plant species: Am. Jour. Botany, v. 36, p. 152-155.

Dolley, C. S., 1889, The botany of the Bahamas: Acad. Nat. Sci. Philadelphia Proc., v. 41, p. 130-134.

Dunham, R. J., 1962, Classification of carbonate rocks according to depositional texture, *in* W. E. Ham, ed., Classification of carbonate rocks—a symposium: AAPG Mem. 1, p. 108-121.

——1965, Vadose pisolite in the Capitan Reef (abs.): AAPG Bull., v. 49, p. 338.

——1969a, Early vadose silt in Townsend mound (reef), New Mexico, *in* G. M. Friedman, ed., Depositional environments in carbonate rocks—a symposium: Soc. Econ. Paleontologists and Mineralogists Spec. Pub. 14, p. 139-181.

——1969b, Vadose pisolite in the Capitan Reef (Permian), New Mexico and Texas, *in* G. M. Friedman, ed., Depositional environments in carbonate rocks—a symposium: Soc. Econ. Paleontologists and Mineralogists Spec. Pub. 14, p. 182-191.

——1971, Meniscus cement, *in* O. P. Bricker, ed., Carbonate cements: Johns Hopkins Univ. Studies in Geology, no. 19, p. 297-300.

Ebanks, W. J., 1967, Recent carbonate sedimentation and diagenesis, Ambergris Cay, British Honduras: Rice Univ., PhD dissert., 189 p.

Ellis, D., 1918, *Phycomycetus* fungi from the English lower coal measures: Royal Soc. Edinburgh Proc., v. 38, p. 130-145.

Evamy, B. D., 1969, The precipitational environment and correlation of some calcite cements deduced from artificial staining: Jour. Sed. Petrology, v. 39, p. 787-821.

Evans, J. W., 1900, Mechanically-formed limestones from Junagarh (Kathiawar) and other localities: Geol. Soc. London Quart. Jour., v. 56, p. 559-583, 588-589.

Fairbridge, R. W., 1950, The geology and geomorphology of Point Peron, Western Australia: Royal Soc. Western Australia Jour., v. 34, p. 35-72.

——1957, The dolomite question, *in* R. J. LeBlanc and J. G. Breeding, eds., Regional aspects of carbonate deposition: Soc. Econ. Paleontologists and Mineralogists Spec. Pub. 5, p. 125-178.

——and Curt Teichert, 1953, Soil horizons and marine bands in the coastal limestones of Western Australia: Royal Soc. New South Wales Jour. and Proc., v. 86, pt. 3, p. 68-86.

Folk, R. L., 1962, Spectral subdivision of limestone types, *in* W. E. Ham, ed., Classification of carbonate rocks—a symposium: AAPG Mem. 1, p. 62-84.

——1965, Some aspects of recrystallization in ancient limestones, *in* L. C. Pray and R. C. Murray, eds., Dolomitization and limestone diagenesis—a symposium: Soc. Econ. Paleontologists and Mineralogists Spec. Pub. 13, p. 14-48.

——1967, Carbonate sediments of Isla Mujeres, Quintana Roo, Mexico, and vicinity, *in* A. E. Weidie, ed., Yucatan field trip guide book, 2d edition: New Orleans Geol. Soc., p. 100-123.

——1969, Toward greater precision in rock-color terminology: Geol. Soc. America Bull., v. 80, p. 725-728.

——and W. C. Ward, 1957, Brazos River bar: a study in the significance of grain size parameters: Jour. Sed. Petrology, v. 27, p. 3-26.

——M. O. Hayes, and R. Shoji, 1962, Carbonate sediments of Isla Mujeres, Quintana Roo, Mexico and vicinity, *in* A. E. Weidie and G. E. Murray, eds., Guidebook, field trip to Peninsula of Yucatan, 1962: New Orleans Geol. Soc., p. 85-100.

Friedman, G. M., 1964, Early diagenesis and lithification in carbonate sediments: Jour. Sed. Petrology, v. 34, p. 777-813.

Gavish, E., and G. M. Friedman, 1969, Progressive diagenesis in Quaternary to late Tertiary carbonate sediments: sequence and time scale: Jour. Sed. Petrology, v. 39, p. 980-1006.

Ginsburg, R. N., E. A. Shinn, and J. H. Schroeder, 1967, Submarine cementation and internal sedimentation within Bermuda reefs (abs.): Geol. Soc. America, Program 1967 Ann. Mtg., p. 78-79; Geol. Soc. America Spec. Paper 115, p. 78-79.

Glover, E. D., and L. C. Pray, 1971, High-magnesium calcite and aragonite cementation within modern subtidal sediment grains, *in* O. P. Bricker, ed., Carbonate cements: Johns Hopkins Univ. Studies in Geology, no. 19, p. 80-87.

Goldsmith, J. R., and D. L. Graf, 1958, Structural and compositional variations in some natural dolomites: Jour. Geology, v. 66, p. 678-693.

Griffiths, J. C., 1967, Scientific method in analysis of sediments: New York, McGraw-Hill, 508 p.

Gross, M. C., 1964, Variations in the O^{18}/L^{16} and C^{13}/C^{12} ratios of diagenetically altered limestones in the Bermuda Islands: Jour. Geology, v. 72, p. 170-194.

Harding, J. L., 1964, Petrology and petrography of the Campeche lithic suite, Yucatan Shelf, Mexico: Texas A&M Univ. Dept. Oceanography and Meteorology, Ref. 64-11T, 140 p.

Hoskin, C. M., 1963, Recent carbonate sedimentation on Alacran Reef, Yucatan, Mexico: U.S. Office Naval Research Foreign Field Research Program Rept. 19 (Natl. Acad. Sci. - Natl. Research Council Pub. 1089), 160 p.

——1971 Aragonite crystals in algal biolithite, Alacran Reef, Mexico, *in* O. P. Bricker, ed., Carbonate cements: Johns Hopkins Univ. Studies in Geology, no. 19, p. 66-67.

Illing, L. V., 1954, Bahaman calcareous sands: AAPG Bull., v. 38, p. 1-95.

Irion, G. and G. Müller, 1968, Mineralogy, petrology, and chemical composition of some calcareous tufa from the Schwabische Alb, Germany, *in* G. Müller and G. M. Friedman, eds., Carbonate sedimentology in central Europe: New York, Springer-Verlag, p. 157-171.

Johnson, D. L., 1967, Caliche on the channel islands: Mineral Information Service, California Div. Mines and Geology, v. 20, p. 151-158.

Johnson, J. H., 1961, Limestone-building algae and algal limestone: Golden, Colorado School of Mines, 197 p.

Kaye, C. A., 1959, Shoreline features and Quaternary shoreline changes in Puerto Rico: U.S. Geol. Survey Prof. Paper 317-B, p. 49-140.

Kidston, R., and W. H. Lang, 1921, Old Red Sandstone plants showing structure from the Rhynie chert bed, Aberdeenshire, Part V: Royal Soc. Edinburgh Trans., v. 52, p. 855-902.

Kindle, E. M., 1923, Range and distribution of certain types of Canadian Pleistocene concretions: Geol. Soc. America Bull., v. 34, p. 609-648.

Kornicker, L. S., 1958, Bahamian limestone crusts: Gulf Coast Assoc. Geol. Socs. Trans., v. 8, p. 167-170.

Krumbein, W. E., 1968, Geomicrobiology and geochemistry of the "Nari-lime-crust" (Israel), *in* G. Müller and G. M. Friedman, eds., Recent developments in carbonate sedimentology in central Europe: New York, Springer-Verlag, p. 138-147.

Land, L. S., 1967, Diagenesis of skeletal carbonates: Jour. Sed. Petrology, v. 37, p. 914-930.

——1971, Submarine lithification of Jamaican reefs, *in* O. P. Bricker, ed., Carbonate cements: Johns Hopkins Univ. Studies in Geology, no. 19, p. 59-62.

——F. T. Mackenzie, and S. J. Gould, 1967, Pleistocene history of Bermuda: Geol. Soc. America Bull., v. 78, p. 993-1006.

Lea, A. M., 1925, Notes on some calcareous insect puparia: Records South Australian Mus., v. 3, p. 35-36.

Leipper, D. F., 1954, Physical oceanography of the Gulf of Mexico, *in* P. S. Galtsoff, ed., Gulf of Mexico, its origin, water, and marine life: U.S. Fish and Wildlife Service Fishery Bull. 89, v. 55.

Logan, B. W., et al, 1969, Late Quaternary carbonate sediments of Yucatán Shelf, Mexico, *in* Carbonate sediments and reefs, Yucatán Shelf, Mexico: AAPG Mem. 11, p. 5-128.

McKee, E. D., and R. C. Gutschick, 1969, History of the Redwall Limestone of northern Arizona: Geol. Soc. America Mem. 144, 726 p.

Meyerhoff, A. A., 1967, Future hydrocarbon provinces of Gulf of Mexico - Caribbean region: Gulf Coast Assoc. Geol. Socs. Trans., v. 17, p. 217-260.

Milliman, J. D., and K. O. Emery, 1968, Sea levels during the past 35,000 years: Science, v. 162, p. 1121-1123.

Müller, G., 1971, "Gravitational" cement: an indicator for the vadose zone of subaerial diagenetic environment, *in* O. P. Bricker, ed., Carbonate cements: Johns Hopkins Univ. Studies in Geology, no. 19, p. 185-188.

Multer, H. C., and J. E. Hoffmeister, 1968, Subaerial laminated crusts of the Florida Keys: Geol. Soc. America Bull., v. 79, p. 183-192.

Murray, G. E., 1961, Geology of the Atlantic and Gulf coastal province of North America: New York, Harper and Bros., 692 p.

——and A. E. Weidie, 1962, Regional geologic summary of Yucatan Peninsula, *in* A. E. Weidie and G. E. Murray, eds., Guidebook, field trip to Peninsula of Yucatan, 1962: New Orleans Geol. Soc., p. 5-51.

Nelson, H. F., 1971, Cementation in the Edwards Formation, Central Texas, *in* O. P. Bricker, ed., Carbonate cements: Johns Hopkins Univ. Studies in Geology, no. 19, p. 155-158.

Newell, N. D., and J. K. Rigby, 1957, Geological studies on the Great

Bahama Bank, *in* R. J. LeBlanc and J. G. Breeding, eds., Regional aspects of carbonate sedimentation: Soc. Econ. Paleontologists and Mineralogists Spec. Pub. 5, p. 15-79.
Northrop, J. I., 1910, A naturalist in the Bahamas: New York, Columbia Univ. Press, 281 p.
Pingitore, N. E., 1970, Diagenesis and porosity modification in *Acropora palmata*, Pleistocene of Barbados, West Indies: Jour. Sed. Petrology, v. 40, p. 712-721.
Porter, C. L., and G. Zebrowski, 1937, Lime-loving molds from Australian sands: Mycologia, v. 29, p. 252-257.
Purdy, E. G., 1968, Carbonate diagenesis—an environmental survey: Estratto da Geologica Romana, v. 7, p. 183-228.
Purser, B. H., 1969, Syn-sedimentary marine lithification of Middle Jurassic limestones of the Paris Basin, France: Sedimentology, v. 12, p. 205-230.
Reeves, C. C., 1970, Origin, classification, and geologic history of caliche on the southern High Plains, Texas and eastern New Mexico: Jour. Geology, v. 78, p. 352-362.
Robinson, R. B., 1967, Diagenesis and porosity development in Recent and Pleistocene oolites from southern Florida and the Bahamas: Jour. Sed. Petrology, v. 37, p. 355-364.
Roehl, P. O., 1967, Stony Mountain (Ordovician) and Interlake (Silurian) facies analogs of Recent low-energy marine and subaerial carbonates, Bahamas: AAPG Bull., v. 51, p. 1979-2032.
Sayles, R. W., 1931, Bermuda during the Ice Age: Am. Acad. Arts Sci. Proc., v. 66, p. 381-467.
Schroeder, J. H., 1969, Experimental dissolution of calcium, magnesium, and strontium from Recent biogenic carbonates: a model of diagenesis: Jour. Sed. Petrology, v. 39, p. 1057-1073.
Seward, A. C., 1898, Fossil plants, V. I: New York, Hafner Publishing Co. (1963), 452 p.
Shinn, E. A., 1969, Submarine lithification of Holocene carbonate sediments in the Persian Gulf: Sedimentology, v. 12, p. 109-144.
Sorby, H. C., 1879, Anniversary address of the President (On the structure and origin of non-calcareous stratified rocks): Geol. Soc. London Quart. Jour., v. 36, p. 46-92.
Stanley, S. M., 1966, Paleoecology and diagenesis of Key Largo limestone, Florida: AAPG Bull., v. 50, p. 1927-1947.
Stehli, F. G., and J. Hower, 1961, Mineralogy and early diagenesis of carbonate sediments: Jour. Sed. Petrology, v. 31, p. 358-371.
Supko, P. R., 1971, "Whisker" crystal cement in a Bahaman rock, *in* O. P. Bricker, ed., Carbonate cements: Johns Hopkins Univ. Studies in Geology, no. 19, p. 143-146.
Swineford, A., A. B. Leonard, and J. C. Frye, 1958, Petrology of the Pliocene pisolitic limestone in the Great Plains: Kansas Geol. Survey Bull. 130, pt. 2, p. 97-116.
Taft, W. H., and J. W. Harbaugh, 1964, Modern carbonate sediments of southern Florida, Bahamas, and Espiritu Santo Island, Baja California: a comparison of their mineralogy and chemistry: Stanford Univ. Pubs. Geol. Sci., v. 8, 133 p.
Taylor, J. C. M., and L. V. Illing, 1969, Holocene intertidal calcium carbonate cementation, Qatar, Persian Gulf: Sedimentology, v. 12, p. 69-107.
Tebbutt, G. E., 1967, Diagenesis of Pleistocene limestone of Ambergris Cay, British Honduras: Rice Univ., PhD dissert., 133 p.
——1975, Paleoecology and diagenesis of Pleistocene limestone on Ambergris Cay, Belize: this volume.

Teichert, C., 1947, Contributions to the geology of Houtman's Abrolhos, Western Australia: Linnean Soc. New South Wales Proc., v. 71, p. 145-196.

Thomas, C. M., 1965, Origin of pisolites (abs.): AAPG Bull., v. 49, p. 360.

Thorp, J., and G. D. Smith, 1949, Higher categories of soil classification: Soil Sci., v. 67, p. 117-126.

Vaughan, T. W., 1914, Preliminary remarks on the geology of the Bahamas, with special reference to the origin of the Bahaman and Floridian oolites: Papers Tortugas Lab., v. 5, pub. 182, p. 47-54.

Ward, W. C., 1967, Shelf islands of northeast Quintana Roo, Mexico, *in* A. E. Weidie, ed., Yucatan field trip guidebook, 2d edition: New Orleans Geol. Soc., p. 124-136.

——1970, Diagenesis of Quaternary eolianites of N.E. Quintana Roo, Mexico: Rice Univ., PhD dissert., 207 p.

——R. L. Folk, and J. L. Wilson, 1970, Blackening of eolianite and caliche adjacent to saline lakes, Isla Mujeres, Quintana Roo, Mexico: Jour. Sed. Petrology, v. 40, p. 548-555.

Webley, D. M., D. J. Eastwood, and C. H. Gimingham, 1952, Development of a soil microflora in relation to plant succession on sand dunes, including the "rhizosphere" flora associated with colonizing species: Jour. Ecology, v. 40, p. 168-178.

Wilson, J. L., 1967, Cyclic and reciprocal sedimentation in Virgilian strata of southern New Mexico: Geol. Soc. America Bull., v. 78, p. 805-818.

——W. C. Ward, and M. J. Brady, 1970, Northeast Yucatan, Mexico—a new area opens for study of carbonate-evaporite sediments: Jour. Sed. Petrology, v. 40, p. 745-749.

Winland, H. D., 1969, Stability of calcium carbonate polymorphs in warm, shallow seawater: Jour. Sed. Petrology, v. 39, p. 1579-1587.

——1971, Non-skeletal deposition of high-Mg calcite in the marine environment and its role in the retention of texture, *in* O. P. Bricker, ed., Carbonate cements: Johns Hopkins Univ. Studies in Geology, no. 19, p. 278-284.

Index

This index consists of two sections, which appear in the following order:

(1) Author Index and

(2) Keyword Index.

The *author index* is arranged alphabetically according to each author's last name. For papers by more than one author, each author's name appears in the index in alphabetical order. The appearance of an author's name followed by the title of an article does not mean that he is the only author of that article. He may be one of two or more authors of the paper whose title follows his name. The author index does *not* show multiple authors in any single listing.

To locate a reference in the *keyword index*, the reader should begin by thinking of the significant words. Then he should look in the index for the keyword entry for each of those words. The reference codes will direct him to the pages.

The columns on the right-hand side of the keyword index give the page number (1 or 3) indicating the nature of the source. The code is:

(1) for phrase from title; and

(3) for phrase from abstract, text, table, figure, or figure caption.

The keyword for each entry is located at the left-hand side of the page. The (>) sign indicates the first word in each title or key phrase. The (<) sign indicates the end of the title or key phrase.

AUTHOR INDEX

KEYWORD INDEX